Fundamentals of Applied Dynamics

Springer Science+Business Media, LLC

Advanced Texts in Physics

This program of advanced texts covers a broad spectrum of topics that are of current and emerging interest in physics. Each book provides a comprehensive and yet accessible introduction to a field at the forefront of modern research. As such, these texts are intended for senior undergraduate and graduate students at the M.S. and Ph.D. levels; however, research scientists seeking an introduction to particular areas of physics will also benefit from the titles in this collection.

Roberto A. Tenenbaum

Fundamentals of Applied Dynamics

With 568 Figures

Springer

Roberto A. Tenenbaum
Departamento de Engenharia Mecânica, EE
Programa de Engenharia Mecânica COPPE
Universidade Federal do Rio de Janeiro
Caixa Postal 68503
Rio de Janeiro, 21945-970 RJ
Brasil

Translated by Elvyn Laura Marshall based on the Portuguese edition, (*Dinâmica*, Editora da UFRJ, Rio de Janeiro, 1997).

Library of Congress Cataloging-in-Publication Data
Tenenbaum, Roberto A.
    Fundamentals of Applied Dynamics / Roberto A. Tenenbaum.
      p. cm.
    Includes bibliographical references and index.
    ISBN 978-1-4419-1844-4   ISBN 978-0-387-21817-5 (eBook)
    DOI 10.1007/978-0-387-21817-5

    1. Dynamics.   I. Title.
  QA845.T37 2003
  531′.11—dc21   2003045454

ISBN 978-1-4419-1844-4   Printed on acid-free paper.

9 8 7 6 5 4 3 2 1   SPIN 10922149

www.springer-ny.com

I dedicate this book to everyone who makes a real effort to improve himself or herself, and especially to Viviane, Isabela, and Miguel.

Preface

When faced with a new textbook on dynamics, a natural question confronts the reader: What is the textbook contribution, if any, relative to the many others already available in the field? With regard to fundamental theory, there is clearly no possible difference, for since Newton, Euler, Lagrange, and D'Alembert there have been no significant developments in the realm of classical mechanics. Nonetheless there has been a generalized and growing dissatisfaction with available textbooks on dynamics. The difficulty encountered by engineering students, or even recent graduates in this area, in correctly analyzing a somewhat more complex mechanical system can be seen as evidence of this dissatisfaction. In an era when engineers face challenges such as modeling a system with several degrees of freedom, designing a mechanical arm, analyzing the stability of an underwater robot, actively controlling the chassis movement of a motor car, or accurately predicting the trajectory of a satellite, as examples, a thorough understanding of dynamics is indispensable.

When confronted with the challenge of a nonconventional problem on classical mechanics, the engineer must not and may not lose himself in a multitude of formulae and methods. In order to safely obtain a solution it is necessary to recognize with accuracy the forces and torques which act upon the system, to identify the number of degrees of freedom with absolute certainty, and to choose appropriate reference coordinates, bases, and axes, describing the motion of the system as a function of the chosen coordinates. To correctly describe the system he or she must master the use of intermediary reference frames. One must also be able to set up the inertia matrices of the system, write a coherent set of equations of motion and kinematics constraints equations, and, finally, solve them or extract relevant information from them. To master all these techniques and consequently be capable of obtaining a reliable result, it is of the utmost importance to have a thorough knowledge of the fundamental concepts of dynamics and, at the same time, to have a solid training in problem-solving methodology.

This book gradually began to take shape as the result of the experience gained over 30 years of teaching — and learning — dynamics and related subjects. It originated from the need for a textbook more in accordance with the methodological unity dictated by the subject and which could simultaneously fulfill the tasks of teaching and training. The reader will find that the text almost always introduces general concepts before introducing specific ones. The author's deliberate choice to do so only appears to present more difficulties at the very beginning. Didactic experience, however, demonstrates the exact opposite: The student, exposed to a concept in its most general form, will rapidly become accustomed to it and will easily master the simplifications which occur in special cases and, most important of all, will not hesitate when faced with more complex situations. In this book, each new concept is introduced along with an illustrative example. Since theory and practice accompany each other, the student is able to implicitly learn useful problem-solving techniques.

It is precisely the methodological approach used in this book, the author believes, that characterizes its contribution, modest though it may be. Although the presentation of concepts is somewhat rigorous, the purpose of this approach is to avoid ambiguities and to develop in the reader the habit of thinking a little more abstractly. Aside from this, several concepts, such as the definition of vector systems, the notion of the angular velocity of a rigid body, or the introduction of the concept of a particle inertia tensor, among others, are presented in a manner considered unusual in basic textbooks of mechanics. The textbook presents a unity within the discipline that is evident to any minimally attentive reader and that is supported by the consistent notation and the methodology used throughout the book. In this manner particle dynamics, system dynamics, and rigid body dynamics, notwithstanding their specificities, are treated uniformly, so that a beginner in the subject will always recognize the principles which permeate the discipline.

The text presents the so-called Newtonian mechanics. Hamilton's, Lagrange's, or Kane's formulations are therefore not discussed here. Experience has shown that a solid basis in Newton-Euler mechanics is a prerequisite for readily mastering the methods of analytic mechanics, thus strengthening the intuition of the future engineer. This is a deliberate choice of the author. This texbook can be seen as a support for an undergraduate first course in dynamics. However, it is intended to prepare engineers to solve simple problems in dynamics and, on the other hand, to create a solid base for a graduate course on analytical mechanics. In this way, graduate students in physics, engineering, and correlate areas will find the text useful.

Instructors will find the text to be reasonably complete, including theory, examples, and problems, covering the essential material to be taught in a two-semester dynamics course, each semester consisting of around 60 hours. Usually the first four chapters can be covered during the first semester and the last four during the second. The natural prerequisites are at least one year of undergraduate-level calculus, one linear algebra course, and a physics course covering the principles of classical mechanics. It it also desirable, but not essential, for the reader to have taken a basic mechanics course, usually offered in all engineering departments, so as to have acquired notions of statics and link analysis.

No textbook, regardless of its excellence, can substitute for the instructor's work in the classroom. It is, naturally, the instructor who must determine the best method to be followed, excluding some topics or adding others according to his or her personal convenience. For example, Section 5.8, which deals with fluids, can be omitted without hindering in any way the understanding of the material that follows. Aside from this, the ideal sequence in a textbook is not always the most adequate one in a classroom. For instance, consider Section 5.7, which covers the conservation principles for mechanical systems. In the text each principle is followed by its respective example, while in the classroom it is more efficient to present a theoretical discussion about all the principles, followed by the set of examples. In this way the student is allowed to decide which principle should be applied in each case. When the student returns to the textbook, however, the direct association between theory and application will always be present. This consideration is also valid for several other topics.

The work of preparing such a textbook would not have been possible without the invaluable help, support, and friendship of many colleagues to whom I am immensely grateful. I would like to thank especially Professor Arthur Palmeira Ripper Neto for reading and commenting on the text, to Professor Antonio Carlos Marques Alvim for helping me to prepare Appendix A, and to Professor Luiz Bevilacqua for his encouragement and optimism. I would like to thank Mrs. Elvyn Marshall, my translator, now a close friend, for her professionalism and sense of humor.

To complete this work, the aid of several students, who gave hours and hours of their time taking care of many details, was essential. Engineer Roberto Seabra dedicated himself with extraordinary competence and determination to the task of transforming my sketches and rough diagrams into final figures stored in computer files. Most of the book's illustrations are his. A tragic accident deprived me of my main collaborator and great friend. Many other students helped me and I am very thankful to all of them.

It would not be possible to conclude without thanking the hundreds of students who, over the last years, dealt with the preliminary versions of the text and helped me improve the book by pointing out an endless number of errors. The remaining ones are my sole responsibility.

Comments, suggestions, and corrections will always be welcome.

Rio de Janeiro   *Roberto A. Tenenbaum*
Spring 2003

*T*o *the Reader*

This book is divided into eight chapters, which are in turn divided into sections, covering the main material in kinematics (Chapter 3), dynamics of particles (Chapter 4), dynamics of systems (Chapter 5), inertia properties (Chapter 6), and dynamics of rigid bodies (Chapters 7 and 8). An introduction to the general principles of dynamics and its general approach (Chapter 1) and a discussion about how to handle forces and torques (Chapter 2) are also given. There are, further, four appendices. Appendix A presents a short review of linear algebra, just to help the reader with vector operations and tensor interpretation. Appendix B shows some linkage modeling, being a complement to Chapter 2. Appendix C gives a reasonably complete table of areas, volumes, centroids, and moments and products of inertia for the most usual geometries. It furnishes a helpful support to Chapter 6. Appendix D reveals the answers for almost all the exercices at the end of each chapter. Last, there is a valuable index.

Each section is identified by two numbers separated by a period, the first number being a reference to the chapter and the second to the section itself. Section 4.7 is therefore the seventh section of Chapter 4. The equations are also identified by two numbers separated by a period, the first number indicating the section and the second indicating sequential numbering within that section. For instance, when Eq. (3.11) is mentioned in the text, a reference is being made to that equation in the same chapter. When a reference must be made to an equation present in a chapter other than the one in which the reference is made, it will consist of three numbers, separated by two periods, where the first number refers to the chapter. As an example, if the reader finds a reference to Eq. (3.3.11) in Chapter 4, a reference is being made to Eq. (3.11) in Section 3.3 of Chapter 3. Figures are also numbered in sequence within each section; when referred to successively in the same section the figure is not reproduced and the reader must search for the page where it was first introduced. When referred to in another section, the figure is then reproduced and in this case is given a new number. Examples are also numbered in sequence within a section. The font used is smaller and the alignment is indented, so that they stand out clearly from the rest of the text. Finally, exercises are given at the end of each chapter. They are organized in series, corresponding to the topic covered in one section or in a group of sections, and are numbered sequentially within the series.

For the English version of this book a set of animations for several of the examples given in the text was prepared. The main purpose of the animations is to give much more information about the motion than that explained in the text. Also, for the examples that deal with nonlinear equations, the numerical integration is provided showing the actual behavior of the particle, system, or body. Since the animations are interactive, the reader may modify parameters or initial conditions to get a deeper insight into the example. Noninteractive video files showing strictly the motion for a prescribed condition are also provided. The animations are available on Springer's Web site at: `www.springeronline.com/038700887X`.

Students must be reminded that reading a textbook or following the corresponding lectures, or both, is not enough for learning dynamics. They must be supported also by the third leg of this structure, that is, working the exercises. A fairly large set of exercises is proposed throughout the book. Working each series by himself at the end of the corresponding sections is the best way to consolidate the material and to verify if it was actually well understood.

The exercises are an important part of the text. Try to work each of them and do not give up if you do not succeed for the first time. Try again and again. And always keep in mind the Zen aphorism: *To know but not to do is not yet knowing.*

Contents

Introduction

The subject called *dynamics* covers a wide range of topics. Even though it possesses a basic theory that is trim and compact, the applications are very numerous and far-reaching. In fact, topics such as the motion of a material particle, draining of a fluid, kinematics of a mechanism, dynamics of a gyroscope, or analysis of a mechnical arm, to mention a few known examples, all belong to the domain of this subject's applications.

This chapter discusses a few introductory topics in the study of dynamics. Section 1.1 presents a very short summary of the history of the subject's origins, briefly summarizing work done before the 17th century, with comments on Galileo's findings, discussing the establishment of the foundations of classical mechanics, dwelling on Newton's formidable work, and analyzing the later contributions of Euler, Lagrange, and D'Alembert, who formalized the mechanics we know today. Section 1.2 introduces the mechanical models, that is, the fundamental concepts employed by dynamics, such as that of force, particle, and body, among many others. Section 1.3 deals with Newton's laws, which will permeate the study of dynamics in its entirety. The aim is to informally introduce the laws, which will then be effectively used in subsequent chapters. Section 1.4 handles the concept of mass center. The objective here is not to enable its practical determination, a subject examined in greater detail in Chapter 6, but to provide only an introduction to this important concept, present throughout the text. Section 1.5 discusses the methodology employed in the resolution of problems on dynamics.

Perhaps the reader will find this treatment premature, which in fact it is, but this approximation will permit a wider panoramic view of dynamics. Finally, Section 1.6 discusses the issue of *notation*. This is an important topic, and in that section the structure common to the entire notational system adopted in this book will be discussed.

1.1 Brief Historical Background

The origin of mechanics goes far into the distant past. From the very beginning, in a continuous effort to conquer the environment, human beings searched for explanations of the origin of phenomena surrounding them. The first phenomena to challenge the human mind must certainly have included free fall, the effort necessary to move objects, and the effect of impact, all of which are of a mechanical nature.

The first more systematic reflections on the motion of bodies and their origin occurred many centuries later, among the Greeks. The Greek architects certainly had enough knowledge about statics to erect safe monuments, although there are few records of such knowledge. Aristotle[1] believed that the concept of *force* involved the idea of something that pulls (or pushes) to maintain a body in motion, an idea known today to be incorrect. In all likelihood, the notion of a force as a causal element in the generation of motion is quite old, its origins probably lying in very primitive concepts that assumed that deities moved the sun, the moon, and the stars. From this point of view, motion needed an agent to produce it. Aristotle therefore defended the idea that a force was necessary for the *maintenance of motion,* or, in other words, that for a body to move at a constant velocity the presence of a force was necessary. The notion of the *variation* of velocity, that is, of acceleration, was only to appear many centuries later, when it was perfectly understood and formulated by Galileo.

Leonardo da Vinci[2] was a man of multiple interests, having left a large number of notes on several questions relating to the context of mechanics of his time. His lack of methodology, however, did not lead

---

[1] Aristotle, Greek philosopher, 384–322 B.C.
[2] Leonardo da Vinci, Italian artist and thinker, 1452–1519.

him to any result worthy of consideration. Although he is considered by some to be the forerunner of Galileo and Newton, statements of his such as "motion is an accident resulting from the inequality between weight and force" or also, "force is the cause of motion; motion is the cause of force" do not appear to lend to his investigations a sufficiently scientific character.

As an architect, he studied the resistance of pillars, beams, and arches. For example, he proposed that the resistance of a beam should be proportional to the area of its cross section. It is suspected, however, that this rule was already well known to the builders of the Parthenon.

The scientific concept of force was apparently introduced by Kepler,[3] who distinguished himself by formulating three laws that govern the movement of the planets around the sun. That was an epoch when cosmogonic conceptions agitated the scientific and also the religious worlds, with the heliocentric conception of Copernicus[4] opposing the geocentric conception of Ptolemy.[5] It is therefore natural that Kepler's attention should have been centered on celestial mechanics.

Galileo[6] made an important contribution to the creation of the modern theory of classical mechanics. Even though historians disagree in their evaluation of his importance in the history of physics, and of mechanics in particular, there is no doubt about his prominence in this field of human knowledge. His most important work, the *Discorsi*,[7] consolidated the knowledge of mechanics at the time. Among other findings, he discovered the parabolic nature of the trajectory of missiles; demonstrated experimentally that the earth's gravitational acceleration is the same for all bodies; conceived and clearly formulated the concept of the *reference frame*, which is still used today in nonrelativistic mechanics; explored with great insight the concept of physical similitude; discovered the laws that govern the motion of the simple pendulum (for small oscillations); and, most important of all, formulated the laws of motion, although in a somewhat imprecise manner. In fact, Newton himself,

---

[3] Johann Kepler, German astronomer, 1571–1630.

[4] Nicolau Copernicus, Polish astronomer, 1473–1543.

[5] Ptolemy, Greek astronomer, second century A.D.

[6] Galileo Galilei, Italian philosopher and mathematician, 1564–1642.

[7] *Discorsi e Dimostrazione Matemat. intorno à due nuove Scienze*, 1638.

naturally in a modest fashion, attributed to Galileo the conception of his first two laws.

It was Galileo who effectively formulated the law of inertia. He understood perfectly that in the absence of applied forces the velocity of a body should remain unchanged. In the *Discorsi,* this law appears as follows: "Whatever the degree is of velocity of an object, it will remain indestructibly imprinted, provided that the external causes of acceleration or deceleration are removed." As to the second law of motion, there is no doubt that it must be credited to Newton. Galileo experimented with the sloping plane and with the motion of projectiles, where the force due to weight was always present as the cause of the motion of bodies. Consequently he did not conceive of forces not proportional to mass and the notion of the *momentum* did not occur to him. Newton would state that the variation of the quantity of motion "is proportional to the applied force and takes place in the direction in which the force is applied."

It was undoubtedly Newton[8] who made the most important contributions to mechanics, and in particular to dynamics, and for this reason is considered the father of classical mechanics. Newton performed a noteworthy revision of the scientific knowledge of his time, consolidating into fundamental laws what had been loosely stated by his predecessors. For example, he showed that Kepler's three laws of planetary motion could be reduced to a single law of universal gravitation and that free-falling bodies were also governed by the same law, thus creating the first and most important synthesis of celestial and terrestrial mechanics.

Newton's most significant contributions in the realm of mechanics are described in the monumental work known as the *Principia,*[9] which brings together in three volumes countless discoveries made over many years. The most important result obtained by Newton was, without question, his second law, known today as the cornerstone of dynamics. Newton's discoveries will be discussed in more detail in Section 1.3.

Euler[10] was another leading figure in the construction of dynamics. He made important contributions in several fields of mathe-

---

[8] Isaac Newton, English scientist, 1642–1727.

[9] *Philosophiæ Naturalis Principia Mathematica,* 1687.

[10] Leonhard Euler, Swiss mathematician, 1707–1783.

matics and physics and was responsible for formulating Newton's second law in its currently most used form, namely, that of the product of the mass and the acceleration being equal to the resultant applied force. Going even further, Euler published this law in 1752, stating that it is equally applicable to a finite or infinite mass, making way for the generalization of the law, which includes fluids as well as rigid bodies. His restless spirit was not satisfied with this finding, which was rigorously not very innovative with respect to Newton. As a result he started to study the problems concerning the motion of the rigid body, which required a more careful approach. In this analysis appeared the six scalars, referred to today as the components of the *inertia tensor*, and the differential equations that govern the rotation of a rigid body about a fixed point, currently known as *Euler's dynamic equations*. It was therefore Euler who developed the concept of the inertial rotation of a body, having published in 1776 laws applicable to any body, or part of a body, rigid or deformable. The laws are as follows:

1. The principle of *momentum*, or of the linear momentum: The total force acting upon a body is equal to the rate of change of the momentum;

2. The principle of *moment of momentum*, or of the angular momentum: The total torque acting upon a body is equal to the rate of change of the angular momentum, where both are measured with respect to the same fixed point.

These laws, known as *Euler's laws of mechanics*, naturally encompass Newton's second law; they are the equations that govern the motion of bodies in general systems and are still used today in so-called Newtonian mechanics.

Classical mechanics was given a new stimulus with the works of D'Alembert[11] and Lagrange.[12] D'Alembert's *Traité de Dynamique* rejects the concept of Newtonian force and also introduces the forces of inertia, reducing, in a way, the problems of dynamics to static situations. D'Alembert also made an attempt, albeit not very successful, to deduce all of mechanics from the laws of collision.

---

[11] Jean Le Rond D'Alembert, French mathematician, 1717–1783.

[12] Joseph-Louis Lagrange, French physicist and mathematician, 1736–1813.

But it was Lagrange who formulated the variational principle, valid for the vast majority of mechanical systems, in his *Méchanique Analitique* (1788). It is known today, curiously, as *D'Alembert's principle*. In a more precise manner, some authors refer to this formulation as the *Lagrangian form of D'Alembert's principle*, thus restoring the real paternity of the dynamical equations within analytical mechanics. It was also Lagrange who introduced *generalized coordinates*.

Analytical mechanics has become an extremely useful and powerful tool for the formulation of the equations of a mechanical system, introducing shortcuts on the way to solving and suppressing linkage forces. But, as Truesdell says:[13] *It cannot be said, from Lagrange's equations, whether a system does or does not have a momentum; Euler's equations at least show this, and this comes from the fact that the integrals of momentum appear naturally in approaches based on Euler's equation. Anyhow, Lagrange's equations are relevant only for certain types of mechanical systems, and are less general than Euler's laws.*

When Newton said about his discoveries that "If I have seen further [than others] it is because I stood upon the shoulders of giants," he was clearly acknowledging the work done by his predecessors and was also describing one important aspect of the evolution of science. Newton's observation reminds us of the Catalan tradition of human towers, whereby very strong individuals form a circle, other such individuals climb upon their shoulders, and so on. The construction of the edifice of science proceeds in a similar fashion, slowly and surely upwards. Each new stage requires another courageous step. (But, unlike the human towers, the tower of scientific knowledge does not occasionally collapse, although it may suffer significant damage due to certain revolutionary discoveries.)

1.2 Mechanical Models

Engineers may be defined as specialists in modeling. In fact, the main task confronting an engineer is that of *solving problems*. This implies a search for the understanding of a usually complex physical reality, start-

---

[13] C. Truesdell, *Essays in the History of Mechanics*, Springer-Verlag, 1968.

ing from simple *models* that approach reality. Models are indispensable tools, for they introduce simplifications that make problems solvable. We are confronted indeed with a difficult task: On the one hand, we must adopt models that are sufficiently complete (and complex) to effectively and fairly closely represent the situation under consideration; on the other hand, we should use models that are simple enough for us to easily reach a solution. The engineer's task is then to discriminate and select, sometimes quite subtly, the most appropriate models for a specific kind of problem, and to evaluate the results that can be expected from these models. It is worth pointing out that technological progress changes our perception of what constitutes an adequate model. In fact, due to the decreasing costs of complex computational tools and the recent availability of numerical simulation, successively more complex models can be adopted. As increasingly more powerful tools become available for their solution, models can become increasingly more sophisticated. Examples of such tools include faster computers, new numerical methods for the integration of equations, software for algebraic manipulation, among others. The fundamental models of mechanics, however, never change.

When the methods associated with a specific theory are used to solve an engineering problem, we are appropriating certain models that are the basis of that theory, whether we are aware of it or not. In this process formal mathematics is constructed on a *deductive basis*. In other words, it is not introduced to us as an *experimental science*, in which results are accepted because they are in accordance with the observations produced by the experiment, but as a structure of *fundamental concepts, axioms, theorems, and inference rules*. Fundamental concepts are defined as notions that are of universal use or based on common sense, and that are therefore accepted without the need for formal definitions. Axioms, on the other hand, are statements of formulae taken to be *true* without the need for *proof*. Let us give an example from Euclidean geometry, a discipline with which the reader is likely to be familiar: the statement that one and only one straight line passes through two points is an axiom, and the notion of a *point* and a *straight line* are fundamental concepts, and therefore undefined. Theorems, on the other hand, are statements or formulae based on the axioms and can be *deduced* using inference rules. The famous Pythagorean theorem, for instance, is a theorem because it can be proved based on the axioms of Euclidean geometry. Finally, rules of inference are elements of mathematical logic that allow theorems to be proved based on axioms and other previously proven theorems.

When dealing with an applied science, such as, for instance, mechanics, we are quite distant from the almost absolute formality of mathematics, but the latter's main elements are still present, as shall be seen. Therefore dynamics, as a branch of physics, or more specifically of classical mechanics, is also regarded as belonging to the realm of applied mathematics. This is so because the subject of mechanics contains a consistent structure of fundamental concepts, principles (axioms), and practical formulae (theorems) that approximate it to pure mathematics, even though it deals with the elements of the physical world, such as bodies and their motion and interactions. The remainder of this section attempts to precisely relate these four categories present in mathematical formalism with their corresponding mechanical equivalents.

In dynamics we will therefore find so-called models, which are nothing more than fundamental concepts accepted without a definition. They are referred to in this manner because they are the result of modeling, or of an idealization of the real physical world or of the world as we see it. Examples of this category include the notion of particles, systems, and force, among others.

We define a *particle* to be a very small body, when compared with the distance it moves. Clearly this definition is not precise, like all others that will follow, and is therefore not formal. It is an approximation of a concept that is, to be more exact, admittedly intuitive. We will make other attempts to approach the concept of a particle. For instance, let us say that a particle is a material point, that is to say, a point with no dimension, but that possesses finite mass. A particle is thus always identified by a point in Euclidean space and is associated with a real number, its mass m. The particle is a fundamental model in classical mechanics, from which principles (axioms) will be formulated, as shall be seen in Section 1.3.

On the other hand, we define an *infinitesimal mass element* to be a body of infinitesimal extent, the mass of which is also infinitesimal.

The difference between a particle and an infinitesimal mass element is subtle but crucial: Both have no dimension, but while the particle's mass is finite, the mass of the infinitesimal mass element is also infinitesimal. Instead of a *mass*, a scalar ρ, named *density*, which is the mass per unit volume, is therefore associated with the infinitesimal mass element; when, in the limit, the volume tends to zero, the element's mass also tends to zero, but the ratio between them tends to a finite value, precisely equal to the density. The model of an infinitesimal mass element will be useful for the modeling of bodies, as shall be seen further ahead.

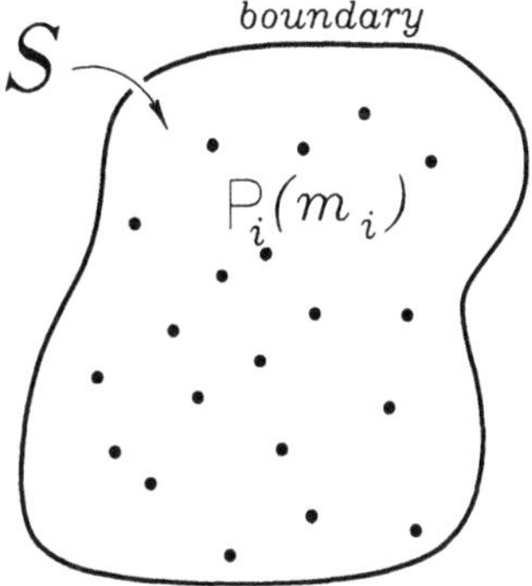

Figure 2.1

A *particle system* consists of a well-defined set of particles. Every system has a *boundary* that distinguishes the particles within from all others which do not belong to the system (see Fig. 2.1). If S is a system consisting of, say, n particles, its mass will be equal to the sum of the masses of the particles belonging to S, as follows:

$$m(S) = \sum_{i=1}^{n} m_i, \qquad (2.1)$$

where m_i is the mass of a generic particle P_i.

The definition above implicitly introduces another model in mechanics, namely the notion of *discrete*. The particle system defined in the preceding paragraph is a discrete system, or in other words, a denumerable one (it contains n elements, where n is an integer). The concept of a discrete system stands in opposition to that of a *continuous system.* A continuous system is also a well-defined system, so it also possesses a boundary, but its elements are infinitesimal mass elements instead of particles and it is not a denumerable system. A continuous system is also called a *body*. The mass of an element is $dm = \rho dV$, where dV is the corresponding volume and ρ is the field of density, which in turn is a function of the element's position (see Fig. 2.2). The mass of body C may therefore be written as an integral over the entire body, as follows:

$$m(C) = \int_C dm = \int_C \rho \, dV. \qquad (2.2)$$

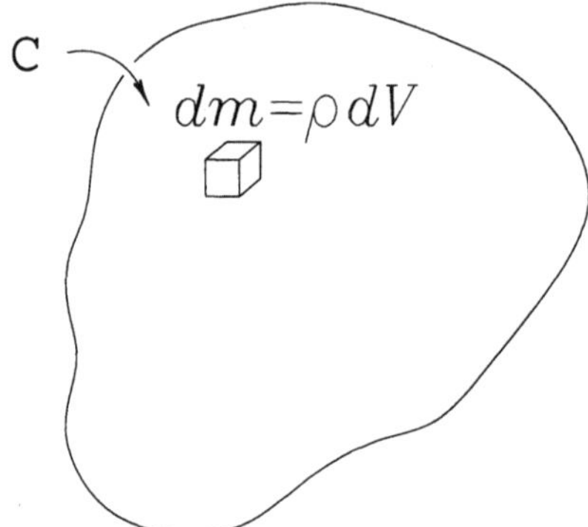

Figure 2.2

The entire subject of dynamics will be developed based on the models of particle, particle systems, and continuous systems (bodies). When we refer to a body, we are in fact referring to any object that satisfies that definition. Fluid flowing inside a pipe, an elastic spring, a deformable beam, a flexible rope, and a stone are all modeled as a *body*. Evidently some distinction should be made between bodies of such widely differing nature, but from the point of view of mechanics all these examples may be considered *bodies* governed by the same equations of motion, as shown in Chapter 5.

Among the examples mentioned in the previous paragraph, one of them (in this case, the stone) is particularly important for dynamics, so the necessary distinction will be made right away. A *rigid body* is defined as a continuous system so that the distance between any two points is time-invariant. Like any model in mechanics, that of a rigid body is an abstraction. A body may be considered rigid if its deformations, or relative motion, can be ignored compared to its global motion (once again we have an inaccurate definition). The dynamics of the rigid body will be studied in Chapter 7. As shall be seen then, major simplifications in the theory will be possible precisely due to the assumption of rigidity, thus justifying its study. Other examples, among those mentioned, are also important. Fluids, for instance, are the subject of *fluid mechanics*, while deformable solids are studied in the field of *elasticity* or in *solid mechanics*. Nevertheless, the bases of all these disciplines are to be found in dynamics.

Since mechanics studies the interactions between bodies and their motion and deals specifically with the relationship between these interactions and the resulting motion, it therefore studies *cause* and *effect*. Motion always originates as a result of a *force* or a *torque*. Force is one of the basic models of mechanics; it can be defined as the interaction between two particles and consists of a vector quantity, which means it has magnitude and direction, and it is the only possible interaction between particles. It can be classified in two different categories: *contact forces* and *field forces*, where the latter are also called *action at a distance* forces. Contact forces, as is clear from the name itself, result from the interaction due to contact between two particles, as is the case in a collision. Field forces exist between particles when there is no mutual contact, as is the case with gravitational forces.

When two bodies interact, torques may also result. Even though, as shall be seen in Chapter 2, a torque may be producd by the *moment* of a force with respect to a given point, the notion of torque will be treated here as a primitive concept. A body may exert a torque upon another body without the intervention of any forces whatsoever, as is the case when the axle of an electric motor activates a hydraulic pump. Forces are normally associated with the generation of translational motion, while torques are normally associated with the generation of rotational motion. The reader should, however, be aware that these ideas are not always true for, as shall be seen in the study of rigid bodies, an off-center force may also produce rotations, and an applied torque may in turn produce a translation of certain points in a body.

The primitive concepts of particle, particle system, mass, infinitesimal mass element, continuous system or body, rigid body, force and torque constitute the infrastructure of mechanics. The theory presented here will totally depend on these fundamental notions. Nonetheless, for the sake of even more completeness and clarity, a few derived concepts of the utmost importance in the study of mechanics will now be introduced. In the chapters that follow we will dwell again, with due care, on these concepts.

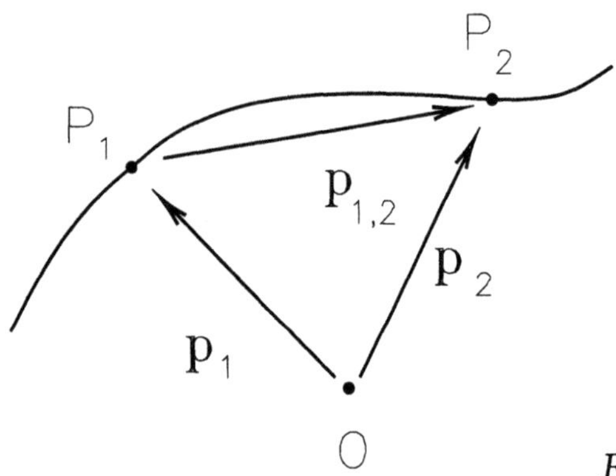

Figure 2.3

When a point moves in space, it is convenient to talk about a *position vector*. If a point P moves between positions P_1 and P_2, given an origin O (see Fig. 2.3), the position vector $\mathbf{p}_1$ and the position vector $\mathbf{p}_2$ determine the position of P, while the vector $\mathbf{p}_{12} = \mathbf{p}_2 - \mathbf{p}_1$ measures the vectorial displacement of the point. (The reader should note that we are referring to a *point*, which means that we may be dealing interchangeably with an individual particle, a particle belonging to a system S, an element in a body, or even a geometric point, without any reference to mass.)

The rate of change in time of the position vector is called its *velocity vector* and the rate of change in time of the velocity vector (therefore the time derivative of the position vector) is called the *acceleration vector*. The concepts of position, velocity, and acceleration will be treated with more care and detail in Chapter 3 and only the concept of position vector may be considered *primitive*, but it is worthwhile introducing it informally here.

A concept in dynamics of the utmost importance is that of the *momentum*. If a particle of mass m is moving in space at velocity $\mathbf{v}$, it may be said to possess a vector property characterized by the product $m\mathbf{v}$, defined as the *momentum vector of the particle*, $\mathbf{G}$. Similarly, when an infinitesimal mass element of mass dm moves at velocity $\mathbf{v}$, its momentum vector $d\mathbf{G}$ is also infinitesimal and is equal to the product $\mathbf{v}dm$. The reader should note that the momentum vector is a multiple of the velocity vector and is therefore always parallel to the latter. Systems and bodies also possess momentum, but these concepts will be introduced at a more appropriate point, in Chapter 5. The concept of the momentum vector of a particle is important because *Newton's second law of motion*, the basic axiom of dynamics, may be formulated in terms of this property, as shall be seen below.

1.3 The Laws of Motion

As has already been mentioned, the foundation of classical mechanics was established by Newton. The *Principia Mathematica*, published in 1687 and consisting of three volumes, examines several areas in mechanics: the motion of bodies, fluid mechanics, the mechanics of the solar system, oscillations, and the propagation of acoustic waves, among other minor topics. Volume I presents the famous axioms or princples that, translated more or less freely, state:

I. *Every body persists in its state of rest or of uniform motion in a straight line unless it is compelled to change that state by forces impressed on it.*

II. *The change of motion is proportional to the driving force applied and occurs along the straight line where this force is exerted.*

III. *To every action there is always an equal and contrary reaction; or, the mutual actions of two bodies upon each other are always equal, and in opposite directions.*

The work of Euler, Lagrange, and D'Alembert followed that of Newton, transforming the laws of motion into their modern formulation. A form of the laws more consistent with the so-called rational mechanics is adopted nowadays, and has made the subject of mechanics more rigorous in the establishment of its axioms, theorems, and fundamental laws.

Newton's first law is also known as the *law of inertia*, for it assigns an *inertial* property to bodies, namely that of *resisting* any change in its state of motion. Formulated in current terms, this law can be stated as follows:

I. *A particle maintains its velocity vector unchanged in an inertial reference frame if the resultant force acting upon it is zero.*

There are several differences between the two formulations. Initially, Newton referred to a body, a somewhat imprecise concept, while the current formulation refers to a particle. The idea of state of rest or uniform motion in a straight line has been entirely substituted by the concept of the invariance of the velocity vector. In fact, if the velocity vector remains constant over time, the state of motion is guaranteed not to change. Aside from this, the velocity vector has been associated with an inertial reference frame, a supposedly primitive concept, or a concept defined in the second law, as shall be seen. Although mentioning inertial reference frames clearly complicates matters, it is absolutely necessary, for without it the law as a whole loses its validity. Finally, the forces impressed on the body have been substituted by the applied resultant force, which is certainly what Newton had in mind.

In mathematical terms, the first law is the following:

$$^{\mathcal{R}}\mathbf{v}^P = \mathbf{v}_0 \qquad \text{if} \quad \mathbf{R} = 0, \tag{3.1}$$

where $^{\mathcal{R}}\mathbf{v}^P$ designates the velocity vector of the particle P in an inertial reference frame $\mathcal{R}$, $\mathbf{v}_0$ is a time-invariant vector, and $\mathbf{R}$ is the resultant applied force.

Newton's second law, also known as the law of change in the momentum, has many modern expressions, which differ in subtle ways. We will adopt the following formulation:

II. *Reference frames exist so that at each instant the time derivative of the momentum vector of a particle is equal to the resultant applied force.*

The second law has also been rewritten to describe a particle. The quantity that expresses the state of motion is now the momentum vector. Therefore, the second law now establishes that the rate of change, with respect to time, of the momentum vector is equal to the resultant of the applied forces. It also states that reference frames (called inertial or Newtonian reference frames) exist in which this relationship is valid. In the meantime, the existence of reference frames for which the derivative of the momentum vector is *not* equal to the resultant force is implied. The second law may, therefore, be taken as a definition of inertial reference frames, even though this is not the most important aspect of the law. All of dynamics is based on the second law, in one way or another. Euler and Navier's equations on fluid mechanics, for instance, originate in Newton's second law.

In mathematical terms, the second law can be expressed as

$$^{\mathcal{R}}\dot{\mathbf{G}}^{P} = \mathbf{R}, \tag{3.2}$$

where $^{\mathcal{R}}\mathbf{G}^{P}$ refers to the momentum vector of the particle P in the inertial reference frame $\mathcal{R}$, the dot above the vector indicates that it is a time derivative in $\mathcal{R}$, and $\mathbf{R}$ is the resultant applied force.

In its more common form, as formulated by Euler, the second law states that the product of the mass of a particle and its acceleration vector in an inertial reference frame is also equal to the resultant applied force

$$m\,^{\mathcal{R}}\mathbf{a}^{P} = \mathbf{R}, \tag{3.3}$$

where m is the mass of the particle P, $^{\mathcal{R}}\mathbf{a}^{P}$ is the acceleration of the particle P in the inertial reference frame $\mathcal{R}$, and $\mathbf{R}$ is the resultant force.

Equations (3.2) and (3.3) are clearly formulations of the same law. Because the momentum vector $^{\mathcal{R}}\mathbf{G}^{P}$ is equal to the product of the mass of the particle and its velocity vector $m\,^{\mathcal{R}}\mathbf{v}^{P}$, assuming that the mass of the particle is constant in time, the derivative of the momentum vector will be equal to $m\,^{\mathcal{R}}\mathbf{a}^{P}$, as stated in Eq. (3.3). Nevertheless, Newton's formulation takes into account systems of variable mass, so that Eq. (3.2) is the expression of an even more general principle than that given by Eq. (3.3). Systems of variable mass will be discussed in Chapter 5.

In its more modern formulation, the third law can also be expressed in terms of particles, as follows:

III.   *The interaction between two particles occurs as the result of two forces; the force particle P exerts on particle Q is vectorially opposite the force Q exerts on P, where both forces act along the straight line containing the two particles.*

Two forces are said to be *vectorially opposite* when they are equal in magnitude and point in opposite directions.

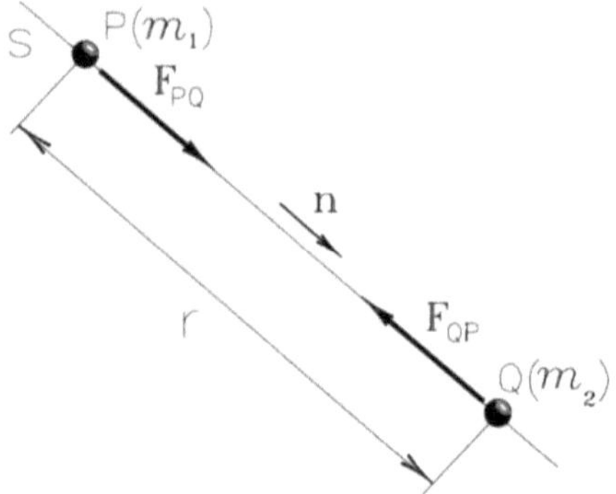

Figure 3.1

Figure 3.1 illustrates this law. Since each particle is a material point, one and only one straight line (s) joins them, which is the support of the forces. Force $\mathbf{F}_{PQ}$ acts upon particle P and another force $\mathbf{F}_{QP}$ upon particle Q, so that the following always holds:

$$\mathbf{F}_{QP} = -\mathbf{F}_{PQ}. \tag{3.4}$$

It is important to emphasize that the third law determines that action and reaction always act upon different particles, also guaranteeing that the forces act on the same line of action.

It is precisely the third law that allows us to generalize the results of the second law, as formulated for particles, to systems consisting of particles and bodies, as shall be seen in Chapter 5.

The *Principia Mathematica* also includes the law of universal gravitation, which is independent from the other three laws, and which states that

IV. *The force of mutual gravitational attraction between two particles in space is proportional to the masses of the particles and inversely proportional to the square of the distance between them.*

In formal terms, let P and Q be two particles of mass m_1 and m_2, respectively, r away from each other (see Fig. 3.1). Then the force Q exerts upon P, $\mathbf{F}_{PQ}$, is given by

$$\mathbf{F}_{PQ} = G\frac{m_1 m_2}{r^2}\mathbf{n}, \tag{3.5}$$

where $\mathbf{n}$ is a unit vector (see Appendix A) on the straight line that joins the particles and G is the universal gravitation constant, the value of which is experimentally obtained and is equal to 6.673×10^{-11} m^3/(kg·s).

The weight of a body is the result of the earth's gravitational force exerted upon it and may be calculated, at least approximately, from the law of universal gravitation. It can be effectively shown that for a small body, which is close to the earth's surface, the center of the particle and the center of the earth can be considered to be particles with the respective masses of the body and the earth. The gravitational acceleration on the surface, g, can then be obtained from the equation

$$g = \frac{GM}{R^2}, \tag{3.6}$$

where G is the universal gravitation constant, $M = 5.976 \times 10^{24}$ kg is the earth's mass, and $R = 6.371 \times 10^6$ m is the average radius of the earth, values that can be obtained indirectly from experiments.

When the above values are substituted into Eq. (3.6), we obtain $g = 9.824$ m/s^2. This value is approximately 0.18% greater than the observed gravitational acceleration. This discrepancy is mainly due to two factors. First, the earth is not an inertial reference frame, and the rotation about its own axis reduces the acceleration of a free-falling body. Moreover, the earth is not a perfect sphere but actually a spheroid flattened at the poles, so acceleration varies according to latitude. Taking into account these effects, the *international equation of gravity* is found to be

$$g = 9.78049\,(1 + 0.0052884\sin^2\gamma - 0.0000059\sin^2 2\gamma), \tag{3.7}$$

where γ is the latitude, in degrees, and g is obtained in m/s^2. This means that g therefore varies between 9.78049 m/s^2 at the equator and 9.83221 m/s^2 at the poles, at sea level.

1.4 Mass Center

The concept of mass center plays a crucial role in the subject of mechanics. This topic will be considered in further detail in Section 6.1, where integration and composition techniques used to determine the mass center of a body will be introduced, along with symmetry properties that are useful in this calculation. Since certain topics to be covered before Section 6.1 require an understanding of the concept of mass center, the purpose of this section is to introduce the concept informally.

Given any system S, discrete or continuous, rigid or deformable, there is always a point in space — let it be called S* — around which the body's mass is evenly distributed. In other words, there will always be a point S* so that, if O is another point the position of which is known, the position vector $\mathbf{p}^*$ of S* with respect to O is the weighted average of all the position vectors of the elements in S (if S is a discrete system, the elements of S are particles, and if S is a body, the elements of S are infinitesimal mass elements). Note that the weighted average is *vectorial* and that the weights present in this average are precisely the masses of the elements of the system.

Therefore, if S is a discrete system, containing n particles P_i, of mass m_i, $i = 1, 2, \ldots, n$, and if $\mathbf{p}_i$ is the position of particle P_i with respect to point O, then the position of the mass center with respect to O is given by

$$\mathbf{p}^* = \frac{1}{m} \sum_{i=1}^{n} \mathbf{p}_i m_i, \tag{4.1}$$

where m is the mass of system S, as defined by Eq. (2.1).

On the other hand, if C is a continuous system (a body) and if $\mathbf{p}$ is the position vector, with respect to point O, of a generic infinitesimal mass element of the body (of mass m), then the position of the mass center with respect to O is

$$\mathbf{p}^* = \frac{1}{m} \int_C \mathbf{p}\, dm, \tag{4.2}$$

where m is now the mass of body C, as defined by Eq. (2.2).

The mass center of a system or body is a point, mathematically defined by Eq. (4.1) or (4.2). It may happen, therefore, that the mass center does not coincide with any point of the body. For example, the geometric center of a homogeneous ring coincides (this is easy to check) with its mass center, this point not actually belonging to the ring.

1.5 Methodology

Here we are going to discuss some comments on the methodology used in the analysis and solution a problem in dynamics.

Using the basic models as a starting point, it is necessary, first, to identify the object under study as a particle, a discrete system of particles, a system involving particles and bodies, a body, rigid body, or a system of rigid bodies. Second, we need to know the reference frame based on which the motion of this object will be observed. Next, we need to define how many coordinates are required to fully characterize the time evolution of the object under study, in other words, to define its motion in the chosen reference frame.

So after defining the model for the object under study, the reference frame, and the coordinates that describe this motion, we move on to the methodology of dynamics itself. The first step, then, will be to identify the forces — and, eventually, the torques — that act on the object. Once this is done, the next step will be to reduce this system of forces and torques to a previously chosen point. When dealing with a particle, the natural point par excellence is the particle itself (although there may be exceptions); when dealing with a system or body, this point may be the mass center, a fixed point on the reference frame, or even another point that becomes more convenient in that specific case. The study of the vector systems — especially, the forces systems — and their reduction to one or more points will be discussed in Chapter 2.

The second step to solve a dynamics problem is the kinematic analysis of the object under study. This consists of expressing, in terms of the chosen coordinates, angular velocities and angular accelerations of the bodies and the intermediary reference frames, whenever applicable, and velocities and accelerations of the particles or points of interest. For this step, the kinematic relations and theorems studied in Chapter 3 will be necessary.

The next step, if the object under study is a particle, is to establish the equations of motion. Chapter 4 discusses the dynamic principles governing the motion of the particle, briefly commented upon in Section 1.3, and also studying other analytical methods, such as the energy method and impulsion method, all derived, as a rule, from Newton's second law. The relations resulting from this procedure consist of differential equations for the coordinates, known as *equations of motion*. The integration of these equations will provide the final solution for the dynamic problem, that is, the time evolution of the coordinates that describe the motion of the particle in the chosen reference frame. Although, in some examples, the solution of the equations of motion is obtained, as a rule, this is not the specific task of dynamics, as it belongs to other branches of applied mathematics, such as differential equations or numerical methods. The animation files provide the numerical solution for several examples of the text.

When the object under study consists of a system of particles, the analytical procedure is the same, the only difference being that the equations and methods used must be generalized for a system. Chapter 5, therefore, discusses this generalization of the dynamics principles for a system of particles, whether discrete or continuous. Here also, having discussed applied forces and analyzed kinematics, the only thing left is to substitute resultant velocities, accelerations, forces and torques in the equations of motion. Integral forms for these equations, such as the energy balance, may be used as an advantage, as we will see later.

Finally, when the object under study is a rigid body or system of rigid bodies, an extra step must be considered. The inertia of a body is a little more complex than the inertia of a particle. In fact, the inertia of a particle is characterized by its *mass*, since a particle only has translational motion, to the extent that a rigid body, showing translational and rotational motion, has, in addition to the translational inertia, a characteristic inertial property of rotation, which is its *inertia tensor*. Hence, for the dynamic analysis of a rigid body, besides considering the system of applied forces, and torques and the study of their kinematics, it will be necessary to find their properties of inertia, before the equations of motion are established. The study of the inertial properties of a system or body will be discussed in Chapter 6. Last, Chapter 7 and Chapter 8 study in detail the analytical methods of the dynamics of the rigid body.

1.6 Notation

Some comments on the notation adopted are necessary. One of the major trumps of any discipline lies precisely in the notation used; this influences the understanding to such an extent that an unsuitable or inaccurate notation can make the topic under discussion unintelligible.

When we choose a notation, we are always faced with, as a rule, an insuperable dilemma. Two fundamental attributes of a notation are, by their own nature, contradictory. On one hand, it must consider as much information as possible so that any possible ambiguities are avoided. On the other hand, simplicity must be one of the main aims. In this text, as explanatory a notation as possible was chosen, simplifying it whenever the context permitted, without raising any possible doubt. Experience has shown that a fuller and more explicit notation can be easily assimilated and helps the reader to understand certain nuances that would become difficult if using a simplified notation. Another point in favor of the adopted notation is that it naturally provides an overview, which this text endeavors to present.

The notation consists of five basic elements, four of which are optional, depending on the case. The first element is the *letter* that indicates a certain quantity. So, if we wish to indicate the mass of a body or particle, we will adopt the notation m. Note that m is in *italics*. Every *scalar* quantity will then be denoted by a letter in italics. When the quantity in question is not, as in the preceding example, scalar but rather a *vectorial* quantity, it is indicated in **bold**. For example, the position vector of a point has been denoted as **p**. So, m indicates mass, a, b, c, r can indicate distances; on the other hand, **p**, **v**, and **a** indicate position, velocity, and acceleration, respectively; **G** indicates momentum, **H** indicates angular momentum, **F** force, and **T** torque, all of them vectorial quantities. A number of other symbols will be presented as the corresponding concepts are being introduced and the reader does not need to worry about memorizing them.

The four other elements that comprise (at most) the notation appear as a subscript (index) or a superscript (exponent) of the principal element. The index usually denotes a *component* or a *numeral*. Hence, if $\mathbf{F}$ is an applied force (a vector), the *component* of the force in the direction of a Cartesian axis, say, axis x, will be denoted as $\mathbf{F}_x$ (a vector) or, when dealing with the *scalar component* of the vector (see Appendix A), as F_x (a scalar). On the other hand, if we are talking about n particles of a system, their masses will be denoted by $m_1, m_2, \ldots, m_n$. The superscript, or exponent, is used to specify to what the quantity in question refers. Thus, $\mathbf{v}^P$ indicates the velocity of the particle P and $\mathbf{G}^C$ indicates the vector momentum of the body C. Note that particles are denoted in Roman-type capitals (P), while bodies are denoted in italics and also in capitals (C). (But, when we write the velocity of P, $\mathbf{v}^P$, the letter P is converted to italics, in the mathematical mode; this should not, however, cause any difficulty.) When we need to clearly indicate the vector position from a point Q to a point P, we will adopt the notation $\mathbf{p}^{P/Q}$, which reads as: position vector of P *with respect to* Q. In this last example a new element appears in the notation; both P and Q have an upper index position, or exponent, in relation to a basic element that indicates the quantity, that is, an element is placed *over* the fraction line, while the other is *under* it. The terminology accompanies the notation; it is the position of P (that which is mentioned) with respect to Q (with respect to that which is mentioned), that is, a vector position from Q to P.

We have now seen four elements, as follows: the quantity (central element of the notation); the lower index, generally indicating either a number among several, or one component; the top index in the numerator, indicating at which point (or body) it is referred; and the top index in the denominator, usually a point (or axis, as will be seen later) with respect to what is being discussed. There is also a fifth element in the notation that is reserved for *reference frames*. So, for example, the velocity of a point P with respect to Q in a reference frame $\mathcal{R}$ will be noted by $^{\mathcal{R}}\mathbf{v}^{P/Q}$. The reference frame, therefore, appears as a top *left* index, in the notation. (The concept of reference frame will be introduced in Chapter 3, but its use in terms of notation can and should be introduced now.)

As we said at the beginning, not every notation requires the use of the five elements. The angular velocity of a body C in a given reference frame $\mathcal{R}$ is indicated by $^{\mathcal{R}}\boldsymbol{\omega}^{C}$, only requiring three elements; the velocity (absolute) of a point P in a reference frame $\mathcal{R}$ is noted by $^{\mathcal{R}}\mathbf{v}^{P}$, also requiring three elements. Now the component in the direction x of the angular momentum vector of a body C with respect to a point O in a reference frame $\mathcal{R}$ will be noted by $^{\mathcal{R}}\mathbf{H}_{x}^{C/O}$, with everything you might need. Note carefully that the text's terminology will say "with respect to" referring to the point O and, on the other hand, will say "in" (or, sometimes, "in relation to") when referring to reference frame $\mathcal{R}$.

Some other notation elements, as well as general simplifications of the notation presented herein, will be introduced throughout the text. But the main idea will always be maintained and the reader must learn it now in order to facilitate the study. It may seem complicated at first, but the consistency of this notation will become quite clear for the reader as it is being presented and even more so when applied, especially facilitating the use.

Exercise Series #1   (Sections 1.1 to 1.6)

P1.1 The critics against Copernicus and Galileo argued that, if the earth moved, then, a heavy body that was dropped from a very high tower should fall to the west of the foot of the tower. Galileo argued, however, that this body would fall slightly to the east. An engineering student, who heard this story, argued that the body would fall directly under the point in which it was dropped. Who is right?

P1.2 Why do you think that Aristotle deduced that force should be in proportion to velocity?

P1.3 Occasionally troublesome paradoxes appear to challenge the formal logic and shake the foundations of the consistency of mathematics. One of the oldest paradoxes ever heard was formulated by Epimenides, a Greek who coined the following immortal words: "All Cretans are liars." Now, as Epimenides himself was from Crete, the statement is true if and only if it is false, and vice versa. There are several mathematical formulations to express the paradox of Epimenides (and his correlates) and such proposals are called *undecidable*. Try also to formulate a logical-mathematical paradox.

P1.4 Explain what the difference is between a system of particles and a body. Why, in a system of particles, is the scalar field ρ not defined?

P1.5 Suppose that a system consists of n particles P_i, with mass m_i, $i = 1, 2, \ldots, n$, and m bodies C_j, with a density ρ_j, $j = 1, 2, \ldots, m$. What will the expression be for the system's mass M?

P1.6 Explain why the earth imprints on every body the same acceleration, independent of the body's mass.

P1.7 What is the advantage of formulating Newton's second law in terms of the time rate of the momentum vector, instead of the broader formulation involving the body's acceleration?

P1.8 Consider the gravitational attraction exerted on a small body by a sphere whose density varies exclusively with the distance to its center. Show that this attraction is equivalent to that exerted by a small sphere of the same mass, located at the center of the original sphere. (This result was demonstrated by Newton, being based on his law of universal gravitation.)

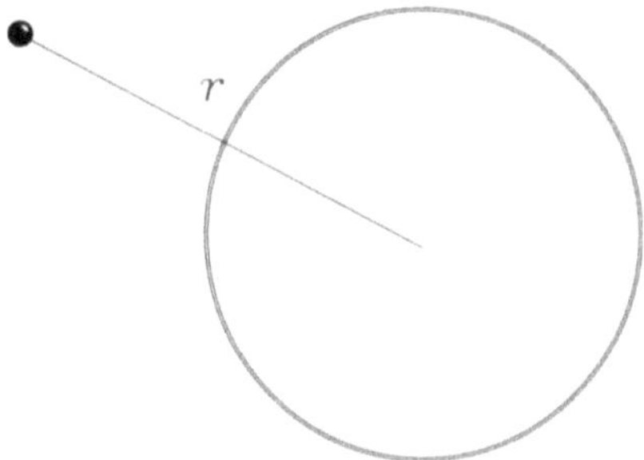

Figure P 1.8

P1.9 The sun's mass is approximately 2×10^{30} kg, the average distance between it and the earth is 1.5×10^{11} m, and the equatorial diameter measures approximately 6.4×10^6 m. Estimate the percentual variation of the weight of a man at the equator between day- and nighttime.

Figure P 1.9

P1.10   Mount Everest, on the boundary between Nepal and Tibet (China), has an altitude of 8848 m and a latitude of approximately 29 degrees. How much do you estimate a man with a mass of 80 kg weighs on reaching the top?

P1.11   Two rings were built from the same wire, one with a radius of r and the other with a radius of $2r$. If the rings are placed on the same plane and touching each other, what will the distance be between the mass center of the set and the tangent point?

P1.12   How far from the center of a homogeneous semicircle is its mass center?

P1.13   Throughout the discussion on methodology, it was repeatedly said that the first step to solving a problem in dynamics is after, of course, characterizing the system to be analyzed, to identify the forces and torques acting on the system. Why is that?

P1.14   How would you denote, from the suggested general notation scheme, the kinetic energy of a body C in a given reference frame $\mathcal{R}$? And how would you denote the resultant moment of a vectors system $\mathcal{V}$ with respect to the axis x_1?

Vectors and Moments

Chapter 2

Vector functions are present in all mechanics. Forces, torques, velocities, angular velocities, accelerations, momenta, and angular momenta are some of the major examples of these kinds of quantities in the subject scope. Among the vector functions, some can be distinguished where the vector is directly associated with a certain point in space of those where this association is not significant. So, for instance, the velocity of a particle is a vector associated with the point in space occupied by it at each instant, while a torque applied to a rigid body is not necessarily associated with any particular point of the body.

This chapter addresses an especially important kind of vector in mechanics: the *moment* vectors. A torque applied to a body is a moment vector; the angular momentum of a body with respect to a given point is also a moment vector. Although torques and angular momenta are different concepts, the vector handling of both is exactly the same and the two functions are discussed together here.

The general purpose of this chapter is the study of *vector systems*. On the one hand, it seeks to give the reader the basic tools to correctly model the forces applied to a mechanical system. On the other, it offers a unified approach to the handling of vectors and their moments, which will make it easier to understand the dynamic properties of a mechanical system — especially the concepts of momentum and angular momentum — facilitating the formulation of equations that govern their motion.

For a systematic and unified approach, Section 2.1 discusses the free, sliding, and bound vector concepts while Section 2.2 defines the moment of a sliding or bound vector with respect to a point or axis, with examples. Section 2.3 introduces the fairly general concept of a vector system, including free and sliding (or bound) vectors and defines the resultant and resultant moment with respect to a point or axis, according to this general approach. The formulation is different from that usually found in the literature and has the advantage of suppressing ambiguities that, for instance, are found when discussing torques applied to a rigid body. Section 2.4 addresses the equivalence of vector systems and the reduction of systems at a given point. It is shown that any vector system can be substituted by a simpler system consisting of just one pair of vectors. Some special systems are also discussed, such as the couple and the null system. Section 2.5 shows the existence of the central axis of a vector system with a nonnull resultant and studies its properties and applications. Section 2.6 specifically discusses the force and torque systems. No attempt has been made to study statics but rather teach the reader how to model the forces and torques acting on a given mechanical system. The contact forces are discussed, paying special attention to the links and phenomenon of friction, field forces, and torques applied to a rigid body.

2.1 Free, Sliding, and Bound Vectors

In a three-dimensional Euclidean space, a given vector can be expressed in three components; if $\mathbf{v}$ is any vector and $\mathbf{n}_1, \mathbf{n}_2, \mathbf{n}_3$ is a basis of orthonormal vectors, the scalar components of $\mathbf{v}$ on this base, $v_j = \mathbf{v} \cdot \mathbf{n}_j$, $j = 1, 2, 3$, where the dot '·' designates *scalar product* (see Appendix A), fully determine the vector $\mathbf{v}$. In its geometric representation, reference is usually made to its elements: magnitude and direction. If the vectors $\mathbf{u}$ and $\mathbf{v}$ are equal, they must have both elements equal and their respective components are necessarily equal on the same basis; in other words,

$$\mathbf{u} = \mathbf{v} \quad \text{if and only if} \quad u_j = v_j, \quad j = 1, 2, 3. \tag{1.1}$$

Furthermore, all algebra for the vectors can be expressed in terms of their components on an arbitrary basis (see Appendix A). Vectors like those described above are called *free vectors*. Examples of free vectors are the angular velocity of a rigid body and a torque applied to a rigid body.

The effect of the action of a force on a rigid body depends on the former's line of action. As discussed in Chapter 1, Newton's third law states, among other things, that, given two particles P and Q, the force exerted, say, by Q over P is a vector associated to the straight line defined by P and Q. In other words, something besides the three components of the vector on a given basis must be specified to fully describe the applied force. So, from a dynamics viewpoint, two forces will be distinguished with the same components — therefore, with equal vectors — and different lines of action. Vectors associated to a certain straight line in the space are called *sliding vectors*. The characterization of a sliding vector requires its components on a given basis and the description of its line of action (the parameters of the equation of this straight line, coordinates of a point on the straight line, or any other form of determination). Two vectorially equal sliding vectors (equality in the usual sense, between free vectors) and associated to the same line of action are called *equivalents* or *equipollents*. Examples of sliding vectors are a force applied on a rigid body and the flow velocity of a fluid in a pipe with a uniform section.

The effect of a force on a deformable body depends, in addition to its line of action, on the point to which the force is applied. Vectors associated to an application point will be called *bound vectors*. To characterize a bound vector one must know its components on a given basis and the coordinates of its application point. Examples of bound vectors are the momentum of a particle and a force applied to the end of a spring.

All vector algebra is defined for free vectors; sums, scalar products, cross products, and other known operations have results exclusively dependent on the respective components of the vectors, thereby constituting free vectors. In other words, whatever the operation between vectors, following the rules of vector algebra, which results in a vector, this will necessarily be a free vector, since it will only depend on the components of the vectors involved in the operation. Algebra of sliding vectors or bound vectors is not, however, prohibited; the operation will only be done as if the vectors are free, and the result must necessarily be a free vector.

Example 1.1   Let us assume that forces $\mathbf{F}_1$ and $\mathbf{F}_2$ and torque $\mathbf{T}$ act on block B, illustrated in Fig. 1.1. $\mathbf{F}_1$ is applied along the axis x.

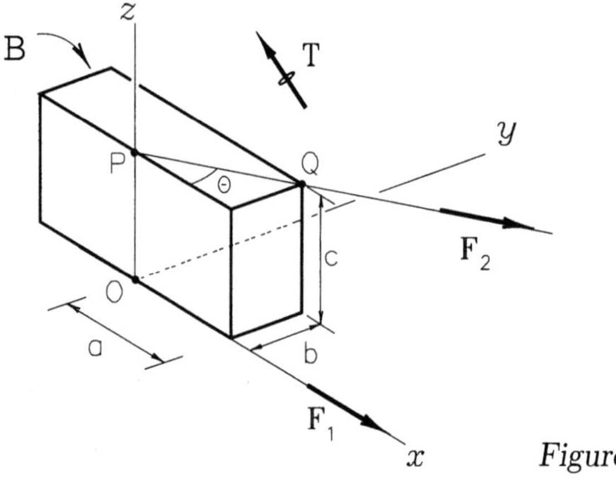

Figure 1.1

If the block can be considered as a rigid body, it makes no difference which is the point of the segment of x inside the block where the force is applied. $\mathbf{F}_1$ is, therefore, a sliding vector associated with axis x and its full characterization is given by its components — $(F_1, 0, 0)$, in the system of Cartesian axes in the figure — and by the axis with which it is associated, in the case x. Force $\mathbf{F}_2$ is also a sliding vector, associated with the straight line that contains points P and Q. It can be fully characterized, for example, by its magnitude, F_2, its direction (from P to Q), and the equation of its line of action: $bx = ay$; $z = c$. Torque $\mathbf{T}$ can be applied at any point of the block; therefore forming a free vector, characterized, for example, by its components (T_1, T_2, T_3). The vector sum $\mathbf{F}_1 + \mathbf{F}_2 = \big((F_1 + F_2\cos\theta), F_2\sin\theta, 0\big)$ is a free vector, not associated, therefore, with any straight line in space. The cross product $\mathbf{p}^{P/O} \times \mathbf{F}_2 = cF_2(-\sin\theta, \cos\theta, 0)$, where $\mathbf{p}^{P/O}$ is the position vector of point P with respect to O, is also a free vector.

2.2 Moments

Let us consider $\mathbf{v}$ as a sliding vector, associated with a straight line r,

O any point in space, and P an arbitrary point on r (see Fig. 2.1). The cross product of vector **p**, position of P with respect to O, with vector **v**, is a free vector, called the *moment of* **v** *with respect to* O

$$\mathbf{M}^{v/O} \rightleftharpoons \mathbf{p} \times \mathbf{v}. \tag{2.1}$$

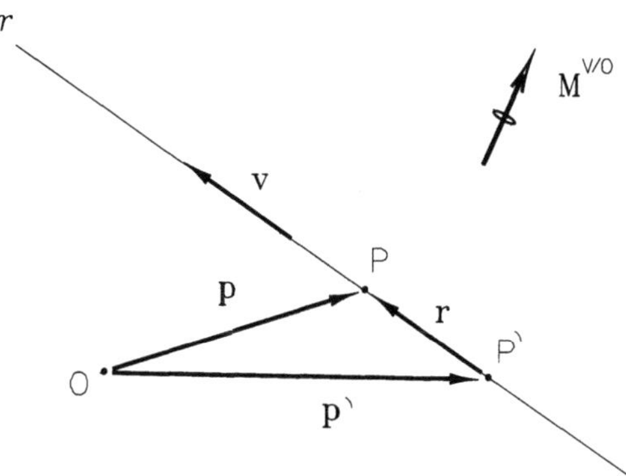

Figure 2.1

Of course, only a sliding vector (or bound vector, a particular case of sliding vector) admits a moment with respect to one point; the position vector **p** is not defined for a free vector. The moment of a vector **v** is always a free vector and orthogonal to **v**. In fact, according to Eq. (2.1), the moment results in an algebraic operation and, as such, does not define a line of action for its result; moreover, as this operation is a cross product, the resulting vector must necessarily be orthogonal to **v** (see Appendix A).

The moment of a vector with respect to a point will be null if the vector is null or if the line of action of the vector contains the point. In fact, product **p** × **v** will be null if one of the vectors is null or if **v** is parallel to **p**.

Lastly, it is worth noting that the moment of a vector with respect to the point O is independent of point P chosen on the line of action of **v**. To check this, one only needs to choose any other point P′ over r and see that $\mathbf{p}' \times \mathbf{v} = \mathbf{p} \times \mathbf{v} - \mathbf{r} \times \mathbf{v} = \mathbf{p} \times \mathbf{v}$, since $\mathbf{r} \times \mathbf{v} = 0$ (see Fig. 2.1).

The physical dimension of the moment vector will always be equal to the physical dimension of the sliding vector that originated it, multiplied by dimension [L], a characteristic of the position vector **p**, that is,

$$\mathrm{Dim}\,[\mathbf{M}^{\mathbf{v}/O}] = \mathrm{Dim}\,[\mathbf{v}] \times [\mathrm{L}]. \tag{2.2}$$

If **F** is a *force*, a sliding or bound vector with dimension $[\mathrm{MLT}^{-2}]$, that is, Newtons (N), in SI units, its moment with respect to a point will be a *torque*, with dimension $[\mathrm{ML}^2\mathrm{T}^{-2}]$, that is, Newtons-meter (Nm), in the same units. If **G** is a *momentum* vector of a particle, a bound vector with dimension $[\mathrm{MLT}^{-1}]$, its moment with respect to a point will be the *angular momentum* vector of the particle with respect to the point, with dimension $[\mathrm{ML}^2\mathrm{T}^{-1}]$.

Given a point O, an axis (straight line) E passing through O and parallel to a certain adimensional unit vector **n** and a sliding vector **v** associated with the line of action r (see Fig. 2.2), the *moment of the vector* **v** *with respect to the axis* E, $\mathbf{M}^{\mathbf{v}/E}$, is defined as the component of the moment of vector **v** with respect to the point O, in the direction of the axis, that is (see Appendix A),

$$\mathbf{M}^{\mathbf{v}/E} \rightleftharpoons \mathbf{M}^{\mathbf{v}/O} \cdot \mathbf{n}\,\mathbf{n}. \tag{2.3}$$

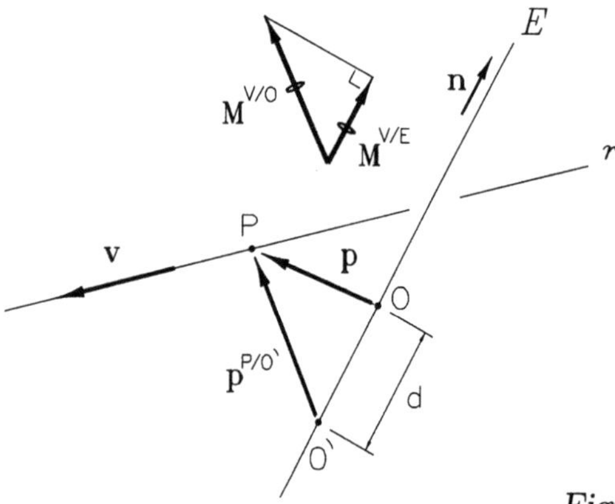

Figure 2.2

The moment of a vector **v** with respect to an axis E is a free vector (result of an algebraic operation) parallel to the axis (its direction is given by the unit vector **n**). The physical dimension of $\mathbf{M}^{\mathbf{v}/E}$ is the same as $\mathbf{M}^{\mathbf{v}/O}$, since **n** is adimensional. Lastly, the moment of a vector with respect to an axis does not depend on the point on the axis chosen for its calculation, which justifies no reference to point O in the notation made for a moment with respect to an axis. In fact, if O' is another point on the axis E (see Fig. 2.2), $\mathbf{M}^{\mathbf{v}/O'} = \mathbf{p}^{P/O'} \times \mathbf{v}$, where $\mathbf{p}^{P/O'}$ is the position vector of point P with respect to point O', as shown. But if d is the distance between points O and O', $\mathbf{p}^{P/O'} = \mathbf{p} + d\mathbf{n}$, so $\mathbf{M}^{\mathbf{v}/O'} \cdot \mathbf{n}\,\mathbf{n} = \mathbf{p} \times \mathbf{v} \cdot \mathbf{n}\,\mathbf{n} + d\mathbf{n} \times \mathbf{v} \cdot \mathbf{n}\,\mathbf{n}$ and, as the mixed product $\mathbf{n} \times \mathbf{v} \cdot \mathbf{n}$ is null, then $\mathbf{M}^{\mathbf{v}/O'} \cdot \mathbf{n}\,\mathbf{n} = \mathbf{M}^{\mathbf{v}/O} \cdot \mathbf{n}\,\mathbf{n} = \mathbf{M}^{\mathbf{v}/E}$ (see Appendix A).

Example 2.1 Consider a particle P, of mass m, moving in the reference system shown in Fig. 2.3. At the instant represented, the position of P is given by the Cartesian coordinates $(s\cos\theta, y_0, s\sin\theta)$; its magnitude velocity v is parallel to straight line r; and a force of magnitude F, directed to point O, acts on the particle. The momentum vector of the particle, defined as $\mathbf{G} = mv\mathbf{n}$, is bound to P.

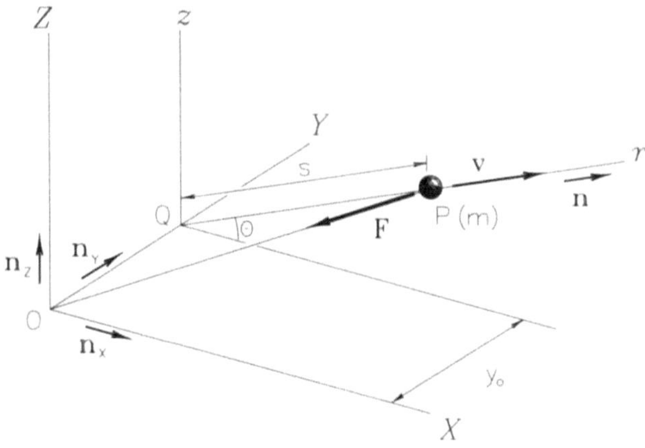

Figure 2.3

The moment of this vector with respect to point O is

$$\mathbf{M}^{\mathbf{G}/O} = \mathbf{p}^{P/O} \times \mathbf{G} = \mathbf{p}^{Q/O} \times \mathbf{G}$$
$$= y_0\mathbf{n}_y \times mv(\cos\theta\mathbf{n}_x + \sin\theta\mathbf{n}_z)$$
$$= mvy_0(\sin\theta\mathbf{n}_x - \cos\theta\mathbf{n}_z).$$

The moment of vector $\mathbf{G}$ with respect to axis X is

$$\mathbf{M}^{\mathbf{G}/X} = \mathbf{M}^{\mathbf{G}/O} \cdot \mathbf{n}_x\,\mathbf{n}_x = mvy_0\sin\theta\mathbf{n}_x,$$

and the moment with respect to axis Z is

$$\mathbf{M}^{G/Z} = \mathbf{M}^{G/O} \cdot \mathbf{n}_z\,\mathbf{n}_z = -mvy_0\cos\theta\mathbf{n}_z.$$

Force F, applied on P, is a bound vector at P. The moment of this force with respect to point O is null, because the support of the force passes through O. The moment of this force with respect to point Q is

$$\mathbf{M}^{F/Q} = \mathbf{p}^{P/Q} \times \mathbf{F}$$
$$= s(\cos\theta\mathbf{n}_x + \sin\theta\mathbf{n}_z)$$
$$\times \frac{-F}{(s^2 + y_0^2)^{1/2}}(s\cos\theta\mathbf{n}_x + y_0\mathbf{n}_y + s\sin\theta\mathbf{n}_z)$$
$$= \frac{Fsy_0}{(s^2 + y_0^2)^{1/2}}(\sin\theta\mathbf{n}_x - \cos\theta\mathbf{n}_z).$$

Note that the moment of $\mathbf{F}$ with respect to Q can also be obtained (and more easily) by

$$\mathbf{M}^{F/Q} = \mathbf{p}^{O/Q} \times \mathbf{F}$$
$$= -y_0\mathbf{n}_y \times \frac{-F}{(s^2 + y_0^2)^{1/2}}(s\cos\theta\mathbf{n}_x + y_0\mathbf{n}_y + s\sin\theta\mathbf{n}_z)$$
$$= \frac{Fsy_0}{(s^2 + y_0^2)^{1/2}}(\sin\theta\mathbf{n}_x - \cos\theta\mathbf{n}_z).$$

The moment of vector $\mathbf{F}$ with respect to the vertical axis z passing through Q is

$$\mathbf{M}^{F/z} = \mathbf{M}^{F/Q} \cdot \mathbf{n}_z\,\mathbf{n}_z = -\frac{Fsy_0}{(s^2 + y_0^2)^{1/2}}\cos\theta\mathbf{n}_z,$$

and the moment of $\mathbf{F}$ with respect to axis Y is

$$\mathbf{M}^{F/Y} = \mathbf{M}^{F/Q} \cdot \mathbf{n}_y\,\mathbf{n}_y = 0.$$

The moment of a vector $\mathbf{v}$ with respect to an axis E passing through a point O will be null if the mixed product $\mathbf{p} \times \mathbf{v} \cdot \mathbf{n}$ is null [see Eqs. (2.1) and (2.3)]. This will happen if $\mathbf{v}$ is null or $\mathbf{p}$ and $\mathbf{v}$ are parallel — and in this case O belongs to the straight line r, the line of action of $\mathbf{v}$, so the axis and straight line are concurrent — or, also, if $\mathbf{v}$ and $\mathbf{n}$ are parallel, meaning that the axis and straight line are parallel (see Fig. 2.2). In short, the moment of a nonnull vector with respect to an axis will be null whenever the vector's line of action and axis are *coplanar*. (There is also another trivial case where the moment vector with respect to an axis vanishes. Which is that?)

The moment of a sliding vector **v** with respect to point O is equal to the vector sum of the moments of the vector with respect to three mutually orthogonal axes that intercept at O. Thus, if $\mathbf{n}_1, \mathbf{n}_2, \mathbf{n}_3$ are orthonormal vectors, parallel to the axes x_1, x_2, x_3, passing through O, then $\mathbf{M}^{v/x_j} = \mathbf{M}^{v/O} \cdot \mathbf{n}_j \, \mathbf{n}_j$, $j = 1, 2, 3$, therefore (see Appendix A),

$$\mathbf{M}^{v/O} = \sum_{j=1}^{3} \mathbf{M}^{v/x_j}. \tag{2.4}$$

Example 2.2   Returning to the preceding example (see Fig. 2.3), the moment of vector **G** with respect to axis Y is null because the **G** line of action intercepts Y at point Q. The moments of vector **F** with respect to axes X, Y, or Z are null because the line of action of **F** intercepts the axes at O. If the angle θ is null, the moment of vector **G** with respect to axis X will also be null since the **G** line of action and axis X will be parallel. In fact, for $\theta = 0$, $\mathbf{M}^{G/O} = -mvy_0\mathbf{n}_z$ and $\mathbf{M}^{G/O} \cdot \mathbf{n}_x = 0$. It is also easy to see, looking at the results of the above example, that $\mathbf{M}^{G/O} = \mathbf{M}^{G/X} + \mathbf{M}^{G/Y} + \mathbf{M}^{G/Z}$, as Eq. (2.4) establishes for any θ value.

2.3 Vector Systems

Consider a set consisting of n sliding vectors of the same physical dimension, $\mathbf{v}_i$, associated with the line of actions r_i, $i = 1, 2, \ldots, n$, respectively, and m free vectors $\mathbf{M}_j$, $j = 1, 2, \ldots, m$, all with the dimension of a moment of a vector from the vector category $\mathbf{v}_i$. A set of vectors defined as such will be called a *vector system*. (Let us not forget that bound vectors are a particular case of sliding vectors and that, therefore, some or even all the vectors $\mathbf{v}_i$ above may consist of bound vectors.)

Example 3.1   The arm shown in Fig. 3.1 is hinged at its end A and can turn freely around the axis x_3. Force F is applied at end B, with components in the three coordinated directions; consider that the vertical force P, the weight of the arm, is applied at point O, mass center of the arm; assume the action of three force components, $\mathbf{F}_1$, $\mathbf{F}_2$, and $\mathbf{F}_3$, on end A, as shown. Lastly, as the arm is free to turn exclusively around the axis x_3, two torque components, $\mathbf{T}_1$ and $\mathbf{T}_2$, shall be applied to it.

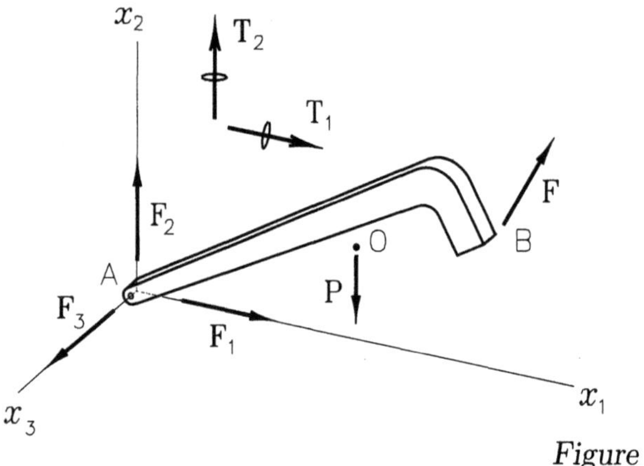

Figure 3.1

This group of seven vectors — five sliding vectors (the forces) and two free vectors (the torques) — forms a vector system, with $n = 5$ and $m = 2$. (If the reader did not clearly understand why these vectors and not others are involved, do not worry: The recognition of the forces and torques applied to a rigid body will be discussed later in this chapter. What matters for now is to recognize that this is a vector system.)

If $\mathcal{V}$ is a vector system consisting of n sliding vectors $\mathbf{v}_i$, $i = 1, 2, \ldots, n$ and m free vectors $\mathbf{M}_j$, $j = 1, 2, \ldots, m$, the vector sum of the n sliding vectors is called *resultant* of the system, that is,

$$\mathbf{R}(\mathcal{V}) \rightleftharpoons \sum_{i=1}^{n} \mathbf{v}_i. \tag{3.1}$$

It is never too late to insist that the resultant of a system, obtained from a usual vector sum, is a free vector, not associated, therefore, with any line of action and, as such, not having defined its moment with respect to any point in the space.

Example 3.2   Figure 3.2 illustrates a system of vectors $\mathcal{V}$ associated to a cube with an edge with a length of $2\,\mathrm{m}$. The vector $\mathbf{v}_1$ with magnitude $5u$ is associated with the axis x_3; the vector $\mathbf{v}_2$ with magnitude $10u$ is associated with the straight line containing A and B; and the vector $\mathbf{v}_3$, with magnitude $15u$, is associated with the straight line containing B and C, with the directions shown, u being a certain physical unit. The system also consists of the free vectors $\mathbf{M}_1$, parallel to axis x_1, with magnitude $20u$m, and $\mathbf{M}_2$, parallel to axis E, with magnitude $30\sqrt{2}u$m, with the directions indicated. The resultant $\mathbf{R}$ of this system will be the free vector

$$\mathbf{R} = \mathbf{v}_1 + \mathbf{v}_2 + \mathbf{v}_3 = 5u(-3\mathbf{n}_1 + 2\mathbf{n}_2 + \mathbf{n}_3).$$

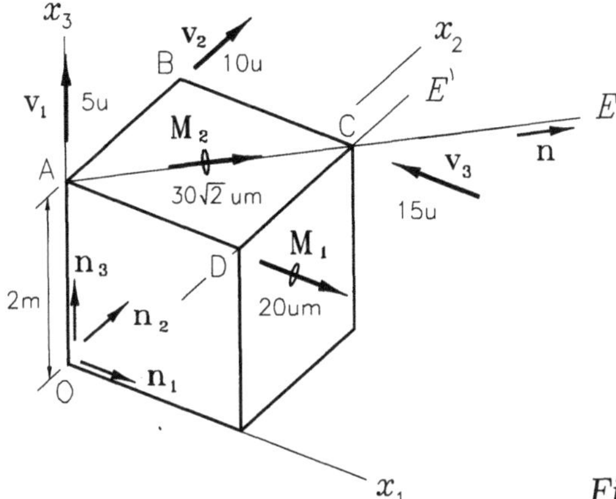

Figure 3.2

The *resultant moment* of a vector system $\mathcal{V}$ with respect to a point O is defined as the vector sum of the moments with respect to O of the sliding vectors of $\mathcal{V}$ with the free vectors of $\mathcal{V}$, that is,

$$\mathbf{M}^{\mathcal{V}/O} \rightleftharpoons \sum_{i=1}^{n} \mathbf{M}^{\mathbf{v}_i/O} + \sum_{j=1}^{m} \mathbf{M}_j. \tag{3.2}$$

It is clear that the resultant moment of a system $\mathcal{V}$ is also a free vector.

The resultant moment of a vector system $\mathcal{V}$ with respect to an axis E, passing through a point O and parallel to an adimensional unit vector $\mathbf{n}$, is defined as the component of the resultant moment of the system at the point, in direction of the axis, that is,

$$\mathbf{M}^{\mathcal{V}/E} \rightleftharpoons \mathbf{M}^{\mathcal{V}/O} \cdot \mathbf{n}\,\mathbf{n}. \tag{3.3}$$

Example 3.3   Returning to the previous example (see Fig. 3.2), the resultant moment of the system with respect to the point O is

$$\mathbf{M}^{\mathcal{V}/O} = \mathbf{M}^{\mathbf{v}_1/O} + \mathbf{M}^{\mathbf{v}_2/O} + \mathbf{M}^{\mathbf{v}_3/O} + \mathbf{M}_1 + \mathbf{M}_2$$

$$= 0 + 2\text{m}\,\mathbf{n}_3 \times 10u\,\mathbf{n}_2 + 2\text{m}\,(\mathbf{n}_2 + \mathbf{n}_3) \times (-15u)\,\mathbf{n}_1$$

$$+\, 20u\text{m}\,\mathbf{n}_1 + 30u\text{m}\,(\mathbf{n}_1 + \mathbf{n}_2)$$

$$= 30u\text{m}\,(\mathbf{n}_1 + \mathbf{n}_3).$$

The resultant moment with respect to the axis x_1 will be $\mathbf{M}^{\mathcal{V}/x_1} = 30u\text{mn}_1$ and the resultant moment with respect to the axis x_2 will be null. The resultant moment of this system with respect to the axis E, which contains A and C vertices, can be directly computed by

$$\mathbf{M}^{\mathcal{V}/E} = \mathbf{M}_1 \cdot \mathbf{n}\,\mathbf{n} + \mathbf{M}_2 = 40\sqrt{2}u\text{m}\,\mathbf{n} = 40u\text{m}\,(\mathbf{n}_1 + \mathbf{n}_2),$$

since the lines of action of the sliding vectors of the system intercept all on axis E.

Once the resultant and resultant moment with respect to any given point O of a vector system $\mathcal{V}$ are known, the resultant moment of a system is determined with respect to any other point. This is guaranteed by a very simple and extremely useful relationship established on what has usually been called the

Moments Transport Theorem. *The resultant moment of a vector system $\mathcal{V}$ with respect to any point O is equal to the vector sum of the resultant moment of the system with respect to a given point O' with the moment, with respect to O, of a sliding vector vectorially equal to the resultant $\mathbf{R}$ of $\mathcal{V}$ and associated with a straight line passing through O', that is,*

$$\mathbf{M}^{\mathcal{V}/O} = \mathbf{M}^{\mathcal{V}/O'} + \mathbf{p}^{O'/O} \times \mathbf{R}. \tag{3.4}$$

The derivation of the theorem is simple, by basing it on the definitions of the resultant moment of a vector system with respect to one point, Eq. (3.2), and resultant of a system, Eq. (3.1). In fact

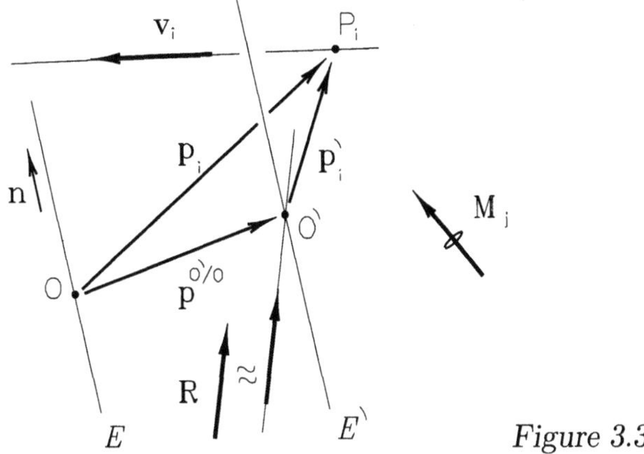

Figure 3.3

(see Fig. 3.3),

$$\mathbf{M}^{\mathcal{V}/O} = \sum_{i=1}^{n} \mathbf{p}_i \times \mathbf{v}_i + \sum_{j=1}^{m} \mathbf{M}_j$$

$$= \sum_{i=1}^{n} \mathbf{p}'_i \times \mathbf{v}_i + \sum_{i=1}^{n} \mathbf{p}^{O'/O} \times \mathbf{v}_i + \sum_{j=1}^{m} \mathbf{M}_j$$

$$= \left(\sum_{i=1}^{n} \mathbf{p}'_i \times \mathbf{v}_i + \sum_{j=1}^{m} \mathbf{M}_j \right) + \mathbf{p}^{O'/O} \times \sum_{j=1}^{m} \mathbf{v}_i$$

$$= \mathbf{M}^{\mathcal{V}/O'} + \mathbf{p}^{O'/O} \times \mathbf{R}. \qquad \blacksquare$$

This result indicates that two free vectors — the resultant, invariant with the point, and a resultant moment, dependent on the chosen point — fully characterize a vector system consisting of an arbitrary number of sliding (or bound) and free vectors.

Equation (3.4) can be extended to resultant moments of a system with respect to different axes. So, if $\mathbf{n}$ is an adimensional unitary vector, defining a direction in space, the resultant moments of a vector system $\mathcal{V}$, with respect to two axes parallel to $\mathbf{n}$, passing through the points O and O' (see Fig. 3.3) are related by

$$\mathbf{M}^{\mathcal{V}/E} = \mathbf{M}^{\mathcal{V}/E'} + \left(\mathbf{p}^{O'/O} \times \mathbf{R} \right) \cdot \mathbf{n}\,\mathbf{n}. \qquad (3.5)$$

Equation (3.5) is the result of projecting Eq. (3.4) in the direction $\mathbf{n}$. The second term on the right can be interpreted as the moment with respect to the axis E of a sliding vector vectorially equal to the resultant of $\mathcal{V}$, whose line of action passes through O$'$. This result is also known as the *parallel axes theorem.*

Example 3.4   Returning to Example 3.2 (see Fig. 3.2), the resultant moment of the system with respect to point A can be obtained, through Eq. (3.4), from

$$\mathbf{M}^{\mathcal{V}/A} = \mathbf{M}^{\mathcal{V}/O} + \mathbf{p}^{O/A} \times \mathbf{R}$$
$$= 30um\,(\mathbf{n}_1 + \mathbf{n}_3) + (-2m)\,\mathbf{n}_3 \times 5u\,(-3\mathbf{n}_1 + 2\mathbf{n}_2 + \mathbf{n}_3)$$
$$= 10um\,(5\,\mathbf{n}_1 + 3\mathbf{n}_2 + 3\,\mathbf{n}_3).$$

The resultant moment of the system with respect to the horizontal axis E', which passes through C and D, is, according to Eq. (3.5),

$$\mathbf{M}^{\mathcal{V}/E'} = \mathbf{M}^{\mathcal{V}/x_2} + \left(\mathbf{p}^{O/D} \times \mathbf{R}\right) \cdot \mathbf{n}_2\,\mathbf{n}_2$$
$$= 0 + \left((-2m)(\mathbf{n}_1 + \mathbf{n}_3) \times 5u\,(-3\,\mathbf{n}_1 + 2\,\mathbf{n}_2 + \mathbf{n}_3)\right) \cdot \mathbf{n}_2\,\mathbf{n}_2$$
$$= 40um\,\mathbf{n}_2.$$

When a vector system consists exclusively of sliding (or bound) vectors, it is called a *simple system.* For a simple system, therefore $m = 0$ and the resultant moment of the system with respect to a point or axis will be the sum of the moments of the sliding vectors comprising the system, with respect to the point or axis.

Some simple systems consist of an infinite number of sliding vectors, each with an infinitesimal magnitude. Systems of this kind are called *distributed systems.* If $d\mathbf{v}$ is a vector of a distributed system $\mathcal{V}$ (see Fig. 3.4), its resultant $\mathbf{R}$ is defined as

$$\mathbf{R} \rightleftharpoons \int_{\mathcal{V}} d\mathbf{v}. \tag{3.6}$$

The resultant moment of a distributed simple system $\mathcal{V}$ with respect to a point O is defined as

$$\mathbf{M}^{\mathcal{V}/O} \rightleftharpoons \int_{\mathcal{V}} \mathbf{p} \times d\mathbf{v}, \tag{3.7}$$

where $\mathbf{p}$ is the position vector with respect to point O, of an arbitrary point on the line of action of $d\mathbf{v}$ (see Fig. 3.4).

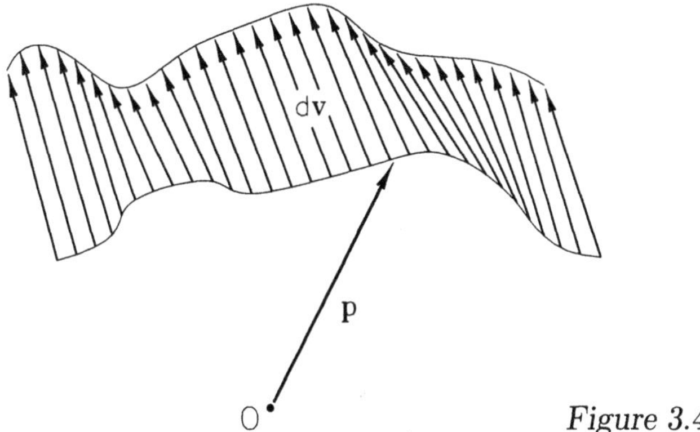

Figure 3.4

The resultant moment of a distributed system with respect to an axis E is defined as in Eq. (3.3), that is, it is the component, in the direction of the axis, of the resultant moment of the system with respect to any point on the same axis.

Example 3.5　Figure 3.5 shows the diagram of a vertical gate, with height a, which holds the water of a tank.

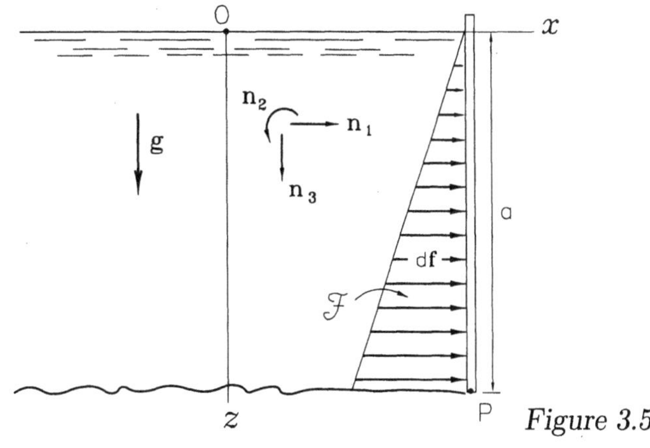

Figure 3.5

The pressure exerted by the fluid on the gate over the atmospheric pressure depends on the depth z, according to the hydrostatic relationship $p(z) = \rho g z$, where p is the pressure, ρ the density of the fluid, and g the gravitational acceleration magnitude. As the pressure varies exclusively with the vertical coordinate on each horizontal surface element, with an area $dA = l\,dz$, where l is the width (uniform) of the gate, an infinitesimal force $d\mathbf{f} = pdA\,\mathbf{n}_1 = \rho glzdz\,\mathbf{n}_1$ will be applied. So a distributed system $\mathcal{F}$, consisting of forces $d\mathbf{f}$ associated to horizontal lines of action (direction x), acts on the gate. The resultant force of the action of the fluid on the gate will be the resultant of this disributed system, given, according to Eq. (3.6), by (assuming ρ and g as constants)

$$\mathbf{R} = \int_0^a d\mathbf{f} = \rho gl \int_0^a zdz\,\mathbf{n}_1 = \frac{1}{2}\rho gla^2\,\mathbf{n}_1.$$

The resultant moment of this system with respect, say, to point P is, according to Eq. (3.7),

$$\mathbf{M}^{\mathcal{F}/P} = \int_0^a \mathbf{p} \times d\mathbf{f} = \rho gl \int_0^a -(a-z)\,\mathbf{n}_3 \times z\,dz\,\mathbf{n}_1 = -\frac{1}{6}\rho gla^3\,\mathbf{n}_2.$$

Example 3.6   Bar B, pivoted on one end at the fixed point O, moves on the plane of the figure with the angle θ varying with time according to the rate $\omega = d\theta/dt$ (see Fig. 3.6).

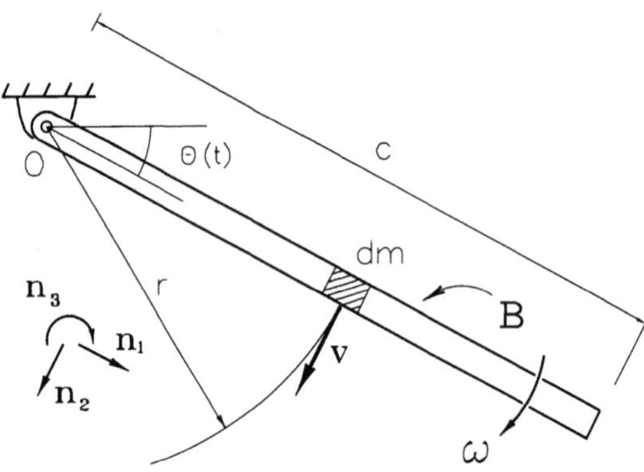

Figure 3.6

The bar is homogeneous with mass m and length c. Each element of B will have a mass $dm = \rho dr$, ρ being its density (mass per unit of length), and the velocity $v = r\omega$ (believe me), in the direction $\mathbf{n}_2$. The momentum vectors of the elements of B, $d\mathbf{G} = \mathbf{v}\,dm$, consist then of a simple distributed system $\mathcal{G}$, whose resultant,

$$\mathbf{R} = \int_B d\mathbf{G} = \int_0^c \rho\omega r\,dr\,\mathbf{n}_2 = \frac{1}{2}\rho\omega c^2\,\mathbf{n}_2 = \frac{1}{2}m\omega c\,\mathbf{n}_2,$$

is the momentum of the bar. Its angular momentum with respect to point O, is the resultant moment of the system $\mathcal{G}$ with respect to O, given by

$$
\begin{aligned}
\mathbf{M}^{\mathcal{V}/O} &= \int_0^c r\,\mathbf{n}_1 \times \rho\omega r\,dr\,\mathbf{n}_2 \\
&= \rho\omega \int_0^c r^2\,dr\,\mathbf{n}_3 \\
&= \frac{1}{3} m\omega c^2\,\mathbf{n}_3.
\end{aligned}
$$

The angular momentum vector of the body with respect to the axis, say, x_3 (axis passing through O, parallel to the unit vector $\mathbf{n}_3$) will be the component, in this direction, of the angular momentum vector with respect to the point O, that is,

$$
\mathbf{M}^{\mathcal{V}/x_3} = \mathbf{M}^{\mathcal{V}/O} \cdot \mathbf{n}_3\,\mathbf{n}_3 = \frac{1}{3} m\omega c^2\,\mathbf{n}_3.
$$

2.4 Equivalent Systems

Two vector systems $\mathcal{V}$ and $\mathcal{V}'$ are said to be *equivalent* if their resultants are equal and if their resultant moments are also equal with respect to some point O, that is,

$$
\mathcal{V} \approx \mathcal{V}' \rightleftharpoons
\begin{cases}
\mathbf{R}(\mathcal{V}) = \mathbf{R}(\mathcal{V}'), \\[2mm]
\mathbf{M}^{\mathcal{V}/O} = \mathbf{M}^{\mathcal{V}'/O}, & \text{for some O,}
\end{cases}
\tag{4.1}
$$

where the symbol '$\approx$' means equivalence.

It is natural that the concept of equivalence is expected to be stronger, such as systems being equivalent with equal resultants and equal resultant moments for *any* point in space. It is easy to see, however, that this is exactly what will happen with systems that fulfill Eq. (4.1); otherwise, let us see: If $\mathcal{V}$ and $\mathcal{V}'$ are equivalent, from the moments transport theorem, Eq. (3.4), then, for any point O$'$

$$
\mathbf{M}^{\mathcal{V}/O'} = \mathbf{M}^{\mathcal{V}/O} + \mathbf{p}^{O/O'} \times \mathbf{R}(\mathcal{V}) = \mathbf{M}^{\mathcal{V}'/O} + \mathbf{p}^{O/O'} \times \mathbf{R}(\mathcal{V}') = \mathbf{M}^{\mathcal{V}'/O'}, \tag{4.2}
$$

as desired, that is, the resultants of the two systems being equal and their resultant moments also being equal for a given point, then the resultant moments will also be equal to each other for any other arbitrarily chosen point.

If $\mathcal{V}$ and $\mathcal{V}'$ are equivalent systems, their moments with respect to any axis E will be equal. In fact, if O is a point on the axis, $\mathbf{M}^{\mathcal{V}/O} = \mathbf{M}^{\mathcal{V}'/O}$ and the component of this relation in the direction of the axis, given by the unit vector $\mathbf{n}$, will be $\mathbf{M}^{\mathcal{V}/O} \cdot \mathbf{n}\,\mathbf{n} = \mathbf{M}^{\mathcal{V}'/O} \cdot \mathbf{n}\,\mathbf{n}$; therefore,

$$\mathbf{M}^{\mathcal{V}/E} = \mathbf{M}^{\mathcal{V}'/E}, \quad \text{for every axis } E \qquad \text{if } \mathcal{V} \approx \mathcal{V}'. \tag{4.3}$$

Example 4.1   Consider the system $\mathcal{V}$, consisting of sliding vectors $\mathbf{u}_1$ and $\mathbf{u}_2$, whose lines of action intercept point A, and by the free vector $\mathbf{M}_A$, orthogonal to the plane of the former (see Fig. 4.1). Also consider the vector system $\mathcal{V}'$, consisting of the sliding vector $\mathbf{v}$, whose line of action intercepts point B and is orthogonal to $\mathbf{n}_3$, with the direction shown, and by the free vector $\mathbf{M}_B$, parallel to $\mathbf{M}_A$. The magnitudes and directions are shown in the figure.

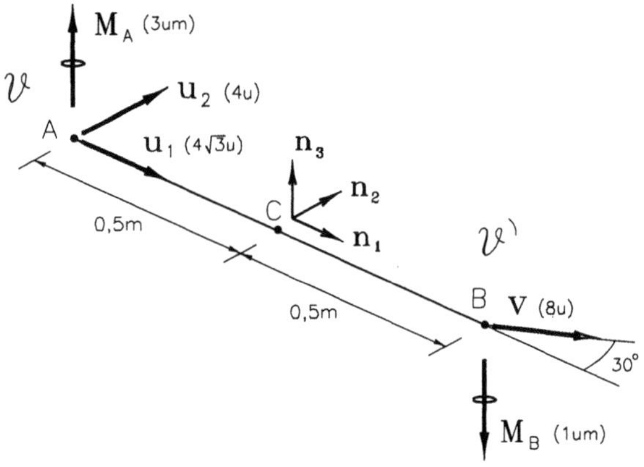

Figure 4.1

The resultant of system $\mathcal{V}$ can be expressed on the basis of $\mathbf{n}_1, \mathbf{n}_2, \mathbf{n}_3$ by

$$\mathbf{R} = \mathbf{u}_1 + \mathbf{u}_2 = 4u\left(\sqrt{3}\mathbf{n}_1 + \mathbf{n}_2\right),$$

and its resultant moment with respect, say, to point B is

$$\mathbf{M}^{\mathcal{V}/B} = \mathbf{p}^{A/B} \times \mathbf{u}_2 + \mathbf{M}_A = -u m\,\mathbf{n}_3.$$

The resultant of system $\mathcal{V}'$ is

$$\mathbf{R}' = \mathbf{v} = 4u\left(\sqrt{3}\mathbf{n}_1 + \mathbf{n}_2\right),$$

and its resultant moment with respect to point B is

$$\mathbf{M}^{\mathcal{V}'/B} = \mathbf{M}_B = -u m\,\mathbf{n}_3.$$

It results then that $\mathcal{V}$ and $\mathcal{V}'$ are equivalent. It is easy to see, for example, that both systems have a null resultant moment with respect to any axis passing through B and parallel to the plane defined by the directions of $\mathbf{n}_1$ and $\mathbf{n}_2$. (Choose any point and calculate the resultant moments of $\mathcal{V}$ and $\mathcal{V}'$ with respect to this point. What conclusion does one reach?)

Every vector system $\mathcal{V}$ has an infinite number of equivalent systems (does the reader agree?). The simplest of them will be, in general, systems consisting of a pair of vectors, one free and the other sliding. Now let us see: Taking a sliding vector equal to the resultant of $\mathcal{V}$ over a line of action passing through a given point Q, and a free vector equal to the resultant moment of $\mathcal{V}$ with respect to Q, we have a new system whose resultant, being equal to the single sliding vector that comprises it, is equal to the resultant of $\mathcal{V}$, and whose resultant moment with respect to point Q is also equal to the resultant moment of $\mathcal{V}$ with respect to Q. It is said, then, that *the system $\mathcal{V}$ was reduced to point* Q. Once a given system of vectors is reduced to point Q, as described in the above procedure, it can be easily reduced to any other point, using the moments transport theorem, Eq. (3.4). In fact, as the resultant is an invariant, one only needs to calculate the new resultant moment based on the previous one, using the theorem, to obtain the new reduction.

Example 4.2 Returning to the previous example (see Fig. 4.1), the system $\mathcal{V}'$ is a reduction of the system $\mathcal{V}$ at point B. The reduction of $\mathcal{V}$ at point C, intermediary between A and B, will consist of a vector equal to $\mathbf{R}$ applied to C and

$$\mathbf{M}^{\mathcal{V}/C} = \mathbf{p}^{A/C} \times \mathbf{u}_2 + \mathbf{M}_A = u m\,\mathbf{n}_3.$$

Note that the same reduction would be obtained from $\mathcal{V}'$, that is,

$$\mathbf{M}^{\mathcal{V}'/C} = \mathbf{p}^{B/C} \times \mathbf{v} + \mathbf{M}_B = u m\,\mathbf{n}_3.$$

In the more general case, as seen above, every vector system can be reduced to an arbitrary point, the reduction consisting of a pair of vectors: a sliding vector (equal to the resultant of the original system) and a free vector (equal to the resultant moment of the original system with respect to the point). Some systems, however, are even more easily reduced, as we will see ahead.

When a system $\mathcal{V}$ has a null resultant and nonnull resultant moment with respect to some point in space, it is called a *couple*. According to the moments transport theorem, Eq. (3.4), the resultant moment of a couple is the same for any point in the space, that is,

$$\mathbf{M}^{\mathcal{V}/O} = \mathbf{M}^{\mathcal{V}/O'} \quad \text{if} \quad \mathbf{R} = 0. \tag{4.4}$$

The *moment of the couple* is then an invariant that characterizes it fully. If $\mathcal{V}$ is a couple consisting of forces and torques, its resultant moment is called the *couple torque*.

Example 4.3   The mechanical system illustrated in Fig. 4.2 consists of a central element of mass $5m$, rigidly connected to four equally spaced spheres, two with mass m each and two with mass $2m$ each, in the configuration shown.

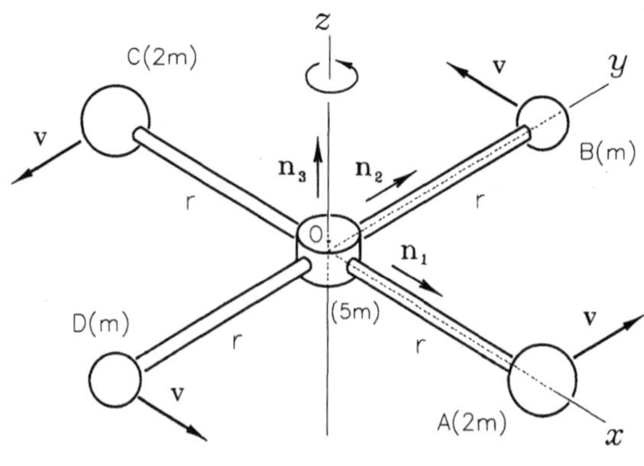

Figure 4.2

The system turns around the axis z at a constant rate so that each of the suspended masses has a velocity of magnitude v. Assuming dimensions so that all elements can be treated as particles, the set of momentum vectors forms a simple vector system with five elements, as follows: $\mathbf{G}_A = 2mv\mathbf{n}_2$, $\mathbf{G}_B = -mv\mathbf{n}_1$, $\mathbf{G}_C = -2mv\mathbf{n}_2$, $\mathbf{G}_D = mv\mathbf{n}_1$, and $\mathbf{G}_O = 0$. The resultant of this sytem is null and the vector system is, therefore, a couple. The resultant moment with respect to point O (the angular momentum of the set of particles with respect to O) is $\mathbf{M}^{\mathcal{V}/O} = 6mvr\mathbf{n}_3$. It is easy to see that the system's resultant moment is the same as for any other point in space.

When a vector system $\mathcal{V}$ has a nonnull resultant and a null resultant moment with respect to a given point O in space, its reduction to that point consists exclusively of a sliding vector vectorially equal to its resultant associated with a line of action passing through the point. It is easy to see that, according to Eq. (3.4), for all points on this support, the resultant moment of the system will also vanish.

Example 4.4   Figure 4.3 illustrates a cylinder floating on a fluid at rest. The system of forces exerted by the fluid on the cylindrical shell is a distributed simple system that, for a vertical cross section, will have the indicated aspect, with the force magnitude varying with depth and its direction always orthogonal to the surface of the cylinder. The lines of action of all components of this system intercept; then the symmetry axis of the cylinder and its resultant moment with respect to this axis will therefore be null. The geometry of the body also guarantees the symmetry of this system of forces in the longitudinal direction, with the consequence that the resultant moment of the system with respect to point O is also null.

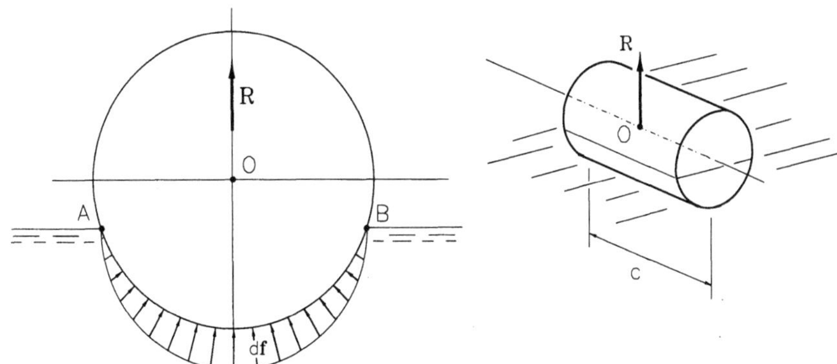

Figure 4.3

The resultant of the system,

$$\mathbf{R} = \int_{-c/2}^{c/2} \int_{A}^{B} d\mathbf{f},$$

will therefore be a vertical vector. The reduction of the system at point O will then consist exclusively of the vector $\mathbf{R}$ associated with the vertical line passing through O, as shown, constituting the *thrust* exerted by the fluid. (Note that the fluid also exerts a distributed force on the cylinder bases, but the symmetry guarantees that these forces cancel each other out and do not contribute to the thrust.)

When a system $\mathcal{V}$ has a null resultant and a null resultant moment with respect to a given point O, it is called a *null·system*. In fact, also according to Eq. (3.4), the resultant moment of a null system will be null for all points, and, consequently, for all axes in the space. As Example 4.4 illustrates, the system of all forces acting on the cylinder bases will constitute a null system.

Example 4.5   Consider the vector system $\mathcal{F}$ consisting of three forces and one torque, applied on a disk, described as follows: two forces, $\mathbf{F}$ and $\mathbf{F}'$, both of a magnitude equal to 5N, exerted by the two ropes, fixed at points B and B$'$ respectively; the weight $\mathbf{P}$ of the disk, vertical and applied on its center, of magnitude 8 N; and the torque $\mathbf{T}$, vertical, applied to the disk, of a magnitude equal to $9\sqrt{3}\,\mathrm{N\,cm}$, in the direction indicated (see Fig. 4.4).

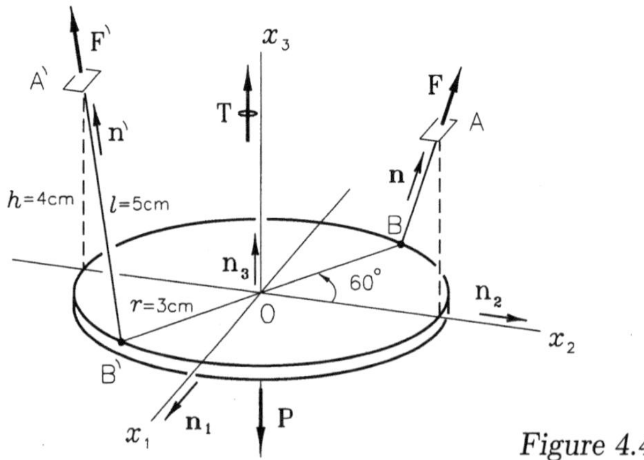

Figure 4.4

Adopting the basis of orthonormal vectors $\mathbf{n}_1, \mathbf{n}_2, \mathbf{n}_3$, the unitary vectors in the directions of the ropes are

$$\mathbf{n} = \frac{1}{10}(3\sqrt{3}\mathbf{n}_1 + 3\mathbf{n}_2 + 8\mathbf{n}_3); \quad \mathbf{n}' = \frac{1}{10}(-3\sqrt{3}\mathbf{n}_1 - 3\mathbf{n}_2 + 8\mathbf{n}_3)$$

and the forces and torques applied to the disk, expressed on the same basis, are

$$\mathbf{F} = \frac{1}{2}(3\sqrt{3}\mathbf{n}_1 + 3\mathbf{n}_2 + 8\mathbf{n}_3)\,\mathrm{N},$$

$$\mathbf{F}' = \frac{1}{2}(-3\sqrt{3}\mathbf{n}_1 - 3\mathbf{n}_2 + 8\mathbf{n}_3)\,\mathrm{N},$$

$$\mathbf{P} = -8\mathbf{n}_3\,\mathrm{N},$$

$$\mathbf{T} = 9\sqrt{3}\mathbf{n}_3\,\mathrm{N\,cm}.$$

The resultant of the system is

$$\mathbf{R} = \mathbf{F} + \mathbf{F}' + \mathbf{P} = 0.$$

The moment of the force $\mathbf{F}$ with respect to point O can be obtained from

$$\mathbf{M}^{\mathbf{F}/O} = \mathbf{p}^{B/O} \times \mathbf{F} = \frac{3}{2}(4\mathbf{n}_1 + 4\sqrt{3}\mathbf{n}_2 - 3\sqrt{3}\mathbf{n}_3)\,\mathrm{N\,cm}.$$

The moment of force $\mathbf{F}'$ with respect to point O is, likewise,

$$\mathbf{M}^{\mathbf{F}'/O} = \mathbf{p}^{B'/O} \times \mathbf{F}' = \frac{3}{2}(-4\mathbf{n}_1 - 4\sqrt{3}\mathbf{n}_2 - 3\sqrt{3}\mathbf{n}_3)\,\mathrm{N\,cm}.$$

The moment of the weight with respect to O is null, of course, due to the symmetry of the disk, and the resultant moment of the system with respect to the same point is

$$\mathbf{M}^{\mathcal{F}/O} = \mathbf{M}^{\mathbf{F}/O} + \mathbf{M}^{\mathbf{F}'/O} + \mathbf{T} = 0.$$

This is, therefore, a null system. It is easy to see that the resultant moment is null with respect to any other point or with respect to any chosen axis.

2.5 Central Axis

The resultant moments of a vector system $\mathcal{V}$ with respect to all points of a line parallel to its resultant are equal to each other. In fact, if O and O' are two points on a line parallel to the resultant $\mathbf{R}$ (see Fig. 5.1), the product $\mathbf{p}^{O/O'} \times \mathbf{R}$ is null, so, from Eq. (3.4), $\mathbf{M}^{\mathcal{V}/O} = \mathbf{M}^{\mathcal{V}/O'}$.

Hence, moving one point parallel to the resultant of the system the resultant moment does not alter. The resultant moment does change, however, when moving the point in an arbitrary direction.

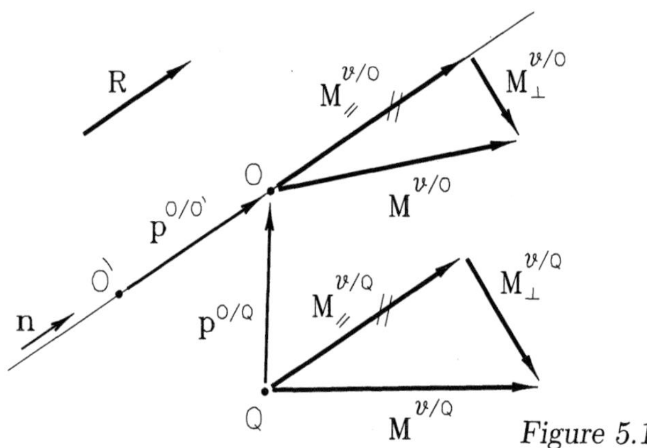

Figure 5.1

Now, if when we move from one point to another we obtain a new resultant moment, it would be useful to consider if there is any particular point for which the resultant moment vanishes. To answer this question, let us take the resultant moment of an arbitrary system $\mathcal{V}$ with respect to a given point O, $\mathbf{M}^{\mathcal{V}/O}$, and let us break it down in the direction of the resultant $\mathbf{R}$ of the system, that is (see Appendix A),

$$\mathbf{M}^{\mathcal{V}/O} = \mathbf{M}_{\parallel}^{\mathcal{V}/O} + \mathbf{M}_{\perp}^{\mathcal{V}/O}, \tag{5.1}$$

where the component of the resultant moment parallel to the resultant is

$$\mathbf{M}_{\parallel}^{\mathcal{V}/O} = \frac{1}{R^2} \mathbf{M}^{\mathcal{V}/O} \cdot \mathbf{R}\,\mathbf{R} \tag{5.2}$$

and the component of the resultant moment orthogonal to the resultant is (see Appendix A)

$$\mathbf{M}_{\perp}^{\mathcal{V}/O} = \frac{1}{R^2} \left(\mathbf{R} \times \mathbf{M}^{\mathcal{V}/O} \right) \times \mathbf{R}. \tag{5.3}$$

If Q is an arbitrary point, the vector difference between the resultant moment of $\mathcal{V}$ with respect to Q and O is, according to the moments transport theorem, given by $\mathbf{p}^{O/Q} \times \mathbf{R}$, a vector orthogonal to $\mathbf{R}$. It is then found that, by changing the point, only the orthogonal component, $\mathbf{M}_{\perp}^{\mathcal{V}/O}$, varies, while the parallel component remains invariant, that is,

$$\mathbf{M}_{\parallel}^{\mathcal{V}/O} = \mathbf{M}_{\parallel}^{\mathcal{V}/Q} = \mathbf{M}_{\parallel}^{\mathcal{V}}. \tag{5.4}$$

The conclusion is that the parallel moment is, like the resultant, an *invariant* of the system $\mathcal{V}$. Therefore, if for a given point P the component of the resultant moment of the system parallel to its resultant is different from zero, there will be no other point in the space with respect to which the resultant moment is null. The answer to the question asked previously is, therefore, negative, that is, it is *untrue*, in the most general case, that there is always a point in the space with respect to which the resultant moment of the system is null. Of course, if for a given point P the parallel moment of the system is null, it will be so for any other point.

As seen above, the parallel moment does not depend on the point, but the orthogonal moment varies with it. It would, then, be worth investigating if there is a point P for which the orthogonal moment vanishes, that is, if there is P so that

$$\mathbf{M}^{\mathcal{V}/P} = \mathbf{M}^{\mathcal{V}}_{/\!/}. \tag{5.5}$$

With this objective, basing ourselves on Eq. (3.4) and substituting Eqs. (5.1), (5.5), (5.4), and (5.3) in succession, we have

$$\mathbf{M}^{\mathcal{V}/O} = \mathbf{M}^{\mathcal{V}/P} + \mathbf{p}^{P/O} \times \mathbf{R};$$

hence,

$$\mathbf{M}^{\mathcal{V}/O}_{/\!/} + \mathbf{M}^{\mathcal{V}/O}_{\perp} = \mathbf{M}^{\mathcal{V}}_{/\!/} + \mathbf{p}^{P/O} \times \mathbf{R};$$

therefore,

$$\mathbf{M}^{\mathcal{V}}_{/\!/} + \frac{1}{R^2}\left(\mathbf{R} \times \mathbf{M}^{\mathcal{V}/O}\right) \times \mathbf{R} = \mathbf{M}^{\mathcal{V}}_{/\!/} + \mathbf{p}^{P/O} \times \mathbf{R}. \tag{5.6}$$

Now note that, after the term $\mathbf{M}^{\mathcal{V}}_{/\!/}$, present in both members, is simplified, we obtain a vector equation that is satisfied for all position vectors $\mathbf{p}^{P/O}$, so that

$$\mathbf{p}^{P/O} = \frac{1}{R^2}\mathbf{R} \times \mathbf{M}^{\mathcal{V}/O} + \lambda\mathbf{R}, \tag{5.7}$$

where λ is an arbitrary real number of dimension $\left[L/\operatorname{Dim}[\mathbf{R}]\right]$.

Equation (5.7) describes a straight line parallel to the resultant passing through the point P*, whose position with respect to point O is given by the vector (see Fig. 5.2)

$$\mathbf{p}^* = \frac{1}{R^2}\mathbf{R} \times \mathbf{M}^{\mathcal{V}/O}. \tag{5.8}$$

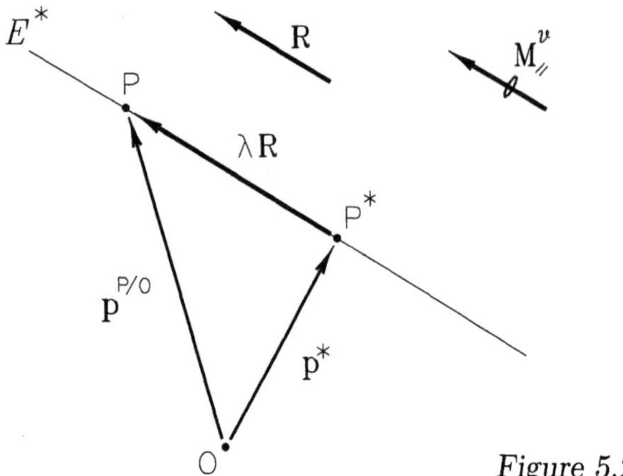

Figure 5.2

This line, the geometric place of the points with respect to which the resultant moment of the system is reduced to the parallel moment, is called the *central axis* of the system. Note that the vector $\mathbf{p}^*$ will exist whenever the resultant of the system is different from zero, that is, the central axis exists for any system that is not a couple or a null system.

The parallel moment, whose magnitude is the least possible among the resultant moments of the system with respect to any point in space, is, for this reason, also called the *minimum moment* of the system and, when the resultant moment with respect to any point O is known, is determined by Eq. (5.2). If the vector system is such that, for any given point O, the resultant moment and resultant of the system are orthogonal, the minimum moment of this system will be null.

Note that P* is the point of the central axis closest to point O. In fact, vector $\mathbf{p}^*$, being orthogonal to $\mathbf{R}$, is perpendicular to the central axis and P* will be the orthogonal projection of O on the axis (see Fig. 5.2).

Example 5.1 Figure 5.3 reproduces the system analyzed in Example 3.2. The parallel moment of this system is, according to Eq. (5.2),

$$\mathbf{M}_{//} = \frac{1}{350u^2}\left[30um(\mathbf{n}_1 + \mathbf{n}_3)\right]$$

$$\cdot \left[5u(-3\mathbf{n}_1 + 2\mathbf{n}_2 + \mathbf{n}_3)\right]\left[5u(-3\mathbf{n}_1 + 2\mathbf{n}_2 + \mathbf{n}_3)\right]$$

$$= -\frac{30}{7}um(-3\mathbf{n}_1 + 2\mathbf{n}_2 + \mathbf{n}_3).$$

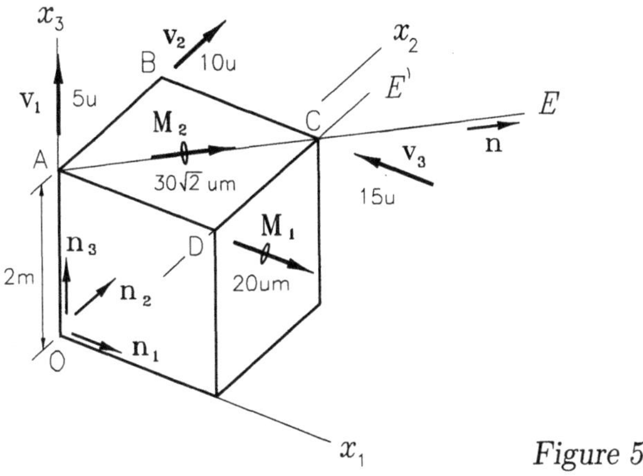

Figure 5.3

The position of the point of the central axis closest to point O is given, according to Eq. (5.8), by the position vector

$$\mathbf{p}^* = \frac{1}{350u^2}\left[5u(-3\mathbf{n}_1 + 2\mathbf{n}_2 + \mathbf{n}_3)\right] \times \left[30um(\mathbf{n}_1 + \mathbf{n}_3)\right]$$
$$= \frac{6}{7}(\mathbf{n}_1 + 2\mathbf{n}_2 - \mathbf{n}_3)\,\text{m}.$$

The central axis of the system is then the straight line given by the equation

$$\mathbf{p} = \frac{6}{7}(\mathbf{n}_1 + 2\mathbf{n}_2 - \mathbf{n}_3)\,\text{m} + \lambda(-3\mathbf{n}_1 + 2\mathbf{n}_2 + \mathbf{n}_3),$$

where $\mathbf{p}$ is the position vector, with respect to point O, of an arbitrary point of the axis and λ is a real number, with dimension [L], that parametrizes the straight line.

Every vector system $\mathcal{V}$ with a nonnull resultant can be reduced to a pair of parallel vectors as follows: A sliding vector equal to the resultant of $\mathcal{V}$, associated to the central axis of the system, and a free vector equal to the parallel moment of the system. In other words, the reduction of any system to an arbitrary point on its central axis consists of exactly two of the system's invariants. When $\mathcal{V}$ is a system of forces, its reduction to an arbitrary point on the central axis forms a *wrench*, the name given to a system formed by a force and a torque parallel to each other. An everyday example of a wrench is the action of a screwdriver. In fact, the action of this tool on a screw consists of a force and a torque, both parallel to the axis of the screw. A wrench is said to be *direct* when the force and torque have the same direction (tightening the screw) and to be *inverse* when the directions are opposite (loosening the screw).

Vector systems whose parallel moment is null can be reduced to a single sliding vector, equal to its resultant and associated to the central axis of the system. This is the case of some particular simple systems, as we will see ahead.

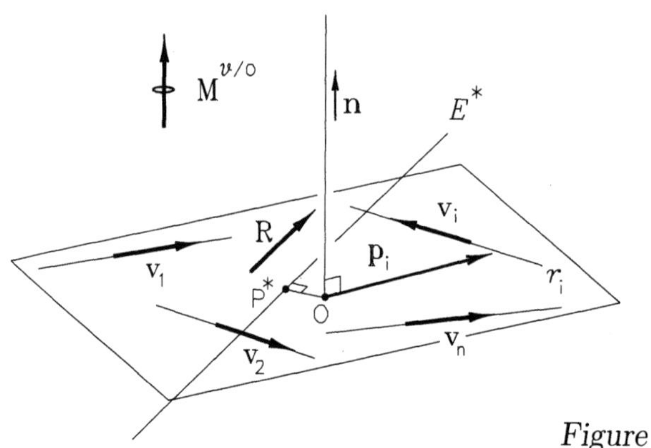

Figure 5.4

A simple system of vectors is called *coplanar* when its lines of action are all contained on the same plane (see Fig. 5.4). On the one hand, the resultant moment of such a system with respect to a point of the plane is necessarily orthogonal to it, since the moment of any of the system's component vectors with respect to a point of the plane is perpendicular to this plane. The resultant of the system, on the other hand, is parallel to the plane, so the parallel moment is null, while the central axis is contained in the plane. The system can, therefore, be reduced to a sliding vector equal to the resultant of the system, associated to the central axis.

Example 5.2   A broad-rimmed hat is laid on a smooth horizontal table. Three lines, fixed to the crown of the hat at points A, B, and C, are pulled horizontally with forces of the same magnitude F, in the directions indicated, skimming the crown of the hat (see Fig. 5.5). We wish to determine a point on the hat rim where a nail must be stuck, so that it does not move. The nail, once it is in place, will prevent the displacement of the point, letting the hat rotate freely around it. The problem is, therefore, to find a point on the rim where the system can be reduced to a single force, with a null resultant moment, thus preventing the hat from rotating.

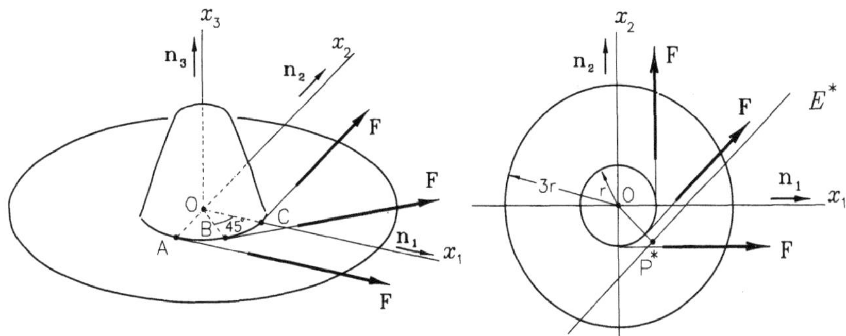

Figure 5.5

The system of forces $\mathcal{F}$ is coplanar and can be reduced to its resultant

$$\mathbf{R} = \frac{2+\sqrt{2}}{2}F(\mathbf{n}_1 + \mathbf{n}_2)$$

applied to a point on the central axis. The resultant moment at point O is

$$\mathbf{M}^{\mathcal{F}/O} = 3Fr\mathbf{n}_3,$$

and a point P* on the central axis can be given by the position vector [see Eq. (5.8)]

$$\mathbf{p}^* = \frac{1}{R^2}\frac{2+\sqrt{2}}{2}F(\mathbf{n}_1 + \mathbf{n}_2) \times 3Fr\mathbf{n}_3$$
$$= \frac{3}{2+\sqrt{2}}r(\mathbf{n}_1 - \mathbf{n}_2)$$
$$= 0.879\,r(\mathbf{n}_1 - \mathbf{n}_2).$$

The central axis will, therefore, be a straight line parallel to $\mathbf{R}$ passing through P*, as shown in the figure. As the resultant moment with respect to any point of E^* is null, the nail, when fixed at any point on this axis, will react with a horizontal force equal to $-\mathbf{R}$, immobilizing the hat.

A simple vector system is called *parallel* when formed by sliding vectors whose line of actions are all parallel to a given straight line. If $\mathbf{n}$ is a unit vector characterizing the direction of the system, its resultant is necessarily parallel to $\mathbf{n}$ and the moment of any of its vectors with respect to an arbitrary point O is orthogonal to **n** (see Fig. 5.6). It then follows that the resultant moment and resultant of a parallel system are always orthogonal, independent of the chosen point, so the minimum moment is null and the system can be reduced to a sliding vector equal to the resultant, having the central axis of the system as line of action.

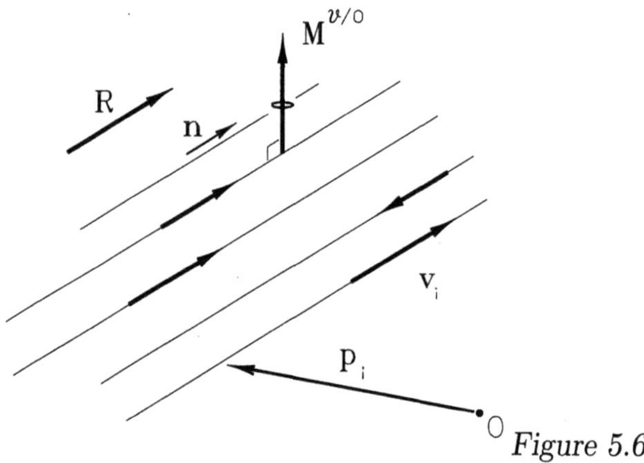

Figure 5.6

Example 5.3   The gravitational force exerted by the earth on a body C close to its surface can, because of the proportions involved, be considered as a parallel distributed system of forces $\mathcal{F}$. The resultant of this system is the weight of the body,

$$\mathbf{P} = \int_C d\mathbf{P} = \int_C \rho g \mathbf{n}\, dV = mg\,\mathbf{n},$$

where ρ is the field of the body's density, g is the magnitude of the gravitational acceleration, **n** is the vertical unitary, pointing to the surface, V is the volume, and m is the mass of the body (see Fig. 5.7). The resultant moment of this system with respect to an arbitrary point O is

$$\mathbf{M}^{\mathcal{F}/O} = \int_C (\mathbf{r} \times \rho g \mathbf{n})\, dV = \int_C \rho \mathbf{r}\, dV \times g\,\mathbf{n},$$

where **r** is the position vector, with respect to point O, of a generic point C. The central axis of this system will be a vertical straight line described, according to Eq. (5.7), by the position vector with respect to point O:

$$\mathbf{p} = \frac{1}{P^2}\mathbf{P} \times \mathbf{M}^{\mathcal{F}/O} + \lambda\mathbf{P}$$

$$= \frac{1}{m}\mathbf{n} \times \left(\int_C \rho\mathbf{r}dV \times \mathbf{n}\right) + \lambda mg\mathbf{n}$$

$$= \frac{1}{m}\int_C \rho\mathbf{r}dV - \frac{1}{m}\int_C \rho\mathbf{r}dV \cdot \mathbf{n}\,\mathbf{n} + \lambda mg\mathbf{n}.$$

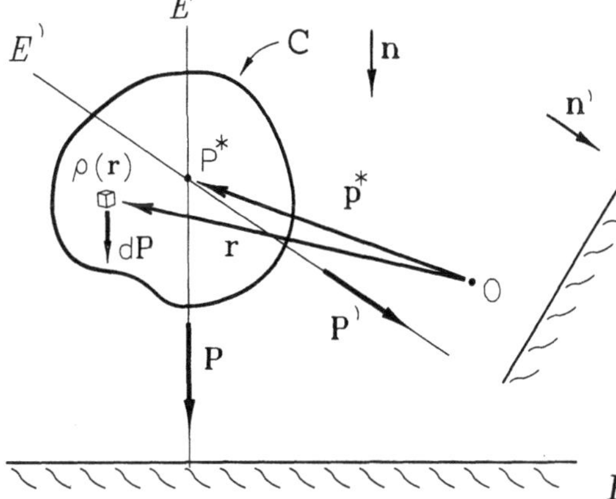

Figure 5.7

Note that the first term of the last line expresses nothing more than the position vector, with respect to O, of the mass center of the body (see Section 1.6),

$$\mathbf{p}^* = \frac{1}{m}\int_C \rho\mathbf{r}dV,$$

while the other two terms are vectors parallel to $\mathbf{n}$ and may be grouped in the form $\beta\mathbf{n}$, where β is an arbitrary scalar. The conclusion, then, is that the central axis of the system of gravitational forces on a body close to the earth's surface is a vertical line that passes through the mass center of the body. Now, modifying the orientation of the body in relation to the earth, only the orientation of the unitary $\mathbf{n}$ ($\mathbf{n}'$) in relation to the body is modified, with the new central axis parallel to $\mathbf{n}'$, passing through the mass center of C (see Fig. 5.7). Now, as the orientation given to the body was arbitrary, the result is that the central axes of all possible configurations will cross each other in the mass center of the body, by which we can, in any case, reduce the gravitational action of the earth on a small body close to its surface, to its weight applied to the mass center of the body.

When the lines of action of a simple vector system all converge at one point, we have a *concurrent system*. The resultant moment of the system with respect to the concurrence point will, naturally, be null, and the central axis of the system will then necessarily pass through the point. Every concurrent system can, therefore, be reduced to a sliding vector equal to its resultant associated to a line of action passing through the concurrence point. (This result is known as *Varignon's theorem*.)

Example 5.4   Consider the system of gravitational forces exerted by a particle P, of mass M, on a homogeneous bar AB, of mass m and length c, in the configuration shown in Fig. 5.8. This is a distributed simple system consisting of the forces of attraction $d\mathbf{F}$ between P and each element of mass $dm = \frac{m}{c} dy$, all with a support passing through P.

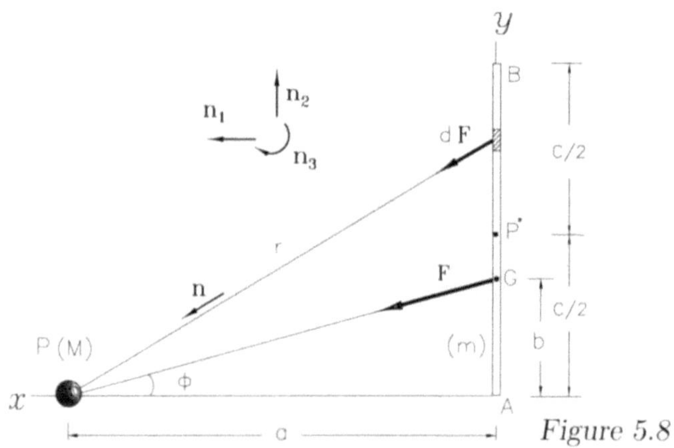

Figure 5.8

Each element of force is, according to the universal gravitational principle, Eq. (1.3.5),

$$d\mathbf{F} = \frac{GM\,dm}{r^2}\,\mathbf{n}$$
$$= \frac{GMm\,dy}{c(a^2 + y^2)^{3/2}}\,(a\mathbf{n}_1 - y\mathbf{n}_2).$$

The resultant gravitational force is then

$$\mathbf{F} = \int_0^c d\mathbf{F} = \frac{GMm}{a(a^2 + c^2)^{1/2}}\left[\mathbf{n}_1 + \left(\frac{a}{c} - \left(1 + \frac{a^2}{c^2}\right)^{1/2}\right)\mathbf{n}_2\right].$$

As the system is concurrent at P, it can be reduced to a gravitational force equal to the resultant calculated above, passing through P. The line of action of this force intercepts the bar at point G, center of gravity of the bar for the gravitational field exerted by particle P. The distance b from this point to the end A of the bar is

$$b = a \tan\phi = a\left[\left(1 + \frac{a^2}{c^2}\right)^{1/2} - \frac{a}{c}\right] = \frac{a^2}{c}\left[\left(1 + \frac{c^2}{a^2}\right)^{1/2} - 1\right].$$

One can see that point G lies between A and P*, mass center of the bar, that is, that $b < c/2$, which is equivalent to

$$\frac{a^2}{c}\left[\left(1 + \frac{c^2}{a^2}\right)^{1/2} - 1\right] < \frac{c}{2},$$

or

$$1 + \frac{c^2}{a^2} < \frac{c^4}{4a^4} + 1 + \frac{c^2}{a^2};$$

therefore,

$$\frac{c^4}{4a^4} > 0,$$

which is always true. The result, then, is that the center of gravity is situated below the mass center of the bar. This result clearly shows that the center of gravity and the mass center of a body are different concepts. The latter depends exclusively on the distribution of the body mass while the former depends also on the nature of the present gravitational field. Of course, as we saw in Example 5.3, both coincide in the case of the earth's gravitational attraction on a body of small dimensions close to its surface. One can also easily see that, in the case under study, G becames closer to P* when the a/c ratio increases. The reduction of the gravitational field at the mass center of the bar will consist of the gravitational force $\mathbf{F}$ applied to P* and a gravitational torque equal to the resultant moment of the system with respect to P*. Using Eq. (3.4), this torque is

$$\mathbf{M}^{\mathcal{F}/P^*} = \mathbf{p}^{G/P^*} \times \mathbf{F} = \frac{GMm(c - 2b)}{2a(a^2 + c^2)^{1/2}}\,\mathbf{n}_3.$$

2.6 Forces and Torques

The first step to be taken to establish the equations that govern the motion of a mechanical system — whether it is a simple particle moving on a plane or a mechanism with multiple interconnected bodies in three-dimensional motion — is to identify the set of forces and torques acting on it. For the sake of simplicity, we will call the system of vectors consisting of forces and torques acting on a mechanical system a *force system*. Once the forces and torques acting on the subject of interest are identified, it is necessary to choose a point to reduce the system; the choice of this point depends on the nature of the encountered force system itself, as well as on kinematic and inertia properties of the body or bodies under study. The general guidelines for choosing the most suitable point for reducing a system, therefore, will not be discussed in this chapter; the matter will be duly discussed later.

Interactions between mechanical elements occur through forces and torques. As discussed in Chapter 1, the concept of force is assumed as primitive in mechanics, the same as in the case of the torque concept. Even though the moment of a force applied with respect to a given point is a torque, applied torques can be considered separately from the existence of force systems that consist of couples with those torques. This is, indeed, the general treatment adopted in Section 2.3 to define vector systems.

The interaction between two particles occurs by means of a force. Thus, given two particles P and Q, Newton's third law states that P exerts on Q a force $\mathbf{F}_{QP}$, associated to the line of action passing through Q and P, while Q exerts on P a force $\mathbf{F}_{PQ}$, also associated to the same line of action (see Fig. 6.1), satisfying the relationship

$$\mathbf{F}_{QP} = -\mathbf{F}_{PQ}. \qquad (6.1)$$

It is convenient to classify the interaction forces in two categories, as follows: *field forces*, or distance action, occurring when the particles are not in contact; and *contact forces*, which are those from direct contact, which only occur when the relative position vector between the particles is null. The former includes the forces of gravita-

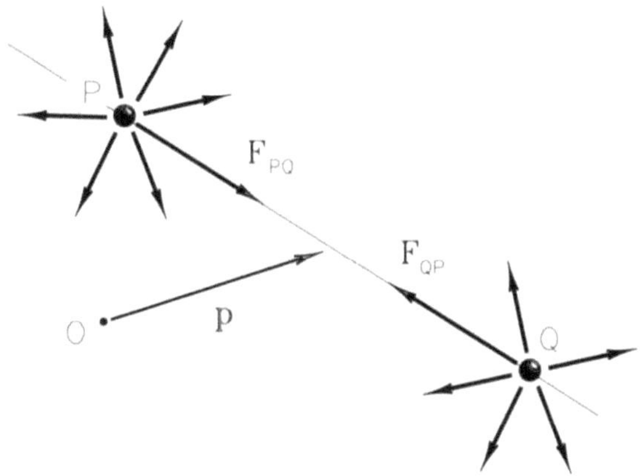

Figure 6.1

tional attraction and electromagnetic fields, among others; in the latter, as examples, are collision and friction forces.

In the mechanical model of a particle, the force system acting on a given particle will always be a concurrent simple system (see Fig. 6.1). Such a system is equivalent, as stated, to a force equal to its resultant applied on the concurrence point. In the case of a particle, therefore, the generally most suitable point for reducing the system is the particle itself.

Given an arbitrary point O, it is trivial that the moments with respect to O of the interaction forces between two particles P and Q satisfy the relationship

$$\mathbf{M}^{\mathbf{F}_{QP}/O} = -\mathbf{M}^{\mathbf{F}_{PQ}/O}, \tag{6.2}$$

that is, the moments with respect to any point are, like the forces, equal and contrary. In fact, the moments result from vectorial products of the same position vector $\mathbf{p}$ (see Fig. 6.1) with equal and opposite force vectors.

Example 6.1 Consider the set of four small spheres of different masses, at rest, laid on a smooth horizontal plane and interconnected by four wires, as shown in Fig. 6.2a. When the horizontal force $\mathbf{F}$ is applied, tractions occur on the wires, each sphere undergoing the forces indicated in Fig. 6.2b. On analyzing it a little more closely, one finds that, besides the forces parallel to the plane, each sphere also undergoes vertical forces, as shown in Fig. 6.2c, for sphere D.

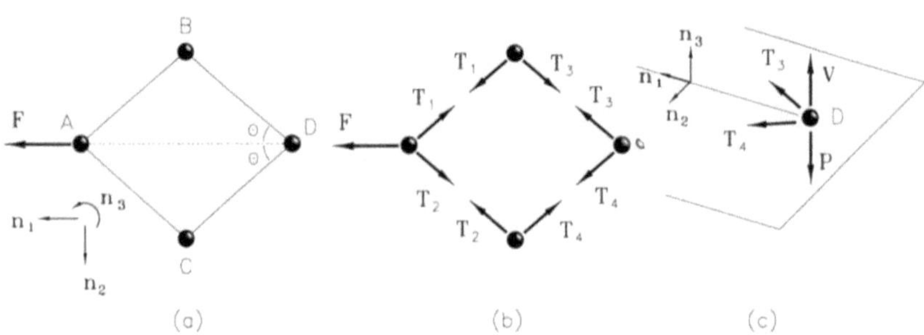

Figure 6.2

The force system acting on D consists of its vertical weight **P**, a field force; the vertical force V, exerted by the smooth plane, a contact force; horizontal tractions **T**$_3$ and **T**$_4$, exerted by the wires, that can be interpreted as contact forces if we include them as elements without mass, but belonging to the system or modeled as field forces exerted by spheres B and C, respectively. The resultant of the force system acting on D is

$$\mathbf{R} = (T_3 + T_4)\cos\theta\,\mathbf{n}_1 + (T_4 - T_3)\sin\theta\,\mathbf{n}_2 + (V - P)\mathbf{n}_3.$$

The system is equivalent, therefore, to **R** applied on D. The reader should not find it hard to analyze the force systems acting on each of the other spheres.

Interaction between two bodies (rigid or otherwise) occurs by means of forces and torques. It therefore requires more careful handling than the interaction between particles, generally involving nonsimple vector systems. When there is interaction without mutual contact, we have a *distance action system*. For example, the gravitational field established between two bodies of arbitrary geometry and whose dimensions are around the same size as the distance between their centers is not generally reducible to a single force. Thus, contrary to what is seen in Example 5.3 — where the system is parallel — and Example 5.4 — where the system is concurrent — the reduction at any point of a body of the gravitational field exerted by another consists of a gravitational force *and* a gravitational torque.

When two bodies have a point or region of their surfaces touching each other, one has a *contact system*. If there is a single point of mutual contact, that is, a single point P of a body C coinciding with

point P$'$ of another body C' (see Fig. 6.3), one has, as in the model of
a particle, a concurrent simple system at the point of contact. There is
not, therefore, in this case, application of a torque between the bodies
(although, of course, there could be a resultant moment with respect to
a point in space other than the point of contact).

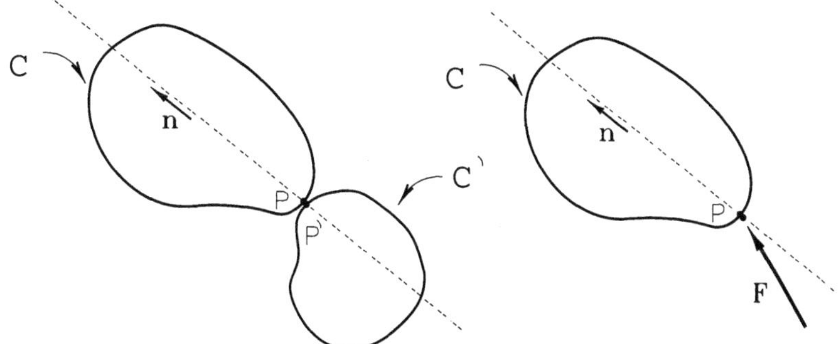

Figure 6.3

When two bodies have a line or region of their surfaces in contact, the interaction occurs through a more general force system. It is
always convenient to model this interaction by reducing this system to
a point P representing this contact (see Fig. 6.4). Reduction will consist, in the most general case, of a force $\mathbf{F}$ (equal to the resultant of
the system) applied to the chosen point and a torque $\mathbf{T}$ (equal to the
resultant moment of the system with respect to the point) that, merely
for convenience and clarity, is also represented as if applied to the point.
By adopting, as usual, an orthonormal basis $\mathbf{n}_1, \mathbf{n}_2, \mathbf{n}_3$ for the decomposition of the vectors, let us say that the contact interaction is modeled
by three mutually orthogonal forces $\mathbf{F}_1, \mathbf{F}_2, \mathbf{F}_3$, and three also mutually
orthogonal torques, $\mathbf{T}_1, \mathbf{T}_2, \mathbf{T}_3$ in the directions of the chosen basis, as
shown in Fig. 6.4.

The contact between two bodies is also called a *link*. The nature of the link will be given by the present force system. When, at the
contact between two bodies, the three force components and the three
torque components are different from zero for an arbitrary orientation of
the base, there is a *rigid link*. This is what happens in the case of welding or fixing. (Although there may be local deformations in the link, we

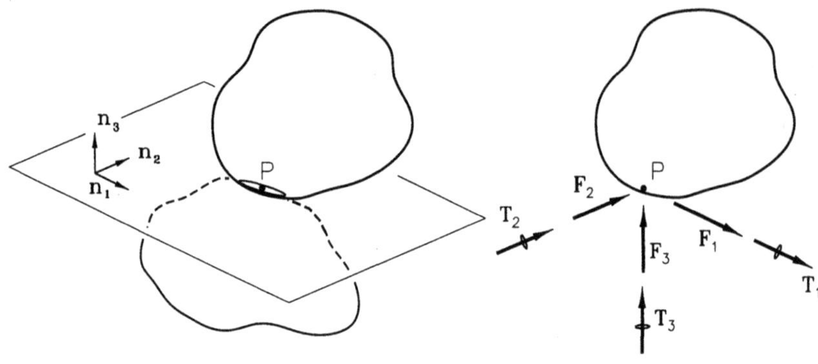

Figure 6.4

will keep the name *rigid link* characterizing the presence of forces and torques in the three directions.) The nature of a link depends on the *kinematic constraints* that the link imposes. We understand a kinematic constraint to be a displacement or a rotation that the presence of the link prevents. Every link can, therefore, be modeled by the displacements and rotations that it does not admit. The rigid link does not admit any relative displacement or rotation between the bodies in contact; hence it is modeled by a resultant force of an arbitrary direction (three components) and a resultant torque also of an arbitrary direction. For each free displacement admitted by a link, the component of the resultant force in the direction of the free displacement is null; for each free rotation admitted by a link, the resultant torque component in the direction of the free rotation is null. Of course, the reduction of the number of force or torque components will depend on the right choice of coordinated directions, that is, if a given direction of movement is free, in order to suppress the respective force or torque component, it is necessary that the direction corresponds to one of the chosen coordinated directions.

Appendix B provides a table of the models usually adopted for the more commonly found links, indicating the respective nonnull components to be considered, at least in principle. Only practice will give the reader confidence to properly identify the relevant components in each case. As a general rule, it is recommended to start by considering *all* six components, then duly eliminating those that correspond to the free displacements and rotations that the link admits. The configurations in Appendix B, far from including all cases, only give the main models adopted. Combinations of these are common; in this case, the components to be considered are the intersection of the sets of components of each link. For example, a ball and socket joint, which has only three force components, mounted on a rectangular slide consisting of two force and three torque components (see Appendix B), results in a link modeled by only two force components.

When one wishes to study the motion of a body C that is bound to other mechanical elements, we start, as already mentioned, by identifying the force system acting on C. Each link must, therefore, be *substituted* by the forces and torques that characterize it. This procedure is called *body isolation*, and the geometric representation of the reductions at the points representing the links is called a *free body diagram*.

Example 6.2   Bar B is linked to the guide A by means of a mechanism that includes a pivot and a runner (see Fig. 6.5a).

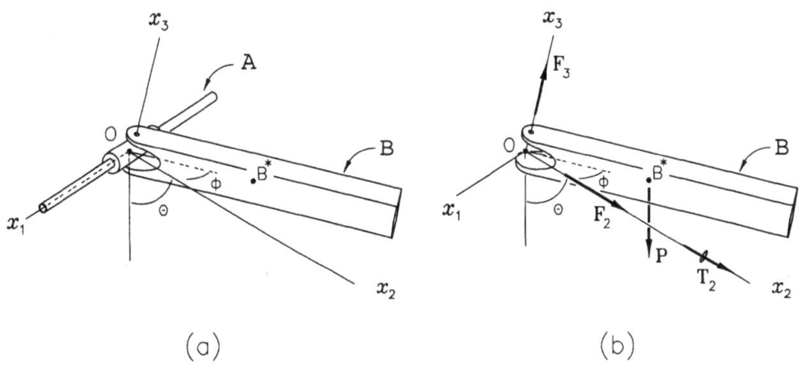

Figure 6.5

Point O, the intersection of the axes of the bar and guide, can be chosen to represent the contact and origin of a system of coordinates for the decomposition of the force and torque vectors involved. This is a compound link; the runner permits free displacement in the direction x_1 and free rotation in the same direction; the pivot admits a free rotation in the direction x_3 (see Appendix B). The force component in the direction x_1 and torque components in the directions x_1 and x_3 will vanish. Taking B, then, as the subject for study, its link with the guide is modeled by a system of forces whose reduction at point O comprises a force applied on O, with components $\mathbf{F}_2$ and $\mathbf{F}_3$ and one torque, $\mathbf{T}_2$. Figure 6.5b illustrates the free body diagram of the bar in which the gravitational action $\mathbf{P}$ is included. Note that the orientation of the axes was chosen to bring to the fore the suppression of the null components ($\mathbf{F}_1$, $\mathbf{T}_1$, and $\mathbf{T}_3$).

2.7 Friction

When two bodies touching at a single point have as a bound force only one component orthogonal to the tangent plane common to their surfaces at the point of contact, this force is called *normal force*, $\mathbf{N}$, and the contact surfaces are said to be *smooth*. When, on the contrary, there are components parallel to the tangent plane, the contact is said to be *with friction* and these components, added up vectorially, form the *friction force*, $\mathbf{F}_a$ (see Fig. 7.1).

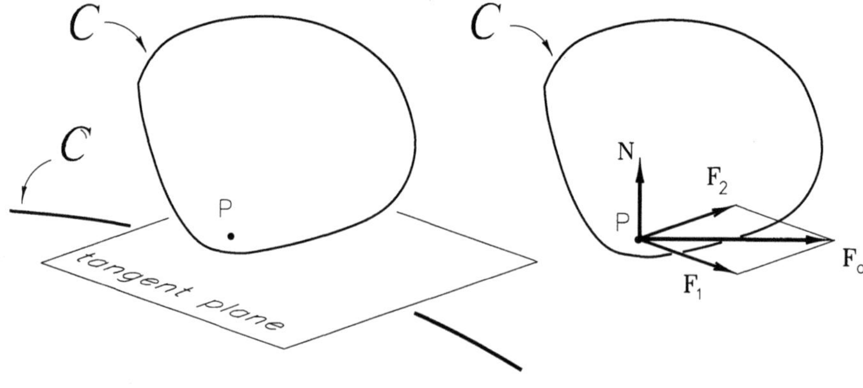

Figure 7.1

When two solid bodies have surfaces touching each other without the presence of a fluid, it is then said that *dry friction* or *Coulomb friction* is present. In this model, the friction component acting on one of the bodies is always a force whose direction is opposite to the relative motion between the surfaces, when sliding occurs, or on its *tendency* of motion where there is none. We understand that this motion tendency is the one to be expected if the contact were smooth, which sometimes is not easy to identify. In order to establish the direction of a tendency to the slide, it is necessary, in most cases, to analyze other links involved, as we will see in the following examples. When the direction of the friction force is unknown, the alternative is to consider two mutually orthogonal force components parallel to the plane tangent to the surface.

In the dry friction model without relative sliding, it is considered that the friction force magnitude has an upper limit depending linearly on the normal component present in the contact, a condition expressed by the inequality

$$|\mathbf{F}_a| \leq \mu |\mathbf{N}|, \tag{7.1}$$

where $\mathbf{F}_a$ is the friction force, $\mathbf{N}$ is the normal force, and the adimensional constant μ, called the *friction coefficient*, expresses, in a simplified form, the complex interaction between two rough surfaces, depending on the material and surface finishing of the bodies in contact. Equation (7.1), consisting of an inequality, merely establishes an upper limit value for the magnitude F_a of the component of the contact force parallel to the tangent plane, a function of the magnitude N of the component orthogonal to the plane. The effective value of the friction force, nonetheless, may only be determined from the dynamic solution of the problem.

Example 7.1 The isosceles triangular plate is at rest, with its vertices A, B, and C lying on the sloping plane π, also having its vertex A pivoted on the fixed pin P, as illustrated in Fig. 7.2a. The action of the pin on the plate consists of the force components $\mathbf{A}_1$ and $\mathbf{A}_2$, parallel to the plane, and there are no torque components (why?). As the pin prevents the vertex A from moving, that is, it inhibits any tendency to sliding, the action of the plane on this support is reduced to the normal $\mathbf{N}_A$. Assuming that the plane is not smooth, the contact on the other two vertices will include normal and friction components. Due to pivoting, the tendency to sliding of the support at B is orthogonal to the edge AB, hence the arbitrated direction for the friction force $\mathbf{F}_B$ (see Fig. 7.2c). Similarly, the friction force $\mathbf{F}_C$ will be orthogonal to the edge AC. Figure 7.2b shows a similar situation, differing, however, by the link at the pin, which is no longer a pivot but a simple contact that we will consider smooth. The action applied by it on the plate can be reduced, then, to force N, parallel to the plane and orthogonal to edge CA, applied at vertex A. The action of the

plane on the vertices can be modeled in this way: On vertex A there is no
tendency to sliding in the direction orthogonal to the edge AC, due to the

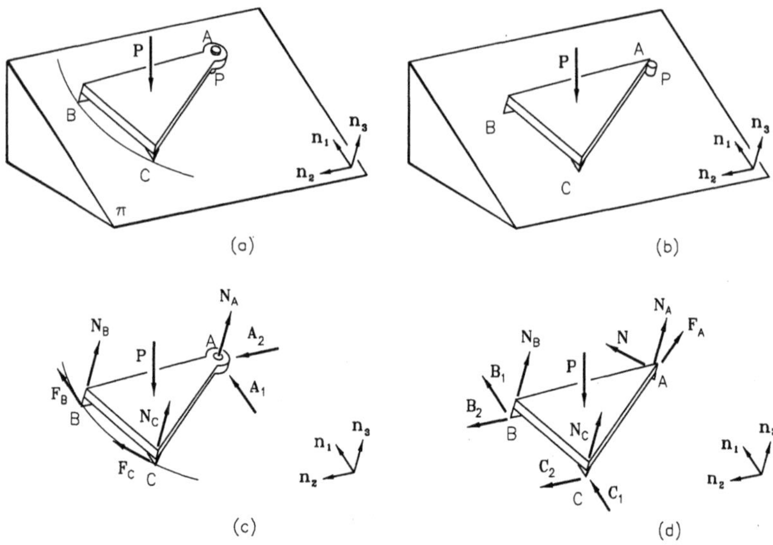

Figure 7.2

presence of the pin, with the result that the contact of the plane is re-
duced to the normal $\mathbf{N}_A$ and to the friction force $\mathbf{F}_A$, parallel to that edge
(see Fig. 7.2d); at vertex B there is a sliding tendency in an unknown direction (at least in principle) and the contact must be modeled with three
components (normal, $\mathbf{N}_B$, and friction, $\mathbf{B}_1$ and $\mathbf{B}_2$); at vertex C, as with
B, we have the normal, $\mathbf{N}_C$, and friction components, $\mathbf{C}_1$ and $\mathbf{C}_2$.

Also with regard to the above example, it is worth noting that,
in both situations studied, the signs chosen for the friction components
are arbitrary. In other words, only the directions to be considered in
each case are part of the links modeling; magnitudes and signs may only
be determined — and not always fully — after a dynamic analysis of
the problem. (Of course, there are situations, such as the weight of a
body or a normal exerted by a simple support, where the sign is known.)
The reader can always infer, using strictly personal guidelines (common
sense, experience, etc.), the direction of an unknown link force; the final
result of the dynamic analysis will indicate, by the sign, if the choice
is right or not. It is recommended, when there is no clear indication

which sign is correct, to infer components in the positive direction of the Cartesian axes adopted; anyhow, the choice will not entail any error.

In the preceding example, it is assumed that there is no sliding of the plate. In principle, therefore, there is not necessarily any relation between the normal and friction components at each link. When the touching surfaces between two bodies have relative motion, it is said that there is dynamic friction; the model, in this case, assumes a linear relationship between the friction and normal components, in the form

$$|\mathbf{F}_a| = \mu'|\mathbf{N}|, \tag{7.2}$$

where μ' is a constant, called the *dynamic friction coefficient*, and is generally dependent on the material and surface finishing of both bodies in contact. If there is sliding, the friction force will always be opposite to the relative motion.

Example 7.2 The homogenous bar B relies on the inclined plane π and pivoted on the pin P, fixed on the plane (see Fig. 7.3). B moves over the plane, turning around P, under the action of gravity. This is reducible, as we have already seen, to its weight $\mathbf{P}$, applied to B^*, the mass center of B. As the bar is homogenous and is fully supported by the plane, the normal force exerted by it can be considered uniformly distributed, as shown in Fig. 7.3b. The resultant of this distribution will be

$$\mathbf{N} = \int_0^c d\mathbf{N}.$$

The friction component exerted by the plane will also be distributed along the bar, having as a resultant

$$\mathbf{F}_a = \int_0^c d\mathbf{F}_a.$$

As there is sliding, each friction force element will have an opposite direction to the motion of the respective point in relation to the plane, resulting in a parallel distribution; on the other hand, sliding guarantees that $|d\mathbf{F}_a| = \mu'|d\mathbf{N}|$, with the result that the distribution, besides parallel, is uniform, as shown in Fig. 7.3b. It is easy to see that the central axis of this distributed system passes through B^*; the action of the plane on the bar can, then, be reduced to $\mathbf{F}_a = \mu'N\mathbf{n}_1$ and $\mathbf{N} = N\mathbf{n}_3$ applied at B^*, as indicated in Fig. 7.3c.

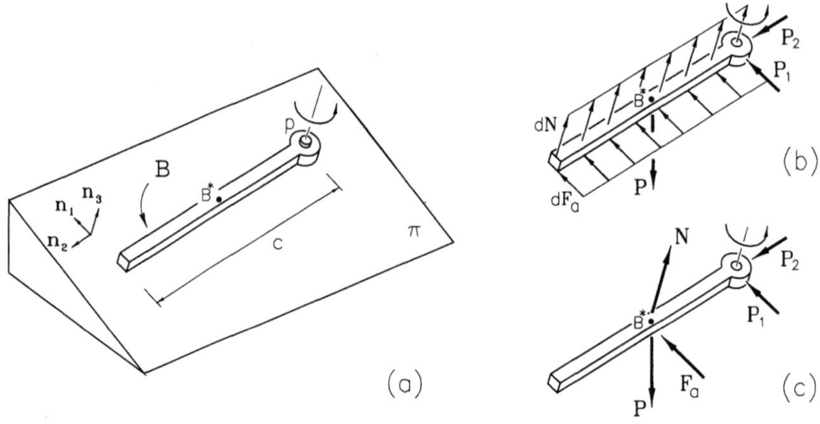

Figure 7.3

When a fluid intervenes in the contact between the bodies (typical case of contact with a lubricant), one then has *viscous friction*. In this model, much more complex than that of dry friction, the friction force depends, essentially, on the relative velocity between the surfaces in contact and the viscosity and thickness of the fluid film between the surfaces. The viscous friction will not be studied here. The following example, illustrating quite a simple case, intends to give only a general idea of the difference in treatment given to the viscous friction model compared to that of dry friction.

Example 7.3   Two flat plates are displaced at a constant relative velocity v, as illustrated in Fig. 7.4, with an oil film of uniform thickness e completely filling the region of mutual contact over area A.

The shearing stress inside the fluid (the dragging force per unit of area exerted by a layer of fluid on its neighboring layer) is given by the ratio

$$\tau = \mu \frac{\partial v_x}{\partial y},$$

where τ is the shearing stress, in the direction of the relative motion, with dimension $[\mathrm{ML}^{-1}\mathrm{T}^{-2}]$, μ is the viscosity of the fluid, with dimension $[\mathrm{ML}^{-1}\mathrm{T}^{-1}]$, and $\partial v_x/\partial y$ is the gradient, in the direction normal to the motion, of the velocity of the fluid. This model of linear relationship between the shearing stress and the velocity gradient is attributed to Newton, and the fluids that satisfy this hypothesis are called *Newtonian fluids*. For

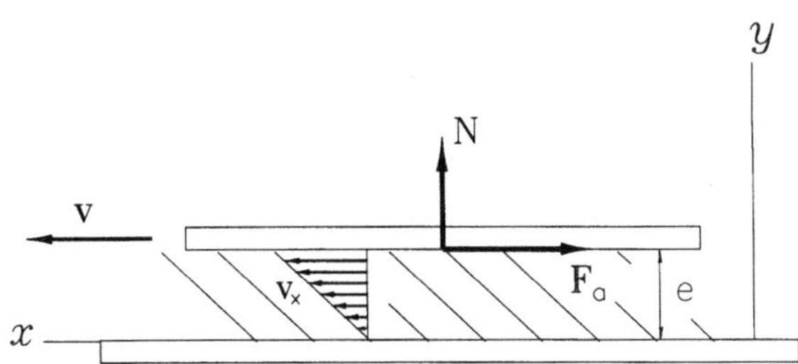

Figure 7.4

a fine film (small e) moving on a steady state (constant v), the velocity profile in the fluid can be assumed to be linear, as shown, and

$$\frac{\partial v_x}{\partial y} = \frac{v}{e}.$$

The magnitude of the friction force applied on the upper plate will then be

$$F_a = \tau A = \mu A \frac{\partial v_x}{\partial y} = \frac{\mu A v}{e}.$$

Note that, in this model, there is no ratio established between the magnitude of the friction force, F_a, and the normal force, N, present between the surfaces in contact.

As mentioned at the beginning of the previous section, the correct modeling of the *force system* acting on an element or a mechanical system whose motion we wish to study is the starting point for a successful solution. If a force or torque component is not considered at this initial stage of analysis, this will give a wrong result; if a component is unduly included, although this would not introduce an error, it will hinder or even make the solution unfeasible, unnecessarily increasing the number of unknown quantities to be determined. It is, therefore, desirable to take special care when modeling the links and distance action systems. A careful study of Appendix B may be of value to the reader.

Example 7.4 Figure 7.5a illustrates a disk welded to a horizontal axis, moving around a second vertical axis in the indicated direction. The link between the axes consists of a joint. The disk relies on the horizontal plane, rolling over it under the action of the torque $\mathbf{T}$, parallel to x_2, as shown, with its center B describing a circular path around point A. Figure 7.5b shows the diagram of the corresponding free body.

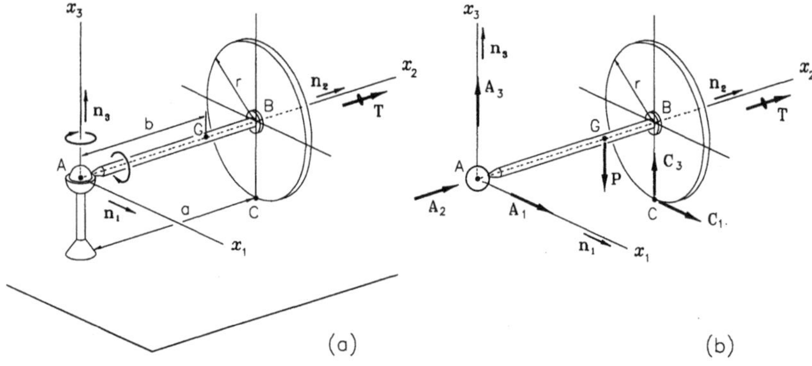

(a)   (b)

Figure 7.5

The forces $\mathbf{C}_3$ (normal) and $\mathbf{C}_1$ (friction) act on the point at which the disk touches the plane. Note that a friction component was not included in the direction of x_2 because, if the contact were smooth, point C would describe a circular path, parallel to the center of the disk. Vertical weight $\mathbf{P}$ acts on point G, the mass center of the set. Three force components, $\mathbf{A}_1$, $\mathbf{A}_2$, and $\mathbf{A}_3$, are applied to point A. As there are no other efforts to be considered, we have a system acting on the set consisting of six forces and one torque, whose resultant is

$$\mathbf{R} = (A_1 + C_1)\mathbf{n}_1 + A_2\mathbf{n}_2 + (A_3 + C_3 - P)\mathbf{n}_3$$

and whose resultant moment with respect, say, to point A is

$$\mathbf{M}^{\mathcal{F}/A} = \mathbf{p}^{C/A} \times (\mathbf{C}_1 + \mathbf{C}_3) + \mathbf{p}^{G/A} \times \mathbf{P} + \mathbf{T}$$
$$= (aC_3 - bP)\mathbf{n}_1 + (T - rC_1)\mathbf{n}_2 - aC_1\mathbf{n}_3.$$

Example 7.5   Plate P, with mass m and mass center P*, is pivoted on O on the fork G that, in its turn, can revolve around the fixed bearing M (see Fig. 7.6a). A light rope, with one end fixed at point A, is being stretched as shown. The Cartesian axes $\{x_1, x_2, x_3\}$ have origin in O, with x_2 aligned with the bearing axis and x_3 orthogonal to the plate.

The orthonormal basis $\mathbf{n}_1, \mathbf{n}_2, \mathbf{n}_3$ is parallel to the chosen coordinated axes. Adopting this basis for the decomposition of the vectors, the force system applied to P will consist of three force components, $\mathbf{F}_1$, $\mathbf{F}_2$, and $\mathbf{F}_3$, exerted by the pivot on O; vertical weight $\mathbf{P}$, applied to P*, traction $\mathbf{F}$, in the direction of the rope, applied to A; and the torque $\mathbf{T}_1$, exerted by the link on O, which permits free rotations in the directions $\mathbf{n}_2$ and $\mathbf{n}_3$ (see Fig. 7.6b).

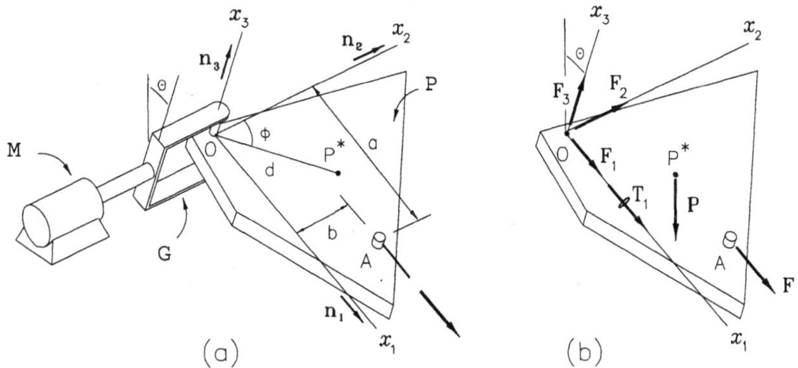

Figure 7.6

The resultant of this system is

$$\mathbf{R} = (F + F_1 + mg\sin\theta)\mathbf{n}_1 + F_2\mathbf{n}_2 + (F_3 - mg\cos\theta)\mathbf{n}_3.$$

Its resultant moment with respect to point O is

$$\begin{aligned}
\mathbf{M}^{\mathcal{F}/O} &= \mathbf{p}^{P^*/O} \times \mathbf{P} + \mathbf{p}^{A/O} \times \mathbf{F} + \mathbf{T}_1 \\
&= (T_1 - mgd\cos\phi\cos\theta)\mathbf{n}_1 + mgd\sin\phi\cos\theta\,\mathbf{n}_2 \\
&\quad - (Fb + mgd\cos\phi\sin\theta)\mathbf{n}_3.
\end{aligned}$$

When a particle is fixed, that is, at rest, it is said that it is in *equilibrium*. The condition for the equilibrium of a particle P is that the resultant of the system of forces acting on P is null, that is,

$$\mathbf{R} = 0 \qquad \text{if P is in equilibrium.} \tag{7.3}$$

In fact, every force system acting on a particle is necessarily a concurrent simple system and, therefore, equivalent to a force equal to its resultant, applied to the particle. Now, Newton's first law establishes exactly that if the resultant of the forces applied to a particle is null, then it will remain at rest or in uniform motion. (Strictly speaking, it will remain at rest or in uniform motion on an inertial reference frame, but this subject involving of the reference frames will be discussed later.) Although both conditions characterize the equilibrium, the equilibrium is usually understood as the condition of rest.

The force system acting on a rigid body is not, as already mentioned, necessarily simple or concurrent. It so happens that stricter conditions must be imposed to characterize its equilibrium. In fact, it is said that a rigid body C is in equilibrium when three of its noncolinear points are fixed. The condition for such a situation to occur is that the force system $\mathcal{F}$ acting on the body is a null system. In other words, a rigid body will be in equilibrium, that is, all its points will be at rest, when the resultant force *and* the resultant moment with respect to some point are both null.

$$\mathbf{R} = 0 \quad \text{and} \quad \mathbf{M}^{\mathcal{F}/O} = 0 \quad \text{if } C \text{ is in equilibrium,} \tag{7.4}$$

where O is an arbitrary point. This result may only be strictly shown further in the text, in Chapter 7, when the dynamics of the rigid body will be studied, but we can assume it as true for this section's requirements.

Example 7.6 Figure 7.7 reproduces the situation studied in Example 4.5

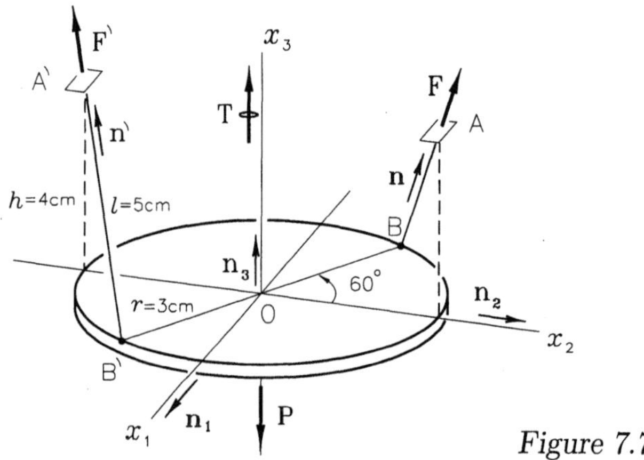

Figure 7.7

(take another look at it). The disk undergoes a force system $\mathcal{F}$, consisting of the forces $\mathbf{P}$, $\mathbf{F}$, and $\mathbf{F}'$ and torque $\mathbf{T}$. As determined in the example, the resultant force $\mathbf{R}$ is null and the resultant moment with respect to point O, $\mathbf{M}^{\mathcal{F}/O}$, is also null. The force system $\mathcal{F}$, therefore, is a null system and the disk is in equilibrium. In fact, the disk would be at rest under the action of the weight and vertical tractions on the ropes if torque $\mathbf{T}$ was not there. By applying the torque, the resultant moment with respect to point O will be compensated by the moment produced by the horizontal components of the tractions on the ropes, while the weight will be counterbalanced by the vertical components of the tractions on the ropes, resulting in a null system.

Example 7.7 Returning now to Example 5.2 (take another look at it), the force system consisting of the three forces of magnitude F applied to the crown of the hat, as illustrated in Fig. 7.8, plus the force

$$\mathbf{P} = -\frac{2+\sqrt{2}}{2}F(\mathbf{n}_1 + \mathbf{n}_2)$$

exerted by the nail stuck in point $\mathbf{P}^* = 0.879\,r\,(1, -1, 0)$ will be a system with both a null resultant and resultant moment, thus guaranteeing the hat's equilibrium.

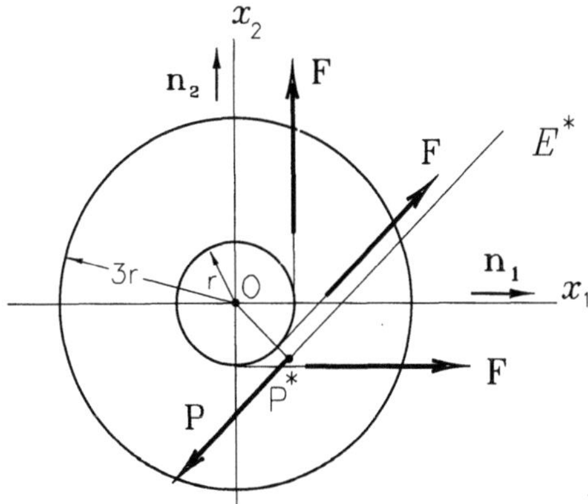

Figure 7.8

Exercise Series #2   (Sections 2.1 to 2.6)

P2.1 A horizontal force of 40N is applied to the arm of the torquimeter, as shown. Determine the magnitude of the moment produced by this force at point O. What is the component of this moment that loosens the screw?

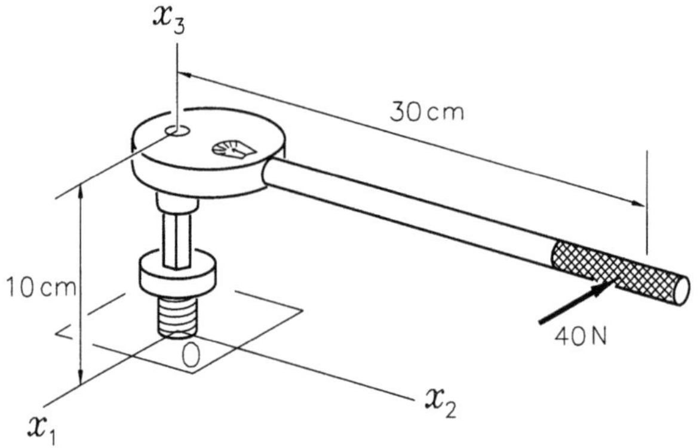

Figure P 2.1

P2.2 In order to drill a wall with the help of a bit, a vertical force of 90N is applied to the arm B and a horizontal force of 60N on handle C, as shown. What is the vertical force that must be applied to the handle so as not to bend the bit at A? What is the reduction of this new system to point A?

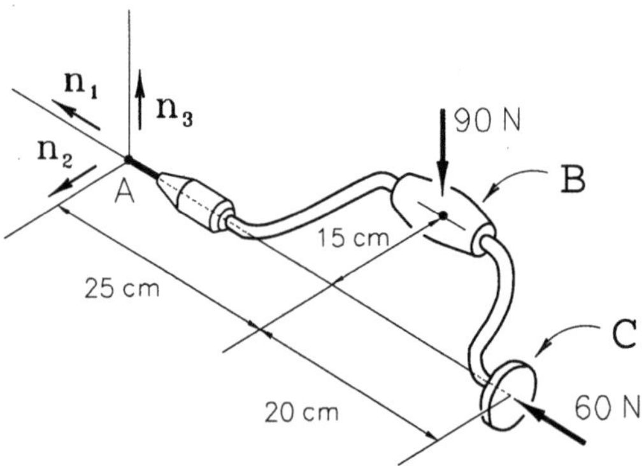

Figure P 2.2

P2.3 A system of forces and torques, consisting of the weight P, the forces on supports A and B and the torques at the input and output axes, equal to 20 N m and 50 N m, respectively, is applied to the gear box. For the box to stay motionless, this system must be null. Determine the forces on supports. If the input torque is increased to 25 N m, it is necessary to screw down support B. Calculate the effort on this screw. (Assuming that there are no losses, the input and output powers always remain equal to each other.)

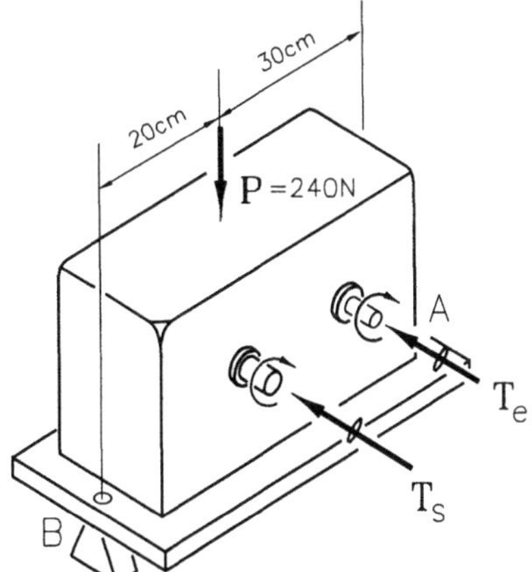

Figure P 2.3

P2.4 A system $\mathcal{S}$ consists of four bound vectors, whose components and respective points of application on a given system of Cartesian axes are $\mathbf{v}_1 = (2, 0, 0)$ in $(0, 1, 2)$; $\mathbf{v}_2 = (0, 1, 1)$ in $(0, 0, 3)$; $\mathbf{v}_3 = (3, -2, 1)$ in $(1, 2, 0)$; $\mathbf{v}_4 = (0, 0, 4)$ in $(-1, 0, 2)$. Determine the magnitude of its resultant, its resultant moment with respect to the origin, and its minimum moment.

P2.5 Referring to the preceding exercise, determine the coordinates of the point closest to the origin where the system can be reduced to a wrench.

P2.6   Consider the system made up of the sliding vectors $\mathbf{v}_i$, $i = 1, \ldots, 4$, associated to the indicated lines of action, and the free vector $\mathbf{M}$. Determine the vector $\mathbf{v}_5$, bound to the vertex A of the cube, with an edge equal to 1 m, whose inclusion in the system permits its reduction to a single nonnull vector. Is there more than one solution?

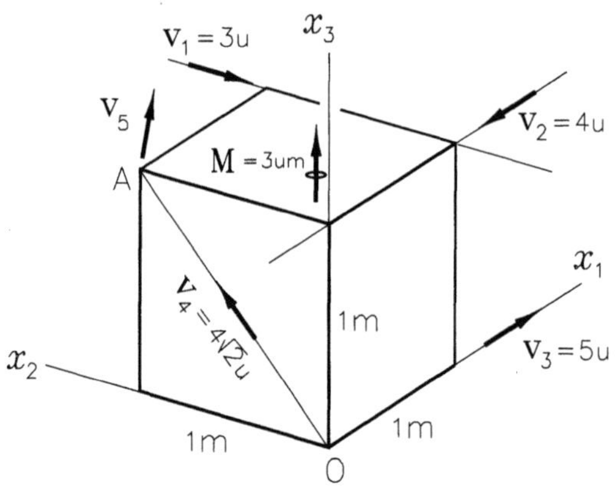

Figure P 2.6

P2.7   Acting on a solid cube with an edge $a = 2\,\text{m}$ are forces $\mathbf{f}_1$, associated to the edge AB, and of 30N magnitude, $\mathbf{f}_2$, associated to the line of action s and of 50N magnitude, $\mathbf{f}_3$, associated to diagonal CF, and $\mathbf{f}_4$, applied to vertex E. Determine the $\mathbf{f}_3$ magnitude and the $\mathbf{f}_4$ components so that the system can be substituted by the action of a screwdriver applied to point O, in the direction of the straight line s. Also determine the efforts to be applied on the screwdriver.

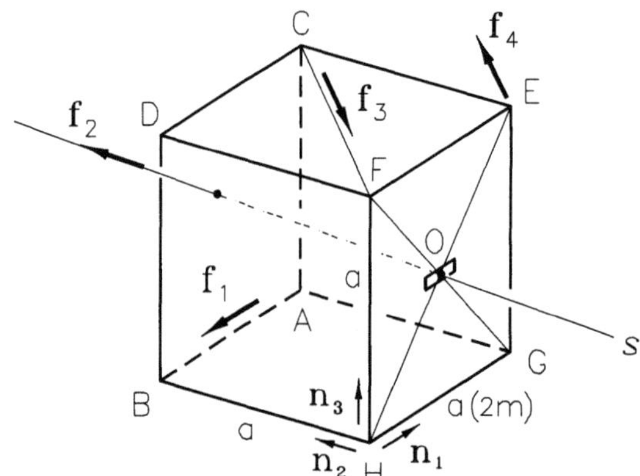

Figure P 2.7

P2.8 The figure illustrates a child's swing in a public square. Among the forces acting on the homogeneous horizontal bar B, consider the subsystem S, comprising the tractions on the four cables and the weight of the bar (200 N). Show that the central axis of this sytem intercepts the axis x_1. If, at a given time, $\mathbf{F}_1 = \mathbf{F}_2 = 100\,\text{N}$, $\mathbf{F}_3 = \mathbf{F}_4 = 120\,\text{N}$, $\beta_1 = 60^0$, $\beta_2 = 30^0$ (children are playing there), determine the x_1 coordinate of the intersecting point.

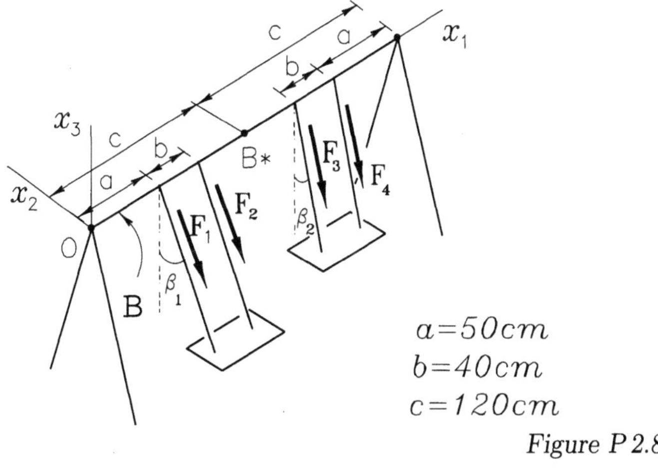

Figure P 2.8

P2.9 A rectangular channel, with width 1, contains water of density ρ, trapped by a cylinder with radius r and mass m, that lies on the rough bottom surface, being kept in position by a horizontal cable, as shown. Determine the traction on the cable to keep the cylinder at rest.

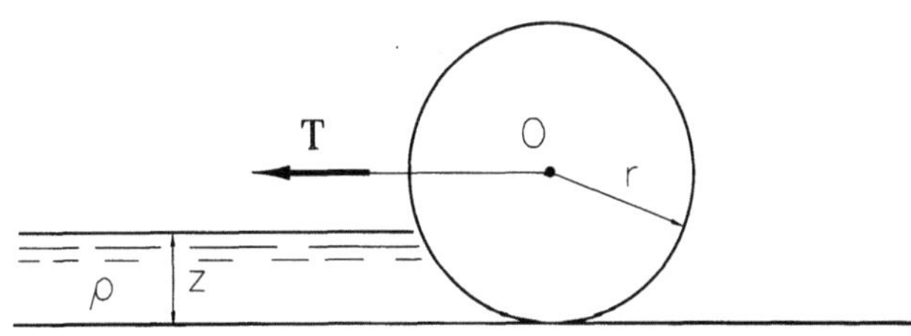

Figure P 2.9

P2.10 Two segments were cut and shaped from a wire with linear density ρ: one rectilinear, with length a, and another in the form of a semicircle with radius r, both fixed in the illustrated configuration. Analyze the gravitational action exerted by the rectilinear segment on the curvilinear one, determining the gravitational torque with respect to point O. Can this system of vectors be reduced to a single force?

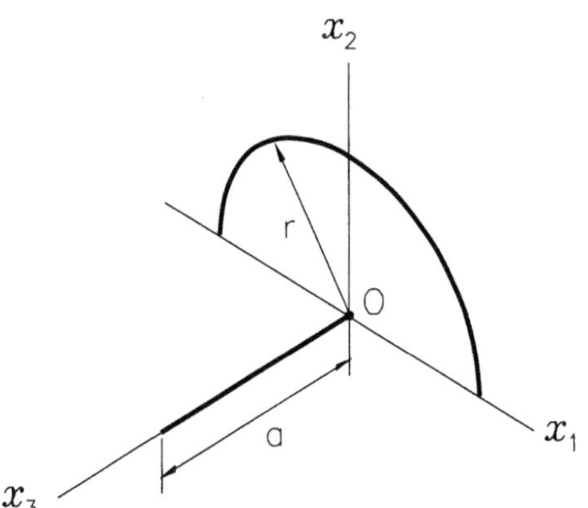

Figure P 2.10

P2.11 Two particles P_1 and P_2, of masses $2m$ and m, respectively, are interconnected by two cords, one with length a and the other $8a$, the latter suspended around a cylinder with a diameter of $2a$, set horizontally on bearings without friction. The system is in equilibrium in the illustrated configuration. Determine the angle θ_1.

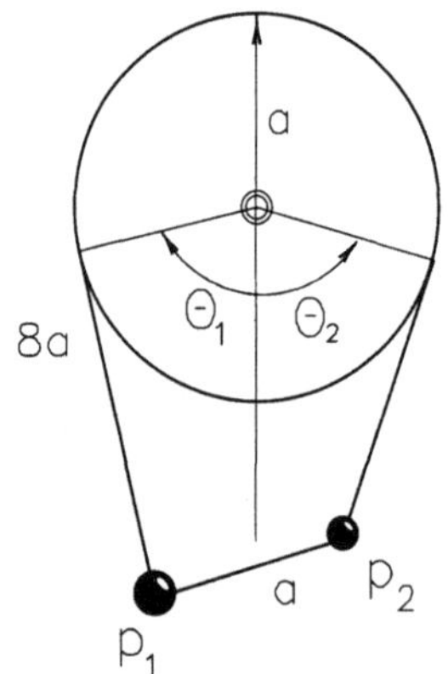

Figure P 2.11

P2.12 The telescopic antenna, of mass m, is pivoted on the point O at the support S and can turn freely around the axis x_1. The antenna is at rest in the vertical position (the friction existing between the elements is sufficient to support its own weight), when a force **F**, with components F_1, F_2, F_3, is applied to the end P, on the Cartesian basis of the figure. Reduce to point G, the antenna's mass center, the system of external forces acting on the antenna at this instant.

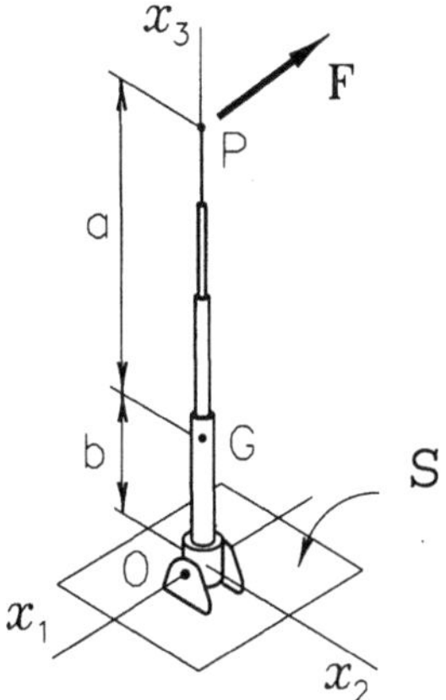

Figure P 2.12

P2.13 The balloon has volume V and specific mass $\rho_0/3$, where ρ_0 is the density of the atmosphere. The AO rope is flexible (does not resist bending), with length a and mass m. Determine the height b of the balloon, the dragging force of the wind, and the force at point O, knowing that the slope at this point is θ_0.

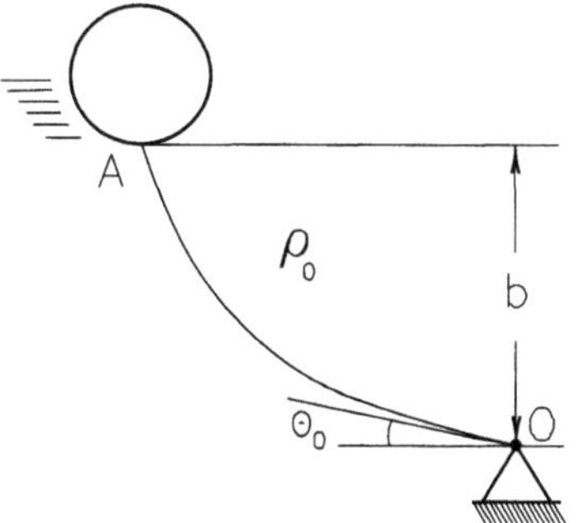

Figure P 2.13

P2.14 Particle P, with a mass of m, is r away from the mass center B*
of the homogeneous bar B, with mass M and length $2a$. By reducing the
system of gravitational forces exerted by P on B at point B*, this results in a
gravitational force and a gravitational torque. Calculate the limit of this pair
when $a/r \longrightarrow 0$.

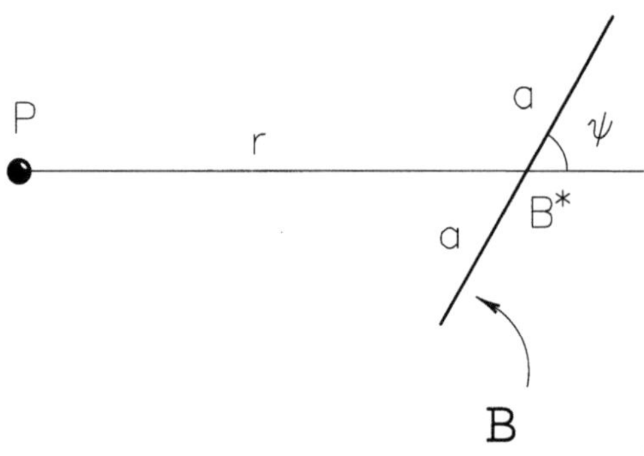

Figure P 2.14

P2.15 To turn the axis, the monkey wrench weighing 30 N would be applied
to the hexagonal end, a horizontal force of 40 N being applied at its end B,
as illustrated. When ascertaining, however, that the monkey wrench was
slightly smaller than necessary, someone suggested that the same result would
be obtained by applying a screwdriver (a wrench, therefore) at some point on
groove A. Determine the r-coordinate of this point.

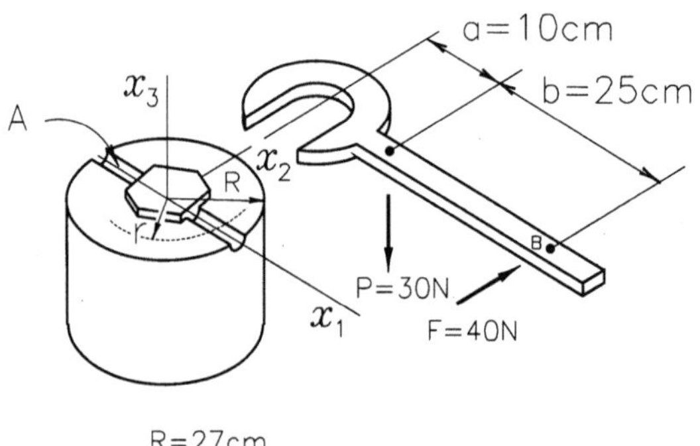

Figure P 2.15

P2.16 The figure is a sketch of a ceiling fan. The rotor/blades set (H) revolves around the symmetry axis, in relation to the motor cover (C), which is fixed on the support (S) by a ball connection that can be tightened to a greater or lesser degree, using a butterfly screw (B). Point G is the mass center of the suspended set. The fan weighs 50 N and is switched off, with its symmetry axis on the horizontal ($\theta = 0$). Tightening the butterfly screw permits the joint to support torques to the maximum limit of 8 N m, guaranteeing, therefore, immobility of the set. When the fan is switched on, the system of forces exerted by the air in motion on the blades is equivalent to a wrench applied at point Q, with resultant $\mathbf{F} = -30\mathbf{n}_1$ N and a resultant torque of $\mathbf{T} = -6\mathbf{n}_1$ N m. Does the cover stay still? Assuming that the tightening moment on the butterfly is proportional to the maximum resistant torque on the joint, how much extra percentage in tightening the butterfly is needed to keep immobility (or, if one concluded that the first tightening is sufficient, what is the percentage of loosening that would still keep this mobility)?

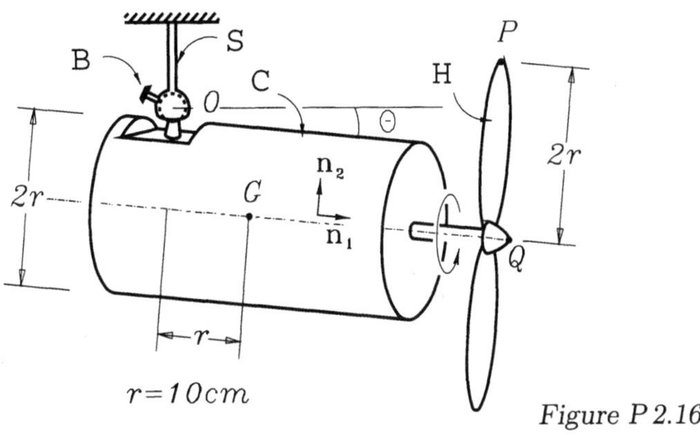

Figure P 2.16

P2.17 The system consists of the sliding vectors $\mathbf{F}_1$, $\mathbf{F}_2$, $\mathbf{F}_3$, and $\mathbf{F}_4$, associated to the straight lines indicated and of magnitudes equal to 5 N, 3 N, $10\sqrt{2}$ N and 2 N, respectively, and the free vectors $\mathbf{M}_1$, of magnitude 10 N m, and $\mathbf{M}_2$, of magnitude 8 N m. The cube in the figure has edges with a length of 2 m. Determine the position vector, with respect to A, of the point of the central axis closest to it.

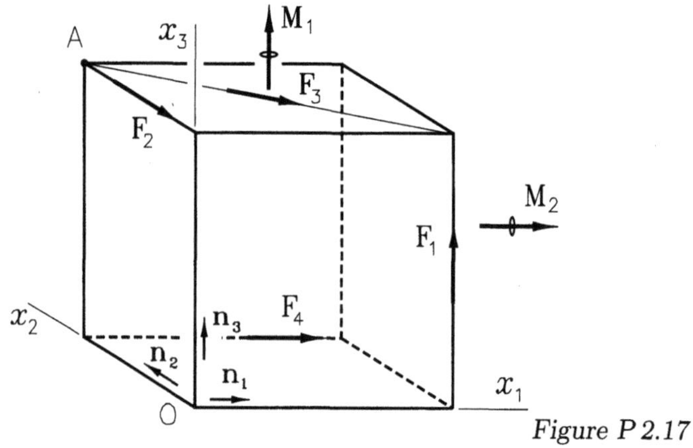

Figure P 2.17

P2.18 Consider the system illustrated in the figure, consisting of two small disks P and Q, of the same mass m each, joined by two rigid light bars of the same length r, at a vertical axis that is revolving in relation to the reference frame $\mathcal{R}$, as indicated. The axes $\{x_1, x_2, x_3\}$ are fixed on the vertical axis and the articulations at O and Q are pivots revolving around the x_3-axis the z-and axis respectively, so that the bars always stay on the plane $x_1 x_2$. Indicate the system of external forces acting on the mechanical system.

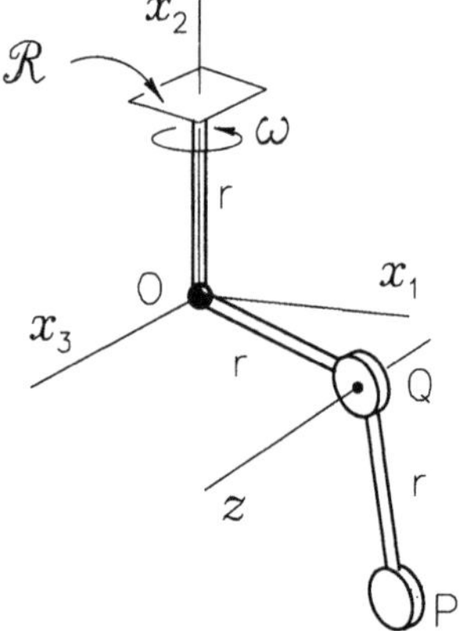

Figure P 2.18

P2.19 Consider the lid, shown in the figure, consisting of a homogeneous rectangular plate of mass m, joined at the support by two hinges, being opened using a cable, passing through a pulley, to which a force of magnitude F is applied. Draw a free body diagram of the plate, indicating all the force and torque components applied. Reduce this system to vertex B.

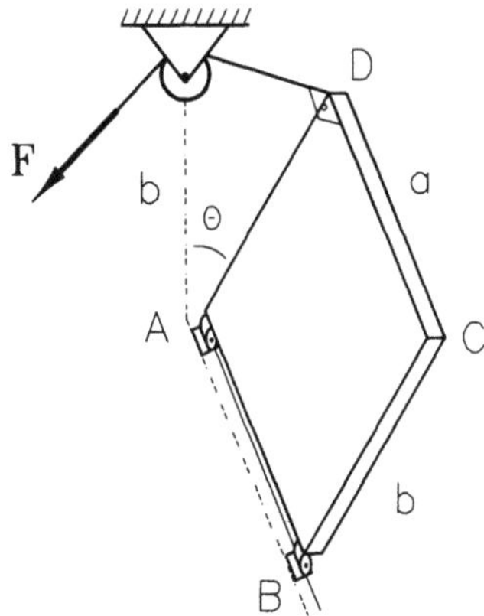

Figure P 2.19

P2.20 A gate, in the shape of a fourth of a cylinder with radius r, is being designed to regulate the level of a canal. The gate shall be able to rotate freely around the axis x, thanks to the pivots at A and B. Specify the relative density d of the building material of the gate, so that it opens when the water level reaches the height $a = \frac{2}{3}r$.

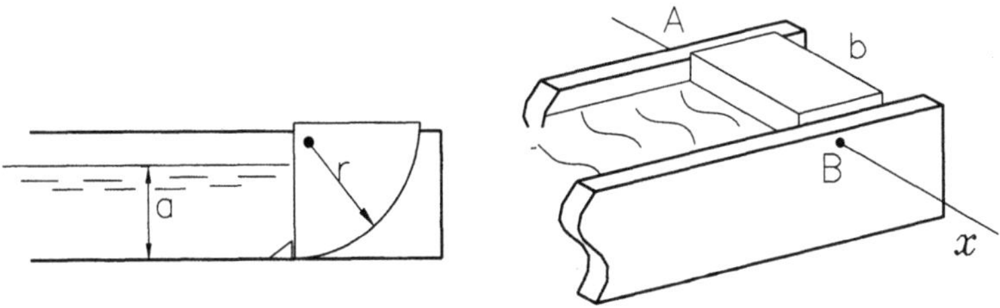

Figure P 2.20

P2.21   The homogeneous bar B, of mass M, is partly lying on the fixed circular base A being pivoted at its end O to a vertical pin fixed on the base, as illustrated. At the other end, Q, a light wire, with length a, is fixed, with a small sphere P, of mass m hanging from it. Consider the mechanical system consisting of B and P and draw a diagram of the external forces applied to it (there is friction between B and A, but the friction on the pin is negligible).

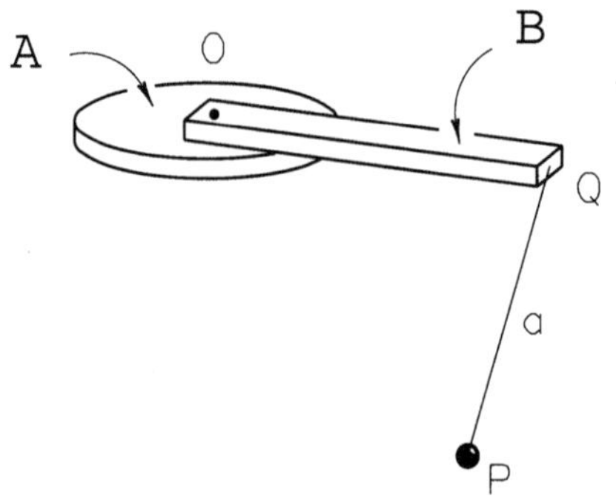

Figure P 2.21

P2.22   The light beam is supported at its ends, subject to the vertical loading $q(x) = Q(1 - \frac{x}{c})$ and to torque $T = \frac{1}{6}Qc^2$, in the direction indicated. Determine the forces V_1 and V_2 on the supports, knowing that the beam is at rest. Now, removing the right support, the force on the left support assumes a new value V_1' and the system is no longer a null system (the beam will move). Can this new system be reduced to a single force applied at one point?

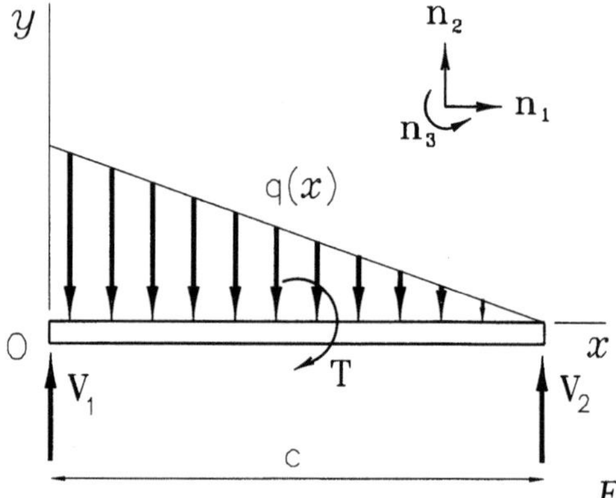

Figure P 2.22

P2.23 Bar B is connected to the guide rod G by a compound link, described below. The axes $\{x_1, x_2, x_3\}$ are fixed to the bar, with x_1 orthogonal to the cursor C and x_2 parallel to its longitudinal axis. The bar can turn freely in relation to the cursor around the axis x_3, while the latter can freely slide with respect to the guide rod, as shown. The axis x_3, however, is built to make a constant angle α, with the guide rod axis. How many force and torque components does this link offer, and how many degrees of freedom does it permit?

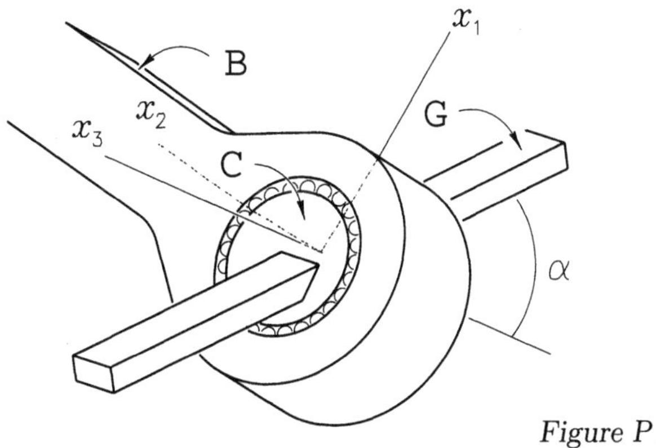

Figure P 2.23

P2.24 A series of n identical blades, with length $2a$ each, are piled on top of each other, as illustrated. What is the maximum balance of each to guarantee the set's equilibrium?

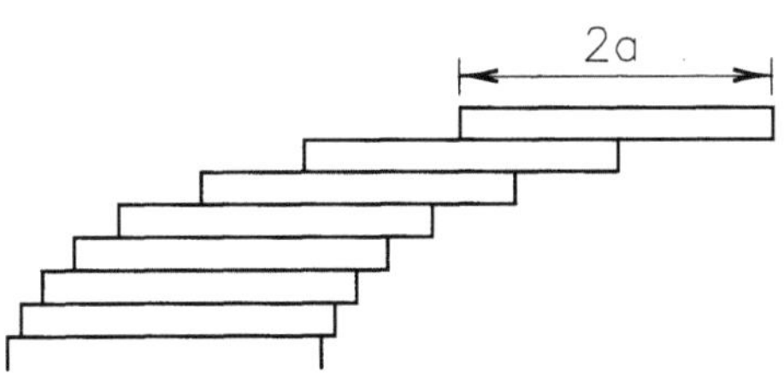

Figure P 2.24

P2.25   Two cylinders with radius r are lying at rest on a horizontal surface, kept together by a rope with length $2r$, joining their centers, and supporting a third homogeneous cylinder, with mass m and radius R, as shown. Assuming that the frictions are negligible, calculate the stress on the rope.

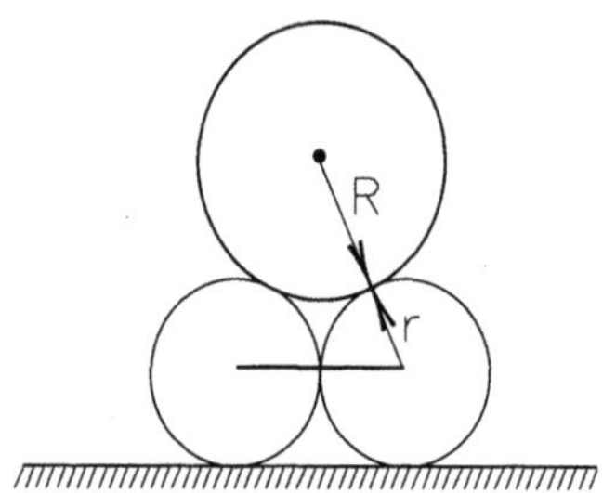

Figure P 2.25

P2.26   A uniform bar with length c can slide inside a cylindrical surface with a radius of r. Determine the maximum angle θ that guarantees the equilibrium of the bar if the friction coefficient at the points of contact is μ.

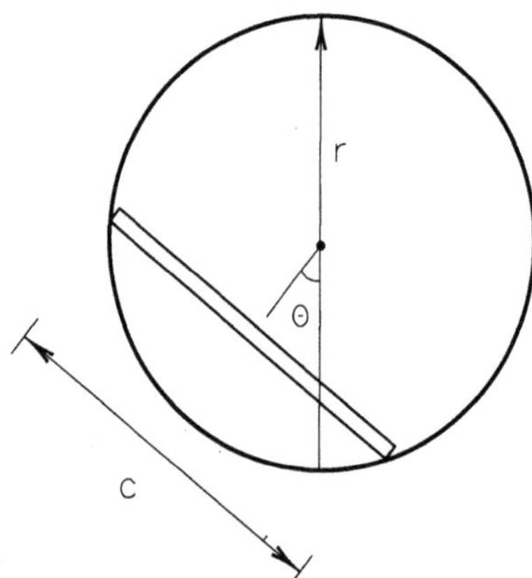

Figure P 2.26

P2.27 Three homogeneous identical balls are lying at rest on a horizontal surface, touching each other and kept together by a rope passing through the common equatorial plane. A fourth ball, with mass of $m = 10$ kg, lies on top of the other three. Calculate the stress on the rope, not considering all frictions.

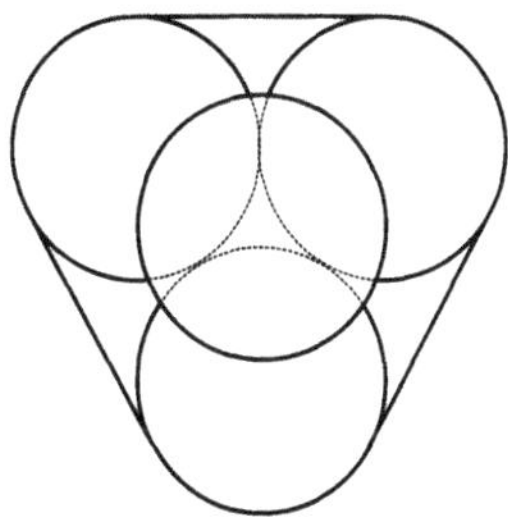

Figure P 2.27

P2.28 A prismatic and homogeneous block of concrete, with density equal to double that of the water, is at rest lying on the bottom of a canal of width a, damming water to a height of $a/2$, as illustrated. The distributed forces exerted by the bottom of the canal on the block have horizontal components (whose resultant is known as *friction force*) and vertical components (whose resultant is called *normal force*). The vertical distributed force is equivalent to a force equal to its resultant applied at a certain point. Determine the coordinate b of this point.

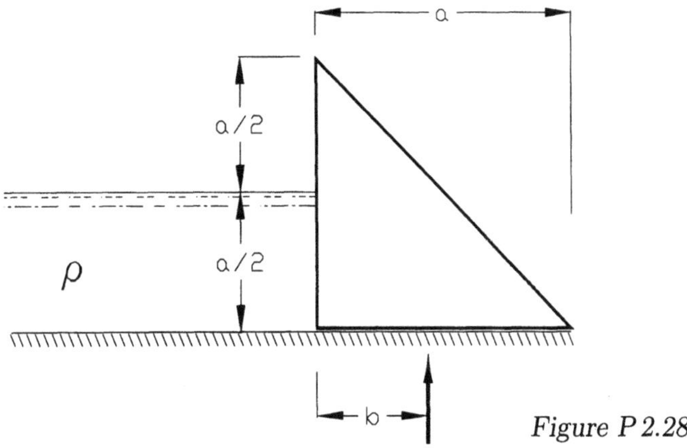

Figure P 2.28

Kinematics

Kinematics plays a fundamental role in the study of dynamics. On the one hand, because it deals precisely with the *motion* of mechanical systems, whose description is the main subject of the discipline; on the other, because, in order to obtain the equations of motion of any system, it will always be necessary to first express velocities and accelerations of its elements as functions of the coordinates that describe the configuration of the system.

The motion of bodies will be discussed here in general terms. Since all motion is relative, the concept of a *reference frame* is introduced right at the beginning and all motion will always be described in relation to one or more reference frames. Special attention, however, will be paid to the relations between the kinematic properties that express the motion of a point or a rigid body when the observer changes the reference frame. The discussion of kinematics will give priority to the study of relative motion between different reference frames; the concept of angular velocity of a rigid body will thereby appear, of course, to be a condition necessary for fully developing the kinematic concepts. The motion of the point will, therefore, be discussed after that of the rigid body, since the kinematic theorems relating velocities and accelerations will result from the relative angular velocities and accelerations with respect to different reference frames.

The differentiation of vector functions is studied in Section 3.1, with emphasis on the dependence of the derivative of a vector on the reference frame where this derivative is calculated. Section 3.2 discusses what is perhaps the most important concept of kinematics: the notion of angular velocity of a rigid body, showing that the angular velocity vector acts as a differential operator between two reference frames. In Section 3.3 a general equation is derived that relates the derivatives of a vector in two reference frames in their relative motion. This relationship will allow the study of the motion in different reference frames, to greatly facilitate the calculation of velocities and accelerations. Section 3.4 introduces the notion of angular acceleration. These first four sections concentrate attention on the relative rotation between rigid bodies since, as already mentioned, all subsequent kinematic relations will be obtained from the general equations deduced therein. Section 3.5 introduces the concepts of position, velocity, and acceleration, which are necessary for describing the motion of the point or particle. In Section 3.6 the so-called kinematic theorems are derived, which relate velocities and accelerations in different reference frames. Section 3.7 introduces the use of *intrinsic coordinates* in the study of the motion of the particle. Then the geometrical and functional relations between the trajectory of a particle and its velocity and acceleration vectors are analyzed in greater detail. In Section 3.8 the motion of the rigid body is studied, first relating the velocities and accelerations of points of a body with its angular velocity and angular accelerations. Next, mention is made of some important categories of motion of a rigid body, such as plane motion and motion with a fixed point. Section 3.9 discusses the *rolling* between two bodies, stating the kinematic premises of this fairly common condition. Last, Section 3.10 addresses the general study of mechanisms, discussing the main procedures of kinematic analysis of multibody systems. Generalized coordinates and holonomicity of a general mechanical system are also discussed in this section.

This chapter is of the utmost importance for understanding almost all the rest of the material in the book. The correct choice of reference frames, bases, and coordinates is fundamental for easy problem-solving and mastering these elements is achieved by studying kinematics in depth. Therefore, it is very important for the reader to clearly understand its content before going ahead. It is, therefore, suggested that the reader, after studying each section, work on the corresponding exercises.

3.1 Differentiation of Vectors and Reference Frames

The differential equations governing the motion of mechanical systems always involve time derivatives of vector functions. For instance, the dynamic equation that expresses Newton's second law for a particle, $m\mathbf{a} = \mathbf{R}$, relates a resultant force vector, $\mathbf{R}$, to an acceleration vector, $\mathbf{a}$, which is, in turn, defined as the second time derivative of a position vector $\mathbf{p}$. In short, to be able to establish equations of motion, it is necessary first to know how to correctly differentiate vectorial functions. On the other hand, the above dynamic equation is only valid when the acceleration is observed from certain reference frames, and not for others. In other words, you must know how to relate the derivatives of a vector in different reference frames. In this section, we will study how to differentiate vector functions and see why their derivatives depend on the observer.

When motion is discussed, a reference frame is implicit. The motion of a point or body can only be described in relation to something else. A *reference frame* will be defined here as a set of non-colinear points, with distances from each other invariant with time. A *rigid body* can be taken as a reference frame, since it fully meets the conditions that define the latter. In practice, the concepts of rigid body and reference frame will be used alternatively. A *plane* can also be adopted as a reference frame, for the same reason. A *line* or *point*, however, does not constitute a reference frame, since it does not fulfill the conditions established in the above definition.

A reference frame may, therefore, be understood as any object to which a system of Cartesian axes can be associated. Nevertheless, the concept of *reference frame* must not be confused with that of *coordinates system*. Different systems of coordinates can be associated with the same reference frame, whenever applicable; but a system of coordinates cannot be fixed simultaneously in two reference frames, unless there is no relative motion between them. Given a reference frame and having chosen a system of Cartesian coordinates, it is usual to adopt a basis of adimensional and mutually orthogonal unit vectors, fixed in the reference frame and parallel to the axes of the system, in which all other vectors are decomposed. It should always be remembered that a vector does not depend on the basis chosen for its decomposition, although its components do, of course, depend on the orientation of the basis.

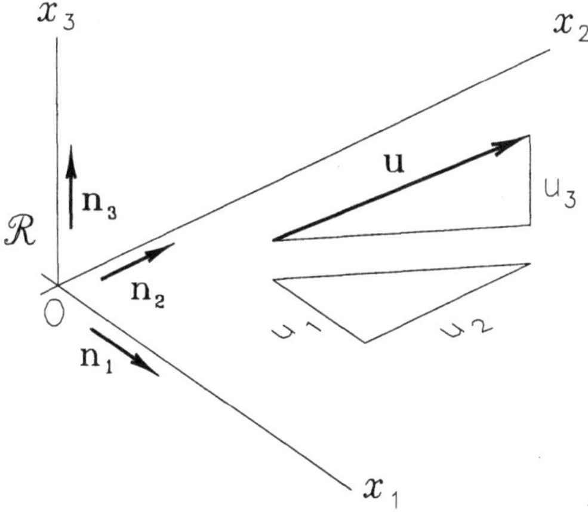

Figure 1.1

Figure 1.1 illustrates a reference frame $\mathcal{R}$, to which a system of Cartesian axes $\{x_1, x_2, x_3\}$ is associated, with origin O, and a basis of orthonormal vectors, $\mathbf{n}_1, \mathbf{n}_2, \mathbf{n}_3$, fixed in $\mathcal{R}$. Any vector $\mathbf{u}$ can be represented by its component in this basis, as (see Appendix A)

$$\mathbf{u} = \sum_{j=1}^{3} u_j \, \mathbf{n}_j = \sum_{j=1}^{3} \mathbf{u} \cdot \mathbf{n}_j \, \mathbf{n}_j. \tag{1.1}$$

If the vector $\mathbf{u}$ varies with time, it is said to be a *vector function*, $\mathbf{u}(t)$, whose scalar components $u_j(t) = \mathbf{u}(t) \cdot \mathbf{n}_j$, $j = 1, 2, 3$, are *scalar functions* of time. As the chosen basis is fixed in $\mathcal{R}$, the time derivatives of the base vectors are null: $d\mathbf{n}_j/dt = 0$, $j = 1, 2, 3$, and the time derivative of the vector $\mathbf{u}$ in $\mathcal{R}$ will be

$$\frac{d\mathbf{u}}{dt} = \sum_{j=1}^{3} \dot{u}_j \mathbf{n}_j, \tag{1.2}$$

where $\dot{u}_j$ is the reduced notation for du_j/dt.

Now let A and B be two different reference frames, moving independently in space, in which the orthonormal basis $\mathbf{a}_1, \mathbf{a}_2, \mathbf{a}_3$ and

$\mathbf{b}_1, \mathbf{b}_2, \mathbf{b}_3$, respectively, are fixed (see Fig. 1.2). An arbitrary vector $\mathbf{u}$ can be expressed alternatively as

$$\mathbf{u} = \sum_{j=1}^{3} \mathbf{u} \cdot \mathbf{a}_j \, \mathbf{a}_j = \sum_{j=1}^{3} \mathbf{u} \cdot \mathbf{b}_j \, \mathbf{b}_j. \tag{1.3}$$

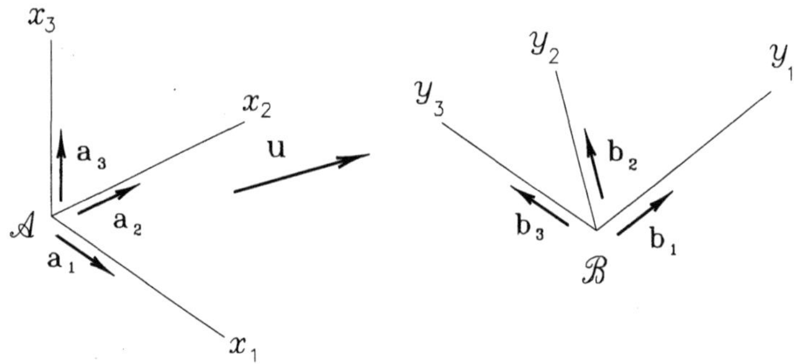

Figure 1.2

Although, in general, $\mathbf{u} \cdot \mathbf{a}_j \neq \mathbf{u} \cdot \mathbf{b}_j$, $j = 1, 2, 3$, both vector sums result in the same vector $\mathbf{u}$. The time derivative of the vector $\mathbf{u}$ will, however, depend on the reference frame from which the vector is observed, since its components on each basis depend on the motion of the respective reference frame. The following example illustrates the point.

Example 1.1   The system of axes $\{x, y\}$, in the plane in Fig. 1.3, rotates around point O in relation to the axes $\{X, Y\}$, according to the function $\theta(t)$ (see Fig. 1.3). The unit vector $\mathbf{n}_1$ is parallel to the axis x, fixed, therefore, in the reference frame with which the system $\{x, y\}$ is associated. In relation then to this reference frame, the time derivative of $\mathbf{n}_1$ will be null: $d\mathbf{n}_1/dt = 0$. Now, by representing $\mathbf{n}_1$ in the basis $\mathbf{a}_1, \mathbf{a}_2$, then $\mathbf{n}_1 = \cos\theta\mathbf{a}_1 + \sin\theta\mathbf{a}_2$ and differentiating with respect to the time in the reference frame associated with the system of axes $\{X, Y\}$ (where the vectors $\mathbf{a}_1$ and $\mathbf{a}_2$ are fixed), then $d\mathbf{n}_1/dt = (-\sin\theta\mathbf{a}_1 + \cos\theta\mathbf{a}_2)\dot{\theta}(t) = \dot{\theta}(t)\mathbf{n}_2$, a different result from the previous one, unless $\theta(t)$ is constant, which only happens if there is no relative motion between the reference frames.

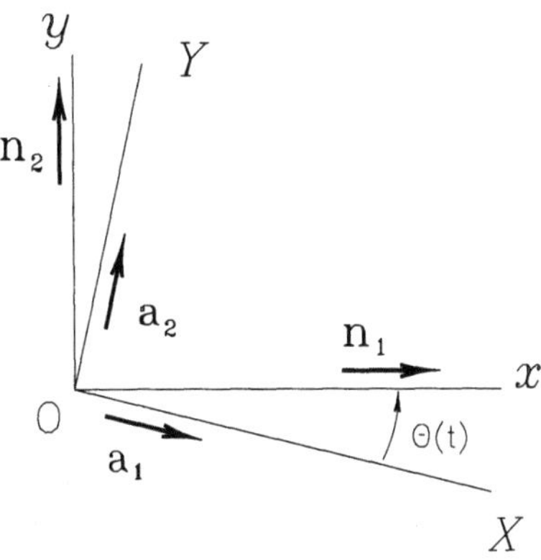

Figure 1.3

It is therefore, clear that the time derivative of a vector must necessarily allude to the reference frame in which it is being computed. The notation $^{\mathcal{R}}d\mathbf{v}/dt$ will be adopted here to designate the *time derivative of the vector* $\mathbf{v}$ *in the reference frame* $\mathcal{R}$, also called the *time rate of* $\mathbf{v}$ *in* $\mathcal{R}$. As the above example illustrates, given an arbitrary vector $\mathbf{v}$, its time derivatives in two reference frames A and B, moving in relation to each other, are generally different, that is,

$$\frac{^{A}d\mathbf{v}}{dt} \neq \frac{^{B}d\mathbf{v}}{dt}, \tag{1.4}$$

and the indication of the reference frame must not be suppressed in the notation of the derivative of a vector, except in a context where a single reference frame is implicit. In this case, the notation $^{\mathcal{R}}d\mathbf{v}/dt$ may be reduced to $d\mathbf{v}/dt$ or even to $\dot{\mathbf{v}}$, for the sake of simplicity. The reader must, however, be sure that there is no ambiguity when choosing the reduced notation, adopting the complete notation whenever there is any doubt regarding the reference frame. The differentiation of scalar functions *is always independent of the reference frame*; therefore, there is no ambiguity in using the abbreviated notation '$\dot{e}$' for the derivative '$\frac{^{\mathcal{R}}d}{dt}e(t)$'.

The differentiation of sums and products of vector functions follows simple working rules. Let $\mathbf{u}(t)$ and $\mathbf{v}(t)$ be vector functions of the variable t and let $e(t)$ be a scalar function of the same variable.

The following relations are fulfilled, under the condition that the vector differentiations are performed in the same reference frame:

$$\frac{^{\mathcal{R}}d}{dt}(\mathbf{u}+\mathbf{v}) = \frac{^{\mathcal{R}}d\mathbf{u}}{dt} + \frac{^{\mathcal{R}}d\mathbf{v}}{dt};$$ (1.5)

$$\frac{^{\mathcal{R}}d}{dt}(e\mathbf{v}) = \dot{e}\mathbf{v} + e\frac{^{\mathcal{R}}d\mathbf{v}}{dt};$$ (1.6)

$$\frac{^{\mathcal{R}}d}{dt}(\mathbf{u}\cdot\mathbf{v}) = \frac{^{\mathcal{R}}d\mathbf{u}}{dt}\cdot\mathbf{v} + \mathbf{u}\cdot\frac{^{\mathcal{R}}d\mathbf{v}}{dt};$$ (1.7)

$$\frac{^{\mathcal{R}}d}{dt}(\mathbf{u}\times\mathbf{v}) = \frac{^{\mathcal{R}}d\mathbf{u}}{dt}\times\mathbf{v} + \mathbf{u}\times\frac{^{\mathcal{R}}d\mathbf{v}}{dt};$$ (1.8)

$$\frac{^{\mathcal{R}}d}{dt}(\mathbf{u}\otimes\mathbf{v}) = \frac{^{\mathcal{R}}d\mathbf{u}}{dt}\otimes\mathbf{v} + \mathbf{u}\otimes\frac{^{\mathcal{R}}d\mathbf{v}}{dt}.$$ (1.9)

(If you still do not know the notation '$\otimes$', which means the *tensorial product* or *dyadic product* between vectors, refer to Appendix A.)

In more general terms, if the vector $\mathbf{p}(t)$ is the result of a multiple product (scalar, vectorial or tensorial product, in any possible order, with the proper brackets) of n vectors $\mathbf{v}_i(t)$, $i = 1, 2, \ldots, n$, that is,

$$\mathbf{p} = \mathbf{v}_1\mathbf{v}_2\cdots\mathbf{v}_n,$$ (1.10)

then the time derivative of $\mathbf{p}$ in a given reference frame can be expressed in terms of the time derivatives, in the same reference frame, of the component vectors, thus

$$\frac{^{\mathcal{R}}d\mathbf{p}}{dt} = \frac{^{\mathcal{R}}d\mathbf{v}_1}{dt}\mathbf{v}_2\cdots\mathbf{v}_n + \cdots + \mathbf{v}_1\mathbf{v}_2\cdots\frac{^{\mathcal{R}}d\mathbf{v}_n}{dt}.$$ (1.11)

Equations (1.5–1.9) and (1.11) may be easily demonstrated by fixing a basis in the reference frame, decomposing the vectors on this basis and then using the differentiation rules of scalar functions.

If $\mathbf{n}_1, \mathbf{n}_2, \mathbf{n}_3$ is an orthonormal basis fixed in the reference frame

$\mathcal{R}$, then

$$
\begin{aligned}
\frac{^{\mathcal{R}}d}{dt}(e\mathbf{v}) &= \frac{^{\mathcal{R}}d}{dt}\left(e\sum_{j=1}^{3}v_j\mathbf{n}_j\right) \\
&= \sum_{j=1}^{3}\frac{^{\mathcal{R}}d}{dt}(ev_j)\mathbf{n}_j \\
&= \sum_{j=1}^{3}(\dot{e}v_j + e\dot{v}_j)\mathbf{n}_j \\
&= \dot{e}\sum_{j=1}^{3}(v_j\mathbf{n}_j) + e\sum_{j=1}^{3}(\dot{v}_j\mathbf{n}_j) \\
&= \dot{e}\mathbf{v} + e\frac{^{\mathcal{R}}d\mathbf{v}}{dt},
\end{aligned}
$$

as Eq. (1.6) illustrates. Note that in the last passage of the above proof Eqs. (1.1) and (1.2) have been used. As an exercise, it is left to the reader to demonstrate the other results.

Let $\mathbf{v}$ be a nonnull vector varying with time in a reference frame $\mathcal{R}$. Its module $v = |\mathbf{v}|$ is defined as (see Appendix A)

$$
v^2 = \mathbf{v} \cdot \mathbf{v}. \tag{1.12}
$$

Differentiating with respect to time in $\mathcal{R}$, then

$$
\begin{aligned}
v\dot{v} &= \frac{1}{2}\left(\frac{^{\mathcal{R}}d\mathbf{v}}{dt}\cdot\mathbf{v} + \mathbf{v}\cdot\frac{^{\mathcal{R}}d\mathbf{v}}{dt}\right) \\
&= \mathbf{v}\cdot\frac{^{\mathcal{R}}d\mathbf{v}}{dt}.
\end{aligned}
\tag{1.13}
$$

This result shows that if a vector $\mathbf{v}$ has a constant module, its time derivative in any reference frame will be either a vector orthogonal to $\mathbf{v}$ or a null vector. In fact, if $\dot{v} = 0$, Eq. (1.13) implies that either $^{\mathcal{R}}d\mathbf{v}/dt$ and $\mathbf{v}$ are orthogonal or $^{\mathcal{R}}d\mathbf{v}/dt = 0$. Consequently, if $\mathbf{v}$ is a nonnull vector fixed in a rigid body C moving in a reference frame $\mathcal{R}$, its module v is constant and

$$
\frac{^{\mathcal{R}}d\mathbf{v}}{dt}\cdot\mathbf{v} = 0, \qquad \mathbf{v} \text{ fixed in } C. \tag{1.14}
$$

Specifically, if $\mathbf{n}_1, \mathbf{n}_2, \mathbf{n}_3$ is an orthonormal basis fixed in a rigid body C that moves in relation to a reference frame $\mathcal{R}$, then

$$\mathbf{n}_1 \cdot \mathbf{n}_1 = \mathbf{n}_2 \cdot \mathbf{n}_2 = \mathbf{n}_3 \cdot \mathbf{n}_3 = 1, \tag{1.15}$$

$$\mathbf{n}_1 \cdot \mathbf{n}_2 = \mathbf{n}_2 \cdot \mathbf{n}_3 = \mathbf{n}_3 \cdot \mathbf{n}_1 = 0. \tag{1.16}$$

Differentiating Eq. (1.15) with respect to time in the reference frame $\mathcal{R}$, using the reduced notation,

$$\dot{\mathbf{n}}_1 \cdot \mathbf{n}_1 = \dot{\mathbf{n}}_2 \cdot \mathbf{n}_2 = \dot{\mathbf{n}}_3 \cdot \mathbf{n}_3 = 0 \tag{1.17}$$

and also differentiating Eq. (1.16), we obtain

$$\begin{aligned}
\dot{\mathbf{n}}_1 \cdot \mathbf{n}_2 &= -\mathbf{n}_1 \cdot \dot{\mathbf{n}}_2, \\
\dot{\mathbf{n}}_2 \cdot \mathbf{n}_3 &= -\mathbf{n}_2 \cdot \dot{\mathbf{n}}_3, \\
\dot{\mathbf{n}}_3 \cdot \mathbf{n}_1 &= -\mathbf{n}_3 \cdot \dot{\mathbf{n}}_1.
\end{aligned} \tag{1.18}$$

If $\mathbf{v}$ is an arbitrary vector fixed in a body C that moves in relation to a reference frame $\mathcal{R}$, Eq. (1.14) guarantees that the time derivative in $\mathcal{R}$ of $\mathbf{v}$ is a vector orthogonal to $\mathbf{v}$ (or a null vector, which is not of interest at the moment).

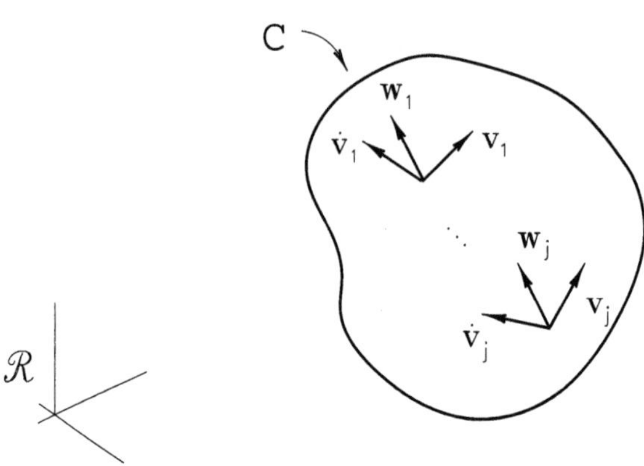

Figure 1.4

Then there is vector $\mathbf{w}$, which depends upon the motion of C in $\mathcal{R}$, that when cross-multiplied by $\mathbf{v}$, results in the vector $\frac{{}^{\mathcal{R}}d\mathbf{v}}{dt}$ (see Fig. 1.4), that is,

$$\frac{{}^{\mathcal{R}}d\mathbf{v}}{dt} = \mathbf{w} \times \mathbf{v}. \tag{1.19}$$

Now consider a set of vectors $\mathbf{v}_1, \mathbf{v}_2, \ldots, \mathbf{v}_n$, fixed in C. Each vector $\mathbf{v}_j$ will have its own time derivative in $\mathcal{R}$ orthogonal to itself and, therefore, equal to the cross product of a vector $\mathbf{w}_j$ with the vector $\mathbf{v}_j$ (see Fig. 1.4). (If the time derivative in $\mathcal{R}$ of a vector $\mathbf{v}_k$ is null, then the vector $\mathbf{w}_k$ will be parallel to $\mathbf{v}_k$.) Although each vector $\mathbf{w}_j$ depends on the motion of C in $\mathcal{R}$, there is no evidence so far of a *relation* between these vectors.

It is possible, however, to demonstrate that the vectors $\mathbf{w}_j$ are all equal to each other, that is, that there is a single vector that, by way of a mere cross product, differentiates any vector fixed in a body moving in a reference frame. This is, perhaps, the most important result of kinematics and is the subject of the following section.

3.2 Angular Velocity of a Rigid Body

Suppose that a rigid body C is moving arbitrarily in space in relation to a reference frame $\mathcal{R}$. Also assume that a certain vector $\mathbf{v}$ is fixed in C, that is,

$$\frac{{}^{C}d\mathbf{v}}{dt} = 0. \tag{2.1}$$

The time derivative of $\mathbf{v}$ in a reference frame $\mathcal{R}$ will depend on how the body C moves in relation to the reference frame. It would be useful to establish a relationship to calculate ${}^{\mathcal{R}}d\mathbf{v}/dt$ based on any parameter that describes the motion of C with respect to $\mathcal{R}$. The following *theorem*, which will be called *Theorem 0*, because it is the most fundamental result of kinematics, establishes this relationship.

> **Theorem 0.** *If a rigid body C moves in relation to a reference frame $\mathcal{R}$, then there is a vector ${}^{\mathcal{R}}\boldsymbol{\omega}^{C}$ so that, for every vector $\mathbf{v}$ fixed in C, its time derivative in the reference frame $\mathcal{R}$ is, at each instant, given by*

$$\frac{{}^{\mathcal{R}}d\mathbf{v}}{dt} = {}^{\mathcal{R}}\boldsymbol{\omega}^{C} \times \mathbf{v}. \tag{2.2}$$

The vector $^{\mathcal{R}}\boldsymbol{\omega}^C$, called the *angular velocity vector of C in $\mathcal{R}$*, is a kinematic property of the body and characterizes, as seen below, the time rate of change of *orientation* of the body in relation to the reference frame. According to Eq. (2.2), the vector $^{\mathcal{R}}\boldsymbol{\omega}^C$ acts as an *operator* that, when cross-multiplied on the left by any vector fixed in the body, yields the time derivative of the vector in the reference frame.

To demonstrate Eq. (2.2), let us take two nonnull vectors $\mathbf{p}$ and $\mathbf{q}$, linearly independent and fixed in C, so that their time derivatives in reference frame $\mathcal{R}$, $\dot{\mathbf{p}}$, and $\dot{\mathbf{q}}$ are nonnull and nonparallel. As $\mathbf{p}$ and $\mathbf{q}$ are fixed in C, their scalar product is a constant and, differentiating with respect to the time in $\mathcal{R}$, gives

$$\dot{\mathbf{p}} \cdot \mathbf{q} + \mathbf{p} \cdot \dot{\mathbf{q}} = 0. \tag{a}$$

Defining the scalar

$$k \rightleftharpoons \dot{\mathbf{p}} \cdot \mathbf{q} = -\mathbf{p} \cdot \dot{\mathbf{q}} \tag{b}$$

and the vector

$$\boldsymbol{\Omega} \rightleftharpoons \dot{\mathbf{p}} \times \dot{\mathbf{q}}, \tag{c}$$

then

$$\boldsymbol{\Omega} \times \mathbf{p} = (\dot{\mathbf{p}} \times \dot{\mathbf{q}}) \times \mathbf{p} = \mathbf{p} \cdot \dot{\mathbf{p}}\, \dot{\mathbf{q}} - \mathbf{p} \cdot \dot{\mathbf{q}}\, \dot{\mathbf{p}} = k\dot{\mathbf{p}}, \tag{d}$$

$$\boldsymbol{\Omega} \times \mathbf{q} = (\dot{\mathbf{p}} \times \dot{\mathbf{q}}) \times \mathbf{q} = \mathbf{q} \cdot \dot{\mathbf{p}}\, \dot{\mathbf{q}} - \mathbf{q} \cdot \dot{\mathbf{q}}\, \dot{\mathbf{p}} = k\dot{\mathbf{q}}. \tag{e}$$

If $k = 0$, the result of Eqs. (d) and (e) is $\boldsymbol{\Omega} \times \mathbf{p} = \boldsymbol{\Omega} \times \mathbf{q} = 0$. Now, by hypothesis, $\mathbf{p} \neq 0$, $\mathbf{q} \neq 0$ and $\boldsymbol{\Omega} \neq 0$. Since $\dot{\mathbf{p}}$, and $\dot{\mathbf{q}}$ are nonnull and nonparallel, it would be necessary for $\boldsymbol{\Omega}$ to be simultaneously parallel to $\mathbf{p}$ and $\mathbf{q}$, which is impossible. Therefore $k \neq 0$ and we can define the vector

$$^{\mathcal{R}}\boldsymbol{\omega}^C \rightleftharpoons \frac{1}{k} \boldsymbol{\Omega}. \tag{f}$$

The result of Eqs. (d), (e), and (f) is that

$$^{\mathcal{R}}\boldsymbol{\omega}^C \times \mathbf{p} = \dot{\mathbf{p}}, \tag{g}$$

$$^{\mathcal{R}}\boldsymbol{\omega}^C \times \mathbf{q} = \dot{\mathbf{q}}. \tag{h}$$

Now let $\mathbf{r}$ be a third vector fixed in C, defined by

$$\mathbf{r} = \mathbf{p} \times \mathbf{q}. \tag{i}$$

Then (see Appendix A),

$$\begin{aligned}
{}^{\mathcal{R}}\boldsymbol{\omega}^{C} \times \mathbf{r} &= {}^{\mathcal{R}}\boldsymbol{\omega}^{C} \times (\mathbf{p} \times \mathbf{q}) \\
&= ({}^{\mathcal{R}}\boldsymbol{\omega}^{C} \times \mathbf{p}) \times \mathbf{q} + \mathbf{p} \times ({}^{\mathcal{R}}\boldsymbol{\omega}^{C} \times \mathbf{q}) \\
&= \dot{\mathbf{p}} \times \mathbf{q} + \mathbf{p} \times \dot{\mathbf{q}} \\
&= \dot{\mathbf{r}}.
\end{aligned} \tag{j}$$

Last, as every vector $\mathbf{v}$ fixed in C can be expressed as a linear combination of the basis formed by $\mathbf{p}$, $\mathbf{q}$, and $\mathbf{r}$, that is,

$$\mathbf{v} = \alpha\mathbf{p} + \beta\mathbf{q} + \gamma\mathbf{r}, \tag{k}$$

where α, β, and γ are constants, then

$$\begin{aligned}
{}^{\mathcal{R}}\boldsymbol{\omega}^{C} \times \mathbf{v} &= {}^{\mathcal{R}}\boldsymbol{\omega}^{C} \times (\alpha\mathbf{p}) + {}^{\mathcal{R}}\boldsymbol{\omega}^{C} \times (\beta\mathbf{q}) + {}^{\mathcal{R}}\boldsymbol{\omega}^{C} \times (\gamma\mathbf{r}) \\
&= \alpha\dot{\mathbf{p}} + \beta\dot{\mathbf{q}} + \gamma\dot{\mathbf{r}} \\
&= \dot{\mathbf{v}}.
\end{aligned}$$
$\blacksquare$

It is worth noting that, whatever are the physical properties of vectors $\mathbf{p}$ and $\mathbf{q}$ used to form the angular velocity vector, then

$$\mathrm{Dim}\,[\boldsymbol{\omega}] = \frac{\mathrm{Dim}\,[\dot{\mathbf{p}}] \times \mathrm{Dim}\,[\dot{\mathbf{q}}]}{\mathrm{Dim}\,[\dot{\mathbf{p}}] \times \mathrm{Dim}\,[\mathbf{q}]}, \tag{2.3}$$

and the angular velocity vector will necessarily have a dimension $[\mathrm{T}^{-1}]$, that is, *radians per second*, in metric units.

Example 2.1   The cube C, with edge a, rotates in relation to the reference frame $\mathcal{R}$, keeping the vertex A fixed in $\mathcal{R}$ (see Fig. 2.1). At a given instant, the time rates in $\mathcal{R}$ of the unit vectors $\mathbf{c}_1$ and $\mathbf{c}_2$, fixed in the cube, are $\dot{\mathbf{c}}_1 = \gamma\mathbf{c}_2 - \beta\mathbf{c}_3$, $\dot{\mathbf{c}}_2 = -\gamma\mathbf{c}_1 + \alpha\mathbf{c}_3$, where α, β, and γ are nonnull real constants. As $\mathbf{c}_1$ and $\mathbf{c}_2$ are linearly independent and their derivatives are not parallel, Eqs. (b), (c), and (f) can be used to calculate the angular velocity vector ${}^{\mathcal{R}}\boldsymbol{\omega}^{C}$:

$$k = \dot{\mathbf{c}}_1 \cdot \mathbf{c}_2 = \gamma,$$

$$\,^{\mathcal{R}}\boldsymbol{\omega}^{C} = \frac{1}{k}\dot{\mathbf{c}}_1 \times \dot{\mathbf{c}}_2 = \alpha\mathbf{c}_1 + \beta\mathbf{c}_2 + \gamma\mathbf{c}_3.$$

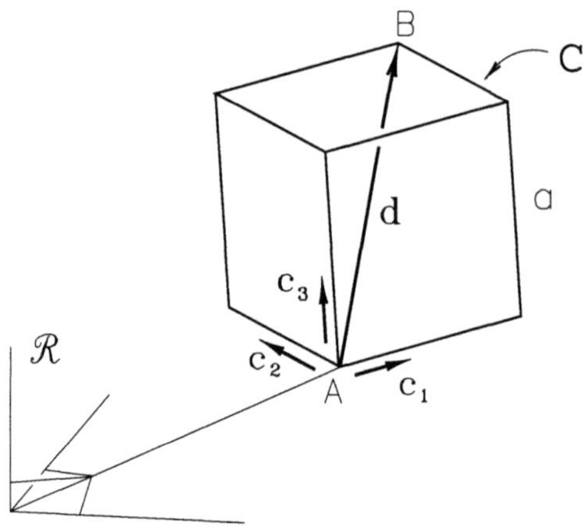

Figure 2.1

Once the angular velocity vector of the cube is known, it is easy to calculate the time derivative of any vector fixed in it using Eq. (2.2). If, for example, $\mathbf{d}$ is the position vector of the vertex B with respect to point A (its module, d, is the diagonal of the cube),

$$\mathbf{d} = a(\mathbf{c}_1 + \mathbf{c}_2 + \mathbf{c}_3),$$

its time derivative in $\mathcal{R}$ will be

$$\dot{\mathbf{d}} = {}^{\mathcal{R}}\boldsymbol{\omega}^C \times \mathbf{d} = a\left[(\beta - \gamma)\mathbf{c}_1 + (\gamma - \alpha)\mathbf{c}_2 + (\alpha - \beta)\mathbf{c}_3\right].$$

Note that

$$\dot{\mathbf{d}} \cdot \mathbf{d} = a^2(\beta - \gamma + \gamma - \alpha + \alpha - \beta) = 0,$$

as would be expected, since $\mathbf{d}$ is a vector fixed in C.

The equation adopted to prove the existence of the angular velocity vector,

$$^{\mathcal{R}}\boldsymbol{\omega}^C = \frac{1}{k}\dot{\mathbf{p}} \times \dot{\mathbf{q}}, \tag{2.4}$$

where

$$k = \dot{\mathbf{p}} \cdot \mathbf{q}, \tag{2.5}$$

although effective in demonstrating the theorem, is rarely used in practice to calculate the angular velocity vector, since it is not usual to know the time derivatives of two vectors fixed in the body in question. What is important is to note that, according to Eqs. (2.4) and (2.5), the angular velocity vector of a rigid body in a given reference frame is *determined* from the time derivatives in the reference frame of two (and only two) independent vectors fixed in the body.

The attentive reader may have asked him or herself, if given a rigid body C moving arbitrarily in a reference frame $\mathcal{R}$, is it always possible, at any instant, to find a pair of vectors $\mathbf{p}$ and $\mathbf{q}$, fixed in C, fulfilling the conditions established in the proof of the theorem. The answer is *yes*, and an example can be given as follows. Suppose C is moving arbitrarily relative to $\mathcal{R}$ and consider the orthonormal basis $\mathbf{n}_1, \mathbf{n}_2, \mathbf{n}_3$ fixed in C, satisfying, therefore, Eq. (1.17). If, for example, $\dot{\mathbf{n}}_1$ and $\dot{\mathbf{n}}_2$ are, at a given instant, not parallel, then $\mathbf{n}_1$ and $\mathbf{n}_2$ can take the place of the vectors $\mathbf{p}$ and $\mathbf{q}$ in constructing the angular velocity vector, which will be ${}^{\mathcal{R}}\boldsymbol{\omega}^C = \frac{\dot{\mathbf{n}}_1 \times \dot{\mathbf{n}}_2}{\dot{\mathbf{n}}_1 \cdot \mathbf{n}_2}$. If, on the contrary, $\dot{\mathbf{n}}_1$ and $\dot{\mathbf{n}}_2$ are, at a given instant, parallel to each other, they will necessarily be parallel to $\mathbf{n}_3$ (unit vector simultaneously orthogonal to $\mathbf{n}_1$ and $\mathbf{n}_2$), which, in turn, is always orthogonal to $\dot{\mathbf{n}}_3$ (see Fig. 2.2). In this case, $\mathbf{n}_3$ and $\mathbf{n}_2$ (or $\mathbf{n}_3$ and $\mathbf{n}_1$) can take the place of the vectors $\mathbf{p}$ and $\mathbf{q}$ in constructing the angular velocity vector, which will be ${}^{\mathcal{R}}\boldsymbol{\omega}^C = \frac{\dot{\mathbf{n}}_2 \times \dot{\mathbf{n}}_3}{\dot{\mathbf{n}}_2 \cdot \mathbf{n}_3}$. In other words, given three linearly independent vectors fixed in a rigid body, it is always possible, at each instant, to find two of these vectors whose time derivatives in a given reference frame are also independent.

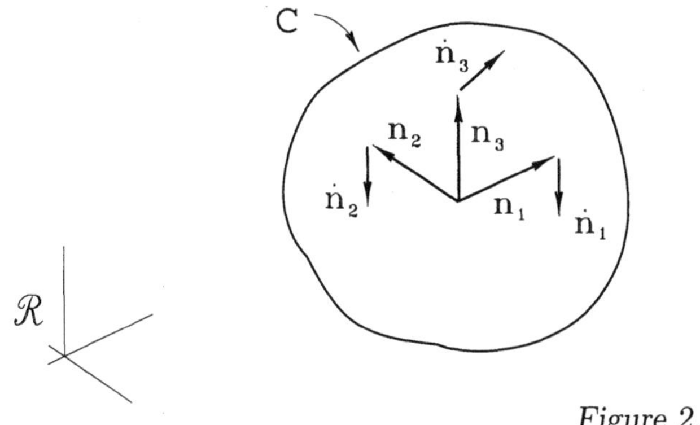

Figure 2.2

Just one last comment regarding the theorem. It has been shown that the angular velocity vector *exists*, ${}^{\mathcal{R}}\boldsymbol{\omega}^C$, which differentiates with respect to $\mathcal{R}$ any and every vector $\mathbf{v}$ fixed in C. Will there be more than one vector with this property? The answer is *no* and this is easy to check. Let us assume that there are two vectors $\boldsymbol{\omega}$ and $\boldsymbol{\omega}'$ so that $\boldsymbol{\omega} \times \mathbf{v} = \boldsymbol{\omega}' \times \mathbf{v} = \dot{\mathbf{v}}$. Then, $(\boldsymbol{\omega} - \boldsymbol{\omega}') \times \mathbf{v} = 0$ and, since $\mathbf{v}$ is arbitrary, $\boldsymbol{\omega} = \boldsymbol{\omega}'$, then *the angular velocity vector is unique*.

When a body B moves in relation to a reference frame A so that, during a certain time interval, there is a unit vector $\mathbf{n}$ fixed simultaneously to B and A, it is said that B has a *simple angular velocity* in A. When this happens, the angular velocity vector of B in A is parallel to the vector $\mathbf{n}$ and may be expressed as

$$^{A}\boldsymbol{\omega}^{B} = \omega\mathbf{n}, \tag{2.6}$$

where

$$\omega = d\theta/dt \tag{2.7}$$

and $\theta(t)$ is the angle formed by two straight lines orthogonal to $\mathbf{n}$, one fixed in B and the other in A, chosen so that θ increases when the line fixed in B rotates in the righthand direction around vector $\mathbf{n}$. Figure 2.3 illustrates the situation.

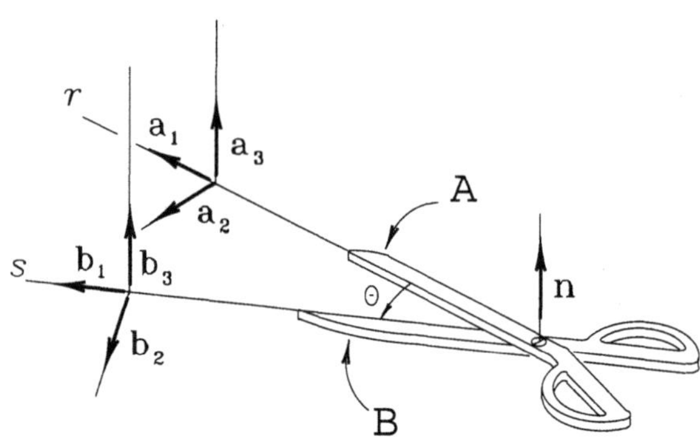

Figure 2.3

The vectors $\mathbf{a}_1, \mathbf{a}_2, \mathbf{a}_3$ form an orthonormal basis fixed in the body A, with $\mathbf{a}_1$ parallel to the line r and $\mathbf{a}_3 = \mathbf{n}$, while the vectors $\mathbf{b}_1, \mathbf{b}_2, \mathbf{b}_3$ form another orthonormal basis, fixed in the body B, with $\mathbf{b}_1$ parallel to the line s and $\mathbf{b}_3 = \mathbf{a}_3 = \mathbf{n}$. The vector $\mathbf{n}$ is fixed in both A and B, thereby configuring a simple angular velocity between B and A. The angular velocity vector of B in A may be broken down in the basis fixed in A: $^A\boldsymbol{\omega}^B = \omega_1\mathbf{a}_1 + \omega_2\mathbf{a}_2 + \omega_3\mathbf{a}_3$, where ω_1, ω_2, and ω_3 are scalar functions of time. The time derivative in A of the vector $\mathbf{n}$, fixed in B, will be $^A d\mathbf{n}/dt = {}^A\boldsymbol{\omega}^B \times \mathbf{n} = \omega_2\mathbf{a}_1 - \omega_1\mathbf{a}_2$. However, as $\mathbf{n}$ is also fixed in A, this derivative will be null and, therefore, $\omega_1 = \omega_2 = 0$. Then $^A\boldsymbol{\omega}^B = \omega_3\mathbf{a}_3 = \omega\mathbf{n}$, as Eq. (2.6) establishes. Now, $\mathbf{b}_1, \mathbf{b}_2, \mathbf{b}_3$ may be expressed in the basis fixed in A: $\mathbf{b}_1 = \cos\theta\mathbf{a}_1 + \sin\theta\mathbf{a}_2$, $\mathbf{b}_2 = -\sin\theta\mathbf{a}_1 + \cos\theta\mathbf{a}_2$, $\mathbf{b}_3 = \mathbf{a}_3$, and differentiating with respect to the time in the reference frame A, yields

$$^A d\mathbf{b}_1/dt = \dot{\theta}(-\sin\theta\mathbf{a}_1 + \cos\theta\mathbf{a}_2) = \dot{\theta}\mathbf{b}_2,$$
$$^A d\mathbf{b}_2/dt = -\dot{\theta}(\cos\theta\mathbf{a}_1 + \sin\theta\mathbf{a}_2) = -\dot{\theta}\mathbf{b}_1,$$
$$^A d\mathbf{b}_3/dt = 0.$$

Now by calculating the same derivatives from Eq. (2.2)

$$^A d\mathbf{b}_1/dt = {}^A\boldsymbol{\omega}^B \times \mathbf{b}_1 = \omega\mathbf{b}_2,$$
$$^A d\mathbf{b}_2/dt = {}^A\boldsymbol{\omega}^B \times \mathbf{b}_2 = -\omega\mathbf{b}_1,$$
$$^A d\mathbf{b}_3/dt = {}^A\boldsymbol{\omega}^B \times \mathbf{b}_3 = 0.$$

When comparing these results, it is evident that $\omega = \dot{\theta}$, as Eq. (2.7) indicates. Also note that the angle θ increases when the line s rotates in relation to line r in the righthand screw direction of $\mathbf{n}$.

Example 2.2 Figure 2.4 illustrates a cylinder C rolling down a ramp R. P, Q, and S are three points fixed to the edge of the cylinder and θ measures the angle between the line passing through P and Q and the edge of the ramp, as illustrated. The vectors $\mathbf{c}_1, \mathbf{c}_2, \mathbf{c}_3$ form an orthonormal basis fixed to C while the vectors $\mathbf{n}_1, \mathbf{n}_2, \mathbf{n}_3$ form an orthonormal basis fixed in R. Vector $\mathbf{c}_3$ is parallel to the cylinder axis and, therefore, is also fixed in R, with $\mathbf{c}_3 = \mathbf{n}_3$, thus characterizing a simple angular velocity between C and R. The angular velocity vector of C in R will therefore be $^R\boldsymbol{\omega}^C = \dot{\theta}\mathbf{c}_3$. If ϕ is the angle formed by the line passing through P and S and the edge of the ramp, it is easy to see that $\phi(t) = \theta(t) + \alpha$, where α is a constant angle. Then $\dot{\phi} = \dot{\theta}$ and the angular velocity vector may be expressed either as $^R\boldsymbol{\omega}^C = \dot{\phi}\mathbf{c}_3$ or even as $^R\boldsymbol{\omega}^C = \dot{\phi}\mathbf{n}_3$. Note that when θ and ϕ increase (cylinder rolling down the ramp), the angular velocity vector has the same direction as $\mathbf{n}_3$.

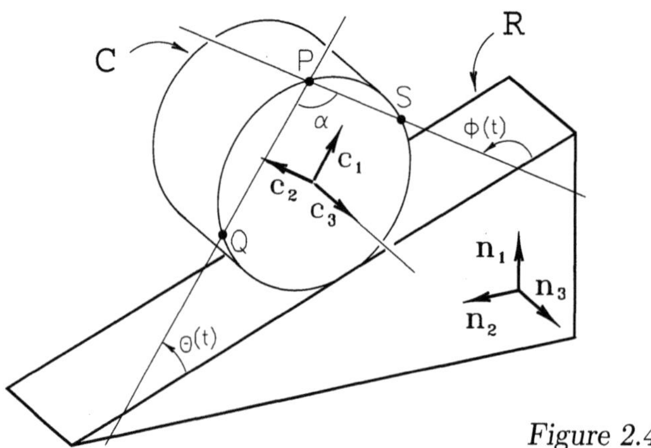

Figure 2.4

3.3 Use of Different Reference Frames

We saw in Section 3.1 that the time derivatives of a vector $\mathbf{v}$ in two reference frames A and B in relative motion are, generally, different, but a general relation has not yet been established between these derivatives, as a result of their relative motion. This can be obtained from Eq. (2.2), deduced in the preceding section, as shown below.

Let us assume that a reference frame B moves in relation to another reference frame A at an angular velocity ${}^{A}\boldsymbol{\omega}^{B}$. In general, the angular velocity vector varies with time, not being fixed in any of the reference frames. Also let three orthonormal vectors $\mathbf{b}_1, \mathbf{b}_2, \mathbf{b}_3$ be fixed in B (see Fig. 3.1). If $\mathbf{u}$ is any vector varying in time in relation to both reference frames, its decomposition on the basis fixed in B will be

$$\mathbf{u} = \sum_{j=1}^{3} u_j, \mathbf{b}_j \tag{3.1}$$

and its time derivative in B will be

$$\frac{{}^{B}d\mathbf{u}}{dt} = \sum_{j=1}^{3} \dot{u}_j, \mathbf{b}_j, \tag{3.2}$$

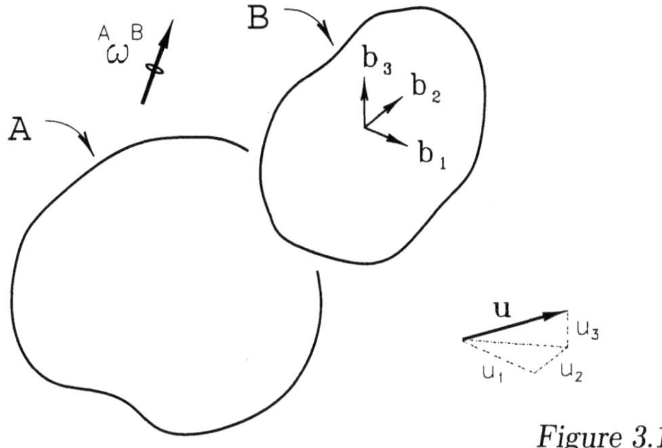

Figure 3.1

since the basis vectors are fixed in B. Now by differentiating Eq. (3.1) in reference frame A, we have

$$\frac{{}^A d\mathbf{u}}{dt} = \sum_{j=1}^{3} \dot{u}_j, \mathbf{b}_j + \sum_{j=1}^{3} u_j \frac{{}^A d}{dt}\mathbf{b}_j. \tag{3.3}$$

One extra term is to be found in Eq. (3.3), if compared with Eq. (3.2), and this is due to the fact that the vectors $\mathbf{b}_j$, $j = 1, 2, 3$, vary in relation to A. Since these vectors are fixed in B, their derivatives in A can be obtained from Eq. (2.2), that is,

$$\frac{{}^A d}{dt}\mathbf{b}_j = {}^A\omega^B \times \mathbf{b}_j, \qquad j = 1, 2, 3. \tag{3.4}$$

By substituting Eqs. (3.2) and (3.4) in Eq. (3.3), then

$$\begin{aligned}
\frac{{}^A d\mathbf{u}}{dt} &= \frac{{}^B d\mathbf{u}}{dt} + \sum_{j=1}^{3} u_j {}^A\omega^B \times \mathbf{b}_j \\
&= \frac{{}^B d\mathbf{u}}{dt} + {}^A\omega^B \times \sum_{j=1}^{3} u_j \mathbf{b}_j,
\end{aligned} \tag{3.5}$$

resulting in the desired relation:

$$\frac{{}^A d\mathbf{u}}{dt} = \frac{{}^B d\mathbf{u}}{dt} + {}^A\omega^B \times \mathbf{u}. \tag{3.6}$$

Equation (3.6) relates the derivatives of any vector **u** with respect to two reference frames moving arbitrarily, in relation to each other. If the vector **u** is fixed in reference frame B, the first term on the right side of the equation vanishes, being reduced to Eq. (2.2). In a more general case, Eq. (3.6) shows that the difference between the derivatives of a vector in two reference frames is given by the term $^A\boldsymbol{\omega}^B \times \mathbf{u}$. In other words, the time rate of change of any vector **u**, in a reference frame A, is equal to the time rate of change of the vector in another reference frame B, added to the derivative, in A, of a vector equal to **u**, but fixed in B.

Example 3.1   Figure 3.2 illustrates part of a gyroscope, consisting of a rotor C, which can rotate freely around its axis of symmetry, fixed to the frame B, which, in its turn, can rotate around the horizontal axis, fixed

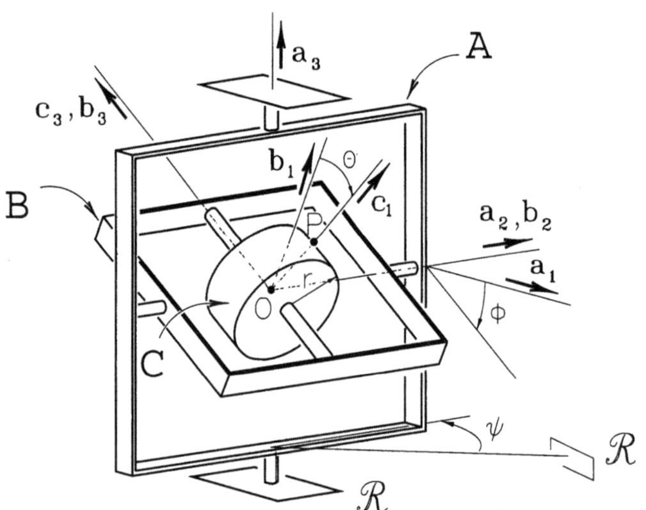

Figure 3.2

to the outer frame A. This, finally, can rotate around the vertical axis, fixed in a reference frame $\mathcal{R}$. The bases of orthonormal vectors: $\mathbf{a}_1, \mathbf{a}_2, \mathbf{a}_3$, $\mathbf{b}_1, \mathbf{b}_2, \mathbf{b}_3$, and $\mathbf{c}_1, \mathbf{c}_2, \mathbf{c}_3$, fixed in A, B, and C, respectively, have their directions indicated. P is a point of the edge of the rotor and the position vector of point P with respect to the center of the rotor, O, is $\mathbf{p} = r\mathbf{c}_1 = r(\cos\theta\mathbf{b}_1 + \sin\theta\mathbf{b}_2)$, where r is the radius of the rotor and θ measures its rotation in relation to B, as indicated. As the vector $\mathbf{c}_3 = \mathbf{b}_3$ is fixed simultaneously in C and B, then there is a simple angular velocity $^B\boldsymbol{\omega}^C =$

$\dot{\theta}\mathbf{b}_3$ and the time derivative of the vector $\mathbf{p}$, fixed in C, in relation to B is, according to Eq. (2.2),

$$\frac{{}^B d\mathbf{p}}{dt} = {}^B\boldsymbol{\omega}^C \times \mathbf{p}$$
$$= \dot{\theta}\mathbf{b}_3 \times r(\cos\theta\mathbf{b}_1 + \sin\theta\mathbf{b}_2)$$
$$= \dot{\theta}r(-\sin\theta\mathbf{b}_1 + \cos\theta\mathbf{b}_2).$$

As the vector $\mathbf{b}_2 = \mathbf{a}_2$ is fixed simultaneously in B and in A, then there is also a simple angular velocity ${}^A\boldsymbol{\omega}^B = \dot{\phi}\mathbf{b}_2$, where ϕ is the angle measuring the rotation of B in relation to A, as shown. The time derivative of the vector $\mathbf{p}$ in relation to the reference frame A can then be obtained, using Eq. (3.6), as

$$\frac{{}^A d\mathbf{p}}{dt} = \frac{{}^B d\mathbf{p}}{dt} + {}^A\boldsymbol{\omega}^B \times \mathbf{p}$$
$$= \dot{\theta}r(-\sin\theta\mathbf{b}_1 + \cos\theta\mathbf{b}_2) + \dot{\phi}\mathbf{b}_2 \times r(\cos\theta\mathbf{b}_1 + \sin\theta\mathbf{b}_2)$$
$$= r(-\dot{\theta}\sin\theta\mathbf{b}_1 + \dot{\theta}\cos\theta\mathbf{b}_2 - \dot{\phi}\cos\theta\mathbf{b}_3).$$

See the corresponding animation.

Note that, for the sake of convenience, in the above example a basis has been adopted, fixed in reference frame B, to calculate the derivative in the reference frame A of vector $\mathbf{p}$, fixed in C. This is a usual procedure and the reader should endeavor always to choose a basis that simplifies the calculation of the derivatives, independent of the reference frame to which it is associated. When necessary, the results can be transferred to another basis, by a simple procedure of vector decomposition.

The relative angular velocities between two reference frames are opposites, that is, given two reference frames A and B in relative motion,

$$ {}^A\boldsymbol{\omega}^B = -{}^B\boldsymbol{\omega}^A. \tag{3.7}$$

In fact, if $\mathbf{u}$ is any nonnull vector, we can use Eq. (3.6) twice and obtain

$$\frac{{}^A d\mathbf{u}}{dt} = \frac{{}^B d\mathbf{u}}{dt} + {}^A\boldsymbol{\omega}^B \times \mathbf{u} = \frac{{}^A d\mathbf{u}}{dt} + {}^B\boldsymbol{\omega}^A \times \mathbf{u} + {}^A\boldsymbol{\omega}^B \times \mathbf{u}, \tag{3.8}$$

so,

$$({}^B\boldsymbol{\omega}^A + {}^A\boldsymbol{\omega}^B) \times \mathbf{u} = 0 \tag{3.9}$$

and, as $\mathbf{u}$ is arbitrary, results in Eq. (3.7).

Example 3.2   With regard to Example 3.1 (see Fig. 3.2), the angular velocity of the frame A in relation to the frame B will be ${}^{B}\boldsymbol{\omega}^{A} = -{}^{A}\boldsymbol{\omega}^{B} = -\dot{\phi}\mathbf{b}_2$ and the angular velocity of the frame B in relation to the rotor C will be ${}^{C}\boldsymbol{\omega}^{B} = -{}^{B}\boldsymbol{\omega}^{C} = -\dot{\theta}\mathbf{b}_3$.

When three or more reference frames move in relation to each other, there is an angular velocity vector that decribes the relative motion of each pair of reference frames (in fact, there are two opposite vectors, as Eq. (3.7) indicates). These vectors keep a relation of additiveness between each other that can be described as follows: Let $\mathcal{R}_1, \mathcal{R}_2, \ldots, \mathcal{R}_N$ be N reference frames moving arbitrarily in space. Their relative angular velocities satisfy the relation

$$\mathcal{R}_1\boldsymbol{\omega}^{\mathcal{R}_N} = {}^{\mathcal{R}_1}\boldsymbol{\omega}^{\mathcal{R}_2} + {}^{\mathcal{R}_2}\boldsymbol{\omega}^{\mathcal{R}_3} + \cdots + {}^{\mathcal{R}_{N-1}}\boldsymbol{\omega}^{\mathcal{R}_N}. \tag{3.10}$$

Specifically, for $N = 3$, a relation very often used is

$$\mathcal{R}_1\boldsymbol{\omega}^{\mathcal{R}_3} = {}^{\mathcal{R}_1}\boldsymbol{\omega}^{\mathcal{R}_2} + {}^{\mathcal{R}_2}\boldsymbol{\omega}^{\mathcal{R}_3}. \tag{3.11}$$

To demonstrate Eq. (3.11), let us take any vector $\mathbf{v}$, fixed in the reference frame $\mathcal{R}_3$. Its derivative in $\mathcal{R}_1$ will, according to Eq. (2.2), be ${}^{\mathcal{R}_1}d\mathbf{v}/dt = {}^{\mathcal{R}_1}\boldsymbol{\omega}^{\mathcal{R}_3} \times \mathbf{v}$. Its derivative in $\mathcal{R}_2$, also obtained from Eq. (2.2), will be ${}^{\mathcal{R}_2}d\mathbf{v}/dt = {}^{\mathcal{R}_2}\boldsymbol{\omega}^{\mathcal{R}_3} \times \mathbf{v}$. Now using Eq. (3.6) to relate the time derivatives of the vector $\mathbf{v}$ in the two reference frames, we have

$$\frac{{}^{\mathcal{R}_1}d\mathbf{v}}{dt} = \frac{{}^{\mathcal{R}_2}d\mathbf{v}}{dt} + {}^{\mathcal{R}_1}\boldsymbol{\omega}^{\mathcal{R}_2} \times \mathbf{v};$$

therefore,

$$\mathcal{R}_1\boldsymbol{\omega}^{\mathcal{R}_3} \times \mathbf{v} = \left({}^{\mathcal{R}_2}\boldsymbol{\omega}^{\mathcal{R}_3} + {}^{\mathcal{R}_1}\boldsymbol{\omega}^{\mathcal{R}_2}\right) \times \mathbf{v},$$

and, as $\mathbf{v}$ is arbitrary, the result is Eq. (3.11). Equation (3.10) can be derived by induction from Eq. (3.11), and will be left to the reader as an exercise.

These results are extremely useful for determining the angular velocity vector of a rigid body moving in a complex manner in a given reference frame by breaking down of this vector in simple angular velocities — generally easy to calculate — whose vector sum is the desired angular velocity vector.

Example 3.3   Returning again to Example 3.1 (see Fig. 3.2), the angular velocity vector of the rotor C in the frame A will be ${}^{A}\boldsymbol{\omega}^{C} = {}^{A}\boldsymbol{\omega}^{B} + {}^{B}\boldsymbol{\omega}^{C} = \dot{\phi}\mathbf{b}_2 + \dot{\theta}\mathbf{b}_3$. If $\psi(t)$ is the angle measuring the rotation of the frame A in the reference frame $\mathcal{R}$, then the simple angular velocity ${}^{\mathcal{R}}\boldsymbol{\omega}^{A} = \dot{\psi}\mathbf{a}_3 = \dot{\psi}(\cos\phi\,\mathbf{b}_1 + \sin\phi\,\mathbf{b}_3)$, and the angular velocity of the rotor in the reference frame $\mathcal{R}$ will be, according to Eq. (3.11),

$$
{}^{\mathcal{R}}\boldsymbol{\omega}^{C} = {}^{\mathcal{R}}\boldsymbol{\omega}^{A} + {}^{A}\boldsymbol{\omega}^{C} = \dot{\psi}\cos\phi\,\mathbf{b}_1 + \dot{\phi}\mathbf{b}_2 + (\dot{\theta} + \dot{\psi}\sin\phi)\mathbf{b}_3.
$$

Although not every motion of a rigid body is, as in the case of the gyroscope rotor in the above example, an addition of simple angular velocities between various reference frames, it is always possible to determine the expression for the angular velocity vector of a body moving arbitrarily in a given reference frame by introducing fictitious reference frames with simple angular velocities with respect to each other, in order to use Eq. (3.10).

Example 3.4   Figure 3.3a illustrates the motion of a coin M rolling over a horizontal plane $\mathcal{R}$, to which the system of Cartesian coordinates $\{X, Y, Z\}$ and the basis of unit vectors $\mathbf{n}_x, \mathbf{n}_y, \mathbf{n}_z$ is associated, as shown.

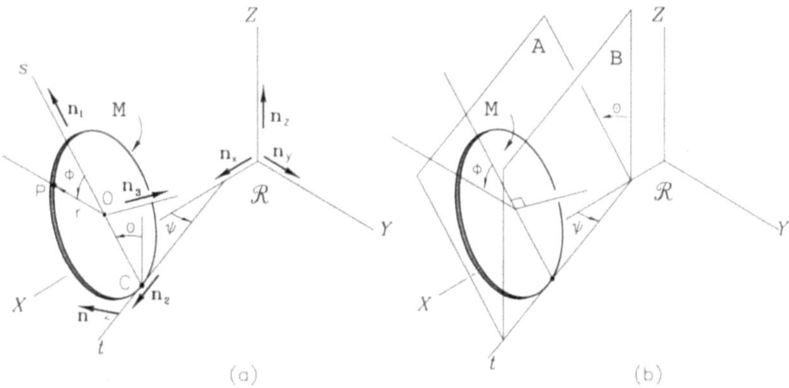

Figure 3.3

Point O is the geometric center of the coin; C is the point on the edge of the coin in contact with the plane, at the instant shown; and P is a fixed point on the surface of the coin. The orientation of the coin in relation to the plane can be described by time functions $\theta(t), \phi(t)$, and $\psi(t)$, where θ measures the angle between the line containing C and O (line s). ϕ is the angle between line s and the line containing P and O. Last, ψ is the angle between the horizontal line touching the edge of the coin at point C (line t) and the Cartesian axis X. These three angles used to describe the orientation of the coin are a variant of what are called *Euler's angles*, which will be discussed later. The basis of orthonormal vectors $\mathbf{n}_1, \mathbf{n}_2, \mathbf{n}_3$, with $\mathbf{n}_1$ parallel to line s, $\mathbf{n}_2$ parallel to line t, and $\mathbf{n}_3$ orthogonal to the plane of the surface of the coin, although convenient to describe the motion, is not in either M or $\mathcal{R}$. The general motion of the coin, with the three angles varying in time, is quite complex. This does not form a simple angular velocity, and to describe it we will introduce two intermediary reference frames (see Fig. 3.3b). So let A be the plane containing the surface of the coin (note that basis $\mathbf{n}_1, \mathbf{n}_2, \mathbf{n}_3$ is fixed in A) and B be the vertical plane containing line t. Vector $\mathbf{n}_3$ is fixed in M and in A and the angular velocity of M at A will be a simple angular velocity, $^{A}\boldsymbol{\omega}^{M} = \dot{\phi}\mathbf{n}_3$. Vector $\mathbf{n}_2$ is fixed in A and in B and configures a simple angular velocity, $^{B}\boldsymbol{\omega}^{A} = \dot{\theta}\mathbf{n}_2$. Last, vector $\mathbf{n}_z$ is fixed in B and in $\mathcal{R}$ and the angular velocity of B in $\mathcal{R}$ is $^{\mathcal{R}}\boldsymbol{\omega}^{B} = \dot{\psi}\mathbf{n}_z$. Using Eq. (3.10), then the angular velocity of the coin in relation to plane $\mathcal{R}$ may be expressed as

$$^{\mathcal{R}}\boldsymbol{\omega}^{M} = {}^{\mathcal{R}}\boldsymbol{\omega}^{B} + {}^{B}\boldsymbol{\omega}^{A} + {}^{A}\boldsymbol{\omega}^{M} = \dot{\psi}\mathbf{n}_z + \dot{\theta}\mathbf{n}_2 + \dot{\phi}\mathbf{n}_3$$

and, by breaking down $\mathbf{n}_z = \cos\theta\,\mathbf{n}_1 + \sin\theta\,\mathbf{n}_3$, then

$$^{\mathcal{R}}\boldsymbol{\omega}^{M} = \dot{\psi}\cos\theta\,\mathbf{n}_1 + \dot{\theta}\mathbf{n}_2 + (\dot{\phi} + \dot{\psi}\sin\theta)\mathbf{n}_3.$$

It is interesting to note the similarity between the results obtained in Examples 3.3 and 3.4. In both, scalar variables θ, ϕ, and ψ describe the *orientation* of the rigid body under study, in relation to a given reference frame. There is, however, neither a "rotation vector" nor an "orientation vector" of the body in the reference frame, whose time derivative in the reference frame is equal to the angular velocity vector of the rigid body in the reference frame. Rotations cannot be modeled as vectors since rotations, in general, are not commutative. Example 3.5 illustrates this point more clearly.

Example 3.5 Consider the rectangular block to which the axes $\{x, y, z\}$ are fixed (see Fig. 3.4a) and whose surface B$'$, opposite to B, is facing the onlooker.

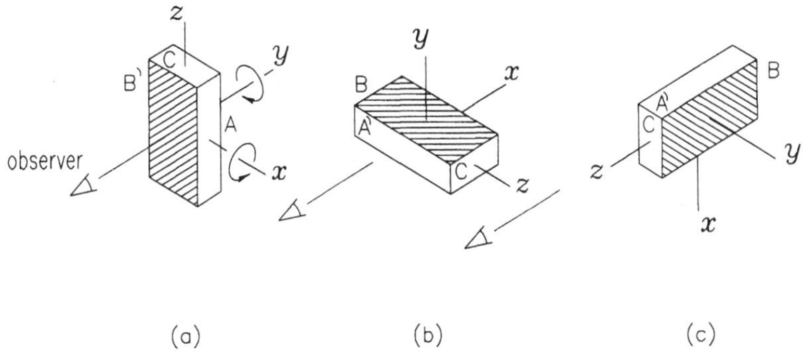

Figure 3.4

After a rotation of $\pi/2$ radians around the x-axis, followed by a rotation of the same angle around the y-axis, both positive, the configuration is that shown in Fig. 3.4b, with the surface A$'$, opposite to A, facing the onlooker. Based now on the same original configuration and proceeding with the same rotations, but in the reverse order, the result is that shown in Fig. 3.4c, with the surface C facing the onlooker. Therefore, here one cannot speak of a "rotation vector" or "orientation vector," since the commutative properties of vector algebra are not fulfilled. The reader will have a better understanding of the idea by reproducing this simple experiment with a matchbox. Also, try repeating the experiment by doing rotations of $\pi/2$ radians around axes fixed to the onlooker. What conclusion is reached?

In general, three scalar functions measuring angles between certain lines in space are needed to determine the *orientation* — also called the *attitude* — of a rigid body in relation to a given reference frame. Examples 3.1 and 3.4 clearly illustrate this fact, describing the body orientation in both using the scalar functions $\theta(t)$, $\phi(t)$, and $\psi(t)$. These functions — called *angular coordinates* of the body — and their time derivatives will be present in the composition of the angular velocity vector of the body in the reference frame, as seen in those examples. The angular velocity vector can be understood, therefore, as a measurement of the time rate of change of the body orientation in the reference frame, even though it is not at all the derivative of any vector, as seen above.

When a rigid body C moves in relation to a reference frame $\mathcal{R}$ so that, at a given interval of time, $^{\mathcal{R}}\boldsymbol{\omega}^{C} = 0$, it is said that in that interval there is *translation* (which, of course, includes the particular case of C being *fixed* in $\mathcal{R}$). When translation occurs, Eq. (3.6) is reduced to

$$\frac{^{\mathcal{R}}d\mathbf{v}}{dt} = \frac{^{C}d\mathbf{v}}{dt} \qquad \text{if} \qquad {^{\mathcal{R}}\boldsymbol{\omega}^{C}} = 0, \tag{3.12}$$

that is, if there is no relative rotation, the time derivative of an arbitrary vector $\mathbf{v}$ is the same in both reference frames. Translational motion is discussed in further detail in Section 3.8.

Returning to Eq. (3.6), it may be conjectured whether the time derivatives of a nonnull vector in two reference frames that have a relative rotational motion may be equal under any circumstance. The equation indicates that this will occur if the vector in question is parallel to the angular velocity vector. In particular, the angular velocity vector itself has equal time derivatives in both reference frames, that is,

$$\frac{^{B}d}{dt}{^{A}\boldsymbol{\omega}^{B}} = \frac{^{A}d}{dt}{^{A}\boldsymbol{\omega}^{B}}, \tag{3.13}$$

as can easily be seen from Eq. (3.6).

3.4 Angular Acceleration

If C is a rigid body moving arbitrarily in a reference frame $\mathcal{R}$, the angular velocity vector of C in $\mathcal{R}$, $^{\mathcal{R}}\boldsymbol{\omega}^{C}$, is, in general, a vector function of time, and therefore is not fixed either in C or in $\mathcal{R}$. Its time derivative in reference frame $\mathcal{R}$ is called the *angular acceleration of body C in the reference frame $\mathcal{R}$*,

$$^{\mathcal{R}}\boldsymbol{\alpha}^{C} \rightleftharpoons \frac{^{\mathcal{R}}d}{dt}{^{\mathcal{R}}\boldsymbol{\omega}^{C}}. \tag{4.1}$$

Note that although it is defined as the time derivative in $\mathcal{R}$ of the angular velocity, Eq. (3.13) shows that the angular acceleration is also equal to the time derivative, in the body, of its angular velocity.

If a rigid body C moves at a simple angular velocity in a given reference frame $\mathcal{R}$, its angular acceleration in this reference frame will be parallel to the angular velocity vector and its scalar component will be the time derivative of the scalar component of the angular velocity vector, that is,

$$^{\mathcal{R}}\boldsymbol{\alpha}^{C} = \dot{\omega}\mathbf{n} = \ddot{\theta}\mathbf{n}, \tag{4.2}$$

where $\omega = \dot{\theta}$ is the scalar component of the simple angular velocity vector and $\mathbf{n}$ is a unit vector simultaneously fixed in C and in $\mathcal{R}$. In fact, to differentiate vector ${}^{\mathcal{R}}\boldsymbol{\omega}^{C}$ in reference frame $\mathcal{R}$, the only requirement is to differentiate its scalar component, since vector $\mathbf{n}$ is constant in $\mathcal{R}$.

Example 4.1 Figure 4.1 reproduces the gyroscope studied in Example 3.1.

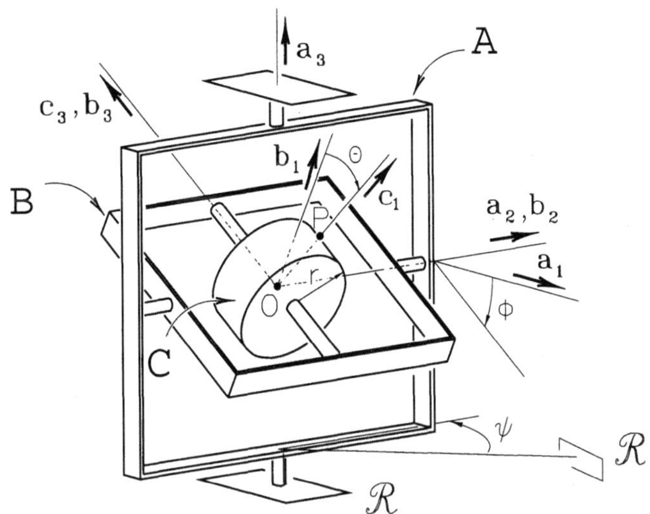

Figure 4.1

The angular acceleration of rotor C in relation to frame B is

$$
{}^{B}\boldsymbol{\alpha}^{C} = \frac{{}^{B}d}{dt}{}^{B}\boldsymbol{\omega}^{C} = \ddot{\theta}\mathbf{b}_3,
$$

and the angular acceleration of rotor C in relation to frame A is

$$
{}^{A}\boldsymbol{\alpha}^{C} = \frac{{}^{A}d}{dt}{}^{A}\boldsymbol{\omega}^{C} = \ddot{\phi}\mathbf{b}_2 + \ddot{\theta}\mathbf{b}_3 + \dot{\theta}\frac{{}^{A}d}{dt}\mathbf{b}_3.
$$

But

$$
\frac{{}^{A}d\mathbf{b}_3}{dt} = {}^{A}\boldsymbol{\omega}^{B} \times \mathbf{b}_3 = \dot{\phi}\mathbf{b}_2 \times \mathbf{b}_3 = \dot{\phi}\mathbf{b}_1,
$$

so

$$
{}^{A}\boldsymbol{\alpha}^{C} = \dot{\theta}\dot{\phi}\mathbf{b}_1 + \ddot{\phi}\mathbf{b}_2 + \ddot{\theta}\mathbf{b}_3.
$$

The angular acceleration of rotor C in reference frame $\mathcal{R}$ (see Example 3.3) is

$$
\begin{aligned}
{}^{\mathcal{R}}\boldsymbol{\alpha}^{C} &= \frac{{}^{\mathcal{R}}d}{dt}{}^{\mathcal{R}}\boldsymbol{\omega}^{C} \\
&= (\ddot{\psi}\cos\phi - \dot{\psi}\dot{\phi}\sin\phi)\mathbf{b}_1 + \ddot{\phi}\mathbf{b}_2 + (\ddot{\theta} + \ddot{\psi}\sin\phi + \dot{\psi}\dot{\phi}\cos\phi)\mathbf{b}_3 \\
&\quad + \dot{\psi}\cos\phi\frac{{}^{\mathcal{R}}d}{dt}\mathbf{b}_1 + \dot{\phi}\frac{{}^{\mathcal{R}}d}{dt}\mathbf{b}_2 + (\dot{\theta} + \dot{\psi}\sin\phi)\frac{{}^{\mathcal{R}}d}{dt}\mathbf{b}_3,
\end{aligned}
$$

where the derivatives of vectors $\mathbf{b}_1, \mathbf{b}_2, \mathbf{b}_3$, fixed in B, in reference frame $\mathcal{R}$, are given as

$$
\frac{{}^{\mathcal{R}}d}{dt}\mathbf{b}_j = {}^{\mathcal{R}}\boldsymbol{\omega}^{B} \times \mathbf{b}_j = (\dot{\psi}\cos\phi\,\mathbf{b}_1 + \dot{\phi}\mathbf{b}_2 + \dot{\psi}\sin\phi\,\mathbf{b}_3) \times \mathbf{b}_j, \quad j = 1, 2, 3.
$$

Working with the cross products, replacing them in the preceding equation, and grouping the terms, we obtain

$$
\begin{aligned}
{}^{\mathcal{R}}\boldsymbol{\alpha}^{C} = {}&(\ddot{\psi}\cos\phi - \dot{\psi}\dot{\phi}\sin\phi + \dot{\theta}\dot{\phi})\mathbf{b}_1 \\
&+ (\ddot{\phi} - \dot{\theta}\dot{\psi}\cos\phi)\mathbf{b}_2 \\
&+ (\ddot{\theta} + \ddot{\psi}\sin\phi + \dot{\psi}\dot{\phi}\cos\phi)\mathbf{b}_3.
\end{aligned}
$$

We suggest that the reader calculates the cross products and checks the result. See the corresponding animation.

The angular accelerations of two reference frames A and B moving in relation to each other are opposite each other, that is,

$$
{}^{A}\boldsymbol{\alpha}^{B} = -{}^{B}\boldsymbol{\alpha}^{A}. \tag{4.3}
$$

In fact, from Eqs. (4.1), (3.7), and (3.13),

$$
{}^{A}\boldsymbol{\alpha}^{B} = \frac{{}^{A}d}{dt}{}^{A}\boldsymbol{\omega}^{B} = \frac{{}^{B}d}{dt}{}^{A}\boldsymbol{\omega}^{B} = -\frac{{}^{B}d}{dt}{}^{B}\boldsymbol{\omega}^{A} = -{}^{B}\boldsymbol{\alpha}^{A}.
$$

Example 4.2   Returning to Example 4.1 (see Fig. 4.1), the angular acceleration of frame B in rotor C is

$$
{}^{C}\boldsymbol{\alpha}^{B} = -{}^{B}\boldsymbol{\alpha}^{C} = -\ddot{\theta}\mathbf{b}_3,
$$

and the angular acceleration of frame A in rotor C is

$$
{}^{C}\boldsymbol{\alpha}^{A} = -{}^{A}\boldsymbol{\alpha}^{C} = -\dot{\theta}\dot{\phi}\mathbf{b}_1 - \ddot{\phi}\mathbf{b}_2 - \ddot{\theta}\mathbf{b}_3.
$$

See the corresponding animation.

Angular acceleration vectors do *not* obey a relation of additiveness similar to Eq. (3.11). In other words, if $\mathcal{R}_1, \mathcal{R}_2$, and $\mathcal{R}_3$ are three different reference frames, then, in general,

$$^{\mathcal{R}_1}\boldsymbol{\alpha}^{\mathcal{R}_3} \neq {}^{\mathcal{R}_1}\boldsymbol{\alpha}^{\mathcal{R}_2} + {}^{\mathcal{R}_2}\boldsymbol{\alpha}^{\mathcal{R}_3}. \tag{4.4}$$

More specifically, by time-differentiating Eq. (3.11) in reference frame $\mathcal{R}_1$, we get

$$\frac{^{\mathcal{R}_1}d}{dt}{}^{\mathcal{R}_1}\boldsymbol{\omega}^{\mathcal{R}_3} = \frac{^{\mathcal{R}_1}d}{dt}{}^{\mathcal{R}_1}\boldsymbol{\omega}^{\mathcal{R}_2} + \frac{^{\mathcal{R}_1}d}{dt}{}^{\mathcal{R}_2}\boldsymbol{\omega}^{\mathcal{R}_3}, \tag{4.5}$$

and, from the definition of angular acceleration and Eq. (3.6),

$$^{\mathcal{R}_1}\boldsymbol{\alpha}^{\mathcal{R}_3} = {}^{\mathcal{R}_1}\boldsymbol{\alpha}^{\mathcal{R}_2} + \frac{^{\mathcal{R}_2}d}{dt}{}^{\mathcal{R}_2}\boldsymbol{\omega}^{\mathcal{R}_3} + {}^{\mathcal{R}_1}\boldsymbol{\omega}^{\mathcal{R}_2} \times {}^{\mathcal{R}_2}\boldsymbol{\omega}^{\mathcal{R}_3}; \tag{4.6}$$

therefore,

$$^{\mathcal{R}_1}\boldsymbol{\alpha}^{\mathcal{R}_3} = {}^{\mathcal{R}_1}\boldsymbol{\alpha}^{\mathcal{R}_2} + {}^{\mathcal{R}_2}\boldsymbol{\alpha}^{\mathcal{R}_3} + {}^{\mathcal{R}_1}\boldsymbol{\omega}^{\mathcal{R}_2} \times {}^{\mathcal{R}_2}\boldsymbol{\omega}^{\mathcal{R}_3}. \tag{4.7}$$

When comparing Eqs. (4.4) and (4.7), it is found that the inequality in Eq. (4.4) will be converted to equality only when $^{\mathcal{R}_1}\boldsymbol{\omega}^{\mathcal{R}_2}$ and $^{\mathcal{R}_2}\boldsymbol{\omega}^{\mathcal{R}_3}$ are parallel (or one of them is null).

Example 4.3   Wire A, in the form of an arc with a quarter of the circumference of a radius R (see Fig. 4.2), rotates around vertical axis z, fixed in laboratory L, according to function $\phi(t)$, which measures the angle between the horizontal segment of the arc and a horizontal line, passing through center O of the circumference and fixed in L. Bar B moves in relation to wire A, having one end fixed at point O and the other sliding along the circular stretch, according to the function $\theta(t)$, which measures the angle between the bar and the horizontal segment of the wire, as shown. The basis of orthonormal vectors $\mathbf{a}_1, \mathbf{a}_2, \mathbf{a}_3$ is fixed in A, with $\mathbf{a}_1$ horizontal and $\mathbf{a}_3$ orthogonal to the plane of the wires, and basis $\mathbf{b}_1, \mathbf{b}_2, \mathbf{b}_3$ fixed in B, with $\mathbf{b}_1$ parallel to the bar and $\mathbf{b}_3 = \mathbf{a}_3$. The angular velocities in question are $^{A}\boldsymbol{\omega}^{B} = \dot{\theta}\mathbf{a}_3$, $^{L}\boldsymbol{\omega}^{A} = -\dot{\phi}\mathbf{a}_2$, and $^{L}\boldsymbol{\omega}^{B} = {}^{L}\boldsymbol{\omega}^{A} + {}^{A}\boldsymbol{\omega}^{B} = -\dot{\phi}\mathbf{a}_2 + \dot{\theta}\mathbf{a}_3$. The angular accelerations of B in A and of A in L can then be obtained from the definition, Eq. (4.1),

$$^{A}\boldsymbol{\alpha}^{B} = \frac{^{A}d}{dt}{}^{A}\boldsymbol{\omega}^{B} = \ddot{\theta}\mathbf{a}_3,$$

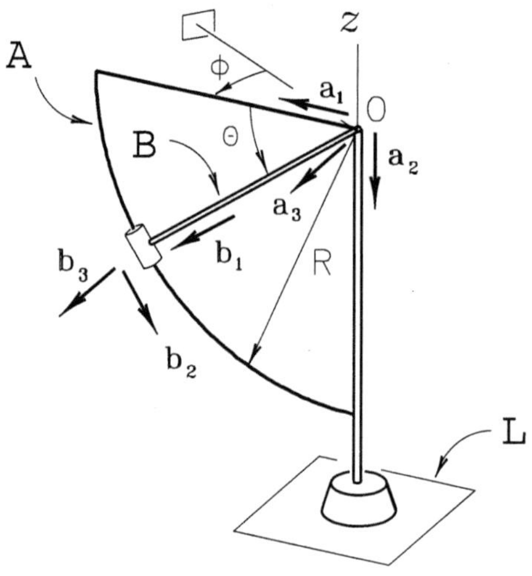

Figure 4.2

$$^{L}\boldsymbol{\alpha}^{A} = \frac{^{L}d}{dt}{}^{L}\boldsymbol{\omega}^{A} = -\ddot{\phi}\mathbf{a}_2.$$

The angular acceleration of B in L can be obtained from Eq. (4.7):

$$^{L}\boldsymbol{\alpha}^{B} = {}^{L}\boldsymbol{\alpha}^{A} + {}^{A}\boldsymbol{\alpha}^{B} + {}^{L}\boldsymbol{\omega}^{A} \times {}^{A}\boldsymbol{\omega}^{B} = -\dot{\phi}\dot{\theta}\mathbf{a}_1 - \ddot{\phi}\mathbf{a}_2 + \ddot{\theta}\mathbf{a}_3.$$

See the corresponding animation.

3.5 Position, Velocity, and Acceleration

The preceding sections concern the relative motion with respect to two
or more reference frames, where the notions of angular velocity and
acceleration play a primary role. The relations between time derivatives
of a vector in different reference frames are also established, expressed
by Eqs. (2.2) and (3.6). We therefore have the tools required for a more
general study of the motion of a point, the subject discussed in this
section. Let P and Q be two points moving independently in relation to
a reference frame $\mathcal{R}$, with which a system of Cartesian axes is associated,
with origin O (see Fig. 5.1). The *position vector of point* P *with respect
to point* Q is the vector whose direction is parallel to the line passing
through P and Q, pointing from Q to P, and whose module is the distance

between P and Q. The notation $\mathbf{p}^{P/Q}$ will be adopted for this vector. The geometric representation of the position vector is usually an oriented line segment, as shown in the figure. If (p_1, p_2, p_3) and (q_1, q_2, q_3) are the Cartesian coordinates of points P and Q, respectively, and $\mathbf{n}_1, \mathbf{n}_2, \mathbf{n}_3$ are orthonormal vectors parallel to the coordinate axes, then

$$\mathbf{p}^{P/Q} = \sum_{j=1}^{3} (p_j - q_j)\mathbf{n}_j. \tag{5.1}$$

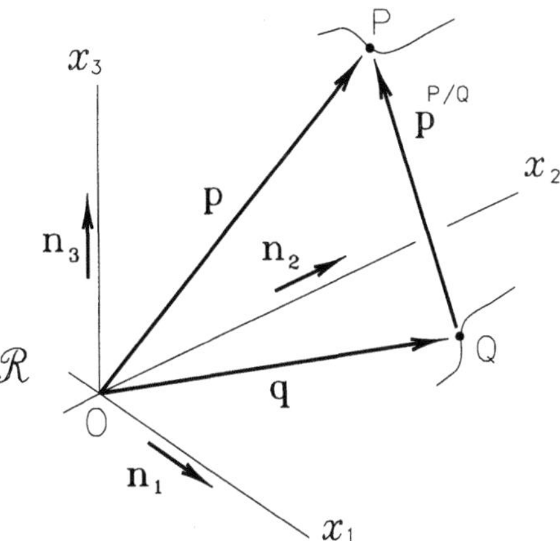

Figure 5.1

In fact, if $\mathbf{p} = \mathbf{p}^{P/O}$ is the position vector of point P with respect to point O and $\mathbf{q} = \mathbf{p}^{Q/O}$ is the position vector of point Q with respect to O, then

$$\mathbf{p} = \sum_{j=1}^{3} p_j \mathbf{n}_j, \qquad \mathbf{q} = \sum_{j=1}^{3} q_j \mathbf{n}_j, \tag{5.2}$$

and the position vector of point P with respect to point Q will be the vector difference between $\mathbf{p}$ and $\mathbf{q}$:

$$\mathbf{p}^{P/Q} = \mathbf{p} - \mathbf{q}. \tag{5.3}$$

When Eq. (5.2) is introduced into Eq. (5.3), the result is Eq. (5.1).

The position vector of point Q with respect to point P is minus the position vector of P with respect to Q, that is,

$$\mathbf{p}^{Q/P} = -\mathbf{p}^{P/Q} = \sum_{j=1}^{3} (q_j - p_j)\mathbf{n}_j. \tag{5.4}$$

Every position vector $\mathbf{p}^{P/Q}$ has a dimension [L], that is, meters, in metric units.

Example 5.1 Figure 5.2 reproduces the system analyzed in Example 4.3, now including a cursor P moving along the bar B and the scalar function $r(t)$, which measures the distance from the cursor to point O.

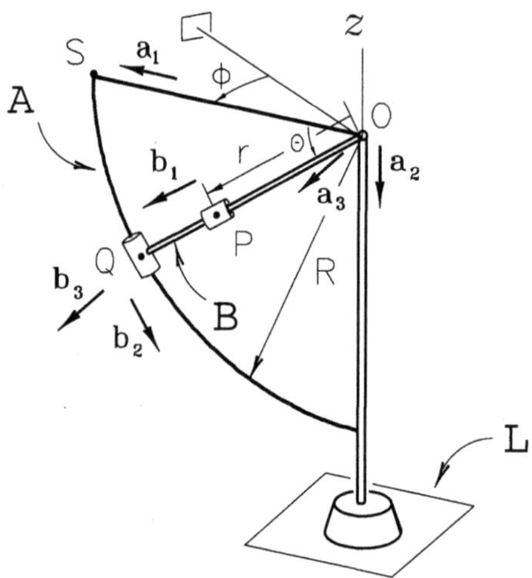

Figure 5.2

The dimensions of the cursor are small when compared to the range of its motion, $0 < r < R$, and may be considered as a particle. Its position vector with respect to point O will be $\mathbf{p}^{P/O} = r\mathbf{b}_1 = r(\cos\theta\mathbf{a}_1 + \sin\theta\mathbf{a}_2)$. The position of point Q, the end of the bar B sliding over the arc of the wire A, with respect to point O is $\mathbf{p}^{Q/O} = R\mathbf{b}_1 = R(\cos\theta\mathbf{a}_1 + \sin\theta\mathbf{a}_2)$. The position of P with respect to point S, fixed in A, is

$$\begin{aligned}
\mathbf{p}^{P/S} &= \mathbf{p}^{P/O} - \mathbf{p}^{S/O} \\
&= r\mathbf{b}_1 - R\mathbf{a}_1 \\
&= (r - R\cos\theta)\mathbf{b}_1 + R\sin\theta\mathbf{b}_2 \\
&= -(R - r\cos\theta)\mathbf{a}_1 + r\sin\theta\mathbf{a}_2.
\end{aligned}$$

See the corresponding animation.

The time derivative in a reference frame $\mathcal{R}$ of the position vector of a point P with respect to another point Q is called the *velocity of* P *relative to* Q *in the reference frame* $\mathcal{R}$:

$$\mathcal{R}\mathbf{v}^{P/Q} \rightleftharpoons \frac{\mathcal{R}d}{dt}\mathbf{p}^{P/Q}. \tag{5.5}$$

Note that the adopted notation refers explicitly to points P and Q of the position vector and to the reference frame where the vector is being differentiated. Every velocity vector has a dimension $[\mathrm{LT}^{-1}]$, that is, m/s, in metric units.

Example 5.2 Referring to the above example (see Fig. 5.2), the velocities of P relative to O, of Q relative to O, and of P relative to S, in reference frame B, are, using the definition, Eq. (5.5),

$$^{B}\mathbf{v}^{P/O} = \dot{r}\mathbf{b}_1, \quad ^{B}\mathbf{v}^{Q/O} = 0, \quad ^{B}\mathbf{v}^{P/S} = (\dot{r} + R\dot{\theta}\sin\theta)\mathbf{b}_1 + R\dot{\theta}\cos\theta\mathbf{b}_2.$$

The relative velocities of the same points in reference frame A are obtained by differentiating, in reference frame A, the respective position vectors:

$$^{A}\mathbf{v}^{P/O} = \dot{r}\mathbf{b}_1 + r\frac{^{A}d}{dt}\mathbf{b}_1 = \dot{r}\mathbf{b}_1 + r\dot{\theta}\mathbf{b}_2;$$

$$^{A}\mathbf{v}^{Q/O} = R\frac{^{A}d}{dt}\mathbf{b}_1 = R\dot{\theta}\mathbf{b}_2;$$

$$^{A}\mathbf{v}^{P/S} = (\dot{r}\cos\theta - r\dot{\theta}\sin\theta)\mathbf{a}_1 + (\dot{r}\sin\theta + r\dot{\theta}\cos\theta)\mathbf{a}_2$$

$$= \dot{r}\mathbf{b}_1 + r\dot{\theta}\mathbf{b}_2.$$

It is worth noting that $^{B}\mathbf{v}^{Q/O} = 0$, because Q and O are both fixed in B and that $^{A}\mathbf{v}^{P/O} = {}^{A}\mathbf{v}^{P/S}$, because S and O are both fixed in A. The first equality is not surprising, since it is to be expected that points fixed to the same body do not have a relative velocity *in the body*. The second equality illustrates the concept of absolute velocity, discussed ahead. See the corresponding animation.

If O is an arbitrary point, fixed in a reference frame $\mathcal{R}$, the velocity of any other point P relative to O in $\mathcal{R}$ is called the *absolute velocity of* P *in* $\mathcal{R}$ or, simply, the *velocity of* P *in* $\mathcal{R}$. Absolute velocity is independent of the choice of point O, fixed in the reference frame, that is,

$$\mathcal{R}\mathbf{v}^{P} \rightleftharpoons \mathcal{R}\mathbf{v}^{P/O}, \qquad \text{for every O fixed in R.} \tag{5.6}$$

In fact, if O and O′ are two different points fixed in $\mathcal{R}$, the positions of any one point P with respect to O and O′ fulfill the relation

$$\mathbf{p}^{P/O} - \mathbf{p}^{P/O'} = \mathbf{p}^{O'/O}. \tag{5.7}$$

Time-differentiating in reference frame $\mathcal{R}$,

$$^{\mathcal{R}}\mathbf{v}^{P/O} - {}^{\mathcal{R}}\mathbf{v}^{P/O'} = 0, \tag{5.8}$$

and once O and O′ are fixed in $\mathcal{R}$, then,

$$^{\mathcal{R}}\mathbf{v}^{P/O} = {}^{\mathcal{R}}\mathbf{v}^{P/O'} = {}^{\mathcal{R}}\mathbf{v}^{P}. \tag{5.9}$$

The velocity of a point P relative to another point Q in a given reference frame $\mathcal{R}$ is equal to the difference of the (absolute) velocities of P and Q in $\mathcal{R}$, that is,

$$^{\mathcal{R}}\mathbf{v}^{P/Q} = {}^{\mathcal{R}}\mathbf{v}^{P} - {}^{\mathcal{R}}\mathbf{v}^{Q}. \tag{5.10}$$

Equation (5.10) results from differentiating Eq. (5.3) in the reference frame $\mathcal{R}$ and from the definitions of relative and absolute velocity.

Example 5.3   Returning again to Example 5.1 (see Fig. 5.2), since O is a point fixed in B, A, and L, then

$$^{B}\mathbf{v}^{P} = {}^{B}\mathbf{v}^{P/O} = \dot{r}\mathbf{b}_1,$$
$$^{B}\mathbf{v}^{Q} = {}^{B}\mathbf{v}^{Q/O} = 0,$$
$$^{A}\mathbf{v}^{P} = {}^{A}\mathbf{v}^{P/O} = \dot{r}\mathbf{b}_1 + r\dot{\theta}\mathbf{b}_2,$$
$$^{A}\mathbf{v}^{Q} = {}^{A}\mathbf{v}^{Q/O} = R\dot{\theta}\mathbf{b}_2.$$

The velocity of cursor P in reference frame L is (in every detail)

$$^{L}\mathbf{v}^{P} = \frac{{}^{L}d}{dt}\mathbf{p}^{P/O}$$
$$= \dot{r}\mathbf{b}_1 + r\frac{{}^{L}d}{dt}\mathbf{b}_1$$
$$= \dot{r}\mathbf{b}_1 + r\,{}^{L}\boldsymbol{\omega}^{B} \times \mathbf{b}_1$$
$$= \dot{r}\mathbf{b}_1 + r(-\dot{\phi}\sin\theta\,\mathbf{b}_1 - \dot{\phi}\cos\theta\,\mathbf{b}_2 + \dot{\theta}\mathbf{b}_3) \times \mathbf{b}_1$$
$$= \dot{r}\mathbf{b}_1 + r\dot{\theta}\mathbf{b}_2 + r\dot{\phi}\cos\theta\,\mathbf{b}_3.$$

Last, it is easy to see that

$$^{A}\mathbf{v}^{P/Q} = {}^{A}\mathbf{v}^{P} - {}^{A}\mathbf{v}^{Q} = \dot{r}\mathbf{b}_1 - (R-r)\dot{\theta}\mathbf{b}_2.$$

See the corresponding animation.

The time derivative in a reference frame $\mathcal{R}$ of the velocity of a point P relative to another point Q in $\mathcal{R}$ is called the *acceleration of* P *relative to* Q *in the reference frame* $\mathcal{R}$:

$$\mathcal{R}\mathbf{a}^{P/Q} \rightleftharpoons \frac{\mathcal{R}d}{dt}\mathcal{R}\mathbf{v}^{P/Q}. \tag{5.11}$$

If O is an arbitrary point fixed in reference frame $\mathcal{R}$, the acceleration of any other point P relative to O in $\mathcal{R}$ is called the *absolute acceleration of* P *in* $\mathcal{R}$ or, simply, the *acceleration of* P *in* $\mathcal{R}$:

$$\mathcal{R}\mathbf{a}^{P} \rightleftharpoons \mathcal{R}\mathbf{a}^{P/O}, \qquad \text{for every O fixed in } \mathcal{R}. \tag{5.12}$$

Likewise with the velocity vectors, if O and O' are any two points fixed in $\mathcal{R}$, the differentiation with respect to time in the reference frame $\mathcal{R}$ of Eq. (5.9) gives

$$\mathcal{R}\mathbf{a}^{P/O} = \mathcal{R}\mathbf{a}^{P/O'} = \mathcal{R}\mathbf{a}^{P}. \tag{5.13}$$

The acceleration of a point P relative to another point Q in a reference frame $\mathcal{R}$ is equal to the difference between the (absolute) accelerations of P and Q in $\mathcal{R}$, that is,

$$\mathcal{R}\mathbf{a}^{P/Q} = \mathcal{R}\mathbf{a}^{P} - \mathcal{R}\mathbf{a}^{Q}. \tag{5.14}$$

This result is an immediate consequence of Eq. (5.10) and the definitions of relative and absolute acceleration.

The time derivative in a reference frame $\mathcal{R}$ of the (absolute) velocity of a point P in $\mathcal{R}$ is equal to the (absolute) acceleration of P in $\mathcal{R}$, that is,

$$\mathcal{R}\mathbf{a}^{P} = \frac{\mathcal{R}d}{dt}\mathcal{R}\mathbf{v}^{P}. \tag{5.15}$$

In fact, if O is a point fixed in $\mathcal{R}$, $\mathcal{R}\mathbf{v}^{P/O} = \mathcal{R}\mathbf{v}^{P}$ and, from Eqs. (5.11) and (5.12), the immediate result is Eq. (5.15).

Example 5.4 Figure 5.3 shows a small cursor C sliding over a guide engraved in the surface of a disk D, which rotates around its axis (horizontal) of symmetry, on a shaft fixed in fork A. The latter, in turn, rotates around the vertical axis z, fixed in the laboratory L, at a simple angular velocity of a constant module ω_0. The basis of orthonormal vectors $\mathbf{a}_1, \mathbf{a}_2, \mathbf{a}_3$ is fixed in A, with $\mathbf{a}_2$ vertical and $\mathbf{a}_3$ parallel to the disk axis; the basis of orthonormal vectors $\mathbf{n}_1, \mathbf{n}_2, \mathbf{n}_3$ is attached to D, with $\mathbf{n}_1$ parallel to the guide and $\mathbf{n}_3$ horizontal. The angular velocity of the disk in relation to the fork is a simple angular velocity and may be expressed as ${}^A\boldsymbol{\omega}^D = \dot{\phi}\mathbf{a}_3$, where ϕ is the indicated angle. The angular velocity of the disk in the laboratory is then ${}^L\boldsymbol{\omega}^D = {}^L\boldsymbol{\omega}^A + {}^A\boldsymbol{\omega}^D = \omega_0\mathbf{a}_2 + \dot{\phi}\mathbf{a}_3$. If $x(t)$ is the coordinate measuring the displacement of the cursor, as illustrated, the velocity of the cursor in D is (point Q is fixed in D)

$$
{}^D\mathbf{v}^C = \frac{{}^D d}{dt}\mathbf{p}^{C/Q} = \dot{x}\mathbf{n}_1 .
$$

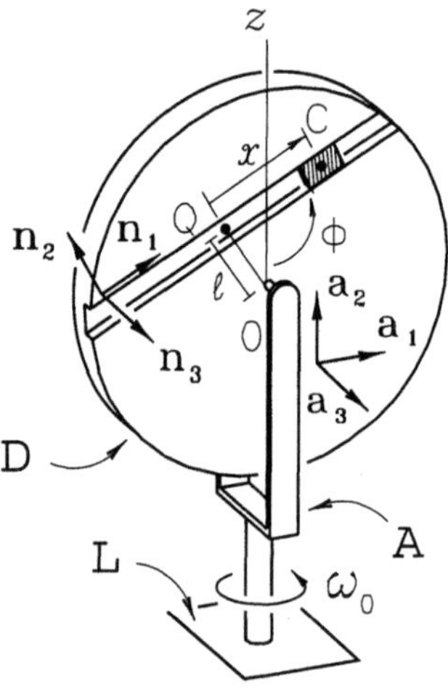

Figure 5.3

The velocities of Q and C in A are

$$
{}^A\mathbf{v}^Q = \frac{{}^A d}{dt}\mathbf{p}^{Q/O} = \frac{{}^A d}{dt}l\mathbf{n}_2 = -l\dot{\phi}\mathbf{n}_1 ,
$$

$$
{}^A\mathbf{v}^C = \frac{{}^A d}{dt}\mathbf{p}^{C/O} = \frac{{}^A d}{dt}(x\mathbf{n}_1 + l\mathbf{n}_2) = (\dot{x} - l\dot{\phi})\mathbf{n}_1 + x\dot{\phi}\mathbf{n}_2 .
$$

The velocity of C relative to Q in A is, therefore,

$$
{}^A\mathbf{v}^{C/Q} = {}^A\mathbf{v}^C - {}^A\mathbf{v}^Q = \dot{x}\mathbf{n}_1 + x\dot{\phi}\mathbf{n}_2 .
$$

The acceleration of C in relation to Q in A is, according to Eq. (5.11),

$$^A\mathbf{a}^{C/Q} = \frac{^Ad}{dt}{}^A\mathbf{v}^{C/Q} = \frac{^Ad}{dt}(\dot{x}\mathbf{n}_1 + x\dot{\phi}\mathbf{n}_2)$$
$$= (\ddot{x} - x\dot{\phi}^2)\mathbf{n}_1 + (x\ddot{\phi} + 2\dot{x}\dot{\phi})\mathbf{n}_2.$$

The (absolute) acceleration of Q in A is, according to Eq. (5.15),

$$^A\mathbf{a}^{Q} = \frac{^Ad}{dt}{}^A\mathbf{v}^{Q} = -l\ddot{\phi}\mathbf{n}_1 - l\dot{\phi}^2\mathbf{n}_2;$$

therefore, the acceleration of C in A is, according to Eq. (5.14),

$$^A\mathbf{a}^{C} = {}^A\mathbf{a}^{C/Q} + {}^A\mathbf{a}^{Q} = (\ddot{x} - l\ddot{\phi} - x\dot{\phi}^2)\mathbf{n}_1 + (x\ddot{\phi} - l\dot{\phi}^2 + 2\dot{x}\dot{\phi})\mathbf{n}_2,$$

or, alternatively, according to Eq. (5.15),

$$^A\mathbf{a}^{C} = \frac{^Ad}{dt}{}^A\mathbf{v}^{C}$$
$$= (\ddot{x} - l\ddot{\phi})\mathbf{n}_1 + (\dot{x} - l\dot{\phi})\dot{\mathbf{n}}_1 + (\dot{x}\dot{\phi} + x\ddot{\phi})\mathbf{n}_2 + x\dot{\phi}\dot{\mathbf{n}}_2$$
$$= (\ddot{x} - l\ddot{\phi} - x\dot{\phi}^2)\mathbf{n}_1 + (x\ddot{\phi} - l\dot{\phi}^2 + 2\dot{x}\dot{\phi})\mathbf{n}_2.$$

In the next section an alternative and more effective form are presented to determine velocity and acceleration vectors, avoiding the need to differentiate vectors. See the corresponding animation.

3.6 Kinematic Theorems

The examples in the preceding section show how to determine velocities and accelerations of points, always obtained from their definitions. In other words, velocities have been determined by differentiating relative position vectors with respect to time, as expressed in Eqs. (5.5) and (5.6). Accelerations have been determined by differentiating, with respect to time, velocity vectors, as expressed in Eqs. (5.11) and (5.15). This procedure can be greatly simplified by using Eqs. (2.2) and (3.6) to obtain relations between velocities and accelerations of different points and of the same point in different reference frames. This section is devoted to establishing and exemplifying the use of these important relations, known as *kinematic theorems*.

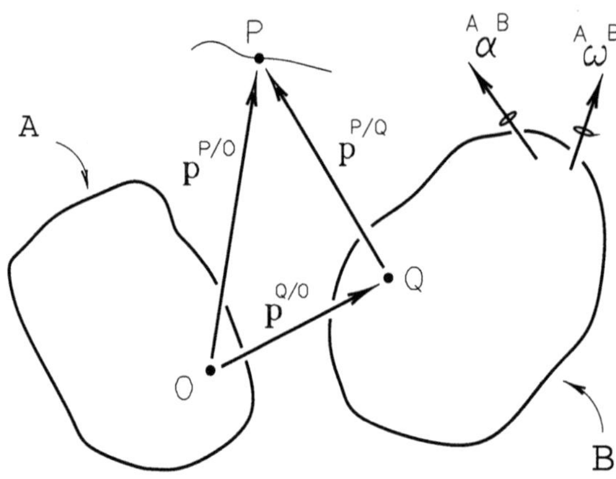

Figure 6.1

Consider A and B as two reference frames moving independently in space and let $^A\omega^B$ and $^A\alpha^B$ be the angular velocity and acceleration of B in A. If Q is a point fixed in reference frame B, O a point fixed in A, and P any point moving arbitrarily in relation to A and B (see Fig. 6.1), the (absolute) velocities of P in A and of P in B are related according to the following theorem:

> **Theorem 1.** *The velocity of a point P in a reference frame A is equal to the vector sum of the velocity of P in another reference frame B with the velocity at A of a point Q fixed in B, plus the cross product of the angular velocity of B in A with the position vector of P with respect to Q, that is,*

$$^A\mathbf{v}^P = {}^B\mathbf{v}^P + {}^A\mathbf{v}^Q + {}^A\omega^B \times \mathbf{p}^{P/Q}. \tag{6.1}$$

The kinematic theorem for velocities relates, therefore, the velocity of a point in two different reference frames moving arbitrarily in relation to each other, showing that the difference between the velocities of the point in the two reference frames depends on their relative angular velocity.

The demonstration is simple when using Eq. (3.6) and the definition of an absolute velocity vector, Eqs. (5.5) and (5.6),

$$
\begin{aligned}
{}^{A}\mathbf{v}^{P} &= \frac{{}^{A}d}{dt}\mathbf{p}^{P/O} \\
&= \frac{{}^{A}d}{dt}\mathbf{p}^{Q/O} + \frac{{}^{A}d}{dt}\mathbf{p}^{P/Q} \\
&= {}^{A}\mathbf{v}^{Q} + \frac{{}^{B}d}{dt}\mathbf{p}^{P/Q} + {}^{A}\boldsymbol{\omega}^{B} \times \mathbf{p}^{P/Q} \\
&= {}^{A}\mathbf{v}^{Q} + {}^{B}\mathbf{v}^{P} + {}^{A}\boldsymbol{\omega}^{B} \times \mathbf{p}^{P/Q}. \qquad\blacksquare
\end{aligned}
$$

Since the velocity of P in relation to Q in reference frame A is the difference between the velocities of P and Q in A [see Eq. (5.10)], Eq. (6.1) can be alternatively expressed as

$$
{}^{A}\mathbf{v}^{P/Q} = {}^{B}\mathbf{v}^{P} + {}^{A}\boldsymbol{\omega}^{B} \times \mathbf{p}^{P/Q}. \tag{6.2}
$$

Example 6.1 Figure 6.2 reproduces the situation studied in Example 5.4. The velocity of point Q in relation to the fork A can now be obtained by using the kinematic theorem, adopting O as the point fixed in reference frame D:

$$
\begin{aligned}
{}^{A}\mathbf{v}^{Q} &= {}^{D}\mathbf{v}^{Q} + {}^{A}\mathbf{v}^{O} + {}^{A}\boldsymbol{\omega}^{D} \times \mathbf{p}^{Q/O} \\
&= 0 + 0 + \dot{\phi}\mathbf{n}_3 \times l\mathbf{n}_2 \\
&= -l\dot{\phi}\mathbf{n}_1.
\end{aligned}
$$

The velocity of the cursor C in relation to fork A, also using Eq. (6.1), is

$$
\begin{aligned}
{}^{A}\mathbf{v}^{C} &= {}^{D}\mathbf{v}^{C} + {}^{A}\mathbf{v}^{O} + {}^{A}\boldsymbol{\omega}^{D} \times \mathbf{p}^{C/O} \\
&= \dot{x}\mathbf{n}_1 + 0 + \dot{\phi}\mathbf{n}_3 \times (x\mathbf{n}_1 + l\mathbf{n}_2) \\
&= (\dot{x} - l\dot{\phi})\mathbf{n}_1 + x\dot{\phi}\mathbf{n}_2,
\end{aligned}
$$

giving the same result as that obtained in Example 5.4. The theorem simplifies the calculation and may be used successively to determine velocities in relation to reference frames where the motion of the point is more complex. Hence, the velocity of the cursor C in the laboratory L is

$$
\begin{aligned}
{}^{L}\mathbf{v}^{C} &= {}^{A}\mathbf{v}^{C} + {}^{L}\mathbf{v}^{O} + {}^{L}\boldsymbol{\omega}^{A} \times \mathbf{p}^{C/O} \\
&= (\dot{x} - l\dot{\phi})\mathbf{n}_1 + x\dot{\phi}\mathbf{n}_2 + 0 + \omega_0\mathbf{a}_2 \times (x\mathbf{n}_1 + l\mathbf{n}_2) \\
&= (\dot{x} - l\dot{\phi})\mathbf{n}_1 + x\dot{\phi}\mathbf{n}_2 - \omega_0(l\cos\phi + x\sin\phi)\mathbf{n}_3.
\end{aligned}
$$

See the corresponding animation.

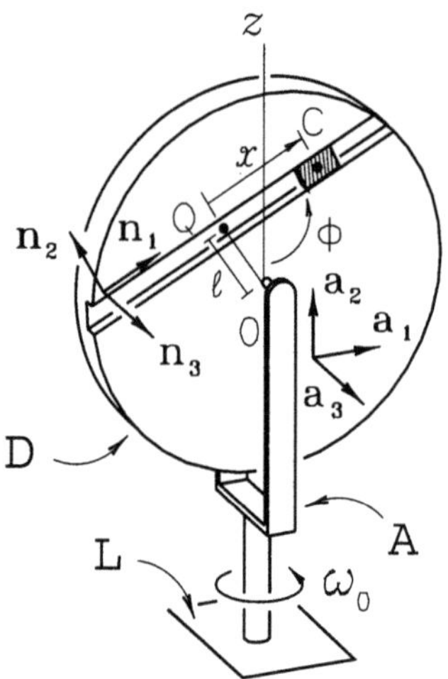

Figure 6.2

As seen from the preceding example, the kinematic theorem
relates the velocities of an arbitrary point P in relation to any two reference frames moving in relation to each other; the difference between
these velocities depends, as seen from Eq. (6.1), on the relative angular velocity between the reference frames (which, as we already know,
is a measure of the time rate of change of their relative orientations)
and on the velocity of any point Q, fixed in one of the reference frames,
in relation to the other. Note that this point Q is, in fact, arbitrary.
When another point Q' is chosen, the terms $^A\mathbf{v}^{Q'}$ and $\mathbf{p}^{P/Q'}$ in Eq. (6.1)
change, of course, but the end result for $^A\mathbf{v}^P$ will necessarily be the
same. This indifference with regard to the choice of point Q, fixed in
reference frame B, is extremely valuable in solving problems, always allowing the choice of a point whose motion in A is known (for example,
a point fixed in both reference frames, in the cases where there is such
a point), simplifying the solution.

A kinematic theorem like the previous one, relating the accelerations of a given point in relation to two reference frames that move
relative to each other would certainly be very valuable. Considering
then the same reference frames A and B and points Q fixed in B and O
fixed in A (see Fig. 6.1), the (absolute) accelerations of P in A and of P
in B will be related according to Theorem 2:

Theorem 2. *The acceleration of a point P in a reference frame A is equal to the vector sum of its acceleration in another reference frame B with another four terms, as follows: acceleration in A of an arbitrary point Q, fixed in B; the double cross product of the angular velocity of B in A with the position vector of P with respect to Q; the cross product of the angular acceleration of B in A with the same position vector; twice the cross product of the angular velocity of B in A with the velocity of P in B, that is,*

$$^A\mathbf{a}^P = {}^B\mathbf{a}^P + {}^A\mathbf{a}^Q + {}^A\boldsymbol{\omega}^B \times ({}^A\boldsymbol{\omega}^B \times \mathbf{p}^{P/Q}) + {}^A\boldsymbol{\alpha}^B \times \mathbf{p}^{P/Q} + 2\,{}^A\boldsymbol{\omega}^B \times {}^B\mathbf{v}^P. \tag{6.3}$$

Equation (6.3) derives from Eq. (6.1) and from the definition of absolute acceleration. In fact, by differentiating both members of Eq. (6.1) in the reference frame A and using Eq. (3.6) to relate the derivatives in the two reference frames, we get

$$
\begin{aligned}
^A\mathbf{a}^P &= \frac{^A d}{dt}{}^A\mathbf{v}^P \\[4pt]
&= \frac{^A d}{dt}{}^B\mathbf{v}^P + \frac{^A d}{dt}{}^A\mathbf{v}^Q + \frac{^A d}{dt}({}^A\boldsymbol{\omega}^B \times \mathbf{p}^{P/Q}) \\[4pt]
&= \frac{^B d}{dt}{}^B\mathbf{v}^P + {}^A\boldsymbol{\omega}^B \times {}^B\mathbf{v}^P + {}^A\mathbf{a}^Q + \frac{^A d}{dt}{}^A\boldsymbol{\omega}^B \times \mathbf{p}^{P/Q} \\[4pt]
&\quad + {}^A\boldsymbol{\omega}^B \times \frac{^A d}{dt}\mathbf{p}^{P/Q} \\[4pt]
&= {}^B\mathbf{a}^P + {}^A\boldsymbol{\omega}^B \times {}^B\mathbf{v}^P + {}^A\mathbf{a}^Q + {}^A\boldsymbol{\alpha}^B \times \mathbf{p}^{P/Q} \\[4pt]
&\quad + {}^A\boldsymbol{\omega}^B \times \left(\frac{^B d}{dt}\mathbf{p}^{P/Q} + {}^A\boldsymbol{\omega}^B \times \mathbf{p}^{P/Q} \right) \\[4pt]
&= {}^B\mathbf{a}^P + {}^A\mathbf{a}^Q + {}^A\boldsymbol{\omega}^B \times ({}^A\boldsymbol{\omega}^B \times \mathbf{p}^{P/Q}) + {}^A\boldsymbol{\alpha}^B \times \mathbf{p}^{P/Q} + 2\,{}^A\boldsymbol{\omega}^B \times {}^B\mathbf{v}^P. \quad \blacksquare
\end{aligned}
$$

It can be seen that the last component, $2\,{}^A\boldsymbol{\omega}^B \times {}^B\mathbf{v}^P$, comes from two different terms, grouped in the final result. This component is called *Coriolis acceleration*.[1]

---

[1] G. Coriolis, French engineer, 1792–1843.

As the acceleration of P relative to Q in reference frame A is given by the difference between the accelerations of P and Q in A [see Eq. (5.14)], Eq. (6.3) can be alternatively expressed as

$$^A\mathbf{a}^{P/Q} = {}^B\mathbf{a}^P + {}^A\boldsymbol{\omega}^B \times ({}^A\boldsymbol{\omega}^B \times \mathbf{p}^{P/Q}) + {}^A\boldsymbol{\alpha}^B \times \mathbf{p}^{P/Q} + 2\,{}^A\boldsymbol{\omega}^B \times {}^B\mathbf{v}^P. \quad (6.4)$$

Example 6.2   Returning to the above example (see Fig. 6.2), the acceleration of cursor C in relation to fork A is, according to Eq. (6.3),

$$^A\mathbf{a}^C = {}^D\mathbf{a}^C + {}^A\mathbf{a}^O + {}^A\boldsymbol{\omega}^D \times ({}^A\boldsymbol{\omega}^D \times \mathbf{p}^{C/O}) + {}^A\boldsymbol{\alpha}^D \times \mathbf{p}^{C/O} + 2\,{}^A\boldsymbol{\omega}^D \times {}^D\mathbf{v}^C,$$

where

$$^D\mathbf{a}^C = \ddot{x}\mathbf{n}_1,$$

$$^A\mathbf{a}^O = 0,$$

$$^A\boldsymbol{\omega}^D \times ({}^A\boldsymbol{\omega}^D \times \mathbf{p}^{C/O}) = \dot{\phi}\mathbf{n}_3 \times \left(\dot{\phi}\mathbf{n}_3 \times (x\mathbf{n}_1 + l\mathbf{n}_2)\right)$$

$$= -\dot{\phi}^2(x\mathbf{n}_1 + l\mathbf{n}_2),$$

$$^A\boldsymbol{\alpha}^D \times \mathbf{p}^{C/O} = \ddot{\phi}\mathbf{n}_3 \times (x\mathbf{n}_1 + l\mathbf{n}_2)$$

$$= \ddot{\phi}(-l\mathbf{n}_1 + x\mathbf{n}_2),$$

$$2\,{}^A\boldsymbol{\omega}^D \times {}^D\mathbf{v}^C = 2\dot{\phi}\mathbf{n}_3 \times \dot{x}\mathbf{n}_1,$$

$$= 2\dot{x}\dot{\phi}\mathbf{n}_2.$$

Bringing the terms together, then

$$^A\mathbf{a}^C = (\ddot{x} - l\ddot{\phi} - x\dot{\phi}^2)\mathbf{n}_1 + (x\ddot{\phi} - l\dot{\phi}^2 + 2\dot{x}\dot{\phi})\mathbf{n}_2,$$

as obtained in Example 5.4. See the corresponding animation.

The kinematic theorem for velocities, Eq. (6.1), has a relatively simple algebraic form, providing geometric interpretation. In fact, it is possible to consider that the velocity of a point P, as seen from a reference frame A, $^A\mathbf{v}^P$, is equal to the velocity of that point with respect to another reference frame B, $^B\mathbf{v}^P$, plus a term equal to the velocity that would have in A a point fixed in B that coincides instantaneously with P, $^A\mathbf{v}^Q + {}^A\boldsymbol{\omega}^B \times \mathbf{p}^{P/Q}$. On the other hand, with regard to the kinematic theorem for accelerations, Eq. (6.3), its algebraic complexity discourages geometric interpretation (particularly the Coriolis term that, as shown in the demonstration of the result, comes from two different parts and does not admit a separate geometric interpretation). Anyhow, with or without geometric interpretations, it is advisable for the reader *always* to adopt the kinematic theorems as safe guides for calculating velocities and accelerations of points in different reference frames.

Example 6.3 The mechanism illustrated in Fig. 6.3 consists of two bars A and B, articulated at the support $\mathcal{R}$ at points O and Q, respectively, and a cursor C, pivoting at the free end of bar A and sliding along bar B, as shown. All the motion occurs on the plane of the figure, and the interdependent coordinates $\theta(t)$, $\phi(t)$, and $r(t)$ conveniently describe the configuration of the mechanism. We wish to determine how the bar B moves, that is, to evaluate $\dot\phi(t)$ and $\ddot\phi(t)$ at the instant when $\theta = 60^0$, $\dot\theta = \omega$, and $\ddot\theta = 0$.

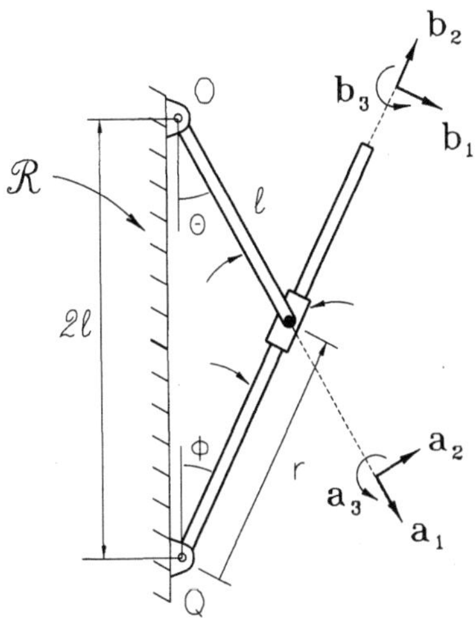

Figure 6.3

The cursor velocity in reference frame $\mathcal{R}$ may be expressed, according to Eq. (6.1), as

$$\begin{aligned}
{}^{\mathcal{R}}\mathbf{v}^C &= {}^{A}\mathbf{v}^C + {}^{\mathcal{R}}\mathbf{v}^O + {}^{\mathcal{R}}\boldsymbol{\omega}^A \times \mathbf{p}^{C/O} \\
&= 0 + 0 + \dot\theta\mathbf{a}_3 \times l\mathbf{a}_1 \\
&= l\dot\theta\mathbf{a}_2.
\end{aligned}$$

Using the same relation but now going through bar B,

$$\begin{aligned}
{}^{\mathcal{R}}\mathbf{v}^C &= {}^{B}\mathbf{v}^C + {}^{\mathcal{R}}\mathbf{v}^Q + {}^{\mathcal{R}}\boldsymbol{\omega}^B \times \mathbf{p}^{C/Q} \\
&= \dot r\mathbf{b}_2 + 0 - \dot\phi\mathbf{b}_3 \times r\mathbf{b}_2 \\
&= r\dot\phi\mathbf{b}_1 + \dot r\mathbf{b}_2.
\end{aligned}$$

By equaling the velocities, the relation is then

$$l\dot\theta\mathbf{a}_2 = r\dot\phi\mathbf{b}_1 + \dot r\mathbf{b}_2.$$

When $\theta = 60^0$ the bars are orthogonal and the bases $\mathbf{a}_1, \mathbf{a}_2, \mathbf{a}_3$ and $\mathbf{b}_1, \mathbf{b}_2, \mathbf{b}_3$ coincide, resulting then in

$$\dot{\phi} = 0 \qquad \text{and} \qquad \dot{r} = \omega l.$$

To calculate $\ddot{\phi}$ and $\ddot{r}$, at the same instant, Eq. (6.3) will be used, going through bar A:

$$
\begin{aligned}
{}^{\mathcal{R}}\mathbf{a}^C &= {}^A\mathbf{a}^C + {}^{\mathcal{R}}\mathbf{a}^O + {}^{\mathcal{R}}\boldsymbol{\omega}^A \times ({}^{\mathcal{R}}\boldsymbol{\omega}^A \times \mathbf{p}^{C/O}) + {}^{\mathcal{R}}\boldsymbol{\alpha}^A \times \mathbf{p}^{C/O} + 2\,{}^{\mathcal{R}}\boldsymbol{\omega}^A \times {}^A\mathbf{v}^C \\
&= 0 + 0 + \dot{\theta}\mathbf{a}_3 \times (\dot{\theta}\mathbf{a}_3 \times l\mathbf{a}_1) + 0 + 0 \\
&= -l\dot{\theta}^2\mathbf{a}_1
\end{aligned}
$$

and through bar B:

$$
\begin{aligned}
{}^{\mathcal{R}}\mathbf{a}^C &= {}^B\mathbf{a}^C + {}^{\mathcal{R}}\mathbf{a}^Q + {}^{\mathcal{R}}\boldsymbol{\omega}^B \times ({}^{\mathcal{R}}\boldsymbol{\omega}^B \times \mathbf{p}^{C/Q}) + {}^{\mathcal{R}}\boldsymbol{\alpha}^B \times \mathbf{p}^{C/Q} + 2\,{}^{\mathcal{R}}\boldsymbol{\omega}^B \times {}^B\mathbf{v}^C \\
&= \ddot{r}\mathbf{b}_2 + 0 - \dot{\phi}\mathbf{b}_3 \times (-\dot{\phi}\mathbf{b}_3 \times r\mathbf{b}_2) - \ddot{\phi}\mathbf{b}_3 \times r\mathbf{b}_2 - 2\dot{\phi}\mathbf{b}_3 \times \dot{r}\mathbf{b}_2 \\
&= (r\ddot{\phi} + 2\dot{r}\dot{\phi})\mathbf{b}_1 + (\ddot{r} - r\dot{\phi}^2)\mathbf{b}_2.
\end{aligned}
$$

Equaling the results in the considered position, then

$$
\begin{aligned}
r\ddot{\phi} + 2\dot{r}\dot{\phi} &= -l\dot{\theta}^2, \\
\ddot{r} - r\dot{\phi}^2 &= 0.
\end{aligned}
$$

So, substituting the velocities calculated above yields

$$\ddot{\phi} = -\frac{l}{r}\dot{\theta}^2 = -\frac{\sqrt{3}}{3}\omega^2 \qquad \text{and} \qquad \ddot{r} = 0.$$

Note that a particularly simple configuration (with $\theta = 60^0$, $\phi = 30^0$, and the mutually orthogonal bars) has been chosen for this example to simplify the algebra involved. The use of kinematic theorems for velocities and accelerations, however, leads to the solution of the problem for a chosen arbitrary configuration. Also, note that it is convenient first to use the kinematic theorem for velocities, since the obtained velocities are always required for determining the accelerations.

To master the use of the kinematic theorems, it is of the utmost importance to gain experience in calculating velocities and accelerations. It is particularly recommended for the reader to work on the corresponding proposed exercises (Series #3) before moving on to study the rest of this chapter.

3.7 Motion of Particles

The motion of a particle P in a reference frame $\mathcal{R}$ is described by its position vector $\mathbf{p}$ with respect to an arbitrary point O fixed in $\mathcal{R}$. Because this vector, expressed as components in any basis, is determined by three scalar functions of time, generally independent to each other, a particle is said to have three *degrees of freedom* if there is no specific constraint to its motion in space. The number of degrees of freedom of a system corresponds to the minimum number of independent coordinates that must be known so that its spatial configuration in relation to a given reference frame is fully determined. In the case of a particle, if $p_1(t), p_2(t), p_3(t)$ are the scalar components of $\mathbf{p}$ in a given basis, these time functions are three coordinates that fully describe the position of P in $\mathcal{R}$. When there is any specific constraint to the motion of P, its number of degrees of freedom is reduced. When, for example, P is restricted in its motion to a given surface $\mathcal{S}$, two coordinates are enough to determine its position in $\mathcal{S}$, and if $\mathbf{n}$ is a unit orthogonal to $\mathcal{S}$ at point P, the kinematic relation $^{\mathcal{R}}\mathbf{v}^P \cdot \mathbf{n} = 0$, establishing that P does not leave the surface, is called the *kinematic constraint.* Each scalar kinematic constraint reduces the number of degrees of freedom of a particle in one unit. If P is confined to move on a line, there will be two kinematic constraints (two velocity components of the particle parallel to the plane orthogonal to the line at the point should be null), and only one coordinate will be sufficient to determine the position of P on the line. Hence, taken as particles, a balloon, a boat, and a train have three, two, and one degrees of freedom, respectively. The kinematic constraints applicable to a particle are of the *holonomic* type. Mechanical systems comprising rigid bodies can be subject to another kind of constraint called *nonholonomic,* in which case the number of coordinates required to describe the system is higher than the number of degrees of freedom. We return to this subject in Section 3.10.

The position vector of P with respect to O is a vector function of time, $\mathbf{p}(t)$, whose first time derivative in a reference frame $\mathcal{R}$ is the velocity of the particle in the reference frame, $^{\mathcal{R}}\mathbf{v}^P$, and whose second time derivative is the acceleration of the particle in the reference frame, $^{\mathcal{R}}\mathbf{a}^P$.

As seen at the beginning of this chapter, the dynamic equations ruling the motion of P in $\mathcal{R}$ involve acceleration vector components. They are, therefore, differential equations where time derivatives of first and second order of the coordinates are present. Depending then on the basis chosen for the breaking down of the acceleration vectors, these equations can become more or less complex, and the choice of a proper basis may facilitate the solution considerably. A notable basis for describing the motion of a particle, called the *intrinsic basis*, will be studied below.

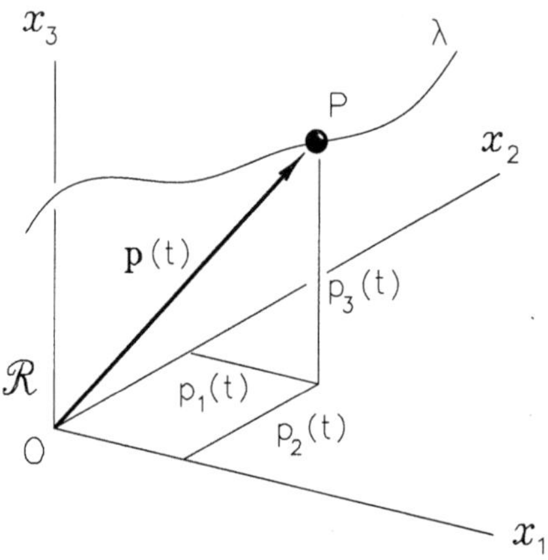

Figure 7.1

If P is a particle moving in a given reference frame $\mathcal{R}$, the geometric place of the successive positions occupied by P over time is its *trajectory* in $\mathcal{R}$ (see Fig. 7.1). The reader should always remember that the trajectory will depend on the reference frame in relation to which the motion of P is observed. The trajectory of P in $\mathcal{R}$ is, mostly, a reverse curve λ. Considering a system of Cartesian axes $\{x_1, x_2, x_3\}$ of origin O, fixed in $\mathcal{R}$, a point P of λ has as coordinates the scalar components of the position vector $\mathbf{p}$ of the point with respect to O (see Fig. 7.1), that is,

$$P : \big(p_1(t), p_2(t), p_3(t)\big). \tag{7.1}$$

Example 7.1 Consider a trajectory λ in the form of a helix with a radius r and pitch a, both constant (see Fig. 7.2). The coordinates (x_P, y_P, z_P) of a general point P can be expressed as $(r\cos\theta, r\sin\theta, z)$; the position vector $\mathbf{p}$ of point P with respect to point O will therefore have the scalar components $p_1 = r\cos\theta$, $p_2 = r\sin\theta$, and $p_3 = z$. Since the helix is a line, only one parameter should be enough to establish P; in fact, it is easy to see that $z = \frac{a}{2\pi}\theta$ and all coordinates of P can be expressed as a function of a single parameter, in the case of the variable θ, P : $(r\cos\theta, r\sin\theta, \frac{a}{2\pi}\theta)$.

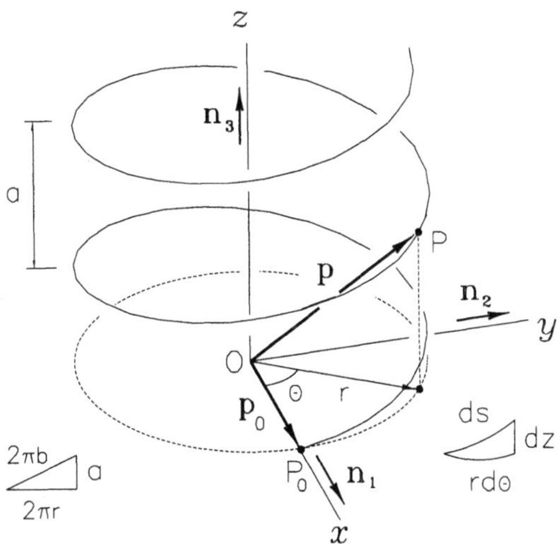

Figure 7.2

Every trajectory can be parametrized according to a conveniently chosen variable. Two different points on the curve must correspond to different values of the parameter. Parameter θ, used in the previous example, is suitable for describing the curve; parameter z, for example, would also be suitable. Of all the possible variables to be chosen to parametrize a given trajectory, one is particularly useful, as discussed ahead.

If P_0 is the position occupied by P at a given instant t_0, that is, $\mathbf{p}_0 = \mathbf{p}(t_0)$, the length s along λ between P_0 and P (see Fig. 7.3) is given by

$$s = \int_{P_0}^{P} ds = \int_{\mathbf{p}_0}^{\mathbf{p}} |d\mathbf{p}|, \qquad (7.2)$$

where $d\mathbf{p}$ is the differential of position vector $\mathbf{p}$, in reference frame $\mathcal{R}$. Note that ds is, by definition, essentially positive. For the sake of simplicity, in this section we omit the indication of reference frame $\mathcal{R}$ in all differentiations. In other words, we assume that $d\mathbf{p} = {}^{\mathcal{R}}d\mathbf{p}$. We also

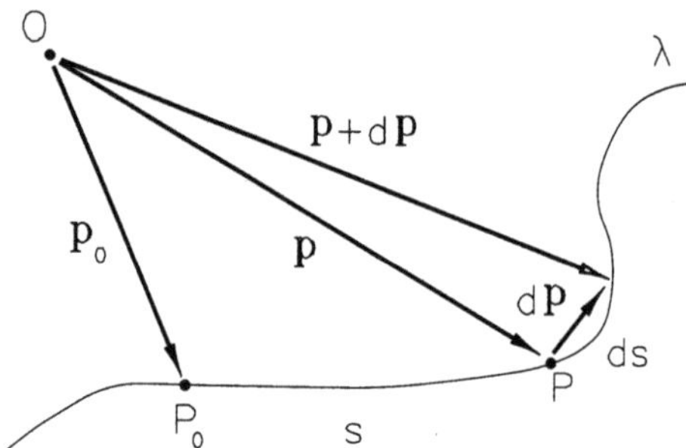

Figure 7.3

adopt the dot over the vector variable as a reduced notation for time derivative in $\mathcal{R}$ of the variable.

Example 7.2   Returning to the previous example (see Fig. 7.2) and adopting $P_0 : (r, 0, 0)$, then

$$ds^2 = |d\mathbf{p}|^2 = \left(r^2 + \left(\frac{a}{2\pi}\right)^2\right)(d\theta)^2;$$

therefore,

$$s = \int_{\mathbf{P_0}}^{\mathbf{P}} |d\mathbf{p}| = \int_0^\theta \left(r^2 + \left(\frac{a}{2\pi}\right)^2\right)^{\frac{1}{2}} d\theta = \left(r^2 + \left(\frac{a}{2\pi}\right)^2\right)^{\frac{1}{2}} \theta.$$

So, if P describes the trajectory, starting at P_0, and gives, for example, two complete rotations around the axis z ($\theta = 4\pi$), the distance covered along the curve is

$$s = 2(4\pi^2 r^2 + a^2)^{\frac{1}{2}}.$$

Line r touching the trajectory λ at point P necessarily has the direction of vector $d\mathbf{p}$ (see Fig. 7.4). A unit vector parallel to r can be defined then as

$$\mathbf{n}_t = \frac{d\mathbf{p}}{|d\mathbf{p}|} = \frac{d\mathbf{p}}{ds} = \frac{\dot{\mathbf{p}}}{\dot{s}}. \qquad (7.3)$$

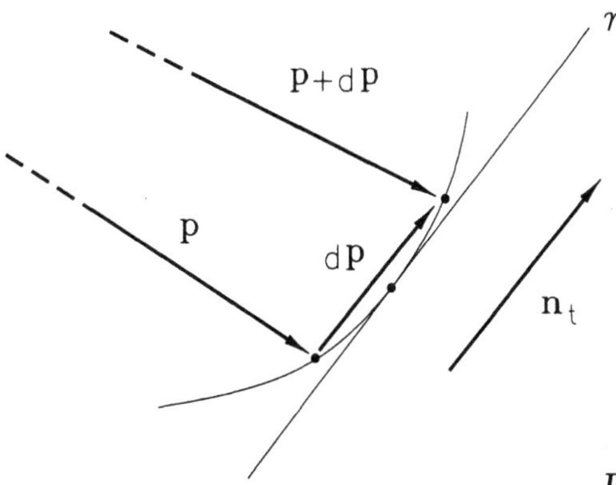

Figure 7.4

Vector $\mathbf{n}_t$, called the *tangent unit vector*, is, in fact, the unit vector that touches the trajectory at the point, pointing in the direction of rising s. Note that at a point where $\dot{s} = 0$, neither $\mathbf{n}_t$ nor the trajectory beyond the point is defined.

If $^{\mathcal{R}}\mathbf{v}^P$ is the (absolute) velocity vector of particle P in reference frame $\mathcal{R}$, then, from Eq. (7.3), we have

$$^{\mathcal{R}}\mathbf{v}^P = \dot{\mathbf{p}} = \dot{s}\mathbf{n}_t, \tag{7.4}$$

that is, *the velocity of P in $\mathcal{R}$ is always tangent to the trajectory* and its module is

$$v = |\mathbf{v}| = \dot{s}. \tag{7.5}$$

Note that $s(t)$ is a monotonic nondecreasing scalar function and, therefore, $\dot{s} \geq 0$, as mentioned above.

Example 7.3   Cursor C moves along guide B according to function $r(t) = r_0 + ut$, where u is a constant velocity, while the guide moves at a simple angular velocity of constant module ω in relation to support A, as illustrated (see Fig. 7.5). An element of arc length of the spiral trajectory is related to the variables $r(t)$, given above, and $\theta(t) = \omega t$ according to $ds^2 = dr^2 + r^2 d\theta^2$. Defining the constant $w^2 = u^2 + r^2\omega^2$, then

$$ds = w\,dt, \qquad \dot{s} = w.$$

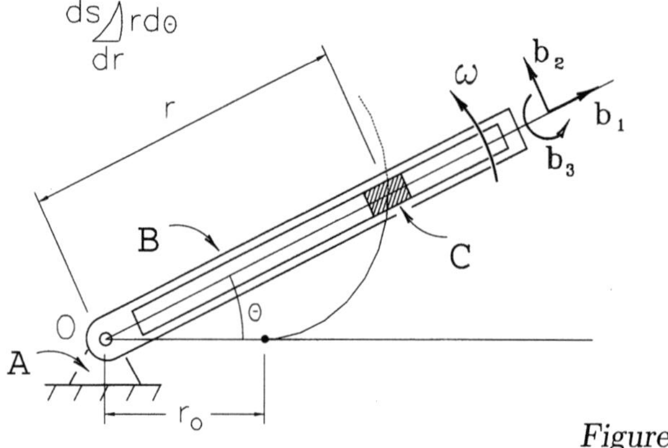

Figure 7.5

By adopting basis $\mathbf{b}_1, \mathbf{b}_2, \mathbf{b}_3$, fixed in B, the position vector of C with respect to pivot O is $\mathbf{p} = r\mathbf{b}_1$ and the velocity of the cursor in A is

$$\mathbf{v} = \dot{\mathbf{p}} = \dot{r}\mathbf{b}_1 + r\dot{\mathbf{b}}_1 = u\mathbf{b}_1 + r\omega\mathbf{b}_2.$$

The module of the velocity vector will then be the speed $v = (u^2 + r^2\omega^2)^{\frac{1}{2}}$, that is, $v = w$. It is therefore immediate that $v = \dot{s}$. The unit vector touching the trajectory at a generic instant is, according to Eq. (7.3),

$$\mathbf{n}_t = \frac{\mathbf{v}}{\dot{s}} = \frac{1}{w}(u\mathbf{b}_1 + r\omega\mathbf{b}_2).$$

We are now going to focus our attention on the neighborhood of a point P, the instantaneous position of a particle that describes a trajectory λ in a reference frame $\mathcal{R}$ (see Fig. 7.6). Let P_1 and P_2 be two positions infinitesimally close to P. The three points therefore define a plane, called the *osculating plane*, that locally contains the curve λ. These three points also define a circumference contained in the osculating plane, which locally approaches the trajectory. Center C of this circumference obviously belongs to the plane and is called the *center of curvature* of the trajectory at the point; the radius of this circumference, ρ, is called the *radius of curvature* of the trajectory at the point (see Fig. 7.6). The infinitesimal arch ds and infinitesimal angle $d\phi$ are naturally related by

$$ds = \rho d\phi. \tag{7.6}$$

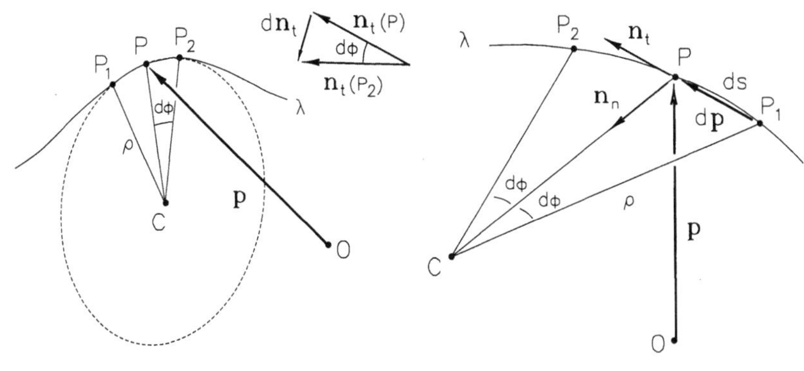

Figure 7.6

Variable ϕ may be expressed as a function of the parameter s, and the rate

$$\kappa = \frac{d\phi}{ds} = \frac{1}{\rho} \tag{7.7}$$

is called the *first curvature* or, simply, the *curvature* of the trajectory at the point. The curvature is a measure of the variation with s of the orientation of the curve in the osculating plane. The orientation of the curve is given by the tangent unit vector, $\mathbf{n}_t$. As its module is constant, the direction of its derivative is necessarily orthogonal to it. In fact, the differential of $\mathbf{n}_t$ between points P and P_2 is $d\mathbf{n}_t$, orthogonal to $\mathbf{n}_t$ and oriented toward the center of curvature C (see Fig. 7.6). So defining the unit vector $\mathbf{n}_n$, orthogonal to $\mathbf{n}_t$ and parallel to the osculating plane, the differential of the tangent unit vector will be $d\mathbf{n}_t = d\phi\,\mathbf{n}_n$, and its time derivative in $\mathcal{R}$ can be expressed as

$$\dot{\mathbf{n}}_t = \frac{d\mathbf{n}_t}{dt} = \dot{\phi}\mathbf{n}_n. \tag{7.8}$$

The rate of change of the tangent unit vector with respect to parameter s, which we write as $\mathbf{n}'_t$, is, according to Eqs. (7.7) and (7.8),

$$\mathbf{n}'_t = \frac{d\mathbf{n}_t}{ds} = \kappa\mathbf{n}_n. \tag{7.9}$$

Vector $\mathbf{n}_n$ is called the *principal normal* of the trajectory at the point. It points to the direction in which the trajectory tends. The rate at which the orientation of the trajectory changes, as a function of displacement

s of the particle, is determined by curvature κ, as shown by Eq. (7.9). It is also worth noting that

$$\dot{\mathbf{n}}_t = \dot{s}\mathbf{n}'_t. \tag{7.10}$$

Also note that, from Eq. (7.7),

$$\rho = \frac{1}{\kappa} = \frac{ds}{d\phi}, \tag{7.11}$$

that is, the curvature is the inverse of the radius of curvature. The higher the rate $d\phi/ds$, the wider the curvature of the trajectory and the smaller its radius of curvature.

Example 7.4   Returning to the helical trajectory discussed in Example 7.1 (also see Example 7.2), the position of a general point P with respect to origin O is (see Fig. 7.7)

$$\mathbf{p} = r\cos\theta\,\mathbf{n}_1 + r\sin\theta\,\mathbf{n}_2 + \frac{a}{2\pi}\theta\,\mathbf{n}_3.$$

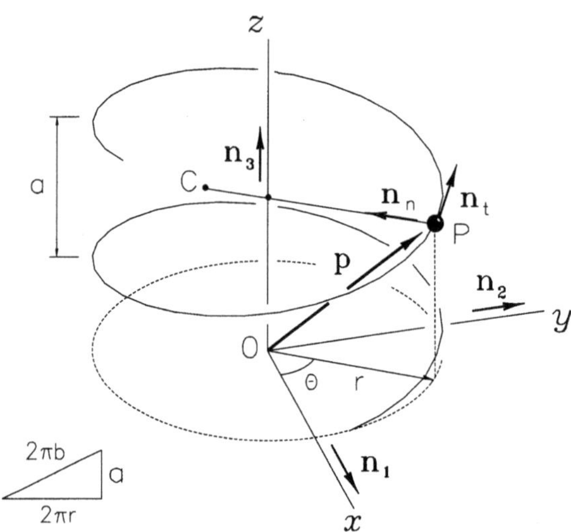

Figure 7.7

So, defining constant $b = \left(r^2 + (a/2\pi)^2\right)^{1/2}$, then $s = b\theta$ and the tangent unit vector, according to Eq. (7.3), is

$$\mathbf{n}_t = \frac{d\mathbf{p}}{ds} = \frac{1}{b}\frac{d\mathbf{p}}{d\theta} = \frac{1}{b}\left(-r\sin\theta\,\mathbf{n}_1 + r\cos\theta\,\mathbf{n}_2 + \frac{a}{2\pi}\mathbf{n}_3\right).$$

By differentiating with respect to time,

$$\dot{\mathbf{n}}_t = -\frac{r\dot{\theta}}{b}(\cos\theta\,\mathbf{n}_1 + \sin\theta\,\mathbf{n}_2).$$

The principal normal unit vector, therefore, is

$$\mathbf{n}_n = \frac{\dot{\mathbf{n}}_t}{|\dot{\mathbf{n}}_t|} = -(\cos\theta\,\mathbf{n}_1 + \sin\theta\,\mathbf{n}_2),$$

a horizontal vector oriented toward the vertical axis z (see Fig. 7.7). The derivative of the tangent unit vector with respect to s can be obtained from Eq. (7.10):

$$\mathbf{n}_t' = \frac{\dot{\mathbf{n}}_t}{\dot{s}} = \frac{\dot{\mathbf{n}}_t}{b\dot{\theta}} = \frac{r}{b^2}\mathbf{n}_n.$$

The curvature, according to Eq. (7.9), is

$$\kappa = \frac{r}{b^2} = \frac{r}{r^2 + \left(\frac{a}{2\pi}\right)^2}.$$

It is then found that the helix has a constant curvature. The radius of curvature, also constant, according to Eq. (7.11), is

$$\rho = r + \frac{a^2}{4\pi^2 r}.$$

Note that the center of curvature C is mobile and does not belong to the z-axis. If $a = 0$, the helix becomes a circumference of radius r, and the center of curvature will then coincide with point O, the center of this circumference.

The acceleration vector of a particle P moving in a trajectory λ in relation to a reference frame $\mathcal{R}$ is parallel to the osculating plane and therefore may be broken down into the directions of the tangent and principal normal directions according to

$$^{\mathcal{R}}\mathbf{a}^P = \mathbf{a}_t + \mathbf{a}_n = a_t\mathbf{n}_t + a_n\mathbf{n}_n. \tag{7.12}$$

Component $\mathbf{a}_t$, being tangent to the trajectory at the point, is called the *tangential acceleration*; component $\mathbf{a}_n$, pointing toward the center of

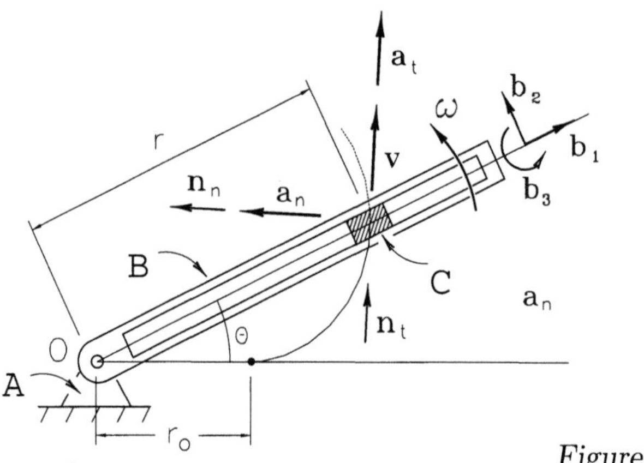

Figure 7.8

curvature, is called the *centripetal acceleration*. In fact, by the definition
of the acceleration vector, then

$$^{\mathcal{R}}\mathbf{a}^P = {^{\mathcal{R}}}\dot{\mathbf{v}}^P = \frac{^{\mathcal{R}}d}{dt}(\dot{s}\mathbf{n}_t) = \ddot{s}\mathbf{n}_t + \dot{s}\dot{\mathbf{n}}_t = \ddot{s}\mathbf{n}_t + \dot{s}^2\kappa\mathbf{n}_n. \tag{7.13}$$

The tangential acceleration is, therefore,

$$\mathbf{a}_t = \ddot{s}\mathbf{n}_t, \tag{7.14}$$

and the centripetal acceleration is

$$\mathbf{a}_n = \dot{s}^2\kappa\mathbf{n}_n = \frac{\dot{s}^2}{\rho}\mathbf{n}_n = \frac{v^2}{\rho}\mathbf{n}_n. \tag{7.15}$$

In short, the acceleration vector of a particle P moving in a trajectory
λ in relation to a reference frame $\mathcal{R}$ is always parallel to the osculating plane at the point. It has a component tangent to the trajectory
with magnitude $\ddot{s}$, that is, the second time derivative of displacement s
along the curve, and another component in the direction of the principal
normal, with magnitude dependent on speed $\dot{s}$, that is, the first time
derivative of s and the radius of curvature, as shown in Eq. (7.15).

Example 7.5   Returning to Example 7.3 (see Fig. 7.8), the acceleration
of cursor C in reference frame A is, by definition,

$$\mathbf{a} = \dot{\mathbf{v}} = u\dot{\mathbf{b}}_1 + \dot{r}\omega\mathbf{b}_2 + r\omega\dot{\mathbf{b}}_2 = -r\omega^2\mathbf{b}_1 + 2u\omega\mathbf{b}_2.$$

By twice differentiating the relation $w^2 = u^2 + r^2\omega^2$, we obtain

$$\ddot{s} = \dot{v} = \dot{w} = \frac{ru\dot{\omega}^2}{w}.$$

Then, the tangential acceleration component, according to Eq. (7.14), is

$$\mathbf{a}_t = \ddot{s}\mathbf{n}_t = \frac{ru\omega^2}{w^2}(u\mathbf{b}_1 + r\omega\mathbf{b}_2).$$

According to Eq. (7.12), the centripetal component of acceleration is

$$\mathbf{a}_n = \mathbf{a} - \mathbf{a}_t$$

$$= -r\omega^2\mathbf{b}_1 + 2u\omega\mathbf{b}_2 - \frac{ru\omega^2}{w^2}(u\mathbf{b}_1 + r\omega\mathbf{b}_2)$$

$$= \frac{w^2 + u^2}{w^2}\omega\left(-r\omega\mathbf{b}_1 + u\mathbf{b}_2\right).$$

The principal normal unit vector, being the unit vector in the corresponding direction, is then

$$\mathbf{n}_n = \frac{1}{w}\left(-r\omega\mathbf{b}_1 + u\mathbf{b}_2\right),$$

and it is immediately ascertained that $\mathbf{n}_t \cdot \mathbf{n}_n = 0$. We can then extract from Eq. (7.15) the value of the radius of curvature, a function of the point,

$$\rho = \frac{w^3}{\omega(w^2 + u^2)} = \frac{(u^2 + r^2\omega^2)^{3/2}}{\omega(2u^2 + r^2\omega^2)}.$$

Defining now a third unit vector

$$\mathbf{n}_b \rightleftharpoons \mathbf{n}_t \times \mathbf{n}_n, \tag{7.16}$$

called *binormal* to the trajectory at the point, we determine an orthonormal basis, $\mathbf{n}_t, \mathbf{n}_n, \mathbf{n}_b$, which accompanies the motion of particle P along trajectory λ (see Fig. 7.9). This set of vectors, called the *intrinsic orthonormal basis*, is fixed in a reference frame $\mathcal{S}$, always with $\mathbf{n}_t$ and $\mathbf{n}_n$ parallel to the osculating plane at the point and with vector $\mathbf{n}_b$ orthogonal to the plane. From the condition of orthogonality, then $\mathbf{n}_b \cdot \mathbf{n}_t = 0$ and differentiating with respect to time in $\mathcal{R}$, $\dot{\mathbf{n}}_b \cdot \mathbf{n}_t + \mathbf{n}_b \cdot \dot{\mathbf{n}}_t = 0$. But, as $\dot{\mathbf{n}}_t = \dot{s}\kappa\mathbf{n}_n$, then $\dot{\mathbf{n}}_b \cdot \mathbf{n}_t = 0$. Therefore, it is concluded that $\dot{\mathbf{n}}_b$ is simultaneously orthogonal to $\mathbf{n}_t$ and $\mathbf{n}_b$, that is, it is parallel to $\mathbf{n}_n$. It may therefore be written as

$$\dot{\mathbf{n}}_b = -\dot{\psi}\mathbf{n}_n, \tag{7.17}$$

where $\dot{\psi}$ may be interpreted as the time derivative of an angle ψ between successive osculating planes so that ψ decreases when the plane rotates in the direction of $\mathbf{n}_t$ (see Fig. 7.9). It is interesting to note that both derivatives of the tangent and binormal unit vectors are parallel to the principal normal unit vector.

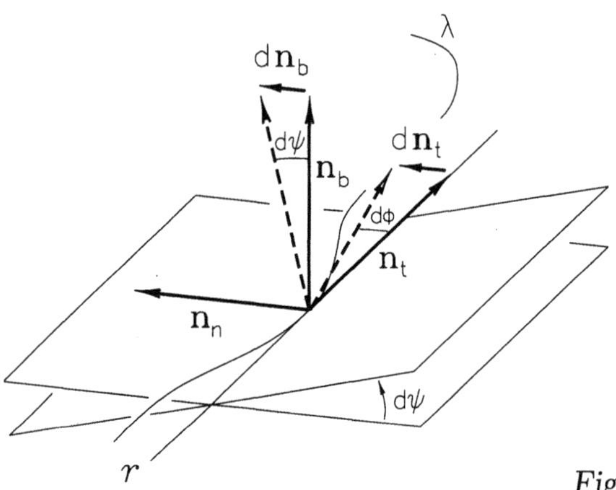

Figure 7.9

Variable ψ can also be expressed as a function of parameter s, defining the rate

$$\tau = \frac{d\psi}{ds} \tag{7.18}$$

as the *second curvature* or *torsion* of the trajectory at the point. Torsion is, therefore, a measure of the variation of the orientation of the osculating plane around the direction of the tangent line.

The reference frame $\mathcal{S}$ where the intrinsic basis is fixed moves in relation to reference frame $\mathcal{R}$ with angular velocity ${}^{\mathcal{R}}\boldsymbol{\omega}^{S}$, which can be determined by using Eq. (2.2). By breaking down ${}^{\mathcal{R}}\boldsymbol{\omega}^{S} = \omega_t\mathbf{n}_t + \omega_n\mathbf{n}_n + \omega_b\mathbf{n}_b$, then $\dot{\mathbf{n}}_t = {}^{\mathcal{R}}\boldsymbol{\omega}^{S} \times \mathbf{n}_t = \omega_b\mathbf{n}_n - \omega_n\mathbf{n}_b = \dot{\phi}\mathbf{n}_n$ and $\dot{\mathbf{n}}_b = {}^{\mathcal{R}}\boldsymbol{\omega}^{S} \times \mathbf{n}_b = \omega_n\mathbf{n}_t - \omega_t\mathbf{n}_n = -\dot{\psi}\mathbf{n}_n$. On comparing the results, we obtain $\omega_t = \dot{\psi}$, $\omega_n = 0$, and $\omega_b = \dot{\phi}$, resulting in

$$ {}^{\mathcal{R}}\boldsymbol{\omega}^{S} = \dot{\psi}\mathbf{n}_t + \dot{\phi}\mathbf{n}_b. \tag{7.19}$$

Also using Eqs. (7.7) and (7.18), we may express the angular velocity of $\mathcal{S}$ in reference frame $\mathcal{R}$ alternatively as

$$^{\mathcal{R}}\boldsymbol{\omega}^{S} = \dot{s}(\tau\mathbf{n}_t + \kappa\mathbf{n}_b). \tag{7.20}$$

It is then easy to determine the time derivative in the reference frame $\mathcal{R}$ of the principal normal vector, that is,

$$\dot{\mathbf{n}}_n = {}^{\mathcal{R}}\boldsymbol{\omega}^{S} \times \mathbf{n}_n = -\dot{\phi}\mathbf{n}_t + \dot{\psi}\mathbf{n}_b = \dot{s}(-\kappa\mathbf{n}_t + \tau\mathbf{n}_b). \tag{7.21}$$

In short, the time derivatives of the component vectors of the intrinsic orthonormal basis in the reference frame $\mathcal{R}$ are

$$\begin{aligned}
\dot{\mathbf{n}}_t &= \dot{\phi}\mathbf{n}_n = \dot{s}\kappa\mathbf{n}_n, \\
\dot{\mathbf{n}}_n &= -\dot{\phi}\mathbf{n}_t + \dot{\psi}\mathbf{n}_b = \dot{s}(-\kappa\mathbf{n}_t + \tau\mathbf{n}_b), \\
\dot{\mathbf{n}}_b &= -\dot{\psi}\mathbf{n}_n = -\dot{s}\tau\mathbf{n}_n.
\end{aligned} \tag{7.22}$$

Of course, the derivatives with respect to s of the intrinsic basis vectors in $\mathcal{R}$ will be

$$\begin{aligned}
\mathbf{n}_t' &= \kappa\mathbf{n}_n, \\
\mathbf{n}_n' &= -\kappa\mathbf{n}_t + \tau\mathbf{n}_b, \\
\mathbf{n}_b' &= -\tau\mathbf{n}_n.
\end{aligned} \tag{7.23}$$

Equations (7.23) are called *Serret Frenet formulas* and express how the vectors of the intrinsic basis vary along the trajectory; Equatons (7.22) express how the same vectors vary as the point (particle) moves forward. The difference is subtle but important. The derivatives with respect to variable s depend exclusively on the *form* of the trajectory, regardless of how the point moves along it; the time derivatives, on the other hand, depend on the motion of the point. So, for example, the curvature of the helix, established in Example 7.4, and its torsion, calculated in Example 7.6 ahead, are independent of the motion of P. Consequently, $\mathbf{n}_t'$, $\mathbf{n}_n'$, and $\mathbf{n}_b'$ will also be independent of time.

Example 7.6 Going even farther back to the helical trajectory of Example 7.1 (also see Examples 7.2 and 7.4), the binormal vector at point P can be established, according to Eq. (7.16), by

$$\begin{aligned}
\mathbf{n}_b &= \frac{1}{b}\left(-r\sin\theta\mathbf{n}_1 + r\cos\theta\mathbf{n}_2 + \frac{a}{2\pi}\mathbf{n}_3\right) \times (-\cos\theta\mathbf{n}_1 - \sin\theta\mathbf{n}_2) \\
&= \frac{1}{b}\left(\frac{a}{2\pi}\sin\theta\mathbf{n}_1 - \frac{a}{2\pi}\cos\theta\mathbf{n}_2 + r\mathbf{n}_3\right).
\end{aligned}$$

The time derivative of this vector is

$$\dot{\mathbf{n}}_b = \frac{\dot{\theta}}{b}\left(\frac{a}{2\pi}\cos\theta\,\mathbf{n}_1 + \frac{a}{2\pi}\sin\theta\,\mathbf{n}_2\right) = -\frac{a}{2\pi b}\dot{\theta}\mathbf{n}_n.$$

The torsion of the trajectory at the point, according to Eqs. (7.17) and (7.18), is then

$$\tau = \frac{\dot{\psi}}{\dot{s}} = \frac{a}{2\pi b^2} = \frac{a}{2\pi r^2 + a^2/2\pi},$$

therefore constant along the curve. The angular velocity of reference frame $\mathcal{S}$, where the intrinsic basis is fixed, is, according to Eq. (7.20),

$$^{\mathcal{R}}\boldsymbol{\omega}^{\mathcal{S}} = \frac{\dot{\theta}}{b}\left(\frac{a}{2\pi}\mathbf{n}_t + r\mathbf{n}_b\right).$$

Note that the time rate of change of the spatial orientation of the intrinsic basis is a function of $\dot{\theta}$, depending, therefore, on time. In other words, the angular velocity of reference frame $\mathcal{S}$ depends on the motion of the particle along the trajectory, characterized by the function $s(t)$. The curvature, κ, and torsion, τ, are, on the other hand, solely functions of the geometric nature of the trajectory and may depend on the point but not on time. (In this specific example of the helix, it was found that both the curvature and torsion are constant, not depending on the point of the curve; this *is not*, however, a valid conclusion for any trajectory.) In short, the time derivatives of the intrinsic units are, according to Eqs. (7.22),

$$\dot{\mathbf{n}}_t = \frac{r}{b}\dot{\theta}\mathbf{n}_n, \qquad \dot{\mathbf{n}}_n = -\frac{r}{b}\dot{\theta}\mathbf{n}_t + \frac{a}{2\pi b}\dot{\theta}\mathbf{n}_b, \qquad \dot{\mathbf{n}}_b = -\frac{a}{2\pi b}\dot{\theta}\mathbf{n}_n.$$

The derivatives with respect to parameter s of the same units are, according to Eqs. (7.23),

$$\mathbf{n}_t' = \frac{r}{b^2}\mathbf{n}_n, \qquad \mathbf{n}_n' = -\frac{r}{b^2}\mathbf{n}_t + \frac{a}{2\pi b^2}\mathbf{n}_b, \qquad \mathbf{n}_b' = -\frac{a}{2\pi b^2}\mathbf{n}_n.$$

Let us now assume that a particle P moves along the helical trajectory, leaving point P_0, according to function $s(t) = \frac{1}{2}a_0 t^2 + u_0 t$, where a_0 and u_0 are scalar constants. As $\theta(t) = s(t)/b$, the velocity of P at an arbitrary instant of the motion is

$$\mathbf{v} = \dot{s}\mathbf{n}_t = \frac{a_0 t + u_0}{b}\left(-r\sin\frac{s}{b}\mathbf{n}_1 + r\cos\frac{s}{b}\mathbf{n}_2 + \frac{a}{2\pi}\mathbf{n}_3\right).$$

The scalar component of the tangential acceleration is $a_t = \ddot{s} = a_o$, therefore constant, and the scalar component of the centripetal acceleration is $a_n = \dot{s}^2\kappa = r(a_0 t + u_0)^2/b^2$, variable in time. The acceleration of P at a general instant is then

$$\mathbf{a}(t) = a_0 \mathbf{n}_t + (a_0 t + u_0)^2 \frac{r}{b^2} \mathbf{n}_n$$

$$= -\left(\frac{a_0 r}{b} \sin \frac{s}{b} + \frac{(a_0 t + u)^2 r}{b^2} \cos \frac{s}{b} \right) \mathbf{n}_1$$

$$+ \left(\frac{a_0 r}{b} \cos \frac{s}{b} - \frac{(a_0 t + u)^2 r}{b^2} \sin \frac{s}{b} \right) \mathbf{n}_2 + \frac{a_0 a}{2\pi b} \mathbf{n}_3.$$

When the trajectory of a particle P moving in a reference frame $\mathcal{R}$ is contained on a plane π fixed in $\mathcal{R}$, it is said that P is describing a *plane motion*. In this case, the osculating plane coincides with plane π, the tangent unit vector and principal normal vector are parallel to the plane, and the velocity and acceleration vectors of P in $\mathcal{R}$ are also parallel to the plane (see Fig. 7.10).

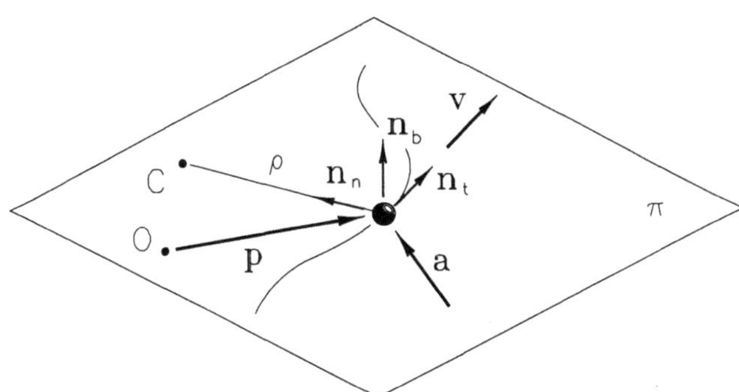

Figure 7.10

Since the trajectory is a flat curve, the center of curvature belongs to the plane and the torsion is null at any point. In fact, if the trajectory is flat, the position vector $\mathbf{p}$ with respect to a point O of the plane is always parallel to the plane and the vector $\mathbf{n}_t$, in the direction of $d\mathbf{p}$, is also parallel. Therefore, the variation of the tangent unit vector is also parallel to the plane, thus guaranteeing that the principal normal vector, $\mathbf{n}_n$, is in the orthogonal direction to $\mathbf{n}_t$ parallel to the plane. Consequently, the velocity and acceleration will also be parallel to the plane. Since the position vector of the center of curvature C with respect to point P has the direction of $\mathbf{n}_n$ and P belongs to π, then C will necessarily belong to π (see Fig. 7.10). Last, as the binormal vector $\mathbf{n}_b$ stays orthogonal to the plane, its time derivative in $\mathcal{R}$ is null and, from Eqs. (7.22), the torsion of the curve is also null.

Example 7.7 Returning again to Example 7.3 (also see Example 7.5), vectors $\mathbf{n}_t$ and $\mathbf{n}_n$ have been expressed as a linear combination of unit vectors $\mathbf{b}_1$ and $\mathbf{b}_2$, parallel to the plane of motion, the same occurring with the velocity and acceleration vectors of the cursor. The time derivative of the tangent unit vector is (check)

$$
\begin{aligned}
\dot{\mathbf{n}}_t &= \frac{d}{dt}\left[\frac{1}{w}(u\mathbf{b}_1 + r\omega\mathbf{b}_2)\right] \\
&= \frac{w^2 + u^2}{w^3}\omega(-r\omega\mathbf{b}_1 + u\mathbf{b}_2) \\
&= \frac{w^2 + u^2}{w^2}\omega\mathbf{n}_n.
\end{aligned}
$$

The curvature of the trajectory is, according to Eqs. (7.22),

$$
\kappa = \frac{w^2 + u^2}{w^3}\omega,
$$

and therefore variable. The radius of curvature is, at each instant,

$$
\rho = \frac{1}{\kappa} = \frac{w^3}{(w^2 + u^2)\omega},
$$

as established in Example 7.5. Since $\mathbf{n}_b = \mathbf{b}_3$, therefore, the torsion is, according to Eqs. (7.22), null and the time derivative of the principal normal vector is, also according to Eqs. (7.22),

$$
\begin{aligned}
\dot{\mathbf{n}}_n &= -\kappa \dot{s}\mathbf{n}_t \\
&= -\frac{w^2 + u^2}{w^2}\omega\mathbf{n}_t \\
&= -\frac{w^2 + u^2}{w^3}\omega(u\mathbf{b}_1 + r\omega\mathbf{b}_2).
\end{aligned}
$$

The reader can check this last result, as an exercise, by directly differentiating the expression for vector $\mathbf{n}_n$, obtained in Example 7.5.

3.8 Rigid Body Motion

The description of the rigid body motion requires more information than for the description of the motion of a particle. Its configuration in a given reference frame can be obtained in different ways. For example, the three components of the position vector of a point of the body with respect to a point fixed in the reference frame and the three angles conveniently chosen to characterize its spatial *orientation* in this reference frame comprise a set of six real scalar functions that can fully describe the position of the rigid body. It is then said that a rigid body, if there is no specific constraint to its motion, has six *degrees of freedom* in space.

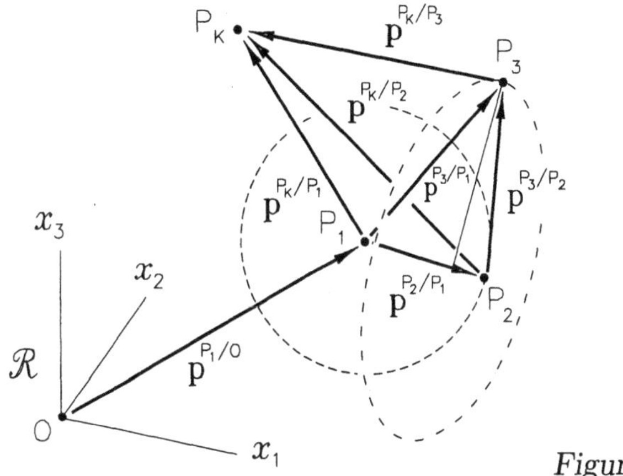

Figure 8.1

Figure 8.1 illustrates a simple way of checking the above statement. Point P_1 belongs to a rigid body (not shown) and, in order to determine its position in reference frame $\mathcal{R}$, we need to have three coordinates, $c_1(t), c_2(t),$ and $c_3(t)$ (for example, the scalar components of the vector $\mathbf{p}^{P_1/O}$, where O is a point fixed in $\mathcal{R}$). If P_2 is another point of the body, the rigid condition requires the distance between P_1 and P_2 to be constant; therefore, with the position of P_1 fixed in relation to $\mathcal{R}$, point P_2 must be on the surface of a sphere with center in P_1; therefore, two coordinates, $c_4(t), c_5(t),$ are enough to determine its position on the surface (latitude and longitude, say). Now if P_3 is another point of the body, its distances to points P_1 and P_2 are also constant. Therefore, if the positions of P_1 and P_2 in relation to $\mathcal{R}$ are also fixed, point P_3 must be on a circumference contained in a plane orthogonal to the line passing through P_1 and P_2, with the center on this line (see Fig. 8.1). Therefore, only one coordinate, $c_6(t)$, is required to calculate its position in relation to $\mathcal{R}$. Last, if P_k is any point of the body, its distances from the three aforementioned points of the body are constant and no extra coordinate is required to calculate its position in relation to $\mathcal{R}$. Six coordinates, therefore, fully determine the configuration of a rigid body in a given reference frame.

Let C be a rigid body moving arbitrarily in relation to a reference frame $\mathcal{R}$ at angular velocity $^{\mathcal{R}}\boldsymbol{\omega}^C$ and angular acceleration $^{R}\boldsymbol{\alpha}^C$. Also, let P and Q be two points fixed in C. The position vector of P with respect to Q, $\mathbf{p}^{P/Q}$, is a vector of a constant module, that is,

$$\frac{^{\mathcal{R}}d}{dt}\mathbf{p}^{P/Q} \cdot \mathbf{p}^{P/Q} = 0. \tag{8.1}$$

As this vector is fixed in C, its time derivative at C is null, that is,

$$\frac{^{C}d}{dt}\mathbf{p}^{P/Q} = {}^{C}\mathbf{v}^{P/Q} = {}^{C}\mathbf{v}^{P} - {}^{C}\mathbf{v}^{Q} = 0. \tag{8.2}$$

The first kinematic theorem, Eq. (6.1), in this case, is then reduced to

$$^{\mathcal{R}}\mathbf{v}^{P} = {}^{\mathcal{R}}\mathbf{v}^{Q} + {}^{\mathcal{R}}\boldsymbol{\omega}^{C} \times \mathbf{p}^{P/Q}. \tag{8.3}$$

In short, Eq. (8.3) relates the velocities, in a given reference frame $\mathcal{R}$, of two points of a rigid body. Since the relative velocity between two points is the difference between their absolute velocities, Eq. (5.10), the above relationship can alternatively be expressed as

$$^{\mathcal{R}}\mathbf{v}^{P/Q} = {}^{\mathcal{R}}\boldsymbol{\omega}^{C} \times \mathbf{p}^{P/Q}, \tag{8.4}$$

that is, the difference between the velocities of two points of a rigid body in a given reference frame is equal to the cross product of the angular velocity vector of the body in the reference frame by the relative position vector between the points.

Example 8.1 Consider the system shown in Fig. 8.2, where disk D, with radius r, rotates around its axis of symmetry z in relation to arm B that, in turn, rotates around horizontal axis y, in relation to bearing A. Orthonormal vectors $\mathbf{b}_1, \mathbf{b}_2, \mathbf{b}_3$ are attached to B, with $\mathbf{b}_2$ parallel to y and $\mathbf{b}_3$ parallel to z. The orthogonality between axes y and z, guaranteed by the connection at O, limits the motion of D in A to two degrees of freedom. The motion of the disk in relation to the bearing is fully described by the angular coordinates $\phi(t)$ and $\theta(t)$, as illustrated.

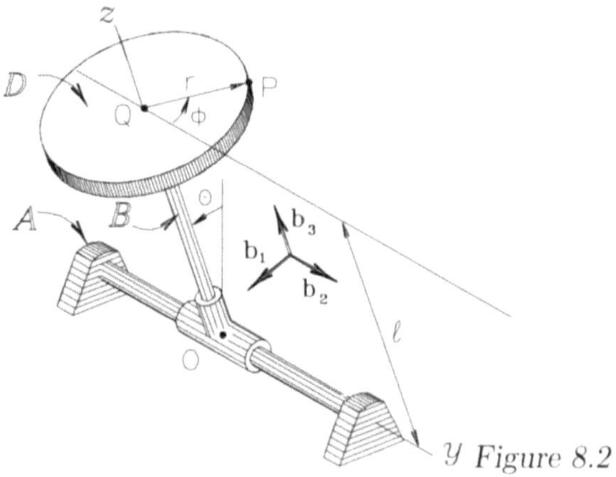

Figure 8.2

The angular velocities in question are

$$^{B}\boldsymbol{\omega}^{D} = \dot{\phi}\mathbf{b}_3, \quad ^{A}\boldsymbol{\omega}^{B} = \dot{\theta}\mathbf{b}_2, \quad ^{A}\boldsymbol{\omega}^{D} = \dot{\theta}\mathbf{b}_2 + \dot{\phi}\mathbf{b}_3.$$

As O and Q are fixed in B, the velocity of point Q in A can be found from Eq. (8.3) to be

$$^{A}\mathbf{v}^{Q} = {}^{A}\mathbf{v}^{O} + {}^{A}\boldsymbol{\omega}^{B} \times \mathbf{p}^{Q/O} = 0 + \dot{\theta}\mathbf{b}_2 \times l\mathbf{b}_3 = l\dot{\theta}\mathbf{b}_1.$$

The velocity in relation to the bearing of point P, fixed on the edge of the disk, can then be calculated, using again Eq. (8.3):

$$^{A}\mathbf{v}^{P} = {}^{A}\mathbf{v}^{Q} + {}^{A}\boldsymbol{\omega}^{D} \times \mathbf{p}^{P/Q}$$
$$= l\dot{\theta}\mathbf{b}_1 + (\dot{\theta}\mathbf{b}_2 + \dot{\phi}\mathbf{b}_3) \times r(-\sin\phi\,\mathbf{b}_1 + \cos\phi\,\mathbf{b}_2)$$
$$= (l\dot{\theta} - r\dot{\phi}\cos\phi)\mathbf{b}_1 - r\dot{\phi}\sin\phi\,\mathbf{b}_2 + r\dot{\theta}\sin\phi\,\mathbf{b}_3.$$

See the corresponding animation.

Returning to the rigid body C moving in a given reference frame $\mathcal{R}$ with angular velocity $^{\mathcal{R}}\boldsymbol{\omega}^C$ and angular acceleration $^{\mathcal{R}}\boldsymbol{\alpha}^C$, we can establish a relation between the accelerations of P and Q in reference frame $\mathcal{R}$ by using the second kinematic theorem, Eq. (6.3). In fact, as P and Q are fixed in C, $^C\mathbf{a}^P = {}^C\mathbf{v}^P = 0$, then

$$^{\mathcal{R}}\mathbf{a}^P = {}^{\mathcal{R}}\mathbf{a}^Q + {}^{\mathcal{R}}\boldsymbol{\omega}^C \times ({}^{\mathcal{R}}\boldsymbol{\omega}^C \times \mathbf{p}^{P/Q}) + {}^{\mathcal{R}}\boldsymbol{\alpha}^C \times \mathbf{p}^{P/Q}. \tag{8.5}$$

Equation (8.5) establishes the relation between the accelerations of two points of a rigid body moving arbitrarily in a given reference frame. By replacing Eq. (5.14) in the above relation, an alternative form is then obtained:

$$^{\mathcal{R}}\mathbf{a}^{P/Q} = {}^{\mathcal{R}}\boldsymbol{\omega}^C \times ({}^{\mathcal{R}}\boldsymbol{\omega}^C \times \mathbf{p}^{P/Q}) + {}^{\mathcal{R}}\boldsymbol{\alpha}^C \times \mathbf{p}^{P/Q}, \tag{8.6}$$

which calculates the relative acceleration between two points of the same rigid body moving in a given reference frame.

Example 8.2   Returning to the previous example (see Fig. 8.2), the angular accelerations in question are

$$^A\boldsymbol{\alpha}^B = \ddot{\theta}\mathbf{b}_2, \qquad ^B\boldsymbol{\alpha}^D = \ddot{\phi}\mathbf{b}_3, \qquad ^A\boldsymbol{\alpha}^D = \dot{\theta}\dot{\phi}\mathbf{b}_1 + \ddot{\theta}\mathbf{b}_2 + \ddot{\phi}\mathbf{b}_3.$$

The acceleration of point Q in A can be obtained from Eq. (8.5), so

$$\begin{aligned}
^A\mathbf{a}^Q &= {}^A\mathbf{a}^O + {}^A\boldsymbol{\omega}^B \times ({}^A\boldsymbol{\omega}^B \times \mathbf{p}^{Q/O}) + {}^A\boldsymbol{\alpha}^B \times \mathbf{p}^{Q/O} \\
&= 0 + \dot{\theta}\mathbf{b}_2 \times (\dot{\theta}\mathbf{b}_2 \times l\mathbf{b}_3) + \ddot{\theta}\mathbf{b}_2 \times l\mathbf{b}_3 \\
&= l\ddot{\theta}\mathbf{b}_1 - l\dot{\theta}^2\mathbf{b}_3.
\end{aligned}$$

The acceleration of point P in reference frame A can now also be obtained from Eq. (8.5),

$$^A\mathbf{a}^P = {}^A\mathbf{a}^Q + {}^A\boldsymbol{\omega}^D \times ({}^A\boldsymbol{\omega}^D \times \mathbf{p}^{P/Q}) + {}^A\boldsymbol{\alpha}^D \times \mathbf{p}^{P/Q},$$

where

$$^A\mathbf{a}^Q = l\ddot{\theta}\mathbf{b}_1 - l\dot{\theta}^2\mathbf{b}_3,$$
$$^A\boldsymbol{\omega}^D \times ({}^A\boldsymbol{\omega}^D \times \mathbf{p}^{P/Q}) = r\left((\dot{\theta}^2 + \dot{\phi}^2)\sin\phi\,\mathbf{b}_1 - \dot{\phi}^2\cos\phi\,\mathbf{b}_2 + \dot{\theta}\dot{\phi}\cos\phi\,\mathbf{b}_3\right),$$
$$^A\boldsymbol{\alpha}^D \times \mathbf{p}^{P/Q} = r(-\ddot{\phi}\cos\phi\,\mathbf{b}_1 - \ddot{\phi}\sin\phi\,\mathbf{b}_2 + (\ddot{\theta}\sin\phi + \dot{\theta}\dot{\phi}\cos\phi)\mathbf{b}_3);$$

therefore,

$$^A\mathbf{a}^P = (l\ddot{\theta} - r\ddot{\phi}\cos\phi + r(\dot{\theta}^2 + \dot{\phi}^2)\sin\phi)\mathbf{b}_1$$
$$- r(\ddot{\phi}\sin\phi + \dot{\phi}^2\cos\phi)\mathbf{b}_2$$
$$+ (r\ddot{\theta}\sin\phi - l\dot{\theta}^2 + 2r\dot{\theta}\dot{\phi}\cos\phi)\mathbf{b}_3.$$

Specifically, if the disk rotates in relation to the arm at an angular velocity of constant module ω_3 and the latter rotates in relation to the bearing at another angular velocity of constant module ω_2, the angular velocity of the disk in relation to the bearing will be $^A\boldsymbol{\omega}^D = \omega_2\mathbf{b}_2 + \omega_3\mathbf{b}_3$ and its angular acceleration in the same reference frame will be $^A\boldsymbol{\alpha}^D = \omega_2\omega_3\mathbf{b}_1$. In this case, the velocity and acceleration of point P in A when, say, $\phi = 0$, are reduced to

$$^A\mathbf{v}^P = (l\omega_2 - r\omega_3)\mathbf{b}_1,$$
$$^A\mathbf{a}^P = -r\omega_3^2\mathbf{b}_2 + (2r\omega_2\omega_3 - l\omega_2^2)\mathbf{b}_3.$$

See the corresponding animation.

When a rigid body C moves in a reference frame $\mathcal{R}$ so that its angular velocity in that reference frame is null for a certain interval of time, it is said that the body is *in translation* in the reference frame in that interval. When this happens, the velocities of all points of the body in the reference frame are equal to each other and the accelerations of all points of the body in the reference frame are also equal to each other. In fact, from Eq. (8.3), if $^\mathcal{R}\boldsymbol{\omega}^C = 0$, then $^\mathcal{R}\mathbf{v}^P = {}^\mathcal{R}\mathbf{v}^Q$ for any points P and Q of the body. On the other hand, as the angular velocity vector is always null, $^\mathcal{R}\boldsymbol{\alpha}^C = 0$ and, from Eq. (8.5), $^\mathcal{R}\mathbf{a}^P = {}^\mathcal{R}\mathbf{a}^Q$ for any points P and Q of the body. The motion of a rigid body that translates in a given reference frame is established by the position vector describing the trajectory of one of its points, since the orientation of the body in $\mathcal{R}$ does not alter and all points of the body move on trajectories parallel to each other. A rigid body with translational motion in a given reference frame, therefore, has three degrees of freedom, provided there is no further constraint against its motion.

Example 8.3 The rectangular block R, shown in Fig. 8.3, hangs from the support S by four bars of the same length c, joined by spherical joints at the suspension points on S and at vertices A, B, C, and D, as illustrated.

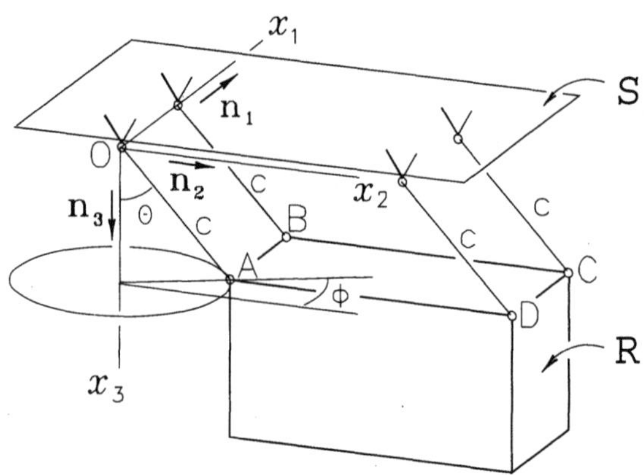

Figure 8.3

The basis of orthonormal vectors $\mathbf{n}_1, \mathbf{n}_2, \mathbf{n}_3$ is parallel to the axes $\{x_1, x_2, x_3\}$, fixed in S. The block moves, keeping its edges parallel to the axes, translating, therefore, about S; the description of its motion will be complete if the trajectory of, say, vertex A is known. As A is liable to move on the surface of the sphere with center at O and radius c, two coordinates, $\theta(t)$ and $\phi(t)$, are sufficient to configure the motion and R will have only two degrees of freedom in S. Note that in this case θ and ϕ are not coordinates that measure the orientation of R on S, but only convenient coordinates to describe the position vector

$$\mathbf{p}^{A/O} = c \left(\sin\theta \sin\phi \, \mathbf{n}_1 + \sin\theta \cos\phi \, \mathbf{n}_2 + \cos\theta \, \mathbf{n}_3 \right).$$

The velocity of any point P of R will then be given by

$$
\begin{aligned}
{}^{S}\mathbf{v}^{P} &= \frac{{}^{S}d}{dt} \mathbf{p}^{A/O} \\
&= c \left((\dot\theta \cos\theta \sin\phi + \dot\phi \sin\theta \cos\phi) \mathbf{n}_1 \right. \\
&\quad \left. + (\dot\theta \cos\theta \cos\phi - \dot\phi \sin\theta \sin\phi) \mathbf{n}_2 - \dot\theta \sin\theta \, \mathbf{n}_3 \right).
\end{aligned}
$$

See the corresponding animation.

When a rigid body C moves at simple angular velocity in a reference frame $\mathcal{R}$, the angular velocity vector and angular acceleration vector are, as shown in Sections 3.2 and 3.4, always parallel to a unit vector $\mathbf{n}$, fixed in C and in $\mathcal{R}$. If, moreover, a point Q of the body moves in a trajectory parallel to the plane orthogonal to $\mathbf{n}$, then all points of the body move in trajectories parallel to this plane and it is said that the body has *plane motion* in the reference frame. In fact, if $^{\mathcal{R}}\mathbf{v}^Q \cdot \mathbf{n} = 0$, the velocity of any point P of C in $\mathcal{R}$ is, from Eq. (8.3), $^{\mathcal{R}}\mathbf{v}^P = {}^{\mathcal{R}}\mathbf{v}^Q + {}^{\mathcal{R}}\boldsymbol{\omega}^C \times \mathbf{p}^{P/Q}$, and, projecting toward $\mathbf{n}$, then $^{\mathcal{R}}\mathbf{v}^P \cdot \mathbf{n} = {}^{\mathcal{R}}\mathbf{v}^Q \cdot \mathbf{n} + \omega\mathbf{n} \times \mathbf{p}^{P/Q} \cdot \mathbf{n}$. Both terms on the right side are null; therefore, whatever P is,

$$^{\mathcal{R}}\mathbf{v}^P \cdot \mathbf{n} = 0 \qquad \text{if } C \text{ has plane motion in } \mathcal{R}. \qquad (8.7)$$

As the velocity of any point stays parallel to the plane orthogonal to $\mathbf{n}$, there will be no acceleration component parallel to $\mathbf{n}$, that is, for every point P of the body,

$$^{\mathcal{R}}\mathbf{a}^P \cdot \mathbf{n} = 0 \qquad \text{if } C \text{ has plane motion in } \mathcal{R}. \qquad (8.8)$$

Example 8.4 The system shown in Fig. 8.4 consists of a disk D, with radius r, rotating around its vertical axis of symmetry in relation to the horizontal arm B, with length R, which, in its turn, rotates around the vertical axis z in relation to support A at angular velocity of constant module Ω.

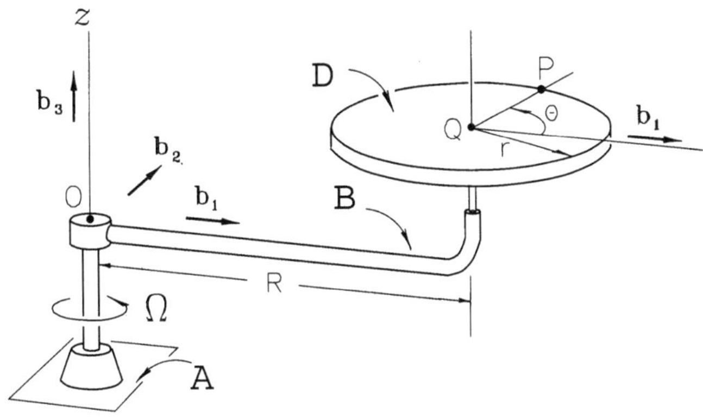

Figure 8.4

The orthonormal basis $\mathbf{b}_1, \mathbf{b}_2, \mathbf{b}_3$ is fixed in B, with $\mathbf{b}_1$ parallel to the axis of the arm and $\mathbf{b}_3$ vertical. As $\mathbf{b}_3$ is fixed simultaneously in A, B, and D, then there is simple angular velocity between the three reference frames, with $^{A}\boldsymbol{\omega}^B = \Omega\mathbf{b}_3$, $^{B}\boldsymbol{\omega}^D = \dot{\theta}\mathbf{b}_3$, and $^{A}\boldsymbol{\omega}^D = (\Omega + \dot{\theta})\mathbf{b}_3$, θ being the angle shown. The angular accelerations are ${}^A\boldsymbol{\alpha}^B = 0$, ${}^B\boldsymbol{\alpha}^D = {}^A\boldsymbol{\alpha}^D = \ddot{\theta}\mathbf{b}_3$ (the crossed term does not appear, since the angular velocities are parallel to each other). Point Q, the center of the disk, describes a circular trajectory with its center on the z-axis; its velocity and acceleration in A are

$$
\begin{aligned}
{}^A\mathbf{v}^Q &= {}^A\mathbf{v}^O + {}^A\boldsymbol{\omega}^B \times \mathbf{p}^{Q/O} \\
&= R\Omega\mathbf{b}_2, \\
{}^A\mathbf{a}^Q &= {}^A\mathbf{a}^O + {}^A\boldsymbol{\omega}^B \times \left({}^A\boldsymbol{\omega}^B \times \mathbf{p}^{Q/O}\right) + {}^A\boldsymbol{\alpha}^B \times \mathbf{p}^{Q/O} \\
&= -R\Omega^2\mathbf{b}_1.
\end{aligned}
$$

Since the disk has a simple angular velocity in A and one of its points (Q) describes a plane trajectory, the disk will have plane motion in A. The velocity and acceleration of point P, fixed to the edge of the disk, in A, are

$$
\begin{aligned}
{}^A\mathbf{v}^P &= {}^A\mathbf{v}^Q + {}^A\boldsymbol{\omega}^D \times \mathbf{p}^{P/Q} \\
&= -r(\Omega + \dot{\theta})\sin\theta\,\mathbf{b}_1 + \left(R\Omega + r(\Omega + \dot{\theta})\cos\theta\right)\mathbf{b}_2, \\
{}^A\mathbf{a}^P &= {}^A\mathbf{a}^Q + {}^A\boldsymbol{\omega}^D \times \left({}^A\boldsymbol{\omega}^D \times \mathbf{p}^{P/Q}\right) + {}^A\boldsymbol{\alpha}^D \times \mathbf{p}^{P/Q} \\
&= -\left(r\ddot{\theta}\sin\theta + R\Omega^2 + r(\Omega + \dot{\theta})^2\cos\theta\right)\mathbf{b}_1 \\
&\quad + r\left(\ddot{\theta}\cos\theta - (\Omega + \dot{\theta})^2\sin\theta\right)\mathbf{b}_2.
\end{aligned}
$$

As the reader can see, no velocity or acceleration components are in the direction of $\mathbf{b}_3$.

Example 8.5 The crankshaft mechanism, illustrated in Fig. 8.5, consists

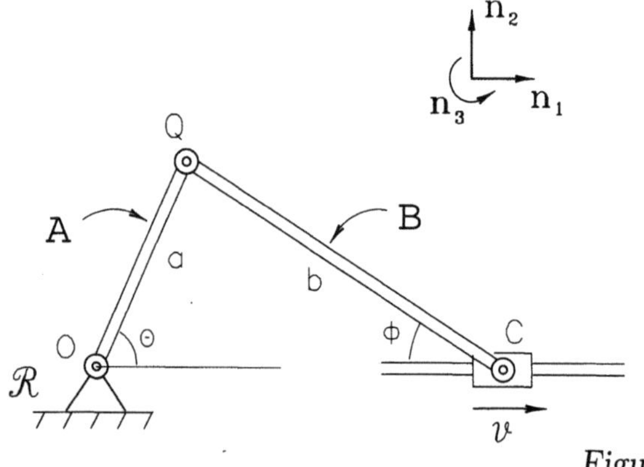

Figure 8.5

of a cursor C, which moves over a straight guide, and two rods, A and B joined at point O, fixed in $\mathcal{R}$, and at pins Q and C, as shown. We wish to find the angular velocity in $\mathcal{R}$ of rod B as a function of the coordinate $\theta(t)$. Since A and B perform a plane motion and $\mathbf{n}_3$ is always perpendicular to the plane of motion, the angular velocities of A and B in $\mathcal{R}$ are simple and parallel to $\mathbf{n}_3$. Point Q, fixed in A and in B, moves on the plane of the figure. The result then is that both bars have a plane motion and

$$\mathcal{R}\boldsymbol{\omega}^A = \dot{\theta}\mathbf{n}_3, \quad \mathcal{R}\boldsymbol{\omega}^B = \omega\mathbf{n}_3, \quad \mathcal{R}\boldsymbol{\alpha}^A = \ddot{\theta}\mathbf{n}_3, \quad \mathcal{R}\boldsymbol{\alpha}^B = \dot{\omega}\mathbf{n}_3,$$

where $\omega(t)$ is the angular velocity to be established. Also,

$$\begin{aligned}
\mathcal{R}\mathbf{v}^Q &= \mathcal{R}\mathbf{v}^O + \mathcal{R}\boldsymbol{\omega}^A \times \mathbf{p}^{Q/O} \\
&= 0 + \dot{\theta}\mathbf{n}_3 \times a(\cos\theta\,\mathbf{n}_1 + \sin\theta\,\mathbf{n}_2) \\
&= a\dot{\theta}(-\sin\theta\,\mathbf{n}_1 + \cos\theta\,\mathbf{n}_2).
\end{aligned}$$

But, on the other hand,

$$\begin{aligned}
\mathcal{R}\mathbf{v}^Q &= \mathcal{R}\mathbf{v}^C + \mathcal{R}\boldsymbol{\omega}^B \times \mathbf{p}^{Q/C} \\
&= v\mathbf{n}_1 + \omega\mathbf{n}_3 \times b(-\cos\phi\,\mathbf{n}_1 + \sin\phi\,\mathbf{n}_2) \\
&= (v - b\omega\sin\phi)\mathbf{n}_1 - b\omega\cos\phi\,\mathbf{n}_2.
\end{aligned}$$

By equaling term by term, we have

$$\omega = -\frac{a}{b}\frac{\cos\theta}{\cos\phi}\,\dot{\theta} = -\frac{a}{b}\frac{\cos\theta}{\sqrt{1 - \left(\frac{a}{b}\sin\theta\right)^2}}\,\dot{\theta}.$$

Note that the angular velocity of bar B is null when $\theta = \pm\pi/2$.

When a rigid body C moves in a reference frame $\mathcal{R}$ in such a way that a point O of C stays fixed in $\mathcal{R}$, it performs a *motion with a fixed point*. When this happens, all points on the line passing through O and parallel, at a given instant, to the angular velocity vector of the body in the reference frame has a null velocity in $\mathcal{R}$ at that instant (see Fig. 8.6).

In fact, if Q is a point of the body so that its position vector with respect to O is parallel to the angular velocity vector, then $\mathcal{R}\boldsymbol{\omega}^C \times \mathbf{p}^{Q/O} = 0$ and, from Eq. (8.3), $\mathcal{R}\mathbf{v}^Q = \mathcal{R}\mathbf{v}^O = 0$ for every point Q on the line. It may then be said that instantaneously everything occurs as if the body rotates around this line, which, for this reason, is called *instantaneous axis of rotation*.

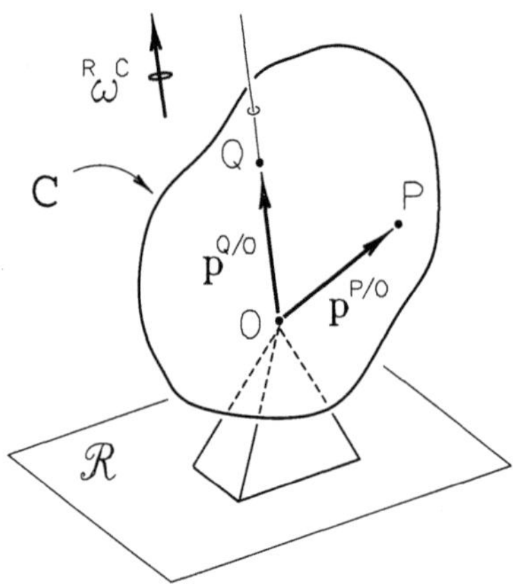

Figure 8.6

In general, however, the angular velocity vector of the body in the reference frame is not fixed either in C or in $\mathcal{R}$ (it is enough that $^{\mathcal{R}}\boldsymbol{\alpha}^C$ is a nonnull vector) and this instantaneous axis of rotation will, therefore, not be fixed in the body or in the reference frame.

The velocity, in a reference frame $\mathcal{R}$, of a point P of a rigid body C moving in $\mathcal{R}$ with a fixed point O, is, according to Eq. (8.3),

$$^{\mathcal{R}}\mathbf{v}^P = {}^{\mathcal{R}}\boldsymbol{\omega}^C \times \mathbf{p}^{P/O}. \tag{8.9}$$

Therefore, the velocity of any point of the body is always orthogonal to the instantaneous axis of rotation. The acceleration of this same point is, according to Eq. (8.5),

$$^{\mathcal{R}}\mathbf{a}^P = {}^{\mathcal{R}}\boldsymbol{\omega}^C \times ({}^{\mathcal{R}}\boldsymbol{\omega}^C \times \mathbf{p}^{P/O}) + {}^{\mathcal{R}}\boldsymbol{\alpha}^C \times \mathbf{p}^{P/O}. \tag{8.10}$$

A rigid body C moving with a point fixed in a given reference frame generally has three degrees of freedom in this reference frame. In fact, the constraint to the motion of the fixed point reduces by three the number of degrees of freedom of the body, dropping, therefore, from six to three. In general, three angular coordinates are required to decribe the motion of a rigid body with a fixed point. When C is a body of revolution, it is particularly convenient to adopt a set of angular coordinates, known as *Euler's angles*, described in the following example.

Example 8.6 Figure 8.7 illustrates the motion of a top C moving freely, supported by a horizontal plane with friction, with its point O staying fixed in the plane. The plane is a reference frame $\mathcal{R}$ in relation to which the motion of the top will be described. The axes $\{x, y, z\}$ are fixed in $\mathcal{R}$, with x and y parallel to the horizontal plane and z vertical. Consider A as a reference frame, not shown, moving at a simple angular velocity in relation to $\mathcal{R}$, around axis z. Axes x_1 and z are fixed in A, and $\psi(t)$ is the angle between x and x_1, as shown. Also consider another reference frame B moving at a simple angular velocity in relation to A, around axis x_1. Axes $\{x_1, x_2, x_3\}$ are fixed in B, and $\theta(t)$ measures the angle between axes z and x_3, as shown. Last, the top moves at a simple angular velocity in relation to B, around axis x_3, in the direction shown. Angle $\phi(t)$, measured on a plane orthogonal to x_3, expresses the rotation of C around x_3. Orthonormal basis $\mathbf{n}_1, \mathbf{n}_2, \mathbf{n}_3$ is fixed in $\mathcal{R}$ and orthonormal basis $\mathbf{b}_1, \mathbf{b}_2, \mathbf{b}_3$ is fixed in B, with the directions indicated.

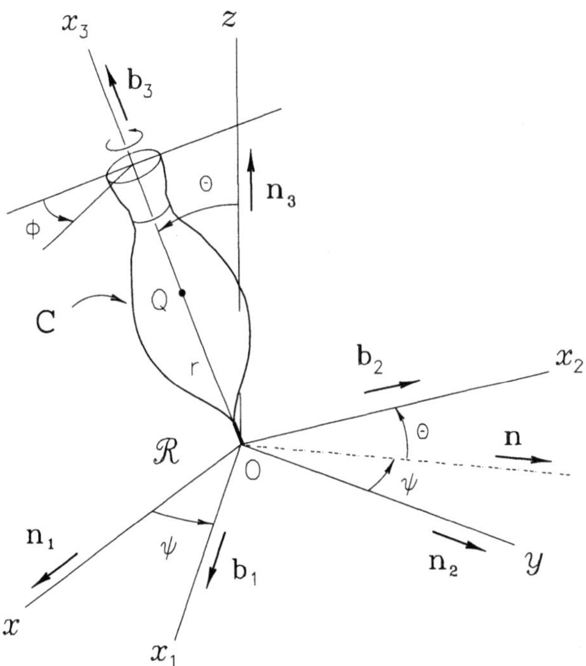

Figure 8.7

The angular velocity of the top in relation to B (rotation of the top around its axis of symmetry), called *spin*, is

$$^{B}\boldsymbol{\omega}^{C} = \dot{\phi}\mathbf{b}_3.$$

The angular velocity of reference frame B in relation to reference frame A, called *nutation*, is

$$^A\boldsymbol{\omega}^B = \dot{\theta}\mathbf{b}_1.$$

The angular velocity of reference frame A in relation to reference frame $\mathcal{R}$, called *precession*, is

$$^\mathcal{R}\boldsymbol{\omega}^A = \dot{\psi}\mathbf{n}_3 = \dot{\psi}(\sin\theta\mathbf{b}_2 + \cos\theta\mathbf{b}_3).$$

The angular velocity vector of the top in $\mathcal{R}$ will therefore be the vector sum of the angular velocities obtained above [see Eq. (3.10)], that is,

$$^\mathcal{R}\boldsymbol{\omega}^C = \dot{\theta}\mathbf{b}_1 + \dot{\psi}\sin\theta\mathbf{b}_2 + (\dot{\phi} + \dot{\psi}\cos\theta)\mathbf{b}_3.$$

Scalar functions $\phi(t)$, $\theta(t)$, and $\psi(t)$, adopted for the description of the angular motion of the top, are called Euler's angles. Although it is not the only possible description for the orientation of a rigid body moving with a fixed point, it is extremely practical and used widely. If Q is a point of the axis of symmetry of the top, being at a distance r from point O (see Fig. 8.7), its velocity in $\mathcal{R}$, according to Eq. (8.9), is

$$^\mathcal{R}\mathbf{v}^Q = {}^\mathcal{R}\boldsymbol{\omega}^C \times \mathbf{p}^{Q/O} = r(\dot{\psi}\sin\theta\mathbf{b}_1 - \dot{\theta}\mathbf{b}_2).$$

The time derivatives in $\mathcal{R}$ of basis vectors $\mathbf{b}_1, \mathbf{b}_2, \mathbf{b}_3$ can be obtained, according to Eq. (2.2), as

$$\begin{aligned}
\dot{\mathbf{b}}_j &= {}^\mathcal{R}\boldsymbol{\omega}^B \times \mathbf{b}_j \\
&= (\dot{\theta}\mathbf{b}_1 + \dot{\psi}\sin\theta\mathbf{b}_2 + \dot{\psi}\cos\theta\mathbf{b}_3) \times \mathbf{b}_j, \qquad j = 1, 2, 3,
\end{aligned}$$

that is,

$$\begin{aligned}
\dot{\mathbf{b}}_1 &= \dot{\psi}(\cos\theta\mathbf{b}_2 - \sin\theta\mathbf{b}_3), \\
\dot{\mathbf{b}}_2 &= -\dot{\psi}\cos\theta\mathbf{b}_1 + \dot{\theta}\mathbf{b}_3, \\
\dot{\mathbf{b}}_3 &= \dot{\psi}\sin\theta\mathbf{b}_1 - \dot{\theta}\mathbf{b}_2.
\end{aligned}$$

The angular acceleration of the top may then be obtained by differentiation:

$$\begin{aligned}
^\mathcal{R}\boldsymbol{\alpha}^C &= \frac{^\mathcal{R}d}{dt}{}^\mathcal{R}\boldsymbol{\omega}^C \\
&= \ddot{\theta}\mathbf{b}_1 + (\ddot{\psi}\sin\theta + \dot{\psi}\dot{\theta}\cos\theta)\mathbf{b}_2 + (\ddot{\phi} + \ddot{\psi}\cos\theta - \dot{\psi}\dot{\theta}\sin\theta)\mathbf{b}_3 \\
&\quad + \dot{\theta}\dot{\mathbf{b}}_1 + \dot{\psi}\sin\theta\dot{\mathbf{b}}_2 + (\dot{\phi} + \dot{\psi}\cos\theta)\dot{\mathbf{b}}_3 \\
&= (\ddot{\theta} + \dot{\phi}\dot{\psi}\sin\theta)\mathbf{b}_1 + (\ddot{\psi}\sin\theta + \dot{\psi}\dot{\theta}\cos\theta - \dot{\phi}\dot{\theta})\mathbf{b}_2 \\
&\quad + (\ddot{\phi} + \ddot{\psi}\cos\theta - \dot{\psi}\dot{\theta}\sin\theta)\mathbf{b}_3.
\end{aligned}$$

Let us now consider the top moving with constant $\theta(t)$, $\dot{\phi}(t)$, and $\dot{\psi}(t)$, that is, a motion with constant spin and precession and null nutation. (This is one of the possible modes of motion of a top, as discussed in Chapter 7.) In this case, then $\dot{\theta} = \ddot{\theta} = \ddot{\psi} = \ddot{\phi} = 0$, and the angular velocity and acceleration of the top are reduced to

$$^{\mathcal{R}}\boldsymbol{\omega}^C = \dot{\psi}\sin\theta\mathbf{b}_2 + (\dot{\phi} + \dot{\psi}\cos\theta)\mathbf{b}_3,$$
$$^{\mathcal{R}}\boldsymbol{\alpha}^C = \dot{\psi}\dot{\phi}\sin\theta\mathbf{b}_1.$$

The acceleration of point Q is, in this case,

$$^{\mathcal{R}}\mathbf{a}^Q = {}^{\mathcal{R}}\mathbf{a}^O + {}^{\mathcal{R}}\boldsymbol{\omega}^C \times ({}^{\mathcal{R}}\boldsymbol{\omega}^C \times \mathbf{p}^{Q/O}) + {}^{\mathcal{R}}\boldsymbol{\alpha}^C \times \mathbf{p}^{Q/O}$$
$$= r\sin\theta\dot{\psi}^2(\cos\theta\mathbf{b}_2 - \sin\theta\mathbf{b}_3)$$
$$= r\sin\theta\dot{\psi}^2\mathbf{n},$$

where $\mathbf{n}$ is the unit vector orthogonal to the plane defined by axes z and x_1, in the indicated direction. In fact, in the condition of constant precession and null nutation, point Q describes a circular trajectory with radius $r\sin\theta$, parallel to the horizontal plane, at constant speed $v = r\sin\theta|\dot{\psi}|$, null tangential acceleration ($\ddot{\psi} = 0$), and centripetal acceleration

$$\mathbf{a}_c = \frac{v^2}{\rho}\mathbf{n} = r\sin\theta\dot{\psi}^2\mathbf{n},$$

as established above. See the corresponding animation.

As expected, the expressions for the angular velocity vector in Examples 3.3, 3.4, and 8.6 are similar. Both the gyroscope in Example 3.3 and the top in the above study perform rigid body motion with a fixed point and, in both cases, the angular coordinates adopted are Euler's angles. In the case of the coin, studied in Example 3.4, the point of contact C has instantaneously zero velocity in relation to the plane, although, as will be seen in the following section, its acceleration is different from zero, therefore instantaneously configuring a rigid body motion with a fixed point.

3.9 Rolling

Let there be two rigid bodies C and C' moving so that, at each instant, there is a point P of the surface of C touching a point P' of the surface of C' (see Fig. 9.1). It is said that there is *rolling* between C and C' if, for some reference frame $\mathcal{R}$,

$$^{\mathcal{R}}\mathbf{v}^P = {}^{\mathcal{R}}\mathbf{v}^{P'},\tag{9.1}$$

or the equivalent,

$$^{\mathcal{R}}\mathbf{v}^{P/P'} = 0.\tag{9.2}$$

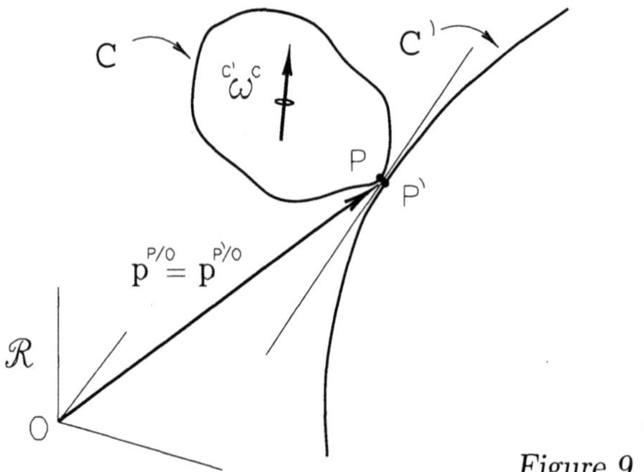

Figure 9.1

By using Eq. (6.1), it is easy to check that if the velocities of P and P' in $\mathcal{R}$ are equal at a given instant in $\mathcal{R}$, their velocities will be equal in any reference frame at that instant. In fact, if $\mathcal{R}'$ is any other reference frame and O is a point fixed in $\mathcal{R}$, then

$$\begin{aligned}
{}^{\mathcal{R}'}\mathbf{v}^P &= {}^{\mathcal{R}}\mathbf{v}^P + {}^{\mathcal{R}'}\mathbf{v}^O + {}^{\mathcal{R}'}\boldsymbol{\omega}^{\mathcal{R}} \times \mathbf{p}^{P/O}, \\
{}^{\mathcal{R}'}\mathbf{v}^{P'} &= {}^{\mathcal{R}}\mathbf{v}^{P'} + {}^{\mathcal{R}'}\mathbf{v}^O + {}^{\mathcal{R}'}\boldsymbol{\omega}^{\mathcal{R}} \times \mathbf{p}^{P'/O},
\end{aligned}\tag{9.3}$$

and subtracting the second from the first equation, then ${}^{\mathcal{R}'}\mathbf{v}^P - {}^{\mathcal{R}'}\mathbf{v}^{P'} = 0$. Specifically, choosing as a reference frame the actual body C', then

$$^{C'}\mathbf{v}^P = {}^{C'}\mathbf{v}^{P'} = 0,\tag{9.4}$$

that is, if a body C rolls over another body C', then at each instant the point P of C in contact with C' has null velocity in C'. Note, however, that rolling sets a condition exclusively for the velocity of the point of contact, nothing being imposed on its acceleration. In fact, when rolling occurs, at each instant a new point of C will be in contact with C', meaning that the acceleration of this point in C' will be different from zero; if the velocity and acceleration are simultaneously null at a given interval of time, we return during this interval to the condition of motion with a fixed point.

Example 9.1   Figure 9.2 illustrates a rolling situation. Cylinder C, with radius r, rolls over the inner surface of the cylindrical ring A, with radius R. In this case, all points of generatrix z are in contact with the ring surface. Let B be the reference frame defined by the plane containing axes Z (the ring axis of symmetry) and z, reference frame to which the Cartesian axes $\{X, Y, Z\}$ and orthonormal basis $\mathbf{n}_1, \mathbf{n}_2, \mathbf{n}_3$ are associated, as illustrated. Then we have

$$^A\boldsymbol{\omega}^B = \dot{\theta}\mathbf{n}_3, \quad {}^B\boldsymbol{\omega}^C = -\dot{\phi}\mathbf{n}_3, \quad {}^A\boldsymbol{\omega}^C = (\dot{\theta} - \dot{\phi})\mathbf{n}_3,$$

where θ and ϕ measure the rotation of B relative to A and the rotation of C relative to B, respectively, as shown.

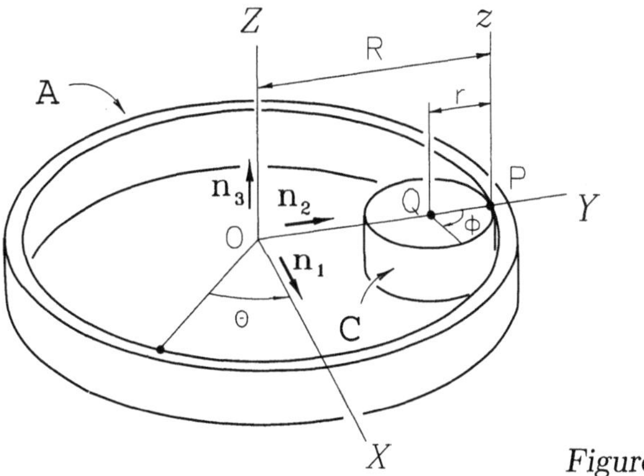

Figure 9.2

The velocity of point Q, fixed in B, in A is

$$^A\mathbf{v}^Q = {}^A\mathbf{v}^O + {}^A\boldsymbol{\omega}^B \times \mathbf{p}^{Q/O} = -(R - r)\dot{\theta}\mathbf{n}_1,$$

and the velocity of the point of contact P, fixed in C, in relation to A, can be obtained from

$$^A\mathbf{v}^P = {}^A\mathbf{v}^Q + {}^A\boldsymbol{\omega}^C \times \mathbf{p}^{P/Q} = (r\dot{\phi} - R\dot{\theta})\mathbf{n}_1.$$

As the rolling condition, Eq. (9.3), must be met, then $^A\mathbf{v}^P = 0$, which establishes the kinematic constraint relating the functions $\dot{\theta}(t)$ and $\dot{\phi}(t)$:

$$\dot{\phi} = \frac{R}{r}\,\dot{\theta}.$$

If the angular velocity vector of body C in relation to body C', $^{C'}\boldsymbol{\omega}^C$, is parallel to the plane touching the contact surfaces of the bodies at point P (see Fig. 9.1), it is said that there is *pure rolling*. If the angular velocity of C in C' is orthogonal to the tangent plane common to the bodies at the point of contact, it is said that there is *pivoting*. In the more general case of rolling between two bodies, the relative angular velocity vector has a pure rolling component and a pivoting component.

Example 9.2   Figure 9.3 shows a circular track A, with inner radius R, in which two identical spheres with a diameter of $2r$ move.

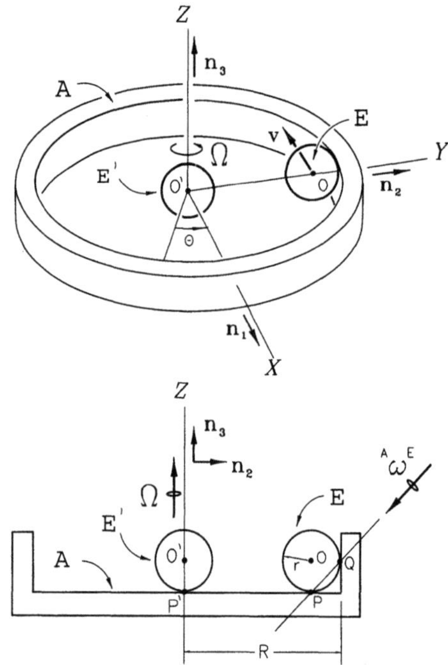

Figure 9.3

Sphere E' has its center O' fixed on the vertical axis Z, rotating at an angular velocity with constant module Ω in relation to the track A. Point

P', fixed in E', is in permanent contact with the horizontal plane of the track. The motion of E' in relation to A is, therefore, purely pivotal, with $^A\boldsymbol{\omega}^{E'} = \Omega\mathbf{n}_3$. Sphere E, on the other hand, describes a more complex motion, with its center O performing a circular trajectory with radius $(R-r)$ on a horizontal plane, with velocity $\mathbf{v}$, and two points, P and Q, in contact with the track. As there is rolling at both points, $^A\mathbf{v}^P = {}^A\mathbf{v}^Q = 0$ and, from Eq. (8.3), $^A\boldsymbol{\omega}^E \times \mathbf{p}^{P/Q} = 0$, then the angular velocity vector of the sphere in relation to the track makes an angle of 45^0 in relation to both the tangent planes at the points of contact, that is, $^A\boldsymbol{\omega}^E = \frac{\sqrt{2}}{2}\omega(\mathbf{n}_2 + \mathbf{n}_3)$. This is, therefore, a case of general rolling, with pure rolling and pivoting components. Now suppose that the center of sphere E moves in relation to the track at constant speed u, that is, $\mathbf{v} = -u\mathbf{n}_1$. From Eq. (8.3), then $^A\boldsymbol{\omega}^E \times \mathbf{p}^{P/O} = {}^A\mathbf{v}^P - {}^A\mathbf{v}^O$, that is, $\frac{\sqrt{2}}{2}\omega(\mathbf{n}_2 + \mathbf{n}_3) \times (-r)\mathbf{n}_3 = -u\mathbf{n}_1$; therefore, $\omega = -\sqrt{2}\,u/r$ and the angular velocity of sphere E is

$$^A\boldsymbol{\omega}^E = -\frac{u}{r}(\mathbf{n}_2 + \mathbf{n}_3).$$

Example 9.3 Consider the motion of the coin M rolling over a horizontal plane $\mathcal{R}$, as described in Example 3.4 (see Fig. 9.4). It is necessary that $^{\mathcal{R}}\mathbf{v}^C = 0$ in order to fulfill the rolling condition. The angular velocity vector of the coin in relation to the reference frame $\mathcal{R}$, determined in the aforementioned example, is

$$^{\mathcal{R}}\boldsymbol{\omega}^M = \omega_1\mathbf{n}_1 + \omega_2\mathbf{n}_2 + \omega_3\mathbf{n}_3,$$

where

$$\omega_1 = \dot{\psi}\cos\theta, \quad \omega_2 = \dot{\theta}, \quad \omega_3 = \dot{\phi} + \dot{\psi}\sin\theta.$$

The time rates of these components are

$$\dot{\omega}_1 = \ddot{\psi}\cos\theta - \dot{\psi}\dot{\theta}\sin\theta, \quad \dot{\omega}_2 = \ddot{\theta}, \quad \dot{\omega}_3 = \ddot{\phi} + \ddot{\psi}\sin\theta + \dot{\psi}\dot{\theta}\cos\theta.$$

The angular velocity of the reference frame A, where the orthonormal basis $\mathbf{n}_1, \mathbf{n}_2, \mathbf{n}_3$ is fixed, is

$$^{\mathcal{R}}\boldsymbol{\omega}^A = \omega_1\mathbf{n}_1 + \omega_2\mathbf{n}_2 + (\omega_3 - \dot{\phi})\mathbf{n}_3,$$

and the time derivatives in $\mathcal{R}$ of the base vectors are (using the simplified notation)

$$\dot{\mathbf{n}}_1 = (\omega_3 - \dot{\phi})\mathbf{n}_2 - \omega_2\mathbf{n}_3,$$

$$\dot{\mathbf{n}}_2 = \omega_1\mathbf{n}_3 - (\omega_3 - \dot{\phi})\mathbf{n}_1,$$

$$\dot{\mathbf{n}}_3 = \omega_2\mathbf{n}_1 - \omega_1\mathbf{n}_2.$$

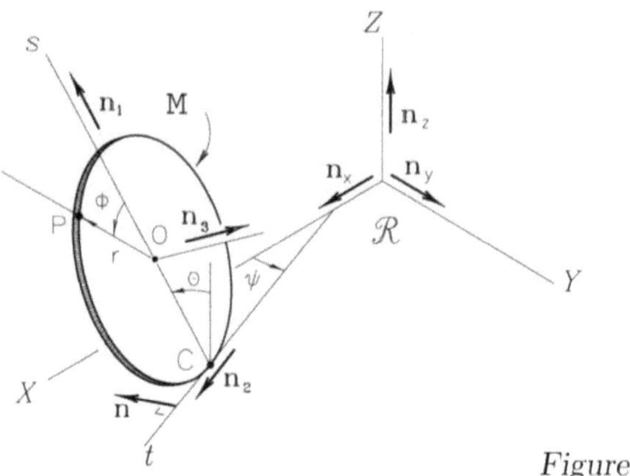

Figure 9.4

The angular acceleration of the coin in reference frame $\mathcal{R}$ is obtained by differentiating the angular velocity vector in $\mathcal{R}$, resulting in

$$^{\mathcal{R}}\boldsymbol{\alpha}^M = \alpha_1 \mathbf{n}_1 + \alpha_2 \mathbf{n}_2 + \alpha_3 \mathbf{n}_3,$$

where

$$\alpha_1 = \dot{\omega}_1 + \dot{\phi}\omega_2, \quad \alpha_2 = \dot{\omega}_2 - \dot{\phi}\omega_1, \quad \alpha_3 = \dot{\omega}_3.$$

The velocity of the center of the coin is obtained from Eq. (8.3):

$$\begin{aligned}
^{\mathcal{R}}\mathbf{v}^O &= {}^{\mathcal{R}}\mathbf{v}^C + {}^{\mathcal{R}}\boldsymbol{\omega}^M \times \mathbf{p}^{O/C} \\
&= 0 + (\omega_1 \mathbf{n}_1 + \omega_2 \mathbf{n}_2 + \omega_3 \mathbf{n}_3) \times r\mathbf{n}_1 \\
&= r(\omega_3 \mathbf{n}_2 - \omega_2 \mathbf{n}_3),
\end{aligned}$$

and its acceleration can be obtained by differentiating the above result in $\mathcal{R}$, that is,

$$\begin{aligned}
^{\mathcal{R}}\mathbf{a}^O &= \frac{^{\mathcal{R}}d}{dt}{}^{\mathcal{R}}\mathbf{v}^O \\
&= r(\dot{\omega}_3 \mathbf{n}_2 - \dot{\omega}_2 \mathbf{n}_3 + \omega_3 \dot{\mathbf{n}}_2 - \omega_2 \dot{\mathbf{n}}_3) \\
&= r\left(-\left(\omega_2^2 + \omega_3(\omega_3 - \dot{\phi})\right)\mathbf{n}_1 + (\dot{\omega}_3 + \omega_1\omega_2)\mathbf{n}_2 \right. \\
&\quad \left. + (-\dot{\omega}_2 + \omega_3\omega_1)\mathbf{n}_3\right).
\end{aligned}$$

Now calculating:

$$\begin{aligned}
^{\mathcal{R}}\boldsymbol{\omega}^M \times ({}^{\mathcal{R}}\boldsymbol{\omega}^M \times \mathbf{p}^{C/O}) &= r[(\omega_2^2 + \omega_3^2)\mathbf{n}_1 - \omega_1\omega_2 \mathbf{n}_2 - \omega_1\omega_3 \mathbf{n}_3]; \\
^{\mathcal{R}}\boldsymbol{\alpha}^M \times \mathbf{p}^{C/O} &= r\left[-\dot{\omega}_3 \mathbf{n}_2 + (\dot{\omega}_2 - \dot{\phi}\omega_1)\mathbf{n}_3\right],
\end{aligned}$$

the acceleration of point C in contact with $\mathcal{R}$ can be determined by Eq. (8.5):

$$
\begin{aligned}
^{\mathcal{R}}\mathbf{a}^C &= {}^{\mathcal{R}}\mathbf{a}^O + {}^{\mathcal{R}}\boldsymbol{\omega}^M \times ({}^{\mathcal{R}}\boldsymbol{\omega}^M \times \mathbf{p}^{C/O}) + {}^{\mathcal{R}}\boldsymbol{\alpha}^M \times \mathbf{p}^{C/O} \\
&= r(\dot{\phi}\omega_3 \mathbf{n}_1 - \dot{\phi}\omega_1 \mathbf{n}_3) \\
&= r\left[(\dot{\phi}^2 + \dot{\phi}\dot{\psi}\sin\theta)\mathbf{n}_1 - \dot{\phi}\dot{\psi}\cos\theta\mathbf{n}_3\right] \\
&= r(\dot{\phi}^2 \mathbf{n}_1 + \dot{\phi}\dot{\psi}\mathbf{n}),
\end{aligned}
$$

where $\mathbf{n}$ is a horizontal unit vector, orthogonal to the line t, as shown. It is interesting to find how the component terms of $^{\mathcal{R}}\mathbf{a}^C$ show various simplifications, in fact annulling the acceleration toward $\mathbf{n}_2$. Actually, there can be no acceleration in this direction; otherwise sliding would occur immediately afterward. Some particular cases illustrate the situation even better. Assuming pure rolling, that is, the coin rolling over the plane with $\dot{\theta} = \dot{\psi} = 0$, then $^{\mathcal{R}}\mathbf{a}^C = r\dot{\phi}^2\mathbf{n}_1$, a centripetal acceleration, which always occurs in the case of pure rolling over a plane. Now when we assume pivoting, that is, the coin moving with $\dot{\theta} = \dot{\phi} = 0$, then $^{\mathcal{R}}\mathbf{a}^C = 0$, meaning that point C stays in contact with the plane throughout the motion, being reduced therefore to the status of motion with a fixed point.

3.10 Mechanical Systems

When a set $\mathcal{S}$ of particles or bodies (or both) moves in relation to a given reference frame $\mathcal{R}$ so that there is a partial or total interdependence between their motions, this is a *mechanical system* or a *mechanism*. If the position vector of each point of $\mathcal{S}$ with respect to a given point O fixed in $\mathcal{R}$ is determined, it is said that the *configuration* of $\mathcal{S}$ in $\mathcal{R}$ is known. The system configuration is determined by means of a set of scalar functions of time, $q_1(t), q_2(t), \ldots, q_r(t)$, called *generalized coordinates* or simply *coordinates* of the system. Every position vector above will therefore be expressed in terms of the chosen coordinates. Each coordinate $q_j(t)$, $j = 1, 2, \ldots, r$, will have a dimension [L] when measuring a distance, or null dimension when measuring an angle.

Example 10.1 Consider a mechanical system $\mathcal{S}$, shown in Fig. 10.1, consisting of a rectangular plate L, with dimensions a and b, whose vertices A and B are confined to move along axes X and Y, respectively, and a small sphere P connected to vertices C and D by two light threads of the same length r.

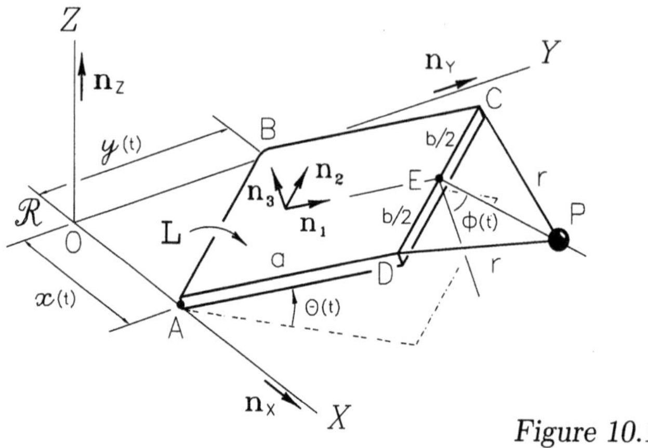

Figure 10.1

The system can therefore be treated as if it were a rigid body and a particle and, assuming that it moves so that the threads stay stretched, its configuration can be fully described by the following set of coordinates: $q_1 = x(t)$, measuring the distance between A and O; $q_2 = y(t)$, measuring the distance between B and O; $q_3 = \theta(t)$, measuring the angle between the plate and the horizontal plane; and $q_4 = \phi(t)$, measuring the angle between the line passing through E and P and the line orthogonal to L passing through E, the midpoint of edge CD. Note that this is not the only possible choice of coordinates (would you have another suggestion?). Note also that, in this case, $r = 4$ and there are two coordinates with the dimension of length and two with the dimension of angle. Any point of S can have its position with respect to O fully established as a function of the chosen coordinates. For example, the position of point D with respect to O may be given as

$$\mathbf{p}^{D/O} = x\mathbf{n}_x + a\mathbf{n}_1 = (x + y\frac{a}{b}\cos\theta)\mathbf{n}_x + x\frac{a}{b}\cos\theta\mathbf{n}_y + a\sin\theta\mathbf{n}_z.$$

See the corresponding animation.

   The number of *mutually independent* generalized coordinates in a mechanical system is known as the *number of degrees of freedom of the system*. A particle, free to move in space, has three degrees of freedom. A rigid body, on the other hand, has, in principle, six degrees of freedom. So, a system S consisting of n particles and m rigid bodies would, in principle, have $3n + 6m$ degrees of freedom. However, it so happens that the elements of a system interact with each other and with the reference frame to which they are associated, resulting in a set of constraints against their motions, called *kinematic constraints*. Every kinematic constraint, therefore, comes from a *link* between two elements of the system or between one of those elements and a body outside the system. To each kinematic constraint corresponds a relation involving the system coordinates and their time rates, so that

$$f(q_1, q_2, \ldots, q_r, \dot{q}_1, \dot{q}_2, \ldots, \dot{q}_r, t) = 0. \tag{10.1}$$

Example 10.2 Returning to the previous example (see Fig. 10.1), the system consists of a particle (P) and a rigid body (L), with, therefore, in principle, $3 + 6 = 9$ degrees of freedom. If $\mathbf{p}$ is the position vector of P with respect to O, its scalar components in the directions of the Cartesian axes would consist of three coordinates of $\mathcal{S}$, so that $q_1 = p_X$, $q_2 = p_Y$, $q_3 = p_Z$. The coordinates, say, of the vertex A: $q_4 = X_A$; $q_5 = Y_A$; $q_6 = Z_A$, and three angles measuring the orientation of the plate in relation to the axes: $q_7 = \alpha_X$; $q_8 = \alpha_Y$; $q_9 = \alpha_Z$, would complete the overall set of coordinates. The system is, however, subject to the following constraints: Vertices A and B of the plate are restricted to moving along the guides fixed to axes X and Y, respectively; and particle P is fixed by nonstretchable threads to vertices C and D of the plate. The following kinematic constraints must, then, be satisfied:

$$^{\mathcal{R}}\mathbf{v}^A \cdot \mathbf{n}_y = 0; \qquad ^{\mathcal{R}}\mathbf{v}^A \cdot \mathbf{n}_z = 0;$$

$$^{\mathcal{R}}\mathbf{v}^B \cdot \mathbf{n}_x = 0; \qquad ^{\mathcal{R}}\mathbf{v}^B \cdot \mathbf{n}_z = 0;$$

$$^{\mathcal{R}}\mathbf{v}^{P/C} \cdot \mathbf{p}^{P/C} = 0; \qquad ^{\mathcal{R}}\mathbf{v}^{P/D} \cdot \mathbf{p}^{P/D} = 0.$$

There is, therefore, a set of six kinematic constraints. When expressing the position and velocity vectors in terms of coordinates $q_1, q_2, \ldots, q_9$, there will be six relations in the form of Eq. (10.1). For example, the first two kinematic constraints above, expressed in terms of the general coordinates, are

$$\dot{q}_5 = 0, \qquad \dot{q}_6 = 0.$$

Note that the effective number of degrees of freedom of the system will be $9 - 6 = 3$. In fact, it is easy to see from the chosen coordinates in

Example 10.1 that, for any configuration of the system, $x^2 + y^2 = b^2$ and, therefore, the y-coordinate can be expressed as a function of the x-coordinate, that is, $y = \sqrt{b^2 - x^2}$ and three coordinates (x, θ, and ϕ) are enough to fully determine the system configuration.

A mechanical system $\mathcal{S}$ consisting of n particles and m rigid bodies, moving in a given reference frame $\mathcal{R}$ and subject to p kinematic constraints in the form of Eq. (10.1), will have

$$l = 3n + 6m - p \qquad (10.2)$$

degrees of freedom. The number of generalized coordinates required to fully describe an arbitrary instantaneous configuration of $\mathcal{S}$ in $\mathcal{R}$ will be $r \geq l$. When a relation of kinematic constraint in the form of Eq. (10.1) *can be integrated*, it is said to be a *holonomic* constraint. When the constraint equation cannot be directly integrated, it is said to be *simple nonholonomic*. A nonholonomic constraint ceases to be considered simple when it cannot be expressed in the form of Eq. (10.1). Holonomic constraints can also be classified as *reonomic* or *escleronomic* to the extent that the function f does or does not depend explicitly on time, respectively. A mechanical system $\mathcal{S}$ is said to be holonomic when all p kinematic constraints are of the holonomic type; otherwise, it will be called nonholonomic. Since they can be directly integrated, holonomic constraint equations result in algebraic expressions interrelating the coordinates. Thus, some coordinates may be expressed as a function of others. In this case, the number of degrees of freedom of the system may equal the number of generalized coordinates adopted for its description, that is, $r = l$. The following examples illustrate better the point.

Example 10.3   Let us consider the mechanical system consisting of disk D and bar B, welded at point Q, which moves in reference frame $\mathcal{R}$ as follows: The bar has its end O linked by a ball and socket to the vertical shaft, fixed in $\mathcal{R}$, and the disk is supported by the horizontal plane, rolling on it (see Fig. 10.2).

Since the disk and bar are welded together, they form a single rigid body and the system would have six degrees of freedom, not taking the kinematic constraints into account. The body (bar plus disk) can be configured by the following set of coordinates: x_O, y_O, z_O, coordinates of point O in the Cartesian system $\{x, y, z\}$, fixed in $\mathcal{R}$, and θ, ϕ, ψ, angles that determine

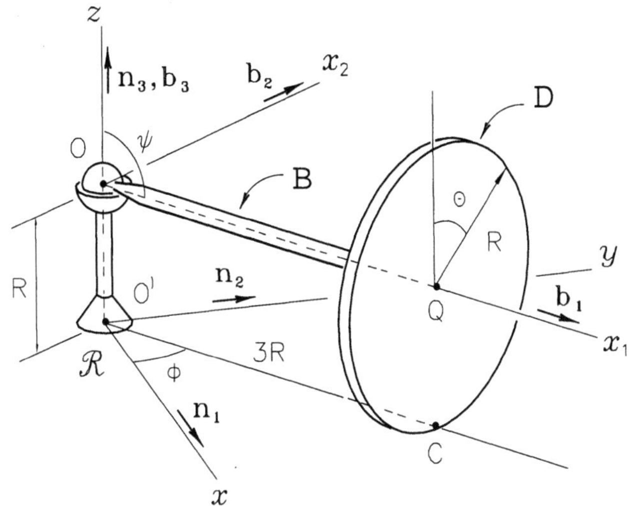

Figure 10.2

the orientation of the body, as shown. The pertinent kinematic constraints
are

$$x_O = 0, \qquad y_O = 0, \qquad z_O = R,$$

ensuring that the point O is fixed in $\mathcal{R}$,

$$\psi = \pi/2,$$

translating the fact that the disk stays in contact with the plane, and

$$^{\mathcal{R}}\mathbf{v}^C = 0,$$

the rolling condition of the disk. Note that the first four kinematic constraints are already in the integrated form, that is, they establish direct
algebraic relations between the coordinates. The original kinematic relations, in the form of Eq. (10.1), would be

$$^{\mathcal{R}}\mathbf{v}^O = 0; \qquad \text{therefore,} \qquad \dot{x}_O = 0, \quad \dot{y}_O = 0, \quad \dot{z}_O = 0,$$

and

$$^{\mathcal{R}}\boldsymbol{\omega}^D \cdot \mathbf{b}_2 = 0; \qquad \text{therefore,} \qquad \dot{\psi} = 0,$$

whose direct integration leads to the previous algebraic equations. The
first four kinematic constraints are, therefore, of the holonomic type and

naturally eliminate the coordinates x_O, y_O, z_O, and ψ. Let us now look at the rolling condition. The angular velocity of the body in $\mathcal{R}$ will be $^{\mathcal{R}}\boldsymbol{\omega}^D = -\dot{\theta}\mathbf{b}_1 + \dot{\phi}\mathbf{b}_3$ and the velocity of the point of contact will then be

$$^{\mathcal{R}}\mathbf{v}^C = {}^{\mathcal{R}}\mathbf{v}^O + {}^{\mathcal{R}}\boldsymbol{\omega}^D \times \mathbf{p}^{C/O} = (3\dot{\phi} - \dot{\theta})R\mathbf{b}_2.$$

To fulfill the rolling condition, then

$$3\dot{\phi} - \dot{\theta} = 0,$$

an equation of the type of Eq. (10.1), which can be directly integrated and solved for θ, resulting in

$$\theta(t) = 3\big(\phi(t) - \phi(0)\big) + \theta(0).$$

This system is, therefore, holonomic, with $n = 0$, $m = 1$, $p = 5$, and $l = 3 \times 0 + 6 \times 1 - 5 = 1$ degree of freedom. The coordinate $\phi(t)$ is sufficient to describe its configuration, provided that the initial configuration is known: $\theta(0)$, $\phi(0)$.

Example 10.4   Let us now consider a new mechanical system derived from the previous one with the introduction of a thin rectangular plate P between the disk and horizontal plane (see Fig. 10.3). Assuming that there is rolling between D and P and that the latter slides over the plane, now the system consists of two rigid bodies ($m = 2$) and the following kinematic constraints:

$$\begin{aligned}
{}^{\mathcal{R}}\mathbf{v}^O &= 0 & \text{(three constraints)}; & \qquad\text{(a)}\\
{}^{\mathcal{R}}\boldsymbol{\omega}^D \cdot \mathbf{b}_2 &= 0 & \text{(one constraint)}; & \qquad\text{(b)}\\
{}^{\mathcal{R}}\mathbf{v}^A \cdot \mathbf{n}_3 &= 0 & \text{(one constraint)}; & \qquad\text{(c)}\\
{}^{\mathcal{R}}\boldsymbol{\omega}^P \cdot \mathbf{n}_1 = {}^{\mathcal{R}}\boldsymbol{\omega}^P \cdot \mathbf{n}_2 &= 0 & \text{(two constraints)}; & \qquad\text{(d)}\\
{}^{\mathcal{R}}\mathbf{v}^C &= {}^{\mathcal{R}}\mathbf{v}^{C'} & \text{(two constraints)}. & \qquad\text{(e)}
\end{aligned}$$

The kinematic constraints (a) and (b) are identical to those in the previous example and therefore holonomic; constraints (c) and (d) translate the fact that the plate stays supported on the horizontal plane, and it is easy to see that they also form holonomic constraints. Let us now look at the nature of constraints (e). We may take as generalized coordinates for this system two coordinates for vertex A of the plate, $x(t)$ and $y(t)$, the angle $\beta(t)$ between the edge illustrated and axis x, and angles $\phi(t)$ and $\theta(t)$, described in the preceding example.

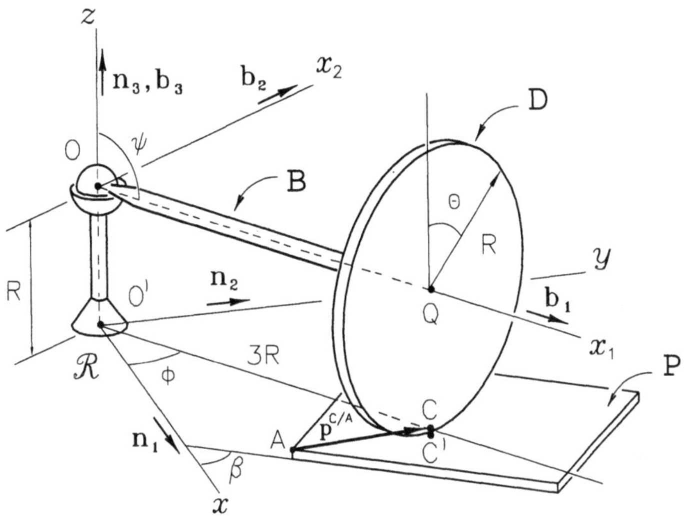

Figure 10.3

The plate angular velocity in $\mathcal{R}$ will be ${}^{\mathcal{R}}\boldsymbol{\omega}^P = \dot{\beta}\mathbf{n}_3$, and the velocity in $\mathcal{R}$ of the vertex A will be ${}^{\mathcal{R}}\mathbf{v}^A = \dot{x}\mathbf{n}_1 + \dot{y}\mathbf{n}_2$. The position relative to A of the point C' of P, instantaneously in contact with D, is

$$\mathbf{p}^{C'/A} = \mathbf{p}^{C/A} = \mathbf{p}^{C/O'} - \mathbf{p}^{A/O'} = (3R\cos\phi - x)\mathbf{n}_1 + (3R\sin\phi - y)\mathbf{n}_2,$$

and the velocity in $\mathcal{R}$ of C' is

$$\begin{aligned}
{}^{\mathcal{R}}\mathbf{v}^{C'} &= {}^{\mathcal{R}}\mathbf{v}^A + {}^{\mathcal{R}}\boldsymbol{\omega}^P \times \mathbf{p}^{C'/A} \\
&= \left(\dot{x} - (3R\sin\phi - y)\dot{\beta}\right)\mathbf{n}_1 + \left(\dot{y} - (3R\cos\phi - x)\dot{\beta}\right)\mathbf{n}_2.
\end{aligned}$$

By substituting it in Eq. (e) (see previous example for the expression of ${}^{\mathcal{R}}\mathbf{v}^C$), then

$$-R(3\dot{\phi} - \dot{\theta})\sin\phi = \dot{x} - (3R\sin\phi - y)\dot{\beta},$$
$$R(3\dot{\phi} - \dot{\theta})\cos\phi = \dot{y} - (3R\cos\phi - x)\dot{\beta}.$$

Note that the two nonlinear differential equations above, involving the coordinates $x(t)$, $y(t)$, $\beta(t)$, $\theta(t)$, and $\phi(t)$, cannot be directly integrated, and it is no longer possible, for example, to find a simple relation between $\theta(t)$ and $\phi(t)$, as in the previous example. There are, therefore, two nonholonomic kinematic constraints and the system will not be holonomic.

Note also that the vector equation (e) resulted, in fact, in only two kinematic constraints, as expected. This is due to the fact that conditions (b), (c), and (d) ensure that there is no vertical component of velocity of the points of contact. The system has, then, $p = 9$ kinematic constraints and $l = 6 \times 2 - 9 = 3$ degrees of freedom. As this is a nonholonomic system, however, to determine a general configuration requires adopting $r = 5$ coordinates (two more, since there are two nonholonomic kinematic constraints), as follows: $\theta(t)$ and $\phi(t)$ to configure D and $x(t)$, $y(t)$, and $\beta(t)$ to configure P.

Now let S be a system consisting of m interconnected rigid bodies, moving in space in relation to a reference frame $\mathcal{R}$. Let C_k be an element of this system, in contact with body C_{k-1} at point Q and with body C_{k+1} at point P (see Fig. 10.4).

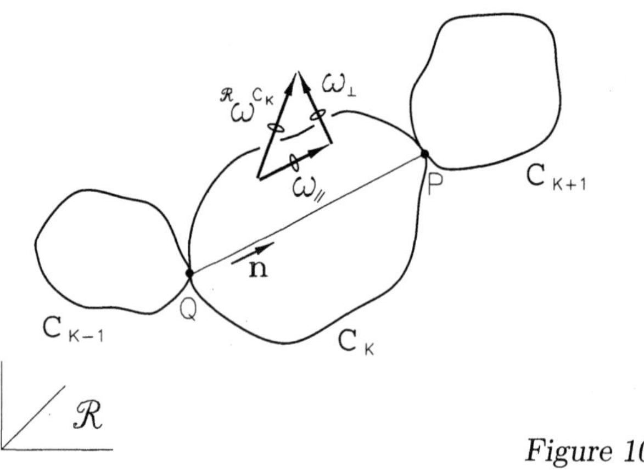

Figure 10.4

Let $\mathbf{n}$ be a unit vector, at each instant parallel to the line containing P and Q, and let $\boldsymbol{\omega}$ be the angular velocity vector of C_k in $\mathcal{R}$. We can break down $\boldsymbol{\omega}$ in the directions parallel and orthogonal to $\mathbf{n}$ (see Appendix A),

$$\boldsymbol{\omega} = \boldsymbol{\omega}_{//} + \boldsymbol{\omega}_\perp, \tag{10.3}$$

where

$$\boldsymbol{\omega}_{//} = \boldsymbol{\omega} \cdot \mathbf{n}\,\mathbf{n} \tag{10.4}$$

and

$$\boldsymbol{\omega}_\perp = \mathbf{n} \times (\boldsymbol{\omega} \times \mathbf{n}). \tag{10.5}$$

The component of the angular velocity vector parallel to the relative position vector between points P and Q, $\boldsymbol{\omega}_{/\!/}$, satisfies, of course, the condition

$$\boldsymbol{\omega}_{/\!/} \times \mathbf{n} = 0, \tag{10.6}$$

while the component of the angular velocity vector orthogonal to the relative position vector, $\boldsymbol{\omega}_{\perp}$, must satisfy the condition

$$\boldsymbol{\omega}_{\perp} \cdot \mathbf{n} = 0. \tag{10.7}$$

Then the result, from Eqs. (8.3), (10.3), and (10.6), is

$$^{\mathcal{R}}\mathbf{v}^{P} = {}^{\mathcal{R}}\mathbf{v}^{Q} + \boldsymbol{\omega}_{\perp} \times \mathbf{p}^{P/Q}. \tag{10.8}$$

The above relation shows that, in a given reference frame, the velocity of a point P of a rigid body is fully established in terms of the velocity of another point Q of the body and of the component of the body angular velocity vector orthogonal to the relative position vector of the points. Equation (10.8) is actually a refinement of Eq. (8.3); as the cross product of the parallel component of the vector angular velocity with the relative position vector is null, this component does not contribute to the relation between the velocities of two points of a body in a given reference frame. The result is quite useful for analyzing mechanisms, where a certain body often acts as a link element between another two. In order to analyze the mechanism, it is not always necessary or sometimes possible to fully determine its motion in the reference frame. In general, to get around the indetermination of the angular velocity vector, Eq. (10.7) is used together with Eq. (10.8).

Example 10.5   Consider the system shown in Fig. 10.5, consisting of a disk D, with radius r, that rotates around the vertical axis Z in relation to support A; a cursor C, which moves along the vertical guide G, fixed in A; and a bar B, with length l, whose ends are connected by ball and socket joints at C and point P, fixed on the edge of D, as shown. The axes $\{X, Y, Z\}$ and orthonormal basis $\mathbf{n}_1, \mathbf{n}_2, \mathbf{n}_3$ are fixed in A, and the unit vector $\mathbf{b}$ is parallel to the axis of bar B. The coordinate $z(t)$, $(l^2 - 4r^2)^{1/2} < z < l$, measures the cursor displacement along the guide, and the coordinate $\theta(t)$ measures the rotation of the disk around the vertical axis.

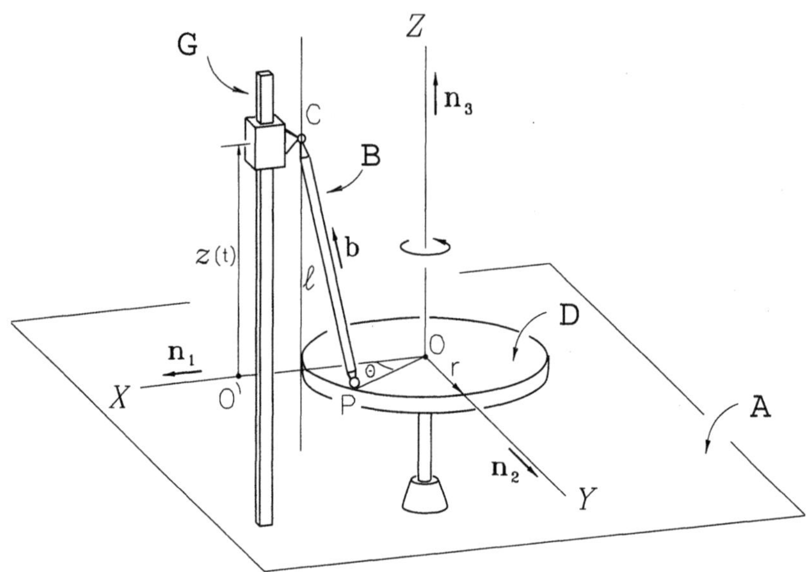

Figure 10.5

The velocities, in A, of points C and P can be expressed as

$$^A\mathbf{v}^C = \dot{z}\mathbf{n}_3,$$
$$^A\mathbf{v}^P = r\dot{\theta}(-\sin\theta\,\mathbf{n}_1 + \cos\theta\,\mathbf{n}_2).$$

The relative position vector between these points is

$$\mathbf{p}^{C/P} = \sum_{j=1}^{3} p_j\mathbf{n}_j,$$

with

$$p_1 = r(1 - \cos\theta), \quad p_2 = -r\sin\theta, \quad p_3 = z.$$

The angular velocity vector of the bar, expressed on the same basis, would
be

$$^A\boldsymbol{\omega}^B = \sum_{j=1}^{3} \omega_j\mathbf{n}_j.$$

The kinematic theorem of velocities for two points of a rigid body, Eq. (8.3)
is, in this case,

$$^A\mathbf{v}^C = {}^A\mathbf{v}^P + {}^A\boldsymbol{\omega}^B \times \mathbf{p}^{C/P}.$$

Then, substituting the components in the above equation, a system of algebraic equations for the scalar components of the angular velocity vector (check) is as follows

$$p_3\omega_2 - p_2\omega_3 = r\sin\theta\,\dot\theta, \tag{a}$$

$$p_1\omega_3 - p_3\omega_1 = -r\cos\theta\,\dot\theta, \tag{b}$$

$$p_2\omega_1 - p_1\omega_2 = \dot z. \tag{c}$$

It is easy to see that the determinant of the matrix of coefficients is null: therefore, the system is indeterminate. In fact, the existing relation between the motion of D and C does not depend on the bar rotation around its axis of symmetry. In other words, the $^A\omega^B$ component in the direction of the unit vector $\mathbf{b}$ is, in fact, indeterminate. Nevertheless, it is possible to compute the component of the angular velocity vector orthogonal to $\mathbf{b}$. Assuming then that

$$\Omega = \sum_{j=1}^{3} \Omega_j \mathbf{n}_j$$

is this component, satisfying, therefore, Eq. (10.7), that is,

$$\Omega \cdot \mathbf{b} = p_1\Omega_1 + p_2\Omega_2 + p_3\Omega_3 = 0, \tag{d}$$

then, according to Eq. (10.8), Ω can take the place of $^A\omega^B$ in the kinematic theorem of velocities and, therefore,

$$p_3\Omega_2 - p_2\Omega_3 = r\sin\theta\,\dot\theta, \tag{e}$$

$$p_1\Omega_3 - p_3\Omega_1 = -r\cos\theta\,\dot\theta, \tag{f}$$

$$p_2\Omega_1 - p_1\Omega_2 = \dot z. \tag{g}$$

There is no doubt that the system of Eqs. (e–g) is also indeterminate but, by including Eq. (d), there is a new system of Eqs. (d–f), the latter determined, whose solution is (check)

$$\Omega_1 = \left((1-\cos\theta)^2 \frac{r^2}{l^2} + \cos\theta\right)\frac{r}{z}\,\dot\theta,$$

$$\Omega_2 = \left(1-(1-\cos\theta)\frac{r^2}{l^2}\right)\frac{r}{z}\sin\theta\,\dot\theta,$$

$$\Omega_3 = (1-\cos\theta)\frac{r^2}{l^2}\,\dot\theta.$$

Specifically, when $\theta = \pi/2$:

$$\Omega_1 = \frac{r^3}{zl^2}\,\dot\theta; \quad \Omega_2 = \left(1 - \frac{r^2}{l^2}\right)\frac{r}{z}\,\dot\theta; \quad \Omega_3 = \frac{r^2}{l^2}\,\dot\theta,$$

and for $\theta = 0$:

$$\Omega_1 = \frac{r}{l}\,\dot\theta; \quad \Omega_2 = \Omega_3 = 0.$$

Last, from Eq. (g),

$$\dot z = -\frac{r^2}{z}\sin\theta\,\dot\theta.$$

In the position $\theta = \pi/2$,

$$\dot z = -\frac{r^2}{z}\,\dot\theta,$$

and for $\theta = 0$,

$$\dot z = 0.$$

The reader should note that the component of the angular velocity vector of the bar parallel to its axis cannot be determined from a kinematic viewpoint. The geometric relation always present between the variables $z(t)$ and $\theta(t)$, $z = [l^2 - 2r^2(1 - \cos\theta)]^{1/2}$, is independent of this component; in fact, when the velocity of end P of the bar is known, which only depends on $\theta(t)$, it is possible to determine the velocity of end C, a function of $z(t)$. See the corresponding animation.

Returning to the system of m rigid bodies, described above (see Fig. 10.4), the angular velocity of C_k in $\mathcal{R}$ can, according to Eq. (10.3), be interpreted as a composition of angular velocities between three reference frames: C_k, $\mathcal{R}$, and a third one, with angular velocity $\boldsymbol{\omega}_\perp$ with respect to $\mathcal{R}$, and in relation to which the angular velocity of C_k is $\boldsymbol{\omega}_{/\!/} = \omega_{/\!/}\mathbf{n}$ (see Fig. 10.6).

The angular acceleration of C_k in $\mathcal{R}$ can then be expressed, according to Eq. (4.6), as

$$^{\mathcal{R}}\boldsymbol{\alpha}^{C_k} = \boldsymbol{\alpha}_\perp + \dot\omega_{/\!/}\mathbf{n} + \boldsymbol{\omega}_\perp \times \boldsymbol{\omega}_{/\!/}, \tag{10.9}$$

where

$$\boldsymbol{\alpha}_\perp = \frac{^{\mathcal{R}}d}{dt}\boldsymbol{\omega}_\perp. \tag{10.10}$$

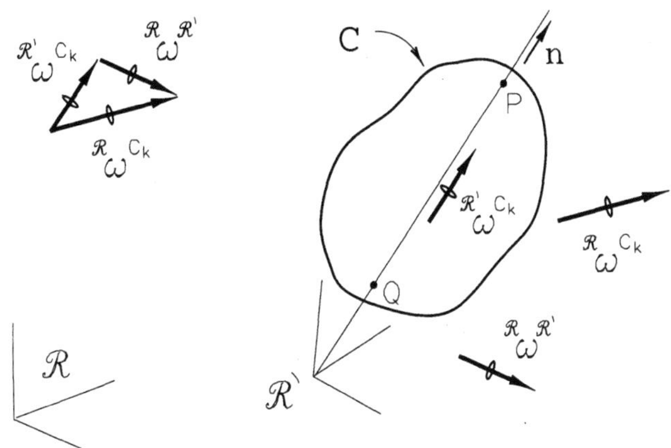

Figure 10.6

Now calculating,

$$^{\mathcal{R}}\boldsymbol{\omega}^{C_k} \times \left(^{\mathcal{R}}\boldsymbol{\omega}^{C_k} \times \mathbf{p}^{P/Q}\right) = \boldsymbol{\omega}_{/\!/} \times \left(\boldsymbol{\omega}_\perp \times \mathbf{p}^{P/Q}\right) + \boldsymbol{\omega}_\perp \times \left(\boldsymbol{\omega}_\perp \times \mathbf{p}^{P/Q}\right),$$

and

$$^{\mathcal{R}}\boldsymbol{\alpha}^{C_k} \times \mathbf{p}^{P/Q} = \boldsymbol{\alpha}_\perp \times \mathbf{p}^{P/Q} + \left(\boldsymbol{\omega}_\perp \times \boldsymbol{\omega}_{/\!/}\right) \times \mathbf{p}^{P/Q}, \tag{10.11}$$

and replacing it in Eq. (8.5), noting that, as $\boldsymbol{\omega}_{/\!/}$ is parallel to $\mathbf{p}^{P/Q}$, then

$$\boldsymbol{\omega}_{/\!/} \times \left(\boldsymbol{\omega}_\perp \times \mathbf{p}^{P/Q}\right) = -\left(\boldsymbol{\omega}_{/\!/} \times \boldsymbol{\omega}_\perp\right) \times \mathbf{p}^{P/Q}, \tag{10.12}$$

we arrive at the relation (check)

$$^{\mathcal{R}}\mathbf{a}^P = {}^{\mathcal{R}}\mathbf{a}^Q + \boldsymbol{\omega}_\perp \times \left(\boldsymbol{\omega}_\perp \times \mathbf{p}^{P/Q}\right) + \boldsymbol{\alpha}_\perp \times \mathbf{p}^{P/Q}. \tag{10.13}$$

Therefore, not only is the component of the angular velocity vector orthogonal to the relative position vector, $\boldsymbol{\omega}_\perp$, enough to determine the relative velocity between two points of a rigid body [see Eq. (10.8)], but its time derivative in the reference frame, $\boldsymbol{\alpha}_\perp$, is also sufficient (jointly, of course, with the vector $\boldsymbol{\omega}_\perp$ itself) to determine the relative acceleration between the points.

It is worth mentioning that vector $\boldsymbol{\alpha}_\perp$ is *not* the component of the angular acceleration of the body orthogonal to the relative position vector between the points; it is solely the angular acceleration in $\mathcal{R}$ of the aforementioned intermediary reference frame, that is, the time derivative in $\mathcal{R}$ of vector $\boldsymbol{\omega}_\perp$, as defined in Eq. (10.10).

Example 10.6   Returning to Example 10.5 (see Fig. 10.5), we now wish to determine the acceleration of the cursor C when it reaches its maximum height, that is, when $\theta = 0$. Then, by adopting the orthogonal component of the vector angular velocity and its time rate at A in the equation of the kinematic theorem for accelerations, as established in Eq. (10.13), we have

$$^A\mathbf{a}^C = {}^A\mathbf{a}^P + \boldsymbol{\Omega} \times (\boldsymbol{\Omega} \times \mathbf{p}^{C/P}) + \dot{\boldsymbol{\Omega}} \times \mathbf{p}^{C/P},$$

where, for $\theta = 0$:

$$^A\mathbf{a}^P = -r\dot{\theta}^2\mathbf{n}_1 + r\ddot{\theta}\mathbf{n}_2;$$

$$\boldsymbol{\Omega} \times (\boldsymbol{\Omega} \times \mathbf{p}^{C/P}) = \Omega_1\mathbf{n}_1 \times (\Omega_1\mathbf{n}_1 \times l\mathbf{n}_3) = -\frac{r^2}{l}\dot{\theta}^2\mathbf{n}_3;$$

$$\dot{\Omega}_1 = \frac{r}{l}\ddot{\theta}; \qquad \dot{\Omega}_2 = \frac{r}{l}\dot{\theta}^2;$$

$$\dot{\boldsymbol{\Omega}} \times \mathbf{p}^{C/P} = l(\dot{\Omega}_2\mathbf{n}_1 - \dot{\Omega}_1\mathbf{n}_2) = r(\dot{\theta}^2\mathbf{n}_1 - \ddot{\theta}\mathbf{n}_2).$$

Then, when $\theta = 0$, the cursor C will have the acceleration

$$^A\mathbf{a}^C = -\frac{r^2}{l}\dot{\theta}^2\mathbf{n}_3.$$

Mechanisms are present throughout the engineering world, from the delicate precision mechanism of a wristwatch to the robust and heavy operating system of a rolling mill. From a strictly kinematic point of view, however, the analytical method is always the same, and we must identify the holonomicity of the system, choose generalized coordinates, establish the pertinent kinematic constraints, determine the angular velocities and accelerations in terms of the selected coordinates, and use the kinematic theorems to determine velocities and accelerations of points. It is worth remembering again that an orthonormal basis should always be chosen to decompose the vectors, which, at least in principle, seems to be most convenient from a vectorial algebra point of view, independent of the reference frame in which it is fixed.

Example 10.7   A mechanism consists of three gears (see Fig. 10.7): A is fixed; B has an axis of symmetry coinciding with that of A and rotates, in relation to the latter, with simple angular velocity of constant module ω_0, in the indicated direction; C engages with A and B, with its axis of symmetry intercepting that of the two others at point O, as illustrated.

The motion of gear C is therefore determined by the motion of gear B in relation to the fixed gear A. The orthonormal basis $\mathbf{n}_1, \mathbf{n}_2, \mathbf{n}_3$, with the directions shown in the figure, is fixed in a reference frame N, coinciding with the plane containing the axes of symmetry of the gears. We now want to determine the angular velocity and acceleration of C in relation to the fixed gear A.

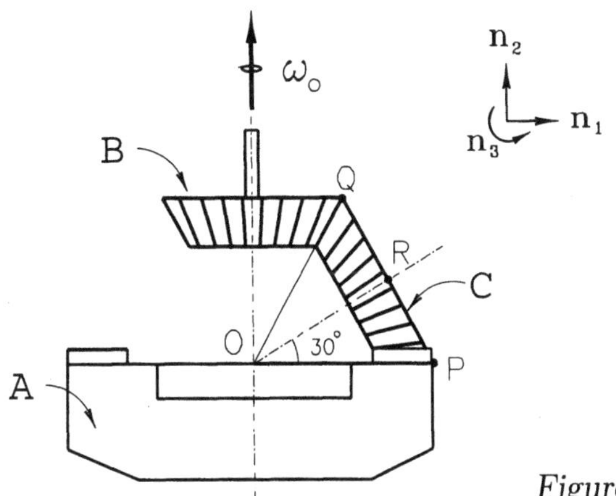

Figure 10.7

Points O and P are fixed in C, O is also fixed in A, and, from the condition of engagement (a rolling condition), $^A\mathbf{v}^P = 0$. Then, from the kinematic theorem,

$$^A\mathbf{v}^P = {}^A\mathbf{v}^O + {}^A\boldsymbol{\omega}^C \times \mathbf{p}^{P/O},$$

the result is that $^A\boldsymbol{\omega}^C$ is parallel to $\mathbf{p}^{P/O}$, that is,

$$^A\boldsymbol{\omega}^C = \omega_1\mathbf{n}_1.$$

Assuming r to be the distance from point Q to the vertical axis (note that, as the geometry is known, r only assumes a scale that will automatically define the other dimensions involved), the velocity in A of the point can also be obtained from the kinematic theorem for velocities

$$^A\mathbf{v}^Q = {}^A\mathbf{v}^O + {}^A\boldsymbol{\omega}^B \times \mathbf{p}^{Q/O}$$

$$= -r\omega_0\mathbf{n}_3.$$

But points O and Q are also fixed in C (Q is one of the points of contact between C and B); therefore,

$$^A\mathbf{v}^Q = {}^A\mathbf{v}^O + {}^A\boldsymbol{\omega}^C \times \mathbf{p}^{Q/O},$$

that is,

$$-r\omega_0\mathbf{n}_3 = \omega_1\mathbf{n}_1 \times r(\mathbf{n}_1 + \sqrt{3}\mathbf{n}_2);$$

therefore,

$$\omega_1 = -\frac{1}{\sqrt{3}}\omega_0.$$

The angular velocity of gear C in relation to A is, therefore,

$$^A\boldsymbol{\omega}^C = -\frac{1}{\sqrt{3}}\omega_0\mathbf{n}_1.$$

Since ω_0 is constant, the angular acceleration of C will depend on the time rate of change of the unit vector $\mathbf{n}_1$ and, to obtain it, we first need to determine the motion in A of reference frame N. Point R, on the axis of symmetry of C, is, of course, fixed in N and C, and its velocity can be computed as

$$^A\mathbf{v}^R = {}^A\mathbf{v}^O + {}^A\boldsymbol{\omega}^C \times \mathbf{p}^{R/O}$$

$$= 0 - \frac{1}{\sqrt{3}}\omega_0\mathbf{n}_1 \times \sqrt{3}r\left(\frac{\sqrt{3}}{2}\mathbf{n}_1 + \frac{1}{2}\mathbf{n}_2\right)$$

$$= -\frac{1}{2}r\omega_0\mathbf{n}_3.$$

Now using the same kinematic theorem for velocities but with O and R fixed in N,

$$^A\mathbf{v}^R = {}^A\mathbf{v}^O + {}^A\boldsymbol{\omega}^N \times \mathbf{p}^{R/O},$$

that is,

$$-\frac{1}{2}r\omega_0\mathbf{n}_3 = \omega_2\mathbf{n}_2 \times \sqrt{3}r\left(\frac{\sqrt{3}}{2}\mathbf{n}_1 + \frac{1}{2}\mathbf{n}_2\right);$$

therefore,

$$\omega_2 = \frac{1}{3}\omega_0.$$

The angular acceleration of C in A can then be established as

$$^A\boldsymbol{\alpha}^C = \frac{^Ad}{dt}{}^A\boldsymbol{\omega}^C$$

$$= -\frac{1}{\sqrt{3}}\omega_0\dot{\mathbf{n}}_1$$

$$= -\frac{1}{\sqrt{3}}\omega_0\,\omega_2\mathbf{n}_2 \times \mathbf{n}_1$$

$$= \frac{1}{3\sqrt{3}}\omega_0^2\,\mathbf{n}_3.$$

See the corresponding animation.

Exercise Series #3   (Sections 3.1 to 3.6)

P3.1 Consider the cube studied in Example 2.1, where the orthonormal basis $\mathbf{c}_1, \mathbf{c}_2, \mathbf{c}_3$ is fixed. Determine the angular velocity vector of the cube in $\mathcal{R}$, at the instant when $\dot{\mathbf{c}}_1 = \mathbf{c}_2 - \mathbf{c}_3$ and $\dot{\mathbf{c}}_2 = -\mathbf{c}_1 + \mathbf{c}_3$. Calculate the time derivative of the vector $\mathbf{c}_3$ at this instant. Ascertain whether it is possible for the cube to move so that, at a given instant, $\dot{\mathbf{c}}_1 = \mathbf{c}_2 + \mathbf{c}_3$ and $\dot{\mathbf{c}}_2 = \mathbf{c}_1 + \mathbf{c}_3$.

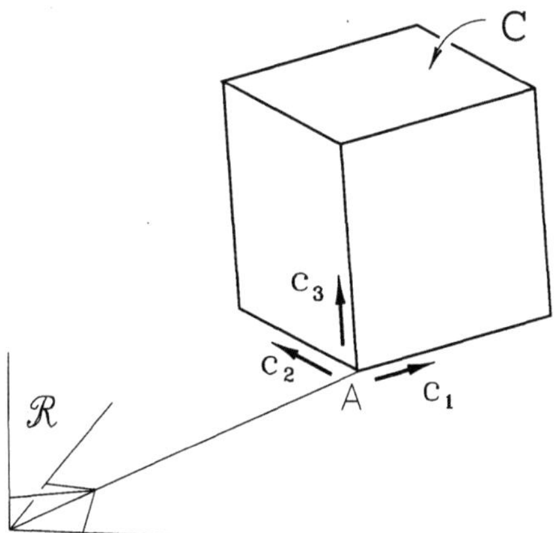

Figure P 3.1

P3.2 The cone C moves in space with its vertex A pinned to fork G, which rotates in $\mathcal{R}$ at a constant module angular velocity Ω. The pinned joint in A allows two movements. One is measured by $\theta(t)$, the angle between axis E and the vertical. The other is measured by $\phi(t)$, the angle between a fixed line at the base of C and the plane containing axes E and x_2. Determine a general relationship for the angular velocity of C in $\mathcal{R}$.

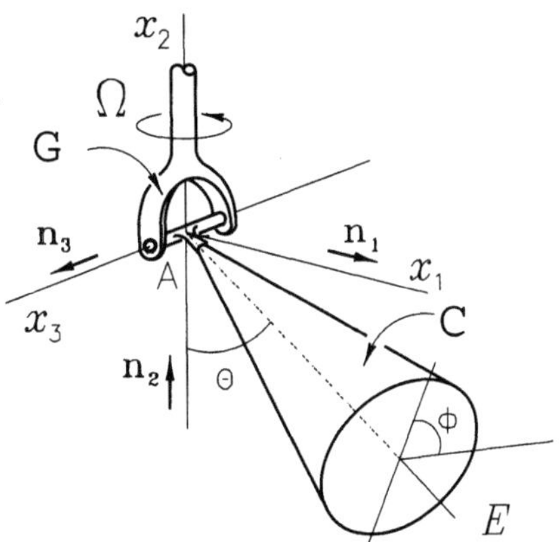

Figure P 3.2

P3.3 For the cone in the previous problem, determine $\dot{\phi}$ as a function of the angle θ, so that the angular acceleration of the cone in $\mathcal{R}$ stays horizontal.

P3.4 Show that if $\mathbf{b}_1, \mathbf{b}_2, \mathbf{b}_3$ is a fixed orthonormal basis on a rigid body B, which moves arbitrarily in a reference frame A, and if $\dot{\mathbf{b}}_j$, $j = 1, 2, 3$, are the time derivatives in A of the basis vectors, the angular velocity vector of B at A may be expressed by

$$^A\boldsymbol{\omega}^B = \dot{\mathbf{b}}_2 \cdot \mathbf{b}_3\, \mathbf{b}_1 + \dot{\mathbf{b}}_3 \cdot \mathbf{b}_1\, \mathbf{b}_2 + \dot{\mathbf{b}}_1 \cdot \mathbf{b}_2\, \mathbf{b}_3.$$

P3.5 The bodies A, B, and C, joined at points P and Q, comprise a mechanism. Given the angular velocities

$$^A\boldsymbol{\omega}^C = \omega_1 \mathbf{a}_1 + \omega_2 \mathbf{a}_2,$$
$$^A\boldsymbol{\omega}^B = \omega_2 \mathbf{a}_2 - \omega_3 \mathbf{a}_3,$$
$$^\mathcal{R}\boldsymbol{\omega}^A = \omega_1 \mathbf{a}_1 + \omega_2 \mathbf{a}_2,$$

where $\mathcal{R}$ is a given reference frame, $\mathbf{a}_1, \mathbf{a}_2, \mathbf{a}_3$ is an orthonormal basis fixed in A, and, $\omega_1, \omega_2, \omega_3$ are constant, find the angular accelerations $^\mathcal{R}\boldsymbol{\alpha}^C$ and $^B\boldsymbol{\alpha}^C$.

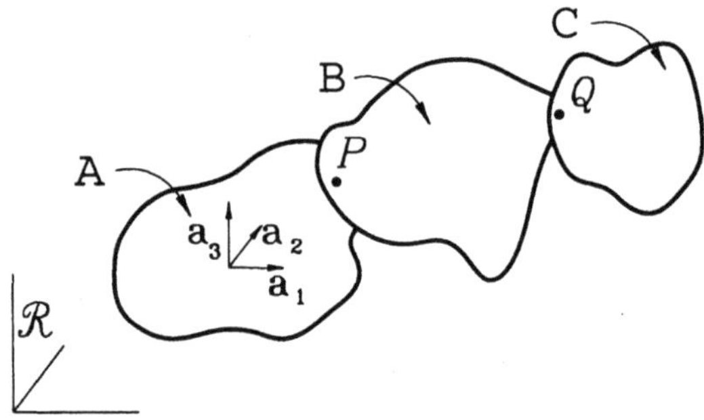

Figure P 3.5

P3.6 Prove that if A, B, C, and D are four bodies moving independently in space, the angular acceleration of D in A can be written as

$$^A\boldsymbol{\alpha}^D = {}^A\boldsymbol{\alpha}^B + {}^B\boldsymbol{\alpha}^C + {}^C\boldsymbol{\alpha}^D + {}^A\boldsymbol{\omega}^B \times {}^B\boldsymbol{\omega}^C + {}^A\boldsymbol{\omega}^B \times {}^C\boldsymbol{\omega}^D + {}^B\boldsymbol{\omega}^C \times {}^C\boldsymbol{\omega}^D.$$

P3.7   Considering the coin M rolling on a horizontal plane $\mathcal{R}$, analyzed in Example 3.4, determine the general relationship for the angular acceleration of M in $\mathcal{R}$, in terms of the functions $\phi(t)$, $\theta(t)$, and $\psi(t)$.

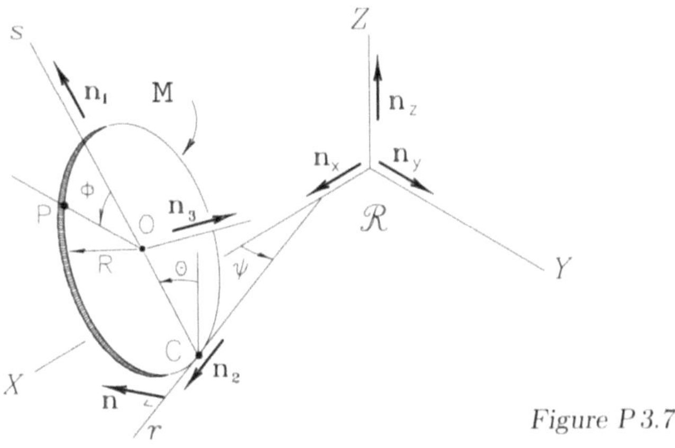

Figure P3.7

P3.8   Show that if a rigid body C moves in relation to a reference frame $\mathcal{R}$, then there exists a linear transformation ${}^{\mathcal{R}}\mathbf{T}^{C}$ that maps a vector $\mathbf{v}$, fixed in the body, into its time derivative in $\mathcal{R}$. That is, for every vector $\mathbf{v}$ fixed in C, its time rate in $\mathcal{R}$, $\dot{\mathbf{v}}$, can be written as

$$\dot{\mathbf{v}} = {}^{\mathcal{R}}\mathbf{T}^{C}\mathbf{v}.$$

Hint: Adopt an orthonormal basis $\mathbf{c}_1, \mathbf{c}_2, \mathbf{c}_3$, fixed in C, in order to decompose the angular velocity vector of the body as

$$ {}^{\mathcal{R}}\boldsymbol{\omega}^{C} = \omega_1 \mathbf{c}_1 + \omega_2 \mathbf{c}_2 + \omega_3 \mathbf{c}_3, $$

and define a tensor $\mathbf{T}$ that, on the same basis, is expressed by the antisymmetric matrix

$$ \mathbf{T} = \begin{pmatrix} 0 & -\omega_3 & \omega_2 \\ \omega_3 & 0 & -\omega_1 \\ -\omega_2 & \omega_1 & 0 \end{pmatrix}. $$

P3.9   Referring to the above exercise, prove that if $\dot{\mathbf{p}}$ and $\dot{\mathbf{q}}$ are the time derivatives in reference frame $\mathcal{R}$ of two linearly independent vectors, fixed in body C, then the tensor ${}^{\mathcal{R}}\mathbf{T}^{C}$ can be written as

$$ {}^{\mathcal{R}}\mathbf{T}^{C} = \frac{1}{\dot{\mathbf{p}} \cdot \mathbf{q}} \left(\dot{\mathbf{q}} \otimes \dot{\mathbf{p}} - \dot{\mathbf{p}} \otimes \dot{\mathbf{q}} \right). $$

Also check that the angular velocity vector of C in $\mathcal{R}$ may also be written as

$$ {}^{\mathcal{R}}\boldsymbol{\omega}^{C} = \frac{1}{2} \left(\frac{\dot{\mathbf{p}} \times \dot{\mathbf{q}}}{\dot{\mathbf{p}} \cdot \mathbf{q}} + \frac{\dot{\mathbf{q}} \times \dot{\mathbf{p}}}{\dot{\mathbf{q}} \cdot \mathbf{p}} \right). $$

P3.10 The support A rotates in relation to the pedestal L at a constant angular velocity of module $\Omega = 4$ rad/s, as illustrated. The engine casing B pivots on the support A and rotates around the pin, coincident with axis x_1 at a constant angular velocity of module $\dot{\theta} = 2$ rad/s. Axes $\{x_1, x_2, x_3\}$ are fixed in B. Disk D rotates in relation to the engine casing B at a constant angular velocity of module $p = 300$ rad/s. Determine the module of the angular velocity $^L\boldsymbol{\omega}^D$ and of the angular acceleration $^L\boldsymbol{\alpha}^D$, at the moment when $\theta = 0$.

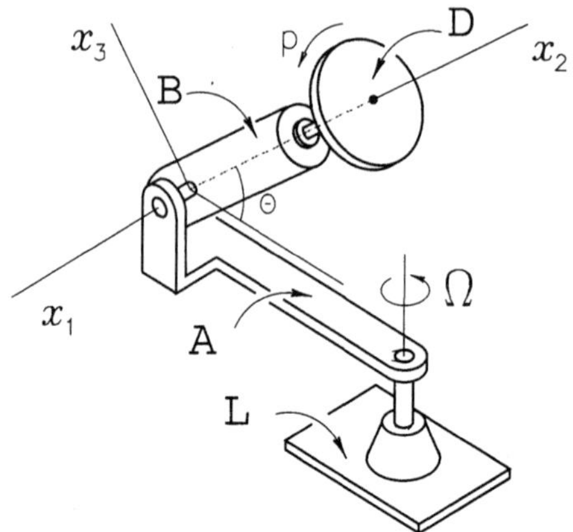

Figure P 3.10

P3.11 Disk D, with radius r, rotates freely around arm B that, in turn, rotates around the fixed vertical axis, at a constant angular velocity of module Ω. The disk rolls on the conical surface, as illustrated. Calculate the modules of the angular velocity and acceleration of the disk in relation to the surface.

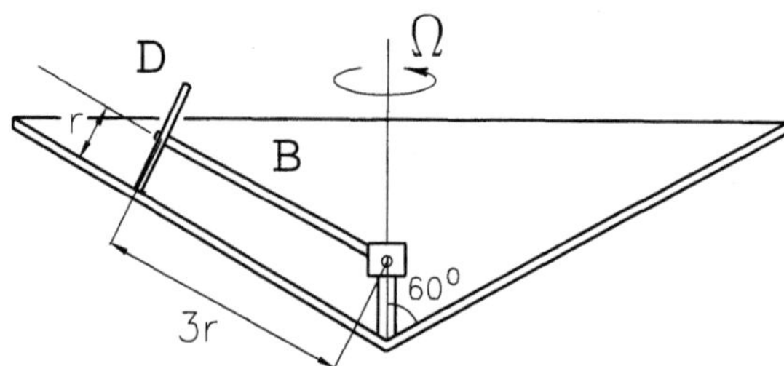

Figure P 3.11

P3.12 Fork G is welded to the bar that can slide lengthwise in relation to the bearing M and can also rotate in relation to the latter around axis x_1. At the moment shown, the velocity of the bar on the bearing is $v = 2$ m/s and the acceleration is $a = 1$ m/s^2, both in the positive direction of the axis. The angular velocity of the bar in relation to the bearing is $\omega = 3$ rad/s, in the direction indicated, and the angular acceleration is null. Bar B, pivoting on the fork, moves in relation to the fork according to the function $\theta(t) = (\pi/2)\cos(\omega t)$. At the moment illustrated, $\theta = \pi/6$ and $\phi = \pi/2$. Calculate the angular velocity and acceleration of B in M at that same instant.

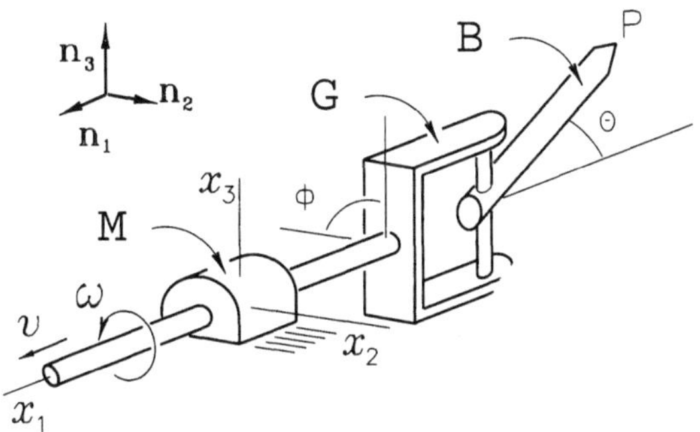

Figure P 3.12

P3.13 A flexible steel tape with length c is being rolled around a fixed drum T, with its end B fixed in T and end A moving to keep the free part of the tape straight. The motion is prescribed by the function $\phi(t)$, which measures the angle between the radii of ends B and C, the last point of contact of the tape with the drum. Find the velocity and acceleration of end A in relation to the drum, for a general position.

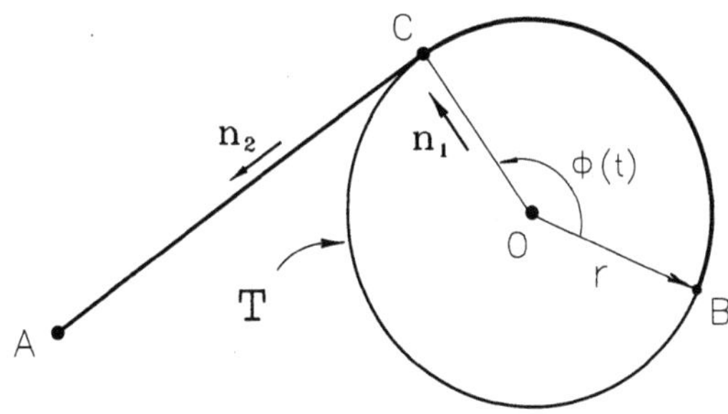

Figure P 3.13

P3.14 Determine the general relationship for the acceleration of cursor P on laboratory L, for the system analyzed in Example 5.1.

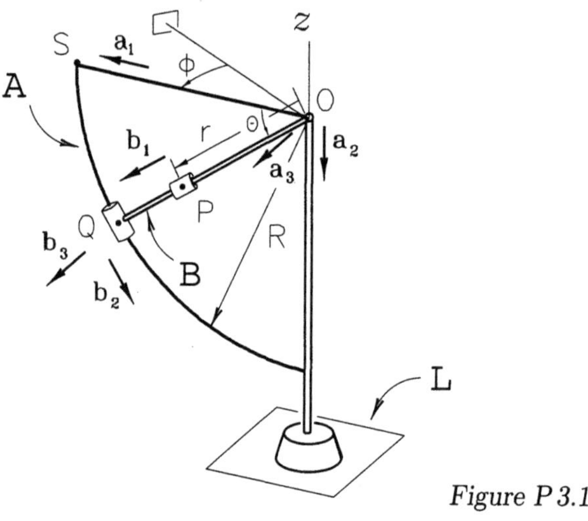

Figure P 3.14

P3.15 Consider the system analyzed in Example 5.4. Suppose that cursor C moves harmonically in the guide according to $x(t) = l\sin(\omega_0 t)$ and that disk D rotates in relation to fork A at the constant rate $\dot{\phi} = 2\omega_0$. Find the acceleration of cursor C in laboratory L, when $x = 0$ ($\dot{x} > 0$), $\phi = \pi$, and when $x = l$, $\phi = \pi/2$.

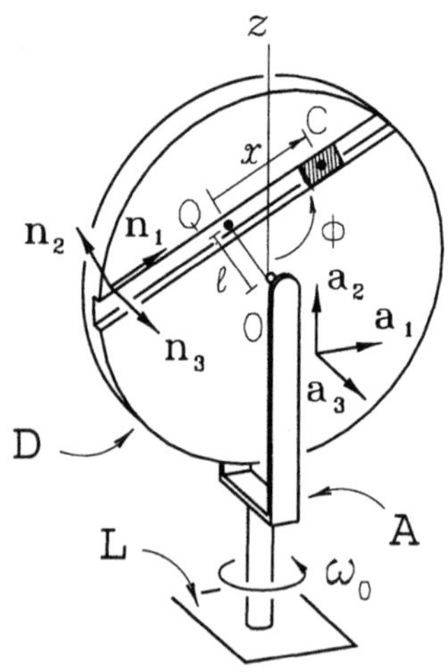

Figure P 3.15

P3.16 The mechanism consists of a cursor C, sliding along bar B, joined by a pivot at point P with arm A that, in its turn, rotates around the vertical axis passing through the point O, in relation to the reference frame $\mathcal{R}$. The system configuration can be described by the functions $\phi(t)$, $\theta(t)$, and $r(t)$, as shown. If $\dot{\phi}(t) = \Omega$, $\theta(t) = \theta_0 \cos(\omega_0 t)$, and $\dot{r}(t) = v$, where Ω, θ_0, ω_0, and v are constants, find the velocity of C in $\mathcal{R}$ and the acceleration of C in A when the bar passes through the vertical position, with positive $\dot{\theta}$.

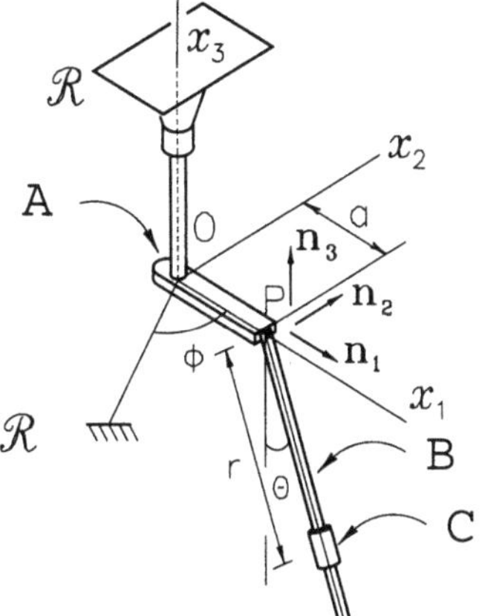

Figure P 3.16

P3.17 The crankshaft A rotates at a constant angular velocity $\dot{\theta} = 12$ rad/s in relation to the support S. By means of a pin at hinge P, bar B is driven to slide through the bushing C, which is pivoted in point Q, as illustrated. Determine the module and direction of the angular velocity and angular acceleration of the bar in relation to the support for positions $\theta = \pi/2$ and $\theta = 0$.

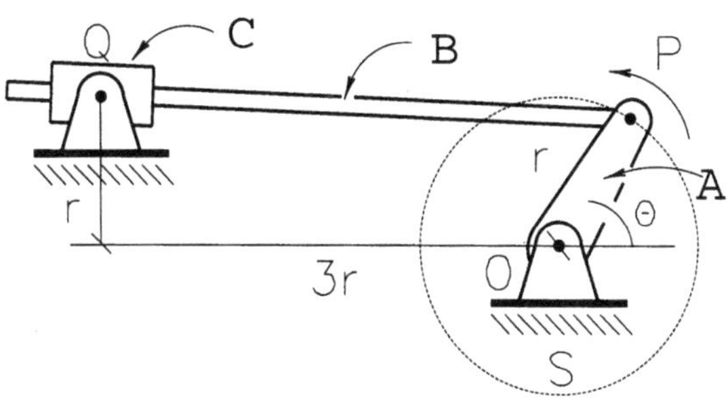

Figure P 3.17

P3.18   The figure illustrates a mechanism consisting of two horizontal guides fixed to the support S, on which cursors A and C slide, with velocities of module u and v, respectively, as illustrated. A telescopic bar joins the cursors. Cursor C is connected by means of a ball and socket. A fork B at the other end of the bar is joined to the last element of the bar, which is connected to cursor A by a pin fixed in A, with an axis orthogonal to x_1. The orthonormal basis $\mathbf{n}_1, \mathbf{n}_2, \mathbf{n}_3$ is fixed in S, while the basis $\mathbf{b}_1, \mathbf{b}_2, \mathbf{b}_3$ is fixed in B, with $\mathbf{b}_1$ in the direction of the pin and $\mathbf{b}_2$ in the direction of the telescopic bar axis. Determine an expression for the angular velocity of the fork on the support, as a function of the velocity of each cursor, at an instant when the position of the center of both cursors, along the respective guides, is equal to z. *Hint:* Write the angular velocity of the fork as a composition of its angular velocity in relation to the cursor A with its angular velocity in relation to S and use the kinematic theorem for velocities, relating the velocities of the center of cursor C in B and in S.

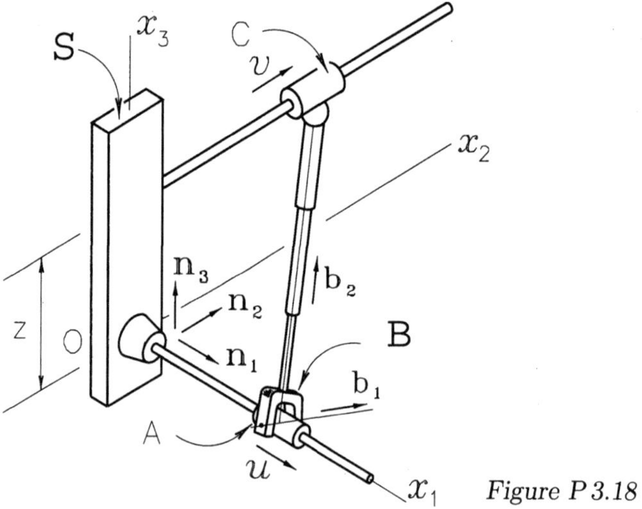

Figure P 3.18

P3.19   Returning to the previous exercise, and for the instant under consideration, find the time rate of change in S of the unit vector $\mathbf{b}_1$. Next, calculate the angular acceleration of the fork in relation to the support, knowing that at that instant $\dot{u} = \dot{v} = a$.

P3.20 The drive mechanism of a windshield wiper consists of three articulated bars. Bar B_1 has its motion prescribed by the function $\theta(t) = \theta_0 - A\cos(2\pi f_0 t)$, where $\theta_0 = 5\pi/12$ rad, $A = \pi/4$ rad, and $f_0 = 1$ Hz. The blade and horizontal bar B_3 are rigidly connected at point Q. Find the velocity of the blade end P at the highest position ($\theta = \pi/2$) and the acceleration, of the same point, at its lowest position ($\theta = \pi/6$).

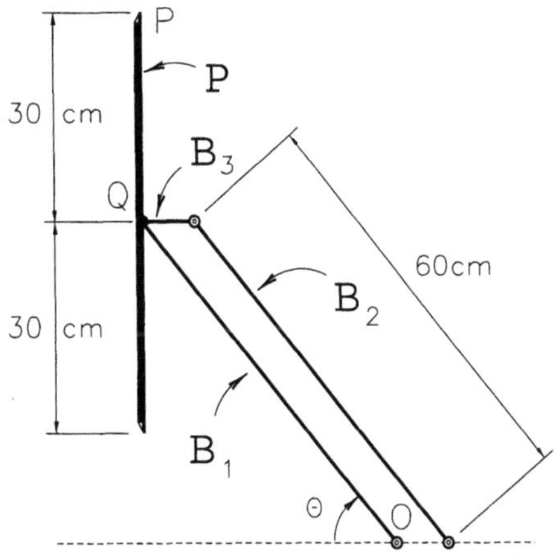

Figure P 3.20

P3.21 Bar B moves inside the cylindrical concavity, staying in the plane of the figure. Its end P slides on the surface, at constant speed v, as shown, while another point B always touches the edge C. Find the angular velocity and acceleration of the bar and the velocity and acceleration of the point of the bar in contact with point C, as a function of the coordinate θ.

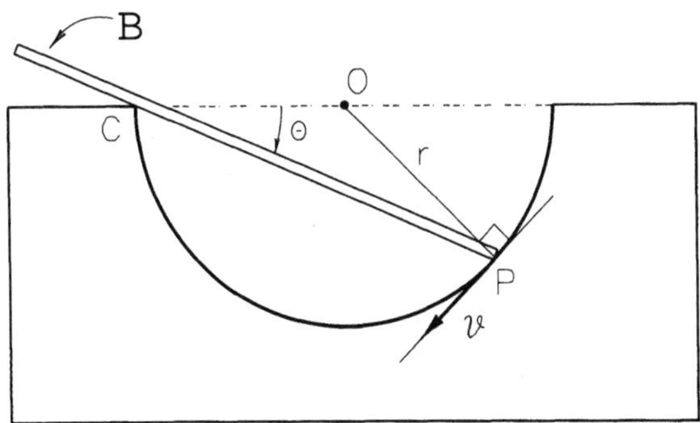

Figure P 3.21

P3.22 The rectangular plate B is hinged at the edge of the fixed plate A, along axis x_2, while the triangular plate C is hinged at the edge of plate B, along axis x_1. The system configuration is described by the angles $\theta(t) = \frac{\pi}{2}\sin(\omega t)$ and $\phi(t) = 2\omega t$. Find the angular acceleration of C in A when $t = 0$ and the velocity in A of vertex P when $t = \pi/2\omega$.

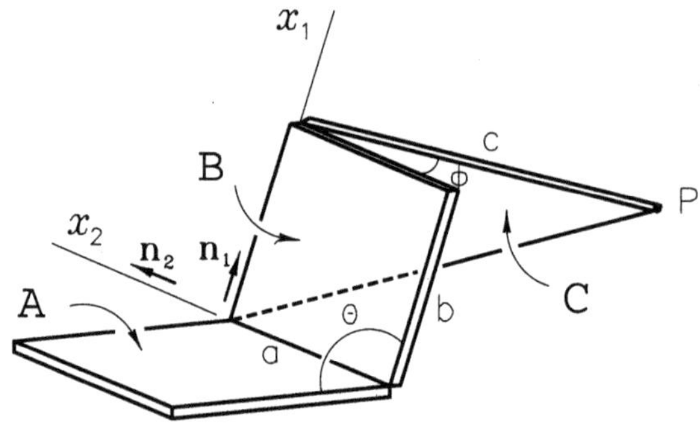

Figure P 3.22

P3.23 The board is hanging by two ropes and swings so that, at a given instant, it is horizontal, with vertex A moving at a velocity $\mathbf{v}^A = v\mathbf{n}_2$, vertex B at a velocity $\mathbf{v}^B = v(-3\mathbf{n}_1 + \mathbf{n}_2)$, and the vertical component of the velocity of vertex C being v, where the basis $\mathbf{n}_1, \mathbf{n}_2, \mathbf{n}_3$, fixed in the board, has the indicated directions and v is constant. Find the angular velocity and acceleration of the board at that instant.

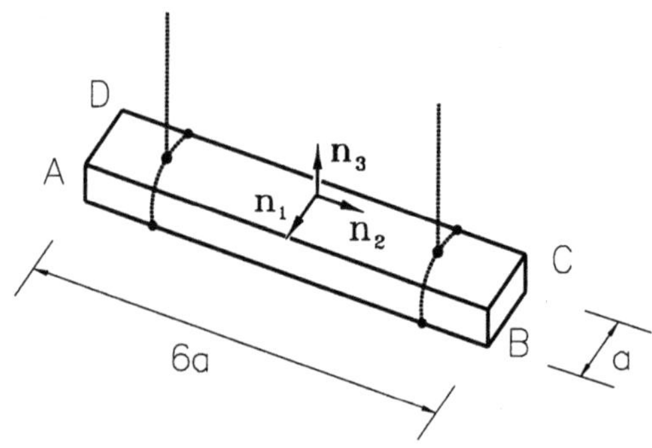

Figure P 3.23

P3.24 This is a sketch of a fan. The rotor-propeller set rotates around the axis of symmetry, in relation to the engine casing C, which, in turn, can rotate in relation to the support S. If the propeller H is rotating at an angular velocity with constant module ω rad/s, in the casing C, in the indicated direction, and this is moving in relation to the support S according to $\theta(t) = (\pi/2)\left(1 - cos(\omega t/10)\right)$ rad, find the acceleration module, in the referential S, of the end P of the propeller, when $\theta = (\pi/2)$ and $\dot{\theta} > 0$, knowing that, at that instant, points P, Q, and G are on the same vertical plane.

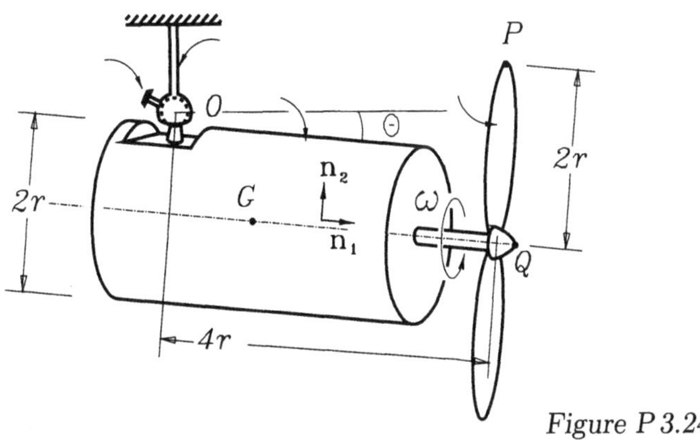

Figure $P\,3.24$

P3.25 Let A and B be two reference frames moving independently and let O be a fixed point in A, Q a fixed point in B, and P a third point moving in relation to both reference frames. Prove that if, at a certain instant, P and Q coincide, the acceleration of P with respect to Q, in reference frame A, may, at that same instant, be expressed by

$$^A\mathbf{a}^{P/Q} = {}^B\mathbf{a}^P + 2\,{}^A\boldsymbol{\omega}^B \times {}^B\mathbf{v}^P.$$

Exercise Series #4   (Section 3.7)

P4.1   The telescopic antenna has its rod A pivoting at point O around a horizontal pin attached to the base B that, in turn, rotates at a simple angular velocity of constant module ω in relation to the reference frame $\mathcal{R}$, as illustrated. The antenna axis slants in relation to the vertical according to the function $\theta(t)$ and the end P moves away from the pivot O according to $r(t)$. The $\mathbf{a}_1, \mathbf{a}_2, \mathbf{a}_3$ basis is fixed in A and the basis $\mathbf{b}_1, \mathbf{b}_2, \mathbf{b}_3$ is fixed in B, as illustrated. Find the velocity of P in $\mathcal{R}$ as a function of coordinates r and θ. Then calculate the tangential acceleration module of P in $\mathcal{R}$ when $r = l$, $\dot{r} = l\omega$, $\ddot{r} = 0$, and, also, $\theta = 30^\circ$, $\dot{\theta} = 0$, $\ddot{\theta} = \omega^2$.

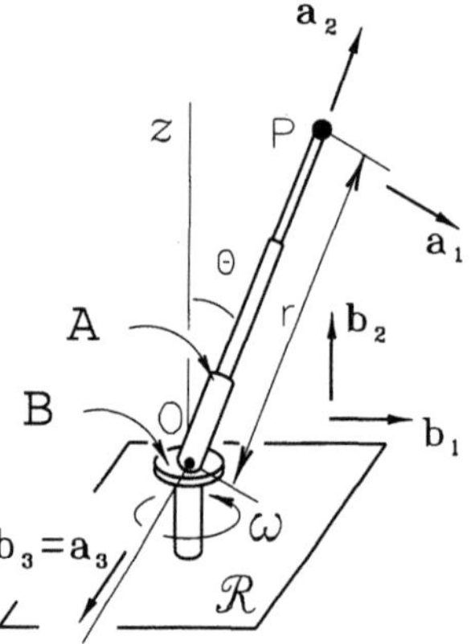

Figure P 4.1

P4.2   The trolley travels along the hyperbolic guide at a constant velocity of module $v = 4$ m/s. Determine its acceleration when it passes by vertex P.

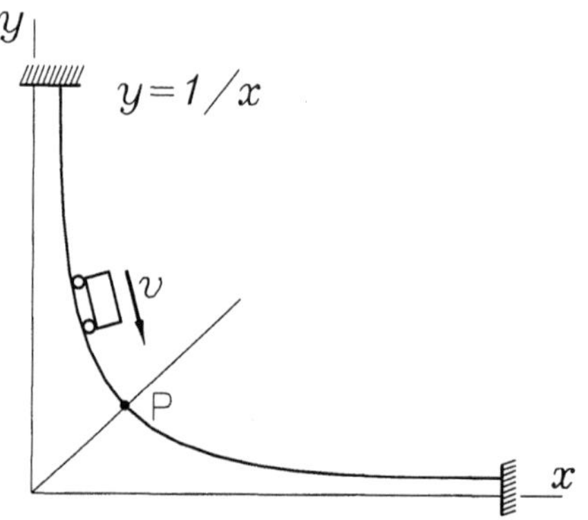

Figure P 4.2

P4.3 A particle moves on the surface of a cylinder defined by the coordinates r, θ, and z, according to $r = a$, $\theta = 2\pi t$, and $z = b\sin(2\pi t)$. Calculate the maximum and minimum values of the radius of curvature of the trajectory.

P4.4 The motion of a particle is described by the cylindrical coordinates $r = a$, $\theta = 2\pi t$, and $z = a\sin^2(2\pi t)$. Determine the modules of the velocity and acceleration of the particle at any instant.

P4.5 The figure shows the system studied in Example 5.4. Now suppose that $x(t) = l\cos(\omega_0 t)$ and $\dot{\phi} = \omega_0$, constant. Calculate the tangential acceleration of the cursor in A at the instant $t = 0$.

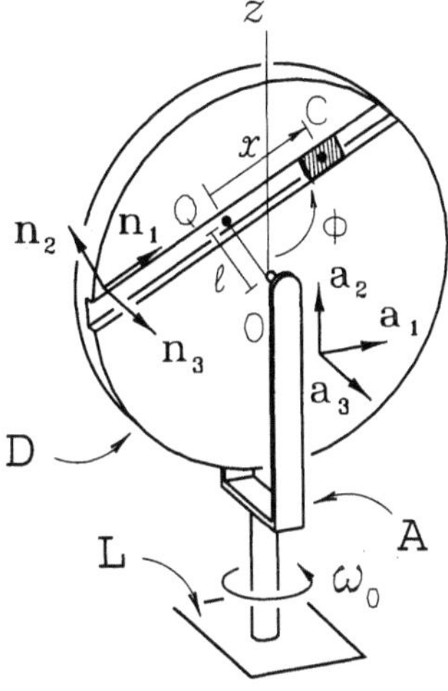

Figure P 4.5

P4.6 With reference to the previous problem, suppose that, at the same instant, $\phi = \pi/2$. Now calculate the tangential acceleration of the cursor in L.

P4.7 A particle performs a periodical motion as a result of the composition of two harmonic motions of the same frequency but different amplitudes and phases, in the directions of two orthogonal axes x and y: $x = A\sin(\omega t + \phi)$; $y = B\sin(\omega t + \psi)$. Determine their trajectory.

P4.8 Find the initial radius of curvature of the trajectory of the plane motion of a particle described by the time functions: $x_1 = 2t$; $x_2 = t^2$, where x_1 and x_2 are the coordinates of the particle in the plane, in meters, and parameter t is the time, in seconds.

P4.9   A disk with radius r rolls on the horizontal surface, with its center moving at constant velocity. A point P, on the edge of the disk, therefore describes a cycloid, as illustrated. Calculate the radius of curvature in any position ϕ.

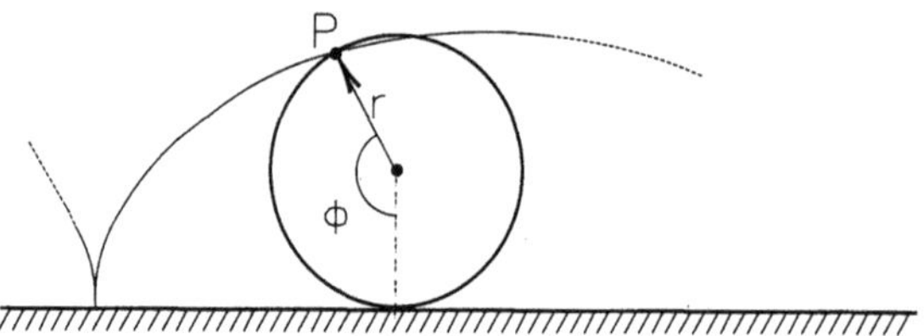

Figure P 4.9

P4.10   With reference to the previous problem, calculate the module and direction of the acceleration of point P, if the disk is 2 m in diameter and makes a complete turn around its own axis of symmetry every $\pi/10$ s.

P4.11   The motion of a particle on the plane is described by the coordinates $x = at$, $y = bt - \frac{1}{2}gt^2$, where x and y are Cartesian coordinates on the plane, and t is the time. Determine the tangential acceleration module of the particle at an instant when its velocity has module v.

P4.12   With reference to the previous exercise, calculate the centripetal acceleration module of the particle, at the same instant.

Exercise Series #5   (Sections 3.8 to 3.10)

P5.1 The stepladder E is initially at rest, supported on points A and B, on the horizontal plane xy, and on points C and D, on the vertical plane yz, as illustrated. At a certain instant, the stepladder starts to move as follows: A remains fixed; B slides on the horizontal plane, in the indicated direction; and C slides on the vertical plane. With the initial velocity of B, v, known, now determine, for that same instant, the module of the stepladder's angular velocity and the velocity of point D.

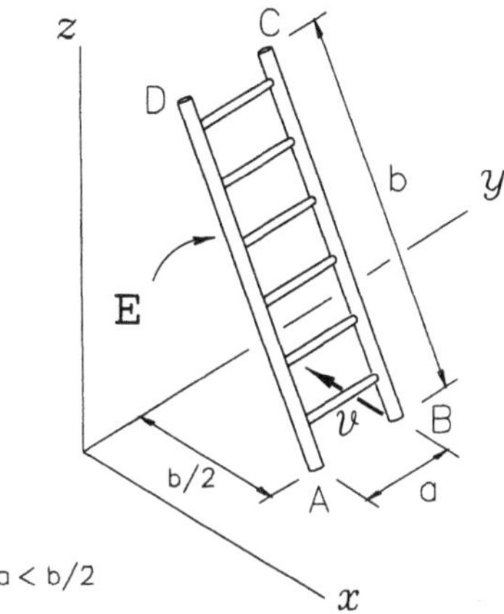

Figure P 5.1

P5.2 With reference to the previous exercise, now assume that the stepladder continues to slide, with point A fixed in position, point B describing a horizontal circle with center A, at a constant module velocity v, and point C always on the vertical plane. Find the velocity of C when it is at a maximum distance from the horizontal plane.

P5.3   Bar B is connected by a pivot, with axis x_1, to disk D, which rotates at a angular velocity of constant module $\Omega = 2$ rad/s, around the vertical axis fixed in A, as illustrated. Determine the acceleration module, in relation to A, of the bar center G, at an instant when $\theta = \pi/3$ rad, $\dot\theta = 4$ rad/s, and $\ddot\theta = 1$ rad/s^2.

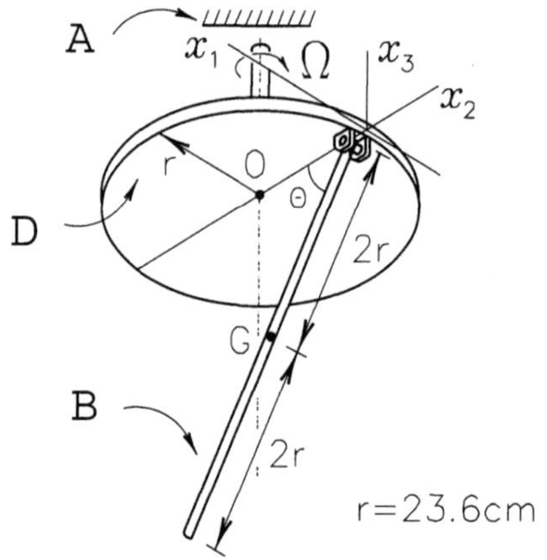

Figure P 5.3

P5.4   The spinning top C moves with its tip in a small depression in bar B (point Q). Bar B rotates at a simple angular velocity of a constant module Ω, in relation to the support A, as illustrated. The motion of the top in B has constant precession and spin with null nutation. Write the general expressions for the angular velocity vector of C in A and for the angular acceleration vector of C in A. Also calculate the velocity of point P, the center of the top, in A, at an instant when $\psi = \pi$, with $\dot\psi > 0$. The basis $\mathbf{b}_1, \mathbf{b}_2, \mathbf{b}_3$ is fixed in B.

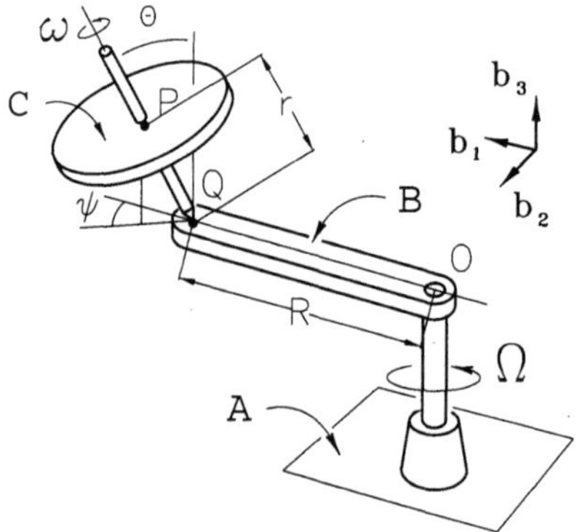

Figure P 5.4

P5.5 Bar B pivots on the fork G and can rotate freely in relation to it around axis x_1. The system of Cartesian axes $\{x_1, x_2, x_3\}$ is fixed in G. The fork rotates in relation to reference frame $\mathcal{R}$ at constant angular velocity Ω, as illustrated. The angle θ, between bar B and vertical axis x_3, is given by $\theta(t) = -\frac{\pi}{2}\cos(\Omega t)$. Determine the relations for the angular velocity and angular acceleration of bar B, in the reference frame $\mathcal{R}$.

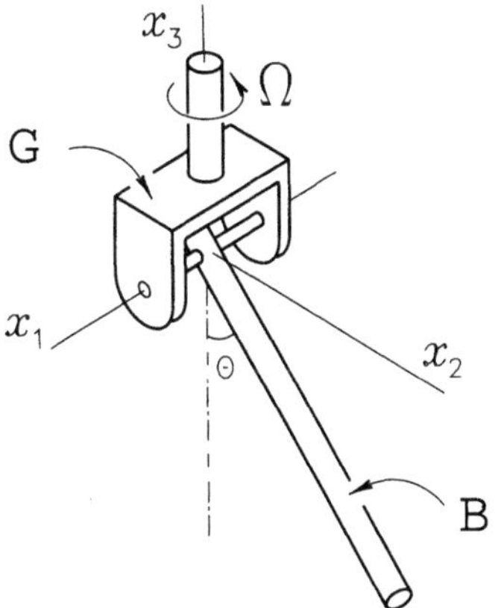

Figure P 5.5

P5.6 The fork G is welded to the bar that can slide lengthwise in relation to the bearing M and can also rotate in relation to the latter in the indicated direction. At the moment shown, the velocity of the bar in the bearing is $v = 2$ m/s, in the positive direction of the x_1-axis. The angular velocity of the bar in relation to the bearing is $\omega = 3$ rad/s, in the indicated direction, when $\phi = \pi/2$. The 20-cm-long bar B pivots on the fork, moving in relation to it according to the function $\theta(t) = (\pi/2)\cos(\omega t)$. At the given moment, $\theta = \pi/6$. Calculate the velocity of end P of the bar, in reference frame M, at this instant.

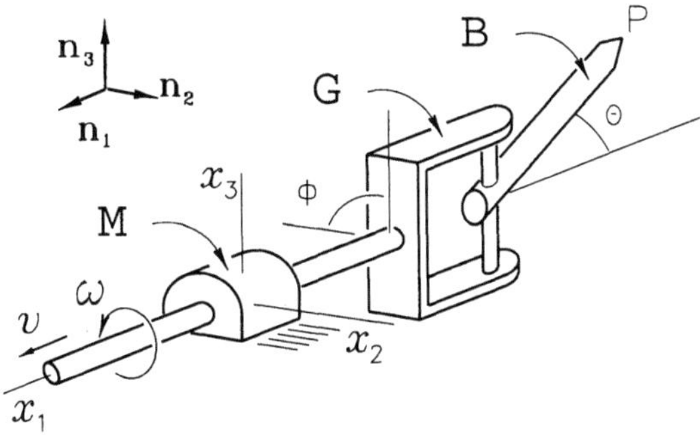

Figure P 5.6

P5.7 Bar B has one end, P, linked to the horizontal plane by means of a ball and socket joint and another end, Q, merely rests on the vertical plane. Assuming that point Q moves on the vertical plane at a velocity of constant module v, find the module of the angular velocity of the bar when Q is at its maximum elevation and at null elevation, immediately before the impact of the bar with the horizontal plane.

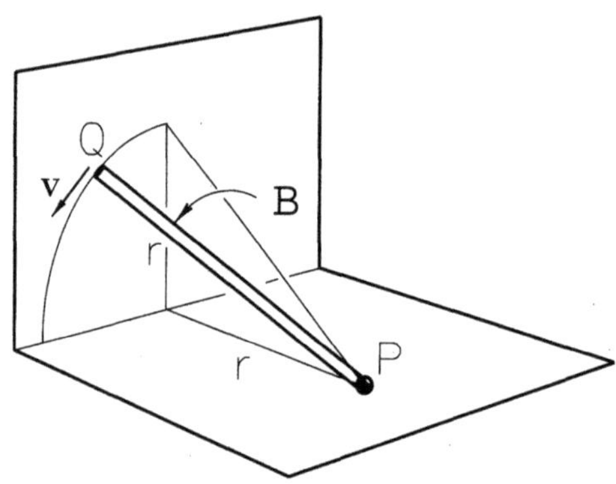

Figure P 5.7

P5.8 With regard to the above problem, what is the module of the angular acceleration of the bar in those two configurations?

P5.9 Cone A rolls over the fixed cone B. Center O', of the base of A, describes a circular motion at a constant module velocity. If the motion is repeated with a frequency of 2 Hz, calculate the angular velocity and acceleration of A.

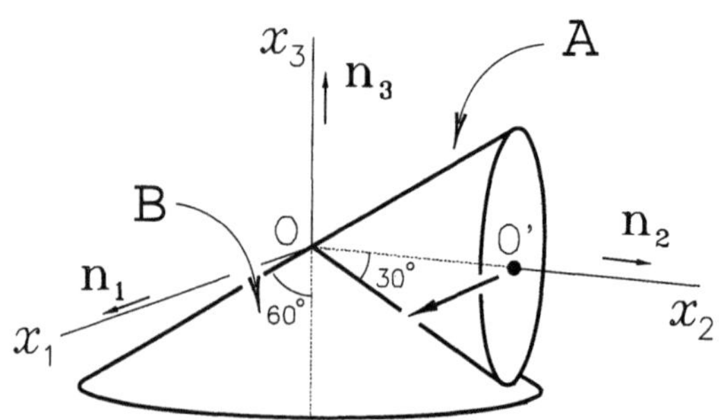

Figure P 5.9

P5.10 The figure illustrates a universal joint, a mechanism that permits torque transmission between two unaligned shafts. Shafts A and B rotate on bearings fixed in a reference frame $\mathcal{R}$ (not shown). Angle β on the vertical plane remains constant. A and B are interconnected by the crosshead C. Points P and Q are common to the fork connected to shaft A and the crosshead, while points R and S are common to the crosshead and the fork connected to shaft B. The orthonormal basis $\mathbf{a}_1, \mathbf{a}_2, \mathbf{a}_3$ is fixed in A, with $\mathbf{a}_1$ parallel to the line PQ and $\mathbf{a}_3$ parallel to the x-axis. The basis $\mathbf{b}_1, \mathbf{b}_2, \mathbf{b}_3$ is fixed in B, with $\mathbf{b}_1$ parallel to the line RS and $\mathbf{b}_3$ parallel to shaft B, as illustrated. Shaft A rotates at a constant angular velocity $\dot{\theta} = \Omega$, where θ is the angle between line PQ and the vertical axis z. Note that the angular velocity of shaft B will be a periodic function of θ, with a period 2π. Find the vector $^{\mathcal{R}}\boldsymbol{\omega}^{B}$ as a function of θ. Then calculate the module of this angular velocity for positions $\theta = 0$ and $\theta = \pi/2$. See the corresponding animation.

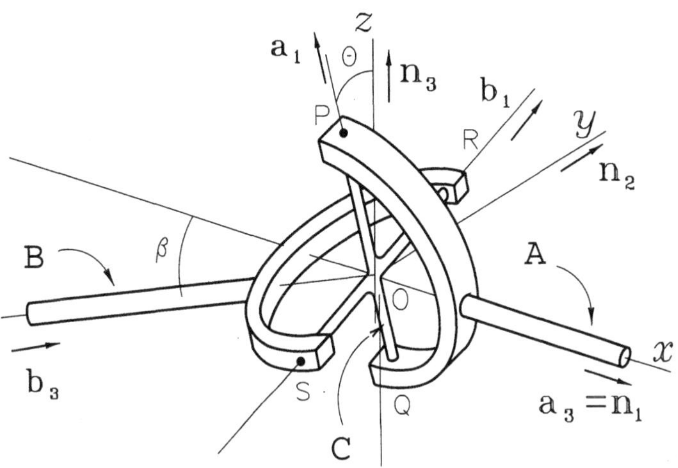

Figure P 5.10

P5.11 With reference to the previous exercise, calculate the module of the crosshead angular velocity for $\theta = 0$ and $\theta = \pi/2$. *Hint:* Use the basis $\mathbf{a}_1, \mathbf{a}_2, \mathbf{a}_3$ to break down all vectors.

P5.12   Disk D rotates in relation to the horizontal arm B, around axis x_2, while it rolls on the horizontal plane A. The arm, in turn, rotates in relation to A, around the x_3-axis, at the constant rate of 30 rpm, in the indicated direction. Find the velocity and acceleration, in A, of point P on the edge of the disk at the instant shown.

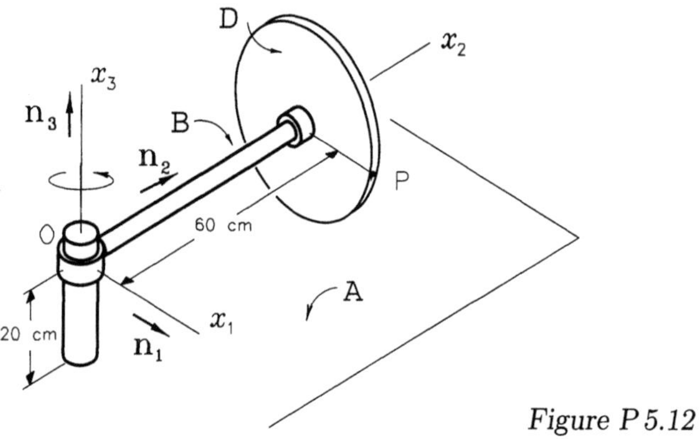

Figure P 5.12

P5.13   The piece of chalk G (a conical stump with a semi-angle β) rolls on the horizontal table M. Is this rolling of G on M pure? Assume that point P, the center of the cone's base, describes a circular trajectory with a constant speed v, in the direction shown. Find the angular velocity and acceleration of G in M.

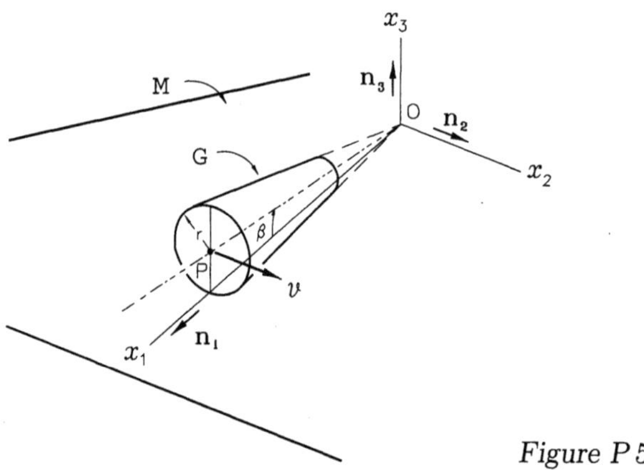

Figure P 5.13

P5.14 The figure illustrates diagrammatically the base frame of a super-market trolley. The wheels A and B can rotate freely around the x_2-axis, in relation to frame M, and roll over the horizontal plane E. Wheels C and D can rotate freely around horizontal axes in relation to the respective sockets that, in turn, can rotate around vertical axes in relation to the frame. At the instant shown, wheels A and B rotate in the same direction in relation to M, at angular velocities of constant module Ω and ω, respectively, as shown. Determine, at that instant, the angular velocity and acceleration of the wheel A and velocity and acceleration of its center (point P), also in relation to E.

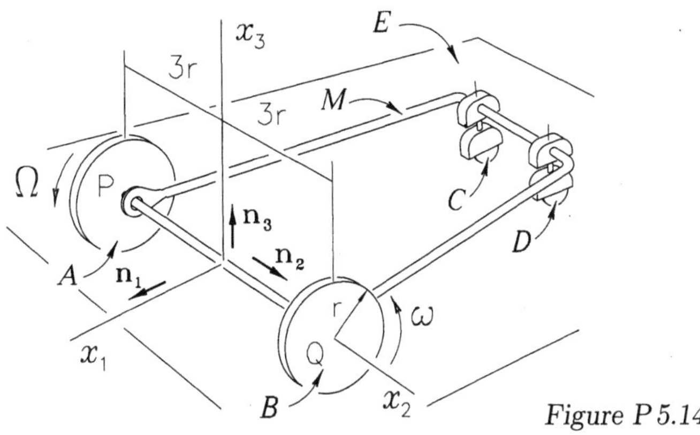

Figure P 5.14

P5.15 Bar B is connected through pin O to the reference frame A. Cursor C slides on B and cursor D slides along the guide welded on A. C and D are linked by a pivot, as shown. Determine the angular velocity of B in A when $x = a/2$, $\dot{x} = u$.

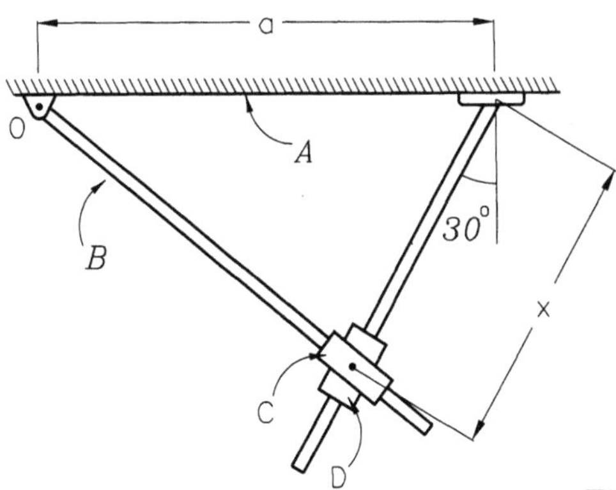

Figure P 5.15

P5.16   The vertical shaft C, with a conical end, is located by a thrust bearing, consisting of a fixed cylindrical base $\mathcal{R}$, on which eight identical balls roll, being kept equidistant by a spacer, not shown. When rotating around its axis of symmetry, the shaft rolls over the balls, while they roll over the plane surface and the cylindrical surface of the base at the same time, as illustrated. When the bearing is designed, it is possible to slightly vary the parameters a, radius of the base, r, radius of the balls, and β, the semi-angle of the cone, thereby altering the points of contact between the bodies in motion. In order to reduce to a minimum the wear of the surface of the conical shaft (harder to change), values are to be found that permit *pure rolling* between the shaft and the balls. Find the proper value for the relation a/r in function of β, to meet this condition.

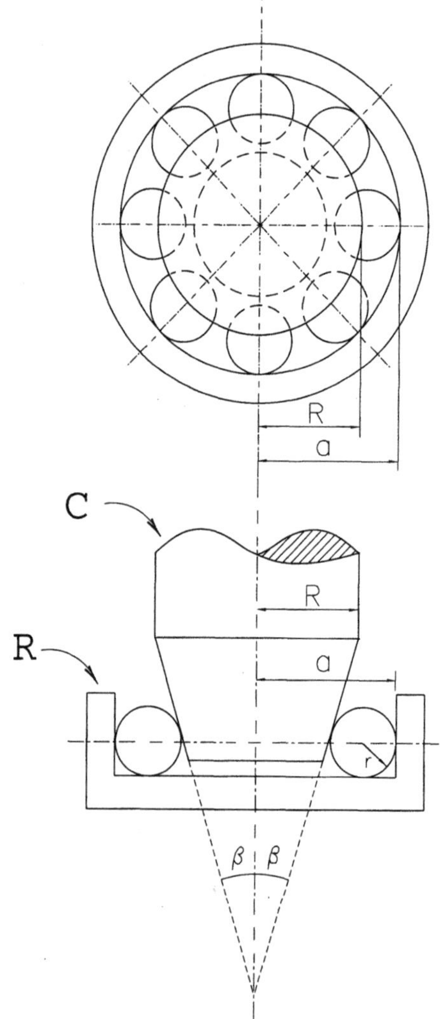

Figure P 5.16

P5.17 The ceiling fan consists of a casing C pivoting around y-axis in relation to the ring A which, in turn, pivots around X-axis in relation to the support S, attached to the ceiling. Inside the casing a motor moves the blade rotor and a small gearbox that moves the shaft on its back. This shaft is connected to arm B, whose end Q pivots on S, around the vertical Z-axis, as illustrated. When the motor starts up, it produces a rotating motion of the blade around the z-axis in relation to the casing, at an angular velocity with constant module Ω, in the direction shown, and a rotating motion of arm B, in the casing, in the same direction, at an angular velocity with constant module ω. The entire set experiences a periodic motion, with the angle θ kept constant. Determine the general expression for the angular velocity of B in S and for the angular velocity of C in S as a function of the angular position $\phi(t)$ (the angle between the X-axis and the direction of arm B). Express the results on the basis $\mathbf{n}_1, \mathbf{n}_2, \mathbf{n}_3$, fixed in S.

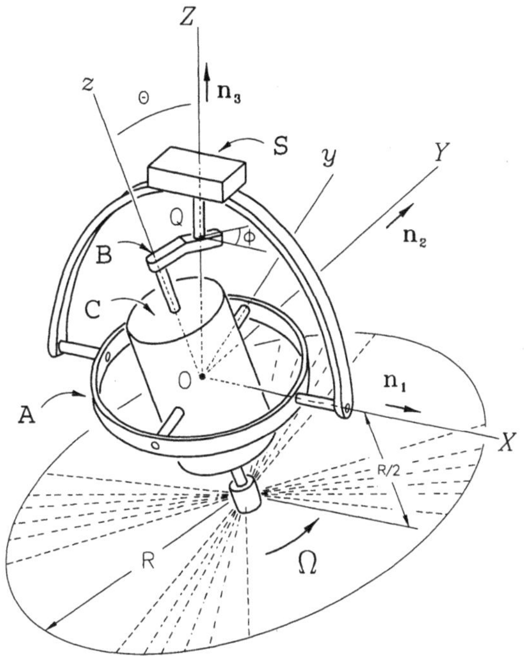

Figure P 5.17

P5.18 With reference to the previous exercise, now consider the fact that the velocity reduction caused by the gearbox is 12:1, that is, $\Omega = 12\omega$. Determine the angular velocity and acceleration of the blade in S at the instant when $\phi = 0$.

P5.19   A planetary velocity conversion system consists of three gears rotating at angular velocities, all parallel to the x_3-axis. The sun gear A, with 24 teeth, is attached to the input shaft of the system and rotates at a constant rate of 200 rpm, in the direction shown, in relation to the internal ring gear B, which is fixed. Gear C, with 18 teeth, engages simultaneously in gear A and in gear B, the latter with 54 teeth. Gear C is free to rotate around a pin attached to support D. Support D is attached to the output shaft. Calculate the angular velocity of the output shaft with respect to the fixed gear B.

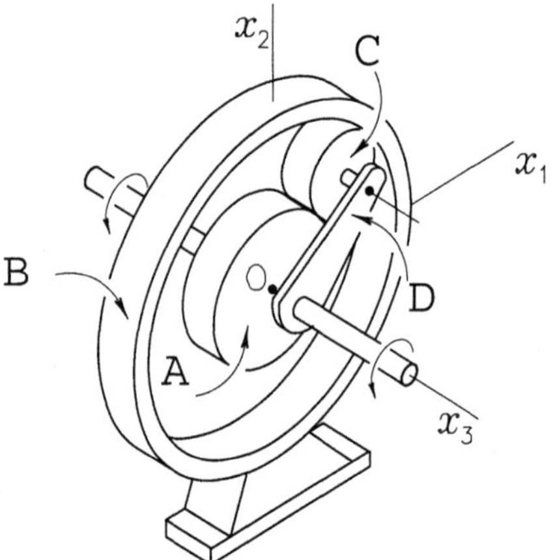

Figure P 5.19

P5.20   The cone C rolls on the conical surface A, with vertex V fixed in A. Point P, the center of the C base, describes a circular trajectory with constant speed v in A, as shown. Determine the angular acceleration of C in A and the acceleration of point Q, fixed to the base of C when it touches the surface A. The orthonormal basis $\mathbf{n}_1, \mathbf{n}_2, \mathbf{n}_3$ is fixed in the plane containing the axes of symmetry of both cones, with $\mathbf{n}_1$ horizontal and radial and $\mathbf{n}_3$ vertical.

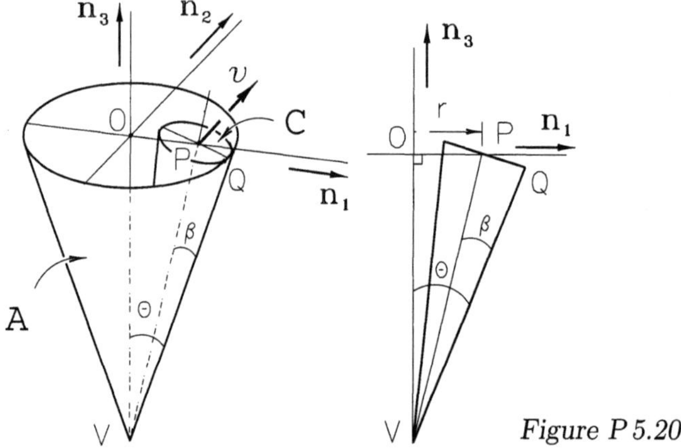

Figure P 5.20

P5.21 The disk rolls on the horizontal plane, rotating freely around the sloping arm, which, in turn, rotates freely around the vertical axis x_3, keeping point O fixed in the support, as shown. The axes $\{x_1, x_2, x_3\}$ are fixed in a reference frame that has, in relation to the horizontal plane, a prescribed angular velocity, $\Omega(t)$. The angle between the sloping arm and the vertical support, θ, remains constant. Find the acceleration of point P of the disk in contact with the plane.

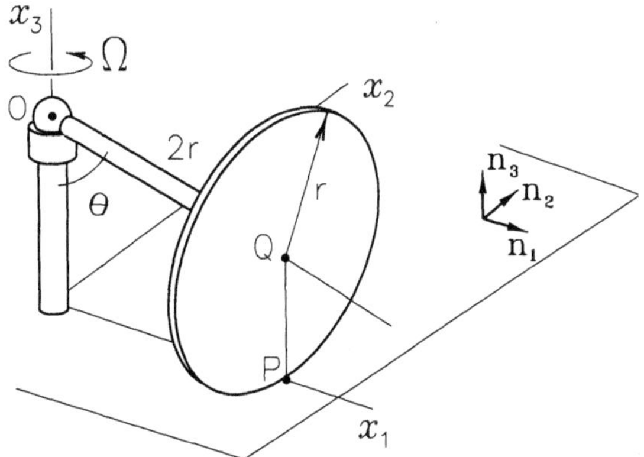

Figure P 5.21

P5.22 The disk D, with radius r, rotates around the sloping arm B, which, in turn, rotates at an angular velocity with constant module Ω, around the vertical shaft fixed in the cylinder of radius r. The two shafts are linked by a ball and socket joint, as illustrated. Knowing that the disk rolls on the cylinder surface, calculate the velocity module, in the cylinder, of point P on the edge of the disk at the instant when it is in a position diametrically opposite of the point of contact.

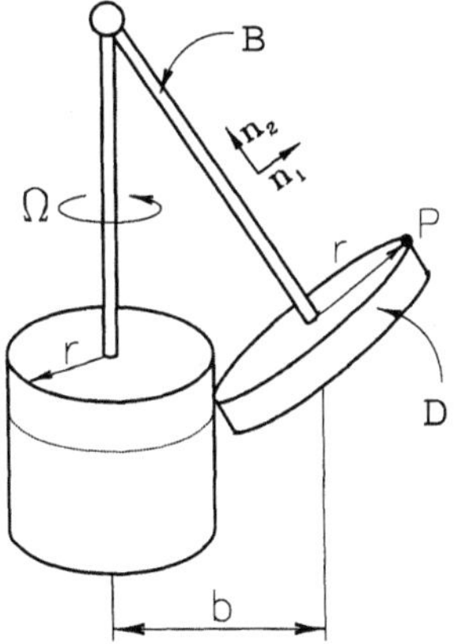

Figure P 5.22

P5.23 The disks D and D' rotate freely around the ends of rod B and roll on the horizontal plane, as shown. Knowing the constant angular velocities ω and ω' of the disks in relation to the vertical plane containing the centers of the disks, calculate the angular velocity and acceleration of D in relation to the horizontal plane.

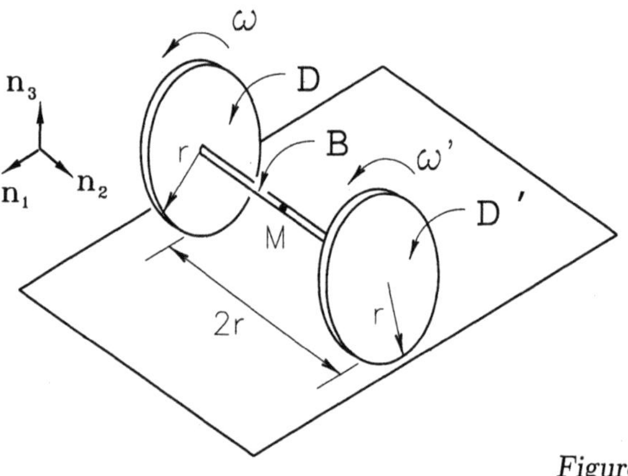

Figure P 5.23

P5.24 The mechanism illustrated consists of three gears. A is fixed, B rotates at a simple angular velocity with constant module ω_0, in relation to A, in the direction indicated, and C is engaged in A and in B, having, therefore, its motion established by the two others (see Example 10.7). Find the angular velocity and acceleration of C in B. See the corresponding animation.

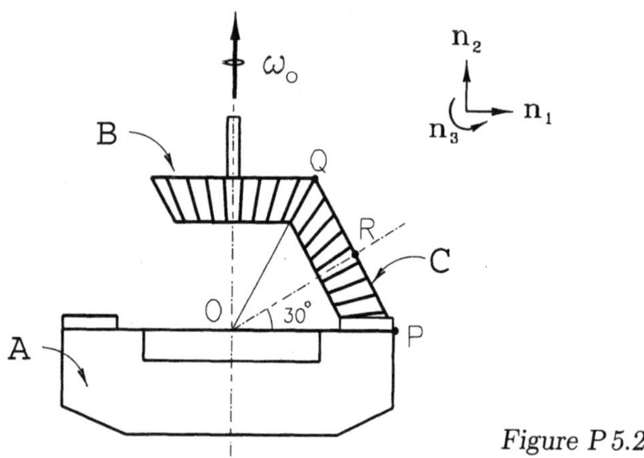

Figure P 5.24

P5.25 Consider the mechanism consisting of the disk D that can rotate freely around its axis of symmetry in relation to the arm B, connected to the trolley C by a ball and socket joint at point O. The trolley, in turn, moves in relation to the table A, along the straight rail, as illustrated. The coordinates $x(t)$, $\theta(t)$, $\phi(t)$, and $\psi(t)$ fully describe the configuration of the system. The orthonormal bases $\mathbf{b}_1, \mathbf{b}_2, \mathbf{b}_3$ and $\mathbf{c}_1, \mathbf{c}_2, \mathbf{c}_3$ are fixed in B and in C, respectively, with the directions shown. Write the general expression, on the basis $\mathbf{c}_1, \mathbf{c}_2, \mathbf{c}_3$, for the angular velocity vector of disk D in the reference frame A. Now assume that $x(t) = \frac{1}{2}at^2 + vt$, $\theta(t) = \omega_0 t$, $\phi(t) = 2\omega_0 t$, and $\psi(t) = \psi_0 \sin(\lambda t)$ and determine the vector angular acceleration of D in A at the instant $t = \pi/2\lambda$. Also find the angular velocity vector of C in relation to D at the same instant.

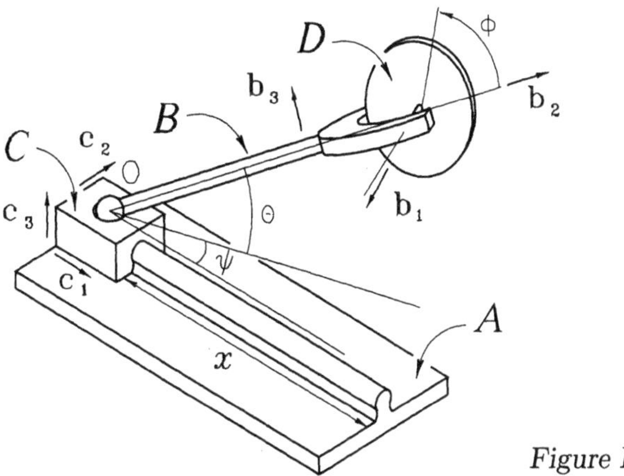

Figure P 5.25

P5.26 The disk D rolls inside the ring A, as illustrated. If the disk center, P, makes a full turn every second, calculate the module of the angular velocity of D in A.

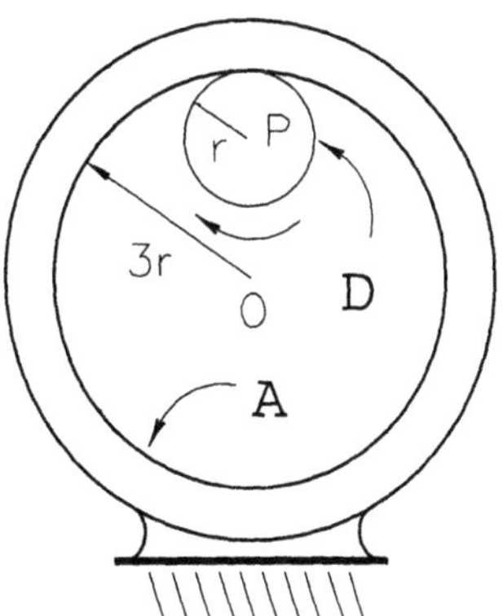

Figure P 5.26

P5.27 The rod B slides on the fixed circular base A, pivoting at its end O on a vertical pin fixed at the center of the basis, as shown. At the other end Q a string with lenght a is fixed with a small ball P hanging from it. How many degrees of freedom does the system have, consisting of B and P, assuming that the wire remains stretched?

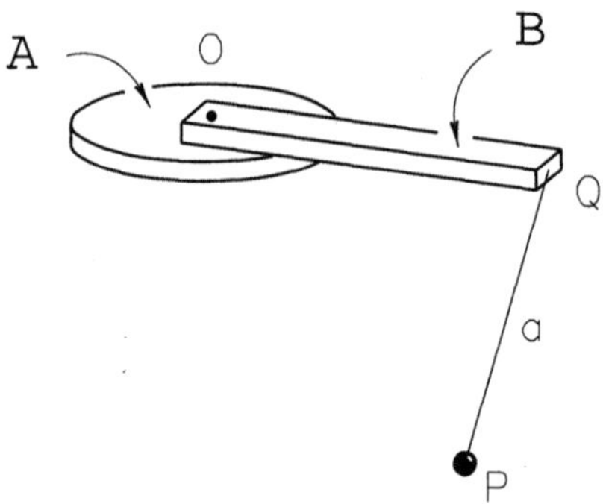

Figure P 5.27

P5.28 Disks D and D' rotate freely around the ends of bar B and roll on the horizontal plane, as illustrated. As the angular velocities ω and ω' of the disks in relation to the vertical plane containing the disk centers are known, find the velocity and acceleration of the midpoint M of bar B in relation to the horizontal plane.

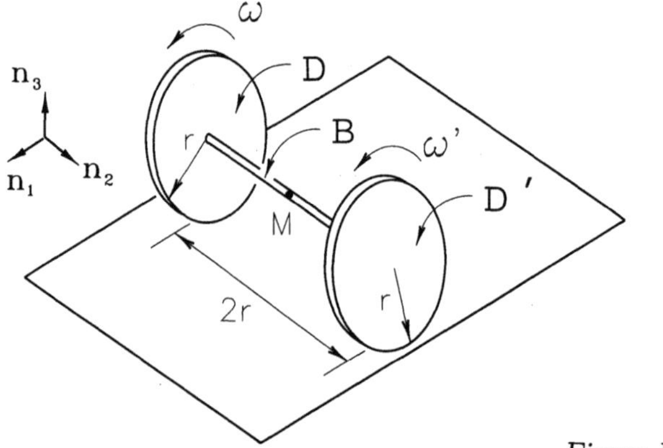

Figure P 5.28

P5.29 The figure illustrates an articulated mechanism, conceived by Leonardo da Vinci, to solve an optical problem. (The purpose of da Vinci's ingenious device, given a candle, convex mirror, and onlooker, was to find which ray of light comes from the flame to reach the onlooker's eye.) Pins P, Q, and O are fixed on a flat horizontal surface (they correspond to the eye, flame, and center of the mirror, respectively, but this is not important when solving the kinematic problem) while pins R and S move in relation to the central rod guide, at velocity v, in the directions shown. If, at the instant shown, the distance between P and R is r, the distance between Q and R is s, and the distance between O and S is a, calculate the angular velocity of the central rod in relation to the plane, at that instant.

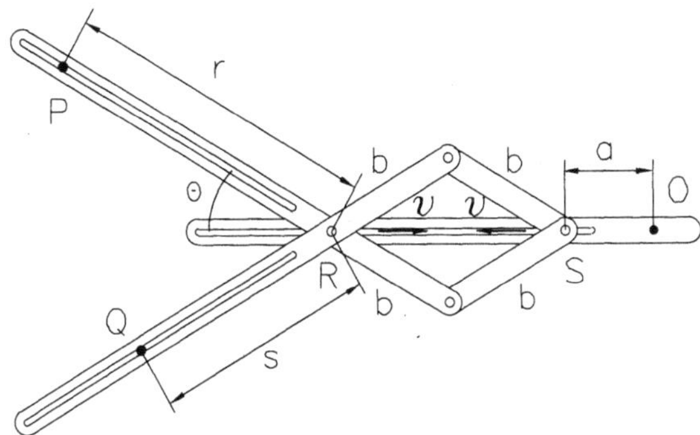

Figure P 5.29

P5.30 The body C consists of a cylinder welded to a rod, with the indicated dimensions. The free end of the rod is linked to the fixed point O by means of a ball and socket joint, and the cylinder rolls on the horizontal plane fixed in a reference frame $\mathcal{R}$, always keeping a point on its bottom edge touching the plane. The motion of C in $\mathcal{R}$ is prescribed and can be described as follows. Consider a reference frame B, defined by the plane containing the vertical axis z and the mass center C^* of C. The orthonormal bases $\mathbf{b}_1, \mathbf{b}_2, \mathbf{b}_3$ and $\mathbf{n}_1, \mathbf{n}_2, \mathbf{n}_3$ are fixed in B, with the directions shown in the figure. B moves in relation to $\mathcal{R}$ at a vertical simple angular velocity with constant module Ω, as illustrated. All points on the x_1-axis, therefore, also have a constant module velocity in the reference frame $\mathcal{R}$. Determine the angular velocity of C in $\mathcal{R}$, the angular acceleration of C in $\mathcal{R}$, the acceleration of the point C^*, and the acceleration of the point of contact between the cylinder and the plane, both in the same reference frame $\mathcal{R}$.

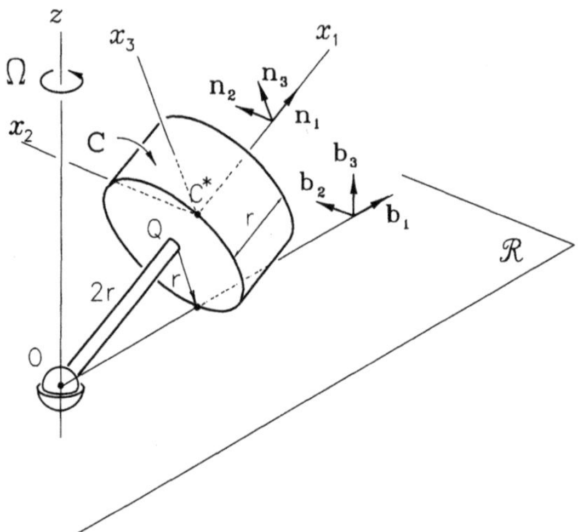

Figure 5.30

P5.31   The figure shows a diagram of the differential system of a motor vehicle, where the torque provided by a shaft E can be transmitted to the drive wheels, permitting them to rotate at different angular velocities in relation to the reference frame fixed in the vehicle. The system operates as follows. The gears A and A' are attached to the axes that drive the wheels. The gears B and B' are engaged with A and A' and free to rotate around pins fixed to the structure C. This structure is attached to the crown gear D, driven, in turn, by the pinion gear driven by shaft E. Note that the gears A, A', and D all rotate around the common axis of symmetry, z, but have partially independent motions. Knowing the angular velocities Ω and Ω', respectively, of the wheels in the reference frame $\mathcal{R}$ and the rate n between the number of teeth of D and E, calculate the angular velocity of shaft E in this same reference frame.

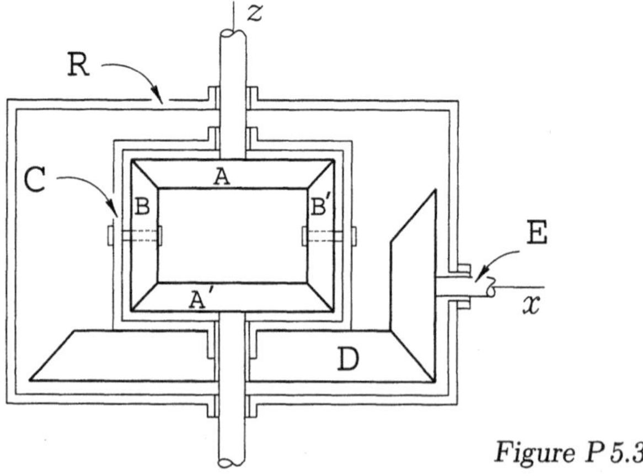

Figure P 5.31

Dynamics of Particles

Chapter 4

This chapter formally discusses the fundamental principles of *dynamics*, applied to the motion of the particle. Although this formulation is not Newton's original one, it is convenient, for the sake of logic and clarity, to formulate the dynamic principles from the point of view of a particle. Once the equations governing the motion of a particle in a given reference frame have been established and the main techniques for solving problems involving its motion have been studied, it is easier and more natural to generalize these principles and techniques for application to systems of particles or bodies in general. This chapter discusses the main methods for analyzing the dynamics of a particle, which must be clearly understood for the study of dynamics of the rigid body and other applications, such as fluid mechanics, solid mechanics, and mechanical vibrations.

Section 4.1 introduces the fundamental concepts of dynamics; momentum vectors, angular momentum with respect to a point, and angular momentum with respect to an axis are defined, as well as the kinetic energy of a particle, in a given reference frame. Section 4.2 is devoted to the study of Newton's second law. Here the lesson is to learn how to establish the equations of motion for a particle and how to choose the basis to help simplify these equations. A thorough analysis of the famous experiment of Foucault's pendulum helps in the discussion on inertial reference frames. Plane motion is discussed in Section 4.3, where evidence is given of the simplifications introduced by this kind of motion.

Section 4.4 deals with an alternative form for the equations of motion, involving the resultant moment with respect to a fixed point and the angular momentum vector of the particle with respect to the point. It is found that the method, in some cases, helps to eliminate undesirable unknowns. Section 4.5 introduces the notion of work of a force applied to a point, with a discussion ensuing on the so-called conservative forces and the calculation of their work based on potential functions, in preparation for the study of energy. In Section 4.6 the dynamic equation is integrated along the displacement of the particle, resulting in the equation that relates work and energy. Mention should be made of the simplicity of this formulation in solving a whole range of problems that do not require integration in the time of the equations of motion. Section 4.7 discusses the problem of impact, presenting the notion of a force's impulse and its applications. The concept of angular impulse is also discussed in that section. Last, in Section 4.8, conservation principles are described. A number of examples illustrate their application in problem-solving, showing the advantage of their use, whenever applicable.

4.1 Dynamic Properties

Let P be a particle with mass m moving in a trajectory — described by the line λ — in a given reference frame $\mathcal{R}$. Let Q also be a point that moves in $\mathcal{R}$ (Q may, for example, be another particle or a point of a rigid body), O a point fixed in $\mathcal{R}$, and $\mathbf{p}$, $\mathbf{q}$, and $\mathbf{p}^{P/Q}$ the position vectors of P with respect to O, Q with respect to O, and P with respect to Q, respectively (see Fig. 1.1). Now adopting the reduced notation for the time rate in $\mathcal{R}$, the (absolute) velocity of the particle P in reference frame $\mathcal{R}$ (see Section 3.5) is

$$^{\mathcal{R}}\mathbf{v}^{P} = \dot{\mathbf{p}} \tag{1.1}$$

and the velocity of P relative to Q in $\mathcal{R}$ is

$$^{\mathcal{R}}\mathbf{v}^{P/Q} = \dot{\mathbf{p}}^{P/Q} = {}^{\mathcal{R}}\mathbf{v}^{P} - {}^{\mathcal{R}}\mathbf{v}^{Q} = \dot{\mathbf{p}} - \dot{\mathbf{q}}. \tag{1.2}$$

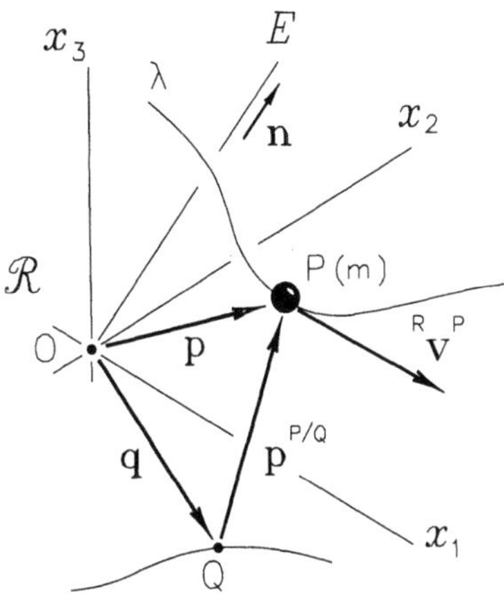

Figure 1.1

The (absolute) acceleration of P in $\mathcal{R}$ (see Section 3.5) is

$$^{\mathcal{R}}\mathbf{a}^{P} = {}^{\mathcal{R}}\dot{\mathbf{v}}^{P} = \ddot{\mathbf{p}} \tag{1.3}$$

and the acceleration of P relative to Q in $\mathcal{R}$ is

$$^{\mathcal{R}}\mathbf{a}^{P/Q} = \ddot{\mathbf{p}}^{P/Q} = {}^{\mathcal{R}}\dot{\mathbf{v}}^{P} - {}^{\mathcal{R}}\dot{\mathbf{v}}^{Q} = \ddot{\mathbf{p}} - \ddot{\mathbf{q}}. \tag{1.4}$$

The *momentum* vector of a particle P in a reference frame $\mathcal{R}$ is a vector *vinculated* to P, defined as the product of its mass m by the velocity of P in $\mathcal{R}$, that is,

$$^{\mathcal{R}}\mathbf{G}^{P} \rightleftharpoons m\,{}^{\mathcal{R}}\mathbf{v}^{P}. \tag{1.5}$$

The time rate in $\mathcal{R}$ of the momentum vector of the particle, assuming its constant mass, using the reduced notation for differentiation in $\mathcal{R}$ will then be

$$^{\mathcal{R}}\dot{\mathbf{G}}^{P} = m\,{}^{\mathcal{R}}\mathbf{a}^{P}. \tag{1.6}$$

It is worth noting that the momentum of a particle, as well as its time rate, depends on the reference frame from which it is being observed.

Example 1.1 The disk D rotates around the vertical axis x_3 in the reference frame $\mathcal{R}$ at a simple angular velocity of constant module ω. Cursor P, with mass m, slides freely over the diametrical guide and the pin Q is fixed to D, as illustrated (see Fig. 1.2). The velocities of the point Q and cursor P in the reference frame D are, respectively,

$$^{D}\mathbf{v}^{Q} = 0, \quad {}^{D}\mathbf{v}^{P} = \dot{x}\mathbf{n}_1.$$

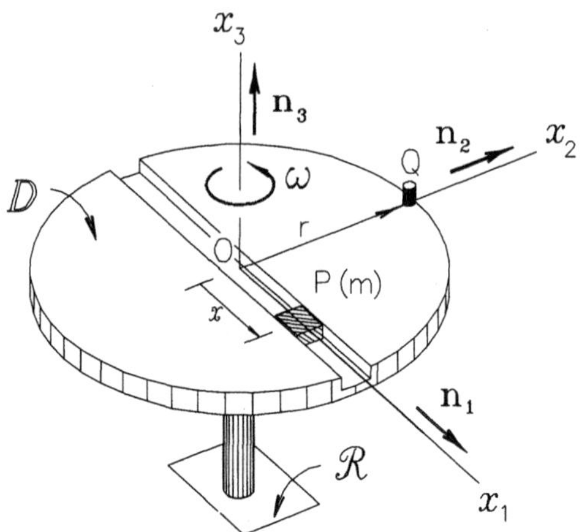

Figure 1.2

The velocities of the point Q and cursor P in the reference frame $\mathcal{R}$ are, respectively,

$$^{\mathcal{R}}\mathbf{v}^Q = -r\omega\mathbf{n}_1, \quad ^{\mathcal{R}}\mathbf{v}^P = \dot{x}\mathbf{n}_1 + x\omega\mathbf{n}_2.$$

The accelerations of P in the reference frames D and $\mathcal{R}$ are, respectively,

$$^D\mathbf{a}^P = {}^D\dot{\mathbf{v}}^P = \ddot{x}\mathbf{n}_1,$$
$$^{\mathcal{R}}\mathbf{a}^P = {}^{\mathcal{R}}\dot{\mathbf{v}}^P = (\ddot{x} - x\omega^2)\mathbf{n}_1 + 2\dot{x}\omega\mathbf{n}_2.$$

The momentum vectors of P in the reference frames D and $\mathcal{R}$ can then be calculated, according to Eq. (1.5):

$$^D\mathbf{G}^P = m{}^D\mathbf{v}^P = m\dot{x}\mathbf{n}_1;$$
$$^{\mathcal{R}}\mathbf{G}^P = m{}^{\mathcal{R}}\mathbf{v}^P = m(\dot{x}\mathbf{n}_1 + x\omega\mathbf{n}_2).$$

The time rates at D and $\mathcal{R}$ of the respective momentum vectors may be obtained from Eq. (1.6):

$$^D\dot{\mathbf{G}}^P = m{}^D\mathbf{a}^P = m\ddot{x}\mathbf{n}_1;$$
$$^{\mathcal{R}}\dot{\mathbf{G}}^P = m{}^{\mathcal{R}}\mathbf{a}^P = m\big((\ddot{x} - x\omega^2)\mathbf{n}_1 + 2\dot{x}\omega\mathbf{n}_2\big).$$

The *angular momentum* vector of a particle P with respect to a point Q in a reference frame $\mathcal{R}$ is defined as the moment, with respect to the point, of the momentum of the particle in the reference frame (see Fig. 1.1), that is,

$$^{\mathcal{R}}\mathbf{H}^{P/Q} \triangleq \mathbf{p}^{P/Q} \times {}^{\mathcal{R}}\mathbf{G}^{P}. \tag{1.7}$$

Substituting Eq. (1.5) in Eq. (1.7) one finds that the alternative equation is then

$$^{\mathcal{R}}\mathbf{H}^{P/Q} = \mathbf{p}^{P/Q} \times m\,{}^{\mathcal{R}}\mathbf{v}^{P}. \tag{1.8}$$

The angular momentum of a particle with respect to a point will be null if the velocity of the particle in the reference frame is null or if the particle and the point coincide or if the velocity vector is parallel to the relative position vector, as stated in Eq. (1.8).

The time rate in $\mathcal{R}$ of the vector $^{\mathcal{R}}\mathbf{H}^{P/Q}$, also using the reduced notation, will be

$$\begin{aligned}
^{\mathcal{R}}\dot{\mathbf{H}}^{P/Q} &= \dot{\mathbf{p}}^{P/Q} \times {}^{\mathcal{R}}\mathbf{G}^{P} + \mathbf{p}^{P/Q} \times {}^{\mathcal{R}}\dot{\mathbf{G}}^{P} \\
&= \left({}^{\mathcal{R}}\mathbf{v}^{P} - {}^{\mathcal{R}}\mathbf{v}^{Q}\right) \times {}^{\mathcal{R}}\mathbf{G}^{P} + \mathbf{p}^{P/Q} \times {}^{\mathcal{R}}\dot{\mathbf{G}}^{P} \\
&= \mathbf{p}^{P/Q} \times {}^{\mathcal{R}}\dot{\mathbf{G}}^{P} - {}^{\mathcal{R}}\mathbf{v}^{Q} \times {}^{\mathcal{R}}\mathbf{G}^{P}
\end{aligned} \tag{1.9}$$

or, using Eqs. (1.5) and (1.6),

$$^{\mathcal{R}}\dot{\mathbf{H}}^{P/Q} = \mathbf{p}^{P/Q} \times m\,{}^{\mathcal{R}}\mathbf{a}^{P} - {}^{\mathcal{R}}\mathbf{v}^{Q} \times m\,{}^{\mathcal{R}}\mathbf{v}^{P}. \tag{1.10}$$

Note that the second term on the right vanishes if Q is fixed in $\mathcal{R}$ or if P and Q have parallel velocities in $\mathcal{R}$.

The angular momentum vector of a particle P with respect to a point O, fixed in the reference frame $\mathcal{R}$, in $\mathcal{R}$, is (see Fig. 1.1)

$$\begin{aligned}
^{\mathcal{R}}\mathbf{H}^{P/O} &= \mathbf{p} \times {}^{\mathcal{R}}\mathbf{G}^{P} \\
&= \mathbf{p} \times m\,{}^{\mathcal{R}}\mathbf{v}^{P},
\end{aligned} \tag{1.11}$$

where $\mathbf{p} = \mathbf{p}^{P/O}$. The time rate in $\mathcal{R}$ of the vector $^{\mathcal{R}}\mathbf{H}^{P/O}$ will then be

$$\begin{aligned}
^{\mathcal{R}}\dot{\mathbf{H}}^{P/O} &= {}^{\mathcal{R}}\mathbf{v}^{P} \times {}^{\mathcal{R}}\mathbf{G}^{P} + \mathbf{p} \times {}^{\mathcal{R}}\dot{\mathbf{G}}^{P} \\
&= \mathbf{p} \times {}^{\mathcal{R}}\dot{\mathbf{G}}^{P}
\end{aligned} \tag{1.12}$$

or

$$\mathcal{R}\dot{\mathbf{H}}^{P/O} = \mathbf{p} \times m\,\mathcal{R}\mathbf{a}^{P}. \tag{1.13}$$

Note that Eq. (1.13) is a particular case of Eq. (1.10).

The angular momentum vector is the moment of a momentum vector with respect to a point; its projection toward an arbitrary unit vector $\mathbf{n}$ will therefore be the moment of this vector with respect to the axis passing through the point and parallel to the unit vector (see Section 2.2). The angular momentum of a particle P with respect to an axis E is then defined in a reference frame $\mathcal{R}$ as

$$\mathcal{R}\mathbf{H}^{P/E} \rightleftharpoons \mathcal{R}\mathbf{H}^{P/O} \cdot \mathbf{n}\,\mathbf{n}, \tag{1.14}$$

where $\mathbf{n}$ is a unit vector parallel to the axis E and O belongs to the axis (see Fig. 1.1). When substituting Eq. (1.11) in Eq. (1.14), then

$$\begin{aligned}
\mathcal{R}\mathbf{H}^{P/E} &= \mathbf{p} \times {}^{\mathcal{R}}\mathbf{G}^{P} \cdot \mathbf{n}\,\mathbf{n} \\
&= \mathbf{p} \times m\,\mathcal{R}\mathbf{v}^{P} \cdot \mathbf{n}\,\mathbf{n}.
\end{aligned} \tag{1.15}$$

The angular momentum vector of a particle P with respect to an axis in a given reference frame does not depend on the point of the chosen axis for its calculation (see Section 2.2) and will be null if the vectors $\mathbf{p}$, $\mathcal{R}\mathbf{v}^{P}$, and $\mathbf{n}$ are coplanar, as illustrated in Eq. (1.15).

If the axis E is fixed in the reference frame, the unit vector $\mathbf{n}$ has a null derivative in this reference frame, and the time rate in $\mathcal{R}$ of the angular momentum of P with respect to E in $\mathcal{R}$ will be

$$\mathcal{R}\dot{\mathbf{H}}^{P/E} = \mathcal{R}\dot{\mathbf{H}}^{P/O} \cdot \mathbf{n}\,\mathbf{n}, \tag{1.16}$$

that is,

$$\mathcal{R}\dot{\mathbf{H}}^{P/E} = \mathbf{p} \times m\,\mathcal{R}\mathbf{a}^{P} \cdot \mathbf{n}\,\mathbf{n}. \tag{1.17}$$

Example 1.2   Returning to the previous example (see Fig. 1.2), the angular momentum vector of the cursor P with respect to the point Q in the reference frame D is, according to Eq. (1.8),

$$^{D}\mathbf{H}^{P/Q} = (x\mathbf{n}_1 - r\mathbf{n}_2) \times m\dot{x}\mathbf{n}_1 = mr\dot{x}\mathbf{n}_3.$$

The angular momentum vector of P with respect to Q in $\mathcal{R}$ will be

$$^{\mathcal{R}}\mathbf{H}^{P/Q} = (x\mathbf{n}_1 - r\mathbf{n}_2) \times m(\dot{x}\mathbf{n}_1 + x\omega\mathbf{n}_2) = m(x^2\omega + r\dot{x})\mathbf{n}_3.$$

The time rate in $\mathcal{R}$ of this vector, according to Eq. (1.10), will be

$$^{\mathcal{R}}\dot{\mathbf{H}}^{P/Q} = (x\mathbf{n}_1 - r\mathbf{n}_2) \times m[(\ddot{x} - x\omega^2)\mathbf{n}_1 + 2\dot{x}\omega\mathbf{n}_2]$$
$$- (-r\omega)\mathbf{n}_1 \times m(\dot{x}\mathbf{n}_1 + x\omega\mathbf{n}_2)$$
$$= m(r\ddot{x} + 2x\dot{x}\omega)\mathbf{n}_3.$$

(Note that the same result would be obtained by directly differentiating $^{\mathcal{R}}\mathbf{H}^{P/Q}$.) The angular momentum vector of the cursor P with respect to the point O in $\mathcal{R}$ is, according to Eq. (1.11),

$$^{\mathcal{R}}\mathbf{H}^{P/O} = x\mathbf{n}_1 \times m(\dot{x}\mathbf{n}_1 + x\omega\mathbf{n}_2) = mx^2\omega\mathbf{n}_3,$$

and its time rate in $\mathcal{R}$, according to Eq. (1.13), will be

$$^{\mathcal{R}}\dot{\mathbf{H}}^{P/O} = x\mathbf{n}_1 \times m\big((\ddot{x} - x\omega^2)\mathbf{n}_1 + 2\dot{x}\omega\mathbf{n}_2\big)$$
$$= 2mx\dot{x}\omega\mathbf{n}_3.$$

The disk moves at a simple angular velocity in $\mathcal{R}$ and, as the points O, Q, and P remain on the same horizontal plane, every angular momentum vector obtained is vertical. In the latter case, therefore, the angular momentum with respect to any axis contained in the disk's plane will be null [see Eq. (1.14)] and the angular momentum with respect to a vertical axis will be equal to the angular momentum with respect to the point of the plane that the axis intercepts. Thus, for example,

$$^{\mathcal{R}}\mathbf{H}^{P/x_3} = {}^{\mathcal{R}}\mathbf{H}^{P/O} = mx^2\omega\mathbf{n}_3.$$

The time rate in $\mathcal{R}$ of this vector, according to Eq. (1.16), will be

$$^{\mathcal{R}}\dot{\mathbf{H}}^{P/x_3} = 2mx\dot{x}\omega\mathbf{n}_3 \cdot \mathbf{n}_3\,\mathbf{n}_3 = 2mx\dot{x}\omega\mathbf{n}_3.$$

The *kinetic energy* of a particle P in a reference frame $\mathcal{R}$ is defined by the expression

$$\mathcal{R}K^P \rightleftharpoons \frac{1}{2}m\,{}^{\mathcal{R}}\mathbf{v}^P \cdot {}^{\mathcal{R}}\mathbf{v}^P. \tag{1.18}$$

If $v = |{}^{\mathcal{R}}\mathbf{v}^P|$ is the module of the velocity vector of the particle in $\mathcal{R}$, its kinetic energy in $\mathcal{R}$ may, therefore, be expressed by

$$\mathcal{R}K^P = \frac{1}{2}mv^2. \tag{1.19}$$

Note that the concept of kinetic energy of a particle retains some similarity with that of momentum, both depending on the mass and velocity of the particle. The kinetic energy, however, is a scalar property, a quadratic function of the module of the velocity vector, while the momentum is a vector property, having the same direction as the velocity vector.

It is also worth noting that the kinetic energy of a particle in a given reference frame may be obtained from the momentum vector of the particle in the reference frame and of its velocity vector in this same reference frame, that is,

$$\mathcal{R}K^P = \frac{1}{2}{}^{\mathcal{R}}\mathbf{G}^P \cdot {}^{\mathcal{R}}\mathbf{v}^P, \tag{1.20}$$

which is evident from the definitions in Eqs. (1.18) and (1.5).

Example 1.3   Once again returning to the configuration described in Example 1.1, the kinetic energy of the cursor P on the disk D is, according to Eq. (1.18),

$$^{D}K^P = \frac{1}{2}m\,{}^{D}\mathbf{v}^P \cdot {}^{D}\mathbf{v}^P = \frac{1}{2}m\dot{x}^2,$$

and the kinetic energy of the cursor in the reference frame $\mathcal{R}$ may be calculated by using Eq. (1.20):

$$\begin{aligned}
\mathcal{R}K^P &= \frac{1}{2}m(\dot{x}\mathbf{n}_1 + x\omega\mathbf{n}_2) \cdot (\dot{x}\mathbf{n}_1 + x\omega\mathbf{n}_2) \\
&= \frac{1}{2}m(\dot{x}^2 + x^2\omega^2).
\end{aligned}$$

The concepts of momentum, angular momentum, and kinetic energy of a particle in a given reference frame are the basis for establishing the equations governing the motion of the particle, as will be seen below.

It is interesting to see that, by projecting the time rate of the momentum vector in the direction of the motion of the particle, that means, performing the dot product

$$
{}^{\mathcal{R}}\dot{\mathbf{G}}^{P} \cdot {}^{\mathcal{R}}d\mathbf{p} = m\,{}^{\mathcal{R}}\mathbf{a}^{P} \cdot {}^{\mathcal{R}}d\mathbf{p}
$$
$$
= m\frac{{}^{\mathcal{R}}d}{dt}{}^{\mathcal{R}}\mathbf{v}^{P} \cdot {}^{\mathcal{R}}d\mathbf{p}
$$
$$
= m\,{}^{\mathcal{R}}d{}^{\mathcal{R}}\mathbf{v}^{P} \cdot \frac{{}^{\mathcal{R}}d}{dt}\mathbf{p} \tag{1.21}
$$
$$
= m\,{}^{\mathcal{R}}\mathbf{v}^{P} \cdot {}^{\mathcal{R}}d{}^{\mathcal{R}}\mathbf{v}^{P}
$$

and, integrating along the trajectory of the particle, then

$$
\int m\,{}^{\mathcal{R}}\mathbf{a}^{P} \cdot {}^{\mathcal{R}}d\mathbf{p} = \int m\,{}^{\mathcal{R}}\mathbf{v}^{P} \cdot {}^{\mathcal{R}}d{}^{\mathcal{R}}\mathbf{v}^{P}
$$
$$
= \frac{1}{2}m\,{}^{\mathcal{R}}\mathbf{v}^{P} \cdot {}^{\mathcal{R}}\mathbf{v}^{P} + C \tag{1.22}
$$
$$
= {}^{\mathcal{R}}K^{P} + C,
$$

where C is a constant, depending on the initial conditions of the motion of P. Equation (1.22) will be especially useful in establishing the balance of energy of the particle, which will be discussed in Section 4.6.

4.2 Newton's Second Law

Let P be a particle with mass m moving in space under the action of a force system $\mathcal{F}$ whose resultant is $\mathbf{R}$ (see Fig. 2.1). As discussed in Chapter 2, every force system acting on a particle is a concurrent simple system, always being equivalent to a force equal to its resultant, applied to the particle itself, the concurrence point. The dynamic law governing the motion of P under the action of this force system — known as *Newton's second law* — may be formulated as follows:

Second Law.   *Reference frames $\mathcal{R}$ exist so that if $^{\mathcal{R}}\mathbf{G}^P$ is the momentum vector of a particle P in $\mathcal{R}$, their time rate in $\mathcal{R}$ is, at each instant, equal to the resultant $\mathbf{R}$ of the force system acting on P, that is,*

$$^{\mathcal{R}}\dot{\mathbf{G}}^P = \mathbf{R}. \tag{2.1}$$

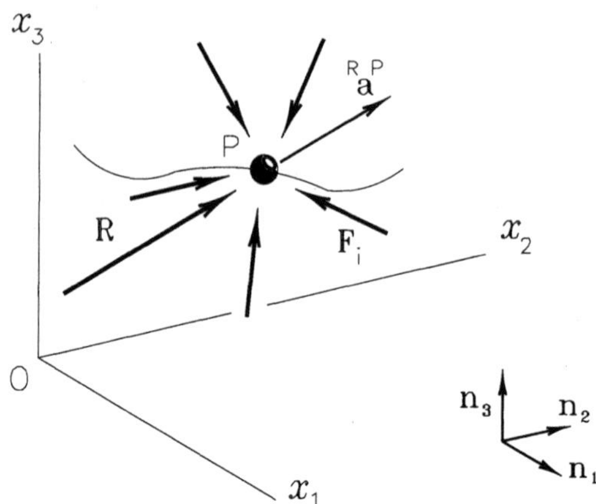

Figure 2.1

Note that, as formulated above, the law establishes the *existence* of reference frames — called *inertial reference frames* — for which the vector equality is valid between the time rate of the momentum and the resultant force, Eq. (2.1). Also note that, in this equation, the reduced notation has been adopted for the time rate; the reader should always bear in mind, however, that it is a differentiation in an inertial reference frame $\mathcal{R}$.

Using Eq. (1.6), we may express the dynamic equation alternatively in a form frequently used,

$$m\,^{\mathcal{R}}\mathbf{a}^P = \mathbf{R}, \tag{2.2}$$

which establishes that the product of the mass of a particle through its acceleration in an inertial reference frame $\mathcal{R}$ is vectorially equal to the resultant of the force system acting on it.

It is usually convenient to break down Eq. (2.2) into three scalar equations obtained from its projection on an orthnormal basis, parallel to a system of Cartesian axes chosen beforehand. So, if $\{x_1, x_2, x_3\}$ is a convenient system of axes for the study of the motion of a particle P, with mass m, and $\mathbf{n}_1, \mathbf{n}_2, \mathbf{n}_3$ is an orthonormal basis parallel to the axes (see Fig. 2.1), then by dot-multiplying Eq. (2.2) sucessively by each basis element:

$$ma_1 = R_1;$$
$$ma_2 = R_2; \qquad (2.3)$$
$$ma_3 = R_3,$$

where a_j and R_j, $j = 1, 2, 3$, are the scalar components of ${}^{\mathcal{R}}\mathbf{a}^P$ and $\mathbf{R}$, respectively, on the chosen basis. Equations (2.3) are usually called *equations of motion* for the particle P.

Equations. (2.3) involve acceleration and forces. Time rates of first and second order of the scalar functions that describe the position of the particle in the reference frame in question are always present in the components of acceleration, as discussed in Chapter 3. For these functions, the *coordinates*, the equations of motion are, therefore, second-order ordinary differential equations, generally not linear. For the components of forces, when unknown, these same equations are, however, simple algebraic equations.

Example 2.1 Figure 2.2 illustrates a curved pipe B, with its axis in the form of a quarter of a circle, turning around the vertical axis z in the reference frame A, at a simple angular velocity of a constant module ω, in the indicated direction. A small sphere P, with mass m, is sliding inside the pipe with negligible friction, under the action of its weight, $m\mathbf{g}$, and force $\mathbf{F}$, applied by the internal contact with the pipe. The basis $\mathbf{n}_1, \mathbf{n}_2, \mathbf{n}_3$, with $\mathbf{n}_1$ directed from O to P and $\mathbf{n}_3$ orthogonal to the plane containing the pipe's axis, and the basis $\mathbf{b}_1, \mathbf{b}_2, \mathbf{b}_3$, fixed in B, with $\mathbf{b}_1$ parallel to the axis z and $\mathbf{b}_3 = \mathbf{n}_3$, are shown in the figure. Ignoring the friction component, in the direction $\mathbf{n}_2$, the resultant of the force system applied to P is

$$\mathbf{R} = (F_1 + mg\sin\theta)\mathbf{n}_1 + mg\cos\theta\,\mathbf{n}_2 + F_3\mathbf{n}_3. \qquad \text{(a)}$$

The acceleration of P in the reference frame A may be obtained from the

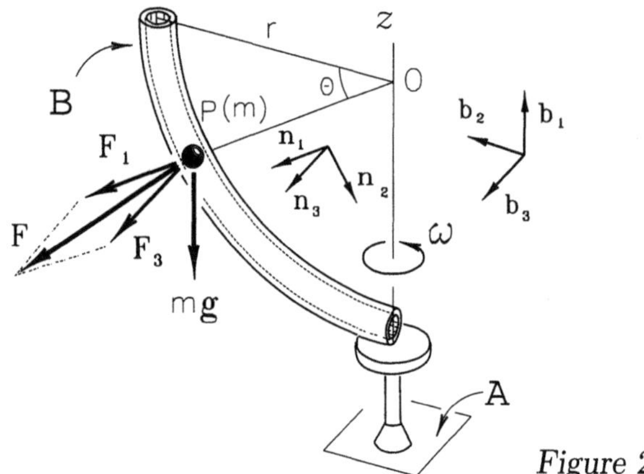

Figure 2.2

kinematic theorem, Eq. (3.6.3),

$$
\begin{aligned}
{}^{A}\mathbf{a}^{P} &= {}^{B}\mathbf{a}^{P} + {}^{A}\mathbf{a}^{O} + {}^{A}\boldsymbol{\omega}^{B} \times ({}^{A}\boldsymbol{\omega}^{B} \times \mathbf{p}^{P/O}) + {}^{A}\boldsymbol{\alpha}^{B} \times \mathbf{p}^{P/O} + 2\,{}^{A}\boldsymbol{\omega}^{B} \times {}^{B}\mathbf{v}^{P} \\
&= r(-\dot{\theta}^2 \mathbf{n}_1 + \ddot{\theta}\mathbf{n}_2) + 0 + \omega\mathbf{b}_1 \times (\omega\mathbf{b}_1 \times r\mathbf{n}_1) + 0 + 2\omega\mathbf{b}_1 \times r\dot{\theta}\mathbf{n}_2 \quad (b) \\
&= r\left[- (\dot{\theta}^2 + \omega^2 \cos^2\theta)\mathbf{n}_1 + (\ddot{\theta} + \omega^2 \cos\theta \sin\theta)\mathbf{n}_2 - 2\omega\dot{\theta} \sin\theta\mathbf{n}_3 \right].
\end{aligned}
$$

Assuming then that the reference frame A is inertial, expressions (a) and (b) may be substituted in Eqs. (2.3) to obtain the following set of equations of motion (check):

$$
\dot{\theta}^2 + \omega^2 \cos^2\theta = -\frac{F_1}{mr} - \frac{g}{r}\sin\theta; \tag{c}
$$

$$
\ddot{\theta} + \omega^2 \cos\theta \sin\theta = \frac{g}{r}\cos\theta; \tag{d}
$$

$$
2\omega\dot{\theta}\sin\theta = -\frac{F_3}{mr}. \tag{e}
$$

Once ω is prescribed, therefore known, the motion of the particle is fully determined by the coordinate $\theta(t)$. Equation (d) is a nonlinear *ordinary second-order differential equation* for the variable $\theta(t)$, of the type

$$
\ddot{\theta}(t) + f(\theta) = 0, \qquad \text{where} \qquad f(\theta) = \left(\omega^2 \sin\theta - \frac{g}{r}\right)\cos\theta,
$$

whose integration would lead to the desired function. The integration of this equation, given the nonlinearity, must be performed numerically. Once $\theta(t)$ is determined and, therefore, $\dot{\theta}(t)$, equations (c) and (e) are *algebraic equations* for the force components F_1 and F_3. See the corresponding animation.

The inertial reference frames, also known as *Newtonian reference frames* — whose existence is necessary for the correct application of the second law — are, in fact, a merely formal concept. The so-called *astronomical reference frame*, fixed in a certain constellation of far-off stars, is usually considered to be the Newtonian reference frame par excellence, in all cases where classic mechanics is applied; it is not, however, suitable for solving commonplace engineering problems, given the complexity in determining accelerations relative to such a reference frame. In practice, only experience can say whether a given adopted reference frame is inertial. In other words, if the motion of a particle, used in the solution of Eq. (2.1) or (2.2), results, for all practical purposes, according to the measurements, then the reference frame adopted for calculating the velocities and accelerations, *in that situation*, is inertial. For the vast majority of engineering applications, the *earth* is an excellent inertial reference frame, that is, the errors from ignoring its motion in outer space are generally less than the errors intrinsic to the observation procedures.

A famous experiment performed by Foucault[1] in 1851, in the *Pantheon*, in Paris, gives a good idea of the question of choice of a reference frame, as we examine in the following example. This experiment, reproduced later by several researchers in different places, is today a demonstrative apparatus in a number of science museums all over the world.

Example 2.2   Figure 2.3 illustrates the earth, T, considered here as a sphere, with center G, turning around its axis $\mathcal{N}$–$\mathcal{S}$ in a reference frame $\mathcal{R}$, making a full turn each sidereal day (86,164 s), with G fixed in $\mathcal{R}$. Earth has, therefore, a simple angular velocity in $\mathcal{R}$, ${}^{\mathcal{R}}\boldsymbol{\omega}^{T} = \omega\mathbf{n}$, where $\mathbf{n}$ is a unit vector in the direction of the earth's polar axis and $\omega = 2\pi/86,164 \approx 7.3 \times 10^{-5}$ rad/s. The Z-axis contains the center G and makes an angle ψ with the polar axis, intercepting the earth's surface at point O; point Q is on the Z-axis, a distant from point O. P is a pendulum with mass m, suspended at Q by a string with length r. The orthonormal basis $\mathbf{n}_1, \mathbf{n}_2, \mathbf{n}_3$ is fixed in T, with directions as shown. Figure 2.4 shows in greater detail the pendulum on the earth's surface, whose position in T may be described by the angles ϕ and θ, as shown.

---

[1] Jean Bernard Leon Foucault, French physicist, 1819–1868.

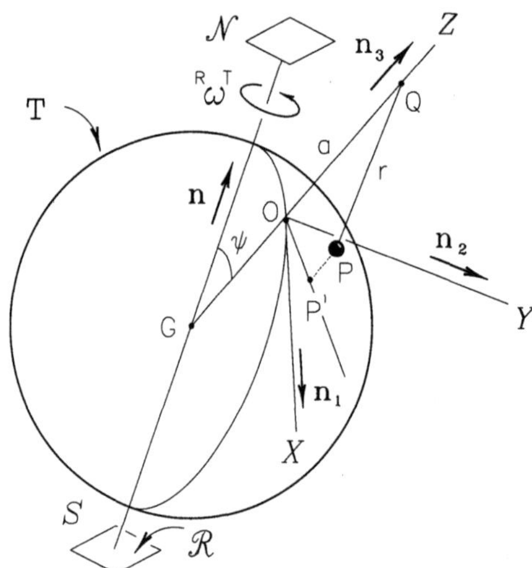

Figure 2.3

The basis $\mathbf{b}_1, \mathbf{b}_2, \mathbf{b}_3$ has $\mathbf{b}_3 = \mathbf{n}_3$, being turned at the angle ϕ, with regard to the basis $\mathbf{n}_1, \mathbf{n}_2, \mathbf{n}_3$; the unit vector $\mathbf{k}$ is oriented from P to Q.

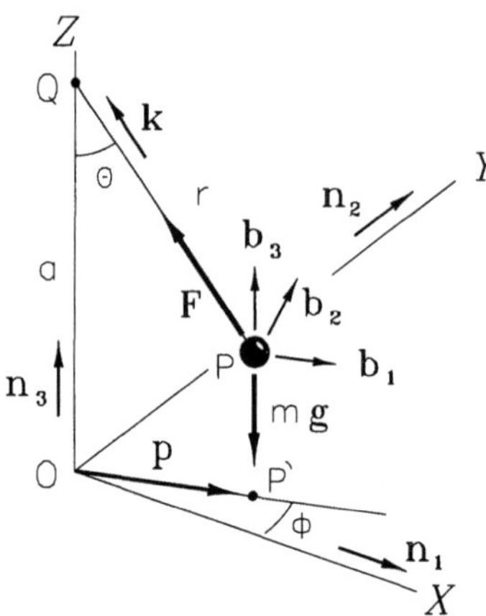

Figure 2.4

We will now study the motion of the pendulum under the action of the earth's gravitational force, $-mg\mathbf{n}_3$, and of the traction on its support, $\mathbf{F} = F\mathbf{k}$, ignoring other effects (the mass of the string, resistance of air, gravitational pull of the moon, etc.), assuming T and, later, $\mathcal{R}$ as inertial reference frames. The resultant force exerted on P is then

$$\mathbf{R} = F\mathbf{k} - mg\mathbf{n}_3 = -F\sin\theta\,\mathbf{b}_1 + (F\cos\theta - mg)\mathbf{b}_3.$$

To determine the acceleration of P in T it is convenient to adopt an intermediary reference frame, B, consisting of the plane containing O, Q, and P, with which the orthonormal basis $\mathbf{b}_1, \mathbf{b}_2, \mathbf{b}_3$ is associated. It is easy to see that ${}^T\boldsymbol{\omega}^B = \dot{\phi}\mathbf{n}_3$ and ${}^T\boldsymbol{\alpha}^B = \ddot{\phi}\mathbf{n}_3$. This gives

$$
\begin{aligned}
{}^T\mathbf{a}^P &= {}^B\mathbf{a}^P + {}^T\mathbf{a}^Q + {}^T\boldsymbol{\omega}^B \times ({}^T\boldsymbol{\omega}^B \times \mathbf{p}^{P/Q}) + {}^T\boldsymbol{\alpha}^B \times \mathbf{p}^{P/Q} + 2\,{}^T\boldsymbol{\omega}^B \times {}^B\mathbf{v}^P \\
&= r\ddot{\theta}(\cos\theta\,\mathbf{b}_1 + \sin\theta\,\mathbf{b}_3) + r\dot{\theta}^2\mathbf{k} + 0 + \dot{\phi}\mathbf{n}_3 \times (\dot{\phi}\mathbf{n}_3 \times (-r)\mathbf{k}) \\
&\quad + \ddot{\phi}\mathbf{n}_3 \times (-r)\mathbf{k} + 2\dot{\phi}\mathbf{n}_3 \times r\dot{\theta}(\cos\theta\,\mathbf{b}_1 + \sin\theta\,\mathbf{b}_3) \\
&= r\Big[(\ddot{\theta}\cos\theta - \dot{\theta}^2\sin\theta - \dot{\phi}^2\sin\theta)\mathbf{b}_1 + (\ddot{\phi}\sin\theta + 2\dot{\phi}\dot{\theta}\cos\theta)\mathbf{b}_2 \\
&\quad + (\ddot{\theta}\sin\theta + \dot{\theta}^2\cos\theta)\mathbf{b}_3\Big].
\end{aligned}
$$

Equation (2.2), expressed in components on the basis $\mathbf{b}_1, \mathbf{b}_2, \mathbf{b}_3$, is as follows:

$$
\ddot{\theta}\cos\theta - (\dot{\theta}^2 + \dot{\phi}^2)\sin\theta = -\frac{F}{mr}\sin\theta, \tag{a}
$$

$$
\ddot{\phi}\sin\theta + 2\dot{\phi}\dot{\theta}\cos\theta = 0, \tag{b}
$$

$$
\ddot{\theta}\sin\theta + \dot{\theta}^2\cos\theta = \frac{F}{mr}\cos\theta - \frac{g}{r}. \tag{c}
$$

Equation (b) is satisfied for $\ddot{\phi} = \dot{\phi} = 0$, that is, B is fixed in T and the trajectory of the pendulum in relation to the earth always stays on the same vertical plane. The system of equations is then reduced to

$$
\ddot{\theta}\cos\theta - \dot{\theta}^2\sin\theta = -\frac{F}{mr}\sin\theta, \tag{d}
$$

$$
\ddot{\theta}\sin\theta + \dot{\theta}^2\cos\theta = \frac{F}{mr}\cos\theta - \frac{g}{r}. \tag{e}
$$

Multiplying both members of Eq. (d) by $\cos\theta$ and of Eq. (e) by $\sin\theta$ and adding the members to each other, then

$$
\ddot{\theta} + \frac{g}{r}\sin\theta = 0,
$$

a nonlinear ordinary differential equation, governing the motion of the pendulum on the vertical plane. The general solution for this equation is fairly arduous, requiring the use of elliptic functions. It may be shown that the solution is periodic with period τ, depending on the initial amplitude θ_0 of the motion, so that

$$
\tau = 4\sqrt{\frac{r}{g}}\int_0^{\frac{\pi}{2}} (1 - k^2\sin^2\beta)^{-1/2}\,d\beta,
$$

where

$$k = \sin\frac{\theta_0}{2}, \qquad \beta = \sin^{-1}\frac{\sin\theta/2}{k}.$$

For small oscillations, however, the differential equation may be linearized. By making the approximation $\sin\theta \approx \theta$, then we obtain

$$\ddot{\theta} + \frac{g}{r}\theta = 0,$$

whose general solution is harmonic, so that

$$\theta(t) = A\cos(\omega_0 t + \varphi),$$

where

$$\omega_0 = \sqrt{\frac{g}{r}}$$

is the natural frequency of the pendulum and the amplitude A and the phase φ will depend on the initial conditions.

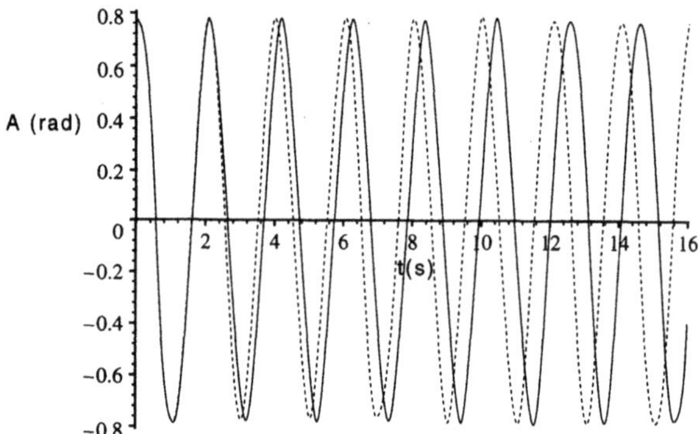

Figure 2.5

For instance, if the pendulum is left at rest with an angular displacement of θ_0, the result is $\varphi = 0$ and $A = \theta_0$. In short, leaving the pendulum with null velocity in relation to the earth and angular displacements ϕ_0 and θ_0, the latter small, the motion expected by equation of motion, taking the earth as an inertial reference frame, consists of

$$\phi(t) = \phi_0, \qquad \theta(t) = \theta_0 \cos\omega_0 t,$$

that is, a slight harmonic motion on a vertical plane in an invariant orientation. Figure 2.5 illustrates the result of the numerical integration of the nonlinearized differential equation, $\ddot{\theta} + \frac{g}{r}\sin\theta = 0$, for $r = 1$ m, showing the periodic behavior (but not harmonic) of the solution, for the initial conditions $\theta_0 = \pi/4$ rad and $\dot{\theta}(0) = 0$. The figure shows $\theta(t)$ (continuous line) for the first 16 seconds, compared with the harmonic solution (broken line) for the same initial conditions. Observe the delay in the nonlinearized equation solution in relation to that of the linearized equation. Experiments conducted with very long pendulums with substantial mass, such as that performed by Foucault, demonstrate, however, that the vertical plane rotates slowly, that is, $\dot{\phi} \neq 0$, the difference between the expected and the observed behavior being a consequence of ignoring the earth's rotation motion around its axis, as will be seen below. So, now taking $\mathcal{R}$ as an inertial reference frame, the acceleration of P in this reference frame shall be calculated, that is,

$$\mathcal{R}\mathbf{a}^P = {}^T\mathbf{a}^P + \mathcal{R}\mathbf{a}^G + \mathcal{R}\boldsymbol{\omega}^T \times (\mathcal{R}\boldsymbol{\omega}^T \times \mathbf{p}^{P/G}) + \mathcal{R}\boldsymbol{\alpha}^T \times \mathbf{p}^{P/G} + 2\,\mathcal{R}\boldsymbol{\omega}^T \times {}^T\mathbf{v}^P.$$

Some simplifying considerations are given here. The average radius of the earth is $R \approx 6.37 \times 10^6$ m and $\omega^2 R \approx 0.0339$ m/s^2, that is, around 0.35% of the gravitational acceleration, which means that the double cross product in the above expression may be ignored. Assuming that the earth's angular velocity is constant in the reference frame $\mathcal{R}$, then $\mathcal{R}\boldsymbol{\alpha}^T = 0$. As, lastly, $\mathcal{R}\mathbf{a}^G = 0$, only the contribution of the acceleration of Coriolis needs to be added to the pendulum's acceleration in relation to the earth:

$$
\begin{aligned}
2\mathcal{R}\boldsymbol{\omega}^T \times {}^T\mathbf{v}^P &= 2\omega\mathbf{n} \times ({}^B\mathbf{v}^P + {}^T\mathbf{v}^Q + {}^T\boldsymbol{\omega}^B \times \mathbf{p}^{P/Q}) \\
&= 2\omega(-\cos\phi\sin\psi\mathbf{b}_1 + \sin\phi\sin\psi\mathbf{b}_2 + \cos\psi\mathbf{b}_3) \\
&\quad \times r(\dot{\theta}\cos\theta\mathbf{b}_1 + \dot{\phi}\sin\theta\mathbf{b}_2 + \dot{\theta}\sin\theta\mathbf{b}_3) \\
&= 2r\omega\big[(\dot{\theta}\sin\psi\sin\phi - \dot{\phi}\cos\psi)\sin\theta\,\mathbf{b}_1 \\
&\quad + (\cos\psi\cos\theta + \sin\psi\cos\phi\sin\theta)\dot{\theta}\,\mathbf{b}_2 \\
&\quad - (\dot{\phi}\cos\phi\sin\theta + \dot{\theta}\sin\phi\cos\theta)\sin\psi\,\mathbf{b}_3\big].
\end{aligned}
$$

Combining the corresponding terms and substituting them in Eq. (2.2), the system of equations governing the pendulum's motion, now assuming $\mathcal{R}$ as an inertial reference frame, is

$$\ddot{\theta}\cos\theta - (\dot{\theta}^2 + \dot{\phi}^2)\sin\theta + 2\omega\sin\theta(\dot{\theta}\sin\psi\sin\phi - \dot{\phi}\cos\psi) = -\frac{F}{mr}\sin\theta, \qquad \text{(f)}$$

$$\ddot{\phi}\sin\theta + 2\dot{\phi}\dot{\theta}\cos\theta + 2\omega\dot{\theta}(\cos\psi\cos\theta + \sin\psi\cos\phi\sin\theta) = 0 \qquad \text{(g)}$$

$$\ddot{\theta}\sin\theta + \dot{\theta}^2\cos\theta - 2\omega\sin\psi(\dot{\phi}\cos\phi\sin\theta + \dot{\theta}\sin\phi\cos\theta) = \frac{F}{mr}\cos\theta - \frac{g}{r}. \qquad \text{(h)}$$

Note that Eqs. (f–h) are reduced to Eqs. (a–c) when $\omega = 0$. Now assuming small amplitudes for the variable $\theta(t)$, the system of equations above may be linearized. So, by approximating $\sin\theta \approx \theta$ and $\cos\theta \approx 1$, ignoring terms of second order in θ and $\dot\theta$ and removing the force F by substituting (f) in (h), then (check)

$$\ddot\phi\theta + 2\dot\theta(\dot\phi + \omega\cos\psi) = 0, \tag{i}$$

$$\ddot\theta + \theta\Big(\frac{g}{r} - \dot\phi^2 - 2\omega\dot\phi\cos\psi\Big) = 0. \tag{j}$$

Equations (i) and (j) are a couple of nonlinear ordinary differential equations (although already linearized in θ) for the coordinates $\theta(t)$ and $\phi(t)$, whose solution will be analyzed below. Essentially, the pendulum's motion will be known if the motion of its projection P$'$ on the horizontal plane is known. So let $\mathbf{p}$ be the position vector of P$'$ with respect to O (see Fig. 2.4),

$$\mathbf{p} = r\sin\theta\,\mathbf{b}_1 \approx r\theta\mathbf{b}_1,$$

whose first and second time rates in the reference frame T are

$$\frac{{}^{T}d}{dt}\mathbf{p} = r(\cos\theta\dot\theta\mathbf{b}_1 + \sin\theta\dot\phi\mathbf{b}_2) \approx r(\dot\theta\mathbf{b}_1 + \dot\phi\theta\mathbf{b}_2),$$

$$\frac{{}^{T}d^2}{dt^2}\mathbf{p} = r\Big[\big((\ddot\theta\cos\theta - (\dot\theta^2 + \dot\phi^2)\sin\theta\big)\mathbf{b}_1 + \big(\ddot\phi\sin\theta + 2\dot\phi\dot\theta\cos\theta\big)\mathbf{b}_2\Big],$$

$$\approx r\Big[(\ddot\theta - \dot\phi^2\theta)\mathbf{b}_1 + (\ddot\phi\theta + 2\dot\phi\dot\theta)\mathbf{b}_2\Big].$$

It is easy to see, by substitution, that, if Eq. (i) and (j) are satisfied, then vector $\mathbf{p}$ satisfies

$$\frac{{}^{T}d^2}{dt^2}\mathbf{p} + 2\omega\cos\psi\mathbf{b}_3 \times \frac{{}^{T}d}{dt}\mathbf{p} + \omega_0^2\mathbf{p} = 0. \tag{k}$$

To verify, substitute $\mathbf{p}$ and its derivatives in Eq. (k) and combine the terms in the directions of $\mathbf{b}_1$ and $\mathbf{b}_2$, comparing them with the expressions present in Eqs. (i) and (j), respectively. To recognize more easily how vector $\mathbf{p}$ behaves, it is convenient to introduce another reference frame A, which moves in relation to the earth at a constant simple angular velocity ${}^{T}\boldsymbol{\omega}^{A} = -\omega\cos\psi\mathbf{b}_3$. So, using Eq. (3.3.6) to convert the time rates in the reference frame T to derivatives in the reference frame A, we then obtain

$$\frac{{}^{A}d}{dt}\mathbf{p} = \frac{{}^{T}d}{dt}\mathbf{p} + {}^{A}\boldsymbol{\omega}^{T} \times \mathbf{p},$$

$$\frac{{}^{A}d^2}{dt^2}\mathbf{p} = \frac{{}^{T}d}{dt}\Big(\frac{{}^{T}d}{dt}\mathbf{p} + {}^{A}\boldsymbol{\omega}^{T} \times \mathbf{p}\Big) + {}^{A}\boldsymbol{\omega}^{T} \times \Big(\frac{{}^{T}d}{dt}\mathbf{p} + {}^{A}\boldsymbol{\omega}^{T} \times \mathbf{p}\Big)$$

$$= \frac{{}^{T}d^2}{dt^2}\mathbf{p} + 2{}^{A}\boldsymbol{\omega}^{T} \times \frac{{}^{T}d}{dt}\mathbf{p} + {}^{A}\boldsymbol{\omega}^{T} \times \big({}^{A}\boldsymbol{\omega}^{T} \times \mathbf{p}\big).$$

Ignoring the double product involving ω^2, and substituting it in Eq. (k), with ${}^A\boldsymbol{\omega}^T = -{}^T\boldsymbol{\omega}^A = \omega\cos\psi\mathbf{b}_3$, then we get

$$\frac{{}^A d^2}{dt^2}\mathbf{p} + \omega_0^2\mathbf{p} = 0. \tag{1}$$

Equation (1) has a known general solution, as follows:

$$\mathbf{p}(t) = \lambda_1\cos\omega_0 t\,\mathbf{a}_1 + \lambda_2\sin\omega_0 t\,\mathbf{a}_2, \tag{m}$$

where $\mathbf{a}_1$ and $\mathbf{a}_2$ are horizontal unit vectors, fixed in A, and λ_1 and λ_2 are constants, all depending on the initial conditions. By leaving the pendulum at rest in relation to the earth, and with a slight separation θ_0 in an arbitrary direction ϕ_0 (see Fig 2.6), then

$$\mathbf{p}(0) = \lambda_1\mathbf{a}_1 = r\theta_0\mathbf{b}_1.$$

Therefore, at the start of the movement,

$$\lambda_1 = r\theta_0; \qquad \mathbf{a}_1 = \mathbf{b}_1.$$

The velocity of the point P' in the earth is

$$\begin{aligned}
{}^T\mathbf{v}^{P'} &= \frac{{}^T d}{dt}\mathbf{p} = \frac{{}^A d}{dt}\mathbf{p} + {}^T\boldsymbol{\omega}^A \times \mathbf{p} \\
&= \omega_0(-\lambda_1\sin\omega_0 t\,\mathbf{a}_1 + \lambda_2\cos\omega_0 t\,\mathbf{a}_2) \\
&\quad - \omega\cos\psi\mathbf{b}_3 \times (\lambda_1\cos\omega_0 t\,\mathbf{a}_1 + \lambda_2\sin\omega_0 t\,\mathbf{a}_2).
\end{aligned}$$

The initial condition of rest in relation to the earth then implies

$$\omega_0\lambda_2\mathbf{a}_2 - \omega\cos\psi\lambda_1\mathbf{b}_3 \times \mathbf{a}_1 = 0;$$

therefore, at the initial instant,

$$\lambda_2 = r\frac{\omega}{\omega_0}\cos\psi\theta_0; \qquad \mathbf{a}_2 = \mathbf{b}_2.$$

The point P' then describes a trajectory on the horizontal plane described by

$$\mathbf{p} = r\theta_0\left(\cos\omega_0 t\,\mathbf{a}_1 + \frac{\omega}{\omega_0}\cos\psi\sin\omega_0 t\,\mathbf{a}_2\right),$$

while the basis $\mathbf{a}_1, \mathbf{a}_2, \mathbf{a}_3$ rotates around the vertical axis in relation to the earth according to $\dot\phi = -\omega\cos\psi$.

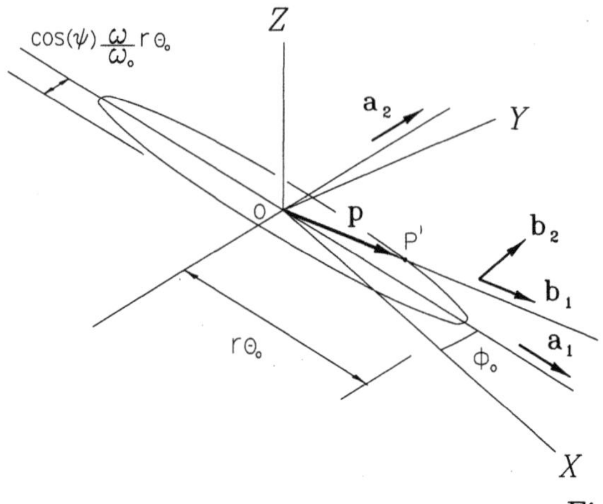

Figure 2.6

The resulting trajectory is, therefore, an ellipse with center at O and semi-axes $r\theta_0$ (the largest) and $r\theta_0 \cos\psi(\omega/\omega_0)$ (the smallest), rotating in a clockwise direction if the experiment is conducted in the northern hemisphere ($\cos\psi > 0$) and in a counterclockwise direction in the southern hemisphere. It is important to note that the eccentricity of the ellipse is very large, with the ratio between its semi-axes (the smallest over the largest) given by $k = (\omega/\omega_0)\cos\psi$, the smallest semi-axis being hard to see with the naked eye. Near the equator, the term in the direction a_2 vanishes and P' describes a straight trajectory fixed on T, as expected when the earth's rotation motion was ignored; in fact, close to the equator, $^{\mathcal{R}}\boldsymbol{\omega}^T$ and $^T\mathbf{v}^P$ are essentially parallel and the contribution of the acceleration of Coriolis vanishes. Although $\mathcal{R}$ does not coincide with the astronomic reference frame, careful experiments have demonstrated a behavior very close to that predicted in the above analysis. This example illustrates how small the error induced by the earth's motion is. For a small pendulum, with a length of say, 1 m, at a latitude of 30^0, then

$$k = \frac{\omega}{\omega_0}\cos\psi = \frac{7.3\times 10^{-5}}{\sqrt{9.8/1}}\cos\frac{\pi}{6} \approx 2.02\times 10^{-5},$$

that is, the smallest axis of the ellipse would be 20 millionths of the largest, while the rotation rate of the axes in relation to the spectator on the earth would be, in module, of the order of

$$\omega\cos\psi = 7.3\times 10^{-5}\times\cos\frac{\pi}{6} \approx 6.3\times 10^{-5}\ \text{rad/s}.$$

See the corresponding animation.

Equation (2.3) comes from Eq. (2.2), through decomposition on a suitable basis. The correct choice of basis may, in most cases, considerably facilitate the resolution of the equation system. There is no general rule to follow when selecting the coordinated directions, and it depends a lot on the experience in solving problems. As a rule of thumb, however, it is normally more effective to adopt an orthonormal basis that, somehow, *accompanies* the motion of the particle in the inertial reference frame. They are usually fixed in an intermediary reference frame, explicit in the context or not. The following example illustrates the use of the intrinsic basis (see Section 3.7) in setting up the equations of motion.

Example 2.3 Consider the motion of a small sphere P, with mass m, smoothly sliding over a rigid wire in the shape of a helix with radius r and constant pitch a (see Fig. 2.7). Let P_0 be P's starting point; O a fixed point on the axis of the helix at the same elevation of P_0; Q a mobile point on the axis of the helix with, at each instant, the same elevation of P (that is, the straight lines passing through P_0 and O and through P and Q are horizontal). The coordinate $\phi(t)$, an angle between the aforementioned straight lines, fully describes the motion of P on the wire. It is convenient, in this case, to choose the basis $\mathbf{n}_1, \mathbf{n}_2, \mathbf{n}_3$, as illustrated, with $\mathbf{n}_1$ horizontal, pointing to the axis of the helix, and $\mathbf{n}_2$ tangent to the curve at point P. This is, of course, a mere inversion of the intrinsic basis for describing the motion of P on the helix, $\mathbf{n}_1$ being the principal normal, $\mathbf{n}_2$ the tangent unit vector, and $\mathbf{n}_3$ opposite the binormal vector. (See Examples 7.1, 7.2, 7.4, and 7.6 in Section 3.7.) As discussed earlier, the displacement along the helix is made by $s(t) = b\phi(t)$, where $b = (r^2 + (a/2\pi)^2)^{1/2}$, and its curvature, constant, is $\kappa = r/b^2$. The acceleration of P may then be expressed by

$$\mathbf{a}^P = \dot{s}^2\kappa\mathbf{n}_1 + \ddot{s}\mathbf{n}_2$$
$$= r\dot{\phi}^2\mathbf{n}_1 + b\ddot{\phi}\mathbf{n}_2.$$

The effect of the wire's contact, represented by two force components, F_1 and F_3 (F_2 does not appear since the friction is ignored) and its weight, acts on P. The resultant of the force system, expressed on the chosen basis, is then

$$\mathbf{R} = F_1\mathbf{n}_1 + mg\sin\theta\mathbf{n}_2 + (F_3 - mg\cos\theta)\mathbf{n}_3,$$

where θ is the slope of the curve, with $\sin\theta = a/2\pi b$, $\cos\theta = r/b$. The

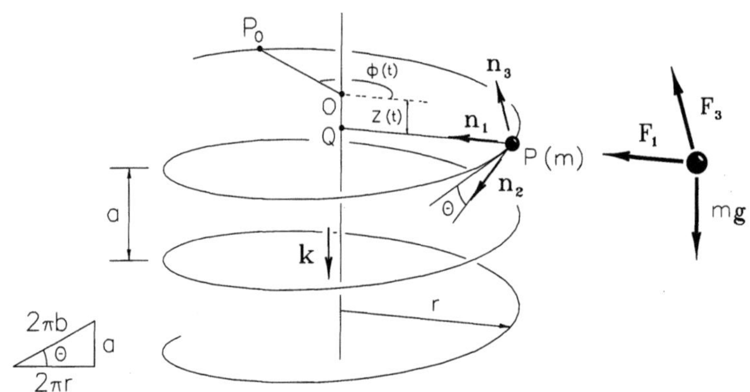

Figure 2.7

equations of motion for P are then

$$mr\dot\phi^2 = F_1,$$

$$mb\ddot\phi = mg\frac{a}{2\pi b},$$

$$0 = F_3 - mg\frac{r}{b}.$$

These expressions enhance the convenience of the chosen coordinated directions. In fact, the equations are uncoupled, which facilitates the solution. The third is an algebraic equation, which immediately determines $F_3 = mg(r/b)$. The second equation provides $\ddot\phi = (ga/2\pi b^2)$, and, directly integrated, results in

$$\phi(t) = \frac{ga}{4\pi b^2}t^2 + At + B,$$

where the constants A and B are determined by the initial conditions. Assuming that the sphere is left from rest in the position $\phi(0) = 0$, then $A = B = 0$ and P's motion is described by

$$\phi(t) = \frac{ga}{4\pi b^2}t^2.$$

Finally, the first equation of motion determines the component F_1:

$$F_1 = mr\dot\phi^2 = \frac{mrg^2a^2}{4\pi^2 b^4}t^2.$$

When the motion of a particle P in an inertial reference frame $\mathcal{R}$ is reproduced identically in a given constant interval of time τ, the motion is said to be *periodic*. For instance, the pendular motions discussed in Example 2.2 (with a short period, in the case of the earth being considered inertial and a long period, assuming $\mathcal{R}$ as an inertial reference frame) are periodic. In some cases, the periodic solution is satisfied only for some particular initial conditions.

Example 2.4   Consider a small sphere P, with mass m, hanging from a string, with length r, at a point Q fixed in an inertial reference frame $\mathcal{R}$ (see Fig. 2.8).

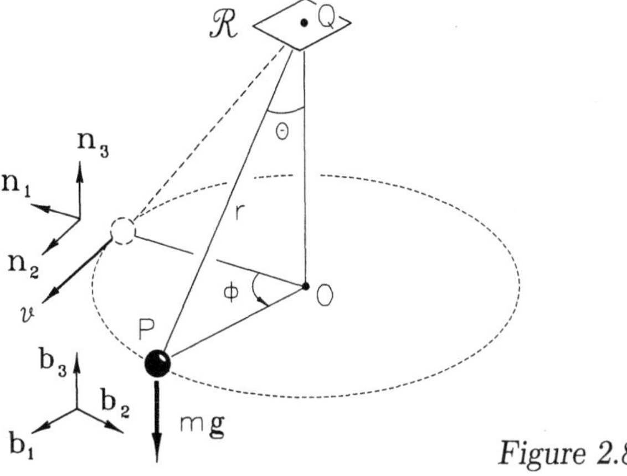

Figure 2.8

The sphere is taken from the vertical position and given an initial velocity of module v, horizontal, in the direction shown. It may be demonstrated that there is an actual value for v so that P stays in a uniform circular motion on the horizontal plane, as shown below. The orthonormal basis n_1, n_2, n_3 is fixed in $\mathcal{R}$, while the basis b_1, b_2, b_3 is fixed in a reference frame B containing the points P, Q, and O and that moves in $\mathcal{R}$ at a simple angular velocity $^{\mathcal{R}}\omega^{B} = \dot{\phi}n_3$. The acceleration of P in B is (check)

$$^{B}\mathbf{a}^{P} = r\ddot{\theta}(\cos\theta\mathbf{b}_1 + \sin\theta\mathbf{b}_3) + r\dot{\theta}^2(-\sin\theta\mathbf{b}_1 + \cos\theta\mathbf{b}_3).$$

The acceleration of P in $\mathcal{R}$ is then (check)

$$^{\mathcal{R}}\mathbf{a}^{P} = {}^{B}\mathbf{a}^{P} + {}^{\mathcal{R}}\mathbf{a}^{Q} + {}^{\mathcal{R}}\omega^{B} \times ({}^{\mathcal{R}}\omega^{B} \times \mathbf{p}^{P/Q}) + {}^{\mathcal{R}}\alpha^{B} \times \mathbf{p}^{P/Q} + 2\,{}^{\mathcal{R}}\omega^{B} \times {}^{B}\mathbf{v}^{P}$$

$$= r(\ddot{\theta}\cos\theta - \dot{\theta}^2\sin\theta - \dot{\phi}^2\sin\theta)\mathbf{b}_1$$

$$+ r(\ddot{\phi}\sin\theta + 2\dot{\phi}\dot{\theta}\cos\theta)\mathbf{b}_2$$

$$+ r(\ddot{\theta}\sin\theta + \dot{\theta}^2\cos\theta)\mathbf{b}_3.$$

The forces acting on P are its weight and the traction **T** exerted by the string, its resultant then being

$$\mathbf{R} = -T\sin\theta\,\mathbf{b}_1 + (T\cos\theta - mg)\mathbf{b}_3.$$

The equations governing the motion of P are therefore

$$\ddot{\theta}\cos\theta - (\dot{\theta}^2 + \dot{\phi}^2)\sin\theta = -\frac{T}{mr}\sin\theta, \qquad (\mathrm{a})$$

$$\ddot{\phi}\sin\theta + 2\dot{\phi}\dot{\theta}\cos\theta = 0, \qquad (\mathrm{b})$$

$$\ddot{\theta}\sin\theta + \dot{\theta}^2\cos\theta = \frac{T}{mr}\cos\theta - \frac{g}{r}. \qquad (\mathrm{c})$$

As it should be, the above equations are the same as those obtained in Example 2.2 when considering the earth as an inertial reference frame. In this case, however, we do not look for a solution for small oscillations but rather a solution with a constant $\theta(t)$. Assuming, then, $\ddot{\theta}(t) = \dot{\theta}(t) = 0$, the equations of motion are reduced to

$$\dot{\phi}^2 = \frac{T}{mr}, \qquad (\mathrm{d})$$

$$\ddot{\phi} = 0, \qquad (\mathrm{e})$$

$$T = \frac{mg}{\cos\theta}. \qquad (\mathrm{f})$$

Equation (e) may be directly integrated, leading to

$$\dot{\phi}(t) = \dot{\phi}(0) = \frac{v}{r\sin\theta}, \qquad \phi(t) = \frac{v}{r\sin\theta}t.$$

So when substituting Eq. (f) in Eq. (d) and solving for v, then

$$v = \sqrt{gr\sin\theta\tan\theta},$$

which therefore establishes the value of the initial velocity required to keep the motion at a prescribed constant angle θ. The motion is therefore periodic, with a period (check: You only need to calculate t for $\phi = 2\pi$)

$$\tau = 2\pi\sqrt{\frac{r\cos\theta}{g}}.$$

See the corresponding animation.

4.3 Plane Motion

When the trajectory of a particle P is kept on a plane fixed in an inertial reference frame, it is said that P has a *plane motion*. This happens if and only if the component of the resultant of the force system acting on P, in the direction orthogonal to the plane, is null. In fact, if π is a fixed plane in an inertial reference frame $\mathcal{R}$, and $\mathbf{n}_1, \mathbf{n}_2, \mathbf{n}_3$ is an orthonormal basis fixed in $\mathcal{R}$, with $\mathbf{n}_1$ and $\mathbf{n}_2$ parallel to π and $\mathbf{n}_3$ orthogonal to the plane (see Fig. 3.1), the condition for P's trajectory to remain contained on the plane is that, at any instant,

$$^{\mathcal{R}}\mathbf{v}^P \cdot \mathbf{n}_3 = 0. \tag{3.1}$$

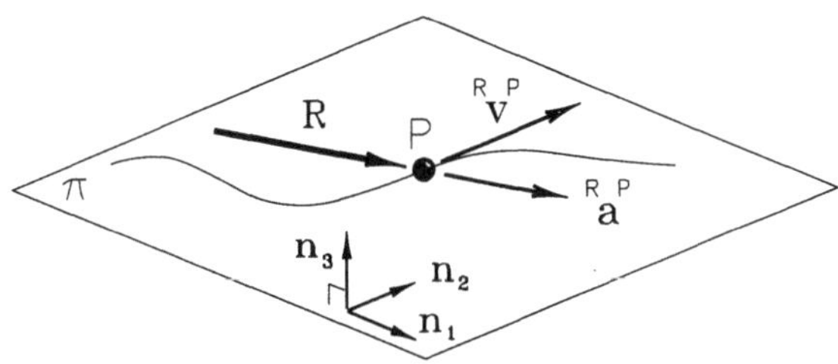

Figure 3.1

Therefore, $a_3 = {}^{\mathcal{R}}\mathbf{a}^P \cdot \mathbf{n}_3 = 0$, the third of Eqs. (2.3) being reduced to

$$R_3 = 0. \tag{3.2}$$

Inversely, if Eq. (3.2) is always satisfied (and $\mathbf{v}_3(t) = 0$ for some t), then $a_3 = 0$ and the trajectory of the particle in $\mathcal{R}$ remains on the plane, satisfying Eq. (3.1).

Example 3.1 Figure 3.2 reproduces the situation examined in Example 1.1 and subsequent examples, including a diagram of the forces acting on the cursor P. The cursor is kept on the plane of the table that, in turn, has a vertical simple angular velocity in $\mathcal{R}$. Assuming, therefore, $\mathcal{R}$ as an inertial reference frame, P has plane motion.

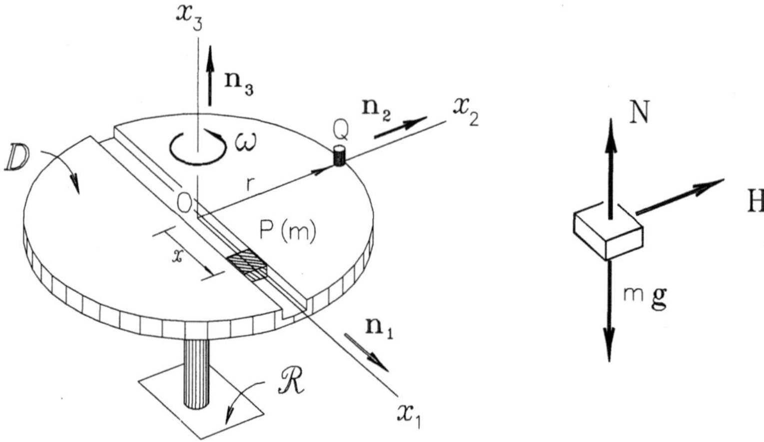

$$\textit{Figure 3.2}$$

Assuming that the effect of friction between the table and cursor is negligible, the forces on the cursor consist of a vertical component, $\mathbf{N}$, and another, horizontal, $\mathbf{H}$, orthogonal to the guide, exerted by the disk and the cursor's weight, $m\mathbf{g}$, as shown (the chosen direction for $\mathbf{H}$ is arbitrary, its true signal being determined, a posteriori, by solving the equations). To meet the condition of a plane motion, Eq. (3.2) must be satisfied, that is,

$$\mathbf{N} - m\mathbf{g} = 0; \qquad \text{therefore,} \qquad \mathbf{N} = mg\mathbf{n}_3.$$

The resultant force on the plane of motion is then

$$\mathbf{R} = H\mathbf{n}_2,$$

and the equations of motion for P are (see Example 1.1)

$$\ddot{x} - \omega^2 x = 0, \tag{a}$$

$$2m\omega\dot{x} = H. \tag{b}$$

So we have an ordinary differential equation for the variable $x(t)$ and an algebraic equation for the component H. Equation (a) has a known general solution, so that

$$x(t) = Ae^{\omega t} + Be^{-\omega t}, \tag{c}$$

where the constants A and B are determined by the initial conditions. The time rate of the variable $x(t)$ is

$$\dot{x}(t) = \omega(Ae^{\omega t} - Be^{-\omega t}). \tag{d}$$

Let us then assume that P is left at rest on the disk in the position $x_0 = r/4$, when $t = 0$. Introducing these conditions in expressions (c) and (d), we then obtain

$$x(0) = A + B = \frac{r}{4},$$

$$\frac{1}{\omega}\dot{x}(0) = A - B = 0,$$

resulting in

$$A = B = \frac{r}{8}.$$

The motion of P in D is therefore described by

$$x(t) = \frac{r}{8}(e^{\omega t} + e^{-\omega t}).$$

Equation (b) provides the solution for the horizontal component H:

$$H(t) = \frac{mr\omega^2}{4}(e^{\omega t} - e^{-\omega t}).$$

See the corresponding animation.

In many situations, even when suitably choosing the coordinated directions, the nonlinearity makes it difficult for a complete solution of the system of equations. Sometimes, however, when wishing to determine a force component, not as a function of time, but as a function of a spatial coordinate, it is not essential to know its time evolution. In the latter case, an option may be made to integrate the equations in the position variable, generally an easier task.

Example 3.2 Consider a pendulum P, with mass m and length r (see Fig 3.3), whose support C can slide over a horizontal guide, fixed in an inertial reference frame $\mathcal{R}$. The support moves with a constant prescribed acceleration a, in the indicated direction, and the pendulum is left in the horizontal position, at rest in relation to the support. We wish to determine the value of the traction T on the pendulum's string, as a function of the angular position θ.

We first must identify the force system acting on P, consisting in this case of its weight, $m\mathbf{g}$, and the action of the string, $T\mathbf{n}_2$, ignoring other minor effects. The resultant of the system is

$$\mathbf{R} = mg\cos\theta\,\mathbf{n}_1 + (T - mg\sin\theta)\mathbf{n}_2.$$

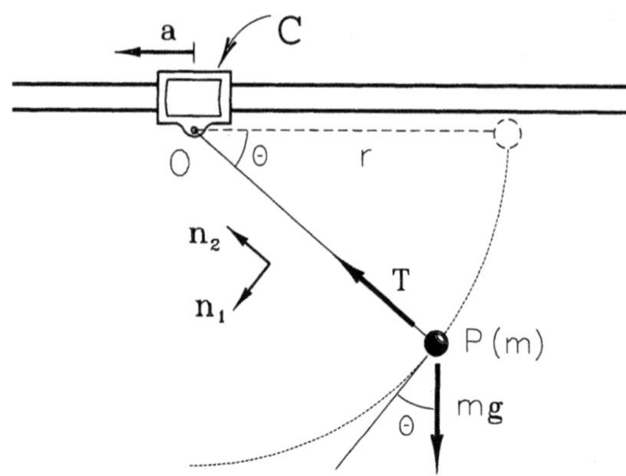

Figure 3.3

The acceleration of P in the reference frame $\mathcal{R}$ may be obtained from the kinematic theorem for accelerations, Eq. (3.6.3), which, as the body C moves in the reference frame $\mathcal{R}$, is reduced to

$$\begin{aligned}
{}^{\mathcal{R}}\mathbf{a}^P &= {}^{C}\mathbf{a}^P + {}^{\mathcal{R}}\mathbf{a}^O \\
&= r\ddot{\theta}\mathbf{n}_1 + r\dot{\theta}^2\mathbf{n}_2 + a(\sin\theta\,\mathbf{n}_1 + \cos\theta\,\mathbf{n}_2) \\
&= (r\ddot{\theta} + a\sin\theta)\mathbf{n}_1 + (r\dot{\theta}^2 + a\cos\theta)\mathbf{n}_2.
\end{aligned}$$

As the pendulum's trajectory and forces involved are parallel to the vertical plane of the figure, the motion is plane, being governed by two dynamic equations in the directions $\mathbf{n}_1$ and $\mathbf{n}_2$:

$$r\ddot{\theta} + a\sin\theta = g\cos\theta, \tag{a}$$

$$r\dot{\theta}^2 + a\cos\theta = \frac{T}{m} - g\sin\theta. \tag{b}$$

It is important to note that in this system of equations, the traction T on the string and the actual position of the particle, described by the scalar function $\theta(t)$, are unknown. In the case of force T, this is an algebraic system of equations; more precisely, due to the fortunate choice of the coordinated directions, we only need Eq. (b) to determine T. From the point of view of determining the angular position θ, there is a second-order nonlinear ordinary differential equation. In fact, reorganizing Eq. (a), we get

$$\ddot{\theta} + \frac{1}{r}(a\sin\theta - g\cos\theta) = 0, \tag{c}$$

which has a simple structure but is difficult to integrate in the time variable. It is worth mentioning that if $a = 0$, we return to the differential equation for the simple pendulum, discussed in Example 2.2, whose harmonic solution, for small oscillations, is not suitable for this case, where the initial condition moves away $\pi/2$ rad from the vertical. As we wish to know $T(\theta)$, the convenient alternative is to integrate Eq. (c) along the angular position, that is,

$$\int \ddot{\theta}\, d\theta + \frac{1}{r} \int (a\sin\theta - g\cos\theta)\, d\theta = C,$$

where C is a constant to be determined. By then calculating the integrals, we obtain

$$\dot{\theta}^2 = \frac{2}{r}(a\cos\theta + g\sin\theta) + 2C.$$

The integration constant must be determined from the given initial condition, $\dot{\theta}(0) = 0$, that is,

$$0 = \frac{2}{r}a + 2C; \qquad \text{therefore,} \qquad C = -\frac{a}{r}.$$

Therefore,

$$\dot{\theta}^2 = \frac{2}{r}\big(a(\cos\theta - 1) + g\sin\theta\big).$$

Now substituting in Eq. (b) and solving for T, we reach the result

$$T = m\big(3g\sin\theta + a(3\cos\theta - 2)\big).$$

On examining the example a little further, we may research the position in which the traction in the string is maximum. Differentiating the above equation with respect to θ, we obtain

$$\frac{dT}{d\theta} = 3m(g\cos\theta - a\sin\theta),$$

and there will be maximum traction when $\theta = \tan^{-1}(g/a)$. Also note that if the pendulum's support is fixed in the inertial reference frame, maximum traction, $T = 3mg$, will occur when $\theta = \pi/2$. See the corresponding animation.

The integration of the equations of motion along the trajectory, that is, along a spatial coordinate, is an extremely useful tool in problem-solving, mainly when wishing to extract information about a certain configuration, once an initial condition is known. In other words, integration along the position variable is generally easier than integration in the time variable. Of course, this procedure will only be satisfactory if we wish to know forces or velocities as a function of positions, and it is not essential to determine the position variable of the particle in function of time. Section 4.6 discusses this subject in further detail.

4.4 Angular Momentum

Let P be a particle, with mass m, moving in an inertial reference frame
$\mathcal{R}$ under the action of a system $\mathcal{F}$ of forces, whose resultant is $\mathbf{R}$ and
whose resultant moment with respect to a point O, fixed in $\mathcal{R}$, is $\mathbf{M}^{\mathcal{F}/O}$
(see Fig. 4.1).

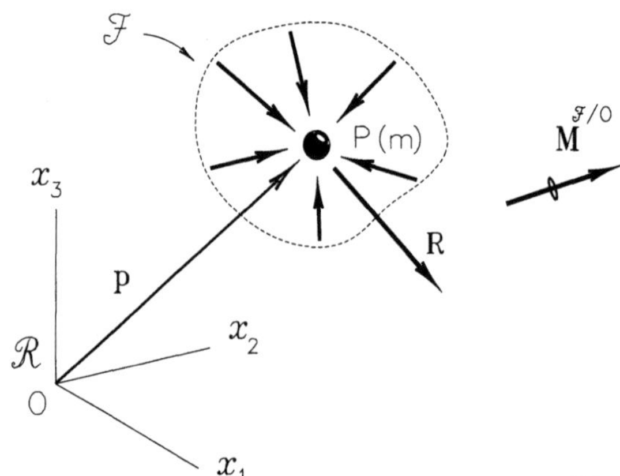

Figure 4.1

As the force system acting on a particle is always concurrent,
the resultant moment may be expressed as (see Section 2.5)

$$\mathbf{M}^{\mathcal{F}/O} = \mathbf{p} \times \mathbf{R}, \tag{4.1}$$

where $\mathbf{p}$ is the position vector of P with respect to O. The resultant of
the force system and the time rate of the momentum of the particle are
both vectors linked to P; the momentum, therefore, with respect to O
of both members of Eq. (2.1) may be calculated, resulting in

$$\mathbf{p} \times {}^{\mathcal{R}}\dot{\mathbf{G}}^P = \mathbf{p} \times \mathbf{R}. \tag{4.2}$$

As, according to Eq. (1.12), $\mathbf{p} \times {}^{\mathcal{R}}\dot{\mathbf{G}}^P = {}^{\mathcal{R}}\dot{\mathbf{H}}^{P/O}$, Eq. (4.2) may be
rewritten as

$${}^{\mathcal{R}}\dot{\mathbf{H}}^{P/O} = \mathbf{M}^{\mathcal{F}/O}. \tag{4.3}$$

Equation (4.3) establishes that the time rate, in an inertial reference
frame, of the angular momentum of a particle with respect to a point

fixed in the reference frame, in this same reference frame, is, at each instant, equal to the resultant moment with respect to the point of the system consisting of all forces acting on the particle. It should be mentioned that Eq. (4.3) is valid only for a point O *fixed* in an inertial reference frame; if Q is an arbitrary point moving in $\mathcal{R}$, $\mathbf{p}^{P/Q}$ is the position vector with respect to Q of the particle P and we calculate the momentum with respect to Q of both members of Eq. (2.1); then $\mathbf{p}^{P/Q} \times {}^{\mathcal{R}}\dot{\mathbf{G}}^P = \mathbf{p}^{P/Q} \times \mathbf{R}$. The second member is the resultant moment with respect to Q, $\mathbf{M}^{\mathcal{F}/Q}$, while the first member, according to Eq. (1.9), will be equal to ${}^{\mathcal{R}}\dot{\mathbf{H}}^{P/Q} + {}^{\mathcal{R}}\mathbf{v}^Q \times {}^{\mathcal{R}}\mathbf{G}^P$, resulting in

$$
{}^{\mathcal{R}}\dot{\mathbf{H}}^{P/Q} + {}^{\mathcal{R}}\mathbf{v}^Q \times {}^{\mathcal{R}}\mathbf{G}^P = \mathbf{M}^{\mathcal{F}/Q}. \tag{4.4}
$$

Equation (4.4), valid for a mobile point Q in the inertial reference frame, is, of course, more general than Eq. (4.3), being reduced to the same form as the latter if Q is fixed in $\mathcal{R}$ or moves at a velocity parallel to that of the particle.

Equation (4.3) was obtained from Eq. (2.1) by cross multiplying with the position vector $\mathbf{p}$; it is, therefore, an alternative form for the equation of motion of a particle although it does not generally bring any new information. Its use will mainly be evident when applied to systems of particles or when establishing the equations of motion for a rigid body, in which cases it will be an independent dynamic equation, as will be shown later.

Example 4.1 Consider the small button B, with mass m, moving in a circular trajectory on the sloping plane, fixed by a string to point O (see Fig. 4.2). Adopting the basis $\mathbf{n}_1, \mathbf{n}_2, \mathbf{n}_3$ as shown, the forces acting on B consist of its weight, $\mathbf{P} = mg(\sin\phi\sin\theta\,\mathbf{n}_1 + \sin\phi\cos\theta\,\mathbf{n}_2 - \cos\phi\,\mathbf{n}_3)$, the traction in the string, $\mathbf{T} = -T\mathbf{n}_1$, the component of the contact with the plane normal to this one, $\mathbf{F}_3 = F_3\mathbf{n}_3$, and the parallel component, or friction, $\mathbf{F}_2 = -F_2\mathbf{n}_2$. The resultant moment of this system $\mathcal{F}$ with respect to point O is

$$
\mathbf{M}^{\mathcal{F}/O} = r\big[(mg\cos\phi - F_3)\mathbf{n}_2 + (mg\sin\phi\cos\theta - F_2)\mathbf{n}_3\big].
$$

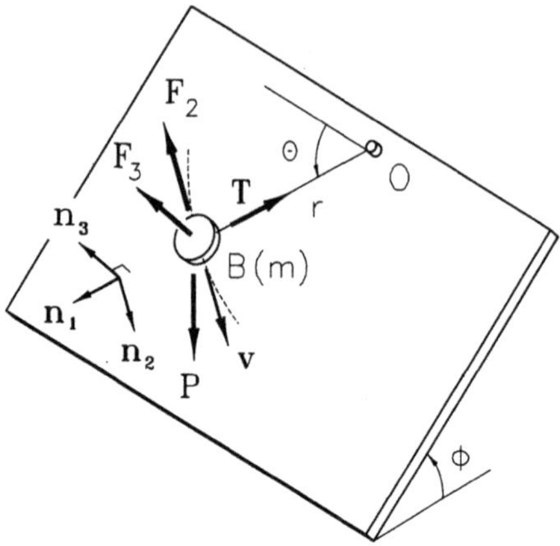

Figure 4.2

B's acceleration in relation to the plane is

$$^{\mathcal{R}}\mathbf{a}^{B} = r(\ddot{\theta}\mathbf{n}_2 - \dot{\theta}^2\mathbf{n}_1),$$

and the derivative of the angular momentum vector with respect to point O in the reference frame fixed on the plane (assumed inertial), according to Eq. (1.13), is

$$^{\mathcal{R}}\dot{\mathbf{H}}^{B/O} = r\mathbf{n}_1 \times mr(\ddot{\theta}\mathbf{n}_2 - \dot{\theta}^2\mathbf{n}_1) = mr^2\ddot{\theta}\mathbf{n}_3.$$

Equation (4.3) then gives the following system of equations governing the button's motion:

$$0 = mg\cos\phi - F_3,$$
$$mr\ddot{\theta} = mg\sin\phi\cos\theta - F_2.$$

The first equation determines $F_3 = mg\cos\phi$. Assuming that the dynamic friction coefficient in the contact between the surfaces is known, μ, then $F_2 = \mu F_3 = \mu mg\cos\phi$ and the second equation gives

$$\ddot{\theta} - \frac{g}{r}(\sin\phi\cos\theta - \mu\cos\phi) = 0,$$

an equation governing the motion of B. Note that by using Eq. (4.3) to substitute Eq. (2.1) suppressed the unknown T, since the traction on the string does not cause momentum with respect to the point O. When wishing to determine it, it is necessary to resort to the component in the direction $\mathbf{n}_1$ of Eq. (2.1). See the corresponding animation.

Choosing a system of Cartesian axes $\{x_1, x_2, x_3\}$ with origin at point O, fixed in the inertial reference frame, the time rate in $\mathcal{R}$ of the angular momentum vector of P with respect to O, in $\mathcal{R}$, may be broken down in the direction of the axes, according to Eq. (1.16), resulting in

$$^{\mathcal{R}}\dot{H}^{P/x_j} = {}^{\mathcal{R}}\dot{\mathbf{H}}^{P/O} \cdot \mathbf{n}_j, \quad j = 1, 2, 3. \tag{4.5}$$

It is worth mentioning that if the axes have a fixed direction in the inertial reference frame, then the scalar components of the time rate of the angular momentum vector are the time rates of the angular momenta with respect to the axes. The resultant moment with respect to the point O, decomposed in the same directions, results in

$$M^{\mathcal{F}/x_j} = \mathbf{M}^{\mathcal{F}/O} \cdot \mathbf{n}_j, \quad j = 1, 2, 3. \tag{4.6}$$

So, when Eq. (4.3) is decomposed in the coordinated directions, then

$$\begin{aligned}
^{\mathcal{R}}\dot{H}^{P/x_1} &= M^{\mathcal{F}/x_1}, \\
^{\mathcal{R}}\dot{H}^{P/x_2} &= M^{\mathcal{F}/x_2}, \\
^{\mathcal{R}}\dot{H}^{P/x_3} &= M^{\mathcal{F}/x_3},
\end{aligned} \tag{4.7}$$

consisting of an alternative form for the equations of motion, also called *equations of motion of the second kind*. In some applications, a combined use of components from Eqs. (2.3) and Eqs. (4.7) is a shortcut in solving the problem.

Example 4.2 Consider the motion of a small sphere P, with mass m, rolling over the inside surface of a cylindrical shell with a vertical axis and radius r (see Fig. 4.3). The sphere is thrown horizontally with a velocity of module v_0 from point A, describing a curvilinear trajectory whose nature is to be determined. O is a point fixed on the vertical axis x_3, at the same level as A; B is a general point in the sphere's trajectory, whose position is given by the coordinates $\phi(t)$ and $z(t)$; Q is the point of the axis that, at each instant, has the same elevation z as the sphere; the angle θ measures the slope of the trajectory; the orthonormal basis $\mathbf{b}_1, \mathbf{b}_2, \mathbf{b}_3$ accompanies the motion of P, with $\mathbf{b}_1$ horizontal and tangent to the cylindrical surface, $\mathbf{b}_2$ orthogonal to the surface, and $\mathbf{b}_3$ vertical. This basis is, therefore, fixed in a reference frame B moving in the shell at a simple angular velocity $\boldsymbol{\omega} = \dot{\phi}\mathbf{b}_3$. The time rates of the vectors of this basis in the reference frame fixed to the shell are

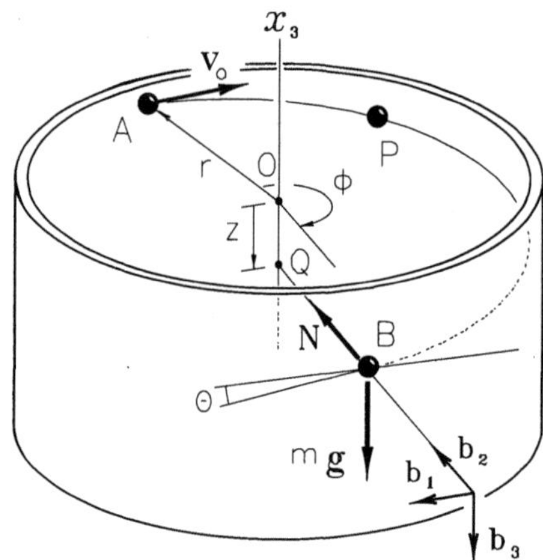

Figure 4.3

$$\dot{\mathbf{b}}_1 = \dot{\phi}\mathbf{b}_2, \quad \dot{\mathbf{b}}_2 = -\dot{\phi}\mathbf{b}_1, \quad \dot{\mathbf{b}}_3 = 0.$$

The definition of the trajectory consists of establishing a working relationship between the coordinates z and ϕ and determining the slope θ in function of one of the coordinates. The forces acting on the sphere are its weight, $mg\mathbf{b}_3$, and the contact force, $\mathbf{N} = N\mathbf{b}_2$. The position of P with respect to point O is described by the vector $\mathbf{p} = -r\mathbf{b}_2 + z\mathbf{b}_3$, and the sphere's velocity and acceleration on the shell will be

$$\mathbf{v}^P = r\dot{\phi}\mathbf{b}_1 + \dot{z}\mathbf{b}_3, \qquad \mathbf{a}^P = r\ddot{\phi}\mathbf{b}_1 + r\dot{\phi}^2\mathbf{b}_2 + \ddot{z}\mathbf{b}_3.$$

The component in the direction of x_3 of Eq. (2.3) will be (considering the shell as an inertial reference frame)

$$m\ddot{z} = mg; \qquad \text{therefore,} \quad \ddot{z} = g. \tag{a}$$

As the acting forces are coplanar with the axis x_3, the resultant moment with respect to this axis is null and the component in this direction of Eq. (4.7) will be (note that the direction in question is fixed in the inertial reference frame)

$$\dot{H}^{P/x_3} = 0.$$

The conclusion then is that H^{P/x_3} remains constant during the motion. In a general position we will have

$$H^{P/x_3} = \mathbf{p}^{P/Q} \times m\mathbf{v}^P \cdot \mathbf{b}_3 = mr^2\dot{\phi}.$$

At point A, $H^{P/x_3} = mrv_0$ and, the values being equal, then

$$\dot{\phi} = \frac{v_0}{r}. \tag{b}$$

Equation (a) may then be expressed as

$$\ddot{z} = \frac{d^2 z}{d\phi^2}\dot{\phi}^2 = \left(\frac{v_0}{r}\right)^2 \frac{d^2 z}{d\phi^2} = g;$$

therefore, taking $\phi = 0$ on point A, we have the intended working relationship

$$z = \frac{gr^2}{2v_0^2}\phi^2.$$

The slope of the trajectory may then be taken from the equation $\tan\theta = dz/r\,d\phi$ (see Fig. 4.3), resulting in

$$\theta = \tan^{-1}\left(\frac{gr}{v_0^2}\phi\right).$$

See the corresponding animation.

4.5 Work and Potentials

Consider P to be a point moving according to a trajectory, described by the curve λ, in a reference frame $\mathcal{R}$ (see Fig. 5.1).

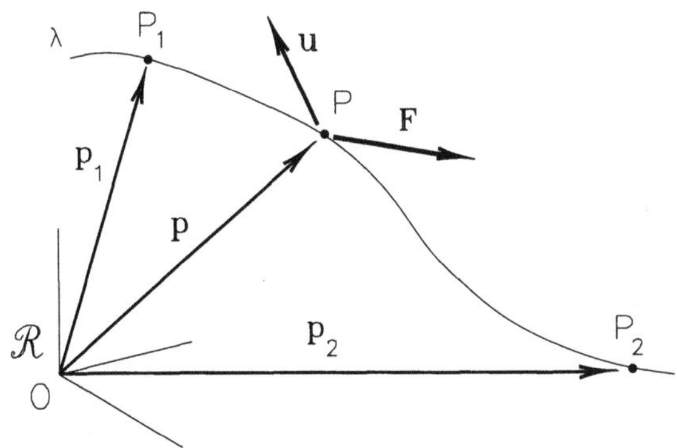

Figure 5.1

The position vector of P with respect to a point O, fixed in $\mathcal{R}$, is a vector function of the time variable, $\mathbf{p}(t)$. Also consider P_1 and P_2 as two points of the trajectory, fixed in $\mathcal{R}$, whose positions with respect to O are given as $\mathbf{p}_1$ and $\mathbf{p}_2$, respectively. If $\mathbf{u}$ is a vector linked to P, it is called *action* of $\mathbf{u}$ between the points P_1 and P_2, in the reference frame $\mathcal{R}$, the scalar defined as

$$^{\mathcal{R}}\mathcal{A}_{12}^{u} \rightleftharpoons \int_{\mathbf{P_1}}^{\mathbf{P_2}} \mathbf{u} \cdot {}^{\mathcal{R}}d\mathbf{p}. \tag{5.1}$$

Note that the action of $\mathbf{u}$ depends on the motion of the point to which it is bound, resulting from its scalar product with the local displacement of the point, between two positions.

If $\mathbf{F}$ is a force linked to a point P moving in a reference frame $\mathcal{R}$ (see Fig. 5.1), the action of $\mathbf{F}$ between two points in the trajectory of P is called the *work* of $\mathbf{F}$ between the points in the reference frame, defined therefore as

$$^{\mathcal{R}}\mathcal{T}_{12}^{F} \rightleftharpoons \int_{\mathbf{P_1}}^{\mathbf{P_2}} \mathbf{F} \cdot {}^{\mathcal{R}}d\mathbf{p}. \tag{5.2}$$

The work of a force is scalar and may give a positive, negative, or null result, whose physical dimension is $[ML^2T^{-2}]$.

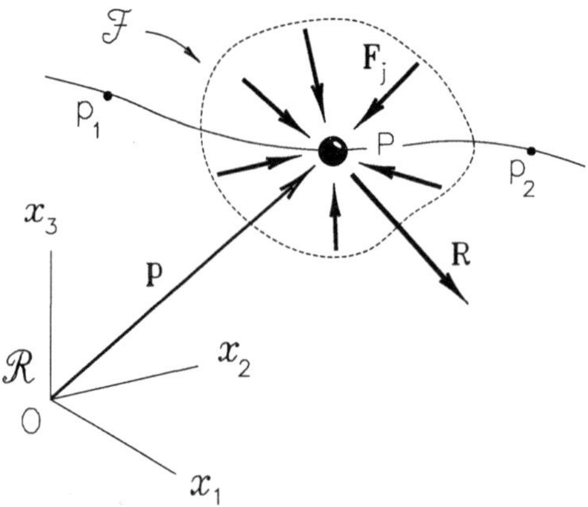

Figure 5.2

If a simple system of forces $\mathcal{F}$, with resultant $\mathbf{R}$, acts on a point P (see Fig. 5.2), the *resultant work* is defined as the sum of the works of the component forces, that is,

$$^{\mathcal{R}}\mathcal{T}_{12}^{\mathcal{F}} \rightleftharpoons \sum_{j=1}^{m} {}^{\mathcal{R}}\mathcal{T}_{12}^{F_j}, \tag{5.3}$$

where $\mathbf{F}_j$, $j = 1, 2, \ldots, m$, are the forces comprising the system $\mathcal{F}$.

The resultant work of a simple system on a point is equal to the work of a force equal to the resultant of the system applied to the point. In fact, substituting Eq. (5.2) in Eq. (5.3), then

$$\begin{aligned}
{}^{\mathcal{R}}\mathcal{T}_{12}^{\mathcal{F}} &= \sum_{j=1}^{m} \int_{\mathbf{p}_1}^{\mathbf{p}_2} \mathbf{F}_j \cdot {}^{\mathcal{R}}d\mathbf{p} \\
&= \int_{\mathbf{p}_1}^{\mathbf{p}_2} \sum_{j=1}^{m} \mathbf{F}_j \cdot {}^{\mathcal{R}}d\mathbf{p} \\
&= \int_{\mathbf{p}_1}^{\mathbf{p}_2} \mathbf{R} \cdot {}^{\mathcal{R}}d\mathbf{p} \\
&= {}^{\mathcal{R}}\mathcal{T}_{12}^{R}.
\end{aligned} \tag{5.4}$$

Example 5.1 The cursor C can slide without friction along the horizontal guide fixed to the support S at the point O and is being pulled by a thread, passing through a mobile axis pulley, to which a constant module force F is applied (see Fig. 5.3). A force system acts on the cursor consisting of its weight, the guide's reactions, and the thread's action. The position of C with respect to O is given by the vector $\mathbf{p} = y\mathbf{n}_2$ and, as the basis $\mathbf{n}_1, \mathbf{n}_2, \mathbf{n}_3$ is fixed on S, ${}^{S}d\mathbf{p} = dy\,\mathbf{n}_2$.

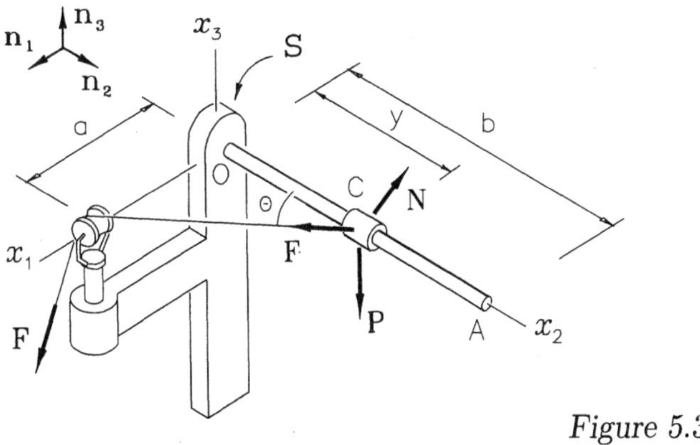

Figure 5.3

The force exerted by the thread may be expressed by

$$\mathbf{F} = F(\sin\theta\,\mathbf{n}_1 - \cos\theta\,\mathbf{n}_2),$$

where

$$\cos\theta = \frac{y}{\sqrt{a^2 + y^2}}, \qquad \sin\theta = \frac{a}{\sqrt{a^2 + y^2}}.$$

The work of this force between points A and O, in the reference frame S, is

$$^S\mathcal{T}_{AO}^F = \int_A^O \mathbf{F} \cdot {}^S d\mathbf{p} = -F \int_b^0 \frac{y\,dy}{\sqrt{a^2 + y^2}} = F\left(\sqrt{a^2 + b^2} - a\right).$$

The weight of the cursor is $\mathbf{P} = -mg\mathbf{n}_3$, with no component in the direction of the cursor's displacement at S. The work of the weight between points A and O, in the reference frame S, will, therefore, be null. Ignoring the friction, the force exerted by the guide on the cursor is $\mathbf{N} = N_1\mathbf{n}_1 + N_3\mathbf{n}_3$ and the scalar product $\mathbf{N}\cdot d\mathbf{p}$ vanishes. The conclusion is that the resulting work between points A and O, at S, is reduced to the work of the force exerted by the thread, calculated above.

If P is a point moving in a reference frame $\mathcal{R}$, its velocity in $\mathcal{R}$ is $^\mathcal{R}\mathbf{v}^P = {}^\mathcal{R}d\mathbf{p}/dt$; therefore, $^\mathcal{R}d\mathbf{p} = {}^\mathcal{R}\mathbf{v}^P\,dt$, and the work of a force $\mathbf{F}$ applied to the point between two positions P_1 and P_2, in the reference frame $\mathcal{R}$, may be expressed alternatively as

$$^\mathcal{R}\mathcal{T}_{12}^F = \int_{t_1}^{t_2} \mathbf{F} \cdot {}^\mathcal{R}\mathbf{v}^P\,dt. \tag{5.5}$$

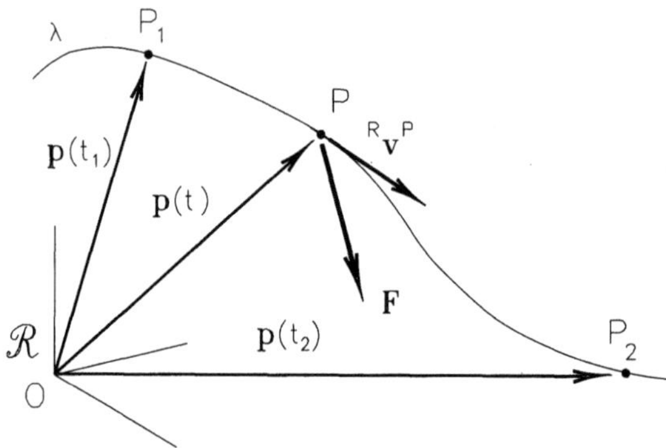

Figure 5.4

Note that the integration now is done in the time, the limits t_1 and t_2 being the instants of the passage of P through the points P_1 and P_2, respectively (see Fig. 5.4).

Example 5.2 The disk D rotates at a simple angular velocity of constant module ω in the support S, with its center O fixed to S (see Fig. 5.5). A constant force $\mathbf{F}$ (F and θ remain invariants in time) is applied to the pin P, fixed to D. We wish to determine the work of $\mathbf{F}$ between the positions (1) and (2), in the reference frame S. The velocity of point P in S may be obtained from

$$
{}^{S}\mathbf{v}^{P} = {}^{S}\mathbf{v}^{O} + {}^{S}\boldsymbol{\omega}^{D} \times \mathbf{p}^{P/O}
$$
$$
= 0 + \omega\mathbf{n}_3 \times r(\sin\phi\,\mathbf{n}_1 - \cos\phi\,\mathbf{n}_2)
$$
$$
= \omega r(\cos\phi\,\mathbf{n}_1 + \sin\phi\,\mathbf{n}_2).
$$

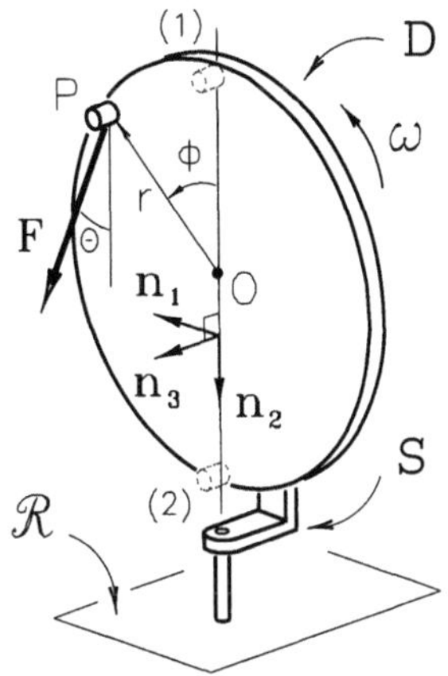

Figure 5.5

The force $\mathbf{F}$, expressed on the chosen basis, is

$$
\mathbf{F} = F(\cos\theta\,\mathbf{n}_2 + \sin\theta\,\mathbf{n}_3).
$$

The desired work is then, according to Eq. (5.5),

$$
{}^{S}\mathcal{T}_{12}^{F} = \int_{t_1}^{t_2} \mathbf{F} \cdot {}^{S}\mathbf{v}^{P}\, dt
$$
$$
= \int_{t_1}^{t_2} F\omega r \cos\theta \sin\phi\, dt
$$
$$
= F\omega r \cos\theta \int_{0}^{\pi/\omega} \sin\omega t\, dt
$$
$$
= 2Fr\cos\theta.
$$

See the corresponding animation.

The work of a force $\mathbf{F}$ applied, either to a particle moving in a reference frame $\mathcal{R}$, as in Example 5.1, or to a point of a body moving in $\mathcal{R}$, as in Example 5.2, depends, as the indicated notation suggests, on the choice of reference frame. In fact, as the velocity of the application point of the force depends on the reference frame, the result of the integration present in Eq. (5.5) will depend on the reference frame involved.

Example 5.3   Consider the system analyzed in the preceding example, but with the support S moving at a simple angular velocity of a constant module Ω in the basis $\mathcal{R}$ (see Fig. 5.6). The velocity of P in $\mathcal{R}$ is

$$
\begin{aligned}
{}^{\mathcal{R}}\mathbf{v}^{P} &= {}^{\mathcal{R}}\mathbf{v}^{O} + {}^{\mathcal{R}}\boldsymbol{\omega}^{D} \times \mathbf{p}^{P/O} \\
&= 0 + (-\Omega\mathbf{n}_2 + \omega\mathbf{n}_3) \times r(\sin\phi\mathbf{n}_1 - \cos\phi\mathbf{n}_2) \\
&= r(\omega\cos\phi\mathbf{n}_1 + \omega\sin\phi\mathbf{n}_2 + \Omega\sin\phi\mathbf{n}_3).
\end{aligned}
$$

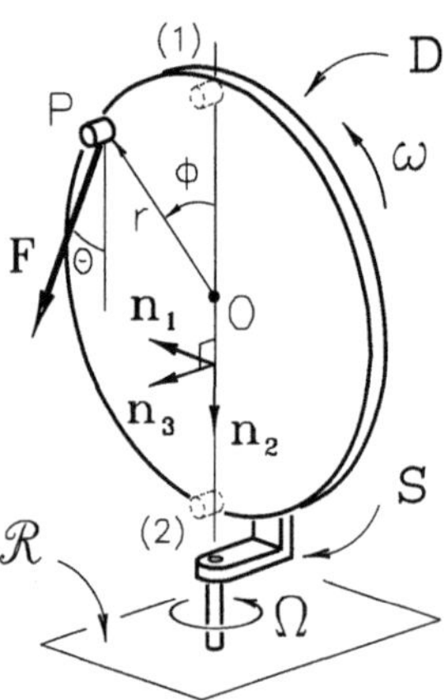

Figure 5.6

The work of the force $\mathbf{F}$ between points (1) and (2) (note that they are exactly the same as in the previous example), in the reference frame $\mathcal{R}$, is then

$$
\begin{aligned}
{}^{\mathcal{R}}\mathcal{T}^{F}_{12} &= \int_{t_1}^{t_2} \mathbf{F} \cdot {}^{\mathcal{R}}\mathbf{v}^{P}\, dt \\
&= \int_{t_1}^{t_2} Fr(\omega\cos\theta\sin\phi + \Omega\sin\theta\sin\phi)\, dt \\
&= Fr(\omega\cos\theta + \Omega\sin\theta) \int_{t_1}^{t_2} \sin(\omega t)\, dt \\
&= 2Fr\left(\cos\theta + \frac{\Omega}{\omega}\sin\theta\right),
\end{aligned}
$$

different, therefore, from the work found in the preceding example. See the corresponding animation.

The work of a force $\mathbf{F}$ applied to a point P depends, as shown in Eqs. (5.2) and (5.5), on the motion of P in the reference frame in question and, of course, on the end points P_1 and P_2, fixed in the reference frame. When, however, $\mathbf{F}$ is such that its work between points P_1 and P_2 *does not depend* on the trajectory of P between those points, the force is said to be *conservative*. Hence, if $\mathbf{F}$ is attached to different points $P, P', P'', \ldots$, moving in different trajectories all passing through the end points P_1 and P_2 (see Fig. 5.7), then $\mathbf{F}$ will be conservative if

$$\int_{\mathbf{P_1}}^{\mathbf{P_2}} \mathbf{F} \cdot {}^{\mathcal{R}}d\mathbf{p} = \int_{\mathbf{P_1}}^{\mathbf{P_2}} \mathbf{F} \cdot {}^{\mathcal{R}}d\mathbf{p}' = \int_{\mathbf{P_1}}^{\mathbf{P_2}} \mathbf{F} \cdot {}^{\mathcal{R}}d\mathbf{p}'' = \cdots = {}^{\mathcal{R}}\mathcal{T}_{12}^{F}. \quad (5.6)$$

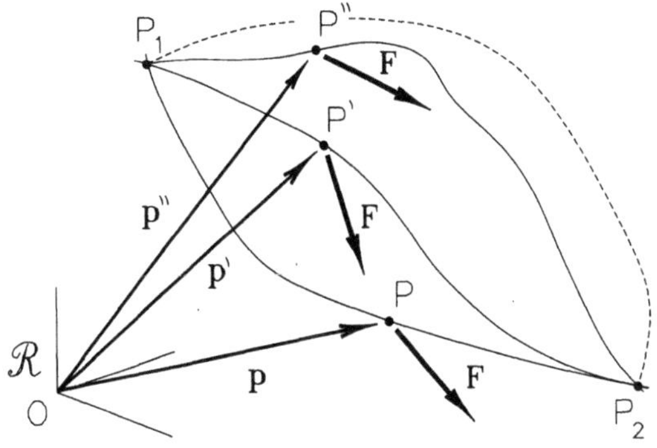

Figure 5.7

Example 5.4 Let us consider a particle P, with mass m, sliding along a curvilinear guide between points A: $(0, 0, c)$ and B: $(a, b, 0)$ (see Fig. 5.8). The guide's geometry may be described by

$$\begin{cases} x = \lambda_x(z), \\ y = \lambda_y(z) \end{cases}$$

satisfying

$$\lambda_x(c) = \lambda_y(c) = 0, \quad \lambda_x(0) = a, \quad \lambda_y(0) = b.$$

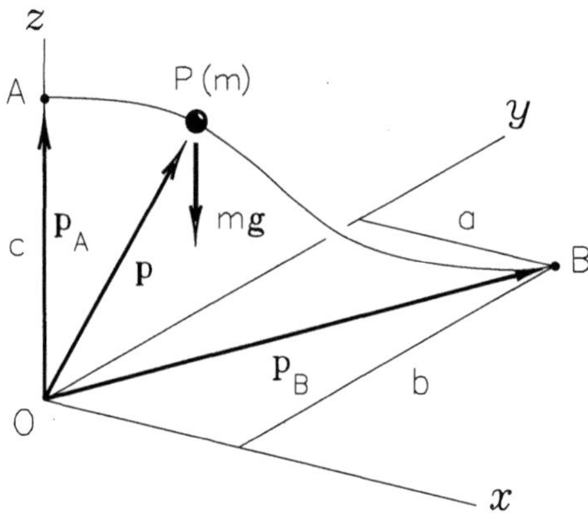

Figure 5.8

Therefore, when moving over the guide, the position of P varies as

$$d\mathbf{p} = (\lambda'_x, \lambda'_y, 1)\, dz,$$

where λ' is the derivative of λ with respect to the its argument (z). The weight force, acting on the particle, is $\mathbf{P} = mg(0,0,-1)$ and its work between the points A and B will be

$$\mathcal{T}^P_{AB} = \int_{\mathbf{P}_A}^{\mathbf{P}_B} \mathbf{P}\cdot d\mathbf{p} = -mg\int_c^0 dz = mgc.$$

Observe that the result of the work of the weight does not depend on the functions $\lambda_x(z)$ and $\lambda_y(z)$, that is, on the guide's format, therefore, on the trajectory described between points A and B, showing then that this is a conservative force.

If $\mathbf{F}$ is a conservative force, then there is a scalar field $\Phi(\mathbf{p})$, called the *potential function* of $\mathbf{F}$, whose gradient is equal to the opposite of the force, that is,

$$\nabla\Phi = -\mathbf{F} \qquad \text{if } \mathbf{F} \text{ is conservative.} \tag{5.7}$$

In fact, if the work of $\mathbf{F}$ does not depend on the trajectory, then there is a function $\Phi(\mathbf{p})$ so that

$$^{\mathcal{R}}\mathcal{T}^F_{12} = \int_{\mathbf{P}_1}^{\mathbf{P}_2} \mathbf{F}\cdot {}^{\mathcal{R}}d\mathbf{p} = \int_{\mathbf{P}_1}^{\mathbf{P}_2} -d\Phi = \Phi(\mathbf{p}_1) - \Phi(\mathbf{p}_2), \tag{5.8}$$

that is, the work of **F** depends only on the values of the function at the two ends of the integration. So, Eq. (5.7) results from the identity $d\Phi(\mathbf{p}) = \nabla\Phi(\mathbf{p}) \cdot d\mathbf{p}$. Note that the negative sign adopted is arbitrary, and resulting from it is the work of the force equal to the first amount less the final of the potential function, as Eq. (5.8) indicates; the reason for this choice will be shown as natural in the following section. It is worth noting that the argument can be inverted: If there is a potential function $\Phi(\mathbf{p})$ whose gradient satisfies Eq. (5.7), then the work of **F** between two points P_1 and P_2 will be equal to $\Phi(\mathbf{p}_1) - \Phi(\mathbf{p}_2)$, independent of the application point's trajectory; therefore, the force is conservative.

Every potential function $\Phi(\mathbf{p})$ has the same physical dimension as the work of a force, that is, $[ML^2T^{-2}]$, also the same as the kinetic energy of a particle.

Example 5.5 Consider the cursor C, with mass m, moving without friction between the ends A and B of a curvilinear guide, under the action of its own weight, of the force exerted by the guide and a linear elastic spring of a constant k and natural length a (see Fig. 5.9). Ignoring the friction, the force exerted by the guide is always orthogonal to the displacement and its work between the points A and B is null.

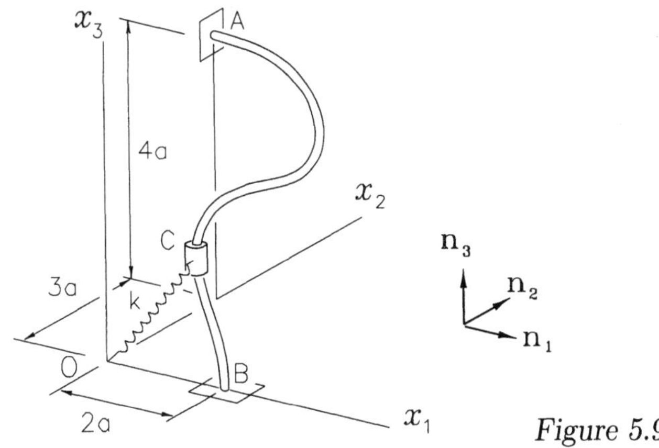

Figure 5.9

As the weight force, $-mg\mathbf{n}_3$, is conservative (see preceding example), it assumes a potential function $\Phi_g(\mathbf{p})$ so that, according to Eq. (5.7),

$$\nabla\Phi_g(\mathbf{p}) = mg\mathbf{n}_3,$$

that is,

$$\frac{\partial \Phi_g}{\partial x_1} = \frac{\partial \Phi_g}{\partial x_2} = 0, \qquad \frac{\partial \Phi_g}{\partial x_3} = mg,$$

therefore,

$$\Phi_g(\mathbf{p}) = mgx_3 + \phi_1,$$

where ϕ_1 is an arbitrary constant establishing a reference level for the potential function. Its value is generally not important since only differences in potentials are used to calculate the work of the force. The work of the weight between A and B is, therefore,

$$\mathcal{T}_{AB}^P = \Phi_g(A) - \Phi_g(B) = mg\big(x_3(A) - x_3(B)\big) = 4mga.$$

If p is the module of the vector $\mathbf{p}$, position of C with respect to O, the force exerted on C by the spring, in a general position, is

$$\mathbf{F}_k = -k\frac{p-a}{p}\mathbf{p}.$$

If this force assumes a potential function $\Phi_k(\mathbf{p})$, it shall, according to Eq. (5.7), satisfy

$$\nabla\Phi_k(\mathbf{p}) = k\frac{p-a}{p}\mathbf{p}.$$

It is easy to check that the function

$$\Phi_k(\mathbf{p}) = \frac{1}{2}k(p-a)^2 + \phi_2,$$

where ϕ_2 is a constant, satisfies the condition (check). The work between points A and B of the spring's force may then be calculated as

$$\mathcal{T}_{AB}^{F_k} = \Phi_k(A) - \Phi_k(B) = \frac{1}{2}k(4a)^2 - \frac{1}{2}ka^2 = \frac{15}{2}ka^2.$$

When a particle P moves in a reference frame $\mathcal{R}$ under the action of a system of forces $\mathcal{F}$ comprising a subsystem $\mathcal{F}_N$, consisting of nonconservative forces, and another subsystem $\mathcal{F}_C$, consisting exclusively of conservative forces, the resultant work between two points may be split in two parts, that is,

$$^{\mathcal{R}}\mathcal{T}_{12}^{\mathcal{F}} = {}^{\mathcal{R}}\mathcal{T}_{12}^{\mathcal{F}_N} + {}^{\mathcal{R}}\mathcal{T}_{12}^{\mathcal{F}_C}. \tag{5.9}$$

Each of the forces $\mathbf{F}_i$, $i = 1, 2, \ldots, m$, belonging to the subsystem $\mathcal{F}_C$, has a potential function $\Phi_i(\mathbf{p})$, naturally satisfying Eq. (5.7). The *potential energy* of P in $\mathcal{R}$ may then be defined as the sum of these potential functions, that is,

$$^{\mathcal{R}}\Phi^P(\mathbf{p}) \rightleftharpoons \sum_{j=1}^{m} \Phi_i(\mathbf{p}). \tag{5.10}$$

Once the potential energy of P has been defined, the resultant work of the conservative subsystem may be defined as the change in the potential energy of particle between both positions, that is,

$$^{\mathcal{R}}\mathcal{T}_{12}^{\mathcal{F}_C} = {}^{\mathcal{R}}\Phi^P(\mathbf{p}_1) - {}^{\mathcal{R}}\Phi^P(\mathbf{p}_2). \tag{5.11}$$

In fact, as $^{\mathcal{R}}\mathcal{T}_{12}^{F_i} = \Phi_i(\mathbf{p}_1) - \Phi_i(\mathbf{p}_2)$, $i = 1, 2, \ldots, m$, totaling the works in the subsystem $\mathcal{F}_C$ and using Eq. (5.10), Eq. (5.11) is reached immediately.

Example 5.6   Returning to the preceding example (see Fig. 5.9), now take the cursor C sliding with friction along the guide. Of the three forces acting on C (weight, spring, and guide), two of them form a conservative subsystem (weight and spring), while the force exerted by the guide will be a nonconservative subsystem. The potential energy of the cursor may be expressed then by (see Example 5.5)

$$^{\mathcal{R}}\Phi^P(\mathbf{p}) = \Phi_g(\mathbf{p}) + \Phi_k(\mathbf{p}) = mgx_3 + \frac{1}{2}k(p - a)^2 + \phi_1 + \phi_2.$$

The resultant work of the conservative forces between the positions A and B will then be

$$^{\mathcal{R}}\mathcal{T}_{AB}^{\mathcal{F}_C} = {}^{\mathcal{R}}\Phi^P(\mathbf{p}_A) - {}^{\mathcal{R}}\Phi^P(\mathbf{p}_B) = 4mga + \frac{15}{2}ka^2.$$

4.6 Work and Energy

If P is a particle with mass m moving in an inertial reference frame $\mathcal{R}$ according to a trajectory λ, under the action of a simple force system $\mathcal{F}$, with resultant $\mathbf{R}$ (see Fig. 6.1), the resultant work performed by the system on the particle between two points P_1 and P_2 of its trajectory is equal to the change in the kinetic energy of P in $\mathcal{R}$, between these points, that is, to the difference between the kinetic energy of the particle at the instant when it passes through P_2 and when it passes through P_1.

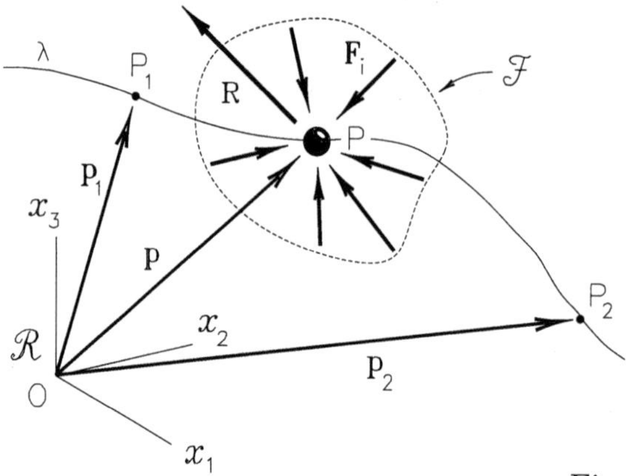

Figure 6.1

In fact, dot-multiplying both members of the dynamic equation for P, Eq. (2.2), by $^{\mathcal{R}}d\mathbf{p}$, $\mathbf{p}$ being the position vector of P with respect to a point O fixed in $\mathcal{R}$, then

$$m\,^{\mathcal{R}}\mathbf{a}^P \cdot {}^{\mathcal{R}}d\mathbf{p} = \mathbf{R} \cdot {}^{\mathcal{R}}d\mathbf{p} \tag{6.1}$$

and, integrating between the positions P_1 and P_2,

$$\int_{\mathbf{p_1}}^{\mathbf{p_2}} m\,^{\mathcal{R}}\mathbf{a}^P \cdot {}^{\mathcal{R}}d\mathbf{p} = \int_{\mathbf{p_1}}^{\mathbf{p_2}} \mathbf{R} \cdot {}^{\mathcal{R}}d\mathbf{p}. \tag{6.2}$$

The left side of the equation above may be identified, according to Eq. (1.22), as the difference $^{\mathcal{R}}K^P(\mathbf{p_2}) - {}^{\mathcal{R}}K^P(\mathbf{p_1})$ (note that the constant of an integration C disappears in the integral defined between the ends $\mathbf{p_1}$ and $\mathbf{p_2}$); the right side, according to Eq. (5.4), consists of the resultant work of the force system, between those points, $^{\mathcal{R}}\mathcal{T}_{12}^{\mathcal{F}}$. The result therefore is that

$$^{\mathcal{R}}K^P(\mathbf{p_2}) - {}^{\mathcal{R}}K^P(\mathbf{p_1}) = {}^{\mathcal{R}}\mathcal{T}_{12}^{\mathcal{F}}. \tag{6.3}$$

We must not forget that the relationship between the change of the kinetic energy and resultant work, expressed by Eq. (6.3), will only be satisfied if $\mathcal{R}$ is an *inertial reference frame*.

Example 6.1   Figure 6.2 reproduces the configuration examined in Example 5.1. Assuming that S is an inertial reference frame and that the cursor leaves from end A at null velocity in S, we wish to determine its velocity when it collides with the support at O.

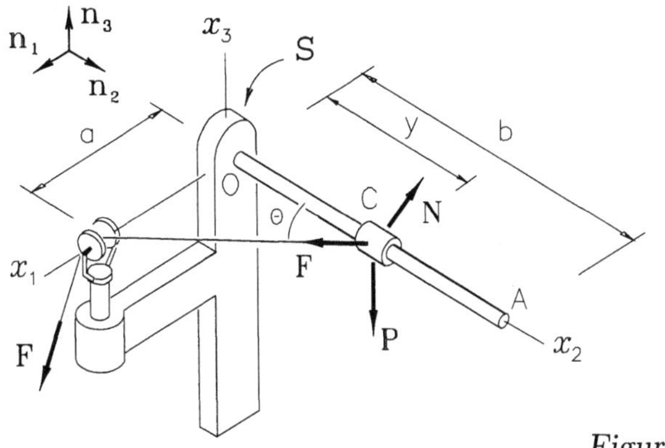

Figure 6.2

The work resulting from the system of every force that acts on the cursor was determined in Example 5.1, as ${}^{S}T^{\mathcal{F}}_{AO} = F(l - a)$, where $l^2 = a^2 + b^2$. As the initial kinetic energy is null, Eq. (6.3) is thereby reduced to

$$\frac{1}{2}mv^2(O) = F(l - a);$$

thus, the desired velocity is

$$v(O) = \sqrt{\frac{2F(l - a)}{m}}.$$

Equation (6.3), relating the resultant work on a particle with the change of its kinetic energy in an inertial reference frame, is an extremely convenient alternative for the equation of motion when wishing to find a velocity as a function of a position, like in the above example, or even a component of force applied to the particle, also as a function of its position in the trajectory. Another important advantage from using Eq. (6.3) lies in the fact that components of forces — generally unknown — that do not carry out work are automatically eliminated, such as those given in Examples 5.1 and 5.5.

Example 6.2 Figure 6.3 reproduces the system studied in Example 2.1. As seen there, the equation governing the motion of the sphere along the pipe, $\ddot{\theta} + (\omega^2 \sin\theta - \frac{g}{r})\cos\theta = 0$, is hard to integrate in the time variable. We may, however, extract valuable information about the behavior of this system, using Eq. (6.3). Assuming that the sphere is left with a null velocity relative to the pipe in the top position ($\theta = 0$), the resulting work up to a general position will be (see Example 2.1)

$$^A\mathcal{T}^{\mathcal{F}}_{0\theta} = {}^A\mathcal{T}^{F_1}_{0\theta} + {}^A\mathcal{T}^{F_3}_{0\theta} + {}^A\mathcal{T}^{P}_{0\theta}. \tag{a}$$

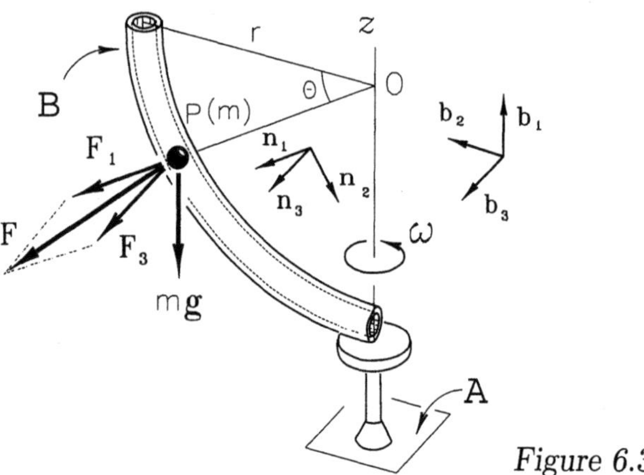

Figure 6.3

The work carried out by the component $\mathbf{F}_1$ is

$$^A\mathcal{T}^{F_1}_{0\theta} = 0, \tag{b}$$

since the velocity of P in A is $^A\mathbf{v}^P = r(\dot\theta\mathbf{n}_2 + \omega\cos\theta\mathbf{n}_3)$, orthogonal, therefore, to the direction of the component $\mathbf{F}_1$. The work done by component $\mathbf{F}_3$ is (see Eq. (e), in Example 2.1)

$$
\begin{aligned}
^A\mathcal{T}^{F_3}_{0\theta} &= \int_0^{t(\theta)} \mathbf{F}_3 \cdot {}^A\mathbf{v}^P \, dt \\
&= -mr^2\omega^2 \int_0^{\theta} 2\cos\theta\sin\theta \, d\theta \\
&= -mr^2\omega^2 \sin^2\theta.
\end{aligned}
\tag{c}
$$

As the work of the weight is a conservative force, it may be expressed directly as (see Example 5.4)

$$^A\mathcal{T}^{P}_{0\theta} = mgr\sin\theta. \tag{d}$$

The kinetic energy in the initial configuration is

$$^AK^P(0) = \frac{1}{2}mr^2\omega^2 \tag{e}$$

and, in a general position, is as follows:

$$^{A}K^{P}(\theta) = \frac{1}{2}mr^{2}(\dot{\theta}^{2} + \omega^{2}\cos^{2}\theta). \tag{f}$$

So, substituting Eqs. (a–f) in Eq. (6.3), then we get

$$mgr\sin\theta - mr^{2}\omega^{2}\sin^{2}\theta = \frac{1}{2}mr^{2}(\dot{\theta}^{2} + \omega^{2}\cos^{2}\theta) - \frac{1}{2}mr^{2}\omega^{2};$$

therefore,

$$\dot{\theta}^{2} = \left(\frac{2g}{r} - \omega^{2}\sin\theta\right)\sin\theta,$$

the result being equivalent to what would be obtained by integrating Eq. (d) in the variable θ, in Example 2.1. Another interesting point is that it is possible for the sphere to stop naturally in the lower position of the pipe. In fact, for this to happen,

$$\dot{\theta}^{2}(\pi/2) = 0,$$

for which it is only necessary that

$$\omega = \sqrt{\frac{2g}{r}}.$$

What happens in this particular case is that the resultant work is

$$^{A}\mathcal{T}_{0\frac{\pi}{2}}^{\mathcal{F}} = -mr^{2}\omega^{2}\sin^{2}\frac{\pi}{2} + mgr\sin\frac{\pi}{2} = -mr(r\omega^{2} - g) = -mgr,$$

negative, therefore, and equal, in module, to the value of the initial kinetic energy, for this value of ω. Consequently, according to Eq. (6.3), the final kinetic energy will be null.

When, in the case of the force system acting on a particle P, moving in an inertial reference frame $\mathcal{R}$, conservative and nonconservative forces compete, only the work performed by the latter needs to be computed by integration along the trajectory, since the potential forces associated with the conservative forces are known. So, if $^{\mathcal{R}}\Phi^{P}(\mathbf{p})$ is the potential energy of the particle, as defined in Eq. (5.10), the substitution of Eq. (5.11) in Eq. (6.3) leads to

$$^{\mathcal{R}}K^{P}(\mathbf{p}_{2}) + {}^{\mathcal{R}}\Phi^{P}(\mathbf{p}_{2}) - {}^{\mathcal{R}}K^{P}(\mathbf{p}_{1}) - {}^{\mathcal{R}}\Phi^{P}(\mathbf{p}_{1}) = {}^{\mathcal{R}}\mathcal{T}_{12}^{\mathcal{F}_{N}}. \tag{6.4}$$

We may then define the *mechanical energy* of the particle P in the reference frame $\mathcal{R}$ as the algebraic sum of its kinetic energy in $\mathcal{R}$ and its potential energy in the same reference frame, that is,

$$^{\mathcal{R}}E^P(\mathbf{p}) \rightleftharpoons {}^{\mathcal{R}}K^P(\mathbf{p}) + {}^{\mathcal{R}}\Phi^P(\mathbf{p}). \tag{6.5}$$

The substitution of Eq. (6.5) in Eq. (6.4) then establishes the following form for the balance of energy of a particle P moving in an inertial reference frame:

$$^{\mathcal{R}}E^P(\mathbf{p}_2) - {}^{\mathcal{R}}E^P(\mathbf{p}_1) = {}^{\mathcal{R}}\mathcal{T}_{12}^{\mathcal{F}_N}, \tag{6.6}$$

that is, the resultant work of the nonconservative forces between two points in an inertial reference frame $\mathcal{R}$ is equal to the change of the mechanical energy of the particle in this reference frame (final energy minus initial energy).

The attentive reader must have perceived the reason in which the potential functions are defined as those whose gradient is equal to *minus* the corresponding force vector, as expressed in Eq. (5.7). In fact, when the resultant work of the conservative forces passes to the other member of the energy balance, Eq. (6.4), it was possible to add the kinetic energy and potential energy of the particle, resulting in Eq. (6.6).

Example 6.3 The cursor C, with mass m, slides along the vertical guide, under the action of its weight and a spring of elastic constant k and natural length a, as shown (see Fig. 6.4), being left at rest with $\theta = 0$. We wish to determine its velocity v when it passes through the position $\theta = \pi/4$. The forces acting on the cursor consist of its weight, $m\mathbf{g}$; the force exerted by the spring, $\mathbf{F}_k = k(l - a)\mathbf{n}$, where $l = a/\cos\theta$ and $\mathbf{n} = -(\cos\theta\mathbf{n}_1 + \sin\theta\mathbf{n}_2)$; the horizontal component of the guide's action, $\mathbf{H}$; and the friction component, $\mathbf{V} = \mu\mathbf{H}$, where μ is the coefficient of dynamic friction between the guide and cursor. As there is no motion in the horizontal direction, the component of the dynamic equation in the directon $\mathbf{n}_1$ is $H - F_k\cos\theta = 0$; therefore,

$$\mathbf{V} = -\mu H\mathbf{n}_2 = -\mu F_k\cos\theta\mathbf{n}_2 = -\mu ka(1 - \cos\theta)\mathbf{n}_2.$$

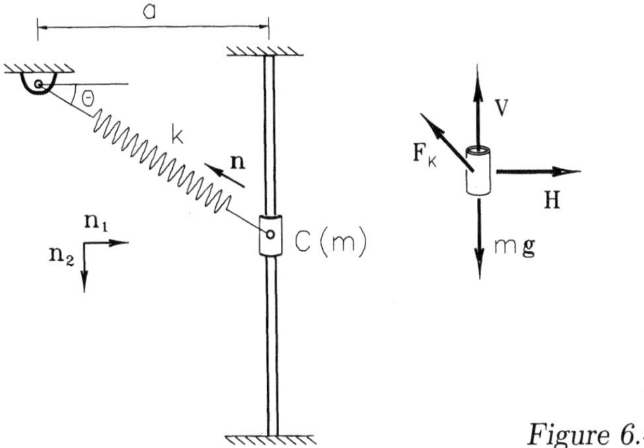

Figure 6.4

The force $\mathbf{H}$ does not work and the forces $\mathbf{F}_k$ and $m\mathbf{g}$ are conservative, with known potential functions. Only the work of the friction force, therefore, should be calculated:

$$\mathcal{T}_{0\frac{\pi}{4}}^{V} = \int_0^{\frac{\pi}{4}} \mathbf{V} \cdot d\mathbf{p}$$

$$= \int_0^{\frac{\pi}{4}} -\mu k a^2 \frac{1 - \cos\theta}{\cos^2\theta}\, d\theta$$

$$= -\mu k a^2 \left[\tan\theta - \ln(\tan\theta + \frac{1}{\cos\theta})\right]_0^{\frac{\pi}{4}}$$

$$= -\mu k a^2 \left(1 - \ln(1 + \sqrt{2})\right).$$

The initial mechanical energy, adopting a null value for the constant of the gravitational potential function, will be

$$E^C(0) = K^C(0) + \Phi^C(0) = 0.$$

The mechanical energy in the required position will be

$$E^C(\pi/4) = K^C(\pi/4) + \Phi^C(\pi/4) = \frac{1}{2}mv^2 - mga + \frac{1}{2}ka^2(\sqrt{2} - 1)^2.$$

So, by substituting these results in Eq. (6.6), the desired velocity is (check)

$$v = \sqrt{a\left(2g - \frac{ka}{m}\left((\sqrt{2} - 1)^2 + 2\mu(1 - \ln(\sqrt{2} + 1))\right)\right)}.$$

See the corresponding animation.

4.7 Impulse and Impact

When a force $\mathbf{F}$ acts on a point P moving in a reference frame $\mathcal{R}$ for a certain time interval (t_1, t_2), it is called *impulse* the vector resulting from the integration of the force along this interval, that is,

$$\mathbf{I}_{12}^F \rightleftharpoons \int_{t_1}^{t_2} \mathbf{F}\, dt. \tag{7.1}$$

The impulse of a force linked to a point P will therefore depend on how the former varies with time and the actual integration interval; it will not, however, explicitly depend on the motion of the point in the referential, as occurs with the *work* of a force. If, for example, P is a point fixed in $\mathcal{R}$, the work performed by any force linked to P will be null, although the impulse of this force can be different from zero, in a given interval. The impulse of a force is a vector quantity with dimension $[\mathrm{MLT}^{-1}]$.

Example 7.1 Figure 7.1 reproduces the system studied in Example 2.4, where a small sphere P, with mass m, moves hanging by a string with length r, describing a circular trajectory on the horizontal plane, with θ constant and velocity of module also constant $v = \sqrt{gr \sin\theta \tan\theta}$. The coordinate ϕ varies linearly, according to $\phi(t) = vt/(r\sin\theta)$. The time interval in which the sphere describes, say, a semicircle will be equal to half the period of motion, that is,

$$t_0 = \frac{\tau}{2} = \pi \sqrt{\frac{r\cos\theta}{g}}.$$

The work done by the weight force in this interval is null because the weight is always orthogonal to the trajectory; the impulse of the weight force, however, is different from zero, that is,

$$\mathbf{I}_{0t_0}^P = \int_0^{t_0} -mg\mathbf{n}_3\, dt = -\pi m \sqrt{gr\cos\theta}\,\mathbf{n}_3.$$

The force exerted by the string on the sphere is

$$\mathbf{T} = T(-\sin\theta\mathbf{b}_1 + \cos\theta\mathbf{b}_3)$$
$$= T(-\sin\theta\cos\phi\mathbf{n}_1 - \sin\theta\sin\phi\mathbf{n}_2 + \cos\theta\mathbf{n}_3),$$

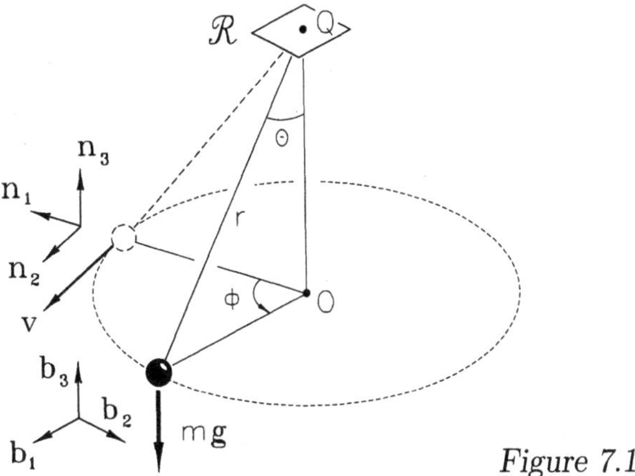

Figure 7.1

with a module $T = mg/\cos\theta$ (see Example 2.4). The work performed on P by the traction on the string is, therefore, null since this force, like the weight, is also orthogonal to the trajectory. The resulting impulse of the force exerted by the string on the sphere, in this same time interval, is

$$
\mathbf{I}_{0t_0}^{T} = \int_0^{t_0} \mathbf{T}\, dt
$$

$$
= \frac{mg}{\cos\theta} \int_0^{t_0} \left(-\sin\theta \cos\phi \mathbf{n}_1 - \sin\theta \sin\phi \mathbf{n}_2 + \cos\theta \mathbf{n}_3 \right) dt
$$

$$
= m\sqrt{gr\cos\theta}\left(-2\tan\theta\,\mathbf{n}_2 + \pi\,\mathbf{n}_3 \right).
$$

Looking at the above result, it is found that the component of the impulse of the force $\mathbf{T}$ in the direction of $\mathbf{n}_1$ vanishes. This occurs because the component of the force in this direction is exactly its opposite during each half of the period in question (see Fig. 7.1), which does not happen with the other two components of force. Note also that the string exerts a force $-\mathbf{T}$ on the fixed point Q, opposite to that exerted on P. Hence the result is that the impulse of this force acting on Q in the interval $(0, t_0)$ is also opposite to the impulse calculated above. The conclusion, therefore, is that the impulse on the fixed point Q, in this interval, is different from zero; the work of $-\mathbf{T}$ on Q will, nevertheless, be null. See the corresponding animation.

If P is a particle moving arbitrarily in a reference frame $\mathcal{R}$ under the action of a force system $\mathcal{F}$, of resultant $\mathbf{R}$, it is called the *resultant impulse*, in a given time interval (t_1, t_2), to the vector sum of the impulses of each of the forces component of $\mathcal{F}$ in that interval, that is,

$$\mathbf{I}_{12}^{\mathcal{F}} \rightleftharpoons \sum_{i=1}^{n} \mathbf{I}_{12}^{F_i} = \sum_{i=1}^{n} \int_{t_1}^{t_2} \mathbf{F}_i \, dt, \qquad (7.2)$$

where the $\mathbf{F}_i$, $i = 1, 2, \ldots, n$, are the forces forming $\mathcal{F}$. Once the system $\mathcal{F}$ can be reduced to a force equal to its resultant applied to P, the resultant impulse in a given interval will be equal to the impulse of the resultant force in that interval. In fact,

$$\mathbf{I}_{12}^{\mathcal{F}} = \sum_{i=1}^{n} \int_{t_1}^{t_2} \mathbf{F}_i \, dt = \int_{t_1}^{t_2} \sum_{i=1}^{n} \mathbf{F}_i \, dt = \int_{t_1}^{t_2} \mathbf{R} \, dt = \mathbf{I}_{12}^{R}. \qquad (7.3)$$

Example 7.2   Returning to the previous example, the resultant of the force system acting on the sphere P is (see Fig. 7.1)

$$\mathbf{R} = -T \sin \theta \mathbf{b}_1 + (T \cos \theta - mg)\mathbf{b}_3 = -T \sin \theta \mathbf{b}_1,$$

and its impulse, in the interval $(0, t_0)$, is simply the vector sum of the impulses due to the weight force and the string's traction, individually, that is,

$$\mathbf{I}_{0t_0}^{R} = \mathbf{I}_{0t_0}^{\mathcal{F}} = \mathbf{I}_{0t_0}^{P} + \mathbf{I}_{0t_0}^{T} = -2m \tan \theta \sqrt{gr \cos \theta} \, \mathbf{n}_2.$$

Note that the vertical component of the impulse due to the string was annulled by the impulse due to the sphere's weight. Do you think that this would happen for any chosen time interval?

If a particle P moves in an inertial reference frame $\mathcal{R}$ under the action of a force system $\mathcal{F}$, with resultant $\mathbf{R}$, the change of the momentum vector of P in $\mathcal{R}$, in a given interval (t_1, t_2), is equal to the resultant impulse in this interval, that is,

$${}^{\mathcal{R}}\mathbf{G}^{P}(t_2) - {}^{\mathcal{R}}\mathbf{G}^{P}(t_1) = \mathbf{I}_{12}^{\mathcal{F}}. \qquad (7.4)$$

In fact, integrating in time the equation of motion of the particle, Eq. (2.1), we have

$$\int_{t_1}^{t_2} {}^{\mathcal{R}}\dot{\mathbf{G}}^{P} \, dt = \int_{t_1}^{t_2} \mathbf{R} \, dt, \qquad (7.5)$$

which immediately leads to Eq. (7.4).

Example 7.3   Returning once again to Example 7.1 (see Fig. 7.1), the momentum vector of P in $\mathcal{R}$ at the instant $t = 0$ is $^{\mathcal{R}}\mathbf{G}^{P}(0) = mv\mathbf{n}_2$ and at the instant $t = t_0$ is $^{\mathcal{R}}\mathbf{G}^{P}(t_0) = -mv\mathbf{n}_2$. The change of its momentum (after minus before) is $-2mv\mathbf{n}_2 = -2m\tan\theta\sqrt{gr\cos\theta}\,\mathbf{n}_2$, equal, therefore, to the resultant impulse in this interval, as determined in Example 7.2.

When a force of relatively great intensity acts on a particle P during a relatively short time interval, it is said that there is an *impact* or *collision*. A force with *relatively great intensity* is understood to be that which, during the time interval in which it acts, has an average module of a much higher value than any of the other forces acting on P; a time interval is understood to be *relatively short* when it is quite small, compared with the time duration of the motion under study.

Example 7.4   The cursor C slides along the sloping straight guide, under the action of its weight, **P**, of the force exerted by the linear spring, $\mathbf{F}_k$, and the force exerted by the guide, the latter conveniently decomposed in $\mathbf{F}_a$, friction force, and **N**, component orthogonal to the motion (see Fig. 7.2).

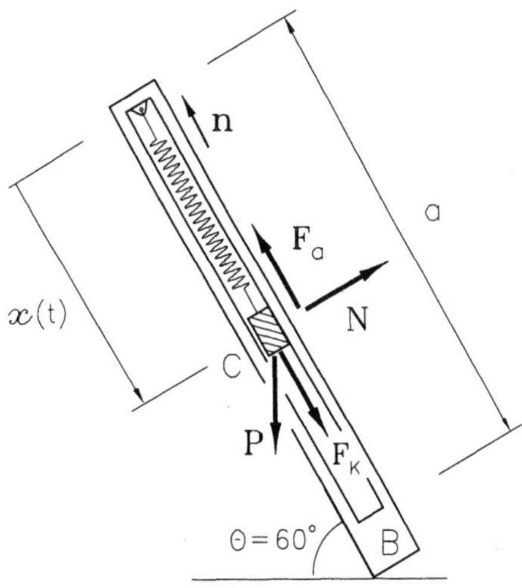

Figure 7.2

At a certain instant, the cursor reaches point B, colliding with the end of the guide. After the impact, the cursor resumes its motion along the guide, in the opposite direction. Just to give an idea of the scope involving the phenomenon, let us imagine the cursor as a steel cylinder with a diameter of 2.5 cm, length 10 cm, and mass 1.5 kg. The duration of the impact between the cursor and stopper is almost exactly equal to the propagation time, there and back, of the wave of compression occurring inside the cursor, after the contact with the end. As this wave propagates in the steel at a velocity of, approximately, 5000 m/s, the total duration of the impact will not be more than 4×10^{-5} s, that is, 40 μs, while a complete cycle of the cursor's motion along the guide occurs in a few seconds. On the other hand, the forces involved before and after the impact are of the order of its weight (14.7 N). Imagining the cursor approaching the guide end at a velocity of, say, 1 m/s, and leaving it, let us assume, with half this velocity, the overall change of its momentum (and, therefore, the impulse undergone), during the collision, will have been 2.25 Ns. Now, a force of impact $\mathbf{F}_I$, to cause this impulse in a time interval of 40 μs, must have an average value of 56,250 N, much greater than the other forces involved. See the corresponding animation.

Impact forces are generally unknown functions of time, which makes it impossible to directly determine their impulse. If the velocity of the particle suffering the impact immediately before and after the phenomenon is known, the impulse may be determined indirectly. In fact, if during the impact a force $\mathbf{F}_I$ has a much higher intensity than the others, the impulse of the impact force will, with a good approximation, be equal to the resulting impulse that, in turn, satisfies Eq. (7.4), that is,

$$\mathbf{I}_{12}^{F_I} \approx \mathbf{I}_{12}^{\mathcal{F}} = m\big(\mathbf{v}(t_2) - \mathbf{v}(t_1)\big). \tag{7.6}$$

Example 7.5   Returning to Example 7.4 (see Fig. 7.2), let us assume that the dynamic friction coefficient is $\mu = 0.2$, that the spring has elastic constant k and a natural length a, and that the cursor is left at rest in the position $x = a/2$. It was also noted that, after the impact with the stopper, the cursor returns, reaching the position $x = 3a/4$ with null velocity. As there is no motion in the orthogonal direction to the guide, $N = P\cos\theta = \frac{1}{2}mg$ and, as it is sliding, $F_a = \mu N = \frac{1}{10}mg$. So, making a balance of energy between the instant t_0, when $x = a/2$, and the instant t_1, when $x = a$, immediately before the impact, as in Eq. (6.6), then

$$E(t_1) - E(t_0) = {}^{\mathcal{R}}\mathcal{T}_{12}^{F_a},$$

where

$$E(t_1) = \frac{1}{2}mv^2(t_1); \qquad E(t_0) = \frac{1}{8}ka^2 + \frac{\sqrt{3}}{4}mga,$$

and

$$^{\mathcal{R}}\mathcal{T}^{F_a}_{01} = \int_0^{a/2} -F_a\,dx = -\frac{1}{20}mga;$$

therefore,

$$v(t_1) = \frac{1}{2}\left[a\left(2\frac{5\sqrt{3}-1}{5}g + \frac{ka}{m}\right)\right]^{\frac{1}{2}}.$$

From the instant t_2, immediately after the collision, until the instant t_3, when the cursor is again at null velocity, the balance of energy expressed by Eq. (6.6) is as follows:

$$\frac{\sqrt{3}}{8}mga + \frac{1}{32}ka^2 - \frac{1}{2}mv^2(t_2) = -\frac{1}{40}mga,$$

resulting in

$$v(t_2) = \frac{1}{2}\left[a\left(\frac{1+5\sqrt{3}}{5}g + \frac{1}{4}\frac{ka}{m}\right)\right]^{\frac{1}{2}}.$$

During the short interval (t_1, t_2) in which the collision occurs, the module of the impact force is much greater, as seen in the preceding example, than that of the other forces involved and their impulse is, for all practical purposes, equal to the resultant impulse. The impulse of the impact force is then, according to Eq. (7.6),

$$\mathbf{I}^{F_I}_{12} = m\big(\mathbf{v}(t_2) - \mathbf{v}(t_1)\big)$$

$$= \frac{m}{2}\left[\left[a\left(\frac{1+5\sqrt{3}}{5}g + \frac{1}{4}\frac{ka}{m}\right)\right]^{\frac{1}{2}}\right.$$

$$\left. + \left[a\left(2\frac{5\sqrt{3}-1}{5}g + \frac{ka}{m}\right)\right]^{\frac{1}{2}}\right]\mathbf{n}.$$

See the corresponding animation.

When a particle P undergoes an impact force in contact with a surface, it is called the *restitution coefficient* to the ratio between the modules of the components orthogonal to the surface, at the point of contact, of the velocities of P immediately after and before the impact, relative to the surface. Thus, if $\mathbf{n}$ is the orthogonal unit vector outside the surface S at the point of contact with P (see Fig. 7.3), $v_1 \cos\theta_1 = -{}^S\mathbf{v}^P(t_1) \cdot \mathbf{n}$ is the module of the orthogonal component to S of the velocity of P in S before the impact, called the *velocity of approach*, and $v_2 \cos\theta_2 = {}^S\mathbf{v}^P(t_2) \cdot \mathbf{n}$ is the orthogonal component after impact, called the *velocity of separation*, then the restitution coefficient, ϵ, is defined as

$$\epsilon \rightleftharpoons \frac{v_2 \cos\theta_2}{v_1 \cos\theta_1}. \tag{7.7}$$

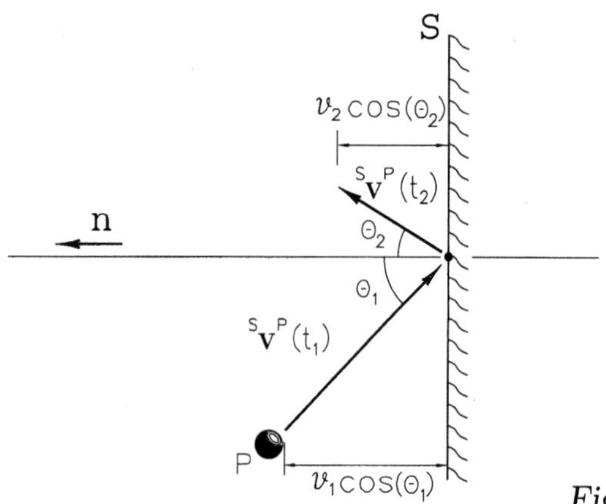

Figure 7.3

When a particle P collides orthogonally with a surface S, the impact is said to be *normal*. In this case, knowing the velocity of P in S immediately before the impact and the restitution coefficient, we may obtain the impulse by the expression (note that $\theta_1 = \theta_2 = 0$)

$$\begin{aligned}
\mathbf{I}_{12}^{F_I} &= m\big(\mathbf{v}(t_2) - \mathbf{v}(t_1)\big) \\
&= m(v_2 + v_1)\mathbf{n} \\
&= mv_1(1 + \epsilon)\mathbf{n}.
\end{aligned} \tag{7.8}$$

The phenomenon of impact is fairly complex, requiring for a more detailed study considerations involving the stresses and strains present in the regions of contact, as well as the propagation of compressive waves in the bodies involved. From the point of view of the model of particle and rigid body, therefore, an overall coefficient, ϵ, is adopted, as in the case of the Coulomb friction phenomenon, to express the effect of the impact. The restitution coefficient, however, depends on both the materials and the geometry of the bodies in question; it also depends on the velocity of the collision. The result, then, is that tables of restitution coefficients cannot be used as often as tables for friction coefficients, which, in practice, restricts their use.

Example 7.6 Returning once more to Example 7.4, now consider that the spring has an elastic constant $k = 8\sqrt{3}mg/a$. The impact between the cursor and stopper is normal and the restitution coefficient is, according to Eq. (7.7),

$$\epsilon = \frac{v(t_2)}{v(t_1)} = \sqrt{\frac{1 + 15\sqrt{3}}{50\sqrt{3} - 2}} \approx 0.552.$$

The impulse of the impact, according to Eq. (7.8), is

$$\begin{aligned}
\mathbf{I}_{12}^{F_I} &= mv(t_1)(1 + \epsilon)\,\mathbf{n} \\
&= 3.218\,m\sqrt{ga}\,\mathbf{n}.
\end{aligned}$$

See the corresponding animation.

When the impact is not normal, it is convenient to decompose the impact force $\mathbf{F}_I$ in a normal component and another of friction (see Fig. 7.4). The impulse due to the impact will, therefore, have a component orthogonal to the surface of contact and another tangent to this, due to the component of friction. When the friction component is very small compared with the normal component, then *smooth impact* occurs; in this case the impulse is a vector parallel to the normal $\mathbf{n}$.

Example 7.7 The small ball B is thrown, moving freely on a vertical plane, until it collides with the horizontal floor, then bouncing on the same original vertical plane, as illustrated (see Fig. 7.5). If v_1, v_2, θ_1, and θ_2 measure the modules and directions, respectively, of the velocity vectors immediately before and after impact, the impulse exerted by the floor will be

$$\mathbf{I}_{12}^{F_I} = \mathbf{I}_{12}^{N} + \mathbf{I}_{12}^{F_a},$$

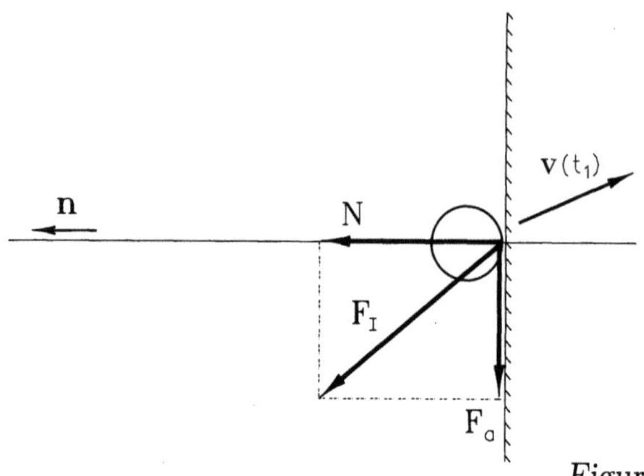

Figure 7.4

where

$$\mathbf{I}_{12}^{N} = \left(\mathbf{G}_2^P - \mathbf{G}_1^P\right) \cdot \mathbf{n}_2\,\mathbf{n}_2 = m(v_2 \cos\theta_2 + v_1 \cos\theta_1)\mathbf{n}_2$$

and

$$\mathbf{I}_{12}^{F_a} = \left(\mathbf{G}_2^P - \mathbf{G}_1^P\right) \cdot \mathbf{n}_1\,\mathbf{n}_1 = m(v_2 \sin\theta_2 - v_1 \sin\theta_1)\mathbf{n}_1.$$

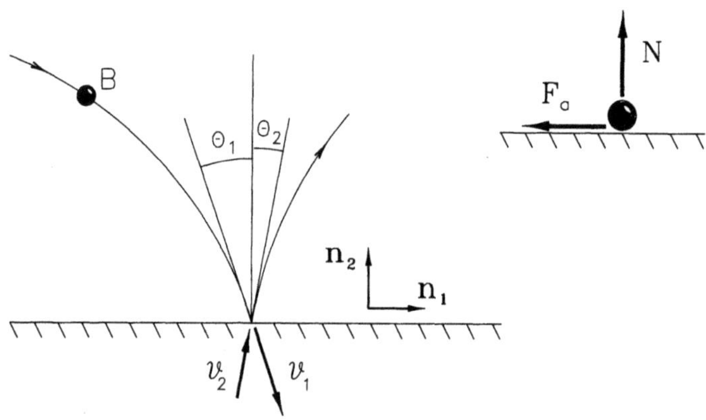

Figure 7.5

Note that the normal impulse component is always positive, toward the outside normal; the friction component, however, will be null if the impact is smooth, giving, in this case, the equation

$$v_2 \sin\theta_2 = v_1 \sin\theta_1. \tag{7.9}$$

It is also worth noting that the phenomenon of impact, even when smooth, involves nonconservative work, resulting, according to Eq. (6.6), in dissipation of the mechanical energy of the particle. In fact, even though Eq. (7.9) is met, that is, that $\mathbf{I}^{F_a} = 0$, the following condition must be satisfied:

$$v_2 \cos \theta_2 \leq v_1 \cos \theta_1, \tag{7.10}$$

which is equivalent to

$$\epsilon \leq 1. \tag{7.11}$$

Equality, in the above equations, expresses a limit condition — never effectively satisfied in practice — called *elastic collision*. In elastic collision, there would be no absorption in the deformation process inherent to the phenomenon of impact. Effectively, elastic collision is assumed when this absorption (a dissipative phenomenon) is very small. In fact, as in the instants immediately before and after impact, the position of the particle is essentially the same, the change in its mechanical energy will be exclusively due to the change in its kinetic energy, with the result that the work performed by the impact force, $\mathbf{F}_I$, will be, according to Eq. (6.6),

$$\mathcal{T}_{12}^{F_I} \approx \frac{1}{2}m(v_2^2 - v_1^2). \tag{7.12}$$

When the collision is normal, $\theta_1 = \theta_2 = 0$ and the work of the impact force may be expressed as

$$\mathcal{T}_{12}^{F_I} \approx \frac{1}{2}mv_1^2(\epsilon^2 - 1). \tag{7.13}$$

Only elastic collision, therefore, keeps the mechanical energy constant; if the contrary holds, the work of the impact force will be negative.

Example 7.8 Returning once again to Example 7.4 (see Fig. 7.2), the cursor's change in mechanical energy, $E^P(t_2) - E^P(t_1)$, due to the impact with the stopper, may be measured by the work of the force of impact that, according to Eq. (7.13), is

$$E^P(t_2) - E^P(t_1) = \frac{1}{2}mv^2(t_1)(\epsilon^2 - 1)$$

$$= -1.54\,mga.$$

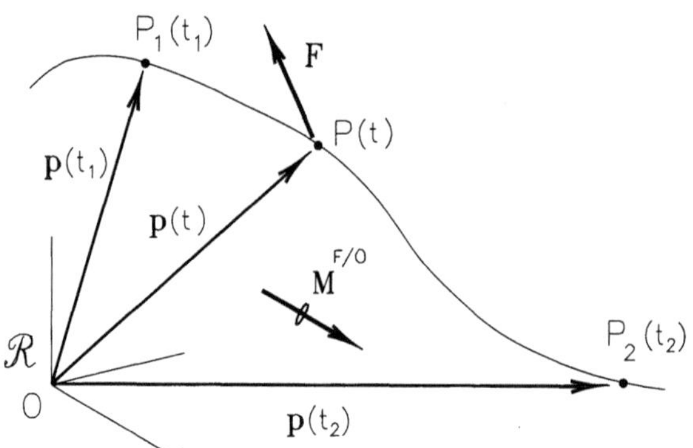

Figure 7.6

If a force **F** acts on a point P moving in a reference frame $\mathcal{R}$ for a certain time interval (t_1, t_2) and its moment with respect to a point O, fixed in $\mathcal{R}$ (see Fig. 7.6), is $\mathbf{M}^{F/O}$, a vector function of time, is called the *angular impulse*, the vector resulting from the integration of the moment during this interval, that is,

$$\mathbf{I}_{12}^{F/O} \rightleftharpoons \int_{t_1}^{t_2} \mathbf{M}^{F/O}\, dt = \int_{t_1}^{t_2} \mathbf{p} \times \mathbf{F}\, dt. \qquad (7.14)$$

The angular impulse of a force bound to a point P with respect to another fixed point O is a vector of dimension $[ML^2T^{-1}]$, depending on how the force (and, consequently, its moment) varies with time.

If P is a particle moving in a reference frame $\mathcal{R}$ under the action of a simple system of forces $\mathcal{F}$, of resultant **R**, it is called the *resultant angular impulse* with respect to a given point O, in a certain time interval (t_1, t_2), to the vector sum of the angular impulses with respect to O of each of the forces forming $\mathcal{F}$ in that interval, that is,

$$\mathbf{I}_{12}^{\mathcal{F}/O} \rightleftharpoons \sum_{i=1}^{n} \int_{t_1}^{t_2} \mathbf{M}_i\, dt, \qquad (7.15)$$

where $\mathbf{M}_i = \mathbf{p} \times \mathbf{F}_i$, $i = 1, 2, \ldots, n$, are the moments, with respect to O, of the forces forming $\mathcal{F}$. It is easy to see, then, that the resultant angular impulse is equal to the angular impulse of the resultant moment vector of the system, that is,

$$\mathbf{I}_{12}^{\mathcal{F}/O} = \sum_{i=1}^{n} \int_{t_1}^{t_2} \mathbf{M}_i\, dt = \int_{t_1}^{t_2} \sum_{i=1}^{n} \mathbf{M}_i\, dt = \int_{t_1}^{t_2} \mathbf{M}^{\mathcal{F}/O}\, dt. \qquad (7.16)$$

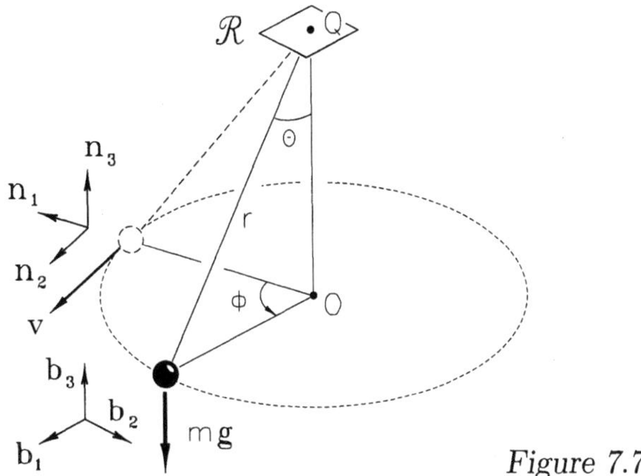

Figure 7.7

If a particle P moves in an inertial reference frame $\mathcal{R}$, under the action of a simple system of forces $\mathcal{F}$, of resultant $\mathbf{R}$, the change in $\mathcal{R}$ of the angular momentum of P with respect to a point O, fixed in $\mathcal{R}$, in a given time interval (t_1, t_2), is equal to the resultant angular impulse acting on P, in this interval, that is,

$$^{\mathcal{R}}\mathbf{H}^{P/O}(t_2) - {}^{\mathcal{R}}\mathbf{H}^{P/O}(t_1) = \mathbf{I}_{12}^{\mathcal{F}/O}. \tag{7.17}$$

In fact, when Eq. (4.3) is integrated in time, then

$$\int_{t_1}^{t_2} {}^{\mathcal{R}}\dot{\mathbf{H}}^{P/O}\, dt = \int_{t_1}^{t_2} \mathbf{M}^{\mathcal{F}/O}\, dt, \tag{7.18}$$

leading immediately to Eq. (7.17).

Example 7.9 Let us go back again to Example 7.1 (see Fig. 7.7), where a small sphere P, with mass m, moves in a circular trajectory on the horizontal plane, at a velocity of a constant module $v = \sqrt{gr \sin\theta \tan\theta}$. The system of forces $\mathcal{F}$ acting on P consists, as already seen, of the traction of the string, $\mathbf{T}$, and its weight, $\mathbf{P} = m\mathbf{g}$. The resultant moment of this system with respect to the fixed point Q is

$$\begin{aligned}
\mathbf{M}^{\mathcal{F}/Q} &= \mathbf{M}^{\mathbf{T}/Q} + \mathbf{M}^{\mathbf{P}/Q} \\
&= 0 + r(\sin\theta \mathbf{b}_1 - \cos\theta \mathbf{b}_3) \times (-mg)\mathbf{b}_3 \\
&= mgr \sin\theta \mathbf{b}_2.
\end{aligned}$$

The string's traction, passing through Q, does not contribute to the resultant moment. The resultant angular impulse in the interval $(0, t_0)$, that is, along a semicircle in the trajectory, is, according to Eqs. (7.14) and (7.16),

$$
\begin{aligned}
\mathbf{I}_{0t_0}^{\mathcal{F}/Q} &= \int_0^{t_0} \mathbf{M}^{\mathcal{F}/Q}\, dt \\
&= \int_0^{t_0} mgr\sin\theta(-\sin\phi\,\mathbf{n}_1 + \cos\phi\,\mathbf{n}_2)\, dt \\
&= -2mvr\cos\theta\,\mathbf{n}_1 .
\end{aligned}
$$

The angular momentum of P with respect to point Q in the inertial reference frame $\mathcal{R}$ is

$$
{}^{\mathcal{R}}\mathbf{H}^{P/Q} = r(\sin\theta\,\mathbf{b}_1 - \cos\theta\,\mathbf{b}_3) \times mv\mathbf{b}_2 = mvr(\cos\theta\,\mathbf{b}_1 + \sin\theta\,\mathbf{b}_3).
$$

At the first instant, $t = 0$, this vector, expressed on the fixed basis in $\mathcal{R}$, is

$$
{}^{\mathcal{R}}\mathbf{H}^{P/Q}(0) = mvr(\cos\theta\,\mathbf{n}_1 + \sin\theta\,\mathbf{n}_3)
$$

and, after turning halfway around, then

$$
{}^{\mathcal{R}}\mathbf{H}^{P/Q}(t_0) = mvr(-\cos\theta\,\mathbf{n}_1 + \sin\theta\,\mathbf{n}_3).
$$

The change of the angular momentum vector in this interval is then

$$
{}^{\mathcal{R}}\mathbf{H}^{P/Q}(t_0) - {}^{\mathcal{R}}\mathbf{H}^{P/Q}(0) = -2mvr\cos\theta\,\mathbf{n}_1 ,
$$

equal, therefore, to the resultant angular impulse in the period, as established by Eq. (7.17). See the corresponding animation.

4.8 Conservation Principles

The equations governing the motion of a particle in an inertial reference frame, studied in the previous sections, are of two kinds. The first kind establishes the equality between the time rate of a vector property of the particle and an acting vector, that may be called the *generation term*. Equations (2.1) and (4.3) fit into this category: The resultant force is the term of generation of momentum of the particle, present in the first equation; the resultant moment is the term of generation of angular momentum of the particle, present in the second. A second class of equations — always arising from the integration of an equation of the first kind — establishes the equality between the change in a property (vector or scalar) between two configurations of the particle and a certain action exerted on it. Equations (6.3), (6.6), (7.4), and (7.17) belong to this category. In fact, Eqs. (6.3) and (6.6) establish that the change in the energy (kinetic or mechanical) of the particle between two positions in its trajectory is equal to the resultant work (total or nonconservative, respectively) done between these positions; Eqs. (7.4) and (7.17) prescribe that the change of the momentum (or angular momentum) of the particle between two instants of its motion is equal to the resultant impulse (or angular impulse) exerted between these instants.

This section is devoted to the study of what are called *conservation principles*, equations derived from the aforementioned when the terms of generation or their integrals vanish. In every equation in this section, t_1 and t_2 represent two arbitrary instants.

When a particle P, with mass m, moves in an inertial reference frame $\mathcal{R}$ under the action of a null force system, its momentum vector is conserved in this reference frame. This result, known as the *principle of conservation of momentum*, comes from Eq. (2.1). In fact, if $\mathcal{F}$ is a null system, its resultant $\mathbf{R}$ is also null, following that $^{\mathcal{R}}\dot{\mathbf{G}}^P = 0$ and, therefore,

$$^{\mathcal{R}}\mathbf{G}^P(t_1) = {}^{\mathcal{R}}\mathbf{G}^P(t_2) \qquad \text{if} \quad \mathbf{R} = 0. \tag{8.1}$$

When a particle is subject to a null force system, it is said to be in *equilibrium*. The momentum of a particle in equilibrium, therefore, satisfies Eq. (8.1). Substituting Eq. (1.5) in Eq. (8.1), then the alternative expression for stating the condition of equilibrium is

$$^{\mathcal{R}}\mathbf{v}^P(t_1) = {}^{\mathcal{R}}\mathbf{v}^P(t_2) \qquad \text{if} \quad \mathbf{R} = 0, \tag{8.2}$$

an equation that expresses the fact that a particle submitted to a null force system moves at a constant velocity in an inertial reference frame, as established by *Newton's first law*.

Example 8.1 The small block P, with mass m, is confined to move on the rail fixed to the sloping plane, as illustrated (see Fig. 8.1), with a coefficient of dynamic friction $\mu = 1/\sqrt{7}$. When adopting the basis $\mathbf{n}_1, \mathbf{n}_2, \mathbf{n}_3$ in the indicated directions, the force exerted by the rail on the block will have three components, $\mathbf{F}_1$, $\mathbf{F}_2$, and $\mathbf{F}_3$, the scalar component of friction being $F_1 = \mu\sqrt{F_2^2 + F_3^2}$ (assuming that the block is sliding). The resultant force on P is then

$$\mathbf{R} = (mg\sin\theta\cos\phi - F_1)\mathbf{n}_1 + (F_2 - mg\sin\theta\sin\phi)\mathbf{n}_2 + (F_3 - mg\cos\theta)\mathbf{n}_3$$

$$= (\frac{\sqrt{2}}{4}mg - F_1)\mathbf{n}_1 + (F_2 - \frac{\sqrt{2}}{4}mg)\mathbf{n}_2 + (F_3 - \frac{\sqrt{3}}{2}mg)\mathbf{n}_3.$$

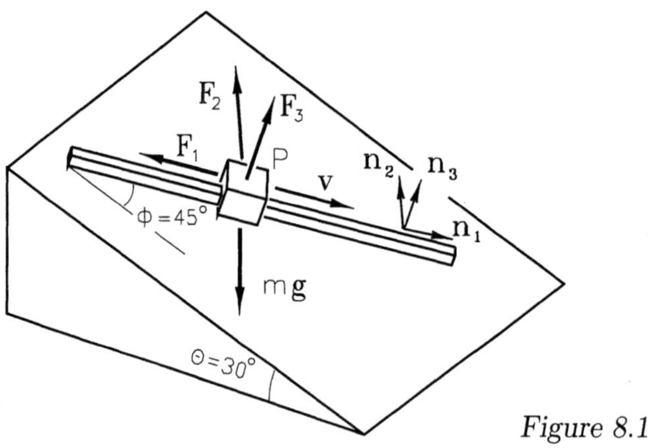

Figure 8.1

As the cursor is confined to move over the rail, its acceleration is $\mathbf{a}^P = a\mathbf{n}_1$, and the equations of motion for the directions $\mathbf{n}_2$ and $\mathbf{n}_3$ are reduced to

$$F_2 = \frac{\sqrt{2}}{4}mg, \qquad F_3 = \frac{\sqrt{3}}{2}mg.$$

The friction component is then

$$F_1 = \frac{1}{\sqrt{7}}\sqrt{\frac{2}{16} + \frac{3}{4}}\, mg = \frac{\sqrt{2}}{4}mg.$$

The resultant force is therefore null, and the block slides over the rail at a constant velocity, conserving its momentum.

A particular case of the condition of equilibrium is *rest*, that is, the situation in which the velocity of the particle in the inertial reference frame is null. In this case, Eq. (8.2) may be expressed as

$$\mathbf{R} = 0 \qquad \text{if} \quad {}^{\mathcal{R}}\mathbf{v}^P(t) = 0, \qquad \text{for every } t, \qquad (8.3)$$

that is, a useful equation for determining possibly unknown components in the system of forces (note that $\mathbf{R} = 0$ is an equation satisfied in any case of equilibrium, not being restricted to the condition of rest). With the adoption of an orthonormal basis $\mathbf{n}_1, \mathbf{n}_2, \mathbf{n}_3$, fixed in $\mathcal{R}$, the condition of equilibrium may be expressed by the scalar equations:

$$\begin{aligned} R_1 &= 0; \\ R_2 &= 0; \\ R_3 &= 0, \end{aligned} \qquad (8.4)$$

where R_j, $j = 1, 2, 3$, are the scalar components of the resultant force on the chosen basis.

Example 8.2 A small block B, with mass m, is at rest, supported on the top of a table and under the action of a linear elastic spring with a constant k and natural length a, as illustrated (see Fig. 8.2). We wish to determine the force exerted by the table top on B. The system $\mathcal{F}$ of forces acting on the block consists of its weight, $-mg\mathbf{n}_2$; the action of the spring, $\mathbf{T} = k\delta\mathbf{n} = \frac{1}{2}ka(-\sqrt{3}\mathbf{n}_1 + \mathbf{n}_2)$; and the force exerted by the table, conveniently resolved into $\mathbf{F}_1$, friction force, and $\mathbf{F}_2$, orthogonal to the top.

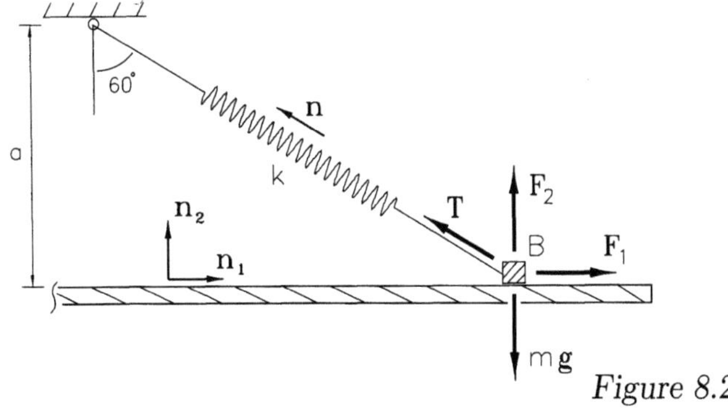

Figure 8.2

As the block is at rest (which implies that the friction coefficient is enough to guarantee the equilibrium), the resultant must be null, leading to the equations

$$F_1 + T_1 = 0,$$

$$F_2 + T_2 - mg = 0,$$

which result in

$$F_1 = \frac{\sqrt{3}}{2}ka,$$

$$F_2 = mg - \frac{1}{2}ka.$$

The force exerted by the table top is then

$$\mathbf{F} = \frac{1}{2}\big(\sqrt{3}ka\mathbf{n}_1 + (2mg - ka)\mathbf{n}_2\big).$$

If, for example, the constant of the spring is $k = mg/2a$, we have

$$F_1 = \frac{\sqrt{3}}{4}mg,$$

$$F_2 = \frac{3}{4}mg,$$

and the coefficient of dry friction present shall satisfy the condition

$$\mu \geq \frac{\sqrt{3}}{3}.$$

If, on the other hand, the friction coefficient is

$$\mu = 1,$$

the elastic constant of the spring may not be greater than

$$k = \frac{2mg}{(1 + \sqrt{3})a}$$

to guarantee the equilibrium (check).

When a particle P, with mass m, moves in an inertial reference frame $\mathcal{R}$ under the action of a force system $\mathcal{F}$ so that its resultant moment with respect to a point O, fixed in $\mathcal{R}$, is null, its angular momentum vector with respect to this point is conserved. This result, called the *principle of conservation of angular momentum*, derives from Eq. (4.3). In fact, if $\mathbf{M}^{\mathcal{F}/O} = 0$, then ${}^{\mathcal{R}}\dot{\mathbf{H}}^{P/O} = 0$ and

$$\mathbf{{}^{\mathcal{R}}H}^{P/O}(t_1) = \mathbf{{}^{\mathcal{R}}H}^{P/O}(t_2) \qquad \text{if} \quad \mathbf{M}^{\mathcal{F}/O} = 0. \tag{8.5}$$

Example 8.3   A small sphere P, with mass m, moves over a smooth, fixed circular platform, under the action of a thread, passing through a small opening in the center O of the platform (see Fig. 8.3). The sphere initially moves with a velocity of constant module v in a circumference of radius r_0, under a traction of module $T_0 = mv^2/r_0$, exerted by the thread. Suddenly, a new force of module $T > T_0$ is applied to the thread. In this condition, both the velocity of the sphere and the radius r will vary. We wish to determine how the tangential component of velocity, v_2, will vary with r. On adopting the mobile basis $\mathbf{n}_1, \mathbf{n}_2, \mathbf{n}_3$, where $\mathbf{n}_3$, being vertical, is fixed in the inertial reference frame, the equation of motion in this direction is reduced to $\mathbf{N} = mg\mathbf{n}_3$, that is, the force exerted by the platform on the sphere is the opposite of its weight (there is no friction).

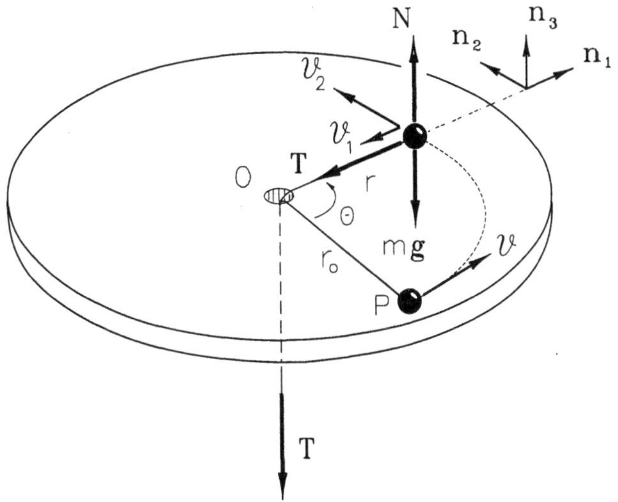

Figure 8.3

As the other force acting on P passes through O, it is concluded that the resultant moment of the force system acting on the sphere with respect to point O is null, thereby satisfying Eq. (8.5). The position vector with respect to point O, at a general instant, is $\mathbf{p} = r\mathbf{n}_1$, and the velocity vector at this same instant, in the inertial reference frame, is

$$^{\mathcal{R}}\mathbf{v}^P = \dot{r}\mathbf{n}_1 + r\dot{\theta}\mathbf{n}_2.$$

The angular momentum vector of P with respect to O, in this reference frame, is then

$$^{\mathcal{R}}\mathbf{H}^{P/O} = \mathbf{p} \times m\,^{\mathcal{R}}\mathbf{v}^P = r\mathbf{n}_1 \times m(\dot{r}\mathbf{n}_1 + r\dot{\theta}\mathbf{n}_2) = mr^2\dot{\theta}\mathbf{n}_3.$$

By making it equal to the initial angular momentum, then

$$mr^2\dot{\theta} = mr_0 v;$$

therefore, the tangential component of the velocity is

$$v_2 = r\dot{\theta} = \frac{r_0}{r}v.$$

See the corresponding animation.

If a particle P moves in an inertial reference frame $\mathcal{R}$ under the action of a force system $\mathcal{F}$ so that its resultant moment with respect to an axis E, fixed in $\mathcal{R}$, is null, its angular momentum vector with respect to this axis is conserved. This condition, which may be called the *principle of conservation of the angular momentum with respect to an axis*, comes from Eq. (4.7) or, equivalently, results from the decomposition, on an arbitrary basis, fixed in an inertial reference frame, of Eq. (8.5). Thus, if x_j is an axis fixed in an inertial reference frame $\mathcal{R}$, so that $M^{\mathcal{F}/x_j} = 0$, then $^{\mathcal{R}}\dot{H}^{P/x_j} = 0$ and

$$^{\mathcal{R}}H^{P/x_j}(t_1) = {}^{\mathcal{R}}H^{P/x_j}(t_2) \qquad \text{if} \quad M^{\mathcal{F}/x_j} = 0. \tag{8.6}$$

Example 8.4   A steel ball B, with mass m, is thrown horizontally with speed v_0 inside a hemispherical bowl with radius r and vertical symmetry axis Z (see Fig. 8.4).

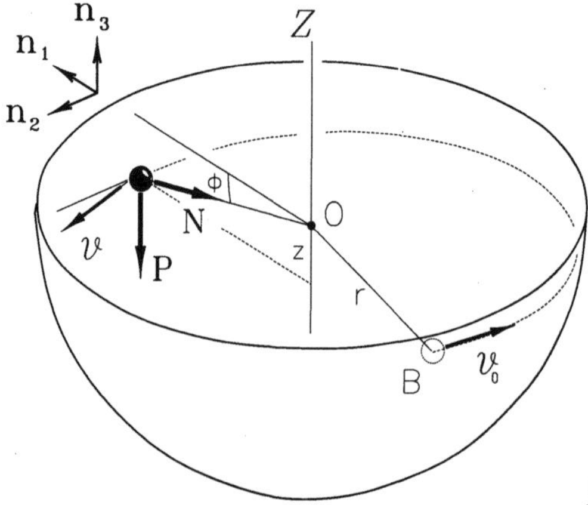

Figure 8.4

The ball describes a trajectory on the spherical surface that may be parametrized by the z-coordinate, measured along the vertical axis, from point O, the center of the hemisphere. The acting forces are the weight of the ball, $\mathbf{P}$, and the force on the smooth contact, $\mathbf{N}$, both coplanar with the Z-axis. It then happens that the resultant moment with respect to this axis is null, therefore conserving the angular momentum of P with respect to Z, that is,

$$H^{P/Z}(z) = H^{P/Z}(0).$$

We have, in the initial condition,

$$H^{P/Z}(0) = mrv_0,$$

and, in a general condition,

$$
\begin{aligned}
H^{P/Z}(z) &= \mathbf{p}^{P/O} \times m\mathbf{v} \cdot \mathbf{n}_3 \\
&= r(\cos\phi\,\mathbf{n}_1 - \sin\phi\,\mathbf{n}_3) \times m(v_1\mathbf{n}_2 + v_2\mathbf{n}_2 + v_3\mathbf{n}_3) \cdot \mathbf{n}_3 \\
&= mrv_2\cos\phi.
\end{aligned}
$$

Therefore,

$$v_2 = v_0/\cos\phi = \frac{r}{\sqrt{r^2 - z^2}}\, v_0,$$

that is, the principle of conservation gives us the component of the ball's velocity, in the direction of the tangent horizontal to the bowl. See the corresponding animation.

When a particle P, with mass m, moves in an inertial reference frame $\mathcal{R}$ under the action of a force system $\mathcal{F}$ so that the resultant work exerted between two positions P_1 and P_2 of its trajectory is null, its kinetic energy is conserved. This *principle of conservation of the kinetic energy* derives immediately from Eq. (6.3). Thus, if $\mathbf{p}_1$ and $\mathbf{p}_2$ are the position vectors with respect to a point O, fixed in $\mathcal{R}$, of the points P_1 and P_2, respectively, then, from Eq. (6.3) we have

$$
{}^{\mathcal{R}}K^P(\mathbf{p}_1) = {}^{\mathcal{R}}K^P(\mathbf{p}_2), \qquad \text{if } {}^{\mathcal{R}}\mathcal{T}^{\mathcal{F}}_{12} = 0. \tag{8.7}
$$

Example 8.5 The cursor C slides, without friction, over the fixed wire A, in the form of an Archimedes spiral on the horizontal plane, described by $r(\theta) = r_0\theta/2\pi$, as illustrated in Fig. 8.5. The cursor passes through the point P_0 $(\theta = 2\pi)$ with speed v_0. The orthonormal basis $\mathbf{b}_1, \mathbf{b}_2, \mathbf{b}_3$ is fixed in a reference frame B moving at a simple angular velocity $\boldsymbol{\omega} = \omega\mathbf{b}_3$ in A (B may be interpreted as the vertical plane containing the axis x_3 and the center of the cursor), with $\mathbf{b}_1$ radial and $\mathbf{b}_3$ vertical. We wish to determine how this angular velocity ω varies with θ. The intrinsic basis $\mathbf{n}_1, \mathbf{n}_2, \mathbf{n}_3$ also being adopted, where $\mathbf{n}_1$ is the tangent unit vector, $\mathbf{n}_2$ is the principal normal, and $\mathbf{n}_3 = \mathbf{b}_3$ is the binormal unit vector (see Section 3.7), the velocity of C in A may be simply expressed as ${}^A\mathbf{v}^P = v\mathbf{n}_1$ and its kinetic energy in a general position of the trajectory will be

$$ {}^A K^C = \frac{1}{2}mv^2. $$

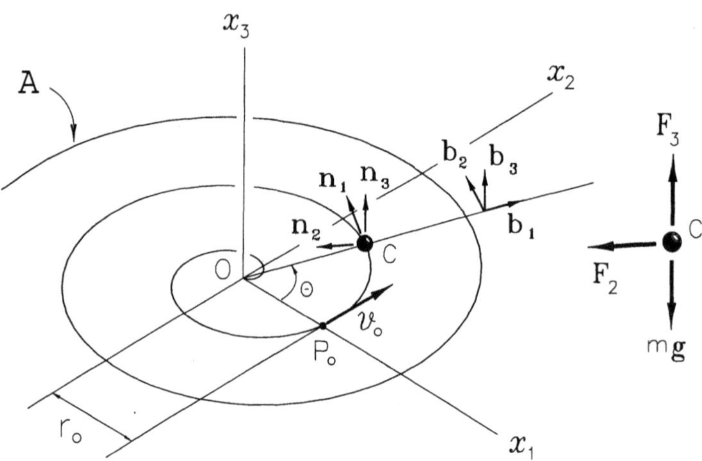

Figure 8.5

The forces acting on C comprise its weight, $-mg\mathbf{n}_3$, and the components $\mathbf{F}_2$ and $\mathbf{F}_3$ exerted by A. There is, therefore, no component of force in the direction of motion, following that the resultant work is null between any two positions of the trajectory and, according to Eq. (8.7), the kinetic energy of C is conserved, that is,

$$ \frac{1}{2}mv^2 = \frac{1}{2}mv_0^2. $$

At any point of the trajectory, therefore, the cursor's velocity will have a module v_0. But the cursor's velocity vector at a general point is

$$\mathbf{v} = \frac{{}^A d}{dt}\mathbf{p} = \frac{{}^A d}{dt}\left(\frac{r_0}{2\pi}\theta\mathbf{b}_1\right) = \frac{r_0}{2\pi}\left(\dot{\theta}\mathbf{b}_1 + \theta\dot{\theta}\mathbf{b}_2\right) = \frac{r_0}{2\pi}\omega(\mathbf{b}_1 + \theta\mathbf{b}_2),$$

and its module will be (assuming $\omega > 0$)

$$v = \frac{r_0}{2\pi}\omega\sqrt{1+\theta^2}.$$

As, in consequence of the principle of conservation of the kinetic energy, the result was $v = v_0$, then the desired equation is

$$\omega(\theta) = \frac{2\pi v_0}{r_0\sqrt{1+\theta^2}}.$$

When a particle P, with mass m, moves in an inertial reference frame $\mathcal{R}$, under the action of a force system $\mathcal{F}$, so that the resultant work of the *nonconservative forces* done between two positions P_1 and P_2 of its trajectory is null, its mechanical energy is conserved. This is called the *principle of conservation of mechanical energy*, being a consequence of Eq. (6.6). In fact, if the resultant work of the nonconservative forces vanishes, then

$$ {}^{\mathcal{R}}E^P(\mathbf{p}_1) = {}^{\mathcal{R}}E^P(\mathbf{p}_2), \qquad \text{if} \quad {}^{\mathcal{R}}\mathcal{T}_{12}^{\mathcal{F}_N} = 0. \tag{8.8}$$

Example 8.6 Returning to Example 8.4 (see Fig. 8.4), we find that there is no resultant work of nonconservative forces. Actually, the force $\mathbf{N}$ does no work at all since it is always orthogonal to the ball's velocity vector and the force $\mathbf{P}$ is conservative. According to Eq. (8.8), therefore, the mechanical energy of the sphere is conserved, that is,

$$E^P(z) = E^P(0);$$

therefore,

$$\frac{1}{2}mv^2 - mgz = \frac{1}{2}mv_0^2 + 0,$$

an equation that, solved for v, gives

$$v(z) = \sqrt{v_0^2 + 2gz}.$$

Note that, when using the principle of conservation of mechanical energy, it was easy to find the velocity of the sphere at a general point in its trajectory, as a function of the variable z, which parametrizes it. It should be mentioned that the principle of conservation of the momentum with respect to the vertical axis provides a component of the velocity vector, while the principle of conservation of mechanical energy informs us which is the module of this same vector. Full determination of the velocity vector, therefore, demands further information. In fact, the kinematic constraint imposed on the ball (there is no velocity component in the radial direction) may be expressed as $\mathbf{v} \cdot \mathbf{p}^{P/O} = 0$, that is,

$$(v_1\mathbf{n}_1 + v_2\mathbf{n}_2 + v_3\mathbf{n}_3) \cdot r(\cos\phi\mathbf{n}_1 - \sin\phi\mathbf{n}_3) = 0.$$

Therefore,

$$v_1 = \frac{z}{\sqrt{r^2 - z^2}}\, v_3,$$

which is the third equation required to determine the components of the velocity vector. See the corresponding animation.

This section has shown, with examples, how the use of the conservation principles can make it easy to solve a number of problems involving the general motion of a particle. The reader should, nevertheless, be very careful when using them. First, there must be clear evidence of the existence of a term of generation (a resultant force or resultant moment) or an integral of one of these terms (a work, say) that vanishes along the motion being studied. Second, it must not be forgotten that the equations expressing the conservation principles are only applicable when the motion is described in an inertial reference frame; otherwise the results obtained will be inconsistent. Last, what information is usually taken concerns the *state of motion* of the particle, that is, velocities as a function of the position; when it is necessary to know the position of the particle as a function of time, it will always require resorting to the equations of motion.

Exercise Series #6 (Sections 4.1 to 4.8)

P6.1 A particle moves over the plane $\{x, y\}$, along the ellipsoid $(x^2/a^2) + (y^2/b^2) = 1$, so that its acceleration is always parallel to the y-axis. Knowing that it passes through the position $(0, b)$ at a velocity v, determine which force is required to maintain this trajectory.

P6.2 A particle rises a plane sloping at 30^0, with a coefficient of dynamic friction $\mu = 0.1$. If, at a certain moment, its velocity is 15 m/s, what is the distance covered and the time taken until it stops?

P6.3 A particle moves on a plane under the action of a force of attraction, which originates from a fixed point O and whose module depends exclusively on the distance r between the particle and the point. What are the module of this force and the trajectory described by the particle, knowing that its velocity varies according to $v = a/r$, where a is a constant?

P6.4 The average radius of the earth's equator is approximately 6377 km. How much should the angular velocity of the earth increase so that a body on its surface, at the equator, does not seem to have any weight?

P6.5 An airplane takes off from an airfield in a straight line, so that the acceleration of Coriolis on it is entirely null. Where is this airfield, and in which direction is the runway?

P6.6 A body on a sloping plane at 30^0 with the horizontal is left at rest. What is its velocity after covering a distance of 2 m, knowing that the friction coefficient is 0.1?

P6.7 A light spring is at rest, compressed at δ_1 by a constant force F. If the direction of the force is suddenly inverted, the spring will be stretched to a maximum deformation δ_2. What is the δ_2/δ_1 ratio?

P6.8 A body is thrown vertically upward from the earth's surface at a velocity of 0.5 km/s. Calculate the maximum height reached by the body. Consider the earth's radius at the throwing point equal to 6375 km, ignore the aerodynamic forces, and approximate the gravity field to an action equivalent to two particles attracted to each other, located at the center of the body and of the earth.

P6.9   The bar B rotates at a simple angular velocity, with a constant module ω, in relation to the inertial reference frame A, as illustrated. The cursor C, with mass m, can slide without friction over the horizontal stretch of the bar B, beginning from rest in B with r essentially null. Calculate the module of the force exerted by the bar as a function of the position r. Is there conservation of the angular momentum of C with respect to the x_3-axis?

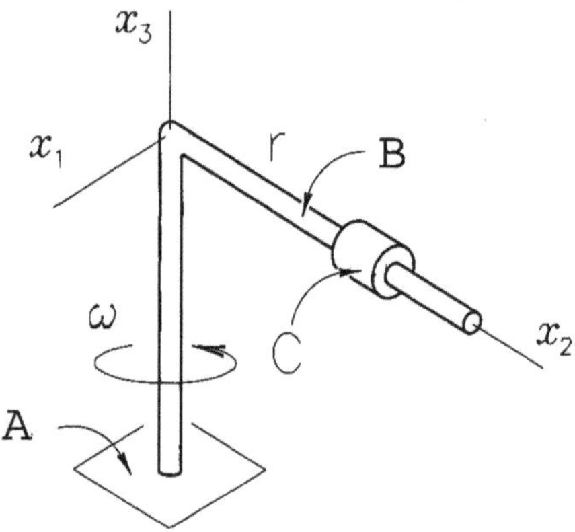

Figure P 6.9

P6.10   A small sphere moves over a hemispherical surface with a radius r, beginning from the top with a horizontal velocity of module v. Determine the angular position θ where the sphere leaves the surface. What should the minimum value of v be so that the sphere leaves the surface, right from the start of its motion? What is the angle θ_M when the sphere leaves the surface, whatever the first condition may be?

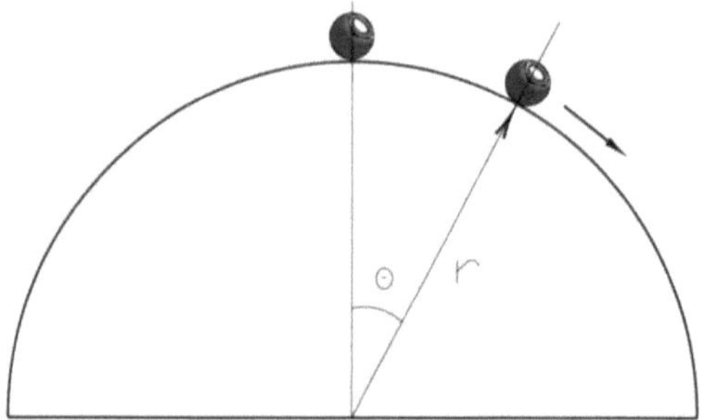

Figure P 6.10

P6.11 The conical rotor B turns around vertical axis z, fixed in an inertial reference frame, at a simple angular velocity of module $w(t)$, as shown. A small sphere P, with mass m, is left from rest at point Q, sliding without friction along the straight guide. Assuming constant angular velocity $\omega = 2\sqrt{g/a}$, determine the position x of the sphere at instant $t = \sqrt{a/4g}$. Calculate, at this instant, the module of the force exerted by the guide in the direction $\mathbf{n}_2$.

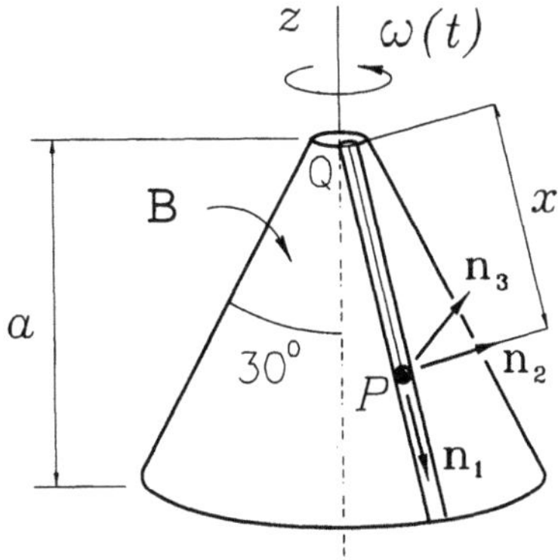

Figure P 6.11

P6.12 Consider the cursor P, sliding without friction over the straight guide, fixed on the disk D that, in turn, rotates at a simple angular velocity of module w in the inertial reference frame $\mathcal{R}$, as described in Example 3.1. Starting from the initial conditions established in that example, determine the instant t when the cursor leaves the guide, knowing that the disk rotates at a constant rate of 12 rpm.

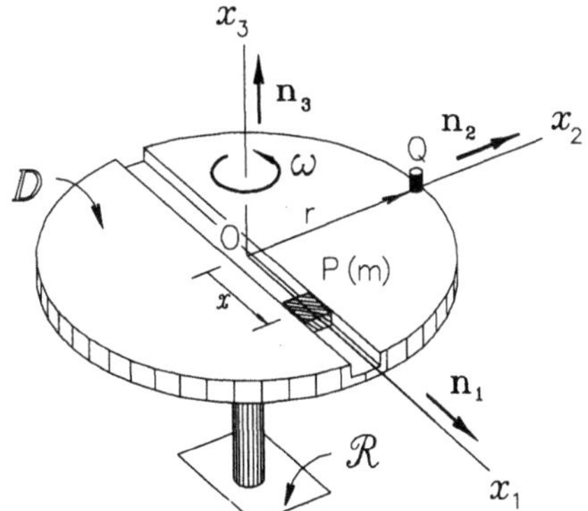

Figure P 6.12

P6.13   A small sphere, with mass m, attached to the end of a light nonstretch string, is left at rest in position (a). Calculate the traction T on the string as a function of the angular position θ.

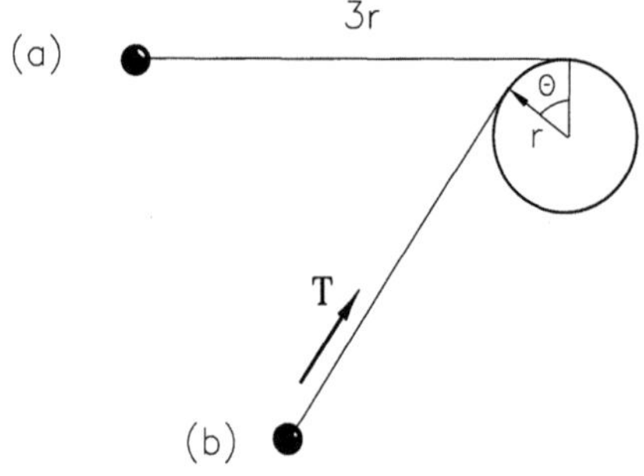

Figure P 6.13

P6.14   Consider the system shown in the figure below, consisting of a cursor P, with mass m, sliding without friction along the straight guide B, which rotates around the wire A, at a simple angular velocity $\dot{\theta} = \omega$, constant. The wire, in turn, moves in the laboratory L, inertial, also at a constant simple angular velocity $\dot{\phi} = \omega$. Given the initial conditions $\theta(0) = r(0) = 0$, determine, as a function of r and θ, the force exerted by the guide on the cursor.

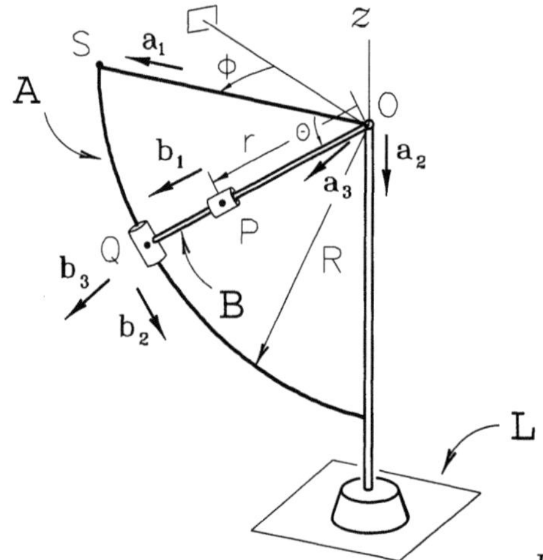

Figure P 6.14

P6.15 A pendulum P, with length r, swings slightly on a vertical plane, with its suspension point, Q, describing a harmonic straight horizontal motion, with a small amplitude, around another point O, described by $a = a_0 \sin pt$, where p is a frequency. The set is at rest when $t = 0$. Determine the function $\theta(t)$ described by the pendulum.

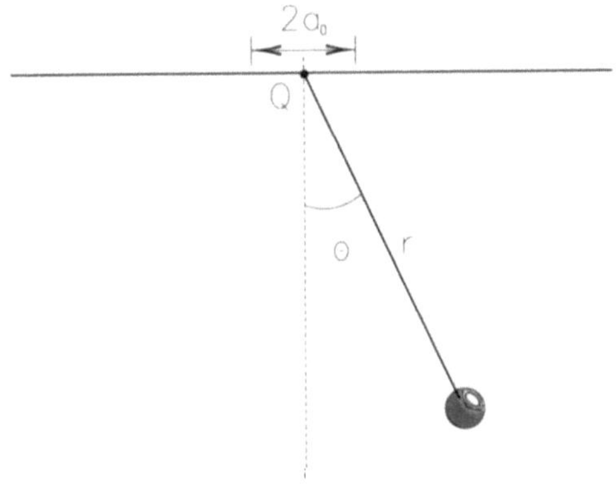

Figure P 6.15

P6.16 A small sphere is left at rest in position (1). The light, flexible string rolls around in the fixed drum as the sphere moves. What is the smallest ratio between the length a of the string and the radius r of the drum so that the string continues stretched when it reaches position (2)?

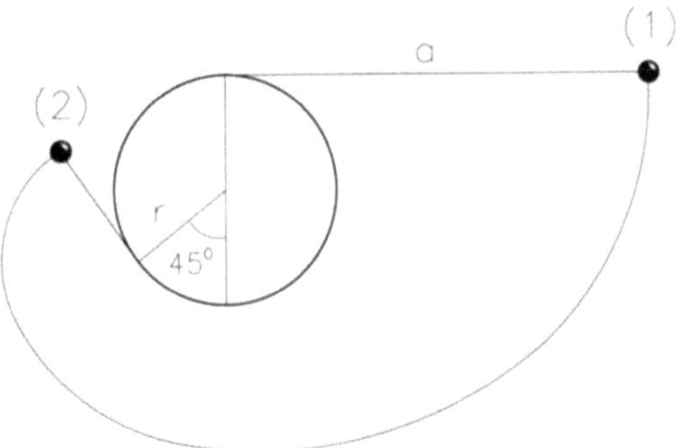

Figure P 6.16

P6.17 A buoy is thrown in free fall, reaching the surface of the water and plunging to the depth $6a$, as shown, returning from this point to the surface. Calculate the work of the pull as the buoy plunges to the point shown in the figure. The density of the water is ρ.

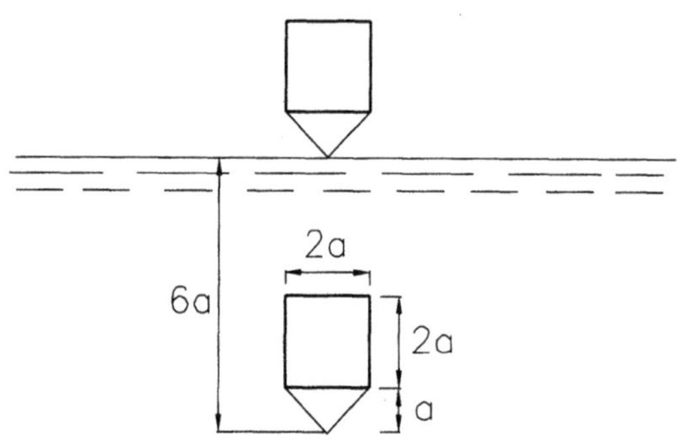

Figure P 6.17

P6.18 A small sphere, with mass m, is attached to the end of a light non-stretch string and is left at rest in position (a). Calculate its velocity in position (b).

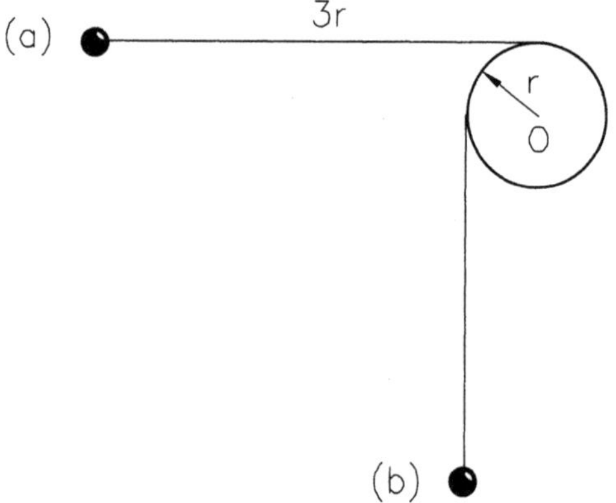

Figure P 6.18

P6.19 A small sphere can move freely inside a symmetric pipe, which is turning at a constant module angular velocity ω around the vertical axis of symmetry y. What shape must the pipe have so that, whatever the initial position in which the sphere is left from rest in the pipe, it stays in the same position?

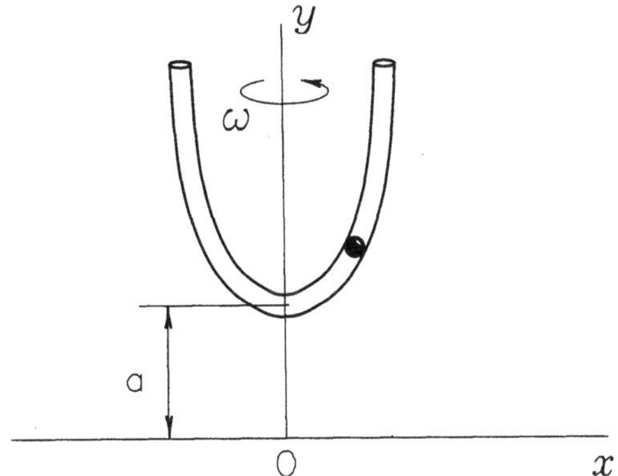

Figure P 6.19

P6.20 A small sphere P, with mass m, is thrown at a horizontal initial velocity v_0 inside a conical surface with a height a and radius of the base R, describing, from there, a trajectory on the surface, as illustrated. If b measures the distance between the initial and current levels of the sphere, determine the vertical component of the sphere's velocity, as a function of the coordinate b.

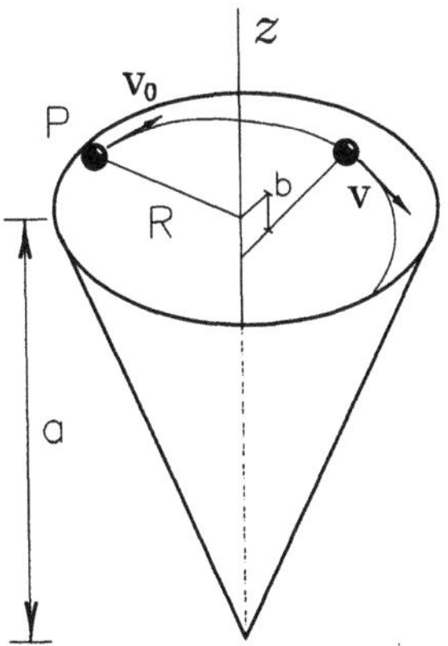

Figure P 6.20

P6.21 The pendulum P is left from rest in position (1). When it moves through position (2), the impact of the string (light and flexible) occurs with the fixed cylinder C, with a radius r. Determine the velocity of P and the traction T on the string, when it reaches position (3).

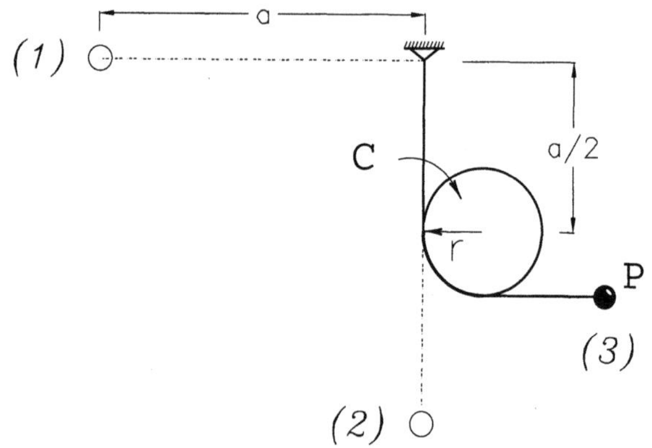

Figure P 6.21

P6.22 The small block P, with mass m, rests on a smooth horizontal plane connected to the fixed pin O, by means of a light linear spring, with elastic constant k and natural length r_0. At the initial instant, the spring's natural length is r_0 and the block has a velocity of module u, on the plane, in the indicated direction. After a time interval, the spring will have another length r and the block another velocity v, with components v_1 and v_2, as illustrated in the figure below. Calculate v as a function of r and of the initial conditions.

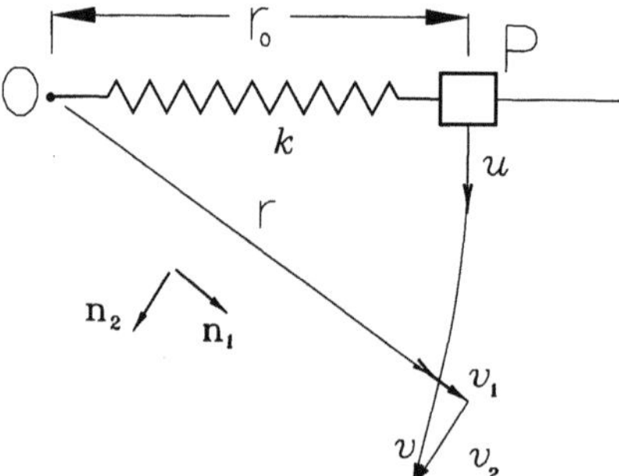

Figure P 6.22

P6.23   A small ball lands obliquely at a velocity v on a fixed horizontal plane and bounces at a velocity $v' = v/4$ and at an angle $\phi = 30^0$ with respect to the normal to the plane. Find the angle θ of landing if the restitution coefficient with the plane is $\epsilon = 1/\sqrt{3}$.

P6.24   A small ball lands obliquely at a velocity v on a fixed horizontal plane and bounces at a velocity $v' = v/\sqrt{2}$. Find the angles of landing and bouncing if the restitution coefficient is $\epsilon = 1/\sqrt{3}$ and the impact is smooth.

P6.25   A simple device for testing restitution coefficients consists of a bar, turning freely on O, which is left at rest in the horizontal position, with a sample of the material to be tested fixed at a distance x from the pivot. After the impact, always in the vertical position, the ângle θ that the bar reaches before causing a new impact is measured. What is the restitution coefficient? What should the distance x be in order to minimize the forces on the swivel, assuming that the test body is much lighter than the test bar?

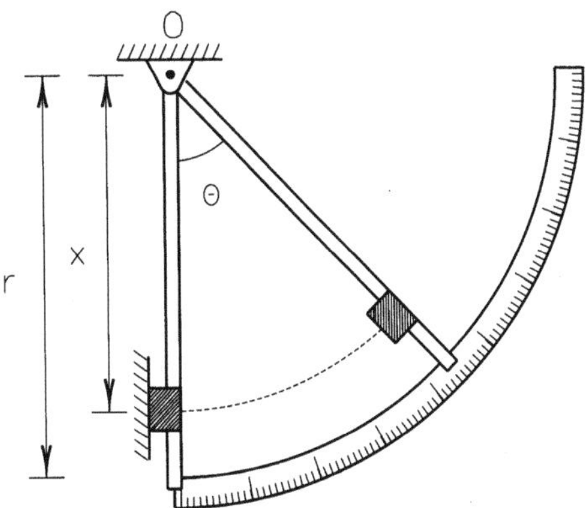

Figure P 6.25

P6.26 The small sphere P can slide freely on bar B and both are at rest in the inertial reference frame A when a vertical angular acceleration with constant module α is applied to B, as shown. So, the initial conditions are: $\omega(0) = 0$, $x(0) = 0$, and $\dot{x}(0) = 0$. Find the horizontal force component applied by B on P, at $t = 0$.

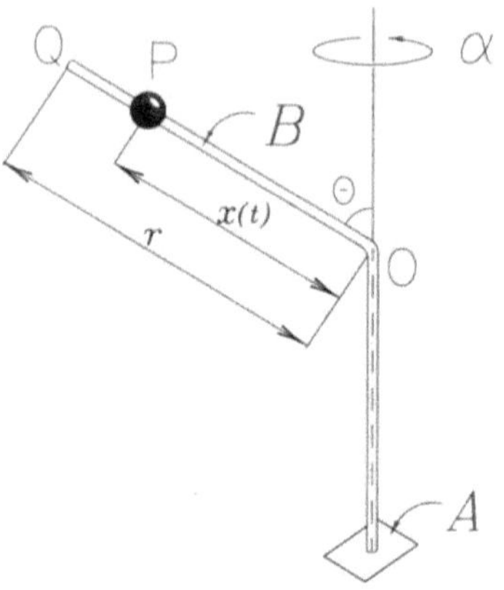

Figure P 6.26

P6.27 Going back to the previous exercise, consider now that $\alpha = 0$ and that the bar rotates at a prescribed angular velocity with constant module Ω. Find the initial horizontal force component applied by B on P for this new situation.

P6.28 For the situation described above, calculate which must be the prescribed value for Ω in order to the sphere arrive to $x = r/2$ with null velocity.

P6.29 Once more concerning to exercise 6.27, verify if the principle of conservation of mechanical energy applies during the motion of P. If the answer is no, compute the change of this energy. What is the explanation for that?

Dynamics of Systems

Chapter 5

The general principles of dynamics are discussed in Chapter 4, and the model of particle adopted in its introduction. Fundamental properties have been defined, such as momentum, angular momentum, and kinetic energy of a particle, and the general laws governing the change of such properties have been formulated; Newton's second law and the balance of mechanical energy are examples of those general laws governing the motion of particles.

This chapter generalizes the concepts and laws discussed in Chapter 4, for discrete and continuous systems of particles. For simplicity's sake, the word *body* will be adopted in reference to a continuous system and the word *system* in reference to a discrete system of particles, that is, a certain numerable set of particles. Here we will see, therefore, how to calculate the momentum, angular momentum, and kinetic energy of a system or body, the equations governing the time rate of these quantities, and their laws of conservation. As discussed below, the mass center of a system or body plays a leading role in the formulation of these equations and several results are greatly simplified when referred to the mass center. Should the reader wish to learn further about the concept of mass center — discussed informally in Section 1.4 — it is recommended to read Section 6.1 before proceeding to study this chapter.

Section 5.1 discusses the dynamic properties of systems and bodies. Definitions are given for momentum, angular momentum with respect to a point, angular momentum with respect to an axis and kinetic energy of a system or body, as well as some important relations between those properties and the mass center motion. Section 5.2 discusses force systems applied on systems and bodies, using Newton's third law to eliminate the forces of interaction between the elements of the system or body itself. In Sections 5.3 and 5.4, Newton's second law is generalized for systems and bodies, respectively, leading to the equations that govern the motion of the mass center. Although these are the most general and important equations in dynamics, they are not enough to fully establish the motion of an arbitrary system. The relationship between the resultant moment of a force system applied on a system or body and the time rate of its angular momentum vector is also discussed. This result, useful, for instance, when analyzing fluid-driven rotary machinery, will prove fundamental for analyzing the motion of the rigid body, as will be seen in Chapter 7. Section 5.5 discusses the more general concept of work done by an arbitrary force system and the potential functions present when at least one portion of the system is conservative. In Section 5.6, then, the generalized law of work and kinetic energy for a system or body is established. It also describes the potential and mechanical energies of systems or bodies, and an analysis made of its overall balance. Section 5.7 discusses the general conservation principles for systems and bodies. Some examples illustrate the simplicity of the methods whenever one or more of these principles is applicable. Last, Section 5.8 gives a simple and very brief introduction to the general approach for handling problems involving the flow of fluids.

5.1 Dynamic Properties

Consider a discrete system S, consisting of N particles P_i, with mass m_i, moving in a reference frame $\mathcal{R}$ with a velocity of $\mathbf{v}_i$, $i = 1, 2, \ldots, N$, (see Fig. 1.1). If $\mathbf{p}_i$ is the position vector with respect to a point O, fixed in $\mathcal{R}$, of the particle P_i and m is the total mass of the system, then the position $\mathbf{p}^*$, with respect to the same point, of the mass center S^* of the system is (see Section 1.4)

$$\mathbf{p}^* = \frac{1}{m} \sum_{i=1}^{N} m_i \mathbf{p}_i. \tag{1.1}$$

The velocity in $\mathcal{R}$ of the mass center, adopting the reduced notation for differentiation in $\mathcal{R}$, is

$$\mathbf{v}^* = {}^{\mathcal{R}}\mathbf{v}^{S^*} = \dot{\mathbf{p}}^* = \frac{1}{m} \sum_{i=1}^{N} m_i \mathbf{v}_i. \tag{1.2}$$

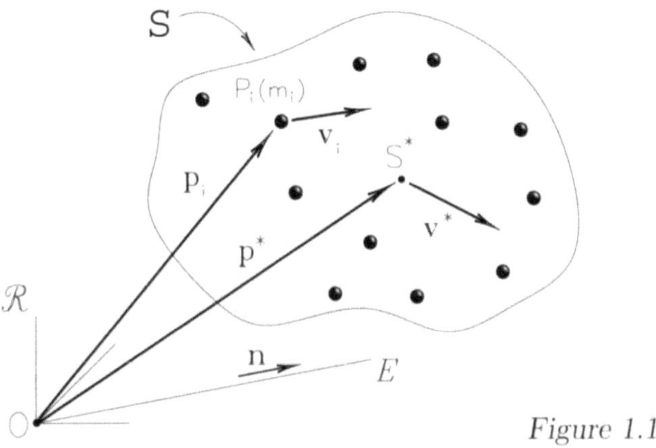

Figure 1.1

The vector $\mathbf{G}_i = m_i \mathbf{v}_i$ is, by definition, the momentum of the particle P_i in the reference frame $\mathcal{R}$. The set of momentum vectors of the particles of S forms, therefore, a simple system $\mathcal{G}$ of vectors $\mathbf{G}_i$, *bound* to the particles P_i. The *resultant* of this system is referred to as the *momentum of the system of particles* in the reference frame $\mathcal{R}$, being therefore given by

$$^{\mathcal{R}}\mathbf{G}^S \rightleftharpoons \sum_{i=1}^{N} \mathbf{G}_i = \sum_{i=1}^{N} m_i \mathbf{v}_i. \tag{1.3}$$

In other words, the momentum in a given reference frame $\mathcal{R}$ of a discrete system of particles S is the sum of the momentum vectors of its components, in the same reference frame.

Given a system S moving in a reference frame $\mathcal{R}$, the *momentum of the mass center* of S is referred to as the momentum vector of a (fictitious) particle with mass m equal to the mass of the system and that moves in $\mathcal{R}$ as its mass center, being, therefore,

$$\mathbf{G}^* \rightleftharpoons m\mathbf{v}^*. \tag{1.4}$$

Since from Eq. (1.2), $m\mathbf{v}^* = \sum_{i=1}^{N} m_i\mathbf{v}_i$, the result of the definitions, Eqs. (1.3) and (1.4), is that

$$^{\mathcal{R}}\mathbf{G}^S = \mathbf{G}^*, \tag{1.5}$$

that is, the momentum of a discrete system of particles in a given reference frame is equal to the momentum of its mass center, in the same reference frame.

Example 1.1   Seven swallows, each with approximately the same mass m, fly in formation, as shown in Fig. 1.2.

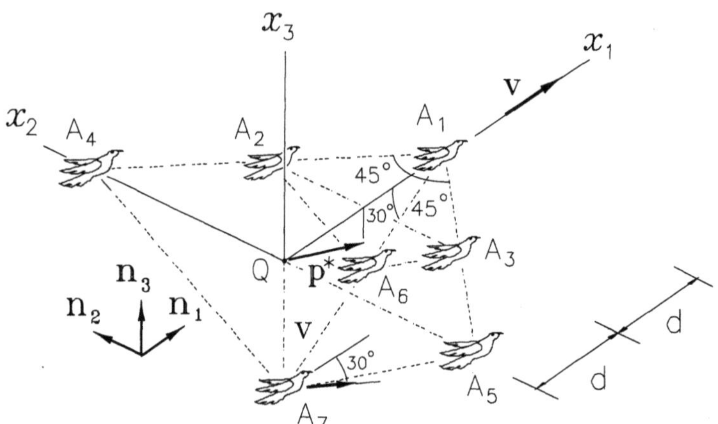

Figure 1.2

Five of them accompany the leader, A_1, gliding with a horizontal velocity v, while A_7 starts to leave the group with a velocity of the same module in the indicated direction. The position of the mass center of the band B

with respect to point Q is, according to Eq. (1.1),

$$\mathbf{p}^* = \frac{1}{7m}[2d\mathbf{n}_1 m + d(\mathbf{n}_1 + \mathbf{n}_2)m + d(\mathbf{n}_1 - \mathbf{n}_2)m + 2d\mathbf{n}_2 m$$

$$- 2d\mathbf{n}_2 m + d(\mathbf{n}_1 - \frac{1}{\sqrt{3}}\mathbf{n}_3)m - d\frac{2}{\sqrt{3}}\mathbf{n}_3 m]$$

$$= \frac{d}{7}(5\mathbf{n}_1 - \sqrt{3}\mathbf{n}_3).$$

The velocity of the mass center can be obtained from Eq. (1.2), which is reduced, with all masses equal, to

$$\mathbf{v}^* = \frac{1}{7}\sum_{i=1}^{7}\mathbf{v}_i = \frac{1}{7}v\left[\left(6 + \frac{\sqrt{3}}{2}\right)\mathbf{n}_1 - \frac{1}{2}\mathbf{n}_3\right].$$

The momentum vector of the band B is then, according to Eqs. (1.4) and (1.5),

$$\mathbf{G}^B = 7m\mathbf{v}^* = mv\left[\left(6 + \frac{\sqrt{3}}{2}\right)\mathbf{n}_1 - \frac{1}{2}\mathbf{n}_3\right].$$

The resultant moment with respect to a point O of the system of momentum vectors $\mathcal{G}$, in a reference frame $\mathcal{R}$, of a system of particles S is called the *angular momentum of the system of particles with respect to the point* in the reference frame $\mathcal{R}$. It is therefore expressed as (see Fig. 1.1)

$$^\mathcal{R}\mathbf{H}^{S/O} \rightleftharpoons \sum_{i=1}^{N}\mathbf{p}_i \times \mathbf{G}_i = \sum_{i=1}^{N}\mathbf{p}_i \times m_i\mathbf{v}_i. \tag{1.6}$$

Of course, as the angular momentum with respect to point O, in $\mathcal{R}$, of the particle P_i is $\mathbf{H}_i^O = \mathbf{p}_i \times \mathbf{G}_i$, the angular momentum vector of a system of particles with respect to a point, in a given reference frame, is equal to the sum of the angular momentum vectors with respect to the point, in the same reference frame, of the particles component of the system.

The resultant moment with respect to an axis E passing through O and parallel to a unit vector $\mathbf{n}$ (see Fig. 1.1) of the vector system $\mathcal{G}$ is referred to as the *angular momentum of the system of particles with respect to the axis*, in the reference frame $\mathcal{R}$. Its expression, therefore, according to Eq. (2.3.3), will be the simple projection in the direction $\mathbf{n}$ of the angular momentum vector with respect to point O, that is,

$$\mathcal{R}\mathbf{H}^{S/E} \rightleftharpoons \mathcal{R}\mathbf{H}^{S/O} \cdot \mathbf{n}\,\mathbf{n}. \tag{1.7}$$

Example 1.2   Returning to the previous example (see Fig. 1.2), the angular momentum vector with respect to point Q of the band of swallows, in the reference frame where the velocities were measured, is, according to Eq. (1.6),

$$
\begin{aligned}
\mathbf{H}^{B/Q} &= \sum_{i=1}^{7} \mathbf{p}_i \times m_i \mathbf{v}_i \\
&= mvd\Big[2\mathbf{n}_1 \times \mathbf{n}_1 + (\mathbf{n}_1 + \mathbf{n}_2) \times \mathbf{n}_1 + (\mathbf{n}_1 - \mathbf{n}_2) \times \mathbf{n}_1 + 2\mathbf{n}_2 \times \mathbf{n}_1 \\
&\quad - 2\mathbf{n}_2 \times \mathbf{n}_1 + (\mathbf{n}_1 - \frac{1}{\sqrt{3}}\mathbf{n}_3) \times \mathbf{n}_1 - \frac{2}{\sqrt{3}}\mathbf{n}_3 \times (\frac{\sqrt{3}}{2}\mathbf{n}_1 - \frac{1}{2}\mathbf{n}_3)\Big] \\
&= -\frac{1+\sqrt{3}}{\sqrt{3}}\,mvd\,\mathbf{n}_2.
\end{aligned}
$$

The angular momentum vector of the set with respect to the x_3-axis is, according to Eq. (1.7),

$$\mathbf{H}^{B/x_3} = \mathbf{H}^{B/Q} \cdot \mathbf{n}_3\,\mathbf{n}_3 = 0.$$

As the reader can see, the angular momentum is null with respect to any other axis passing through Q and orthogonal to $\mathbf{n}_2$. Also note that, even if the swallow A_7 is not leaving the formation, the angular momentum of the system with respect to point Q would be nonnull (check). In other words, even when a system is, as a whole, translating, its angular momentum with respect to an arbitrary point may be different from zero.

When the momentum and angular momentum vectors with respect to a point O in a system S of particles in a given reference frame $\mathcal{R}$ are known, it is easy to establish their angular momentum vector with respect to any other point Q in the same reference frame. In fact, the *moments transport theorem*, Eq. (2.3.4), derived in Section 2.3, establishes that

$$\mathcal{R}\mathbf{H}^{S/Q} = \mathcal{R}\mathbf{H}^{S/O} + \mathbf{p}^{O/Q} \times \mathcal{R}\mathbf{G}^S, \tag{1.8}$$

where $\mathbf{p}^{O/Q}$ is the position vector of O with respect to Q (see Fig. 1.3).

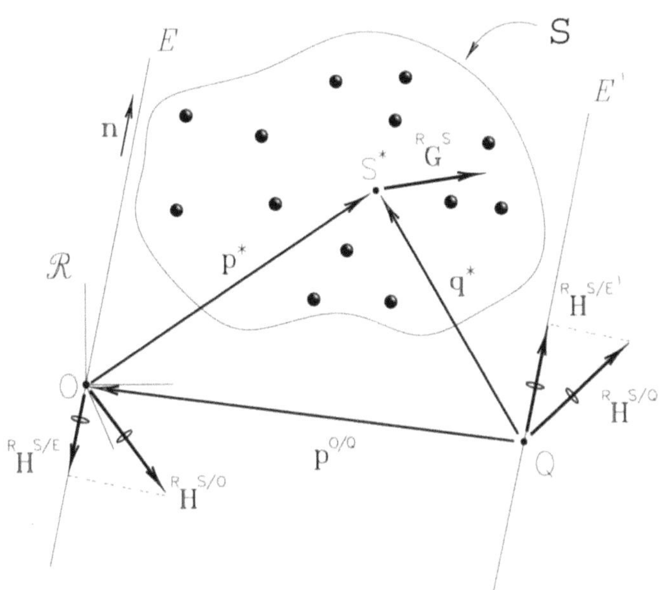

Figure 1.3

Likewise, the angular momentum vector of a system of particles S with respect to two parallel axes E and E' are related, according to Eq. (2.3.5), by (see Fig. 1.3)

$$^{\mathcal{R}}\mathbf{H}^{S/E'} = {}^{\mathcal{R}}\mathbf{H}^{S/E} + \mathbf{p}^{O/Q} \times {}^{\mathcal{R}}\mathbf{G}^{S} \cdot \mathbf{n}\,\mathbf{n}, \tag{1.9}$$

where $\mathbf{n}$ is a unit vector parallel to the axes.

Example 1.3 Figure 1.4 shows part of a mechanism consisting of the base B of a thrust bearing that rotates around the vertical axis z in relation to the support A, with a simple angular velocity of constant module ω. Twelve small identical balls, with mass m each and kept equidistant by a spacer (not shown), roll on the circular track with a velocity of constant module v, as indicated, in relation to the base. When considering the set of balls as a system S of particles P_i, $i = 1, 2, \ldots, 12$, the momentum vector of S in B is, as Eq. (1.3) indicates,

$$^{B}\mathbf{G}^{S} = \sum_{i=1}^{12} m\,^{B}\mathbf{v}^{P_i} = 0,$$

null, therefore, given the symmetry of the system. The angular momentum vector of S with respect to point Q, in the reference frame B, according to

Eq. (1.6), is

$$^{B}\mathbf{H}^{S/Q} = \sum_{i=1}^{12} \mathbf{p}_i \times m\,^{B}\mathbf{v}^{P_i} = 12mvr\mathbf{n}_3.$$

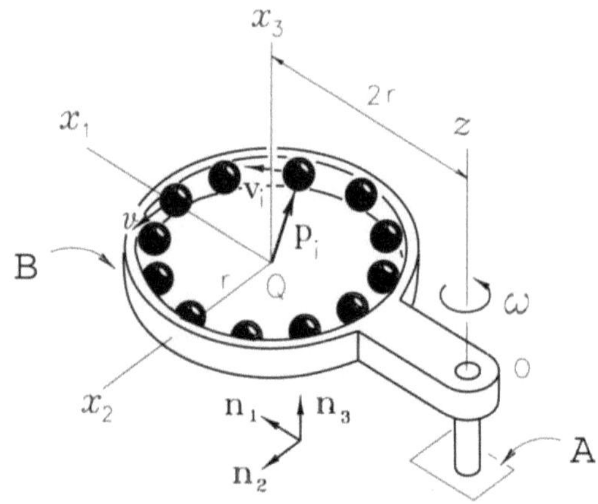

Figure 1.4

The system $\mathcal{G}$ of momentum vectors in the reference frame B consists, therefore, of a *couple*. The angular momentum of S with respect to any point in the reference frame B is then the same (the couple moment). Particularly, the angular momentum vector of S with respect to point O, in B, is, according to Eq. (1.8),

$$^{B}\mathbf{H}^{S/O} = \,^{B}\mathbf{H}^{S/Q} = 12mvr\mathbf{n}_3.$$

The angular momentum vector of S with respect to the x_3-axis in the same reference frame is, as Eq. (1.7) states,

$$^{B}\mathbf{H}^{S/x_3} = \,^{B}\mathbf{H}^{S/Q} \cdot \mathbf{n}_3\,\mathbf{n}_3 = 12mvr\mathbf{n}_3,$$

also being equal to the angular momentum vector with respect to the z-axis, in B. The momentum vector in reference frame A can be calculated using the kinematic theorem to relate velocities in the two reference frames,

that is,

$$^A\mathbf{G}^S = \sum_{i=1}^{12} m\,^A\mathbf{v}^{P_i}$$

$$= m\sum_{i=1}^{12}(^B\mathbf{v}^{P_i} + {}^A\mathbf{v}^Q + {}^A\boldsymbol{\omega}^B \times \mathbf{p}_i)$$

$$= m\Big(0 + \sum_{i=1}^{12} 2\omega r\mathbf{n}_2 + 0\Big)$$

$$= 24m\omega r\mathbf{n}_2.$$

The angular momentum vector of this system with respect to point Q, in the reference frame A, is

$$^A\mathbf{H}^{S/Q} = \sum_{i=1}^{12}\mathbf{p}_i \times m\,^A\mathbf{v}^{P_i}$$

$$= m\sum_{i=1}^{12}\mathbf{p}_i \times \Big(^B\mathbf{v}^{P_i} + {}^A\mathbf{v}^Q + {}^A\boldsymbol{\omega}^B \times \mathbf{p}_i\Big)$$

$$= 12mvr\mathbf{n}_3 + 0 + m\sum_{i=1}^{12}\mathbf{p}_i \times \Big(^A\boldsymbol{\omega}^B \times \mathbf{p}_i\Big)$$

$$= 12mvr\mathbf{n}_3 + m\sum_{i=1}^{12} p_i^2\omega\mathbf{n}_3 - m\sum_{i=1}^{12}\mathbf{p}_i \cdot \omega\mathbf{n}_3\,\mathbf{p}_i$$

$$= 12mr(v + r\omega)\mathbf{n}_3.$$

The angular momentum with respect to point O, in A, is, according to Eq. (1.8),

$$^A\mathbf{H}^{S/O} = {}^A\mathbf{H}^{S/Q} + \mathbf{p}^{Q/O} \times {}^A\mathbf{G}^S$$

$$= 12mr(v + r\omega)\mathbf{n}_3 + 2r\mathbf{n}_1 \times 24m\omega r\mathbf{n}_2$$

$$= 12mr(v + 5\omega r)\,\mathbf{n}_3.$$

See the corresponding animation.

If S is a system consisting of N particles P_i, with mass m_i, moving in a given reference frame $\mathcal{R}$ with velocities $\mathbf{v}_i$, $i = 1, 2, \ldots, N$ (see Fig. 1.5), the *kinetic energy of the system*, in $\mathcal{R}$, is defined as the algebraic sum of the kinetic energies of the constituting particles, that is,

$$^{\mathcal{R}}K^S \rightleftharpoons \sum_{i=1}^{N} {}^{\mathcal{R}}K^{P_i} = \frac{1}{2}\sum_{i=1}^{N} m_i \mathbf{v}_i \cdot \mathbf{v}_i. \tag{1.10}$$

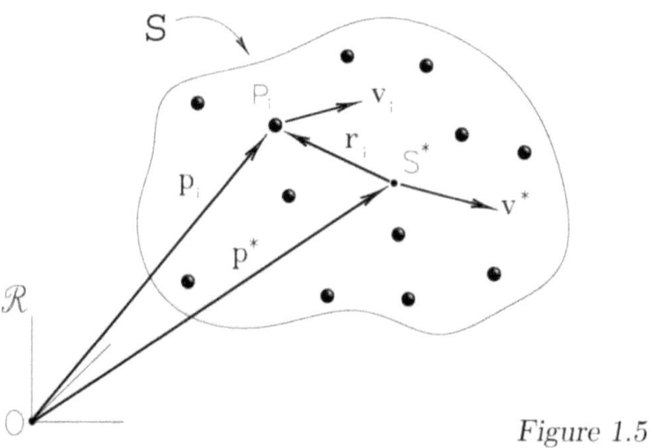

Figure 1.5

If S* is the mass center of S, the *kinetic energy of the mass center* of the system, in a given reference frame $\mathcal{R}$, is defined as the kinetic energy of a particle (fictitious) with the mass equal to the mass of the system and velocity equal to the velocity of the mass center of S, $\mathbf{v}^* = {}^{\mathcal{R}}\mathbf{v}^{S^*}$, that is,

$$^{\mathcal{R}}K^{S^*} \rightleftharpoons \frac{1}{2}m\mathbf{v}^* \cdot \mathbf{v}^*. \tag{1.11}$$

If $\mathbf{r}_i$ is the position vector of the particle P_i with respect to S* (see Fig. 1.5), its time rate in $\mathcal{R}$, $\dot{\mathbf{r}}_i$, is the velocity of P_i relative to S* in $\mathcal{R}$. The kinetic energy of S in $\mathcal{R}$ may then be resolved in two terms, as follows: the kinetic energy of the mass center, as defined in Eq. (1.11); and a *kinetic energy around the mass center*, defined as

$$^{\mathcal{R}}K^{S/S^*} \rightleftharpoons \frac{1}{2}\sum_{i=1}^{N} m_i \dot{\mathbf{r}}_i \cdot \dot{\mathbf{r}}_i. \tag{1.12}$$

In fact, as $\mathbf{v}_i = \mathbf{v}^* + \dot{\mathbf{r}}_i$, then,

$$
\begin{aligned}
{}^{\mathcal{R}}K^S &= \frac{1}{2}\sum_{i=1}^{N} m_i(\mathbf{v}^* + \dot{\mathbf{r}}_i)\cdot(\mathbf{v}^* + \dot{\mathbf{r}}_i) \\
&= \frac{1}{2}\sum_{i=1}^{N} m_i\mathbf{v}^*\cdot\mathbf{v}^* + \frac{1}{2}\sum_{i=1}^{N} m_i\dot{\mathbf{r}}_i\cdot\dot{\mathbf{r}}_i + \mathbf{v}^*\cdot\sum_{i=1}^{N} m_i\dot{\mathbf{r}}_i \\
&= {}^{\mathcal{R}}K^{S^*} + {}^{\mathcal{R}}K^{S/S^*}.
\end{aligned}
\tag{1.13}
$$

Note that the last term in the second line above is null, since it contains the time rate of $\sum_{i=1}^{N} m_i\mathbf{r}_i$, null at every instant t (for more details, see Section 6.1).

Example 1.4 Returning to the preceding example (see Fig. 1.4), the kinetic energy of the system S in the reference frame B, according to Eq. (1.10), is

$$
{}^{B}K^S = \frac{1}{2}m\sum_{i=1}^{12} {}^{B}\mathbf{v}^{P_i}\cdot{}^{B}\mathbf{v}^{P_i} = 6mv^2.
$$

The kinetic energy of the mass center of the system (the point Q) in the reference frame A, according to Eq. (1.11), is

$$
{}^{A}K^Q = \frac{1}{2}(12m){}^{A}\mathbf{v}^Q\cdot{}^{A}\mathbf{v}^Q = 24m\omega^2 r^2.
$$

The time rate in A of the position vector of P_i with respect to the mass center Q is equal to

$$
\dot{\mathbf{p}}_i = \frac{{}^{A}d}{dt}\mathbf{p}_i = \frac{{}^{B}d}{dt}\mathbf{p}_i + {}^{A}\boldsymbol{\omega}^B\times\mathbf{p}_i = {}^{B}\mathbf{v}^{P_i} + {}^{A}\boldsymbol{\omega}^B\times\mathbf{p}_i.
$$

Its squared value is

$$
\dot{\mathbf{p}}_i\cdot\dot{\mathbf{p}}_i = v^2 + ({}^{A}\boldsymbol{\omega}^B\times\mathbf{p}_i)^2 + 2\,{}^{B}\mathbf{v}^{P_i}\cdot({}^{A}\boldsymbol{\omega}^B\times\mathbf{p}_i) = (v + \omega r)^2.
$$

The kinetic energy around the mass center in the reference frame A is then, according to Eq. (1.12),

$$
{}^{A}K^{S/Q} = \frac{1}{2}\sum_{i=1}^{12} m\dot{\mathbf{p}}_i\cdot\dot{\mathbf{p}}_i = 6m(v + \omega r)^2.
$$

The kinetic energy of the system in the reference frame A is, therefore, according to Eq. (1.13), equal to

$$
{}^{A}K^S = {}^{A}K^Q + {}^{A}K^{S/Q} = 6m(v^2 + 2v\omega r + 5\omega^2 r^2).
$$

See the corresponding animation.

Let us now consider a body C, with mass m, moving in a reference frame $\mathcal{R}$ (see Fig. 1.6). Let P be a general point in the body, whose position with respect to a point O, fixed in $\mathcal{R}$, is given by the vector $\mathbf{p} = \mathbf{p}^{P/O}$ and whose velocity in $\mathcal{R}$ is $\mathbf{v} = {}^{\mathcal{R}}\mathbf{v}^P = \dot{\mathbf{p}}$. The position of the mass center C* of C with respect to point O is given by the vector (see Section 1.4)

$$\mathbf{p}^* = \frac{1}{m} \int_C \mathbf{p}\, dm. \tag{1.14}$$

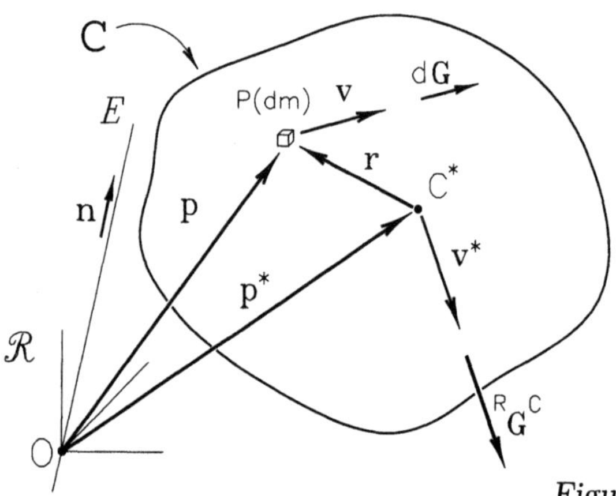

Figure 1.6

The velocity in the reference frame $\mathcal{R}$ of point C* will therefore be

$$\mathbf{v}^* = {}^{\mathcal{R}}\mathbf{v}^{C^*} = \dot{\mathbf{p}}^* = \frac{1}{m} \int_C \mathbf{v}\, dm. \tag{1.15}$$

A mass element of C has momentum in $\mathcal{R}$ given by $d\mathbf{G} = \mathbf{v}\, dm$, and the set of these vectors consists of a simple distributed system of bound vectors $\mathcal{G}$, whose resultant is referred to as the *momentum of the body C*, in the reference frame $\mathcal{R}$, that is,

$$^{\mathcal{R}}\mathbf{G}^C \rightleftharpoons \int_C d\mathbf{G} = \int_C \mathbf{v}\, dm. \tag{1.16}$$

The *momentum of the mass center of a body C* in a reference frame $\mathcal{R}$ is defined as the momentum of a particle (fictitious) with mass equal to that of the body, moving in $\mathcal{R}$ like C*, that is,

$$\mathbf{G}^* \rightleftharpoons m\mathbf{v}^*. \tag{1.17}$$

It immediately results from Eqs. (1.15–1.17) that

$$^{\mathcal{R}}\mathbf{G}^{C} = \mathbf{G}^{*}. \tag{1.18}$$

Note that the momentum vector of a body must necessarily depend on the reference frame in relation to which the velocities are being measured, as indicated in the adopted notation. (The reduced notation, $\mathbf{v}$, $\mathbf{v}^{*}$, $\mathbf{G}^{*}$, etc., must not be used whenever more than one reference frame is involved.)

Example 1.5 The body C, consisting of a rigid homogeneous rod with mass m and length r, is hinged at one end by the peg Q, fixed to the arm B that, in turn, moves with a simple angular velocity of constant module Ω, around the vertical axis x_3, in relation to the support A (see Fig. 1.7). The motion of C in B is described by the angle $\theta(t)$. The coordinated axes $\{x_1, x_2, x_3\}$ are fixed in B and the orthonormal basis $\mathbf{b}_1, \mathbf{b}_2, \mathbf{b}_3$ is parallel to the axes, as shown. The position of point C^*, mass center of C, with respect to the origin O is then given as

$$\mathbf{p}^* = r\mathbf{b}_1 + \frac{r}{2}\sin\theta\,\mathbf{b}_2 - \frac{r}{2}\cos\theta\,\mathbf{b}_3.$$

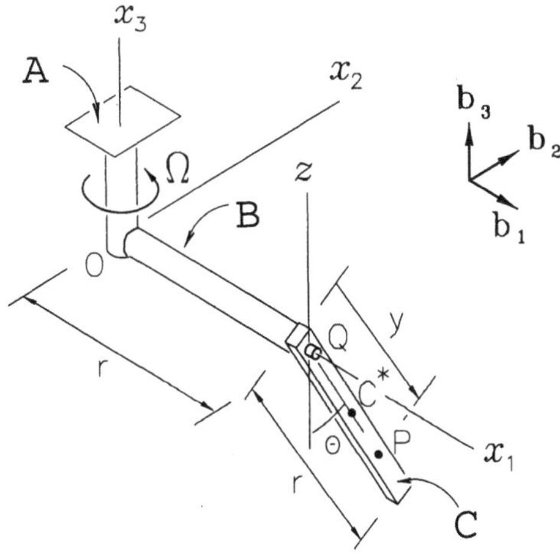

Figure 1.7

The angular velocity of C in B is ${}^B\boldsymbol{\omega}^C = \dot{\theta}\mathbf{b}_1$ and the velocity of point C* in the reference frame B is

$${}^B\mathbf{v}^{C^*} = \dot{\theta}\frac{r}{2}(\cos\theta\mathbf{b}_2 + \sin\theta\mathbf{b}_3).$$

The momentum vector of the body C in the reference frame B is then, according to Eqs. (1.17–1.18),

$${}^B\mathbf{G}^C = m\,{}^B\mathbf{v}^{C^*} = m\frac{r}{2}(\cos\theta\mathbf{b}_2 + \sin\theta\mathbf{b}_3)\dot{\theta}.$$

The velocity of the mass center of the body in the reference frame A may be obtained from

$$\begin{aligned}
{}^A\mathbf{v}^{C^*} &= {}^B\mathbf{v}^{C^*} + {}^A\mathbf{v}^Q + {}^A\boldsymbol{\omega}^B \times \mathbf{p}^{C^*/Q}\\
&= \dot{\theta}\frac{r}{2}(\cos\theta\mathbf{b}_2 + \sin\theta\mathbf{b}_3) + \Omega r\mathbf{b}_2 + \Omega\mathbf{b}_3 \times \frac{r}{2}(\sin\theta\mathbf{b}_2 - \cos\theta\mathbf{b}_3)\\
&= -\Omega\frac{r}{2}\sin\theta\mathbf{b}_1 + (\Omega + \frac{1}{2}\dot{\theta}\cos\theta)r\mathbf{b}_2 + \dot{\theta}\frac{r}{2}\sin\theta\mathbf{b}_3.
\end{aligned}$$

The momentum of the body in the support A is then

$${}^A\mathbf{G}^C = m\,{}^A\mathbf{v}^{C^*} = \frac{1}{2}mr\Big[-\Omega\sin\theta\mathbf{b}_1 + (2\Omega + \dot{\theta}\cos\theta)\mathbf{b}_2 + \dot{\theta}\sin\theta\mathbf{b}_3\Big].$$

As expected, the computed momentum vectors will coincide if $\Omega = 0$. See the corresponding animation.

If C is a body moving in a reference frame $\mathcal{R}$ and $d\mathbf{G} = \mathbf{v}\,dm$ is the momentum in $\mathcal{R}$ of an element of C (see Fig. 1.6), the resultant moment with respect to an arbitrary point O of the distributed system $\mathcal{G}$ of bound vectors $d\mathbf{G}$ is called the *angular momentum of the body* C *with respect to the point* O, in the reference frame $\mathcal{R}$, that is,

$${}^\mathcal{R}\mathbf{H}^{C/O} \rightleftharpoons \int_C \mathbf{p} \times d\mathbf{G} = \int_C \mathbf{p} \times \mathbf{v}\,dm. \tag{1.19}$$

The resultant moment with respect to an axis E, containing the point O and parallel to a given unit vector $\mathbf{n}$ (see Fig. 1.6), of the distributed system $\mathcal{G}$ is called the *angular momentum of the body* C *with respect to the axis* E, in the reference frame $\mathcal{R}$. Its expression is, according to Eq. (2.3.3),

$${}^\mathcal{R}\mathbf{H}^{C/E} \rightleftharpoons {}^\mathcal{R}\mathbf{H}^{C/O} \cdot \mathbf{n}\,\mathbf{n}. \tag{1.20}$$

Example 1.6 Returning to the previous example, let us now find the angular momentum of the rod C with respect to the peg Q, in the reference frame A (see Fig. 1.7). Taking the variable y along the rod, from point Q, and making $dm = (m/r)\,dy$ as mass element associated to the general point P, whose velocity in A is

$$\mathbf{v} = {}^A\mathbf{v}^P = {}^B\mathbf{v}^P + {}^A\mathbf{v}^Q + {}^A\boldsymbol{\omega}^B \times \mathbf{p}^{P/Q}$$
$$= -y\Omega \sin\theta \mathbf{b_1} + (y\dot\theta\cos\theta + \Omega r)\mathbf{b_2} + y\dot\theta\sin\theta\mathbf{b_3},$$

the angular momentum vector is, according to Eq. (1.19),

$${}^A\mathbf{H}^{C/Q} = \int_C \mathbf{p} \times \mathbf{v}\, dm$$
$$= \frac{m}{r}\int_0^r y(\sin\theta\mathbf{b_2} - \cos\theta\mathbf{b_3})$$
$$\times \left(-y\Omega\sin\theta\mathbf{b_1} + (y\dot\theta\cos\theta + \Omega r)\mathbf{b_2} + y\dot\theta\sin\theta\mathbf{b_3}\right) dy$$
$$= \frac{1}{3}mr^2\left((\frac{3}{2}\Omega\cos\theta + \dot\theta)\mathbf{b_1} + \Omega\cos\theta\sin\theta\mathbf{b_2} + \Omega\sin^2\theta\mathbf{b_3}\right).$$

The angular momentum of the rod with respect to the x_1-axis, in the reference frame A, is, according to Eq. (1.20),

$${}^A\mathbf{H}^{C/x_1} = {}^A\mathbf{H}^{C/Q} \cdot \mathbf{b_1}\,\mathbf{b_1} = \frac{1}{3}mr^2\left(\frac{3}{2}\Omega\cos\theta + \dot\theta\right)\mathbf{b_1}.$$

The angular momentum vector of the rod C with respect to the vertical axis z, also passing through Q, in the same reference frame, is then

$${}^A\mathbf{H}^{C/z} = {}^A\mathbf{H}^{C/Q} \cdot \mathbf{b_3}\,\mathbf{b_3} = \frac{1}{3}mr^2\Omega\sin^2\theta\mathbf{b_3}.$$

The reader should be aware of the fact that the angular momentum vector of the rod with respect to a given point is *not* equal to the angular momentum vector of the mass center with respect to the point. In fact, the angular momentum with respect to Q, in A, of a particle (fictitious) with mass m moving as C* would be

$${}^A\mathbf{H}^{C^*/Q} = \mathbf{p}^{C^*/Q} \times m\,{}^A\mathbf{v}^{C^*}$$
$$= \frac{1}{4}mr^2\left((2\Omega\cos\theta + \dot\theta)\mathbf{b_1} + \Omega\cos\theta\sin\theta\mathbf{b_2} + \Omega\sin^2\theta\mathbf{b_3}\right),$$

differing, therefore, from ${}^A\mathbf{H}^{C/Q}$.

As in discrete systems, knowing the momentum vector of a body and its angular momentum vector with respect to a point O, in a given reference frame, it is easy to find its angular momentum with respect to any other point Q, in the same reference frame, resorting to the *moments transport theorem*, Eq. (2.3.4), resulting in

$$^{\mathcal{R}}\mathbf{H}^{C/Q} = {}^{\mathcal{R}}\mathbf{H}^{C/O} + \mathbf{p}^{O/Q} \times {}^{\mathcal{R}}\mathbf{G}^{C}, \tag{1.21}$$

where $\mathbf{p}^{O/Q}$ is the position vector of O with respect to Q (see Fig. 1.8).

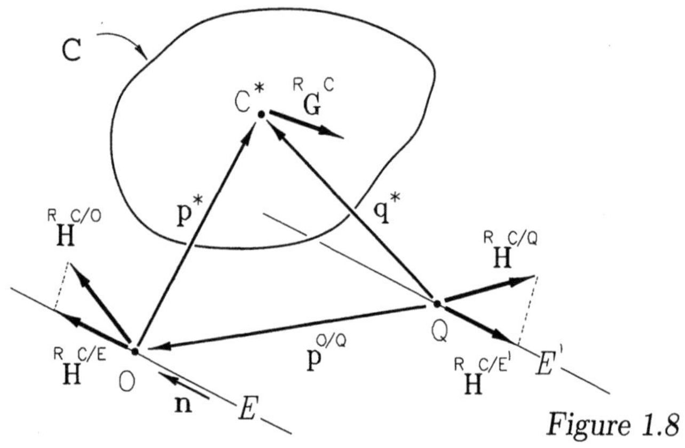

Figure 1.8

As a consequence of the above result, the angular momentum vectors of a body C with respect to two parallel axes E and E' are related, according to Eq. (2.3.5), by (see Fig. 1.8)

$$^{\mathcal{R}}\mathbf{H}^{C/E'} = {}^{\mathcal{R}}\mathbf{H}^{C/E} + \mathbf{p}^{O/Q} \times {}^{\mathcal{R}}\mathbf{G}^{C} \cdot \mathbf{n}\,\mathbf{n}, \tag{1.22}$$

where $\mathbf{n}$ is a unit vector parallel to the axes.

Example 1.7 Returning again to Example 1.5 (see Fig. 1.7), the angular momentum of rod C with respect to point O, in the reference frame A, may be expressed, as in Eq. (1.21), by

$$^{A}\mathbf{H}^{C/O} = {}^{A}\mathbf{H}^{C/Q} + \mathbf{p}^{Q/O} \times {}^{A}\mathbf{G}^{C}$$

$$= \frac{1}{3}mr^{2}\left[(\frac{3}{2}\Omega\cos\theta + \dot{\theta})\mathbf{b}_{1} + \Omega\cos\theta\sin\theta\mathbf{b}_{2} + \Omega\sin^{2}\theta\mathbf{b}_{3}\right]$$

$$+ r\mathbf{b}_{1} \times \frac{1}{2}mr\left[-\Omega\sin\theta\mathbf{b}_{1} + (2\Omega + \dot{\theta}\cos\theta)\mathbf{b}_{2} + \dot{\theta}\sin\theta\mathbf{b}_{3}\right]$$

$$= mr^{2}\left[(\frac{1}{2}\Omega\cos\theta + \frac{1}{3}\dot{\theta})\,\mathbf{b}_{1} + (\frac{1}{3}\Omega\cos\theta - \frac{1}{2}\dot{\theta})\,\sin\theta\,\mathbf{b}_{2} \right.$$

$$\left. + (\Omega(1 + \frac{1}{3}\sin^{2}\theta) + \frac{1}{2}\dot{\theta}\cos\theta)\,\mathbf{b}_{3}\right].$$

The angular momentum of the rod with respect to the x_3-axis is, according to Eq. (1.20),

$$^A\mathbf{H}^{C/x_3} = {}^A\mathbf{H}^{C/O} \cdot \mathbf{b}_3\,\mathbf{b}_3 = mr^2\Big(\Omega(1 + \tfrac{1}{3}\sin^2\theta) + \tfrac{1}{2}\dot\theta\cos\theta\Big)\,\mathbf{b}_3.$$

On the other hand, the same vector may be obtained by using Eq. (1.22). In fact,

$$
\begin{aligned}
^A\mathbf{H}^{C/x_3} &= {}^A\mathbf{H}^{C/z} + \mathbf{p}^{Q/O} \times {}^A\mathbf{G}^C \cdot \mathbf{b}_3\,\mathbf{b}_3 \\
&= \tfrac{1}{3}mr^2\Omega\sin^2\theta\mathbf{b}_3 + r\mathbf{b}_1 \\
&\quad \times \tfrac{1}{2}mr\big[-\Omega\sin\theta\mathbf{b}_1 + (2\Omega + \dot\theta\cos\theta)\mathbf{b}_2 + \dot\theta\sin\theta\mathbf{b}_3\big] \cdot \mathbf{b}_3\,\mathbf{b}_3 \\
&= mr^2\Big(\Omega(1 + \tfrac{1}{3}\sin^2\theta) + \tfrac{1}{2}\dot\theta\cos\theta\Big)\,\mathbf{b}_3.
\end{aligned}
$$

See the corresponding animation.

Examples 1.5 to 1.7 illustrate the calculation of the momentum and angular momentum vectors of a rigid body. Especially when establishing angular momenta, an easier and more practical formalization, preventing, in general, the integration along the body, is discussed in Chapters 6 and 7. For other continuous systems that do not consist of rigid bodies, it is necessary, when establishing angular momentum vectors, to integrate the moments of the momenta, as established in Eq. (1.19).

Example 1.8 The uniform rope C, with mass m and length a, has its end Q fixed to the drum T and is partially stretched while being wound up around the drum (see Fig. 1.9). The motion is governed by the angle $\theta(t)$ with $\theta(0) = 0$ and $\dot\theta(t) = \omega$, constant. The basis $\mathbf{n}_1, \mathbf{n}_2, \mathbf{n}_3$ is fixed to the straight stretch of the rope, as shown.

Now let P be a general point of the rope, s far from point Q, along it. Its position vector with respect to point O, center of the drum, at an arbitrary instant is $\mathbf{p} = (s - r\theta)\mathbf{n}_1 - r\mathbf{n}_2$, and its velocity in relation to T, at the same instant, is

$$\mathbf{v} = \dot{\mathbf{p}} = (s - r\theta)\,\omega\,\mathbf{n}_2.$$

The momentum vector of the rope in the reference frame T may then be

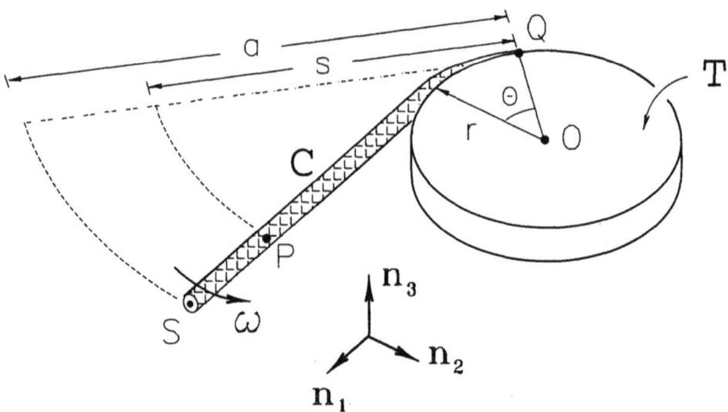

Figure 1.9

calculated, according to Eq. (1.16), as

$$
{}^{T}\mathbf{G}^{C} = \int_{Q}^{S} \mathbf{v}\, dm
$$

$$
= \int_{r\theta}^{a} (s - r\theta)\omega\, \mathbf{n}_2 \, \frac{m}{a}\, ds
$$

$$
= \frac{m\omega}{2a}(a - r\theta)^2\, \mathbf{n}_2.
$$

The angular momentum of the rope with respect to point O in the reference frame T may be obtained from Eq. (1.19) and is equal to

$$
{}^{T}\mathbf{H}^{C/O} = \int_{Q}^{S} \mathbf{p} \times \mathbf{v}\, dm
$$

$$
= \int_{r\theta}^{a} \left((s - r\theta)\mathbf{n}_1 - r\mathbf{n}_2 \right) \times (s - r\theta)\,\omega\, \mathbf{n}_2 \, \frac{m}{a}\, ds
$$

$$
= \frac{m\omega}{a} \int_{r\theta}^{a} (s - r\theta)^2\, ds\, \mathbf{n}_3
$$

$$
= \frac{m\omega}{3a}(a - r\theta)^3\, \mathbf{n}_3.
$$

Note that the momentum and angular momentum vectors vanish when $r\theta = a$, as expected. See the corresponding animation.

If C is a body with mass m moving in a reference frame $\mathcal{R}$, the *kinetic energy of the body* is defined in $\mathcal{R}$ by integrating in the body the kinetic energy of its elements, that is,

$$^{\mathcal{R}}K^C \rightleftharpoons \int_C dK = \int_C \frac{1}{2}\mathbf{v} \cdot \mathbf{v}\, dm, \tag{1.23}$$

where $\mathbf{v}$ is the velocity in $\mathcal{R}$ of a generic point P of the body (see Fig. 1.6).

The *kinetic energy of the mass center of the body*, in a given reference frame $\mathcal{R}$, is defined as the kinetic energy of a (fictitious) particle whose mass is equal to the mass of the body and which moves like its mass center C*, that is,

$$^{\mathcal{R}}K^{C^*} \rightleftharpoons \frac{1}{2}m\mathbf{v}^* \cdot \mathbf{v}^*, \tag{1.24}$$

where $\mathbf{v}^*$ is the velocity of the mass center of the body in the reference frame $\mathcal{R}$.

Also, the *kinetic energy of the body around its mass center*, in a given reference frame $\mathcal{R}$, is defined as the scalar

$$^{\mathcal{R}}K^{C/C^*} \rightleftharpoons \int_C \frac{1}{2}\dot{\mathbf{r}} \cdot \dot{\mathbf{r}}\, dm, \tag{1.25}$$

where $\dot{\mathbf{r}}$ is the time rate in $\mathcal{R}$ of the position vector of a generic point P of C with respect to its mass center, that is, is the velocity of P relative to C* in $\mathcal{R}$ (see Fig. 1.6). In a similar way to discrete systems of particles, the kinetic energy of a body, in a given reference frame $\mathcal{R}$, is equal to the sum of the kinetic energy of its mass center with its kinetic energy around the mass center, in the same reference frame. In fact, from the kinematic relationship $\mathbf{v} = \mathbf{v}^* + \dot{\mathbf{r}}$, then

$$\begin{aligned}
^{\mathcal{R}}K^C &= \int_C \frac{1}{2}(\mathbf{v}^* + \dot{\mathbf{r}}) \cdot (\mathbf{v}^* + \dot{\mathbf{r}})\, dm \\
&= \int_C \frac{1}{2}\mathbf{v}^* \cdot \mathbf{v}^*\, dm + \int_C \frac{1}{2}\dot{\mathbf{r}} \cdot \dot{\mathbf{r}}\, dm + \int_C \mathbf{v}^* \cdot \dot{\mathbf{r}}\, dm \\
&= \frac{1}{2}\mathbf{v}^* \cdot \mathbf{v}^* \int_C dm + \int_C \frac{1}{2}\dot{\mathbf{r}} \cdot \dot{\mathbf{r}}\, dm + \mathbf{v}^* \cdot \int_C \dot{\mathbf{r}}\, dm \\
&= {}^{\mathcal{R}}K^{C^*} + {}^{\mathcal{R}}K^{C/C^*}.
\end{aligned} \tag{1.26}$$

Note that the third term in the third line is null, since it contains the time rate of the vector $\int_C \mathbf{r}\, dm$, null for any instant t (see Section 6.1 for further details).

Example 1.9   Returning to the previous example (see Fig. 1.9), the kinetic energy of the rope C in the reference frame T may, according to Eq. (1.23), be calculated by

$$
\begin{aligned}
{}^{T}K^{C} &= \frac{1}{2}\int_{Q}^{S} \mathbf{v} \cdot \mathbf{v}\, dm \\
&= \frac{m\omega^{2}}{2a}\int_{r\theta}^{a}(s - r\theta)^{2}\, ds \\
&= \frac{m\omega^{2}}{6a}(a - r\theta)^{3}.
\end{aligned}
$$

The initial kinetic energy ($\theta = 0$) is, therefore, $\frac{1}{6}ma^{2}\omega^{2}$ and the final kinetic energy is null. See the corresponding animation.

Example 1.10   Returning again to Example 1.5 (see Fig. 1.7), the kinetic energy of the mass center of the rod C in the reference frame B is, according to Eq. (1.24),

$$
{}^{B}K^{C^{*}} = \frac{1}{2}m\,{}^{B}\mathbf{v}^{C^{*}} \cdot {}^{B}\mathbf{v}^{C^{*}} = \frac{1}{8}mr^{2}\dot{\theta}^{2}.
$$

The position with respect to point C^{*} of a general point P of C is given by the vector $\mathbf{r} = (y - \frac{r}{2})(\sin\theta\mathbf{b}_{2} - \cos\theta\mathbf{b}_{3})$, and its time rate in the reference frame B is

$$
\dot{\mathbf{r}} = \left(y - \frac{r}{2}\right)\dot{\theta}\left(\cos\theta\mathbf{b}_{2} + \sin\theta\mathbf{b}_{3}\right).
$$

The kinetic energy of the rod around its mass center in the reference frame B is, therefore, according to Eq. (1.25),

$$
\begin{aligned}
{}^{B}K^{C/C^{*}} &= \int_{C} \frac{1}{2}\dot{\mathbf{r}} \cdot \dot{\mathbf{r}}\, dm \\
&= \frac{m\dot{\theta}^{2}}{2r}\int_{0}^{r}(y - \frac{r}{2})^{2}\, dy \\
&= \frac{1}{24}mr^{2}\dot{\theta}^{2}.
\end{aligned}
$$

The kinetic energy of the rod in the reference frame B is, according to Eq. (1.26), the sum of the previous ones, that is,

$$
{}^{B}K^{C} = {}^{B}K^{C^{*}} + {}^{B}K^{C/C^{*}} = \frac{1}{6}mr^{2}\dot{\theta}^{2}.
$$

Note that the kinetic energy of the rod was calculated here in the reference frame B. The reader is recommended, as an exercise, to compute the kinetic energy in the reference frame A, where the angular velocity Ω intervenes.

5.2 Force Systems

The systems of forces — a simplified name adopted for systems of vectors consisting of forces, sliding or bound, and torques — were studied in Chapter 2. As discussed therein, those systems may or may not be simple, concentrated (discrete), distributed (continuous), or even consisting of some forces concentrated and others distributed. If such a force system is applied to a mechanical system — whether it is a simple particle, a discrete system of particles, a deformable body, a rigid body, or a mechanism consisting of different elements — it is important to know how to reduce it to a suitable point; the reduction, consisting in general of a resultant force and a resultant torque, is one of the prior requisites necessary for setting up the equations that govern the motion of the mechanical system, as will be seen in the following sections. This section analyzes in a little more detail the forces and torques applied to a mechanical system, paying special attention to the simplifications resulting from the elimination of the so-called internal forces.

Let us first consider a discrete system S, comprising N particles P_i, of mass m_i, which moves in $\mathcal{R}$ with velocity $\mathbf{v}_i$, $i = 1, 2, \ldots, N$ (see Fig. 2.1). Let us assume that the forces $\mathbf{F}_{ij}$, $j = 1, 2, \ldots, m$, applied by particles outside the system S, and the forces $\mathbf{f}_{ik}$, $k = 1, 2, \ldots, N$, applied by the other $N - 1$ particles belonging to S, act on P_i (of course, $\mathbf{f}_{ii} = 0$, since P_i does not apply force on itself).

The set of forces applied to P_i by particles outside the system — hereinafter simply called *external forces* — is a simple concurrent subsystem whose resultant is

$$\mathbf{F}_i = \sum_{j=1}^{m} \mathbf{F}_{ij}. \tag{2.1}$$

The set of forces applied to P_i by particles inside the system — hereinafter simply called *internal forces* — is another simple concurrent subsystem whose resultant is

$$\mathbf{f}_i = \sum_{k=1}^{N} \mathbf{f}_{ik}. \tag{2.2}$$

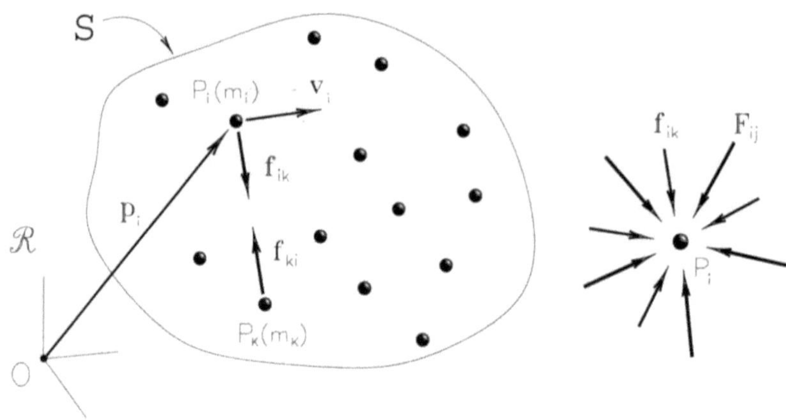

Figure 2.1

The resultant force, therefore, applied on the particle P_i is

$$\mathbf{R}_i = \mathbf{F}_i + \mathbf{f}_i, \tag{2.3}$$

and the resultant of the system $\mathcal{F}$ comprising all forces applied on the system of particles S is

$$\mathbf{R} = \sum_{i=1}^{N} \mathbf{R}_i = \sum_{i=1}^{N} (\mathbf{F}_i + \mathbf{f}_i) = \mathbf{F} + \mathbf{f}, \tag{2.4}$$

being understood that

$$\mathbf{F} \rightleftharpoons \sum_{i=1}^{N} \mathbf{F}_i \tag{2.5}$$

is the *resultant of the external forces applied to the system* and

$$\mathbf{f} \rightleftharpoons \sum_{i=1}^{N} \mathbf{f}_i \tag{2.6}$$

is the *resultant of the internal forces applied to the system.* It is worth mentioning that, when dealing with a discrete system of particles, only forces are applied and the system $\mathcal{F}$ is necessarily a *simple system.*

Now, for each pair of particles P_i and P_k of S there are two forces, $\mathbf{f}_{ik}$ and $\mathbf{f}_{ki}$ (see Fig. 2.1), that, following *Newton's third law,* Eq. (1.3.4), are opposites, that is,

$$\mathbf{f}_{ki} = -\mathbf{f}_{ik}, \qquad i = 1, 2, \ldots, N, \quad k = 1, 2, \ldots, N. \tag{2.7}$$

The result then of Eqs. (2.2), (2.6), and (2.7) is that

$$\mathbf{f} = \sum_{i=1}^{N}\sum_{k=1}^{N} \mathbf{f}_{ik} = 0, \tag{2.8}$$

which means that the sum of all the internal forces applied to the system is null, whatever the case, and, therefore, that, from Eq. (2.4),

$$\mathbf{R} = \mathbf{F}, \tag{2.9}$$

that is, the resultant of all forces acting on a discrete system S of particles is equal to the resultant of the system of external forces acting on it.

Example 2.1 Three small balls, each with the same mass m, stay on a smooth, horizontal flat surface, interconnected by two threads and a linear spring with an elastic constant k and natural length $b/2$, with the configuration shown (see Fig. 2.2).

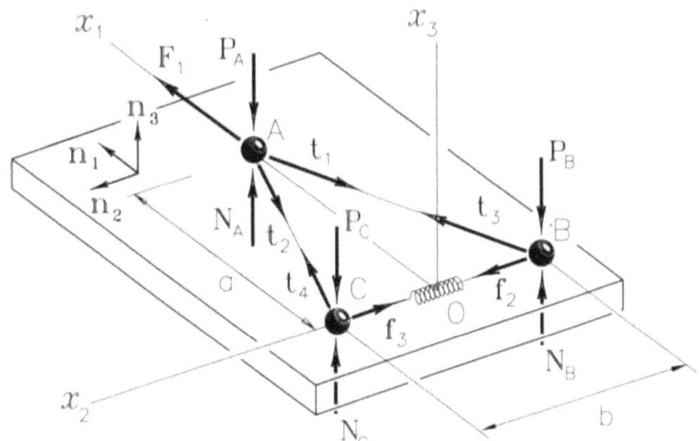

Figure 2.2

The system S, consisting of the three balls, is at rest on the surface when the horizontal force $\mathbf{F}_1$ is applied to ball A. The force system $\mathcal{F}$ acting on S consists of forces $\mathbf{F}_1, \mathbf{t}_1, \mathbf{t}_2, \mathbf{t}_3, \mathbf{t}_4, \mathbf{f}_2, \mathbf{f}_3, \mathbf{P}_A, \mathbf{P}_B, \mathbf{P}_C, \mathbf{N}_A, \mathbf{N}_B$, and $\mathbf{N}_C$, as shown in the figure, where $\mathbf{t}_j$, $j = 1, \ldots, 4$, are the forces applied by the ropes, $\mathbf{f}_2$ and $\mathbf{f}_3$ are the forces applied by the spring, $\mathbf{P}_A, \mathbf{P}_B$, and $\mathbf{P}_C$

are the weights of the balls, and $\mathbf{N}_A, \mathbf{N}_B$, and $\mathbf{N}_C$ are the normal forces applied by the smooth surface. Equation (2.7) guarantees that

$$\mathbf{t}_3 = -\mathbf{t}_1, \quad \mathbf{t}_4 = -\mathbf{t}_2, \quad \mathbf{f}_2 = -\mathbf{f}_3,$$

and the resultant of the forces inside the system is

$$\mathbf{f} = \mathbf{t}_1 + \mathbf{t}_2 + \mathbf{t}_3 + \mathbf{t}_4 + \mathbf{f}_2 + \mathbf{f}_3 = 0,$$

therefore null, as established by Eq. (2.8). The resultant of the external forces is

$$\mathbf{F} = \mathbf{F}_1 + \mathbf{N}_A + \mathbf{N}_B + \mathbf{N}_C - 3mg\mathbf{n}_3.$$

Returning to the internal forces applied to the system S of particles, let us consider the two particles $\mathbf{P}_i$ and $\mathbf{P}_k$, and the two forces $\mathbf{f}_{ik}$ and $\mathbf{f}_{ki}$, of interaction between the former (see Fig. 2.3). Now by computing the moment with respect to an arbitrary point O of these forces, then

$$\mathbf{M}^{\mathbf{f}_{ik}/O} = \mathbf{p}_i \times \mathbf{f}_{ik} = -\mathbf{p}_i \times \mathbf{f}_{ki} = -\mathbf{M}^{\mathbf{f}_{ki}/O}, \qquad (2.10)$$

that is, just as the forces of interaction between two particles of the system are opposite, their moments with respect to an arbitrary point are also opposite.

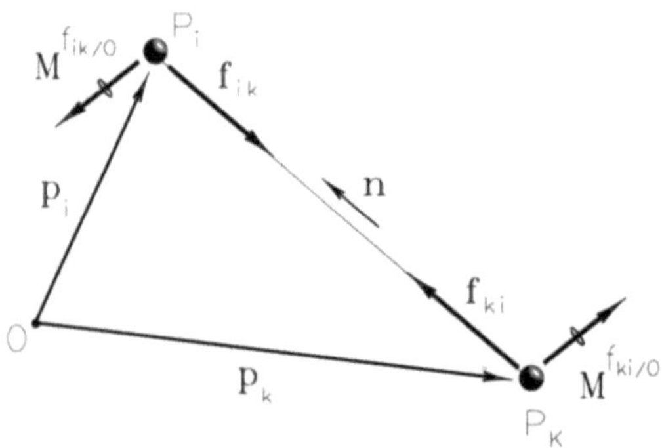

Figure 2.3

It seems convenient, therefore, to fully decompose the force system $\mathcal{F}$ acting on S in two subsystems: the system of external forces, $\mathcal{F}_e$, and the system of internal forces, $\mathcal{F}_i$. Its resultants are defined by Eqs. (2.5) and (2.6), respectively, its vector sum being equal to the resultant of the overall system $\mathcal{F}$, as shown in Eq. (2.4). The resultant moment with respect to point O of the system $\mathcal{F}_i$ is

$$\mathbf{M}^{\mathcal{F}_i/O} = \sum_{i=1}^{N}\sum_{k=1}^{N}\mathbf{M}^{\mathbf{f}_{ik}/O} = 0, \tag{2.11}$$

therefore null, as a consequence of Eq. (2.10).

It can then be said, from Eqs. (2.8) and (2.11), that the system $\mathcal{F}_i$, which comprises all the forces of interaction between the particles of S, is a *null system*. Consequently, the resultant moment with respect to any point O of the system $\mathcal{F}$ is equal to the resultant moment with respect to O of the system of external forces, $\mathcal{F}_e$, that is,

$$\mathbf{M}^{\mathcal{F}/O} = \mathbf{M}^{\mathcal{F}_e/O} = \sum_{i=1}^{N}\mathbf{p}_i \times \mathbf{F}_i. \tag{2.12}$$

Example 2.2 Going back to the previous example (see Fig. 2.2), the resultant moment with respect to point A of the system of internal forces is

$$\mathbf{M}^{\mathcal{F}_i/A} = \mathbf{M}^{\mathbf{t}_1/A} + \mathbf{M}^{\mathbf{t}_2/A} + \mathbf{M}^{\mathbf{t}_3/A} + \mathbf{M}^{\mathbf{t}_4/A} + \mathbf{M}^{\mathbf{f}_2/A} + \mathbf{M}^{\mathbf{f}_3/A}$$

$$= 0 + 0 + 0 + 0 - \frac{1}{2}kab\mathbf{n}_3 + \frac{1}{2}kab\mathbf{n}_3$$

$$= 0.$$

The resultant moment, with respect to the same point, of the system of external forces is

$$\mathbf{M}^{\mathcal{F}_e/A} = \mathbf{M}^{\mathbf{F}_1/A} + \mathbf{M}^{\mathbf{P}_A/A} + \mathbf{M}^{\mathbf{P}_B/A} + \mathbf{M}^{\mathbf{P}_C/A}$$

$$+ \mathbf{M}^{\mathbf{N}_A/A} + \mathbf{M}^{\mathbf{N}_B/A} + \mathbf{M}^{\mathbf{N}_C/A}$$

$$= \frac{b}{2}(N_C - N_B)\mathbf{n}_1 + a(N_C + N_B - 2mg)\mathbf{n}_2.$$

The resultant moment of the overall force system acting on the three balls with respect to the point A is therefore

$$\mathbf{M}^{\mathcal{F}/A} = \mathbf{M}^{\mathcal{F}_e/A} = \frac{b}{2}(N_C - N_B)\mathbf{n}_1 + a(N_C + N_B - 2mg)\mathbf{n}_2.$$

Let us now consider C as a continuous body, with mass m, moving in a reference frame $\mathcal{R}$ under the action of a distributed simple force system $\mathcal{F}$ (see Fig. 2.4). Let P be a point of the body, that means an infinitesimal mass element, with mass dm, whose velocity in $\mathcal{R}$ is **v**. The (infinitesimal) force $d\mathbf{F}$, resultant of the actions performed by particles or bodies outside C, and $d\mathbf{f}$, resultant of the actions performed by all other elements of C, act on P. The (infinitesimal) force resultant on P is, therefore,

$$d\mathbf{R} = d\mathbf{F} + d\mathbf{f}. \tag{2.13}$$

Everything discussed in this section regarding discrete systems has, of course, its counterpart for continuous systems, substituting the sums by corresponding integrals. Thus, the resultant of the system $\mathcal{F}_i$ of internal forces is, as a consequence of the third law, null, that is,

$$\mathbf{f} = \int_C d\mathbf{f} = 0, \tag{2.14}$$

and the resultant moment will be null, with respect to an arbitrary point O in this same system, that is,

$$\mathbf{M}^{\mathcal{F}_i/O} = \int_C \mathbf{p} \times d\mathbf{f} = 0. \tag{2.15}$$

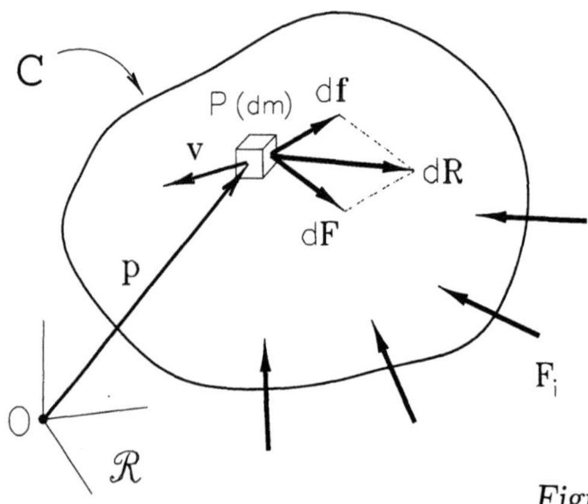

Figure 2.4

The system $\mathcal{F}_i$ is therefore a *null system*. Consequently, the resultant of the overall force system $\mathcal{F}$, acting on C, is equal to the resultant of the system of external forces, $\mathcal{F}_e$,

$$\mathbf{R} = \int_C d\mathbf{R} = \int_C d\mathbf{F} = \mathbf{F}, \qquad (2.16)$$

and the resultant moment with respect to an arbitrary point O of the system $\mathcal{F}$ will, in turn, be equal to the resultant moment with respect to O of the system $\mathcal{F}_e$, that is,

$$\mathbf{M}^{\mathcal{F}/O} = \int_C \mathbf{p} \times d\mathbf{R} = \int_C \mathbf{p} \times d\mathbf{F} = \mathbf{M}^{\mathcal{F}_e/O}. \qquad (2.17)$$

Occasionally the system of external forces acting on a body, although being a simple system, simultaneously comprises concentrated forces, $\mathbf{F}_i$, $i = 1, 2, \ldots, N$, and distributed forces, $d\mathbf{F}$; in the latter case, the calculus of both the resultant and resultant moment with respect to a given point of the system $\mathcal{F}_e$ requires the integration of the distributed forces and the summation of concentrated forces, that is,

$$\mathbf{F} = \sum_{i=1}^{N} \mathbf{F}_i + \int_C d\mathbf{F} \qquad (2.18)$$

and

$$\mathbf{M}^{\mathcal{F}_e/O} = \sum_{i=1}^{N} \mathbf{M}^{\mathbf{F}_i/O} + \int_C \mathbf{p} \times d\mathbf{F}. \qquad (2.19)$$

Example 2.3   A homogeneous cable C is hung by its ends at fixed points A and B, while a pulley, holding a certain load, rolls over the cable (see Fig. 2.5). The external forces acting on C consist of the concentrated forces $\mathbf{F}_A$ and $\mathbf{F}_B$, applied to its ends; the weight force, distributed uniformly along the cable, $d\mathbf{P} = \rho\mathbf{g}\,ds$, where ρ is the mass per unit of length, ds an element of length, and $\mathbf{g}$ the gravitational acceleration; and the distributed force $d\mathbf{F}$, present in the region touching the pulley (other effects, such as those resulting from friction or inertia of the pulley, are not being considered).

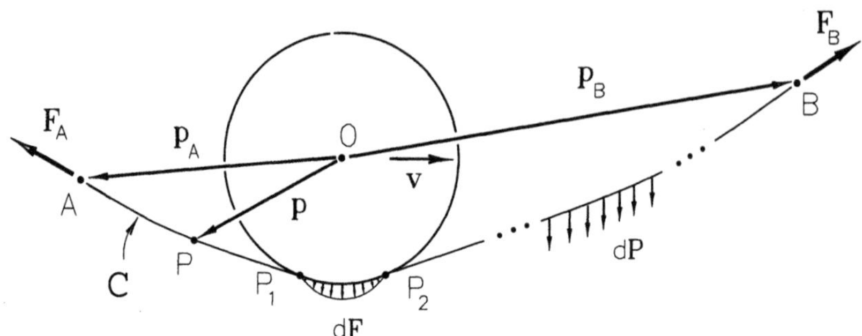

Figure 2.5

At each instant, therefore, the resultant of the aforementioned force system is

$$\mathbf{F} = \mathbf{F}_A + \mathbf{F}_B + m\mathbf{g} + \int_{P_1}^{P_2} d\mathbf{F},$$

where m is the total mass of the cable and P_1 and P_2 are the end contact points of the latter with the pulley, at that instant. The internal forces present between the fibers of the cable, although consisting of a reasonably complex force system, comprise a null system, as guaranteed by Eqs. (2.14) and (2.15). The resultant moment with regard, say, to point O, the center of the pulley, of the acting external forces is

$$\mathbf{M}^{\mathcal{F}e/O} = \mathbf{p}_A \times \mathbf{F}_A + \mathbf{p}_B \times \mathbf{F}_B + \int_A^B \mathbf{p} \times \rho\mathbf{g}\, ds + \int_{P_1}^{P_2} \mathbf{p} \times d\mathbf{F},$$

where $\mathbf{p}$ is the position vector with respect to O of a generic point of the cable.

If C is a rigid body, the system $\mathcal{F}_e$ of external forces acting on C is not necessarily a simple system. Let it then be assumed that, in the most general case, a system $\mathcal{F}_e$ acts on C, consisting of N concentrated forces $\mathbf{F}_i$; a distribution of forces $d\mathbf{F}$, applied to all or part of the body; and a set of M torques $\mathbf{T}_j$ (do not forget that torques are free vectors). The resultant of this system is (see Fig. 2.6)

$$\mathbf{F} = \sum_{i=1}^{N} \mathbf{F}_i + \int_{C'} d\mathbf{F}, \qquad (2.20)$$

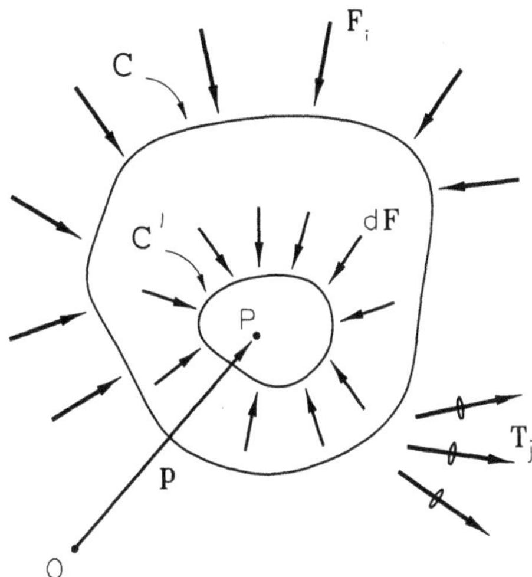

Figure 2.6

where C' is the region of C to which is applied the distributed subsystem (C' may even coincide with C).

The resultant moment of this system with respect to an arbitrary given point O is

$$\mathbf{M}^{\mathcal{F}_e/O} = \sum_{i=1}^{N} \mathbf{p}_i \times \mathbf{F}_i + \sum_{j=1}^{M} \mathbf{T}_j + \int_{C'} \mathbf{p} \times d\mathbf{F}, \qquad (2.21)$$

where $\mathbf{p}$ is the position vector with respect to point O of a generic point of C.

Example 2.4 Figure 2.7 reproduces the system studied in Example 1.5. The system of external forces applied to the rod C consists of a gravitational force distributed along the rod, $d\mathbf{P} = -\frac{m}{r} g\, dy\, \mathbf{b}_3$; a concentrated force applied to the end Q, conveniently broken down into $\mathbf{F}_1$, $\mathbf{F}_2$, and $\mathbf{F}_3$; and a torque due to the coupling on the same point, consisting of the components $\mathbf{T}_2$ and $\mathbf{T}_3$ (see Appendix B). The resultant of this system is, according to Eq. (2.20),

$$\mathbf{F} = \mathbf{F}_1 + \mathbf{F}_2 + \mathbf{F}_3 + \int_0^r d\mathbf{P}$$

$$= F_1\mathbf{b}_1 + F_2\mathbf{b}_2 + (F_3 - mg)\mathbf{b}_3,$$

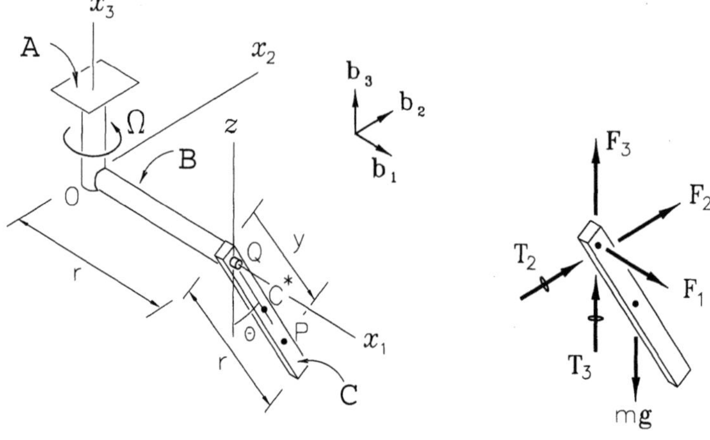

Figure 2.7

and its resultant moment with respect to point Q is, according to Eq. (2.21),

$$\mathbf{M}^{\mathcal{F}_e/Q} = \int_0^r \mathbf{p}^{P/Q} \times d\mathbf{P} + \mathbf{T}_2 + \mathbf{T}_3$$

$$= -\frac{1}{2} mgr \sin\theta \mathbf{b}_1 + T_2 \mathbf{b}_2 + T_3 \mathbf{b}_3.$$

When dealing with a rigid body, the system of internal forces, $\mathcal{F}_i$, may also not be simple, including torques applied reciprocally between its components. If, however, $\mathbf{T}_{jk}$ is the torque that the part C_k of a rigid body C applies on another part C_j of the same body, then the torque $\mathbf{T}_{kj}$ applied by it on the former is opposite to it, that is,

$$\mathbf{T}_{kj} = -\mathbf{T}_{jk}, \tag{2.22}$$

guaranteeing that the system $\mathcal{F}_i$ is null in any case.

The basic aim of the treatment given in this section to the systems of forces acting on the systems of particles or bodies in general was to prepare for the establishment of the equations of motion, the subject of the next two sections. Recognize the fact that the system of internal forces acting on any discrete or continuous system (or body) will allow us to disconsider these forces when analyzing the dynamic equations, as will be shown. When, however, wishing to determine a component of force or torque inside the system or body, this can always be made explicit, cutting the body itself into parts, in order to make the desired component outside the part. The following example illustrates the procedure.

Example 2.5   Returning to the previous example, suppose now that it is of interest to know the forces and torques present in the mass center of the rod C. In this case, it is sufficient to divide the rod in two parts, now considering two bodies, C_1 and C_2, on which new systems of external forces act, as shown in Fig. 2.8.

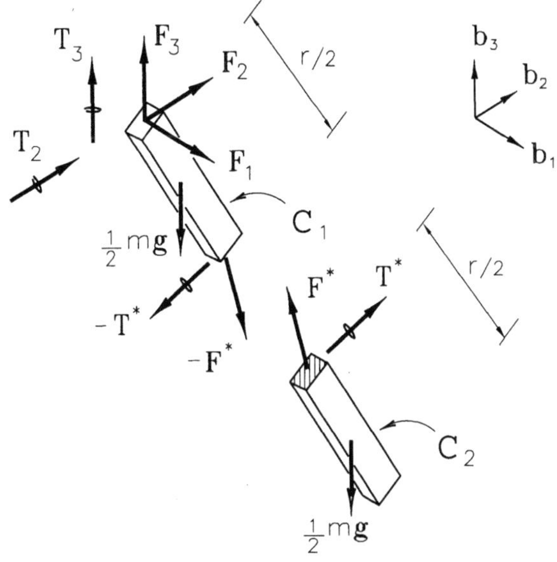

Figure 2.8

Its weight, $\frac{1}{2}m\mathbf{g}$, acts on the lower half of the rod (C_2) and is applied to the respective mass center, and the pair $\{\mathbf{F}^*, \mathbf{T}^*\}$, which is the reduction to the central point of the cut section (coinciding, therefore, with the original mass center of the rod) resulting from the action of C_1 on C_2. The following forces act on the top half of the rod (C_1): its weight, $\frac{1}{2}m\mathbf{g}$, applied to the respective mass center; the force components $\mathbf{F}_1$, $\mathbf{F}_2$, and $\mathbf{F}_3$ and torque components $\mathbf{T}_2$ and $\mathbf{T}_3$, from the link on the point Q; and the pair $\{-\mathbf{F}^*, -\mathbf{T}^*\}$, resulting from the action of C_2 on C_1, reduced to the center of the cut section. Of course, the systems of forces mutually applied by the parts are necessarily opposite, fulfilling Eqs. (2.14), (2.15), and (2.22). (The effective determination of these components of force and torque is only possible after establishing the equations of motion, discussed in the following sections.)

5.3 Equations of Motion

For a clearer explanation, this section will discuss discrete systems only; the equations of motion for bodies will be presented in the next section.

Let P_i, with mass m_i, be a typical particle of a discrete system S, whose motion in an inertial reference frame $\mathcal{R}$ is described by $\mathbf{p}_i(t)$, its position vector with respect to a given point O fixed in $\mathcal{R}$ (see Fig. 3.1). Its velocity in $\mathcal{R}$ is $\mathbf{v}_i = {}^{\mathcal{R}}\mathbf{v}^{P_i} = \dot{\mathbf{p}}_i$, its momentum in $\mathcal{R}$ is $\mathbf{G}_i = m_i\mathbf{v}_i$, and its acceleration in $\mathcal{R}$ is $\mathbf{a}_i = {}^{\mathcal{R}}\mathbf{a}^{P_i} = \dot{\mathbf{v}}_i$.

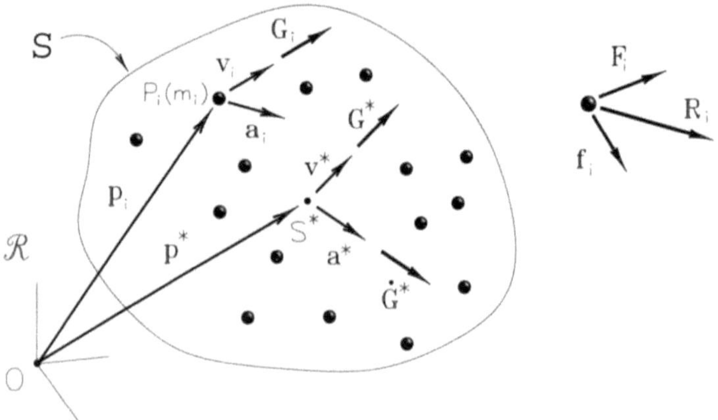

Figure 3.1

If $\mathbf{R}_i = \mathbf{F}_i + \mathbf{f}_i$ is the resultant of the system consisting of all forces acting on P_i, where $\mathbf{F}_i$ is the resultant of the external forces applied to P_i and $\mathbf{f}_i$ is the resultant of the internal forces on the particle (see Section 5.2), then, as a result of Newton's second law, Eq. (4.2.1), we have

$$\dot{\mathbf{G}}_i = \mathbf{R}_i, \tag{3.1}$$

or, from Eq. (4.2.2),

$$m_i\mathbf{a}_i = \mathbf{R}_i. \tag{3.2}$$

It is never too late to remember that the above equations are only valid if the momentum vector, its time rate, and the acceleration vector are being computed in an *inertial reference frame.*

Differentiating Eq. (1.3) with respect to time in $\mathcal{R}$, then $\sum_{i=1}^{N} \dot{\mathbf{G}}_i = {}^{\mathcal{R}}\dot{\mathbf{G}}^S$. On the other hand, we have $\sum_{i=1}^{N} \mathbf{R}_i = \mathbf{F}$ from Eqs. (2.4) and (2.9). When adding Eq. (3.1), over the whole system S, it will then be

$${}^{\mathcal{R}}\dot{\mathbf{G}}^S = \mathbf{F}. \tag{3.3}$$

Equation (3.3) establishes, therefore, the equality between the time rate, in an inertial reference frame $\mathcal{R}$, of the momentum vector in $\mathcal{R}$ of a discrete system of particles S and the resultant of the *external* force system acting on S.

Now remembering that the momentum of a system is identical to the momentum of its mass center, as Eq. (1.5) indicates, Eq. (3.3) may be written in another way, such as

$$\dot{\mathbf{G}}^* = \mathbf{F}. \tag{3.4}$$

Differentiating Eq. (1.2) now with respect to time, then $m\mathbf{a}^* = \sum_{i=1}^{N} m_i \mathbf{a}_i$ and, adding Eq. (3.2) over the whole system S, it gives

$$m\mathbf{a}^* = \mathbf{F}, \tag{3.5}$$

where m is the mass for the system and $\mathbf{a}^*$ the acceleration, in the inertial reference frame $\mathcal{R}$, of the mass center of S (see Fig. 3.1).

Equation (3.3) [or (3.4) or (3.5)] is known as the *first equation of motion of the system* or the *equation of motion of the mass center of the system.*

When an arbitrary system of Cartesian coordinates $\{x_1, x_2, x_3\}$ is chosen, Eq. (3.3) may be written in components, resulting in a set of three scalar equations, so that

$$\begin{aligned} \dot{G}_1^S &= F_1, \\ \dot{G}_2^S &= F_2, \\ \dot{G}_3^S &= F_3, \end{aligned} \tag{3.6}$$

where $\dot{G}_j^S$ and F_j, $j = 1, 2, 3$, are the scalar components, respectively, in the chosen directions of the time rate in the reference frame $\mathcal{R}$ of the momentum vector of the system and (external) resultant force vector. Alternatively, breaking down Eq. (3.5) in the same directions, then

$$
\begin{aligned}
ma_1^* &= F_1, \\
ma_2^* &= F_2, \\
ma_3^* &= F_3,
\end{aligned}
\tag{3.7}
$$

where a_j^*, $j = 1, 2, 3$, are the scalar components of the acceleration of the mass center of the system in the inertial reference frame. Equations (3.7) are known as *equations of motion of the first type.*

As discussed in Section 3.10, a system consisting solely of n particles has, as there are no kinematic constraints, $l = 3n$ degrees of freedom. When a set of p kinematic constraints intervenes, the number of degrees of freedom of the system is reduced to $l = 3n - p$. If the system is holonomic, it will then be necessary to have $r = l$ coordinates $q_1(t), q_2(t), \ldots, q_r(t)$ for its full description. As there are, now, three mutually independent equations of motion, only systems with $r = l \leq 3$ may have their motion totally determined by Eqs. (3.6) or (3.7).

Example 3.1 Let us consider a system S, consisting of three small balls, with the same mass m each, supported on a smooth, horizontal plane fixed to an inertial reference frame, joined by two light rigid rods, with length r (see Fig. 3.2). The Cartesian axes $\{x, y, z\}$, with z vertical, and the basis $\mathbf{n}_1, \mathbf{n}_2, \mathbf{n}_3$, with vectors parallel to the axes, are fixed in the plane. The system is initially at rest, with P_1 coinciding with the origin O, when a constant force $\mathbf{F}_1 = F\mathbf{n}_1$ is applied to it, as shown. Assuming that the three balls always move on the plane, there are three kinematic constraints to be considered:

$$
\mathbf{v}^{P_1} \cdot \mathbf{n}_3 = 0; \quad \mathbf{v}^{P_2} \cdot \mathbf{n}_3 = 0; \quad \mathbf{v}^{P_3} \cdot \mathbf{n}_3 = 0.
$$

The rods connecting P_1 to P_2 and P_3 ensure constant distances between the respective pairs, which means that the component in the direction of the rod of the relative velocity between the particles is null, that is,

$$
\mathbf{v}^{P_2/P_1} \cdot \mathbf{p}^{P_2/P_1} = 0, \quad \mathbf{v}^{P_3/P_1} \cdot \mathbf{p}^{P_3/P_1} = 0.
$$

The holonomic system has, therefore, $l = 3n - p = 3 \times 3 - 5 = 4$ degrees of freedom. By then adopting the coordinates $x(t)$ and $y(t)$, which define the position of P_1 with respect to O, and $\theta(t)$ and $\phi(t)$, which measure the angles formed between the rods and axis x, to describe the configuration of the system at a general instant, the position vectors with respect to O of the three particles are

$$\mathbf{p}_1 = x\mathbf{n}_1 + y\mathbf{n}_2,$$
$$\mathbf{p}_2 = \mathbf{p}_1 - r(\cos\theta\,\mathbf{n}_1 + \sin\theta\,\mathbf{n}_2),$$
$$\mathbf{p}_3 = \mathbf{p}_1 + r(-\cos\phi\,\mathbf{n}_1 + \sin\phi\,\mathbf{n}_2).$$

The position with respect to O of the system's mass center is, therefore,

$$\mathbf{p}^* = x\mathbf{n}_1 + y\mathbf{n}_2 - \frac{r}{3}\Big((\cos\theta + \cos\phi)\mathbf{n}_1 + (\sin\theta - \sin\phi)\mathbf{n}_2\Big).$$

The acceleration of the mass center may then be obtained by differentiating the position vector twice with respect to the time, resulting in

$$\mathbf{a}^* = \ddot{x}\mathbf{n}_1 + \ddot{y}\mathbf{n}_2 + \frac{r}{3}\Big((\ddot{\theta}\sin\theta + \dot{\theta}^2\cos\theta + \ddot{\phi}\sin\phi + \dot{\phi}^2\cos\phi)\mathbf{n}_1$$
$$+ (\ddot{\phi}\cos\phi - \dot{\phi}^2\sin\phi - \ddot{\theta}\cos\theta + \dot{\theta}^2\sin\theta)\mathbf{n}_2\Big).$$

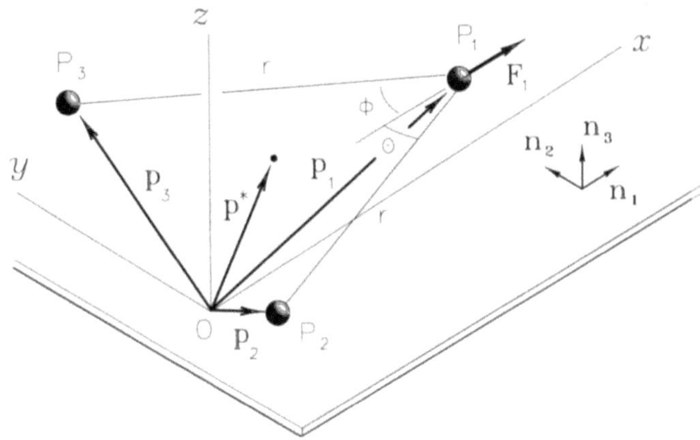

Figure 3.2

The resultant of the system of external forces acting on the system is

$$\mathbf{F} = F\mathbf{n}_1 + (N_1 + N_2 + N_3 - 3mg)\mathbf{n}_3,$$

where N_j, $j = 1, 2, 3$, are forces (normal forces, since the plane is smooth) applied by the surface. When substituting in Eqs. (3.7), then a set of three equations of motion is obtained:

$$\ddot{x} + \frac{r}{3}(\ddot{\theta}\sin\theta + \dot{\theta}^2\cos\theta + \ddot{\phi}\sin\phi + \dot{\phi}^2\cos\phi) = \frac{F}{3m}; \tag{a}$$

$$\ddot{y} + \frac{r}{3}(\ddot{\phi}\cos\phi - \dot{\phi}^2\sin\phi - \ddot{\theta}\cos\theta + \dot{\theta}^2\sin\theta) = 0; \tag{b}$$

$$0 = N_1 + N_2 + N_3 - 3mg. \tag{c}$$

This example points to the fact that Newton's second law, suitable for describing the motion of a single particle (see Section 4.2), is generally insufficient to fully describe the motion of a *system*, only establishing the motion of its mass center. In fact, Eqs. (a) and (b) above are a coupled pair of differential equations for the functions $x(t)$, $y(t)$, $\theta(t)$, and $\phi(t)$, while Eq. (c) is an algebraic equation for the unknown N_1, N_2, and N_3. Of course, if the reader writes the equations of motion in the vertical direction, for each ball, he or she will conclude, assuming that there is no motion in that direction and that the masses are equal, that $N_1 = N_2 = N_3 = mg$; but this information cannot be extracted from Eqs. (3.7), used in the solution of the system as a whole. On the other hand, Eqs. (a) and (b) may be simply written as

$$\ddot{p}_1^* = \frac{F}{3m}, \tag{d}$$

$$\ddot{p}_2^* = 0, \tag{e}$$

whose direct integration provides the motion of the mass center of the system on the plane from the prescribed initial conditions:

$$x(0) = y(0) = 0; \quad \theta(0) = \theta_0; \quad \phi(0) = \phi_0; \quad \dot{x}(0) = \dot{y}(0) = \dot{\theta}(0) = \dot{\phi}(0) = 0,$$

resulting in (check)

$$\mathbf{p}^* = \left(\frac{F}{6m}t^2 - \frac{r}{3}(\cos\theta_0 + \cos\phi_0)\right)\mathbf{n}_1 - \frac{r}{3}(\sin\theta_0 - \sin\phi_0)\mathbf{n}_2.$$

Before finishing the example, it is worth noting that if the ball P_1 is submitted not to the force $\mathbf{F}_1 = F\mathbf{n}_1$ but to an arbitrary force $\mathbf{F}$ with three nonnull components (motion may occur outside the plane), the kinematic constraints would be reduced, consequently increasing the number of degrees of freedom of the system.

Let Q be a point moving arbitrarily in an inertial reference frame $\mathcal{R}$ and P_i a typical particle of a system S that also moves in $\mathcal{R}$ (see Fig. 3.3). The resultant moment with respect to Q of the force system acting on P_i, including the external and internal forces, $\mathbf{M}_i^Q = \mathbf{p}^{P_i/Q} \times \mathbf{R}_i$, is related to the time rate of the angular momentum vector of the particle with respect to Q and with its momentum in $\mathcal{R}$, according to Eq. (4.4.4), by

$$^{\mathcal{R}}\dot{\mathbf{H}}^{P_i/Q} + {}^{\mathcal{R}}\mathbf{v}^Q \times \mathbf{G}_i = \mathbf{M}_i^Q. \qquad (3.8)$$

Equation (3.8) is valid for every particle P_i of S. So, adding for the N particles of the system, then

$$\sum_{i=1}^{N} {}^{\mathcal{R}}\dot{\mathbf{H}}^{P_i/Q} + {}^{\mathcal{R}}\mathbf{v}^Q \times \sum_{i=1}^{N} \mathbf{G}_i = \sum_{i=1}^{N} \mathbf{M}_i^Q. \qquad (3.9)$$

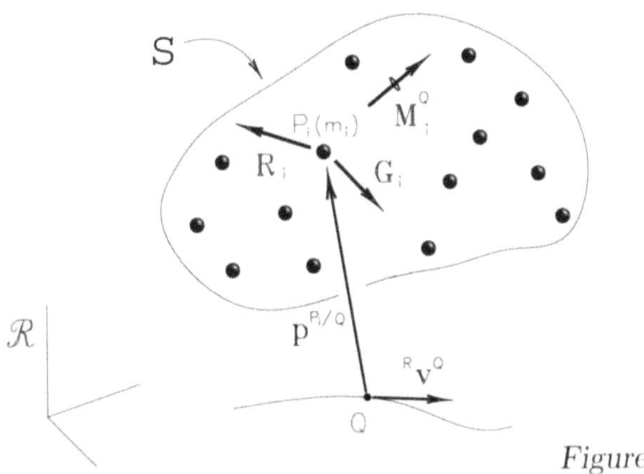

Figure 3.3

The angular momentum vector of the system S with respect to Q in $\mathcal{R}$ is, by definition, $^{\mathcal{R}}\mathbf{H}^{S/Q} = \sum_{i=1}^{N} {}^{\mathcal{R}}\mathbf{H}^{P_i/Q}$ and, therefore, its time rate in $\mathcal{R}$ is

$$^{\mathcal{R}}\dot{\mathbf{H}}^{S/Q} = \sum_{i=1}^{N} {}^{\mathcal{R}}\dot{\mathbf{H}}^{P_i/Q}. \qquad (3.10)$$

On the other hand, Eq. (2.12) guarantees that

$$\mathbf{M}^{\mathcal{F}_e/Q} = \sum_{i=1}^{N} \mathbf{M}_i^Q. \qquad (3.11)$$

Substituting Eqs. (1.3), (3.10), and (3.11) in Eq. (3.9), then

$$^{\mathcal{R}}\dot{\mathbf{H}}^{S/Q} + {}^{\mathcal{R}}\mathbf{v}^{Q} \times {}^{\mathcal{R}}\mathbf{G}^{S} = \mathbf{M}^{\mathcal{F}_e/Q}. \tag{3.12}$$

Equation (3.12), called the *second equation of motion* for the system of particles S, establishes that the time rate in an inertial reference frame $\mathcal{R}$ of the angular momentum vector of a system S with respect to an arbitrary point Q, added vectorially with the cross product between the velocity of Q in $\mathcal{R}$ and the momentum in $\mathcal{R}$ of the system, is equal to the resultant moment with respect to point O of the external forces acting on the system.

Two particular cases of Eq. (3.12) are of special interest: when the chosen point Q coincides with the mass center of the system and when the point Q is fixed in the reference frame $\mathcal{R}$. In both cases, the term $^{\mathcal{R}}\mathbf{v}^{Q} \times {}^{\mathcal{R}}\mathbf{G}^{S}$ vanishes, simplifying the second equation of motion.

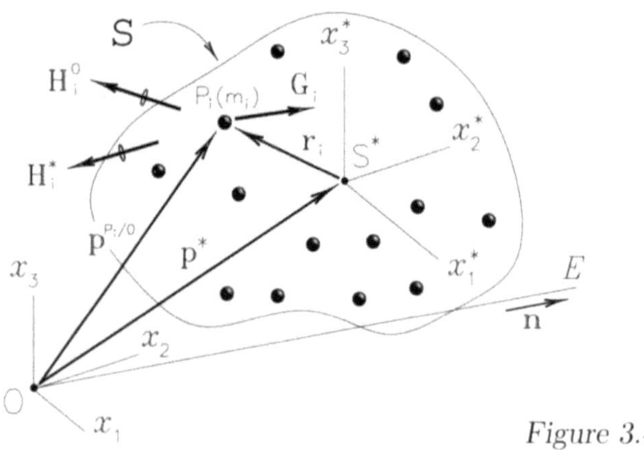

Figure 3.4

Particularly, then, if S^* is the mass center of the system (see Fig. 3.4), the substitution of Q by S^* in Eq. (3.12) leads to

$$^{\mathcal{R}}\dot{\mathbf{H}}^{S/S^*} = \mathbf{M}^{\mathcal{F}_e/S^*}. \tag{3.13}$$

In fact, substituting Q by S^* then, from Eq. (1.4), $^{\mathcal{R}}\mathbf{v}^{S^*} \times {}^{\mathcal{R}}\mathbf{G}^{S} = \mathbf{v}^* \times m\mathbf{v}^* = 0$, therefore resulting in Eq. (3.13).

Equation (3.13), also called the *equation of motion around the mass center*, is a very interesting alternative for Eq. (3.12), given its simplicity. When a system of Cartesian axes $\{x_1^*, x_2^*, x_3^*\}$ is chosen, with origin in S* (see Fig. 3.4), the equation of motion around the mass center will result in a system of three scalar equations, as follows:

$$\dot{H}_1^S = M_1^{\mathcal{F}e};$$
$$\dot{H}_2^S = M_2^{\mathcal{F}e}; \tag{3.14}$$
$$\dot{H}_3^S = M_3^{\mathcal{F}e},$$

where $H_j^S = H^{S/x_j^*}$ and $M_j^{\mathcal{F}e} = M^{\mathcal{F}e/x_j^*}$, $j = 1, 2, 3$. Equations (3.14) are known as *equations of motion of the second type*.

Example 3.2 Two small bodies, with the same mass m each and interconnected by a thread with length $2r$, are at rest on a smooth, horizontal flat surface in the configuration shown in Fig. 3.5a.

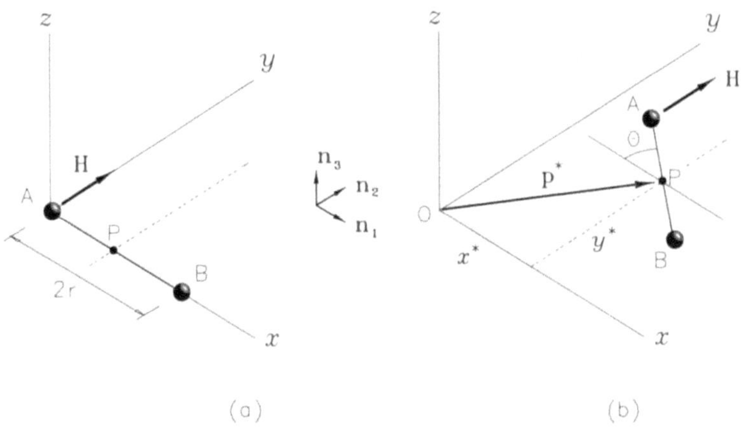

Figure 3.5

In the instant $t = 0$, a horizontal force of constant module H, parallel to axis y is then applied. Figure 3.5b illustrates a general configuration of the system, described by the coordinates $x^*(t)$, $y^*(t)$, and $\theta(t)$, where x^*, y^* are the coordinates of their mass center P and θ is the angle shown. In fact, assuming that the system always moves on the plane (fixed in an inertial reference frame), maintaining the thread straight, then $n = 2$ and $p = 3$, and the system will have $l = 3n - p = 3$ degrees of freedom. The resultant of the acting external forces is

$$\mathbf{F} = H\mathbf{n}_2 + (N_A + N_B - 2mg)\mathbf{n}_3,$$

where N_A and N_B are the forces applied by the smooth plane. The position of the mass center of the system with respect to the origin of the system of coordinates is described by the vector $\mathbf{p}^* = x^*\mathbf{n}_1 + y^*\mathbf{n}_2$ and its acceleration in the inertial reference frame is $\mathbf{a}^* = \ddot{x}^*\mathbf{n}_1 + \ddot{y}^*\mathbf{n}_2$. The first equation of motion, in components, is then

$$2m\ddot{x}^* = 0, \tag{a}$$

$$2m\ddot{y}^* = H, \tag{b}$$

$$0 = N_A + N_B - 2mg. \tag{c}$$

The resultant moment with respect to the mass center of the acting external forces is

$$\mathbf{M}^{\mathcal{F}e/P} = r(N_A - N_B)(\sin\theta\,\mathbf{n}_1 + \cos\theta\,\mathbf{n}_2) - Hr\cos\theta\,\mathbf{n}_3.$$

The angular momentum vector of the system with respect to its mass center is (check)

$$\mathbf{H}^{S/P} = -2mr^2\dot{\theta}\mathbf{n}_3$$

and its time rate in the reference frame is

$$\dot{\mathbf{H}}^{S/P} = -2mr^2\ddot{\theta}\mathbf{n}_3.$$

The second equation of motion of the system may be expressed then by

$$0 = (N_A - N_B)r\sin\theta, \tag{d}$$

$$0 = (N_A - N_B)r\cos\theta, \tag{e}$$

$$2mr^2\ddot{\theta} = Hr\cos\theta. \tag{f}$$

Equations (a) and (b) may be directly integrated and, using the initial conditions $x^*(0) = r$; $y^*(0) = 0$; $\dot{x}^*(0) = \dot{y}^*(0) = 0$, we find the motion of point P:

$$x^*(t) = r; \quad y^*(t) = \frac{H}{4m}t^2.$$

Equations (c) and (d) [or (e)] determine the forces of contact on the plane:

$$N_A = N_B = mg.$$

Last, Eq. (f) may be written as

$$\ddot{\theta} - \frac{H}{2mr}\cos\theta = 0,$$

a nonlinear equation, similar to that of the simple pendulum, and whose integration requires two initial conditions: $\theta(0) = \dot{\theta}(0) = 0$ (see Example 4.2.2, for discussion of the solution of the simple pendulum). Note that this example illustrates a simple situation, forming a system with only three degrees of freedom and involving as unknowns three coordinates (x^*, y^*, θ) and two components of force (N_A, N_B), that may, therefore, be obtained from Eqs. (3.7) and (3.14). See the corresponding animation.

Going back to the general system S, in which P_i is a typical particle, let us now consider O to be a point fixed in the inertial reference frame $\mathcal{R}$ (see Fig. 3.4). Substituting then the moving point Q by the fixed point O in Eq. (3.12), then we have, since $^{\mathcal{R}}\mathbf{v}^O = 0$,

$$^{\mathcal{R}}\dot{\mathbf{H}}^{S/O} = \mathbf{M}^{\mathcal{F}_e/O}. \tag{3.15}$$

Equation (3.15) is one more alternative form to the second equation of motion of a system of particles, providing the same advantage of compactness found in Eq. (3.13).

When choosing an arbitrary system of Cartesian axes with origin at the point O (see Fig. 3.4), Eq. (3.15) resolves into three scalar equations as follows:

$$\begin{aligned} \dot{H}_1^S &= M_1^{\mathcal{F}_e}; \\ \dot{H}_2^S &= M_2^{\mathcal{F}_e}; \\ \dot{H}_3^S &= M_3^{\mathcal{F}_e}, \end{aligned} \tag{3.16}$$

where, now, $H_j^S = H^{S/x_j}$ and $M_j^{\mathcal{F}_e} = M^{\mathcal{F}_e/x_j}$, $j = 1, 2, 3$.

Note that Eqs. (3.14) and (3.16) have an identical form, given the simplification introduced in the notation. The choice between adopting one or the other, or even a composition of them, will depend on the specific convenience for the system under study; nevertheless, only three mutually independent scalar equations will, in any case, be available to solve the problem.

Equations (3.6) [or (3.7)] and (3.14) [or (3.16)] express, in practice, a set of six scalar equations, generally independent of each other, which govern the motion of a system. Consequently, only a system with up to six degrees of freedom will be fully determined by these equations, and if the applied system of external forces is known. A system S, for example, consisting of two particles moving under the action of known forces without kinematic constraints will have $l = 3n - p = 6$ and can be determined; a system consisting of, say, three particles, with two kinematic constraints, will have $l = 3n - p = 7$ and so is undetermined.

If, in particular, a system S moves on a plane π, fixed in an inertial reference frame, two equations of motion of the first type, in directions parallel to the plane, and a single equation of motion of the second type, in the direction normal to the plane, will be available for determining the coordinates that describe the system; the three remaining equations can, however, provide information on unknown force components.

Example 3.3   Two small bodies A and B, with masses m_A and m_B, respectively, slide over a smooth, horizontal flat surface, interconnected by a thread with negligible mass that passes through a peg P, fixed on the plane, as shown (see Fig. 3.6).

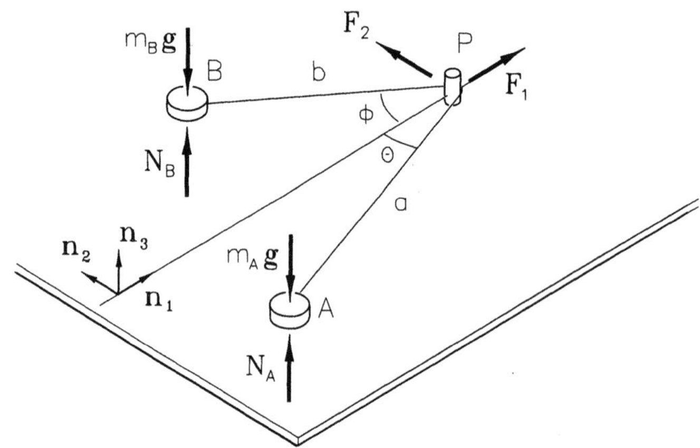

Figure 3.6

The system moves freely on the plane with the thread stretched. Assuming that the friction between the latter and the peg prevents the thread from sliding, the system will then have only two degrees of freedom; the coordinates $\theta(t)$ and $\phi(t)$ describe their configuration adequately. The applied external forces consist of $\mathbf{F}_P$, applied by the peg, conveniently broken down into $F_1\mathbf{n}_1$ and $F_2\mathbf{n}_2$; the weights of the bodies, $-m_A g\mathbf{n}_3$ and $-m_B g\mathbf{n}_3$; and the forces of contact on the surface, $N_A\mathbf{n}_3$ and $N_B\mathbf{n}_3$. The external resultant force will therefore be

$$\mathbf{F} = F_1\mathbf{n}_1 + F_2\mathbf{n}_2 + \big(N_A + N_B - (m_A + m_B)g\big)\mathbf{n}_3.$$

The momentum vector of the system in the reference frame fixed on the surface is

$$\mathbf{G}^S = \mathbf{G}^A + \mathbf{G}^B$$
$$= m_A a \dot{\theta}(\sin\theta\,\mathbf{n}_1 - \cos\theta\,\mathbf{n}_2) + m_B b \dot{\phi}(\sin\phi\,\mathbf{n}_1 + \cos\phi\,\mathbf{n}_2).$$

Differentiating with respect to time and substituting in Eq. (3.6), one has the three components of the first equation of motion for the system:

$$m_A a(\ddot{\theta}\sin\theta + \dot{\theta}^2\cos\theta) + m_B b(\ddot{\phi}\sin\phi + \dot{\phi}^2\cos\phi) = F_1; \tag{a}$$
$$m_A a(-\ddot{\theta}\cos\theta + \dot{\theta}^2\sin\theta) + m_B b(\ddot{\phi}\cos\phi - \dot{\phi}^2\sin\phi) = F_2; \tag{b}$$
$$N_A + N_B - (m_A + m_B)g = 0. \tag{c}$$

Note that, in this set of equations, the coordinates $\theta(t)$ and $\phi(t)$ and the components of force $F_1, F_2, N_A,$ and N_B are unknown. Equations (a–c) are, therefore, not enough to establish the motion of the system. The resultant moment of the system of applied external forces with respect to point P (fixed) is

$$\mathbf{M}^{\mathcal{F}_{e}/P} = a\,(N_A - m_A g)\,(-\sin\theta\,\mathbf{n}_1 + \cos\theta\,\mathbf{n}_2)$$
$$+ b\,(N_B - m_B g)\,(\sin\phi\,\mathbf{n}_1 + \cos\phi\,\mathbf{n}_2).$$

The angular momentum vector of the system with respect to point P is

$$\mathbf{H}^{S/P} = \mathbf{H}^{A/P} + \mathbf{H}^{B/P} = (m_A a^2 \dot{\theta} - m_B b^2 \dot{\phi})\mathbf{n}_3.$$

Differentiating with respect to time and substituting in Eq. (3.15), then the three components of the second equation of motion for the system are

$$0 = -(N_A - m_A g)a\sin\theta + (N_B - m_B g)b\sin\phi, \tag{d}$$
$$0 = (N_A - m_A g)a\cos\theta + (N_B - m_B g)b\cos\phi, \tag{e}$$
$$m_A a^2\ddot{\theta} - m_B b^2\ddot{\phi} = 0. \tag{f}$$

Observe now carefully the set of Eqs. (a–f). Although the number of unknowns (six) is equal to the number of equations, the solution is not necessarily guaranteed. In fact, as $N_A = m_A g$ and $N_B = m_B g$ (which can be easily checked by using Newton's second law in the direction $\mathbf{n}_3$ for A and B separately), Eqs. (c), (d), and (e) are identically null. There then remain Eqs. (a) and (b) (first type, in directions parallel to the plane of motion) and Eq. (f) (second type, in direction orthogonal to the plane), insufficient, in principle, to fully determine $\phi(t)$, $\theta(t)$, F_1, and F_2. Investigating the available equations a little further, it can be seen that a possible solution for Eq. (f) is given by

$$\theta(t) = \omega_A t + \theta_0, \quad \phi(t) = \omega_B t + \phi_0,$$

where ω_A, ω_B, θ_0, and ϕ_0 are constants depending on the initial conditions of the motion. Note that this solution prescribes a uniform circular motion on the plane for both bodies. The components of force applied by the peg can then be established from Eqs. (a) and (b), resulting in

$$F_1 = m_A a \omega_A^2 \cos\theta + m_B b \omega_B^2 \cos\phi,$$
$$F_2 = m_A a \omega_A^2 \sin\theta - m_B b \omega_B^2 \sin\phi.$$

Example 3.4   Consider the double pendulum consisting of two small identical bodies A and B, with mass m each, hanging from threads with the same length r, as shown (see Fig. 3.7). Assuming the earth as an inertial reference frame and abandoning the system from rest on a vertical plane, with arbitrary $\theta(0) = \theta_0$ and $\phi(0) = \phi_0$, the set will move, remaining on the plane. The system, therefore, has two degrees of freedom, described by the coordinates $\theta(t)$ and $\phi(t)$. The applied external forces are the weights and the traction in the top thread, passing through O. The reduction of this force system to point O is given by its resultant,

$$\mathbf{F} = (2mg - T\cos\theta)\mathbf{n}_1 - T\sin\theta\,\mathbf{n}_2,$$

and its resultant moment with respect to O,

$$\mathbf{M}^{\mathcal{F}_e/O} = -mgr(2\sin\theta + \sin\phi)\mathbf{n}_3.$$

The momentum vector of the system, in a general position, is

$$\begin{aligned}
\mathbf{G}^S &= \mathbf{G}^A + \mathbf{G}^B \\
&= mr\big(-(2\dot\theta\sin\theta + \dot\phi\sin\phi)\mathbf{n}_1 + (2\dot\theta\cos\theta + \dot\phi\cos\phi)\mathbf{n}_2\big).
\end{aligned}$$

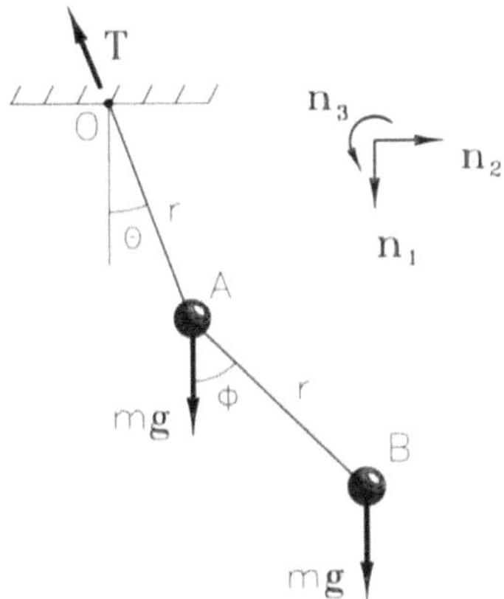

Figure 3.7

Differentiating with respect to time and substituting in Eq. (3.6) results in two components for the first equation of motion:

$$\ddot{\phi}\sin\phi + \dot{\phi}^2\cos\phi + 2(\ddot{\theta}\sin\theta + \dot{\theta}^2\cos\theta) = \frac{T}{mr}\cos\theta - \frac{2g}{r}; \qquad (a)$$

$$\ddot{\phi}\cos\phi - \dot{\phi}^2\sin\phi + 2(\ddot{\theta}\cos\theta - \dot{\theta}^2\sin\theta) = -\frac{T}{mr}\sin\theta. \qquad (b)$$

Note that the equation in the direction of $\mathbf{n}_3$ is identically null. Now calculating the angular momentum vector of the system with respect to point O, then

$$\mathbf{H}^{S/O} = \mathbf{H}^{A/O} + \mathbf{H}^{B/O}$$
$$= mr^2\left[\left(2 + \cos(\phi - \theta)\right)\dot{\theta} + \left(1 + \cos(\phi - \theta)\right)\dot{\phi}\right]\mathbf{n}_3.$$

Differentiating with respect to time and substituting in Eq. (3.16), then

$$\begin{aligned}\left(2 + \cos(\phi - \theta)\right)\ddot{\theta} & \\ + \left(1 + \cos(\phi - \theta)\right)\ddot{\phi} - \sin(\phi - \theta)(\dot{\phi}^2 - \dot{\theta}^2) &= -\frac{g}{r}(2\sin\theta + \sin\phi).\end{aligned} \qquad (c)$$

When removing T from Eqs. (a) and (b) and joining Eq. (c), this gives a nonlinear system with two coupled differential equations for the coordinates $\theta(t)$ and $\phi(t)$. This system is strongly nonlinear and extremely sensitive to the initial conditions. Indeed, for small oscillations, the equations may be linearized (as in the case of the simple pendulum, studied in Chapter 4), resulting in periodic behavior. However, when given arbitrary initial conditions, the behavior of the solution is hard to forecast. This behavior is called *chaotic motion*. See the corresponding animation.

5.4 Continuous Systems

In Section 5.3 the equations of motion are established for discrete systems of particles. As noted in Section 5.1, the relations established for discrete systems always have a counterpart for continuous systems, it being only necessary for this to substitute the masses m_i, of the constituent particles, by the mass $dm = \rho\, dV$, of an infinitesimal mass element, and the sums for the n particles of the discrete system by integrals in the body. The equations present in this section may all be derived by identical procedures to those used for the derivation of their counterpart for discrete systems and, for this reason, are left to the reader.

Let us then consider a body C, with mass m, moving arbitrarily in an inertial reference frame $\mathcal{R}$. The reader should note carefully that, when the term *body* is used, we are referring without distinction to a rigid body, a deformable solid, or a fluid portion in motion. Also let O be a point fixed in $\mathcal{R}$, Q a point moving arbitrarily in $\mathcal{R}$, and C* the mass center of C (see Fig. 4.1).

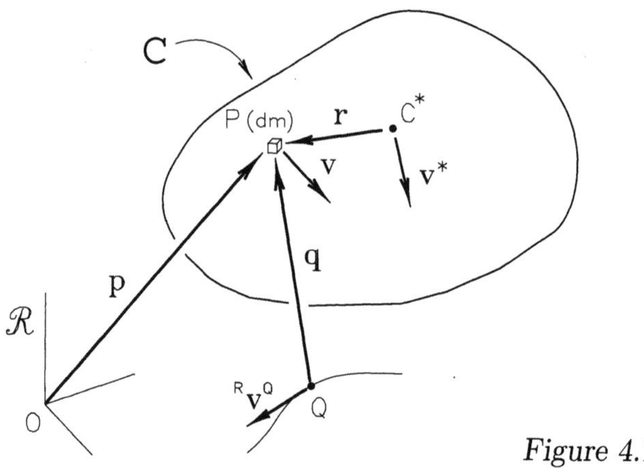

Figure 4.1

Let $\mathcal{F}_e$ be the system of external forces acting on C, whose resultant is $\mathbf{F}$ and whose resultant moments with respect to the points C*, O, and Q are $\mathbf{M}^{\mathcal{F}_e/C^*}$, $\mathbf{M}^{\mathcal{F}_e/O}$, and $\mathbf{M}^{\mathcal{F}_e/Q}$, respectively. The first equation of motion for the body C is then given by

$$^{\mathcal{R}}\dot{\mathbf{G}}^C = \mathbf{F}, \tag{4.1}$$

that is, the time rate, in an inertial reference frame $\mathcal{R}$, of the momentum vector in $\mathcal{R}$ of a body C is, at each instant, equal to the resultant of the system of applied external forces. An interesting alternative form of expressing this result is to consider a particle (fictitious) with mass equal to the mass m of the body and with movement identical to that of its mass center, resulting in

$$\dot{\mathbf{G}}^* = \mathbf{F}, \tag{4.2}$$

or, also, the acceleration in $\mathcal{R}$ of the mass center being $\mathbf{a}^*$,

$$m\mathbf{a}^* = \mathbf{F}. \tag{4.3}$$

Arbitrating a system of Cartesian axes to break down the vectors, Eq. (4.1) assumes the form of three scalar equations:

$$\begin{aligned}
\dot{G}_1^C &= F_1; \\
\dot{G}_2^C &= F_2; \\
\dot{G}_3^C &= F_3.
\end{aligned} \tag{4.4}$$

Alternatively, by resolving Eq. (4.3) into the same directions, this gives the set of three scalar equations, called equations of motion of the first type for a body, as follows:

$$\begin{aligned}
ma_1^* &= F_1; \\
ma_2^* &= F_2; \\
ma_3^* &= F_3.
\end{aligned} \tag{4.5}$$

Now taking moment with respect to point Q, mobile, the second equation of motion for a body C moving in an inertial reference frame $\mathcal{R}$ assumes the general form

$$^{\mathcal{R}}\dot{\mathbf{H}}^{C/Q} + {}^{\mathcal{R}}\mathbf{v}^{Q} \times {}^{\mathcal{R}}\mathbf{G}^{C} = \mathbf{M}^{\mathcal{F}_e/Q}. \tag{4.6}$$

Just as happens to discrete systems, the general form for the second equation of motion involves two terms on the left side of the equation: one consisting of the time rate in the inertial reference frame of the angular momentum of the body with respect to the point; the other consisting of the cross product of the velocity of the point in the reference frame by the momentum vector of the body, in the same reference frame. Also here, two simplified forms are useful for the second equation of motion of a body, as happens when the adopted point Q is the mass center itself of the system (and, in this case, $\mathbf{v}^*$ is parallel to ${}^{\mathcal{R}}\mathbf{G}^C$, canceling out the second term), or when Q is fixed in $\mathcal{R}$ (in this case, canceling out ${}^{\mathcal{R}}\mathbf{v}^Q$).

For the mass center, then, the second equation of motion is reduced to

$$ {}^{\mathcal{R}}\dot{\mathbf{H}}^{C/C^*} = \mathbf{M}^{\mathcal{F}_e/C^*}. \tag{4.7} $$

On the other hand, if the force system is reduced to a point O fixed in the inertial reference frame, the second equation of motion will assume the alternative form

$$ {}^{\mathcal{R}}\dot{\mathbf{H}}^{C/O} = \mathbf{M}^{\mathcal{F}_e/O}. \tag{4.8} $$

Finally, the decomposition in three mutually orthogonal directions, from either Eq. (4.7) or Eq. (4.8), leads to three scalar equations, consisting on the equations of motion of the second kind for a body, so that:

$$ \begin{aligned} \dot{H}_1^C &= M_1^{\mathcal{F}_e}; \\ \dot{H}_2^C &= M_2^{\mathcal{F}_e}; \\ \dot{H}_3^C &= M_3^{\mathcal{F}_e}, \end{aligned} \tag{4.9} $$

where $\dot{H}_j^C$, $j = 1, 2, 3$, are the scalar components of the time rate of the angular momentum vector of the body with respect to the mass center or to the chosen fixed point and $M_j^{\mathcal{F}_e}$, $j = 1, 2, 3$, are the scalar components of the resultant moment with respect to the coordinated axes, with origin in the mass center or in the fixed point, respectively. Note that the decomposition of Eq. (4.7) or Eq. (4.8) provides three mutually independent equations.

As an illustration, a possible alternative is given below for the demonstration, step by step, say, of Eq. (4.8). Other equations above may be derived by adopting a similar procedure.

$$ {}^{\mathcal{R}}\dot{\mathbf{H}}^{C/O} = \frac{{}^{\mathcal{R}}d}{dt}{}^{\mathcal{R}}\mathbf{H}^{C/O} $$

$$= \frac{^{\mathcal{R}}d}{dt} \int_C \mathbf{p} \times \mathbf{v}\, dm \qquad \text{[from Eq. (1.19)]}$$

$$= \int_C \mathbf{v} \times \mathbf{v}\, dm + \int_C \mathbf{p} \times \mathbf{a}\, dm \quad \text{[der. of the prod.]}$$

$$= \int_C \mathbf{p} \times d\mathbf{R} \qquad \text{[second law]}$$

$$= \mathbf{M}^{\mathcal{F}_e/O}. \qquad \text{[from Eq. (2.17)]}$$

The reader will have no difficulty in deriving the other results in this section, following the same steps taken to obtain the valid relations for discrete systems.

Let the reader's attention now turn to the general form that the equations of motion take for discrete or continuous systems. For both kinds of systems, there are two types of equations: one equation that governs the motion of its mass center (Eq. (3.3) for discrete systems and Eq. (4.1) for continuous systems); and an equation governing the motion of the system around its mass center (Eq. (3.13) for discrete systems and Eq. (4.7) for continuous systems). This pair of equations has a special characteristic: On one side of the equations are the time rates, in an inertial reference frame, of the resultant and resultant moment of a system of momentum vectors, this pair consisting of a reduction to the mass center of the system $\mathcal{G}$ of momentum vectors; on the other side are the resultant and the resultant moment of the system $\mathcal{F}_e$ of external forces, this latter pair consisting of a reduction in the mass center of the force system. The first equation of motion, therefore, equals the time rate of the resultant of the system of momentum vectors to the resultant of the system of external forces; the second equation of motion, in turn, equals the time rate of the resultant moment with respect to the mass center of the system of momentum vectors to the resultant moment with respect to the mass center of the system of applied external forces. For the equations of motion of the second type, it is worth remembering that there is always the alternative of reducing the systems to a point fixed in the inertial reference frame, resulting in Eq. (3.15) for discrete systems and in Eq. (4.8) for continuous systems.

Example 4.1   A homogeneous rod B, with mass m and length $2a$, moves on the vertical plane xy, having its end O hanging by a known vertical force V, and the other end Q supported, by means of a roller, on a horizontal plane (see Fig. 4.2). The linear coordinate $x(t)$ of point Q, and the angular coordinate $\theta(t)$, indicating the slope of the rod with respect to the vertical, are enough to describe the system (two degrees of freedom). Let us establish the equations of motion for the rod, in terms of the chosen coordinates and the unknown force N applied to the roller by the horizontal plane. The system of external forces acting on the rod consists of its weight, $m\mathbf{g}$, applied to point C, mass center of the rod, the vertical force $\mathbf{V}$, applied to O, and the vertical force $\mathbf{N}$, applied to Q. The resultant of this system is, therefore,

$$\mathbf{F} = (V + N - mg)\mathbf{n}_2,$$

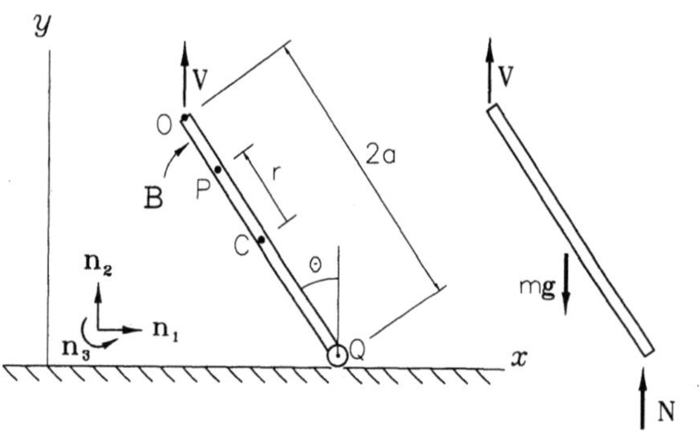

Figure 4.2

and its resultant moment with respect to C is

$$\mathbf{M}^{\mathcal{F}e/C} = (N - V)a\sin\theta\,\mathbf{n}_3.$$

The velocity of the mass center is given by

$$\mathbf{v}^C = \mathbf{v}^Q + \boldsymbol{\omega}^B \times \mathbf{p}^{C/Q}$$
$$= \dot{x}\mathbf{n}_1 + \dot{\theta}\mathbf{n}_3 \times a(-\sin\theta\,\mathbf{n}_1 + \cos\theta\,\mathbf{n}_2)$$
$$= (\dot{x} - a\dot{\theta}\cos\theta)\mathbf{n}_1 - a\dot{\theta}\sin\theta\,\mathbf{n}_2.$$

By differentiating with respect to time, the acceleration of point C can be computed as

$$\mathbf{a}^C = \left(\ddot{x} - a(\ddot{\theta}\cos\theta - \dot{\theta}^2\sin\theta)\right)\mathbf{n}_1 - a\left(\ddot{\theta}\sin\theta + \dot{\theta}^2\cos\theta\right)\mathbf{n}_2.$$

The position of a generic point P of the rod, distant r from the mass center, with respect to the latter, is given by

$$\mathbf{p} = r(-\sin\theta\,\mathbf{n}_1 + \cos\theta\,\mathbf{n}_2),$$

and its velocity is

$$\begin{aligned}
\mathbf{v} &= \mathbf{v}^C + \boldsymbol{\omega}^B \times \mathbf{p} \\
&= (\dot{x} - a\dot{\theta}\cos\theta)\mathbf{n}_1 - a\dot{\theta}\sin\theta\,\mathbf{n}_2 + \dot{\theta}\mathbf{n}_3 \times r(-\sin\theta\,\mathbf{n}_1 + \cos\theta\,\mathbf{n}_2) \\
&= \big(\dot{x} - (a+r)\dot{\theta}\cos\theta\big)\mathbf{n}_1 - (a+r)\dot{\theta}\sin\theta\,\mathbf{n}_2.
\end{aligned}$$

An expression can then be established for the angular momentum vector of the rod with respect to its mass center, given by

$$\begin{aligned}
\mathbf{H}^{B/C} &= \int_B \mathbf{p} \times \mathbf{v}\, dm \\
&= \int_{-a}^{a} r(-\sin\theta\,\mathbf{n}_1 + \cos\theta\,\mathbf{n}_2) \\
&\qquad \times \big((\dot{x} - (a+r)\dot{\theta}\cos\theta)\mathbf{n}_1 - (a+r)\dot{\theta}\sin\theta\,\mathbf{n}_2\big)\frac{m}{2a}\,dr \\
&= \frac{m}{2a}\int_{-a}^{a}\big((a+r)\dot{\theta} - \dot{x}\cos\theta\big)r\,dr\,\mathbf{n}_3 \\
&= \frac{1}{3}ma^2\dot{\theta}\,\mathbf{n}_3.
\end{aligned}$$

Its time rate is then

$$\dot{\mathbf{H}}^{B/C} = \frac{1}{3}ma^2\ddot{\theta}\,\mathbf{n}_3.$$

The equations of motion for the system are then, according to Eqs. (4.5) and (4.9),

$$\ddot{x} - a(\ddot{\theta}\cos\theta - \dot{\theta}^2\sin\theta) = 0, \tag{a}$$

$$a(\ddot{\theta}\sin\theta + \dot{\theta}^2\cos\theta) = g - \frac{V+N}{m}, \tag{b}$$

$$\ddot{\theta} = \frac{3\sin\theta}{ma}(N - V), \tag{c}$$

which consist of a coupled system of equations for $x(t)$, $\theta(t)$, and $N(t)$. The elimination of N from Eqs. (b) and (c) leads to a nonlinear differential equation for $\theta(t)$, which shall require numerical integration. Once $\theta(t)$ is established, Eq. (a) can be solved for $x(t)$ and Eq. (c) for $N(t)$. Also note that, since it is a plane motion, only two components of the equation of motion of the mass center and a single component of the equation of motion around the mass center are of interest, in the case, the rest being identically null. See the corresponding animation.

348   5. *Dynamics of Systems*

Example 4.2   Consider an uncompressible fluid, with density ρ, that flows in a steady state with uniform velocity v in a piece of pipe consisting of a curve at 90^0, on the horizontal plane, having one end rigidly flanged and the other coupled to an expansion joint (see Fig. 4.3a). We wish to determine the forces in the flange, assuming that the forces in the joint and the weight of the fluid are negligible compared to the other forces involved. It is convenient, in this case, to adopt the fluid contained in the curved stretch of pipe and the tubular curve itself as a study system, in order to internalize in the system the distributed forces of interaction between the inside wall of the pipe and the fluid. This methodological alternative will help to establish the forces on the flange, as required, without further difficulties, since the solid part of the system is a rigid body fixed in the inertial reference frame and therefore does not contribute to the momenta.

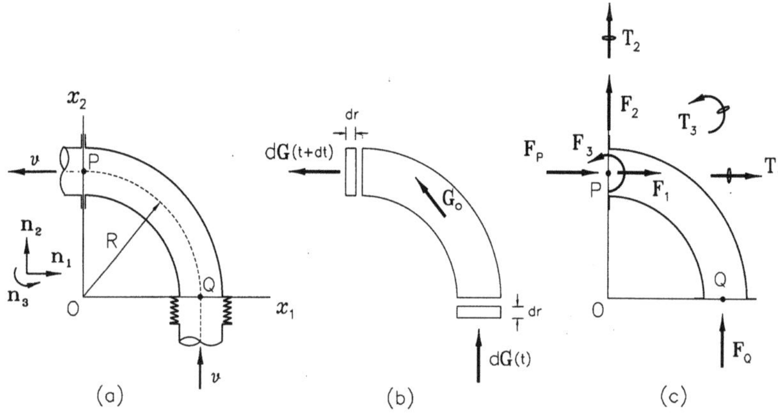

Figure 4.3

In order to establish the time rate of the momentum vector of the system, two subsequent instants will be considered (see Fig. 4.3b). First, a portion of fluid occupies the curve and an element of the joint, with mass $dm = \rho A\, dr = \rho A v\, dt$, where A is the cross section of the pipe. The momentum vector of the body C (fluid plus pipeline) is then given by

$$\mathbf{G}(t) = \mathbf{G}_0 + \rho A v^2\, dt\, \mathbf{n}_2,$$

where $\mathbf{G}_0$ is the momentum of the portion of fluid inside the curve (the pipeline itself is at rest, not contributing, therefore, to $\mathbf{G}$). Shortly afterward, the vector momentum of the system is (see Fig. 4.3b)

$$\mathbf{G}(t + dt) = \mathbf{G}_0 - \rho A v^2\, dt\, \mathbf{n}_1.$$

The time rate of the momentum vector of the body is, therefore,

$$\dot{\mathbf{G}} = \frac{\mathbf{G}(t+dt) - \mathbf{G}(t)}{dt} = -\rho A v^2 \left(\mathbf{n}_1 + \mathbf{n}_2\right).$$

Point O, the curvature center of the pipe, is fixed in the inertial reference frame, being convenient for calculating the angular momentum vector of the system. In fact, as the system is in steady state, v is constant and the angular momentum vector of the body C with respect to point O is constant in time. It results, then, that

$$\dot{\mathbf{H}}^{C/O} = 0.$$

By reducing to point P, in the middle of the flange, the forces applied to it, as it is a rigid link, consist of three force components, $\mathbf{F}_1, \mathbf{F}_2, \mathbf{F}_3$ and three torque components, $\mathbf{T}_1, \mathbf{T}_2, \mathbf{T}_3$. The distributed forces due to pressure p in the fluid also may be considered uniform in the stretch considered. Then, by reducing the pressure in the outlet section to point P and the pressure distribution in the ingoing section to point Q, the local resultants are $\mathbf{F}_P = pA\mathbf{n}_1$ and $\mathbf{F}_Q = pA\mathbf{n}_2$ (see Fig. 4.3c), respectively. Then, reducing the system of all applied external forces to point O, the resultant is

$$\mathbf{F} = (F_1 + pA)\mathbf{n}_1 + (F_2 + pA)\mathbf{n}_2 + F_3\mathbf{n}_3,$$

and the resultant moment is

$$\mathbf{M}^{\mathcal{F}_e/O} = (T_1 + RF_3)\mathbf{n}_1 + T_2\mathbf{n}_2 + (T_3 - RF_1)\mathbf{n}_3.$$

The equations of motion for this system are then, according to Eqs. (4.5) and (4.9),

$$-\rho A v^2 = F_1 + pA,$$
$$-\rho A v^2 = F_2 + pA,$$
$$0 = F_3,$$
$$0 = T_1 + RF_3,$$
$$0 = T_2,$$
$$0 = T_3 - RF_1.$$

By solving the algebraic system for the scalar components of force and torque applied to the flange, then

$$\mathbf{F} = -(p + \rho v^2)A\left(\mathbf{n}_1 + \mathbf{n}_2\right), \quad \mathbf{T} = -(p + \rho v^2)RA\,\mathbf{n}_3.$$

5.5 Work and Potentials

Section 4.5 discussed the concept of the work performed by a force that acts on a point P moving in a given reference frame $\mathcal{R}$. Equations (4.5.2) and (4.5.5) are two alternative ways to calculate the work done by the force between two positions or two instants. It was also shown that the resultant work of a simple concurrent system in a point is equal to the work done by a force equal to the resultant of the system applied on the point, as expressed in Eq. (4.5.4). These concepts are satisfactory for the study of the motion of a particle, on which a simple and concurrent force system always acts. When dealing with systems of particles or bodies, however, this approach is incomplete, since more general systems of forces may be present. This section is devoted to the generalization of the ideas relating to the work applied by a force system and its potential functions.

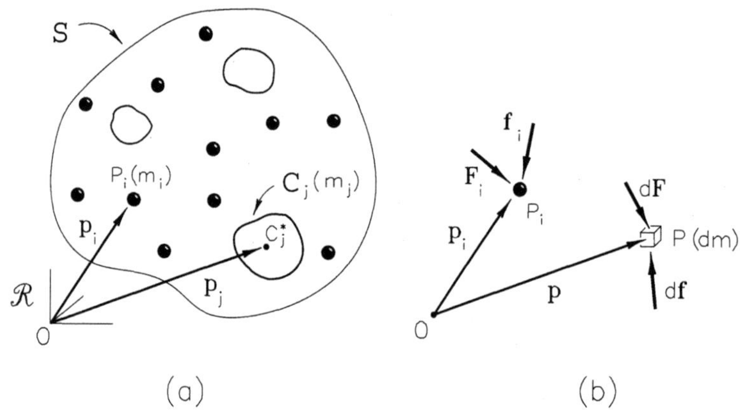

Figure 5.1

Consider, then, S, a *general mechanical system*, consisting of p particles P_i, with mass m_i, $i = 1, 2, \ldots, p$, and q bodies C_j, with mass m_j, $j = 1, 2, \ldots, q$ (see Fig. 5.1a).

Let $\mathcal{F}$ be the system of all forces acting on S, consisting of a subsystem of internal forces, $\mathcal{F}_i$, and a subsystem of external forces, $\mathcal{F}_e$. Let us now be restricted here to the consideration of a simple system of forces, that is, those in which there are no applied torques (see Section 2.3). The study of the work performed by a torque is discussed in Chapter 7, which is devoted to the dynamics of the rigid body. Let us, then, assume that $\mathcal{F}$ comprises K external concentrated forces $\mathbf{F}_k$, L internal concentrated forces $\mathbf{f}_l$, a field of external distributed forces $d\mathbf{F}$, and a field of internal distributed forces $d\mathbf{f}$. The reduction of this force system to a general point O will consist, then, of a force equal to its resultant,

$$\mathbf{R} = \mathbf{F} + \mathbf{f}$$
$$= \sum_{k=1}^{K} \mathbf{F}_k + \int_S d\mathbf{F} + \sum_{l=1}^{L} \mathbf{f}_l + \int_S d\mathbf{f}, \tag{5.1}$$

applied to O and a torque equal to its resultant moment with respect to O,

$$\mathbf{M}^{\mathcal{F}/O} = \mathbf{M}^{\mathcal{F}_e/O} + \mathbf{M}^{\mathcal{F}_i/O}$$
$$= \sum_{k=1}^{K} \mathbf{p}_k \times \mathbf{F}_k + \int_S \mathbf{p} \times d\mathbf{F} + \sum_{l=1}^{L} \mathbf{p}_l \times \mathbf{f}_l + \int_S \mathbf{p} \times d\mathbf{f}, \tag{5.2}$$

where $\mathbf{p}$ are position vectors, with respect to O, of the corresponding points of an application of the forces (see Fig. 5.1b).

As shown in Section 5.2, $\mathbf{f} = 0$ and $\mathbf{M}^{\mathcal{F}_i/O} = 0$. The reduction to O is then restricted to

$$\mathbf{R} = \mathbf{F} = \sum_{k=1}^{K} \mathbf{F}_k + \int_S d\mathbf{F},$$
$$\mathbf{M}^{\mathcal{F}/O} = \mathbf{M}^{\mathcal{F}_e/O} = \sum_{k=1}^{K} \mathbf{p}_k \times \mathbf{F}_k + \int_S \mathbf{p} \times d\mathbf{F}. \tag{5.3}$$

Although the system $\mathcal{F}_i$ does not contribute to the reduction to a point of the system of applied forces, it will generally contribute to the work done on the system S. In fact, the total work of $\mathcal{F}$ in the reference frame $\mathcal{R}$ between two configurations $S(t_1)$ and $S(t_2)$ of the system (see Fig. 5.2) will be the sum of the work done by each component of force, the decomposition being natural:

$$^{\mathcal{R}}\mathcal{T}_{12}^{\mathcal{F}} = {}^{\mathcal{R}}\mathcal{T}_{12}^{\mathcal{F}_e} + {}^{\mathcal{R}}\mathcal{T}_{12}^{\mathcal{F}_i}, \tag{5.4}$$

that is, the *resultant work* of the force system $\mathcal{F}$, in a given reference frame $\mathcal{R}$, between two arbitrary configurations of the system, can be broken down into the works performed by the subsystems $\mathcal{F}_e$ and $\mathcal{F}_i$.

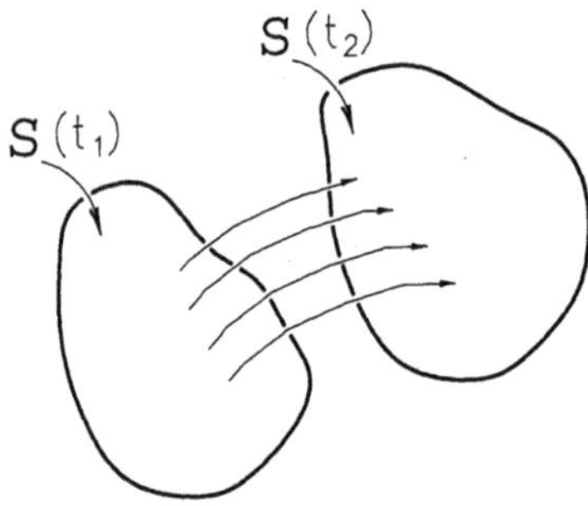

Figure 5.2

The *resultant work of the external forces* on a system S between two arbitrary configurations, in a given reference frame $\mathcal{R}$, is defined, naturally, as the algebraic sum of the works performed by their components, between these configurations, in the same reference frame, then being given by

$$
{}^{\mathcal{R}}\mathcal{T}_{12}^{\mathcal{F}_e} = \sum_{k=1}^{K} \int_1^2 \mathbf{F}_k \cdot {}^{\mathcal{R}}d\mathbf{p}_k + \int_S \int_1^2 d\mathbf{F} \cdot {}^{\mathcal{R}}d\mathbf{p}, \tag{5.5}
$$

where $\mathbf{p}$ is the position vector with respect to a point O, fixed in $\mathcal{R}$, of the point of application of the corresponding force (concentrated or distributed). Note that, for the sake of simplicity, the work of the distributed forces was expressed as an integral throughout the system; of course, the portion of this integral in the region where there are no forces of this nature applied will be identically null.

The *resultant work of the internal forces*, on a system S between two arbitrary configurations in a given reference frame $\mathcal{R}$ is defined, in turn, as the sum of the works, in $\mathcal{R}$, of its components, being equal to

$$
{}^{\mathcal{R}}\mathcal{T}_{12}^{\mathcal{F}_i} = \sum_{l=1}^{L} \int_1^2 \mathbf{f}_l \cdot {}^{\mathcal{R}}d\mathbf{p}_l + \int_S \int_1^2 d\mathbf{f} \cdot {}^{\mathcal{R}}d\mathbf{p}. \tag{5.6}
$$

It is worth noting that if P_i and P_j are any two particles components of S, then, although their mutual interaction consists of a pair of opposite forces, their local displacements, $d\mathbf{p}_i$ and $d\mathbf{p}_j$, are, in general, independent or, at least, different (see Fig. 5.3a), and, therefore, the contribution of these forces for the resultant work of the internal forces is not necessarily null, that is,

$$^{\mathcal{R}}\mathcal{T}_{12}^{\mathbf{f}_{ij}} + \,^{\mathcal{R}}\mathcal{T}_{12}^{\mathbf{f}_{ji}} \neq 0, \qquad \text{in general.} \tag{5.7}$$

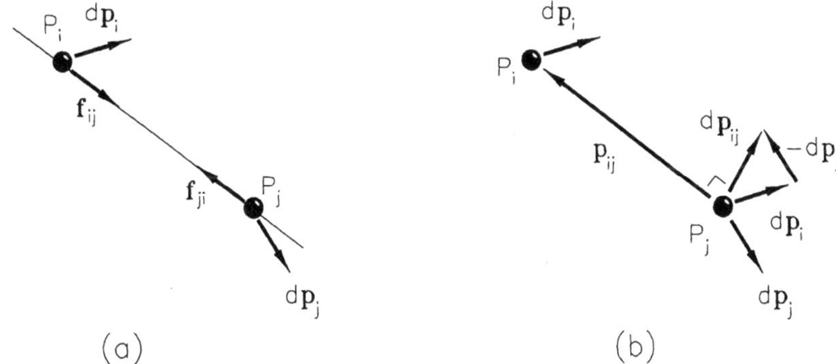

Figure 5.3

An important exception to the inequality just described occurs when two particles of a system S move in a reference frame $\mathcal{R}$ in such a way that their relative distance remains invariant with time. In fact, if the distance between P_i and P_j stays fixed, their relative position vector, $\mathbf{p}_{ij} = \mathbf{p}_i - \mathbf{p}_j$, has a constant module and $^{\mathcal{R}}d\mathbf{p}_{ij} = \,^{\mathcal{R}}d\mathbf{p}_i - \,^{\mathcal{R}}d\mathbf{p}_j$ is a vector necessarily orthogonal to $\mathbf{p}_{ij}$ (see Fig. 5.3b). The result, then, is that the joint contribution of the forces of interaction between the particles under consideration are

$$\begin{aligned}
^{\mathcal{R}}\mathcal{T}_{12}^{\mathbf{f}_{ij}} + \,^{\mathcal{R}}\mathcal{T}_{12}^{\mathbf{f}_{ji}} &= \int_1^2 \mathbf{f}_{ij} \cdot \,^{\mathcal{R}}d\mathbf{p}_i + \int_1^2 \mathbf{f}_{ji} \cdot \,^{\mathcal{R}}d\mathbf{p}_j \\
&= \int_1^2 \mathbf{f}_{ij} \cdot (^{\mathcal{R}}d\mathbf{p}_i - \,^{\mathcal{R}}d\mathbf{p}_j) \\
&= \int_1^2 \mathbf{f}_{ij} \cdot \,^{\mathcal{R}}d\mathbf{p}_{ij} \\
&= 0
\end{aligned} \tag{5.8}$$

if P_i and P_j keep a constant distance.

Just as the contribution for the total work of a given force, component of the system $\mathcal{F}$, is null when the point on which it is applied is fixed in the reference frame or moves in it with a velocity orthogonal to the applied force, Eq. (5.8) guarantees that the joint contribution, for the resultant work, of the interacting forces between elements that keep a constant mutual distance is null. This result is especially important when analyzing the rigid body; in fact, the condition of rigidity will ensure that the contribution of the whole system of internal forces for the resultant work is null.

When some of the forces, either internal or external, acting on the system are *conservative* (see Section 4.5), its contribution for the resultant work may be obtained directly by the change of its potential function, as established in Eq. (4.5.8). When there is a set of conservative forces, internal or external, this will consist of a subsystem $\mathcal{F}_c$, which admits a potential energy, such as defined in Eq. (4.5.10), and the contribution of this subsystem for the resultant work will be equal to the change of this potential energy, as established in Eq. (4.5.11).

Example 5.1   Consider the system S consisting of four small bodies with the same mass m each and connected by threads with length a each and a linear spring, with elastic constant k and natural length a. The system is at rest on a smooth, horizontal plane, in the configuration shown in Fig. 5.4a, when the horizontal force $\mathbf{F}$, which remains constant in time, is applied to the body A.

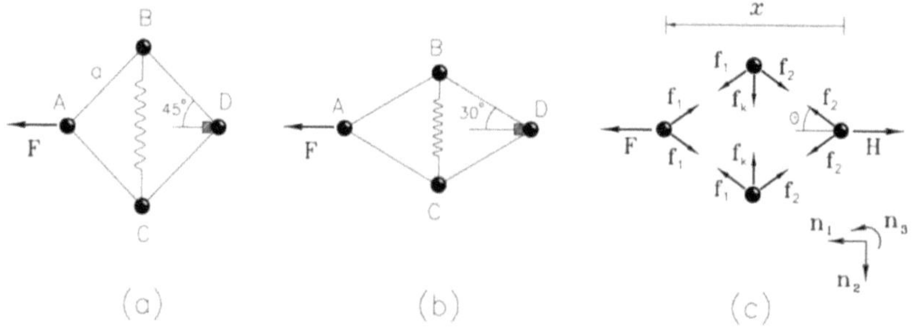

Figure 5.4

A catch, fixed on the plane, prevents the motion of body D and, after an interval of time, the system will be found in the configuration shown in Fig. 5.4b. Figure 5.4c shows all external and internal forces applied on the system (the weights and the normal forces applied by the plane are not of interest, since there is no motion outside the plane). The unknown force **H** is applied by the catch on D. Given its symmetry, the system has only one degree of freedom. It is convenient, nevertheless, to describe it by the coordinates $x(t)$, which measures the distance between D and A, and $\theta(t)$, which measures the angle shown. The infinitesimal displacements of the bodies are

$$d\mathbf{p}^A = dx\,\mathbf{n}_1,$$
$$d\mathbf{p}^B = -a\,d\theta(\sin\theta\,\mathbf{n}_1 + \cos\theta\,\mathbf{n}_2),$$
$$d\mathbf{p}^C = a\,d\theta(-\sin\theta\,\mathbf{n}_1 + \cos\theta\,\mathbf{n}_2),$$
$$d\mathbf{p}^D = 0.$$

Since D is fixed, the work of **H** is null and the resultant work of the external forces will be obtained by calculating the work of the force **F**, that is,

$$\mathcal{T}_{12}^{\mathcal{F}e} = \mathcal{T}_{12}^F = \int_1^2 \mathbf{F}\cdot d\mathbf{p}^A = \int_{a\sqrt{2}}^{a\sqrt{3}} F\,dx = \left(\sqrt{3}-\sqrt{2}\right)Fa.$$

Assuming that the threads remain stretched, their contribution to the resultant work of the internal forces is null, as established by Eq. (5.8). Computing the work done by the forces applied by the spring on B and C, that is, the resultant work of the internal forces, we have

$$\mathcal{T}_{12}^{\mathcal{F}i} = \int_1^2 f_k\mathbf{n}_2\cdot d\mathbf{p}^B + \int_1^2 -f_k\mathbf{n}_2\cdot d\mathbf{p}^C$$
$$= -2ka^2\int_{\pi/4}^{\pi/6}(2\sin\theta - 1)\cos\theta\,d\theta$$
$$= \frac{3-2\sqrt{2}}{2}ka^2.$$

It is interesting to note that the variation of the potential function of the spring between the two configurations under consideration is (see Section 4.5)

$$\Phi_k(b) - \Phi_k(a) = \frac{1}{2}k\delta_b^2 - \frac{1}{2}k\delta_a^2 = 0 - \frac{1}{2}k(a\sqrt{2}-a)^2 = -\frac{3-2\sqrt{2}}{2}ka^2,$$

as would be expected. The total resultant work done over the system, in that interval, is, therefore,

$$\mathcal{T}_{12}^{\mathcal{F}} = a\left(\left(\sqrt{3}-\sqrt{2}\right)F + \frac{3-2\sqrt{2}}{2}ka\right).$$

Example 5.2   The homogeneous rod B, with mass m and length a, is supported on the horizontal plane, revolving at its end O around a fixed vertical peg on the plane (see Fig. 5.5a). The orthonormal bases $\mathbf{n}_1, \mathbf{n}_2, \mathbf{n}_3$ and $\mathbf{b}_1, \mathbf{b}_2, \mathbf{b}_3$ are fixed, respectively, on the plane and rod, which is initially at rest, aligned with axis x_2 ($\theta = 0$), when the constant force $\mathbf{F}_Q = \frac{1}{\sqrt{2}}F_Q(\mathbf{n}_1 - \mathbf{n}_2)$ is applied to the end Q. Assuming that μ is the dynamic friction coefficient between the rod and the plane, then we wish to establish the resultant work from the original configuration until the instant when the rod is aligned with axis x_1. As this is a rigid body, the internal forces do not contribute to the resultant work, as established in Eq. (5.8), that is,

$$\mathcal{T}_{12}^{\mathcal{F}_i} = 0.$$

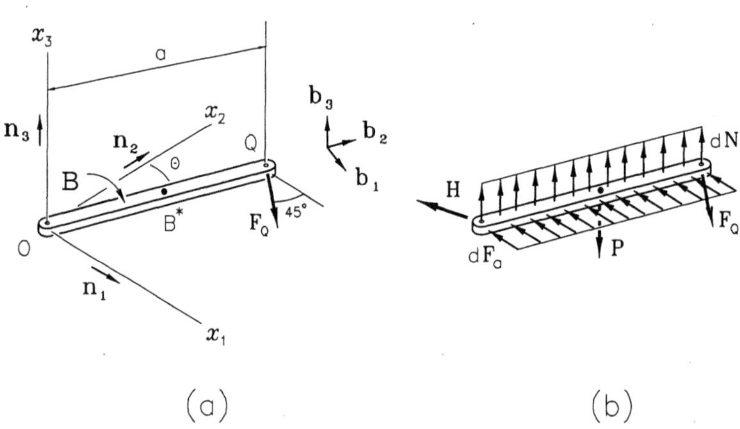

Figure 5.5

The system of external forces, $\mathcal{F}_e$, consists of $\mathbf{F}_Q$, concentrated, applied on Q; the weight $\mathbf{P} = -mg\mathbf{n}_3$ if treated as a concentrated force applied to the mass center of the rod, or $d\mathbf{P} = -g\,dm\mathbf{n}_3$ if treated as a distributed force along the whole rod; the normal distributed force, applied by the plane, $d\mathbf{N} = -d\mathbf{P}$; the concentrated force $\mathbf{H}$, applied by the pivot on O; and, lastly, the friction force, also distributed, applied by the plane, $d\mathbf{F}_a = -\mu\,dN\mathbf{b}_1$ (see Fig. 5.5b). The weight and the normal force do not work, as they are orthogonal to the direction of motion of the respective points of application. Nor does the force $\mathbf{H}$ work as it is applied on a fixed point in the reference frame. Taking r as an integration variable along the rod, the resultant work is then, according to Eqs. (5.4) and (5.5),

$$\mathcal{T}_{12}^{\mathcal{F}} = \int_1^2 \left(\mathbf{F}_Q \cdot d\mathbf{p}^Q + \int_B d\mathbf{F}_a \cdot d\mathbf{p} \right)$$

$$= \int_0^{\pi/2} \left(\frac{F_Q}{\sqrt{2}} (\mathbf{n}_1 - \mathbf{n}_2) \cdot a\, d\theta \mathbf{b}_1 + \int_0^a -\mu \frac{m}{a} g\, dr \mathbf{b}_1 \cdot r\, d\theta \mathbf{b}_1 \right)$$

$$= \int_0^{\pi/2} \left(\frac{F_Q a}{\sqrt{2}} (\cos\theta + \sin\theta) - \frac{\mu m g}{a} \int_0^a r\, dr \right) d\theta$$

$$= \left(\sqrt{2} F_Q - \frac{\pi}{4} \mu m g \right) a.$$

Note that, from the result obtained, it is assumed that $F_Q \geq (\pi/4\sqrt{2})\,\mu m g$; otherwise there is no movement. See the corresponding animation.

If S is a general mechanical system, like that defined at the beginning of this section, and S* is its mass center, whose position with respect to a point O, fixed in the reference frame $\mathcal{R}$, is given by the position vector $\mathbf{p}^*$, the local displacement of a point P_k of S where a concentrated force $\mathbf{F}_k$ is applied (or of a general point P of S where a distributed force $d\mathbf{F}$ is applied), in a reference frame $\mathcal{R}$, it can be broken down, respectively, into

$$\mathcal{R} d\mathbf{p}_k = \mathcal{R} d\mathbf{p}^* + \mathcal{R} d\mathbf{r}_k, \qquad \mathcal{R} d\mathbf{p} = \mathcal{R} d\mathbf{p}^* + \mathcal{R} d\mathbf{r}, \tag{5.9}$$

where $\mathbf{r}_k$ (or $\mathbf{r}$) is the position vector with respect to the mass center, as shown in Fig. 5.6.

When substituting Eq. (5.9) in Eq. (5.5), then we obtain

$$\mathcal{R}\mathcal{T}_{12}^{\mathcal{F}_e} = \int_1^2 \left(\sum_{k=1}^K \mathbf{F}_k \cdot \left(\mathcal{R} d\mathbf{p}^* + \mathcal{R} d\mathbf{r}_k \right) + \int_S d\mathbf{F} \cdot \left(\mathcal{R} d\mathbf{p}^* + \mathcal{R} d\mathbf{r} \right) \right)$$

$$= \int_1^2 \mathbf{F} \cdot \mathcal{R} d\mathbf{p}^* + \int_1^2 \left(\sum_{k=1}^K \mathbf{F}_k \cdot \mathcal{R} d\mathbf{r}_k + \int_S d\mathbf{F} \cdot \mathcal{R} d\mathbf{r} \right). \tag{5.10}$$

The first term of Eq. (5.10) may be interpreted as the *resultant work of the external forces on the mass center* in the reference frame $\mathcal{R}$, that is, it is the work done by the external resultant force, $\mathbf{F}$, as if applied, in the interval, on the mass center given, therefore, by

$$\mathcal{R}\mathcal{T}_{12}^{\mathbf{F}} \rightleftharpoons \int_1^2 \mathbf{F} \cdot \mathcal{R} d\mathbf{p}^*. \tag{5.11}$$

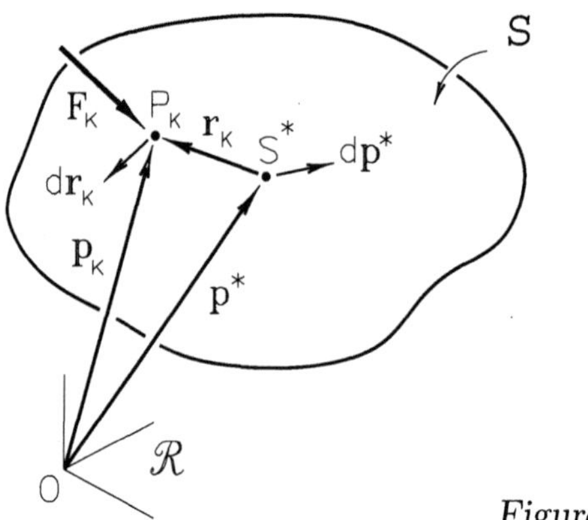

Figure 5.6

The second term of Eq. (5.10) may, in turn, be interpreted as the *resultant work of the external forces around the mass center* in the reference frame $\mathcal{R}$, that is, it is the work done by the external forces, as if the motion of the system were to happen around the mass center, being then given by

$$^{\mathcal{R}}\mathcal{T}_{12}^{\mathcal{F}_e/S^*} \rightleftharpoons \int_{1}^{2} \left(\sum_{k=1}^{K} \mathbf{F}_k \cdot {}^{\mathcal{R}}d\mathbf{r}_k + \int_{S} d\mathbf{F} \cdot {}^{\mathcal{R}}d\mathbf{r} \right). \qquad (5.12)$$

By then substituting Eqs. (5.11) and (5.12) in Eq. (5.10), we see that the resultant work applied by external forces is broken down as

$$^{\mathcal{R}}\mathcal{T}_{12}^{\mathcal{F}_e} = {}^{\mathcal{R}}\mathcal{T}_{12}^{\mathbf{F}} + {}^{\mathcal{R}}\mathcal{T}_{12}^{\mathcal{F}_e/S^*}. \qquad (5.13)$$

Similarly, the *resultant work of the internal forces on the mass center* in reference frame $\mathcal{R}$ may be defined as

$$^{\mathcal{R}}\mathcal{T}_{12}^{\mathbf{f}} \rightleftharpoons \int_{1}^{2} \mathbf{f} \cdot {}^{\mathcal{R}}d\mathbf{p}^* = 0, \qquad (5.14)$$

this work always being null, since $\mathbf{f} = 0$, in any case. There remains, however, the *resultant work of the internal forces around the mass center*, in the reference frame $\mathcal{R}$, that is,

$$^{\mathcal{R}}\mathcal{T}_{12}^{\mathcal{F}_i/S^*} \rightleftharpoons \int_{1}^{2} \left(\sum_{l=1}^{L} \mathbf{f}_l \cdot {}^{\mathcal{R}}d\mathbf{r}_l + \int_{S} d\mathbf{f} \cdot {}^{\mathcal{R}}d\mathbf{r} \right), \qquad (5.15)$$

and this may be *different from zero*, although the resultant of the internal forces vanishes. The result, then, is that

$$\mathcal{R}\mathcal{T}_{12}^{\mathcal{F}_i} = \mathcal{R}\mathcal{T}_{12}^{\mathcal{F}_i/S^*}. \tag{5.16}$$

In short, the resultant work on a system S between two arbitrary configurations, in a given reference frame, will always be the sum of the resultant work of the external forces with the resultant work of the internal forces, as shown in Eq. (5.4). The resultant work of the external forces may be calculated by adding up the works of each of the acting external forces, each obtained by integrating along the trajectory the dot product of the force by the displacement of its point of action, as Eq. (5.5) expresses, or as the resultant work on the mass center plus the resultant work around the mass center, as shown in Eqs. (5.11–5.13), where the displacements now refer to that of the mass center itself and of the application points of the forces relative to the former.

On the other hand, the resultant work of the internal forces may be calculated alternatively by Eq. (5.6), directly, or calculating the work of the internal forces around the mass center, as indicated by Eqs. (5.14–5.16).

Example 5.3   Returning to Example 5.2 (see Fig. 5.5), the resultant work on rod B is now going to be established by using the decomposition of the mass center. As the forces $\mathbf{P}$, $d\mathbf{N}$, and $\mathbf{H}$ do not work, they do not need to be included in the calculation. The external resultant force, not including those that do not work, is then

$$\mathbf{F} = \frac{F_Q}{\sqrt{2}}\left((\cos\theta + \sin\theta)\mathbf{b}_1 + (\sin\theta - \cos\theta)\mathbf{b}_2\right) + \int_0^a -\frac{\mu mg}{a}\,dr\,\mathbf{b}_1.$$

The infinitesimal displacements of the mass center and of a general point in relation to the mass center are, respectively,

$$d\mathbf{p}^* = \frac{a}{2}\,d\theta\,\mathbf{b}_1, \qquad d\mathbf{r} = \left(r - \frac{a}{2}\right)d\theta\,\mathbf{b}_1.$$

The resultant work of the external forces on the mass center is then, according to Eq. (5.11),

$$
^{\mathcal{R}}\mathcal{T}_{12}^{\mathbf{F}} = \int_1^2 \mathbf{F} \cdot {}^{\mathcal{R}} d\mathbf{p}^*
$$

$$
= \int_0^{\frac{\pi}{2}} \left[\frac{F_Q}{\sqrt{2}} \left((\cos\theta + \sin\theta)\mathbf{b}_1 + (\sin\theta - \cos\theta)\mathbf{b}_2 \right) \right.
$$

$$
\left. - \int_0^a \frac{\mu mg}{a}\, dr\, \mathbf{b}_1 \right] \cdot \frac{a}{2}\, d\theta\, \mathbf{b}_1
$$

$$
= \left(\frac{F_Q}{\sqrt{2}} - \frac{\pi}{4}\mu mg \right) a.
$$

The resultant work of the external forces around the mass center is, in turn, according to Eq. (5.12),

$$
^{\mathcal{R}}\mathcal{T}_{12}^{\mathcal{F}_e/S^*} = \int_1^2 \left(\sum_{k=1}^K \mathbf{F}_k \cdot d\mathbf{r}_k + \int_S d\mathbf{F} \cdot d\mathbf{r} \right)
$$

$$
= \int_0^{\frac{\pi}{2}} \frac{F_Q}{\sqrt{2}} \left((\cos\theta + \sin\theta)\mathbf{b}_1 + (\sin\theta - \cos\theta)\mathbf{b}_2 \right) \cdot \frac{a}{2}\, d\theta\, \mathbf{b}_1
$$

$$
+ \int_0^{\frac{\pi}{2}} \int_0^a -\frac{\mu mg}{a}\, dr\, \mathbf{b}_1 \cdot \left(r - \frac{a}{2} \right) d\theta\, \mathbf{b}_1
$$

$$
= \frac{F_Q\, a}{\sqrt{2}}.
$$

The resultant work of the external forces is, therefore, according to Eq. (5.13),

$$
^{\mathcal{R}}\mathcal{T}_{12}^{\mathcal{F}_e} = {}^{\mathcal{R}}\mathcal{T}_{12}^{\mathbf{F}} + {}^{\mathcal{R}}\mathcal{T}_{12}^{\mathcal{F}_e/S^*}
$$

$$
= \left(\sqrt{2}F_Q - \frac{\pi}{4}\mu mg \right) a,
$$

as obtained in the previous example. As this is a rigid body, the resultant work of the internal forces will be null and Eqs. (5.14–5.16) will be identically null.

5.6 Work and Energy

If S is a general mechanical system, as defined in Section 5.5, moving in an inertial reference frame $\mathcal{R}$ under the action of a force system $\mathcal{F}$, the resultant work on S, in $\mathcal{R}$, between two configurations of the system, S_1 and S_2, will be equal to the change in its kinetic energy, in $\mathcal{R}$, between these two states, that is,

$$^{\mathcal{R}}K^S(2) - {}^{\mathcal{R}}K^S(1) = {}^{\mathcal{R}}T_{12}^{\mathcal{F}}. \tag{6.1}$$

Equation (6.1), therefore, establishes the relationship between the resultant work and the change in the kinetic energy of a system, resembling the law expressed in Eq. (4.6.3) for a particle, and generalized now for a general mechanical system. The proof of the result follows, essentially, the same steps for deducing Eq. (4.6.3), adding and integrating now to the whole system.

As a starting point, let us take the equation of motion valid for each one of the p particles of S, $m_i\mathbf{a}_i = \mathbf{R}_i$, $i = 1, 2, \ldots, p$, where $\mathbf{a}_i$ is the acceleration, in the inertial reference frame, of P_i and $\mathbf{R}_i$ is the resultant force applied to the particle [see Eq. (3.2)], and also to a general element of each one of the q bodies of S, $dm_j\mathbf{a}_j = d\mathbf{R}_j$, $j = 1, 2, \ldots, q$, where $\mathbf{a}_j$ is the acceleration, in the inertial reference frame, of an element of the body C_j and $d\mathbf{R}_j$ is the resultant force applied to the element (see Fig. 6.1).

Taking then the equations of motion above, dot-multiplying with the respective local displacements (${}^{\mathcal{R}}d\mathbf{p}_i$ and ${}^{\mathcal{R}}d\mathbf{p}_j$), integrating along the trajectory of each particle or element between the two configurations considered, and, finally, adding for the whole system S, then we obtain

$$\sum_{i=1}^{p} \int_1^2 m_i\mathbf{a}_i \cdot {}^{\mathcal{R}}d\mathbf{p}_i + \sum_{j=1}^{q} \int_{C_j} \int_1^2 \mathbf{a}_j \cdot {}^{\mathcal{R}}d\mathbf{p}_j\, dm_j$$
$$= \sum_{i=1}^{p} \int_1^2 \mathbf{R}_i \cdot {}^{\mathcal{R}}d\mathbf{p}_i + \sum_{j=1}^{q} \int_{C_j} \int_1^2 d\mathbf{R}_j \cdot {}^{\mathcal{R}}d\mathbf{p}_j. \tag{6.2}$$

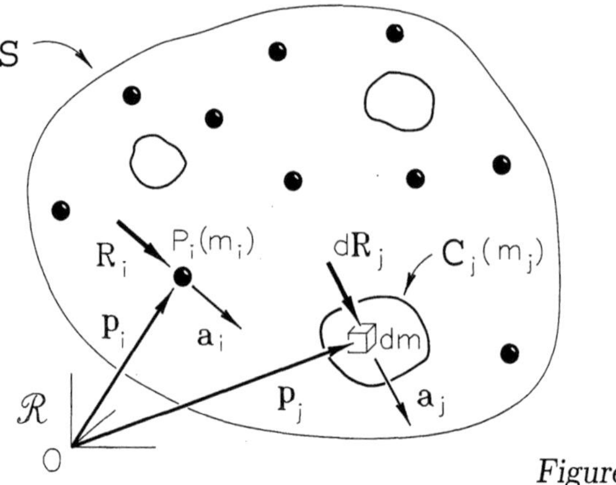

Figure 6.1

Although the above equation may seem complex at first glance, it involves only concepts already studied. In fact,

$$\sum_{i=1}^{p} \int_{1}^{2} m_i \mathbf{a}_i \cdot {}^{\mathcal{R}} d\mathbf{p}_i = \sum_{i=1}^{p} \int_{1}^{2} m_i \mathbf{v}_i \cdot {}^{\mathcal{R}} d\mathbf{v}_i$$
$$= \sum_{i=1}^{p} \frac{1}{2} m_i \mathbf{v}_i \cdot \mathbf{v}_i \bigg|_{1}^{2}.$$

Likewise,

$$\sum_{j=1}^{q} \int_{C_j} \int_{1}^{2} \mathbf{a}_j \cdot {}^{\mathcal{R}} d\mathbf{p}_j \, dm_j = \sum_{j=1}^{q} \int_{C_j} \int_{1}^{2} \mathbf{v}_j \cdot {}^{\mathcal{R}} d\mathbf{v}_j \, dm_j$$
$$= \sum_{j=1}^{q} \frac{1}{2} m_j \mathbf{v}_j \cdot \mathbf{v}_j \bigg|_{1}^{2}.$$

Therefore, the term on the left of Eq. (6.2) corresponds, according to Eqs. (1.10) and (1.23), to the difference between the kinetic energy of the system in the two configurations. On the other hand, according to Eqs. (5.5) and (5.6),

$$\sum_{i=1}^{p} \int_{1}^{2} \mathbf{R}_i \cdot {}^{\mathcal{R}} d\mathbf{p}_i$$

is the work done by all concentrated internal and external forces, applied on the system in the interval, while

$$\sum_{j=1}^{q} \int_{C_j} \int_{1}^{2} d\mathbf{R}_j \cdot {}^{\mathcal{R}}d\mathbf{p}_j$$

is the work done by all distributed internal and external forces applied on the system, in the same interval. The term on the right of Eq. (6.2) is then the total resultant work of the force system acting on the system in the interval, thus concluding the proof of Eq. (6.1).

Example 6.1 Figure 6.2 reproduces the system analyzed in Example 5.2. Assuming that the rod B leaves from rest aligned to axis x_2, we wish to compute the velocity of the point Q when crossing axis x_1. If $\omega(\theta)$ is the module of the angular velocity of the rod in a general position θ, the kinetic energy of B is, according to Eq. (1.23),

$$K^B(\theta) = \int_B \frac{1}{2}\mathbf{v} \cdot \mathbf{v}\, dm$$

$$= \frac{1}{2} \int_0^a \left(\omega(\theta)r\right)^2 \frac{m}{a}\, dr$$

$$= \frac{1}{6}ma^2\omega^2.$$

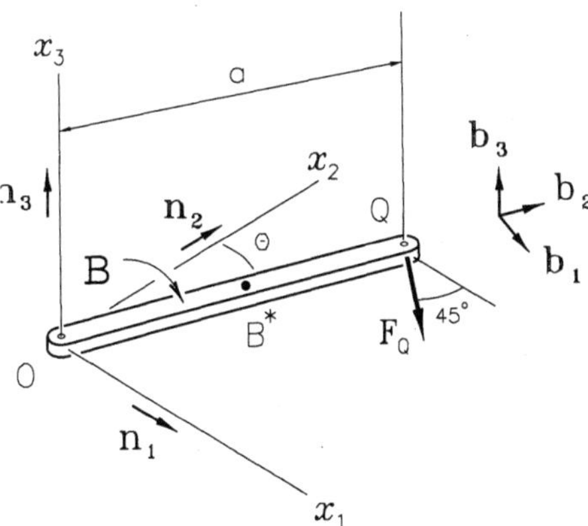

Figure 6.2

The kinetic energy change between the positions $\theta = 0$ and $\theta = \pi/2$ is then

$$K^B(\pi/2) - K^B(0) = \frac{1}{6}ma^2\omega_2^2,$$

where $\omega_2 = \omega(\pi/2)$. From Eq. (6.1), then, we have the relationship to find the angular velocity in the configuration desired (see Example 5.2 for the computation of the resultant work),

$$\frac{1}{6}ma^2\omega_2^2 = \left(\sqrt{2}F_Q - \frac{\pi}{4}\mu mg\right)a.$$

The desired velocity will therefore be

$$v^Q = a\omega_2 = \left(\frac{6\sqrt{2}F_Q a}{m} - \frac{3\pi\mu ag}{2}\right)^{\frac{1}{2}}.$$

This example illustrates the facility of obtaining the solution from Eq. (6.1) whenever we wish to calculate a *velocity*. See the corresponding animation.

When conservative forces are among the forces acting on a system, their contribution to the resultant work may be calculated by the change in its potential function, as stated in Eq. (4.5.8). Should there be more than one conservative force, their joint contribution for the resultant work may be calculated by the change in the sum of the corresponding potential functions — called the potential energy of the system — as stated in Eq. (4.5.11). Breaking down, then, the resultant work on the system between two configurations into a work done by the conservative forces, including internal and external forces, and a work done by the nonconservative forces, also with external and internal forces, as expressed by Eq. (4.5.9), the work energy relationship, Eq. (6.1), may be stated alternatively by

$$^{\mathcal{R}}K^S(2) + {}^{\mathcal{R}}\Phi^S(2) - {}^{\mathcal{R}}K^S(1) - {}^{\mathcal{R}}\Phi^S(1) = {}^{\mathcal{R}}T_{12}^{\mathcal{F}_N}. \tag{6.3}$$

When defining the *mechanical energy of the system*, in a reference frame $\mathcal{R}$, like the algebraic sum, at each instant of its kinetic energy and potential energy in $\mathcal{R}$, that is,

$$^{\mathcal{R}}E^S = {}^{\mathcal{R}}K^S + {}^{\mathcal{R}}\Phi^S, \tag{6.4}$$

the balance of energy of the system may also be given by

$$^{\mathcal{R}}E^S(2) - {}^{\mathcal{R}}E^S(1) = {}^{\mathcal{R}}T_{12}^{\mathcal{F}_N}. \tag{6.5}$$

It is evident that Eq. (6.5) is valid only when $\mathcal{R}$ is an inertial reference frame.

Example 6.2 Going back to Example 5.1 (see Fig. 6.3), we now wish to establish the velocity of body A in configuration (b), recalling that the system was in rest in configuration (a). The internal forces contributing to the resultant work between the two conservative configurations are those applied by the linear spring on the bodies B and C and whose change in the potential function, established in the aforementioned example, is

$$\Phi_k(b) - \Phi_k(a) = -\frac{3 - 2\sqrt{2}}{2}ka^2.$$

The only external force acting on the plane of motion is the constant and also conservative force **F**. The change in its potential function in the interval considered is

$$\Phi_F(b) - \Phi_F(a) = -F\big(x(b) - x(a)\big) = -(\sqrt{3} - \sqrt{2})\,Fa.$$

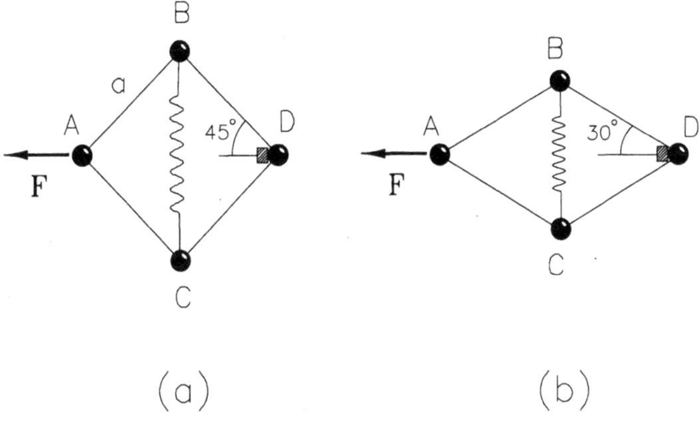

Figure 6.3

The change in the potential energy of the system S is then

$$^{\mathcal{R}}\Phi^S(b) - {}^{\mathcal{R}}\Phi^S(a) = -\Big((\sqrt{3} - \sqrt{2})F + \frac{3 - 2\sqrt{2}}{2}ka\Big)\,a.$$

The initial kinetic energy is null and the kinetic energy in configuration (b) is (check the kinematic relation between v_B, v_C, and v_A)

$$^{\mathcal{R}}K^S(b) = \sum_{i=1}^{4}\frac{1}{2}m_i v_i^2$$

$$= \frac{1}{2}m(v_A^2 + v_B^2 + v_C^2)$$

$$= \frac{3}{2}mv_A^2.$$

Then by substituting in Eq. (6.3) and solving for v_A, we would get

$$v_A = \left[\frac{2a}{3m} \left((\sqrt{3} - \sqrt{2})F + \left(\frac{3}{2} - \sqrt{2} \right) ka \right) \right]^{\frac{1}{2}}.$$

Now going back to the equation of motion for the mass center of a system, Eq. (3.5), or for a body, Eq. (4.3), dot-multiplying both members by the local displacement of the mass center in the inertial reference frame, $^{\mathcal{R}}d\mathbf{p}^*$, and integrating along the trajectory of the mass center of the system or body in the reference frame, we will have

$$\int_1^2 m\mathbf{a}^* \cdot {}^{\mathcal{R}}d\mathbf{p}^* = \int_1^2 \mathbf{F} \cdot {}^{\mathcal{R}}d\mathbf{p}^*. \tag{6.6}$$

Again, it is easy to identify the term on the left, from the Eqs. (4.1.22) and (1.11), as the change between two configurations of the kinetic energy of the mass center, $^{\mathcal{R}}K^{S^*}(2) - {}^{\mathcal{R}}K^{S^*}(1)$; the term on the right corresponds, according to Eq. (5.11), to the resultant work of the external forces over the mass center, in the inertial reference frame, it being therefore demonstrated that

$$^{\mathcal{R}}K^{S^*}(2) - {}^{\mathcal{R}}K^{S^*}(1) = {}^{\mathcal{R}}\mathcal{T}_{12}^{\mathbf{F}}. \tag{6.7}$$

In short, Eq. (6.7) establishes that the change of the kinetic energy of the mass center of a system or body between any two configurations, in an inertial reference frame, is equal to the resultant work of the external forces on the mass center.

Looking at Eqs. (6.1) and (6.7) together, it is natural to find that the terms of the latter consist, on both sides, of parts of the former. In fact, from Eq. (1.13), which establishes a decomposition for the kinetic energy of a system, and from Eqs. (5.4), (5.13), and (5.16), which give a decomposition for the overall resultant work applied on the system, and subtracting Eq. (6.7) from Eq. (6.1), the result is

$$^{\mathcal{R}}K^{S/S^*}(2) - {}^{\mathcal{R}}K^{S/S^*}(1) = {}^{\mathcal{R}}\mathcal{T}_{12}^{\mathcal{F}_e/S^*} + {}^{\mathcal{R}}\mathcal{T}_{12}^{\mathcal{F}_i/S^*}. \tag{6.8}$$

Equation (6.8), therefore, expresses the equality between the change of the kinetic energy of a system around its mass center, in an inertial reference frame, between two arbitrary configurations of the system, and the resultant work around the mass center, in the same reference frame and between the same configurations, the external and internal forces contributing to that.

5.7 Conservation Principles

In the study we did on the dynamic behavior of systems and bodies, a set of equations governing their motion in an inertial reference frame was established; the equations of motion of the first and second kinds and the energy balances comprise the fundamental relationship that help to determine the time evolution of mechanical systems.

In the same way as happens for a particle, canceling the generation term of any of those equations, there is a *principle of conservation*, that is, the corresponding property is conserved.

In this section we will examine the principles of conservation applicable to discrete systems of particles and continuous bodies. For the sake of simplicity, we will call S a general, discrete or continuous mechanical system, or even a combination of them.

The principles of conservation prove to be extremely useful in problem-solving. Whenever one or more of these principles apply, using them greatly simplifies the analysis of the mechanical system. The reader shall therefore always be alert to identify possible generation terms (resultant force, resultant moment, resultant moment with respect to axis, resultant work, nonconservative resultant work, etc.) that vanish in a given situation.

The principles of conservation for systems are, essentially, the same as those studied for a particle. The major differences are precisely due to the fact that, in some cases, only *external* forces and torques, when null, will guarantee the corresponding principle of conservation, as will be seen below.

If the external resultant force applied to a system S is null during a certain interval, the momentum vector of the system is conserved in an inertial reference frame $\mathcal{R}$, that is,

$$^{\mathcal{R}}\mathbf{G}^{S} = \mathbf{G}_0 \qquad \text{if} \qquad \mathbf{F} = 0. \tag{7.1}$$

Equation (7.1) expresses, therefore, the *momentum principle of conservation*, where $\mathbf{G}_0$ is a constant vector. It is easy to check that Eq. (7.1) derives immediately from Eq. (3.3) when $\mathbf{F} = 0$.

Example 7.1   Two snooker balls, B_1 and B_2, with masses m_1 and m_2, roll with negligible friction over a table at the indicated velocities (see Fig. 7.1), colliding at the point O. We wish to determine their new velocities immediately after the impact. The collision, as we know, is an event of short duration (see Section 4.7) and only internal forces intervene in this interval. The external resultant force is, therefore, null and Eq. (7.1) is satisfied. The momentum of the system before the collision is

$$^{\mathcal{R}}\mathbf{G}^{S}(1) = m_1\mathbf{v}_1 + m_2\mathbf{v}_2$$

and, after the impact, is

$$^{\mathcal{R}}\mathbf{G}^{S}(2) = m_1\mathbf{v}_1' + m_2\mathbf{v}_2'.$$

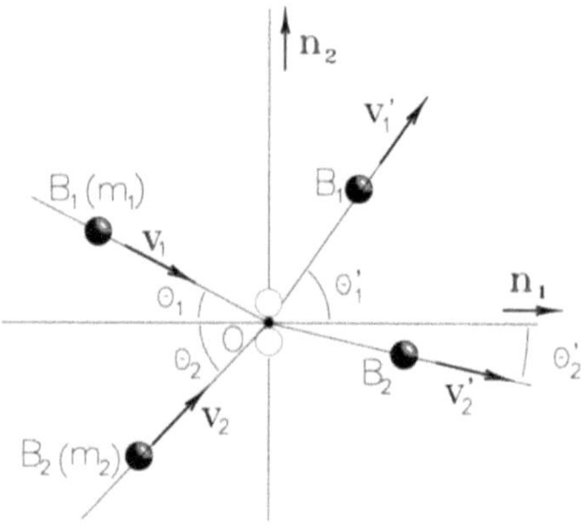

Figure 7.1

From the principle of conservation we then have

$$m_1v_1'\cos\theta_1' + m_2v_2'\cos\theta_2' = m_1v_1\cos\theta_1 + m_2v_2\cos\theta_2, \tag{a}$$

$$m_1v_1'\sin\theta_1' - m_2v_2'\sin\theta_2' = -m_1v_1\sin\theta_1 + m_2v_2\sin\theta_2. \tag{b}$$

Note that this pair of equations is insufficient to determine the four unknowns, as follows: v_1'; v_2'; θ_1'; and θ_2' (the snooker game is not easy). If the impact is *smooth*, and there are no interacting forces in direction $\mathbf{n}_1$, the component of the momentum vector of each ball for itself is conserved in direction $\mathbf{n}_1$, which results in the following additional relations (this is not exactly what happens with snooker balls):

$$m_1v_1'\cos\theta_1' = m_1v_1\cos\theta_1; \tag{c}$$

$$m_2v_2'\cos\theta_2' = m_2v_2\cos\theta_2. \tag{d}$$

Note that Eq. (a) is dependent on Eqs. (c) and (d) and that the set of three mutually independent equations (b–d) is also insufficient for a complete solution. Now, if the restitution coefficient, ϵ, is known (which would be rare in the case of snooker balls), we will then have a fourth relation (see Section 4.7),

$$v_1' \sin \theta_1' + v_2' \sin \theta_2' = \epsilon(v_1 \sin \theta_1 + v_2 \sin \theta_2),$$

that will help determine the two desired velocities (four components). See the corresponding animation.

When the momentum of a system is conserved in an inertial reference frame, in a given interval, the velocity vector of the mass center of the system in the reference frame also remains constant, that is,

$$\mathbf{v}^* = \mathbf{v}_0^* \qquad \text{if} \qquad \mathbf{F} = 0. \tag{7.2}$$

Equation (7.2) results directly from Eqs. (1.4), (1.5), and (7.1), $\mathbf{v}_0^*$ being a constant vector.

Example 7.2   Two identical homogeneous rods can turn freely around the common end O and are supported on a smooth, horizontal flat surface (see Fig. 7.2). Initially, the rods are set out orthogonally, with B_2 at rest and B_1 turning with an angular velocity ω_0, as shown. We wish to establish the velocity of the mass center of the set at the instant when the rods overlap.

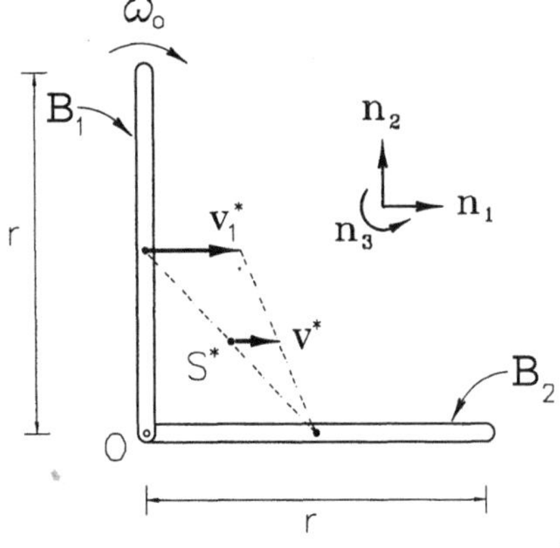

Figure 7.2

The velocity of the mass center in the initial condition is, according to Eq. (1.2),

$$\mathbf{v}^* = \frac{1}{2}(\mathbf{v}_1^* + \mathbf{v}_2^*) = \frac{1}{2}\left(\frac{1}{2}r\omega_0\mathbf{n}_1 + 0\right) = \frac{1}{4}r\omega_0\mathbf{n}_1.$$

Now, since the system of acting external forces is null (weights and normal counterbalance, since there is no motion outside the plane), the momentum of the system is conserved and, according to Eq. (7.2), the velocity of the mass center does not alter; the desired velocity is then that calculated above. (The reader may conjecture how the whole set will be moving at the moment under study; well, this is another story, involving other principles of conservation, and the problem will be left as an exercise. See the corresponding animation.)

Equation (7.1) expresses a vectorial principle of conservation. When, nevertheless, a scalar component of the external resultant force in a fixed direction in the inertial reference frame is null in a given interval, the corresponding scalar component of the momentum vector of the system will be conserved in this reference frame, that is,

$$^{\mathcal{R}}G_j^S = G_{0j} \qquad \text{if} \qquad F_j = 0. \tag{7.3}$$

Equation (7.3), which expresses the *principle of conservation of the momentum in a given direction*, is no less than the projection of Eq. (7.1) in the direction of the unit vector $\mathbf{n}_j$.

Example 7.3   A pendulum P, with mass m and length r, is hung on a cursor C, with mass m_0, which may slide without friction on a horizontal guide (see Fig. 7.3). The set is left, at rest (the guide is an inertial reference frame), in the position shown. We wish to calculate the velocities at the instant when the pendulum support passes through the vertical position. The acting external forces are the weight of the pendulum, the weight of the cursor, and the force (normal) applied by the guide on the latter. As all forces are vertical, the scalar component of the external resultant force on the system in the horizontal direction is null, the horizontal component of the momentum vector therefore being conserved. Since the initial condition is at rest and, in the final condition, both velocities are horizontal, Eq. (7.3) is simply expressed by

$$m_0 v_C - m v_P = 0.$$

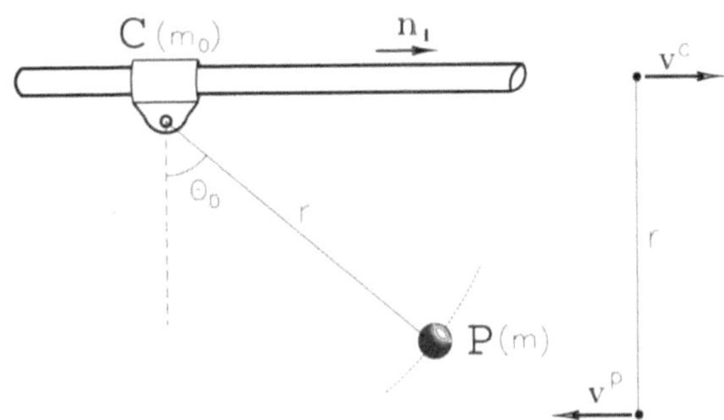

Figure 7.3

The preceding relation is not enough to determine the two unknowns (note that there is no conservation of the momentum vector component in the vertical direction). We will return to this problem soon. See the corresponding animation.

If the resultant moment, with respect to the mass center of a system S, of the acting external forces is null in a certain interval, the angular momentum vector of the system with respect to the mass center is conserved in an inertial reference frame $\mathcal{R}$, that is,

$$^{\mathcal{R}}\mathbf{H}^{S/S^*} = \mathbf{H}_0^* \qquad \text{if} \qquad \mathbf{M}^{\mathcal{F}_e/S^*} = 0. \tag{7.4}$$

Equation (7.4) expresses the *principle of conservation of the angular momentum with respect to the mass center*, where $\mathbf{H}_0^*$ is a constant vector. It is easy to see that Eq. (7.4) immediately results from Eq. (3.13) when the resultant moment of the external forces vanishes in a given interval.

Example 7.4 A pencil L (not yet sharpened), with mass m, is thrown into space as shown in Fig. 7.4a. Initially, its mass center O has a velocity on the plane of the figure, with module v_0 and slope θ_0 with the vertical, and its angular velocity, simple and of module ω_0, has a direction orthogonal to the figure's plane, as shown. What will the angular velocity ω of the pencil and the velocity v of its mass center be at the instant when the latter is horizontal (see Fig. 4.7b)? Assuming the aerodynamic forces as negligible, the only applied external force during the motion is its weight $\mathbf{P} = mg\mathbf{n}_1$. Therefore, the horizontal component of the momentum vector of the pencil is conserved and, consequently, the horizontal component of the velocity of the point O, thereby resulting in that, according to Eq. (7.2),

$$v^O = v_0 \sin \theta_0.$$

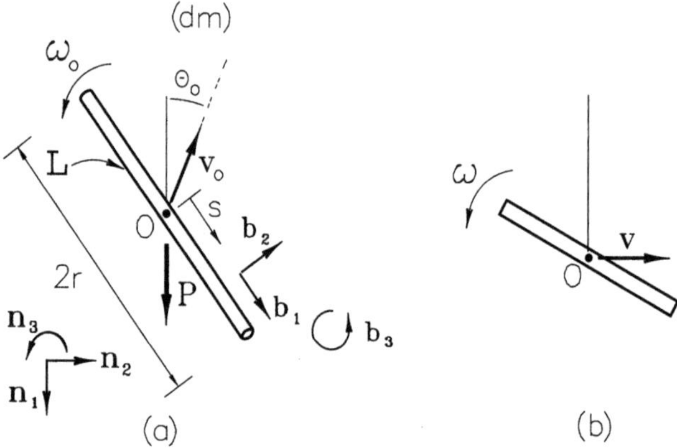

Figure 7.4

The external resultant torque with respect to O is, of course, null, thus conserving the angular momentum of the pencil with respect to its mass center, as stated in Eq. (7.4). At a general instant, this vector is as follows:

$$^{\mathcal{R}}\mathbf{H}^{L/O} = \int_L \mathbf{p} \times \mathbf{v}\, dm$$

$$= \int_{-r}^{r} s\mathbf{b}_1 \times (\mathbf{v}^O + s\omega\mathbf{b}_2)\, \frac{m}{2r}\, ds$$

$$= \frac{m}{2r} \left[\int_{-r}^{r} s\, ds\, \mathbf{b}_1 \times \mathbf{v}^O + \omega \int_{-r}^{r} s^2\, ds\, \mathbf{b}_3 \right]$$

$$= \frac{1}{3} mr^2 \omega\, \mathbf{b}_3.$$

(Note that the term multiplying the velocity of the mass center vanishes automatically.) Equaling the final to the initial angular momentum, then we simply have

$$\omega = \omega_0.$$

In conclusion, the conservation principles applicable to the system guarantee that the pencil continues revolving with a constant angular velocity and with its center describing a trajectory (parabolic, but demonstrating this simple fact is up to the reader) on the vertical plane, so that the horizontal component of its velocity vector remains constant.

If the resultant moment with respect to a point O, fixed in an inertial reference frame $\mathcal{R}$, of the external forces acting on a mechanical system S is null in a certain interval, the angular momentum vector of the system with respect to O is conserved in $\mathcal{R}$, that is,

$$^{\mathcal{R}}\mathbf{H}^{S/O} = \mathbf{H}_0^O \qquad \text{if} \qquad \mathbf{M}^{\mathcal{F}_e/O} = 0. \tag{7.5}$$

Equation (7.5) expresses the *principle of conservation of the angular momentum with respect to a fixed point*, where $\mathbf{H}_0^O$ is a constant vector. It is easy to verify that Eq. (7.5) results from Eq. (3.15) when the resultant moment with respect to O is null.

Example 7.5   Consider a pair of barbells H, consisting of two small balls, with mass m each, connected rigidly by a light rod with length $3r$, moving over a smooth, horizontal flat surface with velocity v, as shown in Fig. 7.5a.

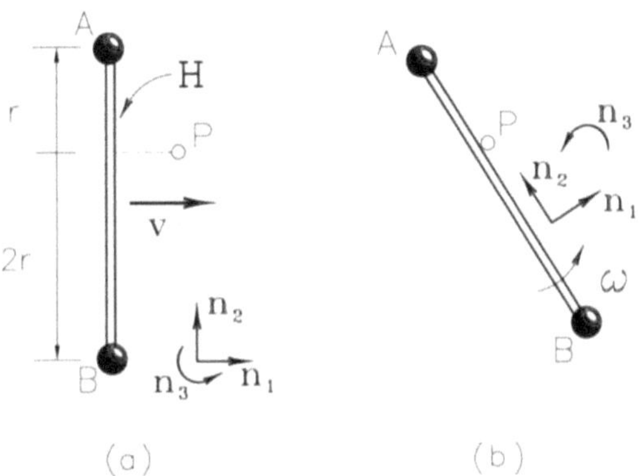

Figure 7.5

At a certain instant, the set collides with a peg, P, fixed to the surface; a fitting device (not shown) ensures that, after the collision, the rod and peg remain together, the barbells revolving around the point P, as shown in Fig. 7.5b. We want to determine the angular velocity of the barbells immediately after the impact. Now, as the only noncounterbalanced external force is that applied by the peg, the resultant moment with respect to P is null, meaning then that we may use Eq. (7.5), referring to the point P. The angular momentum of the barbells with respect to P before the impact is

$$^{\mathcal{R}}\mathbf{H}^{H/P}(1) = r\mathbf{n}_2 \times mv\mathbf{n}_1 - 2r\mathbf{n}_2 \times mv\mathbf{n}_1$$

$$= mrv\mathbf{n}_3.$$

After the collision, the angular momentum is

$$^{\mathcal{R}}\mathbf{H}^{H/P}(2) = r\mathbf{n}_2 \times (-mr\omega)\mathbf{n}_1 - 2r\mathbf{n}_2 \times 2mr\omega\mathbf{n}_1$$

$$= 5mr^2\omega\mathbf{n}_3,$$

therefore,

$$\omega = \frac{v}{5r}.$$

See the corresponding animation.

When the resultant moment with respect to an axis E, fixed in an inertial reference frame $\mathcal{R}$, of the external forces acting on a mechanical system S is null in a certain interval, the angular momentum vector of the system with respect to E is conserved in $\mathcal{R}$, that is,

$$^{\mathcal{R}}\mathbf{H}^{S/E} = \mathbf{H}_0^E \qquad \text{if} \qquad \mathbf{M}^{\mathcal{F}_e/E} = 0. \tag{7.6}$$

Equation (7.6) expresses the *principle of conservation of the angular momentum with respect to a fixed axis*, where $\mathbf{H}_0^E$ is a constant vector. Equation (7.6) may be interpreted as an immediate consequence of Eqs. (3.16), for a given direction, when the corresponding component of the resultant moment vanishes, or, identically, as the projection of Eq. (7.5) in a given direction. Note that Eq. (7.6) is also valid for an axis E^*, containing the mass center of the system and with a fixed orientation in the inertial reference frame.

Example 7.6   Let us consider a system S consisting of a small disk with mass m, sliding over a horizontal table without friction, and a small cylinder, with mass $3m$, hanging from a rope, with length $4r$, passing through the hole in the center of the table, and whose opposite end is fixed to the disk (see Fig. 7.6). In the first phase of the motion of the system, the cylinder is immobile, lying on a support, while the disk describes a circular motion around the center of the table, with a radius $2r$ and velocity of constant module v, as shown in Fig. 7.6a. The support under the cylinder is then suddenly removed, thus altering the motion of the system. We want to establish the velocities v for the disk, and v' for the cylinder, at the instant when the configuration of the system is that indicated in Fig. 7.6b. Assuming that there is no friction, the external forces acting in the system are the weight of the bodies, the normal force applied by the table on the disk, and the force applied by the table on the rope, at point O.

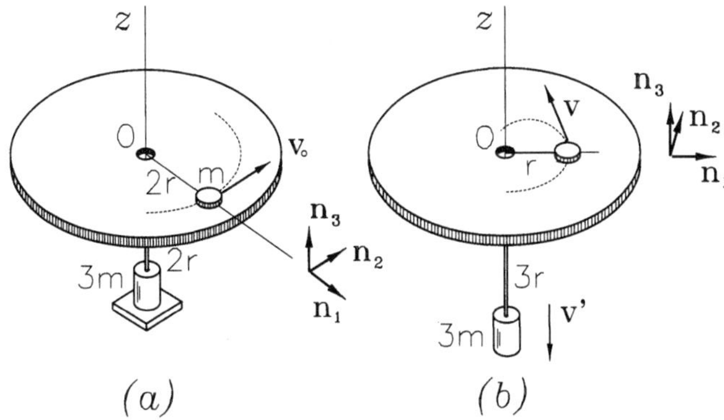

Figure 7.6

The resultant moment with respect to the z-axis is, therefore, null, so the angular momentum of the system with respect to this axis is conserved, as established in Eq. (7.6). In the beginning of the movement, we have

$$^{\mathcal{R}}\mathbf{H}^{S/z}(1) = \left(2r\mathbf{n}_1 \times mv_0\mathbf{n}_2\right) \cdot \mathbf{n}_3 \, \mathbf{n}_3 + 0 = 2mrv_0\mathbf{n}_3,$$

and, in the final configuration, breaking down the velocity of the disk in the basis of the figure, we have

$$^{\mathcal{R}}\mathbf{H}^{S/z}(2) = \left(r\mathbf{n}_1 \times m(v_1\mathbf{n}_1 + v_2\mathbf{n}_2)\right) \cdot \mathbf{n}_3 \, \mathbf{n}_3 + 0 = mrv_2\mathbf{n}_3.$$

The component of the velocity of the disk orthogonal to the rope may then be determined by making the above equations equal, resulting in

$$v_2 = 2v_0.$$

Note that the velocity component of the disk in the direction of the rope is, in module, equal to the velocity v' of the cylinder, since we assumed that the rope is unstretchable; establishing it, however, will depend on the application of another principle of conservation, as shown latter. See the corresponding animation.

When the resultant work of all external and internal forces applied to a system S in a given interval is null, its kinetic energy is conserved in an inertial reference frame, in this interval, that is,

$$^{\mathcal{R}}K^{S}(2) = {}^{\mathcal{R}}K^{S}(1) \qquad \text{if} \qquad {}^{\mathcal{R}}\mathcal{T}_{12}^{\mathcal{F}} = 0. \tag{7.7}$$

Equation (7.7) expresses, therefore, the *principle of conservation of the kinetic energy* of a system, deriving directly from Eq. (6.1) when the resultant work is null. This is an extremely useful and easy-to-use equation for finding velocities.

Example 7.7   Two small bodies A and B with the same mass m slide freely over a smooth horizontal flat surface, connected by a thread, passing by a peg P, fixed to the surface (see Fig. 7.7). In the beginning, the velocities are $2v$ and v, respectively, as shown. We wish to compute the new velocities immediately before there is an impact between the bodies.

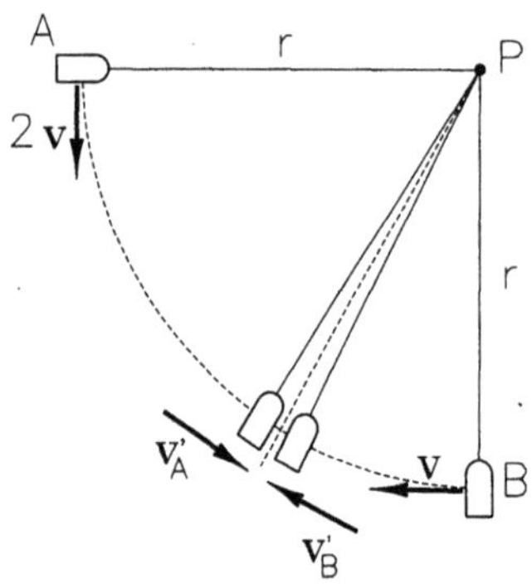

Figure 7.7

The weights and (normal) forces applied by the plane are counterbalanced, meaning that their overall contribution to the resultant work is null. Assuming that the friction between the thread and the peg is enough to prevent the relative sliding, the result is that the force applied on it does not work. As the thread is nonstretchable, there will be no contribution from the internal forces to the resultant work (the traction in the thread is always orthogonal to the velocity, for both bodies). Equation (7.7), therefore, is applicable, which, in the case, is as follows:

$$\frac{1}{2}mv_A'^2 + \frac{1}{2}mv_B'^2 = \frac{1}{2}mv_A^2 + \frac{1}{2}mv_B^2,$$

that is,

$$v_A'^2 + v_B'^2 = 5v^2. \tag{a}$$

As the only noncounterbalanced external force passes by P, the angular momentum of the system is therefore conserved with respect to P and, from Eq. (7.5), we have

$$mrv_A' - mrv_B' = mrv_A - mrv_B;$$

therefore,

$$v_A' - v_B' = v. \tag{b}$$

The pair of Eqs. (a, b) provides the solution for the problem. The reader should note that, after the impact, Eq. (7.5) will continue to prevail; but the forces present during the collision between the bodies do work (with an unknown value, at least in principle), making the use of Eq. (7.7) unfeasible after contact.

When the resultant work of the nonconservative forces — external and internal — applied to a system S is null in a given interval, its mechanical energy is conserved in an inertial reference frame, in that interval, that is,

$$^{\mathcal{R}}E^S(2) = {}^{\mathcal{R}}E^S(1) \qquad \text{if} \qquad {}^{\mathcal{R}}\mathcal{T}_{12}^{\mathcal{F}_N} = 0. \tag{7.8}$$

Equation (7.8) expresses the *principle of conservation of the mechanical energy* of a system, deriving directly from Eq. (6.5) in the case of the nonconservative work being null.

Example 7.8 Going back to Example 7.3 (see Fig. 7.3), where the weight of the cursor and force applied on it by the guide do not work, since they are orthogonal to the velocity of C in the inertial reference frame; nor does the traction in the rope work since the distance between C and P remains invariant (see Section 4.7) and the weight of the pendulum is a conservative force. As there are no other contributions for the resultant work to consider, Eq. (7.8) then applies. The mechanical energy of the system in the first configuration will be null, taking from there the reference for the gravitational potential energy; the mechanical energy of the system in the final position (rope in the vertical) is then

$$^{\mathcal{R}}E^S(2) = \frac{1}{2}\left(mv_P{}^2 + m_0 v_C{}^2\right) - mgr(1 - \cos\theta_0),$$

resulting in, according to Eq. (7.8),

$$v_P{}^2 + \frac{m_0}{m}v_C{}^2 = 2gr(1 - \cos\theta_0).$$

There are now two independent relations for establishing the desired velocities.

Example 7.9   Now returning to Example 7.6 (see Fig. 7.6), it is easy to see that the work of the nonconservative forces between the two configurations under study is null, the mechanical energy being, therefore, conserved. In the initial condition, this energy is (taking this position as reference for the gravitational potential energy)

$$^{\mathcal{R}}E^S(1) = {}^{\mathcal{R}}K^S(1) + {}^{\mathcal{R}}\Phi^S(1)$$
$$= \frac{1}{2}mv_0^2.$$

In the final condition, we will have

$$^{\mathcal{R}}E^S(2) = {}^{\mathcal{R}}K^S(2) + {}^{\mathcal{R}}\Phi^S(2)$$
$$= \frac{1}{2}m(v_1^2 + v_2^2) + \frac{1}{2}3mv_1^2 - 3mgr.$$

When the energies are equal, remembering that, as it was obtained beforehand, $v_2 = 2v_0$, then, for the final velocity of the cylinder, we have

$$v' = v_1 = \sqrt{\frac{3}{2}gr - \frac{3}{4}v_0^2}$$

and, for the final velocity of the disk,

$$v = \sqrt{\frac{3}{2}gr + \frac{13}{4}v_0^2}.$$

Collisions usually involve nonconservative forces, for which reason Eq. (7.8) cannot be used when a collision occurs between the elements of a system or between an element of the system and a body outside it.

Example 7.10 Returning to Example 7.5 (see Fig. 7.5), the mechanical energy of the barbells, before the collision, is

$$^{\mathcal{R}}E^{H}(1) = {}^{\mathcal{R}}K^{H}(1) = 2 \cdot \frac{1}{2}mv^2 = mv^2.$$

After the collision, this energy will be

$$\begin{aligned}
^{\mathcal{R}}E^{H}(2) &= {}^{\mathcal{R}}K^{H}(2) \\
&= \frac{1}{2}m\big((r\omega)^2 + (2r\omega)^2\big) \\
&= \frac{1}{10}mv^2.
\end{aligned}$$

The mechanical energy of the system has therefore been reduced by 90%, due to the impact.

5.8 Fluids

One of the important applications in this chapter is found to be the study of *fluid dynamics*. Although this section offers only a very brief introduction to this subject, it is interesting to discuss here the principal methods of an overall analysis for the flow of fluids and check how, with slight alterations, the general equations for systems apply to the motion of fluids, leading to the solution of a whole important class of engineering problems.

When analyzing the motion of a fluid, it is convenient to consider a fixed region in space — more precisely, fixed in an inertial reference frame — for study purposes. This region, called the *control volume*, differs from the concept of a *system*, precisely because it is considered as fixed. (In fact, mobile control volumes may be considered, making the corrections due to the motion of the respective reference frame, but this topic will not be discussed here.)

This analytical method, also known as *Euler's description* of a system, consists, therefore, of fixing a region through which a stream of fluid flows. Let us imagine a water pressure pump, pumping water from a cistern to a raised tank. Now, it is especially inconvenient for us to analyze the *system*, consisting of a portion (how much?) of water in the cistern, another portion (how much?) of water in the tank, and the water actually flowing inside the pump and pipes. If, nevertheless, we take the water inside the pump as the control volume, it is possible, at each instant, to make balances of, for example, momentum and energy, thus assessing forces, flows, and consumed power.

Let us then consider a stream of flowing fluid and a region to be studied, indicated by the initials CV (control volume) in Fig 8.1.

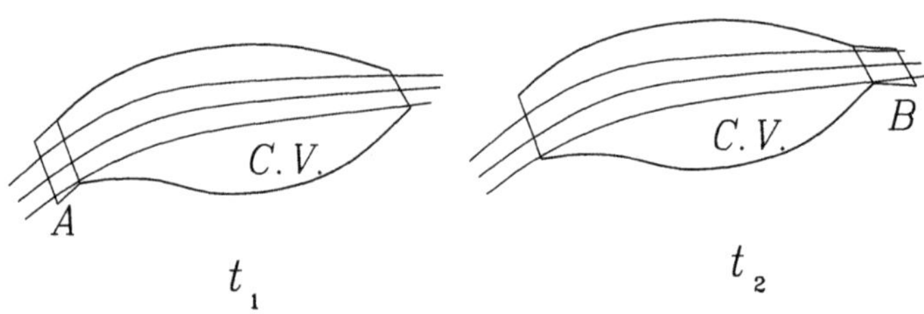

Figure 8.1

Let us take a continuous system that consists of the fluid occupying the regions A and CV at an instant t_1 and the regions CV and B at an instant t_2. Let us now assume that a certain *property P* of the system (for example, the mass of the system, the momentum of the system, or the kinetic energy of the system) whatever the subject under study. This property may be expressed throughout the system as

$$P(t) = \int_V p\,dV, \tag{8.1}$$

where V is the sum of the regions A and CV at the instant t_1 and is the sum of the regions CV and B at the instant t_2. The term p, in the integral, is the *specific property* corresponding to P (p is a density if P

is the mass, is the specific kinetic energy if P is the kinetic energy of the system, etc.).

The time rate of the property P for the system is

$$\frac{dP}{dt} = \frac{d}{dt} \int_V p \, dV. \tag{8.2}$$

This time rate may be calculated, as the limit, when $\Delta t = t_2 - t_1$ approaches zero, of

$$\frac{dP}{dt} = \lim_{\Delta t \to 0} \frac{1}{\Delta t} \left(\int_{VC+B} p \, dV - \int_{VC+A} p \, dV \right). \tag{8.3}$$

Now, the term involving the limit of the integral in the control volume, the latter being fixed, is equal to the partial derivative of the property P with respect to time, that is,

$$\lim_{\Delta t \to 0} \frac{1}{\Delta t} \int_{VC} \left(p(t_2) - p(t_1) \right) dV = \frac{\partial}{\partial t} \int_{VC} p \, dV. \tag{8.4}$$

In the rest of the region, (A+B), reduced in the limit to the boundary of the control volume, the time rate of the product $p \, dV$ is due to the variation of dV, that is, to the time rate of the volume in this boundary. The time rate of the volume on the boundary is equal to the velocity of the fluid flow in the direction orthogonal to the boundary, multiplied by the area of flow, that is, $dV/dt = \mathbf{v} \cdot d\mathbf{A}$, where $d\mathbf{A}$ is the product of the element of an area, dA, by the outside normal unit vector $\mathbf{n}$ (see Fig. 8.2). Therefore,

$$\lim_{\Delta t \to 0} \frac{1}{\Delta t} \left(\int_B p \, dV - \int_A p \, dV \right) = \int_\delta p\mathbf{v} \cdot d\mathbf{A}, \tag{8.5}$$

where δ indicates the boundary of the control volume. This term is also called the *flux of the property* P through the boundary of the control volume.

Joining these terms, then, leads to the equation for the time rate of the property P of a system, in terms of Euler's description, also known as the *transport theorem* or also as *Reynolds' theorem*,

$$\frac{dP}{dt} = \frac{\partial}{\partial t} \int_{VC} p \, dV + \int_\delta p\mathbf{v} \cdot d\mathbf{A}. \tag{8.6}$$

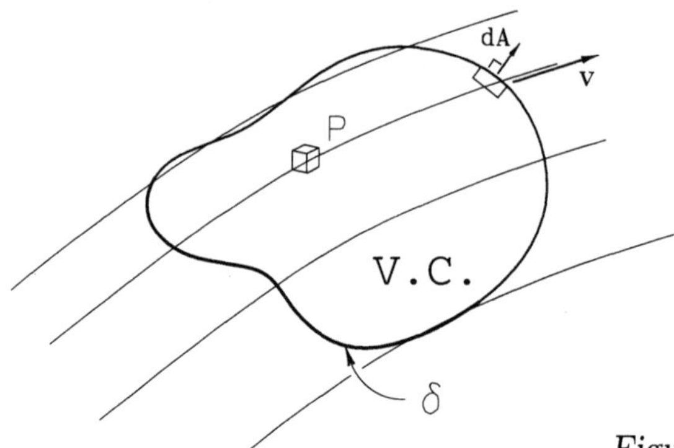

Figure 8.2

Equation (8.6) is the relation that we need to study a system that is in continuous motion. It establishes that the time rate of any property inside a system is always equal to the time rate of the property inside an arbitrary control volume (a partial derivative with respect to time) *plus* the net flow (which is what goes out less what comes in) of this property out of the control volume. Note that the net flow is guaranteed by the integral in the term on the right of Eq. (8.5): At the points where the flow goes *outward* from the control volume, the dot product between the velocity vector and outside normal force is positive, while at the points where the flow goes *inward* to the control volume, the scalar product will be negative. Let us now consider some applications of the transport theorem.

If P is the *mass* m of a system (p will therefore be its density, ρ), the principle of conservation of the mass — implicit in the concept of system — is expressed by

$$\frac{dm}{dt} = \frac{d}{dt} \int_V \rho \, dV = 0. \tag{8.7}$$

Translating in terms of a control volume, that is, substituting Eq. (8.6) in Eq. (8.7), then

$$\frac{\partial}{\partial t} \int_{VC} \rho \, dV + \int_\delta \rho \mathbf{v} \cdot d\mathbf{A} = 0. \tag{8.8}$$

Equation (8.8) establishes that the time rate of the mass inside an arbitrary control volume plus the flux of mass outside the control volume must be zero. If, for example, we have a compressible flow (a gas), there may be an accumulation of mass inside the control volume, strictly compensated by a negative flux of mass (inflowing more than outflowing).

If P is the momentum $\mathbf{G}$ of a system (p will, therefore, be the specific momentum, $\rho\mathbf{v}$), Eq. (4.1) states that

$$\frac{d\mathbf{G}}{dt} = \frac{d}{dt}\int_V \rho\mathbf{v}\,dV = \mathbf{F}, \tag{8.9}$$

that is, the time rate of the momentum (always in an inertial reference frame) is equal to the outside resultant force. In terms of Euler's description, that is, substituting Eq. (8.6) in Eq. (8.9), it results in

$$\frac{\partial}{\partial t}\int_{VC} \rho\mathbf{v}\,dV + \int_\delta \rho\mathbf{v}\,\mathbf{v}\cdot d\mathbf{A} = \mathbf{F}, \tag{8.10}$$

that is, that the time rate, in an arbitrary control volume, of the momentum vector plus the flux of momentum is equal to the resultant of the external forces applied to this control volume.

If P is the angular momentum $\mathbf{H}^{S/O}$ of a system with respect to a point fixed in an inertial reference frame (p will, therefore, be the specific angular momentum with respect to the point, $\rho\mathbf{p}\times\mathbf{v}$, $\mathbf{p}$ being the position vector with respect to the point), Eq. (4.8) states that

$$\frac{d\mathbf{H}^{S/O}}{dt} = \frac{d}{dt}\int_V \rho\mathbf{p}\times\mathbf{v}\,dV = \mathbf{M}^{\mathcal{F}_e/O}, \tag{8.11}$$

that is, the time rate of the angular momentum of a system with respect to a point fixed in the inertial reference frame is equal to the resultant moment of the external forces with respect to the point. In terms of Euler's description, that is, substituting Eq. (8.6) in Eq. (8.11), the result is

$$\frac{\partial}{\partial t}\int_{VC} \rho\mathbf{p}\times\mathbf{v}\,dV + \int_\delta \rho\mathbf{p}\times\mathbf{v}\,\mathbf{v}\cdot d\mathbf{A} = \mathbf{M}^{\mathcal{F}_e/O}, \tag{8.12}$$

that is, that the time rate, in an arbitrary control volume, of the angular momentum vector with respect to a fixed point plus the flux of angular momentum, with respect to the point, is equal to the resultant moment of the system of external forces with respect to the point, applied to the control volume.

It is never too late to remember that Eq. (4.7) may be used instead of Eq. (4.8), resulting in

$$\frac{\partial}{\partial t} \int_{VC} \rho \mathbf{r} \times \mathbf{v} \, dV + \int_{\delta} \rho \mathbf{r} \times \mathbf{v} \, \mathbf{v} \cdot d\mathbf{A} = \mathbf{M}^{\mathcal{F}_e/P^*}, \qquad (8.13)$$

where the position vector $\mathbf{r}$ is relative, now, to the mass center of the control volume, P^*, and the resultant moment is also calculated with respect to this point.

When, in a flow, the properties remain invariant with time, it is said that there is a *steady state*. The steady state is the condition for which the flow machines are usually designed. Hence, hydraulic pumps, fans, turbines, and many other items of equipment that use fluids usually operate in a steady state. When the properties are *not* invariant in time, it is said that there is a *transient state*. So, a fan, a hydraulic pump, and a turbine, when started up, necessarily pass through a transient state before achieving the steady state. When there is a steady state, the partial derivative with respect to the time of a property inside the control volume is null. In a steady state, therefore, Eq. (8.8) is reduced to

$$\int_{\delta} \rho \mathbf{v} \cdot d\mathbf{A} = 0. \qquad (8.14)$$

Also in a steady state, Eq. (8.10) is reduced to

$$\int_{\delta} \rho \mathbf{v} \, \mathbf{v} \cdot d\mathbf{A} = \mathbf{F}. \qquad (8.15)$$

Also in a steady state, Eq. (8.12) is as follows:

$$\int_{\delta} \rho \mathbf{p} \times \mathbf{v} \, \mathbf{v} \cdot d\mathbf{A} = \mathbf{M}^{\mathcal{F}_e/O}. \qquad (8.16)$$

And, last, for a steady state, Eq. (8.13) results in

$$\int_{\delta} \rho \mathbf{r} \times \mathbf{v} \, \mathbf{v} \cdot d\mathbf{A} = \mathbf{M}^{\mathcal{F}_e/P^*}. \qquad (8.17)$$

Example 8.1 Figure 8.3 illustrates the situation analyzed in Example 4.2 (read it again). Let us see how it can be solved in light of the concepts introduced in this section. Taking the curve as a control volume, the boundary to be considered involves the inflow and outflow. As this is a steady state, then, from Eq. (8.14),

$$\int_\delta \rho \mathbf{v} \cdot d\mathbf{A} = \rho A(v_s - v_e) = 0;$$

therefore, the inflow velocity, v_e, and outflow velocity, v_s, are equal in module, as assumed in Example 4.2, that is,

$$v_s = v_e = v.$$

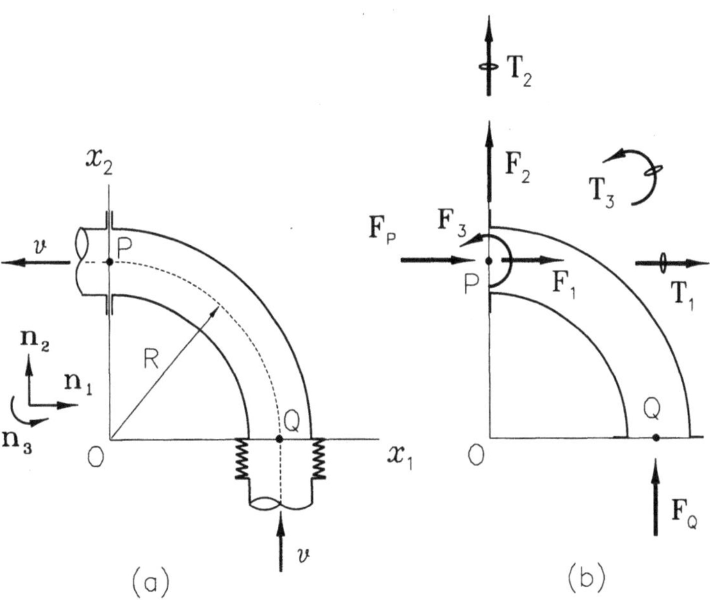

Figure 8.3

Now analyzing the momentum, we have

$$\int_\delta \rho \mathbf{v}\, \mathbf{v} \cdot d\mathbf{A} = \rho v A(\mathbf{v}_s - \mathbf{v}_e)$$
$$= -\rho v^2 A(\mathbf{n}_1 + \mathbf{n}_2).$$

Therefore, Eq. (8.15) results in

$$-\rho v^2 A(\mathbf{n}_1 + \mathbf{n}_2) = (F_1 + pA)\mathbf{n}_1 + (F_2 + pA)\mathbf{n}_2 + F_3\mathbf{n}_3. \tag{a}$$

Note that the time rate of the momentum of the system, calculated in Example 4.2, is, in fact, the flux of momentum outward from the control volume. Last,

$$\int_\delta \rho \mathbf{p} \times \mathbf{v}\, \mathbf{v} \cdot d\mathbf{A} = \rho R v A(v - v) = 0;$$

therefore, from Eq. (8.16), we get

$$0 = (T_1 + RF_3)\mathbf{n}_1 + T_2\mathbf{n}_2 + (T_3 - RF_1)\mathbf{n}_3. \tag{b}$$

Equations (a) and (b) are enough to determine the forces in the flange. Solving then for the unknown $\mathbf{F}$ and $\mathbf{T}$, we obtain (check)

$$\mathbf{F} = -(p + \rho v^2)A\,(\mathbf{n}_1 + \mathbf{n}_2), \quad \mathbf{T} = -(p + \rho v^2)RA\,\mathbf{n}_3.$$

In fact, the results obtained here are exactly the same as those obtained in Example 4.2. Note, however, that, by adopting the equations for fluids discussed herein, the solution has proven to be easier and much more straightforward.

Exercise Series #7 (Sections 5.1 to 5.8)

P7.1 A chain, with length a and mass m, is lying in a pile on the floor. A winch then starts to hoist up an end with a prescribed constant vertical velocity of module v, applying on the end link a force F that will depend, naturally, on the distance x from the floor, as shown. Find $F(x)$.

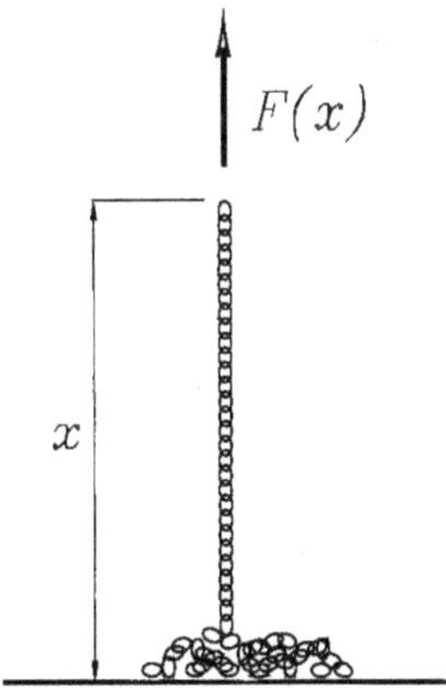

Figure P 7.1

P7.2 The system, consisting of three identical balls with the same mass m and connected by two threads, with the same length a, is at rest in the configuration shown lying on a smooth, horizontal plane, when the force F is applied to the central body, as shown. Find the initial acceleration of this body.

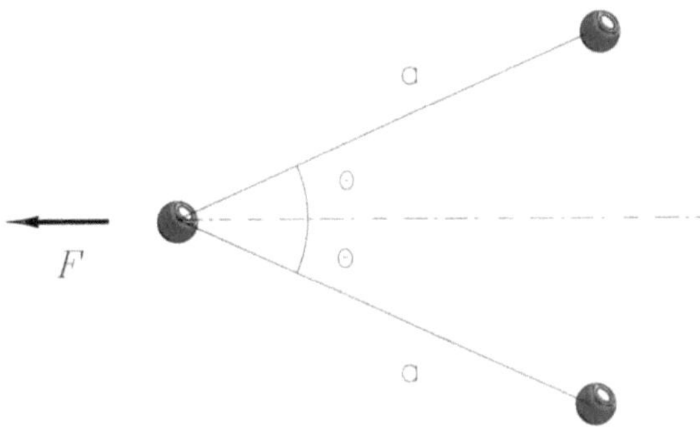

Figure P 7.2

P7.3   The set of four balls, with the same mass m, is interconnected by unstretchable threads, being at rest on a smooth, horizontal plane in the configuration shown, when a force with constant module F is suddenly applied. Find the initial acceleration of ball A.

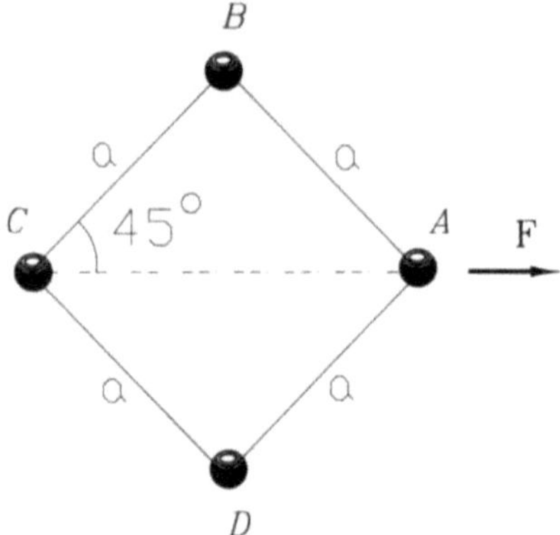

Figure P 7.3

P7.4   A bicycle chain rests on a triangular prism, with its center coinciding with the top end of the prism, as shown. Find the horizontal acceleration to be applied to the prism in order to ensure that the chain does not move in relation to the prism.

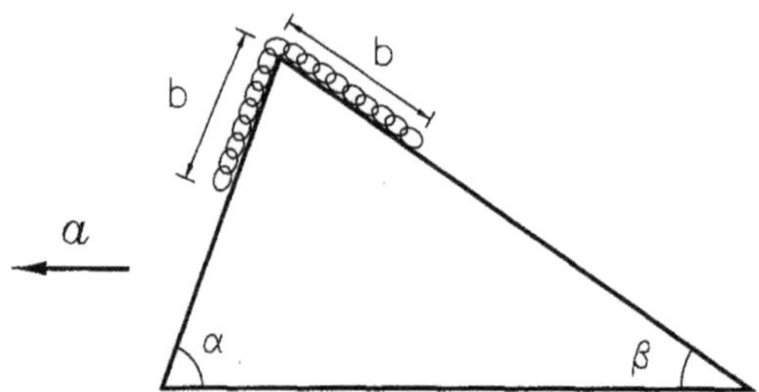

Figure P 7.4

P7.5 A pendulum consists of two masses, m and $4m$, interconnected by a light r-long rod, revolving around a horizontal axis passing through point O. Choose the distance x that minimizes the period of the pendulum's swinging with small oscillations.

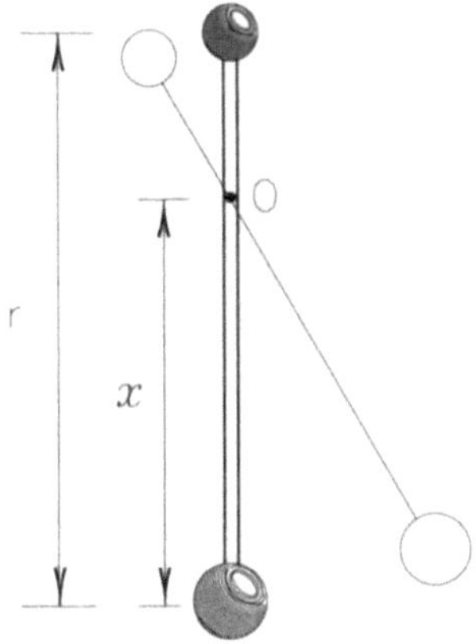

Figure P 7.5

P7.6 Two small repair wagons on a railroad, with masses $8m$ and $10m$, move on parallel rails and in opposite directions, driven by operators with masses m and $1.5\,m$, respectively. When passing each other, the operators change wagon, jumping at the same time from one wagon to the other, orthogonally to the line direction. Find the velocities v_1 and v_2 of the wagons soon after the operators have settled if the modules of the velocities immediately before the change are $u_1 = 2u$ (the lighter) and $u_2 = 3u$, respectively.

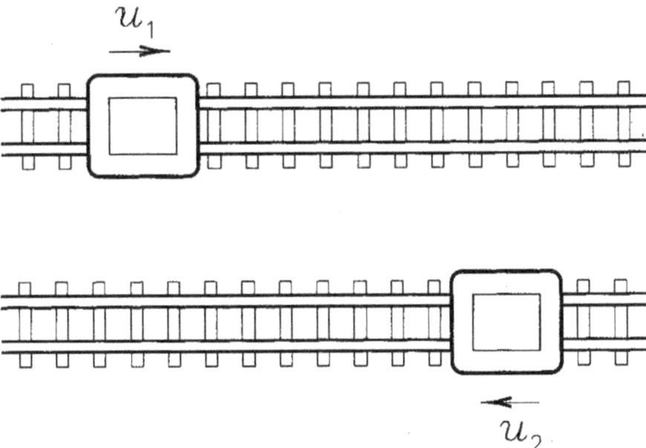

Figure P 7.6

P7.7 The vane B, shaped as an arc of a circle, is revolving freely with a angular velocity ω around the vertical axis z, in relation to the reference frame A, when a small ball is inserted in the top end at null initial velocity in relation to A. After the impact between the ball and the vane, the former descends until it reaches the bottom end, leaving the vane. Find the angular velocity of the vane at this instant.

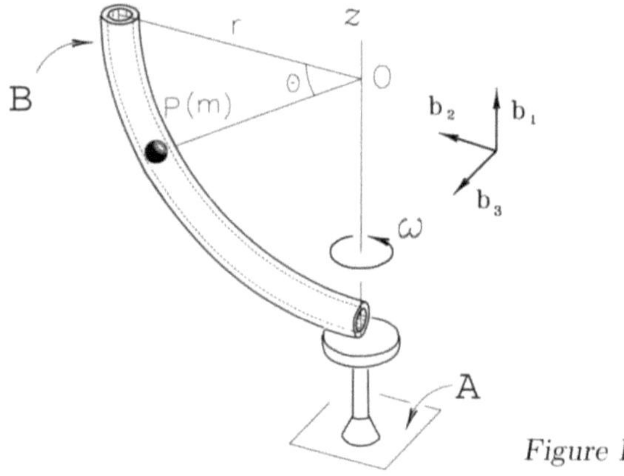

Figure P 7.7

P7.8 A chain, with length a and mass m, is lying in a pile on the floor. A winch then starts to hoist up an end with a prescribed constant vertical velocity of module v, applying on the end link a force F that will, naturally, depend on the distance x from the floor, as indicated. As it is being raised, each new link lifted collides with the one before, resulting in dissipation of energy. Calculate the overall loss of energy in the hoisting process (see P 7.1).

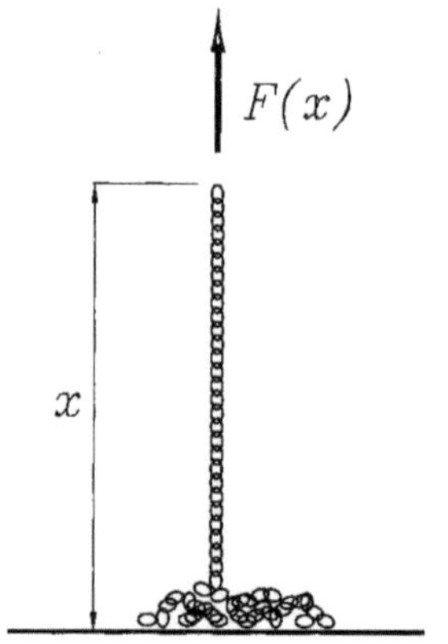

Figure P 7.8

P7.9 The matchbox is lying over one end of rod AB, at rest in the vertical, as shown. After a light horizontal push, the end B starts to slide, with negligible friction, over the horizontal plane. Assuming that the mass of the matchbox is much smaller than that of the rod, find the angle θ of slope of the rod in relation to the vertical at the moment when the matchbox leaves end A.

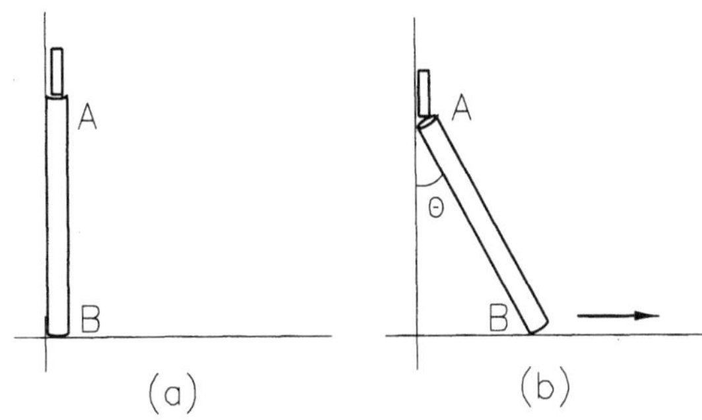

Figure P 7.9

P7.10 Cursors A and B, with masses m_1 and m_2, respectively, are interconnected by a light linear elastic spring, with constant k and natural length a. If the set is left at rest with the spring stretched to $2a$, find the velocity of B when the spring is relaxed, assuming that the friction is negligible in both guides. Also calculate the amplitude of the motion of A.

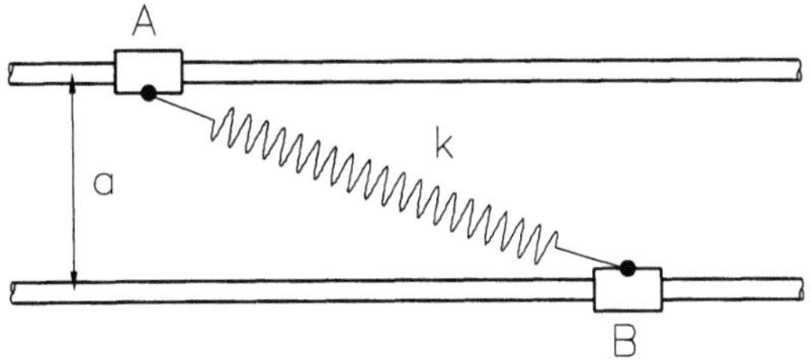

Figure P 7.10

P7.11 In the previous problem, there is a periodic motion of the system. Find this period.

P7.12   Two identical snooker balls A and B collide face on, in experiment (a). B is initially at rest and A has velocity v_0 before and $\frac{1}{4}v_0$ after the collision. In experiment (b), both have a velocity of module v, in orthogonal directions, before the collision. Find the final velocities of each ball and the respective directions on the plane, for experiment (b), assuming a smooth collision.

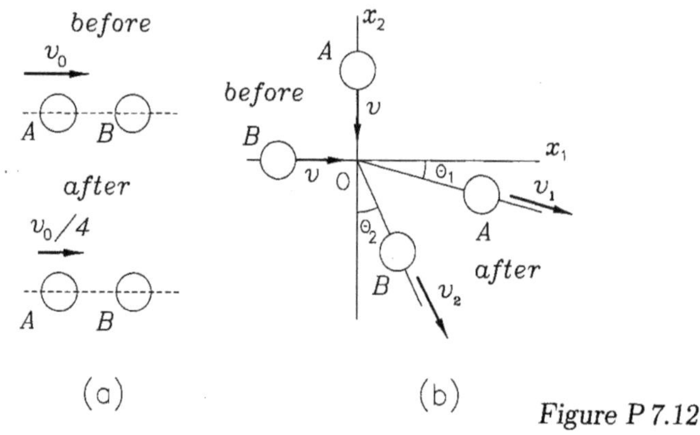

Figure P 7.12

P7.13   Two small balls with mass m each are interconnected by an unstretchable rope with length c and are left at rest in configuration (1). After a free fall from height a, the rope collides with a fixed horizontal rod, with radius r, and starts to roll around it. Find the traction in the free stretch of the rope when in configuration (2).

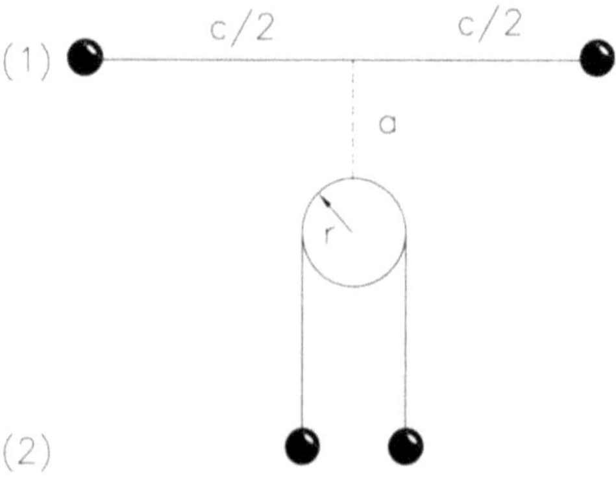

Figure P 7.13

P7.14 Analyze the system shown in the figure, consisting of four small balls, A, B, C, and D, each with the same mass m, and D interconnected by threads with a length $2r$ to the other three. At the instant $t = t_1$, the system is configured as follows: A, B, and C are lying on the smooth surface of a table occupying the vertices of an equilateral triangle and at a distance r from the center hole, while D is hanging by the three threads, as shown. At this same instant, A, B, and C have velocities of module v, in directions orthogonal to the respective radii, while D is at rest. Find the velocity of D at the instant $t = t_2$, when it has descended $r/2$.

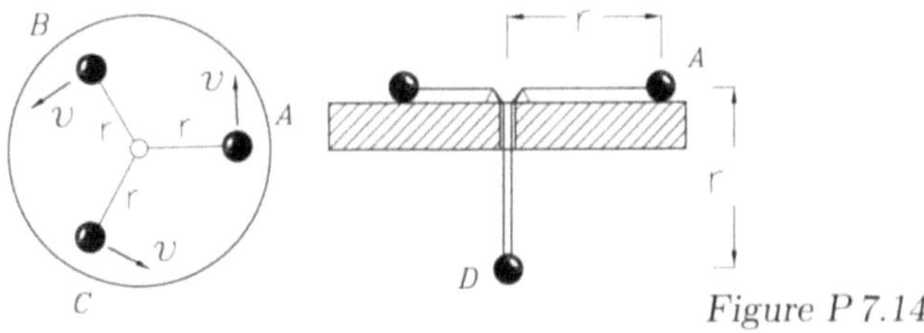

Figure P 7.14

P7.15 A body with a mass of 10 kg drops freely and, when reaching the point P, of elevation a, breaks into three parts A, B, and C, with masses of 2, 4, and 4 kg, respectively. Once known the directions of the velocities of the fragments (see figure) and the velocity module of A are known, $v_a = 12$ m/s, immediately after the fracture, find the vertical velocity of the body immediately beforehand.

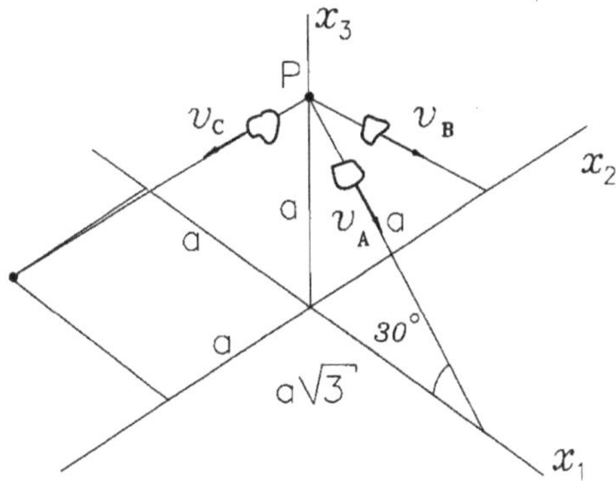

Figure P 7.15

P7.16 Consider the system consisting of four small bodies with the same mass m, joined by light rods with length a and connected by two identical linear springs, with elastic constant k and natural length a. The system is left from rest in configuration (a) on a smooth, horizontal plane. Find the velocity of the body B in configuration (b).

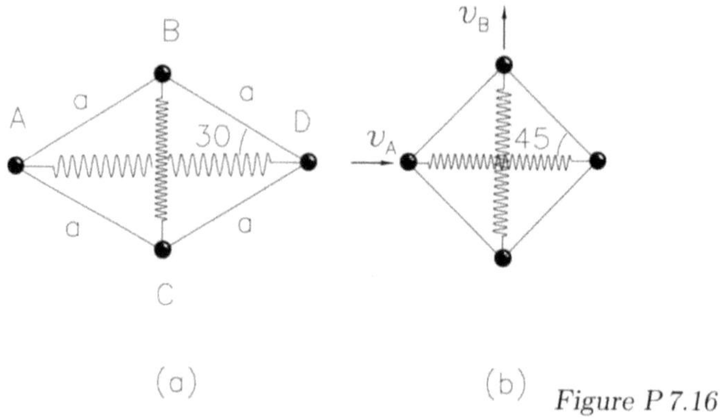

Figure P 7.16

P7.17 Three balls, with the same mass m, are connected by two light bars, and the angle θ may vary freely. The set is sliding over a smooth, horizontal plane with a constant velocity v, in the configuration shown in (a), with $\theta = 30^0$. The middle ball collides at a certain instant with the fixed rivet C, coming to an abrupt stop after the collision, as shown in (b). Calculate the velocities v_1 and v_2 of the other two balls immediately after the collision.

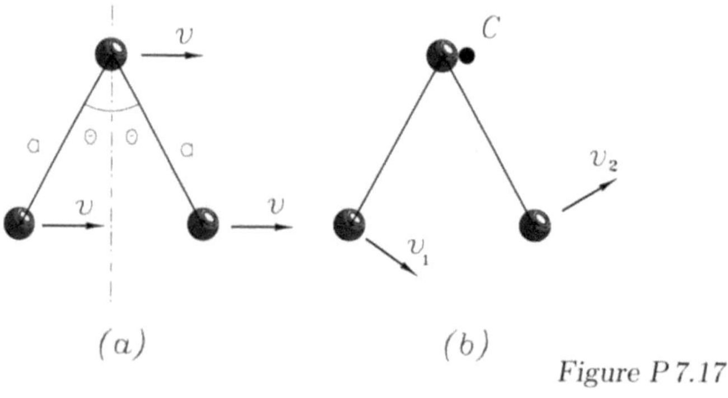

Figure P 7.17

P7.18 Two identical balls B and C are at rest on a horizontal plane, when a third ball A, identical to the other two and moving at a velocity $v_0 = 10$ m/s, collides with ball B. A series of collisions then occurs between the balls until they move away. Calculate the final velocities of each ball, knowing that the coefficient of restitution is equal to 0.4 in all impacts.

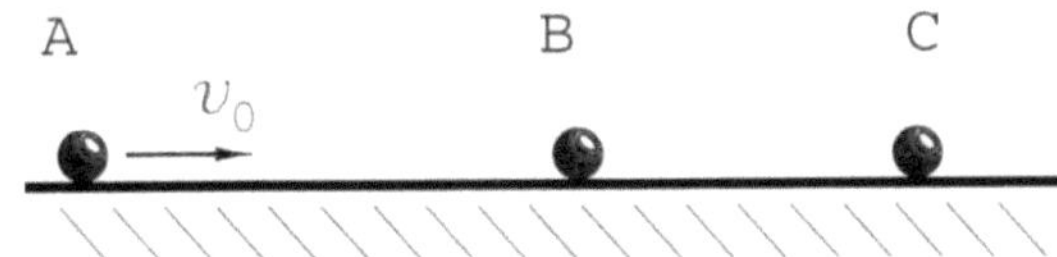

Figure P 7.18

P7.19 Three balls, with masses m_1, m_2, and m_3, are aligned in a smooth, straight conduit, at a certain distance from each other. A flick on the first ball (m_1) makes it collide with the second that, in turn, collides with the third in perfectly elastic collisions. For which mass of the ball in the middle is the velocity of the third ball maximum?

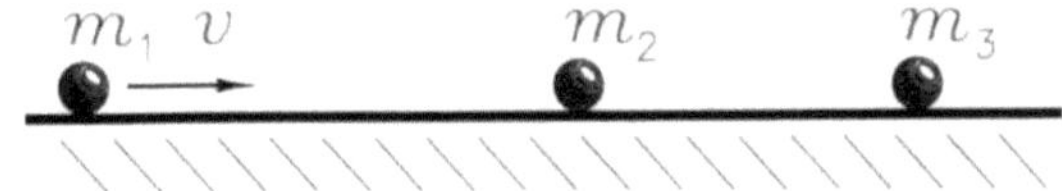

Figure P 7.19

P7.20   Two identical balls collide front on with a coefficient of restitution $\epsilon = 0.3$. What is the relation between the velocities so that one of the balls stops after the collision?

P7.21   Three identical homogeneous rods, each with mass m, are at rest on a smooth, horizontal plane, freely interconnected at point O, in the configuration shown. Calculate the acceleration of O at the instant when force F is applied.

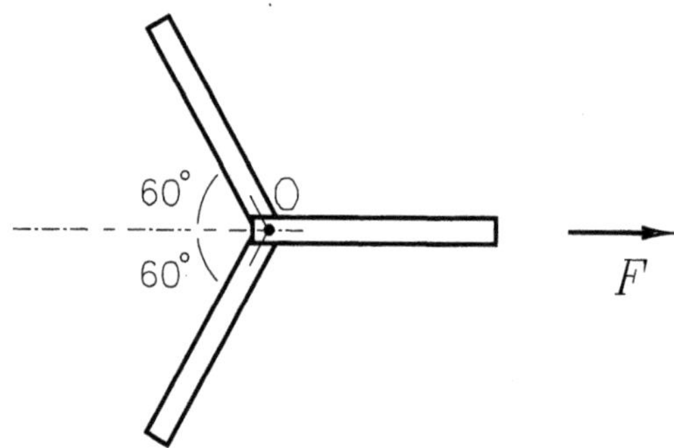

Figure P 7.21

P7.22   A channel has inlet and outlet sections equal to 0.02 m^2, with a deviation of 60^0. There is a steady-state stream of water, entering the channel at velocity $v_1 = 2$ m/s. What is the horizontal component of the resultant force that the water exerts on the channel walls?

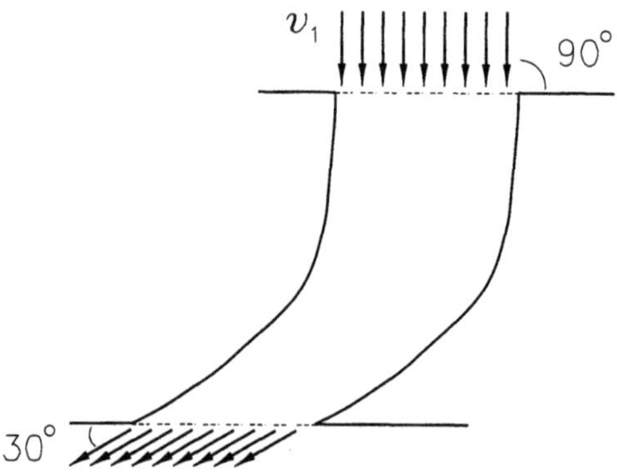

Figure P 7.22

P7.23 A jet of water on the blade of a turbine is diverted by it. If the volummetric flow of water is Q, the density of water is ρ, the inflow and outflow velocities are v_1 and v_2, the latter making the angle indicated with the horizontal, find the resultant force applied by the water on the blade if this is not moving and if the blade moves with a constant velocity of v_0, in the direction of the jet flow on it.

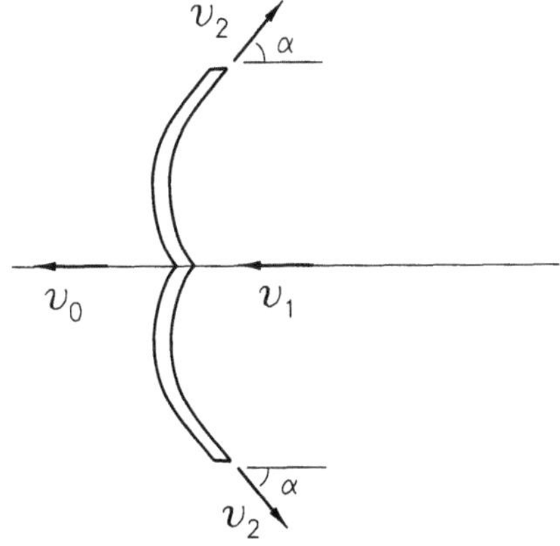

Figure P 7.23

P7.24 A jet of water is sprayed at velocity v from a nozzle with a rectangular crosssection, with unit width and height e, on a flat plate, making an angle θ, as shown. Calculate the flows in the two branches of water (the inflow and outflow velocities are equal).

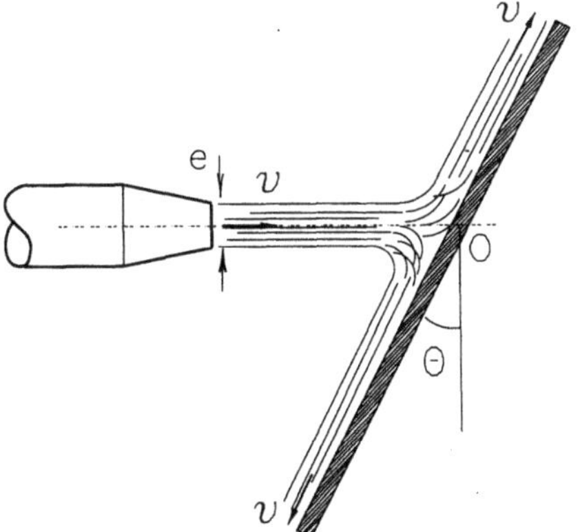

Figure P 7.24

P7.25 For the jet of water, with density ρ in the preceding problem, find the force and torque that must be applied to O to keep the plate immobile.

P7.26 A rigid body consists of a regular hexagon whose sides consist of light
rods and the vertices are small identical balls each with mass m. The body
moves on the vertical plane, traveling with a horizontal velocity v_0, as shown
in the left, when ball A collides with the end of a fixed table. After the impact,
ball A remains fixed and the body turns around this point until ball B collides,
in turn, with the surface of the table top, as shown in the right. Calculate the
angular velocity of the body immediately before this second impact and the
percentage of energy loss due to the first impact.

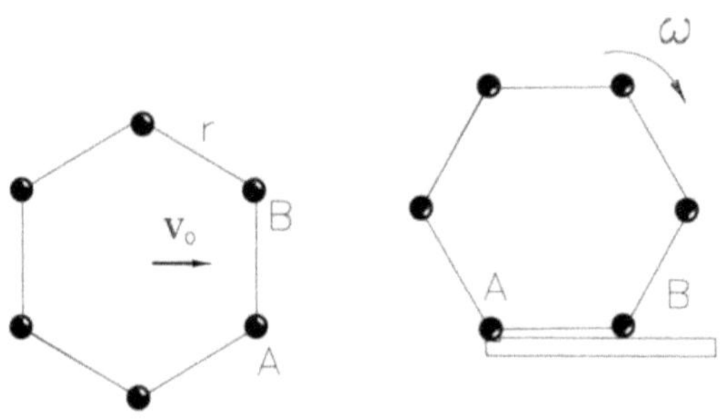

Figure P 7.26

P7.27 Three slender homogeneous rods are at rest on a vertical plane, in
the configuration illustrated. The first rod, with mass m_1, is joined to the
fixed point and to the second rod, with mass m_2, which has its other end
joined to the third rod, with mass m_3, which, in turn, is under the action of
the horizontal force **F**. Calculate the angles α, β, and γ.

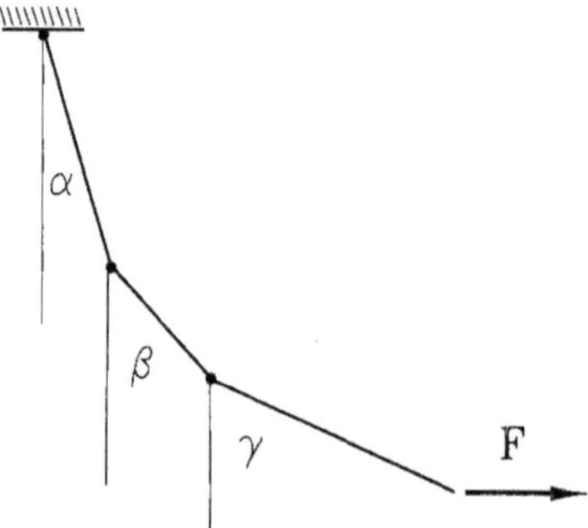

Figure P 7.27

P7.28 The system consists of six light joined rods, four with length $2a$ and two with length a each, as shown in the figure. Seven small balls each with mass m are fixed at the joining points so that two cursors are fixed at the ends A and B, each with mass $2m$, which may slide without friction on the horizontal guide. The system is at rest with $\theta = 0$ when vertical force $F = 10\,mg$ is applied to ball C, as shown. Find the velocity of C when $\theta = \pi/3$.

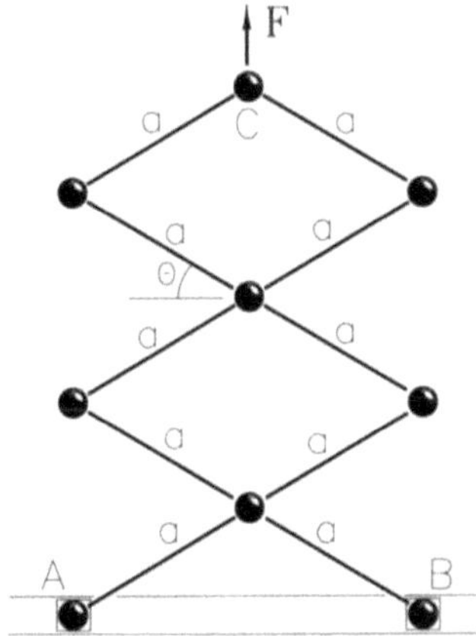

Figure P 7.28

P7.29 A flexible cable with length c is left at rest on the perfectly smooth surface of a hemisphere with radius r, in the position shown. Determine the initial acceleration of end A.

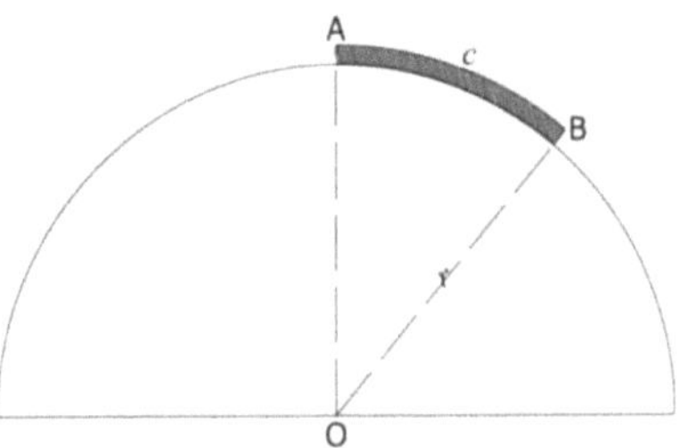

Figure P 7.29

*I*nertia

Chapter 6

This chapter is a bridge between the general principles of dynamics, discussed in Chapters 4 and 5, and their application to the study of the motion of the rigid body, discussed in Chapter 7. The main purpose is to give a standard consistent treatment to the concept of rotational inertia of a body, which is simultaneously streamlined from the theoretical viewpoint, and operational from the practical viewpoint. To cover both aspects, a tensorial approach was chosen. The reader with little knowledge of the tensorial language will find a small but comprehensive summary in Appendix A on the topics of linear algebra necessary and, the author believes, sufficient to understand this chapter.

The concept of mass center, discussed in a more or less informal way in Chapter 1, is now discussed in detail in Section 6.1 and with due care. This is a concept of vital importance, useful in studying the inertia properties, when obtaining the equations of motion for a particles system, as seen in Chapter 5, and in the dynamic analysis of the rigid body, in Chapter 7. Section 6.1 helps the reader establish the position of the mass center and centroid of systems and bodies. Appendix C presents a table with the position of the centroid for the most common geometric figures.

The rest of the chapter is devoted to the study of the inertia properties of a system. The concepts of inertia tensor — and its matrix representation in a system of Cartesian coordinates, the inertia matrix — inertia vectors, and moments and products of inertia are first established for a particle P with respect to a point O, in Section 6.2. The reason for this unusual approach is to facilitate the comprehension of the geometric relations present in those properties, and to make the concept of inertia tensor of the mass center of a body or system natural. In Section 6.3 these concepts are then generalized for discrete systems of particles and continuous bodies; in the former case, the inertia tensor is seen as a sum of the inertia tensors of the particles involved, the same being valid for inertia vectors and moments and products of inertia; in the second, the inertia tensor is presented in the same way as an integration in the body of the inertia tensor of a mass element. In the next section, the inertia properties resolved in Cartesian coordinates are studied, paying attention to the calculation of moments and products of inertia, either by integration or by using tables that lead the composition of the inertia matrix to the adopted coordinate system. The tables of moments and products of inertia for the most common geometries are presented in Appendix C as a support to this section.

Section 6.5 discusses the transposition of axes, indicating what to do to establish the inertia properties of a body with respect to a point, when its inertia properties are known with respect to another point. Here the importance of mass center is again shown, which, as will be highlighted, facilitates this transposition. Lastly, Section 6.6 discusses the important theme of the so-called principal directions of inertia, which diagonalize the inertia matrix, facilitating operations with the inertia tensor. The principal directions of inertia will be of the utmost importance for studying the dynamics of the rigid body, since its adoption substantially simplifies the equations of motion, as will be demonstrated in Chapter 7.

6.1 Mass and Mass Center

Let us consider, first, S as a system consisting of N particles P_i, with mass m_i, $i = 1, 2, \ldots, N$, and position, with respect to a given point O, $\mathbf{p}_i = \mathbf{p}^{P_i/O}$ (see Fig. 1.1). The mass m of the system S is the sum of the masses of the particles involved, that is,

$$m = \sum_{i=1}^{N} m_i, \tag{1.1}$$

and its *mass center* is defined as the point S*, whose position with respect to point O, $\mathbf{p}^* = \mathbf{p}^{S^*/O}$, is given by

$$\mathbf{p}^* = \frac{1}{m} \sum_{i=1}^{N} \mathbf{p}_i m_i. \tag{1.2}$$

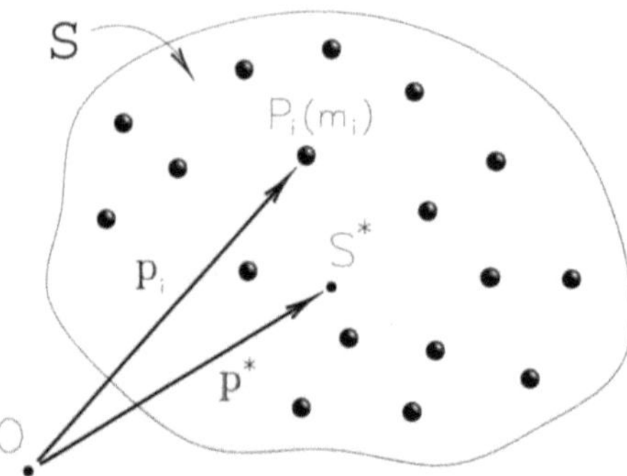

Figure 1.1

The position vector of mass center is, therefore, none other than a vectorial weighted average of the spatial mass distribution of the system.

Example 1.1   Consider the system consisting of four particles P_i, $i = 1, 2, 3, 4$, with the same mass $m_i = 2$ kg and a particle P_5, with mass $m_5 = 8$ kg, with the configuration shown in Fig. 1.2. The mass of the system is $m = 4m_1 + m_5 = 16$ kg and the position of its mass center with respect to the origin of the system of axes is, according to Eq. (1.2),

$$\mathbf{p}^* = \frac{1}{m} \sum_{i=1}^{5} m_i \mathbf{p}_i = \frac{a}{16}(2\mathbf{n}_1 + 2\mathbf{n}_2 - 2\mathbf{n}_1 + 2\mathbf{n}_3 + 0) = \frac{a}{8}(\mathbf{n}_2 + \mathbf{n}_3).$$

Note that the coordinate of the mass center in the direction of axis x_1 is null, due to the symmetry of the system with respect to the plane $x_2\,x_3$.

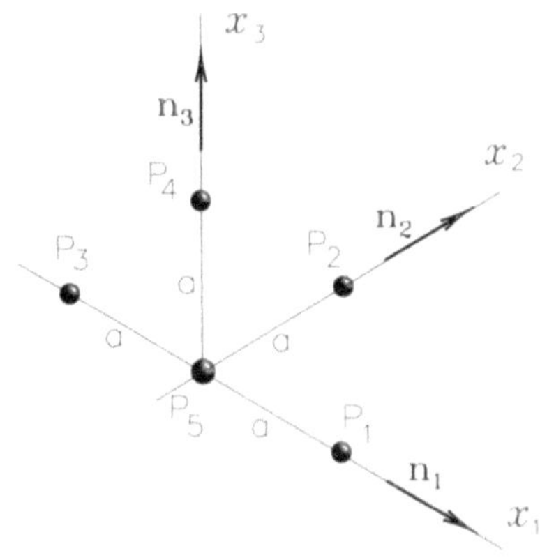

Figure 1.2

Now let C be a continuous body with volume V, inside of which a scalar field called the *density* of the body, $\rho(\mathbf{p})$, where $\mathbf{p}$ is the position vector of a general point of the body with respect to a given point O (see Fig. 1.3), is defined.

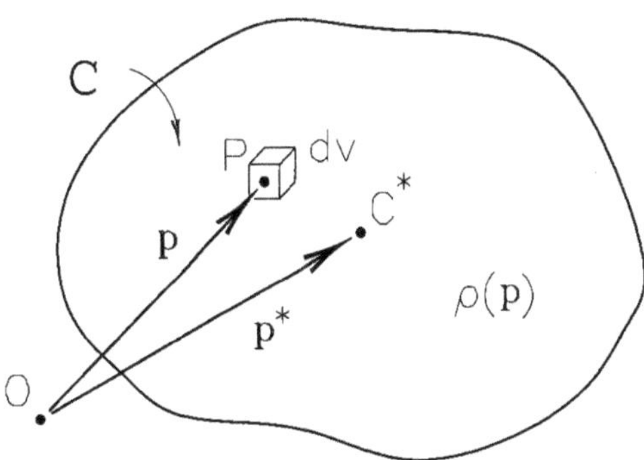

Figure 1.3

The mass of the body is given by integrating this field over the body volume, that is,

$$m = \int_C dm = \int_C \rho \, dV, \tag{1.3}$$

and the mass center of the body is defined as point C*, whose position regarding O, $\mathbf{p}^* = \mathbf{p}^{C^*/O}$, is given by

$$\mathbf{p}^* = \frac{1}{m} \int_C \mathbf{p} \, dm = \frac{1}{m} \int_C \rho \, \mathbf{p} \, dV. \tag{1.4}$$

Equation (1.4) is the continuous version of the weighted average expressed in Eq. (1.2).

When the density of a body C is a uniform field, that is, when ρ does not depend on the position, the body is said to be *homogeneous* and the expression for the position of its mass center is reduced to

$$\mathbf{p}^* = \frac{1}{V} \int_C \mathbf{p}\, dV. \tag{1.5}$$

The position of the mass center of a homogeneous body, as shown in the above expression, is a property solely of the *geometry* of the body, in this case called the *centroid* of the body. The centroids of the most common geometric figures are given in Appendix C.

When the body consists essentially of a line, as in the case of a wire, where only one dimension matters, the ρ, present in Eq. (1.4), is a mass by unit of length and, when the body is also homogeneous, the integration expressed by Eq. (1.5) is performed along this line, with a linear integration element, ds. Section C.1 presents a table of centroids of the most common lines.

When the body is essentially flat, as in the case of a plate, where one dimension does not matter, the ρ, present in Eq. (1.4), is a mass by unit of area and, when the body is also homogeneous, the integration expressed by Eq. (1.5) is performed in a plane region, occupied by the body, with a two-dimensional element of integration, dA. Section C.2 presents a table of centroids of the most common sections or areas.

When a plane cuts a body into two parts so that one is the mirror image of the other, it is said to be a *plane of symmetry*. If the body is homogeneous, this is enough for there to be geometric symmetry; if the body is not homogeneous, the symmetry must necessarily include the mass distribution. When a body has a plane of symmetry, its mass center belongs to the plane. If it has two or more planes of symmetry, its mass center must be in its intersection. In fact, if $\mathbf{n}$ is a unit vector orthogonal to plane π, of symmetry for body C (see Fig. 1.4), the scalar component in direction $\mathbf{n}$ of the position vector of the mass center of the body with respect to a point O, belonging to π, is

$$\mathbf{p}^* \cdot \mathbf{n} = \frac{1}{m} \int_C \mathbf{p} \cdot \mathbf{n}\, dm$$

$$= \frac{1}{m} \int_{C^-} \mathbf{p} \cdot \mathbf{n}\, dm + \frac{1}{m} \int_{C^+} \mathbf{p} \cdot \mathbf{n}\, dm \qquad (1.6)$$

$$= 0.$$

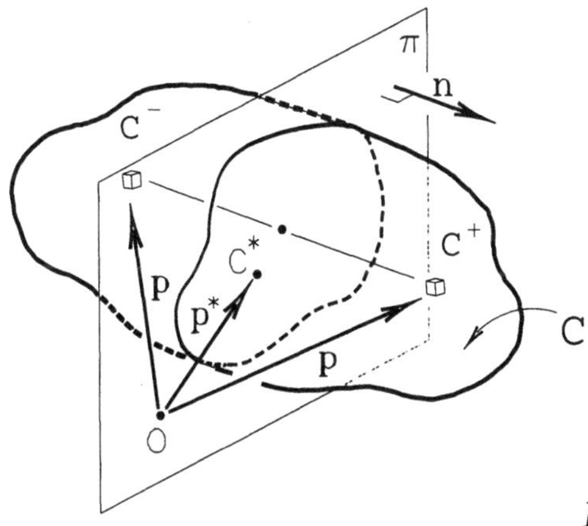

Figure 1.4

Note that the integration was broken down into two regions, C^- and C^+, according to the plane of symmetry. As for each element dm in C^-, with $\mathbf{p} \cdot \mathbf{n} < 0$, there is another identical element in C^+, with $\mathbf{p} \cdot \mathbf{n} > 0$, the integrals in both regions will have the same module and opposite signals, canceling each other out. In other words, Eq. (1.6) guarantees that the mass center C^* belongs to the plane of symmetry.

In the most general case, the determination of the position of the mass center or centroid of a body requires calculating the integral in Eq. (1.4), with the simplification introduced by Eq. (1.6), when there is symmetry.

Example 1.2 A straight circular cone, with height a and base radius R (see Fig. 1.5), has a density varying linearly with elevation z, according to the relationship $\rho = \rho_0(1 + 2z/a)$. The cone mass can be evaluated using Eq. (1.3), adopting a homogeneous element of integration in the shape of a disk, with volume $dV = \pi r^2\, dz = \pi(R/a)^2 z^2\, dz$, that is,

$$m = \pi \rho_0 \frac{R^2}{a^2} \int_0^a \left(1 + 2\frac{z}{a}\right) z^2\, dz = \frac{5}{6} \pi \rho_0 R^2 a.$$

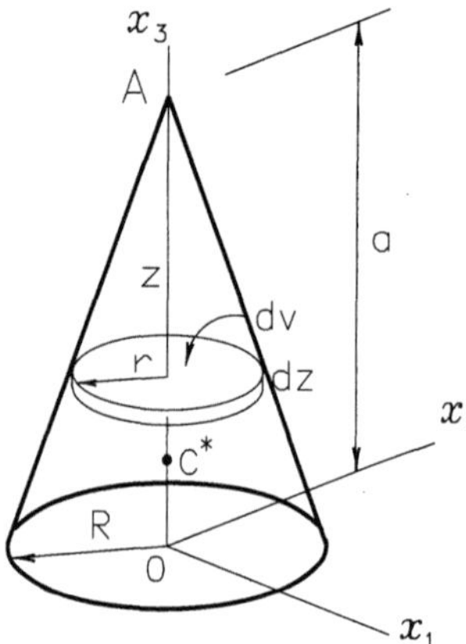

Figure 1.5

As every vertical plane passing through A is of symmetry for the cone, its mass center, C^*, necessarily belongs to the axis of symmetry x_3, that is, if $\mathbf{p}^* = (x_1^*, x_2^*, x_3^*)$ is the position of the mass center of the cone with regard to the origin of the axes, then, by Eq. (1.6), $x_1^* = x_2^* = 0$. The coordinate x_3^* may be obtained from Eq. (1.4), adopting the same element of integration

$$x_3^* = a - \frac{1}{m} \int_0^a \rho z \, dV = a - \frac{\pi \rho_0 R^2}{ma^2} \int_0^a \left(1 + 2\frac{z}{a}\right) z^3 \, dz = \frac{11}{50} a.$$

Note that, as Appendix C indicates, the vertical coordinate of the centroid of the straight circular cone is $x_3^* = a/4$, differing substantially from the vertical coordinate of the mass center, established above.

Establishing the mass center of a body with complex geometry may, sometimes, be simplified by breaking it down into smaller parts whose mass centers are known or easy to obtain; the mass center of the whole body may then be calculated using Eq. (1.2), as if it were a discrete system of particles whose masses are equal to the masses of the parts of the body and whose positions coincide with those of the respective mass centers of these parts. In fact, by arbitrarily breaking a body C, of mass m, down into N parts $C_1, C_2, \ldots, C_N$, of masses $m_1, m_2, \ldots, m_N$, respectively, and calling $\mathbf{p}_i^*$ the position vector, with respect to a given point O, of the mass center of part C_i (see Fig. 1.6), then from Eq. (1.4)

$$\mathbf{p}_i^* = \frac{1}{m_i} \int_{C_i} \mathbf{p} \, dm, \qquad i = 1, 2, \ldots, N. \tag{1.7}$$

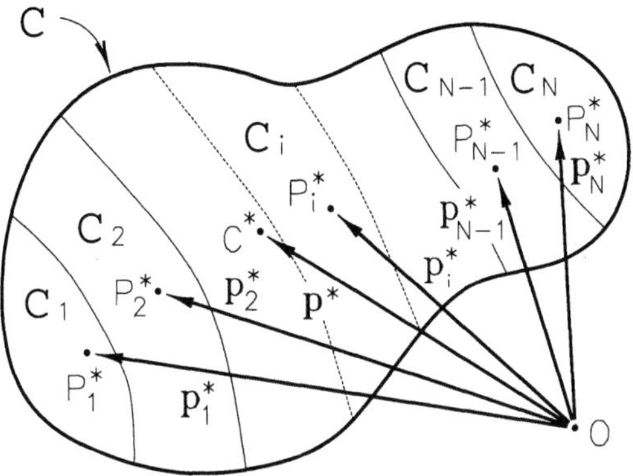

Figure 1.6

Now by multiplying both members of the equation by m_i and adding to i, we have

$$\sum_{i=1}^{N}\mathbf{p}_i^* m_i = \sum_{i=1}^{N}\int_{C_i}\mathbf{p}\,dm = \int_C \mathbf{p}\,dm = m\mathbf{p}^*,$$

which, when divided by $m = m_1 + m_2 + \cdots + m_N$, results in Eq. (1.2).

Example 1.3   Consider the metal piece shown in Fig. 1.7, milled from a homogeneous block, with the indicated dimensions.

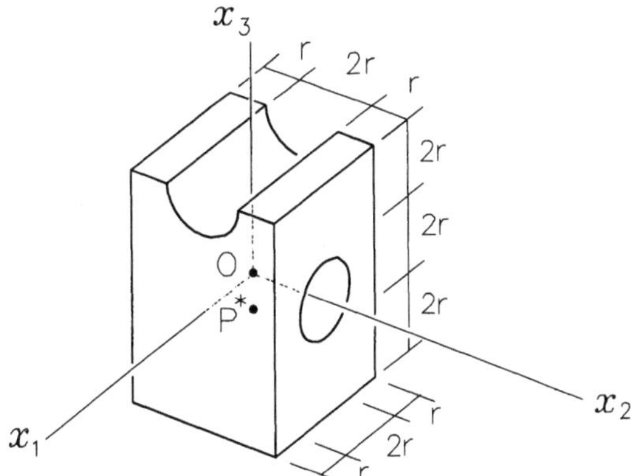

Figure 1.7

To establish the coordinates of its centroid, $\mathrm{P}^*\colon (x_1^*, x_2^*, x_3^*)$, let us first observe that the planes $x_2 x_3$ and $x_3 x_1$ are of symmetry for the piece; therefore,

$$x_1^* = x_2^* = 0.$$

Coordinate x_3^* may be easily obtained by breaking the body down into three elements: C_1, consisting of a parallelpiped with a volume $V_1 = 96r^3$ and centroid P_1^*: $(0,0,0)$; C_2, consisting of a cylinder with a volume $V_2 = -4\pi r^3$ and centroid P_2^*: $(0,0,0)$; and C_3, consisting of a half-cylinder with volume $V_3 = -2\pi r^3$ and centroid P_3^*: $(0,0,z)$, where $z = (3 - 4/3\pi)r$ (see Appendix C). The volume of the body is

$$V = V_1 + V_2 + V_3 = (96 - 6\pi)r^3.$$

As the masses are proportional to the respective volumes, the use of Eq. (1.2) to establish the coordinate x_3^* results in

$$\begin{aligned}
x_3^* &= \frac{1}{V}(V_1 \mathbf{p}_{1_3}^* + V_2 \mathbf{p}_{2_3}^* + V_3 \mathbf{p}_{3_3}^*) \\
&= \frac{1}{(96 - 6\pi)r^3}\left(0 + 0 - 2\pi r^3\left(3 - \frac{4}{3\pi}\right)r\right) \\
&= -0.21r.
\end{aligned}$$

The centroid of the piece is, therefore, P^*: $(0,0,-0.21r)$.

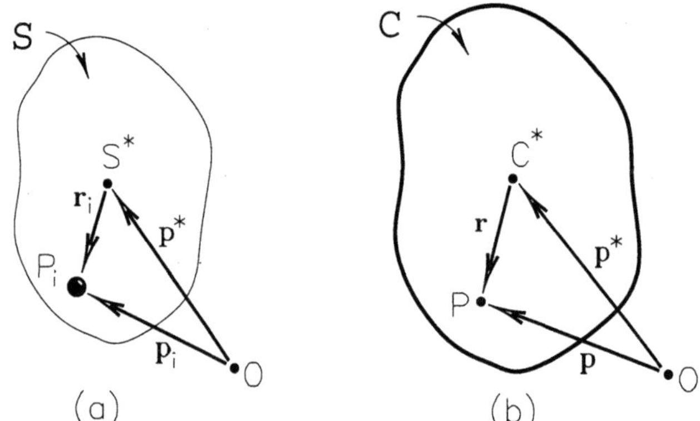

Figure 1.8

    If S^* is the mass center of a system S of particles and $\mathbf{r}_i$ is the position vector of any particle P_i, of mass m_i, with respect to S^* (see Fig. 1.8a), then

$$\sum_{i=1}^{N} \mathbf{r}_i m_i = 0. \tag{1.8}$$

In fact, substituting the relationship $\mathbf{p}_i = \mathbf{p}^* + \mathbf{r}_i$ in Eq. (1.2) gives

$$\mathbf{p}^* = \frac{1}{m} \sum_{i=1}^{N} (\mathbf{p}^* + \mathbf{r}_i) m_i = \mathbf{p}^* + \frac{1}{m} \sum_{i=1}^{N} \mathbf{r}_i m_i,$$

fulfilling Eq. (1.8). This result of particular interest, as will be seen later, establishes the simple geometric fact that the position vector of the mass center of a system with respect to the actual mass center is null.

If C^* is the mass center of a body C and $\mathbf{r}$ is the position vector of an arbitrary point P of C with respect to C^* (see Fig. 1.8b), then

$$\int_C \mathbf{r}\, dm = 0. \tag{1.9}$$

The demonstration is similar to that of the previous equation and will be left to the reader. By differentiating Eqs. (1.8) and (1.9) with respect to time, then

$$\sum_{i=1}^{N} \dot{\mathbf{r}}_i m_i = 0, \tag{1.10}$$

$$\int_C \dot{\mathbf{r}}\, dm = 0, \tag{1.11}$$

and, by differentiating again, then

$$\sum_{i=1}^{N} \ddot{\mathbf{r}}_i m_i = 0, \tag{1.12}$$

$$\int_C \ddot{\mathbf{r}}\, dm = 0. \tag{1.13}$$

Equations (1.8–1.13) were used in the deduction from equations of motion to discrete systems of particles and continuous bodies, studied in Chapter 5. It is mainly thanks to these relationships, which greatly simplify the dynamic equations, that the mass center plays a leading role in the dynamic analysis of systems and bodies.

6.2 Inertia Properties of a Particle

The *mass* is the inertia property of translation of a body; it indicates, as Newton's second law states, the resistance that the body offers to accelerate under the action of a given resultant force. So, when under identical resultant forces, the mass centers of two bodies of different masses will accelerate with different intensities, inversely proportional to the respective masses. The *inertia property of translation* of a body is, therefore, usually a constant scalar property of the body.

We know that the action of a system of forces on a given body may always be reduced to a chosen point, this reduction consisting of a resultant force and a resultant torque with respect to the point. This torque is related, as seen in Chapter 5, to the time rate of the angular momentum of the body with respect to the point. There must be, therefore, an associated *inertia property of rotation* that measures the resistance that the body offers to, say, change its angular velocity. Take another look, for instance, at Example 4.1, Section 5.4. Note that the time rate of the angular momentum vector of rod B with respect to point C, in module, resulted from the product of a constant, $\frac{1}{3}ma^2$, multiplied by $\ddot{\theta}$, which measures the angular acceleration of the rod in the reference frame. This constant is, as seen later in this chapter, the *moment of inertia* of the rod with respect to point C. It is, then, a measure of the inertia of the body rotation around the axis passing through C and orthogonal to the plane of figure (see Fig. 5.4.2).

Nevertheless, what happens is that, depending on the mass distribution of the body (a permanent distribution, when a rigid body), its inertia of rotation for each axis passing through the chosen point will be different. This means that neither a simple scalar — nor a vector — will be sufficient to characterize this property. In fact, the rotational inertia of a body is a *tensorial* property. The inertia tensor of a body with respect to a given point, therefore, collects all information relating to the resistance offered by the body in modifying its state of angular motion, whatever the direction of the resultant torque applied. As in the aforementioned example, the inertia tensor will always have a dimension of mass multiplied by a quadratic distance.

To simplify the approach of the concept of inertia tensor, in this section we will study the inertia properties of a particle; it is easier, once the fundamental concepts are understood, to then generalize for discrete systems and bodies. (If the reader is not familiar with algebraic operations involving tensors and matrices, it is advisable to read Appendix A before proceeding to study this chapter. The appendix summarizes the tensor and matrix algebra required to understand the following material.)

Let P be a particle with mass m and O a given point, not coinciding with P. Also, let $\mathbf{p}$ be the position vector of P with respect to O and let $\mathbf{a}$ and $\mathbf{b}$ be any two adimensional unit vectors; axes E_a and E_b contain point O and are parallel to $\mathbf{a}$ and $\mathbf{b}$, respectively (see Fig. 2.1).

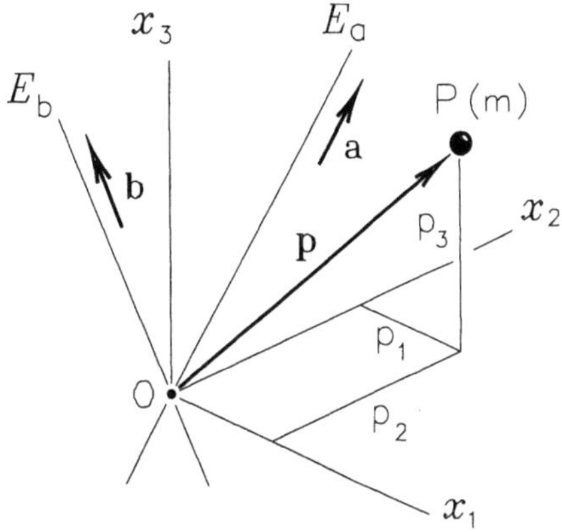

Figure 2.1

The *inertia tensor* of the particle P with respect to point O is defined by

$$\mathbb{I}^{P/O} \rightleftharpoons (p^2\mathbb{1} - \mathbf{p}\otimes\mathbf{p})m, \tag{2.1}$$

where p is the module of vector $\mathbf{p}$, $\mathbb{1}$ is the identity tensor, and $\mathbf{p}\otimes\mathbf{p}$ is the tensorial product of $\mathbf{p}$ by $\mathbf{p}$ (see Appendix A).

Note that the inertia tensor is the resultant of the sum of two symmetric tensors. In fact, $mp^2\mathbb{1}$ is symmetrical since it is a multiple of the identity tensor and $-m\mathbf{p}\otimes\mathbf{p}$ is symmetric by construction. The inertia tensor is, therefore, a *symmetric tensor*.

In the same way as the mass of a particle P is a measure of its inertia of translation, the inertia tensor of a particle with respect to a given point O is a measure of the inertia of P to move *around* point O. This notion will be discussed later.

The *inertia vector* of a particle P with respect to a point O for a given direction **a** is defined as the result of the product of the inertia tensor of P with respect to O with the unitary **a**, that is,

$$\mathbf{I}_a^{P/O} \rightleftharpoons \mathbb{II}^{P/O} \cdot \mathbf{a}. \tag{2.2}$$

As, by definition, $\mathbb{1} \cdot \mathbf{a} = \mathbf{a}$ and $\mathbf{p} \otimes \mathbf{p} \cdot \mathbf{a} = \mathbf{p} \cdot \mathbf{a}\,\mathbf{p}$ (see Appendix A), the inertia vector can be expressed as

$$\mathbf{I}_a^{P/O} = (p^2 \mathbf{a} - \mathbf{p} \cdot \mathbf{a}\,\mathbf{p})m, \tag{2.3}$$

or also by using the identity for the double vector product (see, again, Appendix A),

$$\mathbf{I}_a^{P/O} = \mathbf{p} \times (\mathbf{a} \times \mathbf{p})m. \tag{2.4}$$

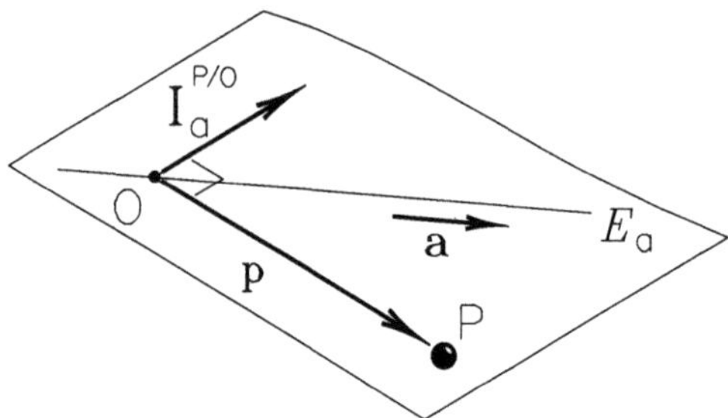

Figure 2.2

Note that $\mathbf{I}_a^{P/O}$ is orthogonal to **p** and parallel to the plane defined by **p** and **a**, not being, however, necessarily parallel to **a** (see Fig. 2.2). Also note that, while the inertia tensor is a function solely of point O (and, of course, of the mass of the particle), the inertia vector also depends on the direction chosen.

Example 2.1 The particle P, of mass m, is distant d from point O, the origin of the system of axes $\{x_1, x_2, x_3\}$, in which the orthonormal basis $\mathbf{n}_1, \mathbf{n}_2, \mathbf{n}_3$ is fixed. In this system, the position vector of P with respect to O is $\mathbf{p} = \frac{d}{\sqrt{2}}(\mathbf{n}_1 + \mathbf{n}_2)$ (see Fig. 2.3). The inertia tensor of P with respect to O is, according to Eq. (2.1),

$$\mathbb{I}^{P/O} = md^2\left(\mathbb{1} - \frac{1}{2}(\mathbf{n}_1 + \mathbf{n}_2)\otimes(\mathbf{n}_1 + \mathbf{n}_2)\right).$$

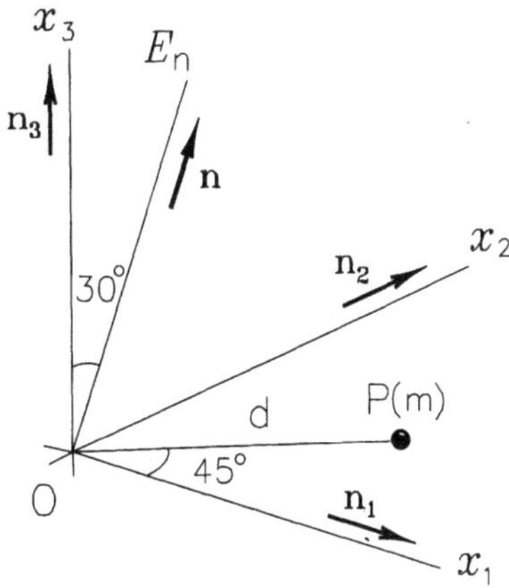

Figure 2.3

The inertia vector of P with respect to O for the direction $\mathbf{n}_1$ is, according to Eq. (2.3),

$$\mathbf{I}_{n_1}^{P/O} = \mathbb{I}^{P/O}\cdot\mathbf{n}_1 = md^2\left(\mathbf{n}_1 - \frac{1}{2}(\mathbf{n}_1 + \mathbf{n}_2)\right) = \frac{1}{2}md^2(\mathbf{n}_1 - \mathbf{n}_2).$$

Likewise, the inertia vectors of P with respect to O for the directions $\mathbf{n}_2$ and $\mathbf{n}_3$ are

$$\mathbf{I}_{n_2}^{P/O} = \mathbb{I}^{P/O}\cdot\mathbf{n}_2 = \frac{1}{2}md^2(\mathbf{n}_2 - \mathbf{n}_1),$$

$$\mathbf{I}_{n_3}^{P/O} = \mathbb{I}^{P/O}\cdot\mathbf{n}_3 = md^2\mathbf{n}_3.$$

The inertia vector of P with respect to O for the direction $\mathbf{n}$ (the axis E_n is on plane x_2x_3) may be obtained from Eq. (2.4):

$$\mathbf{I}_n^{P/O} = \frac{d}{\sqrt{2}}(\mathbf{n}_1 + \mathbf{n}_2)\times\left(\frac{1}{2}(\mathbf{n}_2 + \sqrt{3}\mathbf{n}_3)\times\frac{d}{\sqrt{2}}(\mathbf{n}_1 + \mathbf{n}_2)\right)m$$

$$= \frac{1}{4}md^2(-\mathbf{n}_1 + \mathbf{n}_2 + 2\sqrt{3}\mathbf{n}_3).$$

It is interesting to note that the calculated inertia vectors, with the exception of $\mathbf{I}_{n_3}^{P/O}$, *are not parallel* to the respective unit vectors; the general conditions in which this parallelism may occur will be discussed later in this chapter.

The *moment of inertia* of the particle P with respect to point O for direction $\mathbf{a}$ is defined as the scalar component, in the direction of $\mathbf{a}$, of the inertia vector for the same direction (see Fig. 2.2), that is,

$$I_{aa}^{P/O} \rightleftharpoons \mathbf{I}_a^{P/O} \cdot \mathbf{a}. \tag{2.5}$$

The moment of inertia is a scalar property and may be obtained directly from the inertia tensor of the particle with respect to the point, according to

$$I_{aa}^{P/O} = \mathbf{a} \cdot \mathbf{II}^{P/O} \cdot \mathbf{a}. \tag{2.6}$$

The above relationship results directly from the definitions, Eqs. (2.2) and (2.5), and from the fact that the inertia tensor is symmetric. Returning to Eq. (2.5) and introducing Eq. (2.3), we have

$$\begin{aligned} I_{aa}^{P/O} &= \left(p^2 - (\mathbf{p} \cdot \mathbf{a})^2\right) m \\ &= md^2, \end{aligned} \tag{2.7}$$

where d is the distance between P and the axis E_a, passing through O and parallel to $\mathbf{a}$ (see Fig. 2.4).

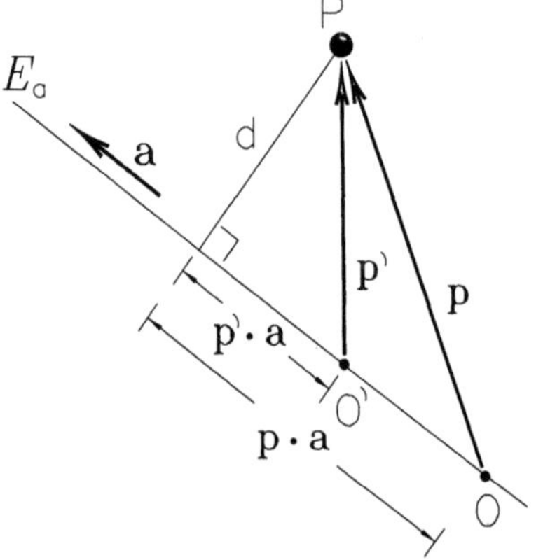

Figure 2.4

Note that the moment of inertia of a particle with respect to a point O for a given direction is a nonnegative scalar ($m > 0$, $d^2 \geq 0$).

Given any other point O' on E_a, the moment of inertia of P with respect to O' for direction $\mathbf{a}$ would be (see Fig. 2.4) $I_{aa}^{P/O'} = \left(p'^2 - (\mathbf{p}' \cdot \mathbf{a})^2 \right) m = md^2$, that is,

$$I_{aa}^{P/O'} = I_{aa}^{P/O} = I_{E_a}^{P},\tag{2.8}$$

whenever O and O' are on the same axis E_a. The result then from Eqs. (2.7) and (2.8) is that $I_{E_a}^{P}$ is the *moment of inertia of* P *with respect to the axis* E_a, being equal to the product of the mass of the particle by its quadratic distance to the axis. Here it is worth mentioning about the adopted notation: The point is indicated in the top index when there is no express reference to the axis; when, on the contrary, the axis is indicated in the bottom index, it is unnecessary to indicate the point, since it is indifferent.

The *product of inertia* of the particle P with respect to the point O, for two arbitrary directions $\mathbf{a}$ and $\mathbf{b}$, is defined as the scalar component, toward $\mathbf{b}$, of the inertia vector of P with respect to O for the direction $\mathbf{a}$ (see Fig. 2.1), that is,

$$I_{ab}^{P/O} \rightleftharpoons \mathbf{I}_{a}^{P/O} \cdot \mathbf{b}.\tag{2.9}$$

Note that, as the inertia tensor is symmetric, $I_{ab}^{P/O} = (\mathbb{I}^{P/O} \cdot \mathbf{a}) \cdot \mathbf{b} = (\mathbb{I}^{P/O} \cdot \mathbf{b}) \cdot \mathbf{a} = I_{ba}^{P/O}$, that is, the product of inertia *is independent* of the order of the unit vectors $\mathbf{a}$ and $\mathbf{b}$, which means

$$I_{ab}^{P/O} = I_{ba}^{P/O}.\tag{2.10}$$

Note also that, except when $\mathbf{a} = \mathbf{b}$ (and, in this case, the product of inertia is reduced to a moment of inertia), the product of inertia of P with respect to another point O' for the directions $\mathbf{a}$ and $\mathbf{b}$ will differ generally from that with respect to O, for the same directions. Now substituting Eq. (2.3) in Eq. (2.9), the following expression for the product of inertia is obtained:

$$I_{ab}^{P/O} = \left(p^2 \mathbf{a} \cdot \mathbf{b} - (\mathbf{p} \cdot \mathbf{a})(\mathbf{p} \cdot \mathbf{b}) \right) m.\tag{2.11}$$

Depending, therefore, on the directions of the unit vectors $\mathbf{a}$ and $\mathbf{b}$, the product of inertia may be positive, negative, or null.

We should note, lastly, that inertia tensors, inertia vectors, moments of inertia, and products of inertia of a particle with respect to a point and for any directions are all properties with the same physical dimension, $[ML^2]$.

Example 2.2   Returning to the previous example (see Fig. 2.3), the moment of inertia of P with respect to the axis x_1 is, according to Eq. (2.5),

$$I_{x_1}^P = \frac{1}{2}md^2(\mathbf{n}_1 - \mathbf{n}_2)\cdot\mathbf{n}_1 = \frac{1}{2}md^2.$$

Likewise, the moment of inertia of P with respect to the axis E_n is

$$I_{E_n}^P = \frac{1}{4}md^2(-\mathbf{n}_1 + \mathbf{n}_2 + 2\sqrt{3}\mathbf{n}_3)\cdot\frac{1}{2}(\mathbf{n}_2 + \sqrt{3}\mathbf{n}_3) = \frac{7}{8}md^2.$$

Using, as an alternative, Eq. (2.7), we have

$$I_{E_n}^P = \left(p^2 - (\mathbf{p}\cdot\mathbf{n})^2\right)m = \frac{7}{8}md^2.$$

The product of inertia of P with respect to O for directions $\mathbf{n}$ and $\mathbf{n}_3$ may be obtained using Eq. (2.11), by

$$I_{nn_3}^{P/O} = \left(p^2\mathbf{n}\cdot\mathbf{n}_3 - (\mathbf{p}\cdot\mathbf{n})(\mathbf{p}\cdot\mathbf{n}_3)\right)m = \left(d^2\frac{\sqrt{3}}{2} - 0\right)m = \frac{\sqrt{3}}{2}md^2.$$

Alternatively, since the inertia vector for direction $\mathbf{n}$ has already been calculated, the same product of inertia may be obtained using Eq. (2.9),

$$I_{nn_3}^{P/O} = \frac{1}{4}md^2(-\mathbf{n}_1 + \mathbf{n}_2 + 2\sqrt{3}\mathbf{n}_3)\cdot\mathbf{n}_3 = \frac{\sqrt{3}}{2}md^2.$$

Given a particle P, of mass m, and a system of Cartesian axes $\{x_1, x_2, x_3\}$ with origin O, the moment of inertia of P with respect to the coordinate axis x_j is, according to Eq. (2.7),

$$I_{jj}^{P/O} = I_{x_j}^P = (p^2 - p_j^2)m, \qquad j = 1, 2, 3, \tag{2.12}$$

where $p_j = \mathbf{p}\cdot\mathbf{n}_j$ is the scalar component of the position vector of P with regard to O in direction $\mathbf{n}_j$ (see Fig. 2.5). Therefore, recalling that $p^2 = p_1^2 + p_2^2 + p_3^2$, the moments of inertia with respect to the three coordinate axes will be

$$\begin{aligned}
I_{11}^{P/O} &= (p_2^2 + p_3^2)m, \\
I_{22}^{P/O} &= (p_3^2 + p_1^2)m, \\
I_{33}^{P/O} &= (p_1^2 + p_2^2)m.
\end{aligned} \tag{2.13}$$

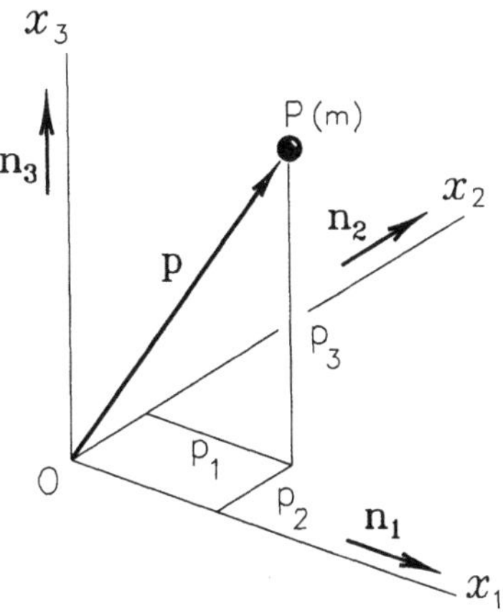

Figure 2.5

The product of inertia of P with respect to O for coordinate directions x_j and x_k is, according to Eq. (2.11),

$$I_{jk}^{P/O} = \left(p^2 \mathbf{n}_j \cdot \mathbf{n}_k - (\mathbf{p} \cdot \mathbf{n}_j)(\mathbf{p} \cdot \mathbf{n}_k)\right) m. \tag{2.14}$$

But $\mathbf{n}_j \cdot \mathbf{n}_k = 0$, for $j \neq k$, and $\mathbf{p} \cdot \mathbf{n}_j = p_j$. Of course, if $j = k$, we fall back on Eq. (2.12). Therefore, the products of inertia for the coordinate directions are

$$I_{12}^{P/O} = I_{21}^{P/O} = -p_1 p_2 \, m,$$
$$I_{23}^{P/O} = I_{32}^{P/O} = -p_2 p_3 \, m, \tag{2.15}$$
$$I_{31}^{P/O} = I_{13}^{P/O} = -p_3 p_1 \, m.$$

The inertia tensor of a particle P with respect to a point O may, in a system of Cartesian coordinates, be expressed by a symmetric matrix called the *inertia matrix*, so that

$$\mathbf{II}^{P/O} = \begin{pmatrix} I_{11} & I_{12} & I_{13} \\ I_{21} & I_{22} & I_{23} \\ I_{31} & I_{32} & I_{33} \end{pmatrix}, \tag{2.16}$$

where components I_{jj} and I_{jk}, $j, k = 1, 2, 3$, are the moments and products of inertia of P with respect to O for the coordinate directions, respectively, in a reduced notation. In fact, Eq. (2.6) provides

$$I_{jj}^{P/O} = \mathbf{n}_j \cdot \mathbf{II}^{P/O} \cdot \mathbf{n}_j, \qquad j = 1, 2, 3, \tag{2.17}$$

and, likewise, from Eqs. (2.2) and (2.9), we have

$$I_{jk}^{P/O} = \mathbf{n}_j \cdot \mathbf{II}^{P/O} \cdot \mathbf{n}_k, \qquad j, k = 1, 2, 3. \tag{2.18}$$

Example 2.3   Returning again to Example 2.1 (see Fig. 2.3), the moments of inertia of P with respect to the coordinate axes, according to Eq. (2.13), are

$$I_{11}^{P/O} = \frac{1}{2}md^2, \quad I_{22}^{P/O} = \frac{1}{2}md^2, \quad I_{33}^{P/O} = md^2.$$

Note that the moment of inertia with respect to axis x_3 is equal to the sum of the two other moments of inertia (why?). The products of inertia with respect to the coordinate axes, according to Eq. (2.15), are

$$I_{12}^{P/O} = -\frac{d}{\sqrt{2}}\frac{d}{\sqrt{2}}m = -\frac{1}{2}md^2, \quad I_{23}^{P/O} = I_{31}^{P/O} = 0.$$

Note that all products of inertia involving the direction x_3 (there are four of them) are null, since vector **p** does not have a component in this direction. The inertia tensor of P with respect to O may then be expressed in the system of Cartesian coordinates of the figure by the matrix

$$\mathbf{II}^{P/O} = \frac{1}{2}md^2 \begin{pmatrix} 1 & -1 & 0 \\ -1 & 1 & 0 \\ 0 & 0 & 2 \end{pmatrix}.$$

The moment of inertia with respect to axis E_n may then be calculated by

$$I_{E_n}^P = \left(0, \frac{1}{2}, \frac{\sqrt{3}}{2}\right) \cdot \frac{1}{2}md^2 \begin{pmatrix} 1 & -1 & 0 \\ -1 & 1 & 0 \\ 0 & 0 & 2 \end{pmatrix} \cdot \begin{pmatrix} 0 \\ \frac{1}{2} \\ \frac{\sqrt{3}}{2} \end{pmatrix} = \frac{7}{8}md^2.$$

The product of inertia for directions **n** and $\mathbf{n}_3$ may be calculated by

$$I_{nn_3}^{P/O} = \left(0, \frac{1}{2}, \frac{\sqrt{3}}{2}\right) \cdot \frac{1}{2}md^2 \begin{pmatrix} 1 & -1 & 0 \\ -1 & 1 & 0 \\ 0 & 0 & 2 \end{pmatrix} \cdot \begin{pmatrix} 0 \\ 0 \\ 1 \end{pmatrix} = \frac{\sqrt{3}}{2}md^2,$$

as was obtained in Example 2.2.

6.3 Inertia Properties of Systems and Bodies

The studies of inertia properties in the preceding section may easily be extended to discrete or continuous systems of particles, as will be seen below.

Let us first consider S as a system consisting of N particles. Let P_i be a general particle of S, of mass m_i, whose position with respect to a given point O is described by vector $\mathbf{p}_i$ (see Fig. 3.1).

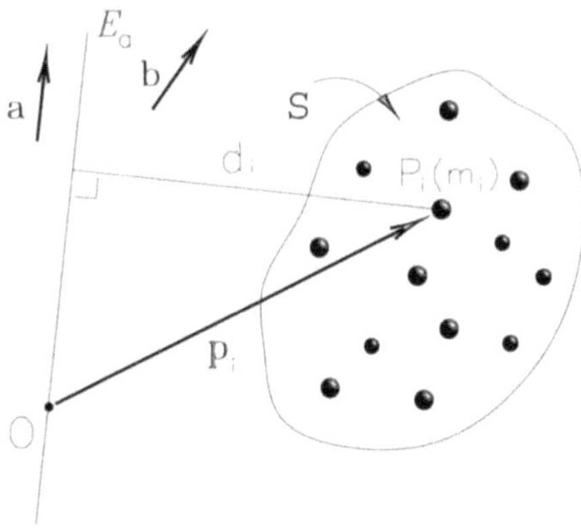

Figure 3.1

The *inertia tensor* of S with respect to point O is defined as the sum of the inertia tensors of the particles belonging to S with respect to the point, that is,

$$\mathbb{II}^{S/O} \rightleftharpoons \sum_{i=1}^{N} (p_i^2 \mathbb{1} - \mathbf{p}_i \otimes \mathbf{p}_i) m_i. \tag{3.1}$$

Being the resultant of the sum of N symmetric tensors, the inertia tensor of a system S is also a symmetric tensor.

Now consider a continuous body C, with volume V and mass m, and let P be a general point of C, whose position with respect to a given point O is described by the position vector $\mathbf{p}$ (see Fig. 3.2).

The *inertia tensor* of C with respect to point O is defined as the integral in the body of the inertia tensor of an infinitesimal element of mass dm with respect to the point, that is,

$$\mathbb{II}^{C/O} \rightleftharpoons \int_C (p^2 \mathbb{1} - \mathbf{p} \otimes \mathbf{p}) \, dm. \tag{3.2}$$

It is easy to see that, as in the way of a discrete system of particles, the inertia tensor of a body C with respect to a point O is also a symmetric tensor.

The *inertia vector* of S or of C with respect to point O for direction $\mathbf{a}$ is defined as the product of the inertia tensor of S or of C with respect to O by unit vector $\mathbf{a}$ (see Figs. 3.1 and 3.2), that is,

$$\mathbf{I}_a^{S/O} \rightleftharpoons \mathbb{II}^{S/O} \cdot \mathbf{a}, \tag{3.3}$$

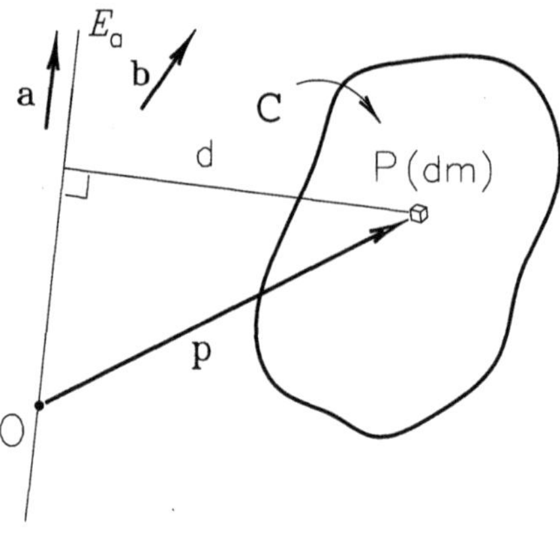

Figure 3.2

$$\mathbf{I}_a^{C/O} \rightleftharpoons \mathbb{II}^{C/O} \cdot \mathbf{a}. \tag{3.4}$$

From the product, we obtain, in a similar way of Eq. (2.3),

$$\mathbf{I}_a^{S/O} = \sum_{i=1}^{N} (p_i^2 \mathbf{a} - \mathbf{p}_i \cdot \mathbf{a}\,\mathbf{p}_i)\, m_i, \tag{3.5}$$

$$\mathbf{I}_a^{C/O} = \int_C (p^2 \mathbf{a} - \mathbf{p} \cdot \mathbf{a}\,\mathbf{p})\, dm, \tag{3.6}$$

or, also using the relationship for the double vector product, we have the alternative expressions

$$\mathbf{I}_a^{S/O} = \sum_{i=1}^{N} \mathbf{p}_i \times (\mathbf{a} \times \mathbf{p}_i) m_i, \tag{3.7}$$

$$\mathbf{I}_a^{C/O} = \int_C \mathbf{p} \times (\mathbf{a} \times \mathbf{p})\, dm. \tag{3.8}$$

The inertia vector of a system of particles with respect to a given point for a certain direction is, therefore, the sum of the inertia vectors of the particles of the system with regard to that point and for that direction. Likewise, the inertia vector of a body C with respect to a point for a given direction is the result of the integration in the body of the respective inertia vector of its mass element. Although each component of the sum or integral is orthogonal to the respective position vector, the resultant vector will have a general direction in space.

Example 3.1 Consider a system S consisting of particles P_1, P_2, P_3, and P_4, each of mass m, and P_5, of mass $2m$, arranged as shown in Fig. 3.3. The inertia vector of S with respect to P_5 for the direction of the unit vector $\mathbf{n}$, according to Eq. (3.5), is

$$\mathbf{I}_n^{S/P_5} = \sum_{i=1}^{5} (p_i^2 \mathbf{n} - \mathbf{p}_i \cdot \mathbf{n}\,\mathbf{p}_i)\, m_i$$

$$= 4ma^2 \left(\frac{1}{2}\mathbf{n}_2 + \frac{\sqrt{3}}{2}\mathbf{n}_3\right) - \left(\frac{1}{2}a^2\mathbf{n}_2 + \frac{\sqrt{3}}{2}a^2\mathbf{n}_3\right) m$$

$$= \frac{3}{2}ma^2(\mathbf{n}_2 + \sqrt{3}\,\mathbf{n}_3).$$

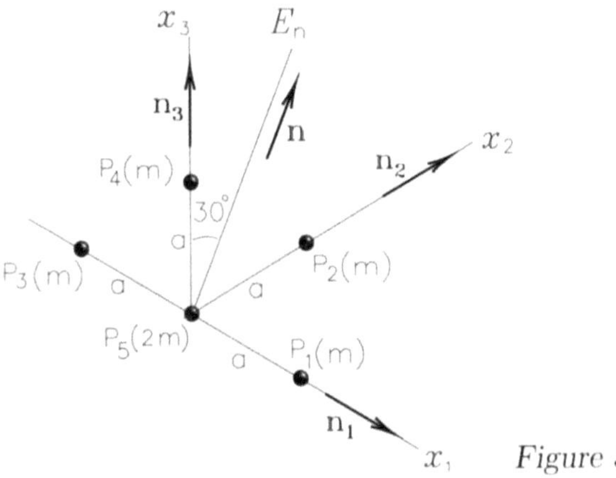

Figure 3.3

Note that only particles P_1 to P_4 contribute to the term on the left of the expression of the inertia vector, $(p_i^2\mathbf{n})$, since $p_5 = 0$. Particles P_5, P_1, and P_3 do not contribute to the term on the right, whose position vectors do not have a component in the direction $\mathbf{n}$. Also note that the inertia vector obtained is parallel to the corresponding unit vector.

Example 3.2 The inertia tensor of the homogeneous disk D, of mass m and radius R, with respect to its center O (see Fig. 3.4), according to Eq. (3.2), is

$$\mathbb{I}^{D/O} = \int_D (p^2 \mathbb{1} - \mathbf{p} \otimes \mathbf{p})\, dm.$$

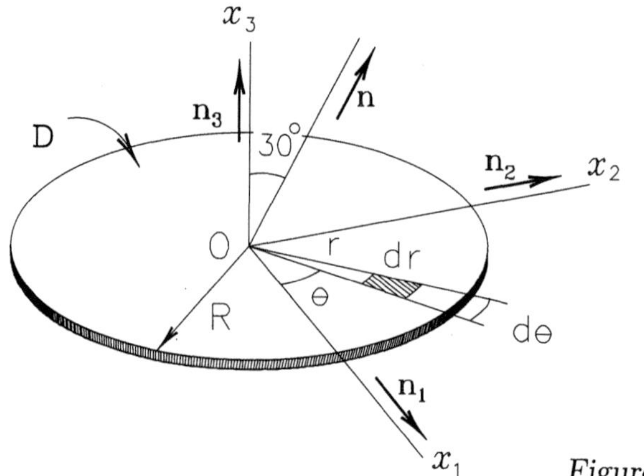

Figure 3.4

Taking the integration element, with mass $dm = (m/\pi R^2)r\,dr\,d\theta$, and the position vector $\mathbf{p} = r(\cos\theta\,\mathbf{n}_1 + \sin\theta\,\mathbf{n}_2)$, with $p^2 = r^2$ and remembering that

$$\mathbf{p} \otimes \mathbf{p} = r^2\big[\cos^2\theta(\mathbf{n}_1 \otimes \mathbf{n}_1) + \sin^2\theta(\mathbf{n}_2 \otimes \mathbf{n}_2)$$
$$+ \cos\theta\sin\theta(\mathbf{n}_1 \otimes \mathbf{n}_2) + \sin\theta\cos\theta(\mathbf{n}_2 \otimes \mathbf{n}_1)\big],$$

we have

$$\mathbf{II}^{D/O} = \frac{m}{\pi R^2}\int_0^{2\pi}\int_0^{R}\left(r^2\mathbb{1} - \mathbf{p}\otimes\mathbf{p}\right)r\,dr\,d\theta$$

$$= \frac{m}{\pi R^2}\int_0^{R} r^3\,dr\int_0^{2\pi}\Big[\mathbb{1} - \big(\cos^2\theta\,\mathbf{n}_1\otimes\mathbf{n}_1 + \sin^2\theta\,\mathbf{n}_2\otimes\mathbf{n}_2$$

$$+ \cos\theta\sin\theta\,(\mathbf{n}_1\otimes\mathbf{n}_2 + \mathbf{n}_2\otimes\mathbf{n}_1)\big)\Big]\,d\theta$$

$$= \frac{1}{4}mR^2(2\mathbb{1} - \mathbf{n}_1\otimes\mathbf{n}_1 - \mathbf{n}_2\otimes\mathbf{n}_2).$$

The inertia vector of the disk with respect to point O for the direction of $\mathbf{n} = \frac{1}{2}(\mathbf{n}_2 + \sqrt{3}\,\mathbf{n}_3)$ is, according to Eq. (3.4),

$$\mathbf{I}_n^{D/O} = \mathbf{II}^{D/O}\cdot\mathbf{n}$$

$$= \frac{1}{4}mR^2\left[(\mathbf{n}_2 + \sqrt{3}\,\mathbf{n}_3) - 0 - \frac{1}{2}\mathbf{n}_2\right]$$

$$= \frac{1}{8}mR^2(\mathbf{n}_2 + 2\sqrt{3}\,\mathbf{n}_3).$$

The inertia vector for the direction $\mathbf{n}_1$ is

$$\mathbf{I}_{n_1}^{D/O} = \mathbf{II}^{D/O}\cdot\mathbf{n}_1 = \frac{1}{4}mR^2(2\mathbf{n}_1 - \mathbf{n}_1 - 0) = \frac{1}{4}mR^2\mathbf{n}_1,$$

and the inertia vector for the direction $\mathbf{n}_3$ is

$$\mathbf{I}_{n_3}^{D/O} = \mathbf{II}^{D/O} \cdot \mathbf{n}_3 = \frac{1}{4}mR^2(2\mathbf{n}_3 - 0 - 0) = \frac{1}{2}mR^2\mathbf{n}_3.$$

It is interesting to note that the inertia vectors with respect to point O for the directions $\mathbf{n}_1$ and $\mathbf{n}_3$ are parallel to the respective directions. Try to calculate the inertia vector for direction $\mathbf{n}_2$. What did you find? Note that the inertia vector associated with direction $\mathbf{n}$ is *not* parallel to $\mathbf{n}$.

The projection of the inertia vector of a system S or body C with respect to a point O for the direction $\mathbf{a}$ in the same direction is the *moment of inertia* of the system or body with respect to the point for that direction, that is,

$$I_{aa}^{S/O} \rightleftharpoons \mathbf{I}_{a}^{S/O} \cdot \mathbf{a} = \mathbf{a} \cdot \mathbf{II}^{S/O} \cdot \mathbf{a}, \tag{3.9}$$

$$I_{aa}^{C/O} \rightleftharpoons \mathbf{I}_{a}^{C/O} \cdot \mathbf{a} = \mathbf{a} \cdot \mathbf{II}^{C/O} \cdot \mathbf{a}. \tag{3.10}$$

Replacing Eq. (3.5) in Eq. (3.9) gives

$$\begin{aligned}
I_{aa}^{S/O} &= \sum_{i=1}^{N} \left(p_i^2 - (\mathbf{p}_i \cdot \mathbf{a})^2 \right) m_i \\
&= \sum_{i=1}^{N} m_i d_i^2,
\end{aligned} \tag{3.11}$$

and, likewise, replacing Eq. (3.6) in Eq. (3.10),

$$\begin{aligned}
I_{aa}^{C/O} &= \int_{C} \left(p^2 - (\mathbf{p} \cdot \mathbf{a})^2 \right) dm \\
&= \int_{C} d^2\, dm.
\end{aligned} \tag{3.12}$$

The moment of inertia of a system S (a body C) with respect to a point O for a direction $\mathbf{a}$ is, therefore, a nonnegative scalar, equal to the sum of the products of the masses of the particles (of the mass elements) by their respective quadratic distances to the axis containing the point, and is parallel to unit vector $\mathbf{a}$ (see Figs. 3.1 and 3.2). Equations (3.11) and (3.12) are also evidence that the moment of inertia of a system or body depends solely on its mass distribution around the axis containing the point and parallel to the given direction. The result is that, in the same way as with the moment of inertia for a particle, the moments of inertia of a system or body with respect to two different points of the same axis are equal. That is why we may call them *moment of inertia with respect to the axis*, that is,

$$I^S_{E_a} = I^{S/O}_{aa}, \qquad \text{O belonging to } E_a, \tag{3.13}$$

$$I^C_{E_a} = I^{C/O}_{aa}, \qquad \text{O belonging to } E_a. \tag{3.14}$$

The projection of the inertia vector of a system S or a body C, with respect to a given point O for a direction $\mathbf{a}$, over another different direction, defined by the unit vector $\mathbf{b}$, is the *product of inertia* of S or C with respect to O for the directions $\mathbf{a}$ and $\mathbf{b}$, that is,

$$I^{S/O}_{ab} \rightleftharpoons \mathbf{I}^{S/O}_{a} \cdot \mathbf{b} = \mathbf{a} \cdot \mathbf{II}^{S/O} \cdot \mathbf{b}, \tag{3.15}$$

$$I^{C/O}_{ab} \rightleftharpoons \mathbf{I}^{C/O}_{a} \cdot \mathbf{b} = \mathbf{a} \cdot \mathbf{II}^{C/O} \cdot \mathbf{b}. \tag{3.16}$$

Replacing Eq. (3.5) in Eq. (3.15) and Eq. (3.6) in Eq. (3.16), we have

$$I^{S/O}_{ab} = \sum_{i=1}^{N} \left(p_i^2 \mathbf{a} \cdot \mathbf{b} - (\mathbf{p}_i \cdot \mathbf{a})(\mathbf{p}_i \cdot \mathbf{b}) \right) m_i, \tag{3.17}$$

$$I^{C/O}_{ab} = \int_C \left(p^2 \mathbf{a} \cdot \mathbf{b} - (\mathbf{p} \cdot \mathbf{a})(\mathbf{p} \cdot \mathbf{b}) \right) dm. \tag{3.18}$$

The above expressions show that the product of inertia of a system or body with respect to a point for two given directions is a scalar equal to the algebraic sum of the corresponding products of inertia of the particles or mass elements involved. It is also important to note that, given the symmetry of the inertia tensor, the order of unit vectors in the composition of the products of inertia is irrelevant, that is,

$$I^{S/O}_{ab} = \mathbf{a} \cdot \mathbf{II}^{S/O} \cdot \mathbf{b} = \mathbf{b} \cdot \mathbf{II}^{S/O} \cdot \mathbf{a} = I^{S/O}_{ba}, \tag{3.19}$$

$$I^{C/O}_{ab} = \mathbf{a} \cdot \mathbf{II}^{C/O} \cdot \mathbf{b} = \mathbf{b} \cdot \mathbf{II}^{C/O} \cdot \mathbf{a} = I^{C/O}_{ba}. \tag{3.20}$$

Example 3.3 Returning to the system studied in Example 3.1 (see Fig. 3.3), the moment of inertia of S with respect to axis E_n, passing through the origin and parallel to $\mathbf{n}$, may be obtained from Eq. (3.11),

$$I_{E_n}^S = \sum_{i=1}^{5} m_i d_i^2 = a^2 m + \frac{3}{4} a^2 m + a^2 m + \frac{1}{4} a^2 m + 0 = 3ma^2.$$

The product of inertia of S with respect to the origin for directions $\mathbf{n}$ and $\mathbf{n}_2$ may be calculated from Eq. (3.17),

$$I_{nn_2}^{S/P_5} = \mathbf{I}_n^{S/P_5} \cdot \mathbf{n}_2 = \frac{3}{2} ma^2 (\mathbf{n}_2 + \sqrt{3}\,\mathbf{n}_3) \cdot \mathbf{n}_2 = \frac{3}{2} ma^2.$$

The product of inertia for directions $\mathbf{n}$ and $\mathbf{n}_1$ results in

$$I_{nn_1}^{S/P_5} = \frac{3}{2} ma^2 (\mathbf{n}_2 + \sqrt{3}\,\mathbf{n}_3) \cdot \mathbf{n}_1 = 0.$$

This last result is due to the symmetry of the system with respect to the plane $x_2\, x_3$, as seen ahead.

Example 3.4 Returning to Example 3.2, the moment of inertia of the disk with respect to point O for direction $\mathbf{n}$ (see Fig. 3.4), according to Eq. (3.10), is

$$I_{nn}^{D/O} = \mathbf{I}_n^{D/O} \cdot \mathbf{n} = \frac{1}{8} mR^2 (\mathbf{n}_2 + 2\sqrt{3}\,\mathbf{n}_3) \cdot \frac{1}{2} (\mathbf{n}_2 + \sqrt{3}\,\mathbf{n}_3) = \frac{7}{16} mR^2.$$

The moments of inertia of the disk with respect to axes x_3 and x_1 may also be obtained from Eq. (3.10), considering what is expressed by Eq. (3.14):

$$I_{x_3}^D = \mathbf{I}_{n_3}^{D/O} \cdot \mathbf{n}_3 = \frac{1}{2} mR^2;$$

$$I_{x_1}^D = \mathbf{I}_{n_1}^{D/O} \cdot \mathbf{n}_1 = \frac{1}{4} mR^2.$$

The product of inertia of the disk with respect to point O for directions $\mathbf{n}$ and $\mathbf{n}_3$, according to Eq. (3.16), is

$$I_{nn_3}^{D/O} = \mathbf{I}_n^{D/O} \cdot \mathbf{n}_3 = \frac{1}{8} mR^2 (\mathbf{n}_2 + 2\sqrt{3}\,\mathbf{n}_3) \cdot \mathbf{n}_3 = \frac{\sqrt{3}}{4} mR^2.$$

As Eq. (3.20) shows, this product of inertia may otherwise be calculated by

$$I_{n_3 n}^{D/O} = \mathbf{I}_{n_3}^{D/O} \cdot \mathbf{n} = \frac{1}{2} mR^2 \mathbf{n}_3 \cdot \frac{1}{2} (\mathbf{n}_2 + \sqrt{3}\,\mathbf{n}_3) = \frac{\sqrt{3}}{4} mR^2.$$

It is also easy to see that

$$I_{n_1 n_3}^{D/O} = I_{n_3 n_1}^{D/O} = 0.$$

Try calculating the products of inertia $I_{n_1 n_2}^{D/O}$ and $I_{n_2 n_3}^{D/O}$. What conclusion does the reader draw from the results?

When a system S of particles has a plane of symmetry, the inertia vector of S with respect to a point of the plane for the direction orthogonal to this plane is parallel to this direction and its module is the moment of inertia of S with respect to the axis orthogonal to the plane containing the point. Moreover, every product of inertia of S with respect to a point of the plane for two directions, one being orthogonal to the plane and the other parallel to that one, will be null. In fact, if π is a plane of symmetry for S, O is any point of π, $\mathbf{n}$ is a unit vector orthogonal to the plane, and E is the axis parallel to $\mathbf{n}$ passing through O (see Fig. 3.5), for every particle P_i^+, with mass m_i, on one side of the plane, there will be another one, P_i^-, of the same mass, satisfying the symmetry condition. If point Q_i is the orthogonal projection of P_i^+ and P_i^- on π, and $\mathbf{r}_i$ is the position vector of Q_i with respect to O, the position vectors of the particles in question can be expressed by $\mathbf{p}_i^+ = \mathbf{r}_i + \lambda\mathbf{n}$ and $\mathbf{p}_i^- = \mathbf{r}_i - \lambda\mathbf{n}$, where $\lambda = \mathbf{p}_i^+ \cdot \mathbf{n} \geq 0$. So, $\mathbf{n} \times \mathbf{p}_i^+ = \mathbf{n} \times \mathbf{p}_i^- = \mathbf{n} \times \mathbf{r}_i$.

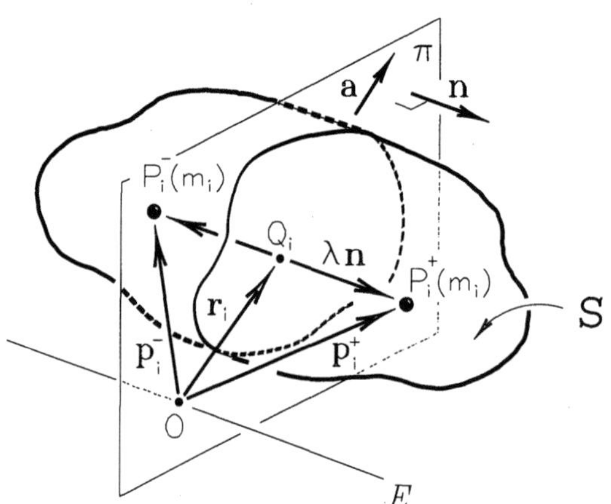

Figure 3.5

Now calculating the double products:

$$\mathbf{p}_i^+ \times (\mathbf{n} \times \mathbf{p}_i^+) = \mathbf{r}_i \times (\mathbf{n} \times \mathbf{r}_i) + \lambda\mathbf{n} \times (\mathbf{n} \times \mathbf{r}_i);$$
$$\mathbf{p}_i^- \times (\mathbf{n} \times \mathbf{p}_i^-) = \mathbf{r}_i \times (\mathbf{n} \times \mathbf{r}_i) - \lambda\mathbf{n} \times (\mathbf{n} \times \mathbf{r}_i),$$

and adding to the pair of particles, we have $2\mathbf{r}_i \times (\mathbf{n} \times \mathbf{r}_i)$, a vector parallel to $\mathbf{n}$ (and in the same direction as the latter), since $\mathbf{r}_i$ and $\mathbf{n}$ are orthogonal. Of course, since a symmetric system with respect to plane π consists of pairs of opposite particles, the vector sum in Eq. (3.7) must necessarily be a vector with the same direction as $\mathbf{n}$. In other words, the inertia vector will, in fact, be orthogonal to the plane. Now, since the moment of inertia is, by definition, the projection of the inertia vector in the direction of the unit vector, then, of course,

$$\mathbf{I}_n^{S/O} = I_{nn}^{S/O}\mathbf{n} \tag{3.21}$$

if O belongs to the symmetry plane and $\mathbf{n}$ is orthogonal to the plane. In turn, the products of inertia relative to direction $\mathbf{n}$ and any other parallel to the plane of symmetry will be null since the projection of the inertia vector in any direction orthogonal to $\mathbf{n}$ will be null (see Fig. 3.5), that is,

$$I_{an}^{S/O} = I_{na}^{S/O} = 0 \tag{3.22}$$

whenever O is on the plane of symmetry, with $\mathbf{a}$ parallel and $\mathbf{n}$ orthogonal to the plane. As we will see in Section 6.6, whenever a system admits a plane of symmetry, the direction orthogonal to the plane will be called a *principal direction of inertia*. The inertia vector associated with such direction is, therefore, parallel to it and its module, the moment of inertia for that direction, is called the *principal moment of inertia*.

Example 3.5 Returning to the system S in Example 3.1 (see Fig. 3.3), it is easy to see that plane $x_2\,x_3$ is a plane of symmetry for S. The inertia vector with regard to the origin for direction $\mathbf{n}_1$ is

$$\mathbf{I}_{n_1}^{S/P_5} = \sum_{i=1}^{5}(p_i^2\mathbf{n}_1 - \mathbf{p}_i \cdot \mathbf{n}_1\,\mathbf{p}_i)m_i = 4a^2 m\mathbf{n}_1 - 2a^2 m\mathbf{n}_1 = 2ma^2\mathbf{n}_1,$$

a parallel vector, therefore, to $\mathbf{n}_1$ and whose module is equal to the moment of inertia of S with respect to axis x_1,

$$I_{x_1}^{S} = \mathbf{I}_{n_1}^{S/P_5} \cdot \mathbf{n}_1 = 2ma^2.$$

On the other hand, every product of inertia involving the direction $\mathbf{n}_1$ and any other direction $\mathbf{a}$, parallel to plane $x_2\,x_3$, will be null, since $\mathbf{n}_1 \cdot \mathbf{a} = 0$.

When a continuous body C admits a plane of symmetry, the inertia vector of C with respect to a point O of the plane for the direction $\mathbf{n}$, orthogonal to the plane (see Fig. 3.6), is parallel to $\mathbf{n}$ and its module is the moment of inertia of C with respect to an axis E parallel to $\mathbf{n}$ passing through O, that is,

$$\mathbf{I}_n^{C/O} = I_E^C \mathbf{n}. \tag{3.23}$$

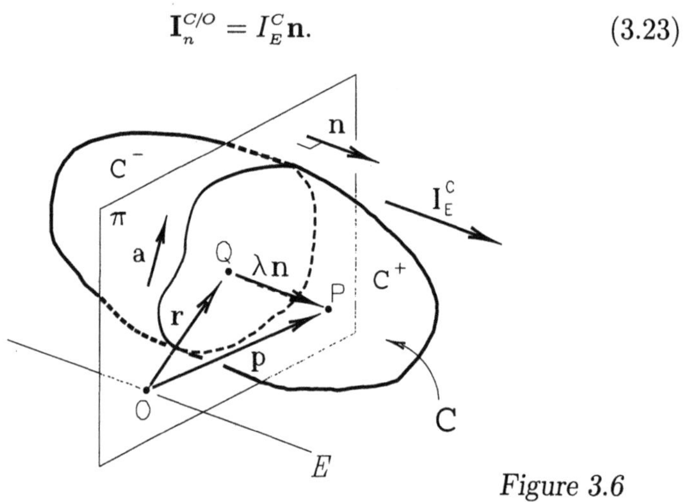

Figure 3.6

In fact, as the plane π splits the body into two symmetrical regions C^+ and C^-, the integrals $\int_{C^+} \mathbf{p} \times (\mathbf{n} \times \mathbf{p})\,dm$ and $\int_{C^-} \mathbf{p} \times (\mathbf{n} \times \mathbf{p})\,dm$ will have equal components in direction $\mathbf{n}$ and opposite components in directions parallel to the plane (see the argument for the case of discrete systems), its sum resulting in a vector parallel to $\mathbf{n}$. As the moment of inertia is the projection of the inertia vector in the direction in question, its value will be the vector module itself. For the same reason, if $\mathbf{a}$ is a vector parallel to π, the product of inertia of the body with respect to O for directions $\mathbf{n}$ and $\mathbf{a}$ will necessarily be null, that is,

$$I_{an}^{C/O} = I_{na}^{C/O} = 0. \tag{3.24}$$

Example 3.6 Returning once again to the disk studied in Example 3.2 (see Fig. 3.4), the three coordinate planes are of symmetry, for which reason the inertia vectors for the directions of the axes are parallel to the respective directions. It was found, in that example, that $\mathbf{I}_{n_1}^{D/O} = \frac{1}{4}mR^2\mathbf{n}_1$, parallel, therefore, to $\mathbf{n}_1$, whose module, $\frac{1}{4}mR^2$, is equal to the moment of inertia with respect to the axis x_1, as seen in Example 3.4. The inertia vectors with respect to point O for the other two coordinate directions fulfill the same condition of being parallel to the respective directions. This is not the same, however, in the case of the inertia vector for direction **n**, which has been shown not to be parallel to **n**. In fact, the plane orthogonal to unit **n** containing point O is not of symmetry for the disk.

This section showed that the properties of symmetry result in special conditions for the inertia vectors associated with directions orthogonal to the planes of symmetry. The above example, in particular, illustrates a situation in which the three coordinate planes are of symmetry to the body, with the result that the three inertia vectors associated with the coordinate directions are parallel to them. We will see in Section 6.6 that symmetry is a sufficient but not a necessary condition to ensure parallelism between the inertia vector and the corresponding direction. In other words, we will see that it is possible to find directions for which this parallelism occurs, even when the body does not present any symmetry whatsoever.

6.4 Cartesian Coordinates

Inertia tensors and vectors and moments and products of inertia are handled more easily when using a system of axes. The results below concern the representation of the inertia properties of a continuous body C on a Cartesian base; these relations will be especially valuable in the study of the motion of the rigid body, discussed in Chapter 7. The reader should not find it hard to obtain the equivalent expressions for discrete systems of particles, merely replacing integrals by sums.

So, let C be a continuous body with mass m and volume V, $\{x_1, x_2, x_3\}$ a system of Cartesian axes with origin O (which may or may not belong to the body) and $\mathbf{n}_1, \mathbf{n}_2, \mathbf{n}_3$ an orthonormal basis associated with the axes (see Fig. 4.1). If P is a general point of C, with coordinates (p_1, p_2, p_3), the position vector of P with respect to O will be $\mathbf{p} = p_1\mathbf{n}_1 + p_2\mathbf{n}_2 + p_3\mathbf{n}_3$.

The moment of inertia of C with respect to each coordinate

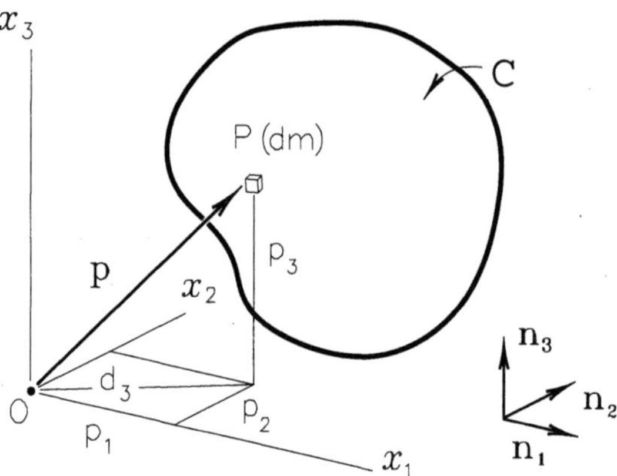

Figure 4.1

axis x_j, $j = 1, 2, 3$, will be, according to Eq. (3.12),

$$I_{x_j}^C = \int_C \left(p^2 - (\mathbf{p} \cdot \mathbf{n}_j)^2 \right) dm$$

$$= \int_C (p_1^2 + p_2^2 + p_3^2 - p_j^2)\, dm, \quad j = 1, 2, 3, \tag{4.1}$$

resulting, then, in the following expressions for the moments of inertia $I_{jj}^{C/O} = I_{x_j}^C$:

$$I_{11}^{C/O} = \int_C (p_2^2 + p_3^2)\, dm;$$

$$I_{22}^{C/O} = \int_C (p_3^2 + p_1^2)\, dm; \tag{4.2}$$

$$I_{33}^{C/O} = \int_C (p_1^2 + p_2^2)\, dm.$$

It is easy to see that the moment of inertia of the body with respect to any coordinate axis is equal to the integral of the product of the mass elements by the respective quadratic distances to the axis. For example, $p_1^2 + p_2^2 = d_3^2$, where d_3 is the distance from point P to axis x_3 (see Fig. 4.1).

The product of inertia of C with respect to the origin of the axis system for two different coordinate directions, $\mathbf{n}_j, \mathbf{n}_k$, $j, k = 1, 2, 3$, or in other words, the product of inertia of C with respect to axes x_j, x_k, is, according to Eq. (3.18),

$$I^{C/O}_{n_j n_k} = \int_C \left(p^2 \mathbf{n}_j \cdot \mathbf{n}_k - (\mathbf{p} \cdot \mathbf{n}_j)(\mathbf{p} \cdot \mathbf{n}_k)\right) dm$$
$$= -\int_C p_j p_k \, dm, \qquad j, k = 1, 2, 3, \tag{4.3}$$

resulting, then, in the following expressions for the products of inertia $I^{C/O}_{jk}$:

$$I^{C/O}_{12} = I^{C/O}_{21} = -\int_C p_1 p_2 \, dm;$$
$$I^{C/O}_{23} = I^{C/O}_{32} = -\int_C p_2 p_3 \, dm; \tag{4.4}$$
$$I^{C/O}_{31} = I^{C/O}_{13} = -\int_C p_3 p_1 \, dm.$$

Example 4.1 Consider the homogeneous wire A, with mass m and shape of a quarter of a circumference of radius r (see Fig. 4.2). Taking a mass element $dm = (2/\pi)\, m\, d\theta$ and the position vector of a general point of the wire with respect to its center O, $\mathbf{p} = (r\cos\theta, r\sin\theta, 0)$, the moment of inertia of the wire with respect to axis x_1 is, according to Eq. (4.2),

$$I^A_{x_1} = \frac{2}{\pi} mr^2 \int_0^{\frac{\pi}{2}} \sin^2\theta \, d\theta = \frac{1}{2} mr^2.$$

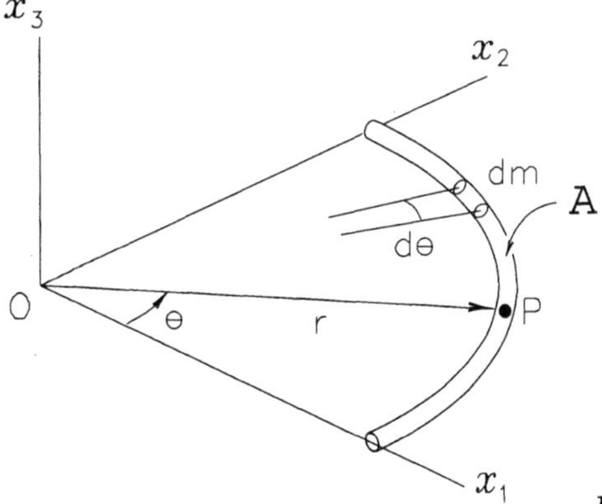

Figure 4.2

Likewise, the moments of inertia with respect to the other two coordinate axes are

$$I_{x_2}^A = \frac{2}{\pi} mr^2 \int_0^{\frac{\pi}{2}} \cos^2 \theta \, d\theta = \frac{1}{2} mr^2,$$

$$I_{x_3}^A = \frac{2}{\pi} mr^2 \int_0^{\frac{\pi}{2}} (\sin^2 \theta + \cos^2 \theta) \, d\theta = mr^2.$$

It is interesting to observe that, in this example, $I_{x_3}^A = I_{x_1}^A + I_{x_2}^A$. The product of inertia of the wire with respect to point O for the directions of axes x_1 and x_2 is, according to Eq. (4.4), given by

$$I_{12}^{A/O} = -\frac{2}{\pi} mr^2 \int_0^{\frac{\pi}{2}} \sin \theta \cos \theta \, d\theta = -\frac{1}{\pi} mr^2.$$

The products of inertia with respect to the other two pairs of axes will be null, since they involve the coordinate $p_3 = 0$; this result was already to be expected, since the plane $x_1 x_2$ is of symmetry for the wire.

The above example also illustrates a couple of properties common to the plane figures: The products of inertia involving a direction orthogonal to the plane of the figure are null, and the moment of inertia with respect to an axis orthogonal to the plane is equal to the sum of the moments of inertia with respect to any two orthogonal axes in the plane that intercept the first.

So, if C is a flat body, O is any point of the plane, and $\{x_1, x_2, x_3\}$ is a system of Cartesian axes with origin O, x_3 being perpendicular to the plane (see Fig. 4.3), the position vector $\mathbf{p}$ of an element of mass dm with respect to point O will have $p_3 = 0$. Integrals $\int_C p_j p_3 \, dm$, $j = 1, 2$, will, of course, be null and, therefore, the corresponding products of inertia will also be null:

$$I_{13}^{C/O} = I_{31}^{C/O} = I_{23}^{C/O} = I_{32}^{C/O} = 0; \qquad C \text{ and O in the plane } x_1 x_2. \quad (4.5)$$

On the other hand, the quadratic distance of element dm to axis x_3 is $p^2 = p_1^2 + p_2^2$, equal, therefore, to the sum of the quadratic distances of the element to axes x_1 and x_2, resulting in

$$I_{33}^{C/O} = I_{11}^{C/O} + I_{22}^{C/O}, \qquad C \text{ and O in the plane } x_1 x_2. \quad (4.6)$$

The inertia tensor of a body C with respect to a point O may, in a system of Cartesian coordinates with origin in O, be expressed by a

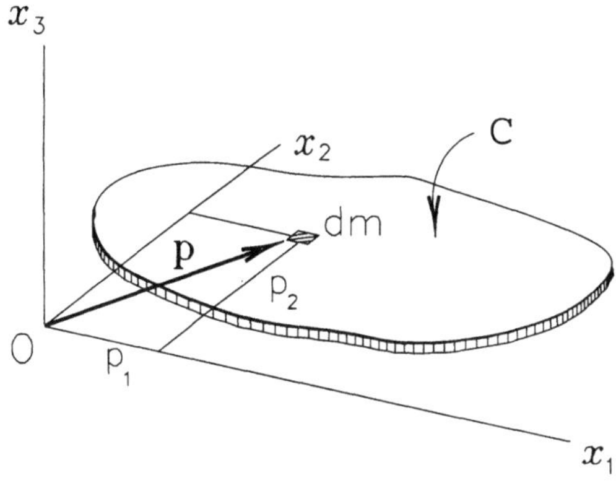

Figure 4.3

symmetric matrix — called the *inertia matrix* — whose main diagonal consists of the moments of inertia of C with respect to the three coordinate axes, the rest of its elements consisting of the products of inertia with respect to O, for the same axes, that is,

$$\mathbf{II}^{C/O} = \begin{pmatrix} I_{11}^O & I_{12}^O & I_{13}^O \\ I_{21}^O & I_{22}^O & I_{23}^O \\ I_{31}^O & I_{32}^O & I_{33}^O \end{pmatrix}. \tag{4.7}$$

Observe that, for the sake of simplicity, a reduced notation was adopted for the elements of the matrix, omitting the body C. This means that $I_{jj}^O \equiv I_{jj}^{C/O}$ and $I_{jk}^O \equiv I_{jk}^{C/O}$. Should more than one body be involved, however, it is advisable to make the corresponding distinction. Equation (4.7) results directly from Eqs. (3.10) and (3.16). In fact, given the orthogonality of the Cartesian axis, the moments and products of inertia for the coordinate directions will be such that

$$I_{jj}^{C/O} = [\mathbf{n}_j] \cdot \begin{pmatrix} I_{11}^O & I_{12}^O & I_{13}^O \\ I_{21}^O & I_{22}^O & I_{23}^O \\ I_{31}^O & I_{32}^O & I_{33}^O \end{pmatrix} \cdot [\mathbf{n}_j]^T = I_{jj}^O, \qquad j = 1, 2, 3,$$

$$\tag{4.8}$$

$$I_{jk}^{C/O} = [\mathbf{n}_j] \cdot \begin{pmatrix} I_{11}^O & I_{12}^O & I_{13}^O \\ I_{21}^O & I_{22}^O & I_{23}^O \\ I_{31}^O & I_{32}^O & I_{33}^O \end{pmatrix} \cdot [\mathbf{n}_k]^T = I_{jk}^O, \qquad j, k = 1, 2, 3.$$

Once the inertia tensor of a body with respect to a point is expressed in a system of Cartesian coordinates through a symmetric matrix — whose elements are the moments and products of inertia of the body for these same axes — the inertia vectors of the body with respect to the point for the coordinate directions also acquire a simple expression, so that

$$\mathbf{I}_{n_j}^{C/O} = \mathbb{II}^{C/O} \cdot \mathbf{n}_j = \left(I_{1j}^O, I_{2j}^O, I_{3j}^O \right), \qquad j = 1, 2, 3, \tag{4.9}$$

that is:

$$\mathbf{I}_1^{C/O} = I_{11}^O \mathbf{n}_1 + I_{21}^O \mathbf{n}_2 + I_{31}^O \mathbf{n}_3;$$
$$\mathbf{I}_2^{C/O} = I_{12}^O \mathbf{n}_1 + I_{22}^O \mathbf{n}_2 + I_{32}^O \mathbf{n}_3; \tag{4.10}$$
$$\mathbf{I}_3^{C/O} = I_{13}^O \mathbf{n}_1 + I_{23}^O \mathbf{n}_2 + I_{33}^O \mathbf{n}_3.$$

Equations (4.10) come from the multiplication of the inertia tensor for each of the base unit vectors, as the reader can easily see.

Example 4.2   Returning to the wire studied in the preceding example (see Fig. 4.2), the encountered moments and products of inertia with respect to the coordinate axes were

$$I_{11}^O = I_{22}^O = \frac{1}{2}mr^2, \quad I_{33}^O = mr^2, \quad I_{12}^O = -\frac{1}{\pi}mr^2, \quad I_{23}^O = I_{31}^O = 0.$$

The inertia tensor with respect to point O may, therefore, be expressed in the system of axes adopted by the matrix

$$\mathbb{II}^{A/O} = mr^2 \begin{pmatrix} 1/2 & -1/\pi & 0 \\ -1/\pi & 1/2 & 0 \\ 0 & 0 & 1 \end{pmatrix}.$$

The inertia vector for, say, direction $\mathbf{n}_1$ may then be easily calculated by

$$\mathbf{I}_1^{A/O} = mr^2 \begin{pmatrix} 1/2 & -1/\pi & 0 \\ -1/\pi & 1/2 & 0 \\ 0 & 0 & 1 \end{pmatrix} \cdot \begin{pmatrix} 1 \\ 0 \\ 0 \end{pmatrix} = mr^2 \left(\frac{1}{2}\mathbf{n}_1 - \frac{1}{\pi}\mathbf{n}_2 \right).$$

or, otherwise, using Eq. (4.10),

$$\mathbf{I}_1^{A/O} = \frac{1}{2}mr^2 \mathbf{n}_1 + \left(-\frac{1}{\pi}mr^2 \right) \mathbf{n}_2 + 0\mathbf{n}_3 = mr^2 \left(\frac{1}{2}\mathbf{n}_1 - \frac{1}{\pi}\mathbf{n}_2 \right).$$

The mere observation of the inertia matrix provides interesting information. For example, the fact that the last line (or column) has only one nonnull element is evidence that the inertia vector for direction $\mathbf{n}_3$ has this same orientation, which was to be expected, as it is a flat figure. In fact,

$$\mathbf{I}_3^{A/O} = mr^2 \begin{pmatrix} 1/2 & -1/\pi & 0 \\ -1/\pi & 1/2 & 0 \\ 0 & 0 & 1 \end{pmatrix} \cdot \begin{pmatrix} 0 \\ 0 \\ 1 \end{pmatrix} = mr^2 \mathbf{n}_3.$$

In principle, establishing the moments and products of inertia of a body C with regard to three mutually orthogonal axes with origin in a given point O requires computing the integrals present in Eqs. (4.2) and (4.4). When the body is homogeneous, the moments and products of inertia may be expressed as a function of its mass m and its geometric dimensions, with the advantage that the result may be tabulated for the more common geometries. Appendix C presents the formulas for moments and products of inertia with respect to the mass center and another notable point for a reasonable broad set of usual geometries. When a body is nonhomogeneous or, even when homogeneous, has its own geometry not found in the available tables, its moments and products of inertia must be obtained by integration, according to Eqs. (4.2) and (4.4), after adopting a convenient system of coordinates; if there is symmetry, Eq. (3.24) can be used, thereby saving the computing of one or more defined integrals.

Example 4.3   Consider a homogeneous prism P with a right-angled triangular base and mass m, for which the moments and products of inertia with respect to point O are to be found (see Fig. 4.4). Its volume is $V = a^3$ and its density is $\rho = m/a^3$. Taking an element of mass $dm = \rho\, dV = \frac{m}{a^3}\, dx\, dy\, dz$, the moment of inertia with respect to axis x can be computed, according to Eq. (4.2), by

$$I_{xx}^{P/O} = \frac{m}{a^3} \int_{-a}^{a} \int_{0}^{a} \int_{0}^{a-y} (y^2 + z^2)\, dx\, dy\, dz = \frac{1}{2}ma^2.$$

By symmetry, the moments of inertia with respect to axes x and y are equal; therefore,

$$I_{yy}^{P/O} = \frac{1}{2}ma^2.$$

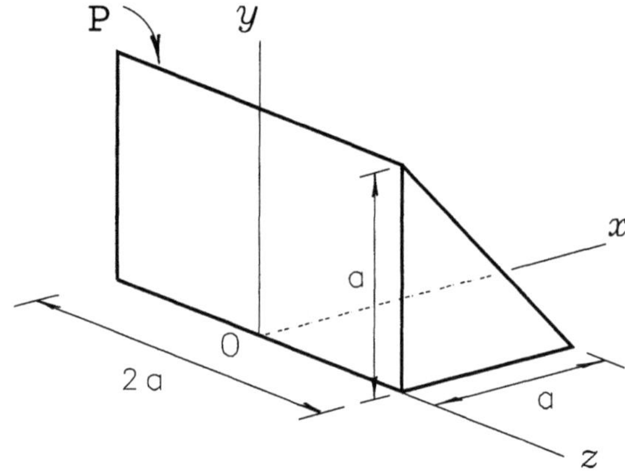

Figure 4.4

The moment of inertia with respect to axis z, also according to Eq. (4.2), is

$$I_{zz}^{P/O} = \frac{m}{a^3} \int_{-a}^{a} \int_{0}^{a} \int_{0}^{a-y} (x^2 + y^2)\, dx\, dy\, dz = \frac{1}{3} ma^2.$$

Since plane xy is of symmetry for the prism, according to Eq. (3.24), the products of inertia involving the coordinate z orthogonal to the plane are null, that is,

$$I_{zx}^{P/O} = I_{zy}^{P/O} = 0.$$

Finally, the product of inertia for axes x and y can, according to Eq. (4.4), be calculated by

$$I_{xy}^{P/O} = -\frac{m}{a^3} \int_{-a}^{a} \int_{0}^{a} \int_{0}^{a-y} xy\, dx\, dy\, dz = -\frac{1}{12} ma^2.$$

The inertia tensor of the prism with respect to point O can, therefore, be expressed by the matrix

$$\mathrm{II}^{P/O} = ma^2 \begin{pmatrix} 1/2 & -1/12 & 0 \\ -1/12 & 1/2 & 0 \\ 0 & 0 & 1/3 \end{pmatrix}.$$

When a body is flat and homogeneous, it is usual to define *area moments and products of inertia* for the coordinate directions. If ρ is the mass per unit (uniform) of area of body C and $\{x_1, x_2, x_3\}$ is a system of Cartesian axes with origin in a point O of the plane of the figure, with x_3 orthogonal to this plane (see Fig. 4.5), its moments of inertia with respect to the coordinate axes are

$$I_{jj}^{C/O} = \int_C d_j^2 \, dm = \rho \int_C d_j^2 \, dA, \qquad j = 1, 2, 3, \tag{4.11}$$

where the integral on the surface of the flat body is defined as the *area moment of inertia* with respect to axis x_j:

$$J_{jj}^{C/O} \rightleftharpoons \int_C d_j^2 \, dA, \qquad j = 1, 2, 3. \tag{4.12}$$

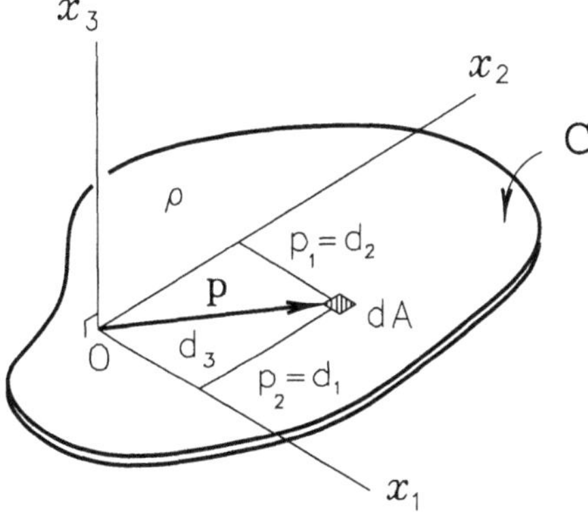

Figure 4.5

Likewise, the products of inertia with respect to a point O of the plane for the coordinate directions are

$$I_{jk}^{C/O} = -\int_C p_j p_k \, dm = -\rho \int_C p_j p_k \, dA, \qquad j, k = 1, 2, 3, \tag{4.13}$$

where the integral on the surface of the flat body is defined as the *area product of inertia* with respect to point O for axes x_j and x_k:

$$J_{jk}^{C/O} \rightleftharpoons -\int_C p_j p_k \, dA, \qquad j, k = 1, 2, 3. \tag{4.14}$$

In short, moments of inertia, products of inertia, area moments of inertia, and area products of inertia are merely related by

$$I_{jj}^{C/O} = \rho J_{jj}^{C/O}, \qquad I_{jk}^{C/O} = \rho J_{jk}^{C/O}, \qquad j, k = 1, 2, 3, \qquad (4.15)$$

where ρ is the mass per unit of area, uniform throughout the body, with dimension $[ML^{-2}]$.

Area moments and products of inertia of a flat homogeneous body have the dimension $[L^4]$. Being purely geometric definitions, moments and products of inertia may be given for a cross section of a body. This concept is particularly important in deformable solid mechanics, but will not be discussed here. Section C.2 of Appendix C gives a table of inertia properties of area for the more usual geometries. Properties of inertia of bodies with an essentially flat geometry, such as plates, sheets, etc., may be established by using this table or a similar one.

It is vitally important to note that for flat bodies, $p_3 = 0$ and the area products of inertia involving direction x_3 are all null, that is,

$$J_{13}^{C/O} = J_{31}^{C/O} = J_{23}^{C/O} = J_{32}^{C/O} = 0. \qquad (4.16)$$

For this reason only the area product of inertia $J_{12}^{C/O}$ is assigned to the plane figures in Appendix C. Also note that, since the area moment of inertia with respect to axis x_3 is equal to the sum of the moments of inertia of area with respect to axes x_1 and x_2, that is,

$$J_{33}^{C/O} = J_{11}^{C/O} + J_{22}^{C/O}, \qquad (4.17)$$

nor is the former found in the table of plane figures in Appendix C.

Example 4.4   Consider the thin, homogeneous semi-elliptical plate P, with surface density $\rho = 2$ kg/m^2 and semi-axes $a = 30$ cm and $b = 20$ cm (see Fig. 4.6). Appendix C provides the following area moments and products of inertia with respect to point A:

$$J_{11}^{P/A} = \frac{\pi}{8}ab^3; \qquad J_{22}^{P/A} = \frac{\pi}{8}a^3b; \qquad J_{12}^{P/A} = 0.$$

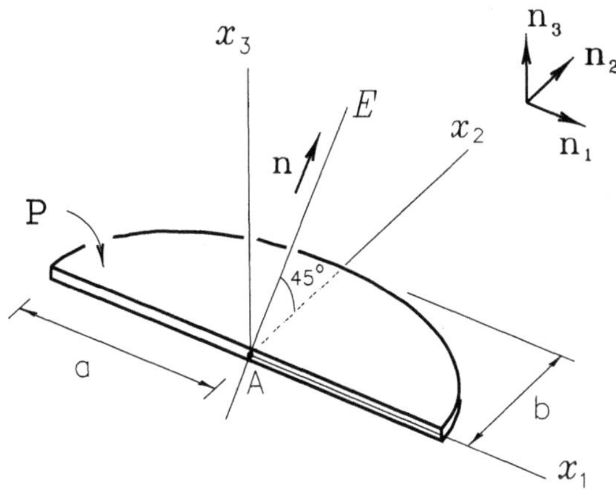

Figure 4.6

The other elements that comprise the inertia tensor with respect to point A are

$$J_{33}^{P/A} = J_{11}^{P/A} + J_{22}^{P/A} = \frac{\pi}{8}ab(b^2 + a^2), \quad J_{23}^{P/A} = J_{31}^{P/A} = 0.$$

The inertia tensor of the plate with respect to point A is then

$$\mathbf{II}^{P/A} = \rho \begin{pmatrix} J_{11}^{P/A} & J_{12}^{P/A} & J_{13}^{P/A} \\ J_{21}^{P/A} & J_{22}^{P/A} & J_{23}^{P/A} \\ J_{31}^{P/A} & J_{32}^{P/A} & J_{33}^{P/A} \end{pmatrix} = \rho\frac{\pi}{8}ab \begin{pmatrix} b^2 & 0 & 0 \\ 0 & a^2 & 0 \\ 0 & 0 & b^2 + a^2 \end{pmatrix}.$$

Now if E is an axis passing through A and parallel to $\mathbf{n} = \frac{1}{\sqrt{2}}(\mathbf{n_2} + \mathbf{n_3})$, the moment of inertia of the plate with respect to E is

$$I_E^P = \frac{1}{\sqrt{2}}(0, 1, 1) \cdot \rho\frac{\pi}{8}ab \begin{pmatrix} b^2 & 0 & 0 \\ 0 & a^2 & 0 \\ 0 & 0 & b^2 + a^2 \end{pmatrix} \cdot \frac{1}{\sqrt{2}} \begin{pmatrix} 0 \\ 1 \\ 1 \end{pmatrix}$$

$$= \frac{\pi}{16}\rho ab(2a^2 + b^2)$$

$$= 5.18 \times 10^{-3} \text{ kg m}^2.$$

6.5 Transfer of Axes

The tables of inertia properties in Appendix C give expressions for moments and products of inertia, for the most common geometries, with respect to one or, at most, two notable points. It would be extremely useful to extend the coverage of these tables to any other point in space without resorting to new integrating procedures. This is, in fact, possible, with the help of the relationships established below, which are commonly called *transfer of inertia properties.*

Let C be a continuous body with mass m and volume V, O an arbitrary point (that may or may not belong to C), origin of a system of Cartesian axes $\{x_1, x_2, x_3\}$, and C* the mass center of C, origin of another system of axes $\{x_1^*, x_2^*, x_3^*\}$ parallel to the first system (see Fig. 5.1). The inertia tensor of C with respect to O is equal to the sum of the inertia tensor of C with respect to C* with the inertia tensor of the mass center with respect to point O, that is,

$$\mathrm{II}^{C/O} = \mathrm{II}^{C/C^*} + \mathrm{II}^{C^*/O}. \tag{5.1}$$

The inertia tensor $\mathrm{II}^{C^*/O}$, called the *inertia tensor of the mass center with respect to point* O, is actually the inertia tensor of a (fictitious) particle, whose mass is the same as that of body C and its position in space coincides with that of point C*.

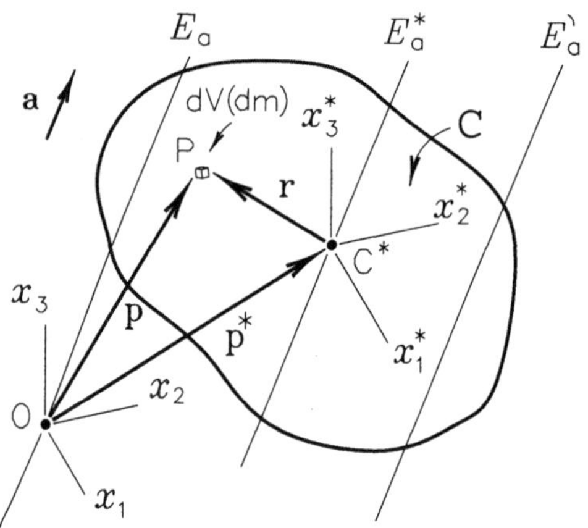

Figure 5.1

Before deriving Eq. (5.1), let us see how this result is reflected in the relationship between inertia vectors. Thus, by dot-multiplying Eq. (5.1) by an arbitrary unit vector **a**, then from Eqs. (2.2) and (3.4), we have

$$\mathbf{I}_a^{C/O} = \mathbf{I}_a^{C/C^*} + \mathbf{I}_a^{C^*/O}, \tag{5.2}$$

that is, the inertia vector of the body with respect to point O for a given direction is equal to the sum of the inertia vector of the body with respect to its mass center with the inertia vector of the mass center with respect to point O, both for the same direction $\mathbf{a}$. In fact, expressing the position vector of a generic point P of the body according to $\mathbf{p} = \mathbf{p}^* + \mathbf{r}$ (see Fig. 5.1), using Eq. (3.8), and expanding the double vector product $\mathbf{p} \times (\mathbf{a} \times \mathbf{p})$, we get

$$\mathbf{I}_a^{C/O} = \int_C (\mathbf{p}^* + \mathbf{r}) \times \left(\mathbf{a} \times (\mathbf{p}^* + \mathbf{r})\right) dm$$

$$= \mathbf{p}^* \times (\mathbf{a} \times \mathbf{p}^*)m + \mathbf{p}^* \times \left(\mathbf{a} \times \int_C \mathbf{r}\, dm\right)$$

$$+ \int_C \mathbf{r}\, dm \times (\mathbf{a} \times \mathbf{p}^*) + \int_C \mathbf{r} \times (\mathbf{a} \times \mathbf{r})\, dm.$$

As, from Eq. (3.8),

$$\int_C \mathbf{r} \times (\mathbf{a} \times \mathbf{r})\, dm = \mathbf{I}_a^{C/C^*},$$

and Eq. (2.4),

$$\mathbf{p}^* \times (\mathbf{a} \times \mathbf{p}^*)m = \mathbf{I}_a^{C^*/O},$$

as well as in Eq. (1.9), the terms involving the integral $\int_C \mathbf{r}\, dm$ are null, the validity of Eq. (5.2) is confirmed. Now, if the product of a tensor T_1 by an arbitrary unit vector $\mathbf{a}$ is always equal to the product of another tensor T_2 by the same unit vector, whichever it is, then the two tensors are necessarily equal. In other words, demonstrating Eq. (5.2) for an arbitrary direction, the validity of Eq. (5.1) is automatically established.

Equations (5.1) and (5.2) may be interpreted as if the inertia of rotation of a body around any point O were equivalent to its inertia of rotation around its mass center, plus the inertia of rotation of a (fictitious) particle, with the mass equal to that of the body and concentrated in the mass center, with respect to the given point.

Properties of transfer of axes that relate moments and products of inertia also result from these relations. Thus, by projecting Eq. (5.2) in the same direction $\mathbf{a}$ (see Fig. 5.1), according to Eqs. (2.5), (2.8), and (3.10), we have

$$I_{E_a}^C = I_{E_a^*}^C + I_{E_a}^{C^*}. \tag{5.3}$$

Equation (5.3) relates the moments of inertia of a body with respect to two parallel axes, *one of them passing through the mass center of the body*. This last constraint cannot be dismissed; by wishing to relate the moments of inertia of a body with respect to any two parallel axes, Eq. (5.3) should be used twice; otherwise we find $I^C_{E'_a} = I^C_{E^*_a} + I^{C^*}_{E'_a}$, where $I^C_{E^*_a} = I^C_{E_a} - I^{C^*}_{E_a}$ (see Fig. 5.1), resulting then in

$$I^C_{E'_a} = I^C_{E_a} + I^{C^*}_{E'_a} - I^{C^*}_{E_a}. \tag{5.4}$$

Finally, a relation of transfer of axes for the products of inertia may be obtained by dot-multiplying Eq. (5.2) by a unit vector **b**, different from **a**, to obtain

$$I^{C/O}_{ab} = I^{C/C^*}_{ab} + I^{C^*/O}_{ab}. \tag{5.5}$$

Expressed in components, in Cartesian axes, Eq. (5.1) assumes the matrix form

$$\begin{pmatrix} I^O_{11} & I^O_{12} & I^O_{13} \\ I^O_{21} & I^O_{22} & I^O_{23} \\ I^O_{31} & I^O_{32} & I^O_{33} \end{pmatrix} = \begin{pmatrix} I^*_{11} & I^*_{12} & I^*_{13} \\ I^*_{21} & I^*_{22} & I^*_{23} \\ I^*_{31} & I^*_{32} & I^*_{33} \end{pmatrix} + \begin{pmatrix} I^{C^*/O}_{11} & I^{C^*/O}_{12} & I^{C^*/O}_{13} \\ I^{C^*/O}_{21} & I^{C^*/O}_{22} & I^{C^*/O}_{23} \\ I^{C^*/O}_{31} & I^{C^*/O}_{32} & I^{C^*/O}_{33} \end{pmatrix}, \tag{5.6}$$

where I^O_{jj} and I^O_{jk} are moments and products of inertia, respectively, of the body with respect to point O, for the coordinate directions, I^*_{jj} and I^*_{jk} are the inertia properties of the body with respect to its mass center, C^*, for directions parallel to the former, and $I^{C^*/O}_{jj}$ and $I^{C^*/O}_{jk}$ are the moments and products of inertia of the mass center with respect to point O, for the same directions.

The moments of inertia for the coordinate directions satisfy the relations

$$I^O_{11} = I^*_{11} + m(p^{*2}_2 + p^{*2}_3),$$
$$I^O_{22} = I^*_{22} + m(p^{*2}_3 + p^{*2}_1), \tag{5.7}$$
$$I^O_{33} = I^*_{33} + m(p^{*2}_1 + p^{*2}_2).$$

Equations (5.7) come from the substitution of Eq. (2.13) in Eq. (5.3). Now replacing Eq. (2.15) in Eq. (5.5), we get the relations of transfer between parallel Cartesian axes for the products of inertia:

$$I^O_{12} = I^*_{12} - mp^*_1 p^*_2,$$
$$I^O_{23} = I^*_{23} - mp^*_2 p^*_3, \tag{5.8}$$
$$I^O_{31} = I^*_{31} - mp^*_3 p^*_1.$$

Example 5.1   Consider the homogeneous beam V, of mass m and dimensions a, b, and c, as shown in Fig. 5.2. The table in Appendix C gives the following expressions for the moments of inertia relative to point O:

$$I_{11}^{O} = \frac{1}{12}m(b^2 + c^2), \quad I_{22}^{O} = \frac{1}{12}m(c^2 + a^2), \quad I_{33}^{O} = \frac{1}{12}m(a^2 + b^2),$$

with all products of inertia null, due to the symmetry.

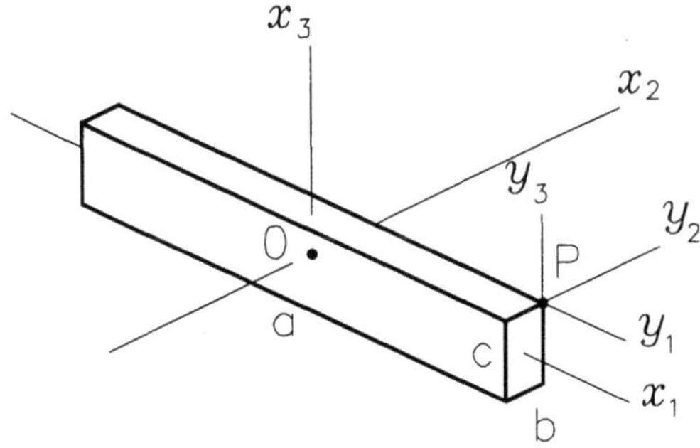

Figure 5.2

The inertia tensor, therefore, assumes the form of a diagonal matrix:

$$\mathbf{II}^{V/O} = \frac{1}{12}m \begin{pmatrix} b^2 + c^2 & 0 & 0 \\ 0 & c^2 + a^2 & 0 \\ 0 & 0 & a^2 + b^2 \end{pmatrix}.$$

To establish the moments and products of inertia with respect to axes parallel to the former, with origin in vertex P, we will use Eqs. (5.7) and (5.8):

$$I_{11}^{P} = \frac{1}{12}m(b^2 + c^2) + m\left(\frac{b^2}{4} + \frac{c^2}{4}\right) = \frac{1}{3}m(b^2 + c^2);$$

$$I_{22}^{P} = \frac{1}{12}m(c^2 + a^2) + m\left(\frac{c^2}{4} + \frac{a^2}{4}\right) = \frac{1}{3}m(c^2 + a^2);$$

$$I_{33}^{P} = \frac{1}{12}m(a^2 + b^2) + m\left(\frac{a^2}{4} + \frac{b^2}{4}\right) = \frac{1}{3}m(a^2 + b^2);$$

$$I_{12}^{P} = 0 - m\frac{a}{2}\frac{b}{2} = -\frac{1}{4}mab;$$

$$I_{23}^{P} = 0 - m\frac{b}{2}\frac{c}{2} = -\frac{1}{4}mbc;$$

$$I_{31}^{P} = 0 - m\frac{c}{2}\frac{a}{2} = -\frac{1}{4}mca.$$

The inertia tensor of the beam with respect to point P is then

$$\mathbb{II}^{V/P} = \frac{1}{12} m \begin{pmatrix} 4(b^2 + c^2) & -3ab & -3ac \\ -3ab & 4(c^2 + a^2) & -3bc \\ -3ac & -3bc & 4(a^2 + b^2) \end{pmatrix}.$$

Tensors, vectors, moments, and products of inertia are additive properties, that is, established by sums or integrals; so, if a body C can be arbitrarily divided into N parts, $C_1, C_2, \ldots, C_N$ (see Fig. 5.3), the inertia properties of the body with respect to a given point O will be the sum of the respective inertia properties of its parts, with respect to the point.

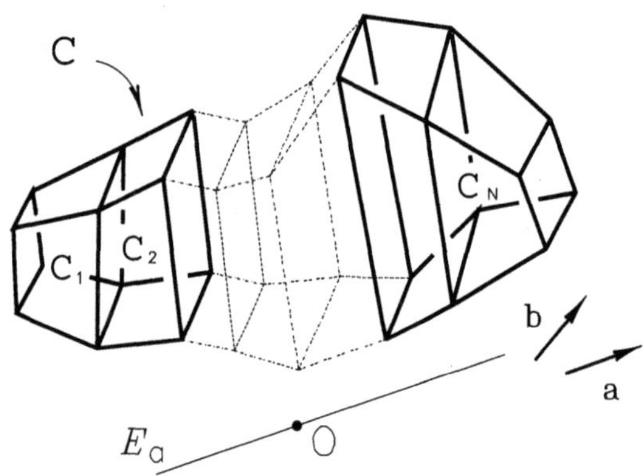

Figure 5.3

In fact, the inertia tensor of C with respect to O, being

$$\begin{aligned}
\mathbb{II}^{C/O} &= \int_C (p^2 \mathbb{1} - \mathbf{p} \otimes \mathbf{p})\, dm \\
&= \sum_{i=1}^{N} \int_{C_i} (p^2 \mathbb{1} - \mathbf{p} \otimes \mathbf{p})\, dm \\
&= \sum_{i=1}^{N} \mathbb{II}^{C_i/O},
\end{aligned} \tag{5.9}$$

is equal to the sum of the inertia tensors of each of the parts C_i, with respect to O. Now let $\mathbf{a}$ and $\mathbf{b}$ be two unit vectors, defining two different directions in space. The projection of Eq. (5.9) in direction **a**, according to Eq. (3.4), produces an additive relation for inertia vectors, that is,

$$\mathbf{I}_a^{C/O} = \sum_{i=1}^{N} \mathbf{I}_a^{C_i/O}. \tag{5.10}$$

By repeating the dot product with unit vector **a**, then, according to Eqs. (3.10) and (3.14), we have

$$I_{E_a}^{C} = \sum_{i=1}^{N} I_{E_a}^{C_i}, \tag{5.11}$$

where E_a is an axis parallel to **a**, passing through O. Finally, by projecting Eq. (5.10) toward **b**, we obtain the relation of additiveness for the products of inertia,

$$I_{ab}^{C/O} = \sum_{i=1}^{N} I_{ab}^{C_i/O}. \tag{5.12}$$

Example 5.2 Figure 5.4 illustrates a body consisting of a homogeneous rod B, of mass m and length r, welded to an also homogeneous wheel A, of mass $2m$ and radius r, supported by a horizontal plane.

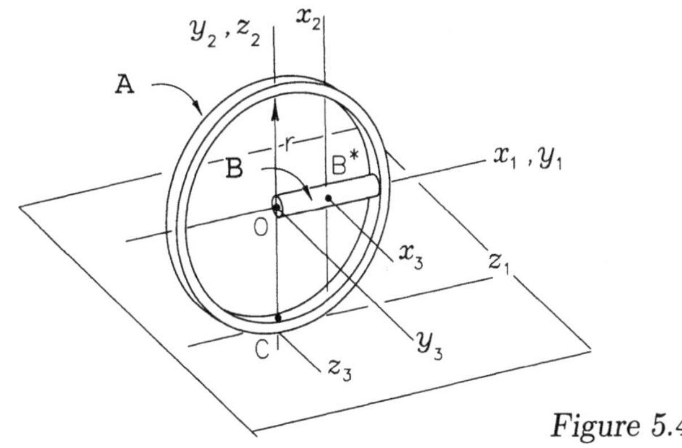

Figure 5.4

Axes $\{x_1, x_2, x_3\}$, $\{y_1, y_2, y_3\}$, and $\{z_1, z_2, z_3\}$ are parallel to each other, with origin in the center of the rod, B^*, in the middle of the wheel, O, and on point C, contact of the wheel with the horizontal plane, respectively. We wish to establish the inertia tensor of the solid with respect to the point of contact C. The tables in Appendix C and the properties of symmetry give us

$$I_{11}^{B/B^*} = 0, \qquad\qquad I_{11}^{A/O} = I_{22}^{A/O} = mr^2,$$

$$I_{22}^{B/B^*} = I_{33}^{B/B^*} = \tfrac{1}{12}mr^2, \qquad I_{33}^{A/O} = 2mr^2,$$

$$I_{jk}^{B/B^*} = 0, \quad j,k = 1,2,3, \qquad I_{jk}^{A/O} = 0, \quad j,k = 1,2,3.$$

Transferring the moments and products of inertia of the rod from axes $\{x_1, x_2, x_3\}$ to axes $\{z_1, z_2, z_3\}$, and the moments and products of inertia of the wheel from axes $\{y_1, y_2, y_3\}$ also to axes $\{z_1, z_2, z_3\}$, using for this purpose Eqs. (5.7) and (5.8), then

$$I_{11}^{B/C} = 0 + mr^2 = mr^2, \qquad\qquad I_{11}^{A/C} = mr^2 + 2mr^2 = 3mr^2,$$

$$I_{22}^{B/C} = \tfrac{1}{12}mr^2 + \tfrac{1}{4}mr^2 = \tfrac{1}{3}mr^2, \quad I_{22}^{A/C} = mr^2 + 0 = mr^2,$$

$$I_{33}^{B/C} = \tfrac{1}{12}mr^2 + \tfrac{5}{4}mr^2 = \tfrac{4}{3}mr^2, \quad I_{33}^{A/C} = 2mr^2 + 2mr^2 = 4mr^2,$$

$$I_{12}^{B/C} = 0 - \tfrac{1}{2}mr^2 = -\tfrac{1}{2}mr^2, \qquad I_{12}^{A/C} = 0 + 0 = 0,$$

$$I_{23}^{B/C} = 0 - 0 = 0, \qquad\qquad I_{23}^{A/C} = 0 + 0 = 0,$$

$$I_{31}^{B/C} = 0 - 0 = 0, \qquad\qquad I_{31}^{A/C} = 0 + 0 = 0.$$

The inertia tensor of the body with respect to point C is, therefore, the sum of the component tensors, that is,

$$\mathbf{I}^{(A+B)/C} = mr^2 \begin{pmatrix} 1 & -1/2 & 0 \\ -1/2 & 1/3 & 0 \\ 0 & 0 & 4/3 \end{pmatrix} + mr^2 \begin{pmatrix} 3 & 0 & 0 \\ 0 & 1 & 0 \\ 0 & 0 & 4 \end{pmatrix}$$

$$= mr^2 \begin{pmatrix} 4 & -1/2 & 0 \\ -1/2 & 4/3 & 0 \\ 0 & 0 & 16/3 \end{pmatrix}.$$

6.6 Principal Directions of Inertia

The preceding section showed how to establish the inertia properties of a body (or discrete system of particles) with respect to a point when its inertia properties with respect to another point are known. These results are important because the inertia tensor of a body (or system) is a function solely of its distribution of mass around the point. The rotation of the coordinate axes around the point does not raise problems, since the tensor does not depend on the orientation of the axes (although the components of the inertia matrix do). The inertia tensor with respect to the point is established after having arbitrated a convenient orientation from the geometric viewpoint and computed the moments and products of inertia for those chosen axes; moments and products of inertia with respect to any other directions are then very easily calculated using Eqs. (3.10) and (3.16) or Eqs. (3.9) and (3.15), for discrete systems.

Let us then consider a body C and an arbitrary point O, and let $\mathbf{II}^{C/O}$ be the inertia tensor of C with respect to O. Given an axis E passing through O, the moment of inertia of the body with respect to the axis is given, as was shown, by $I_E^C = \mathbf{I}_n^{C/O} \cdot \mathbf{n} = \mathbf{n} \cdot \mathbf{II}^{C/O} \cdot \mathbf{n}$, where $\mathbf{n}$ is a unit vector parallel to the axis. For each arbitrated $\mathbf{n}$, the moment of inertia, a nonnegative scalar, will have its own value, and it will be natural to expect maximum and minimum values for one or more particular directions. On the other hand, we saw that the inertia vector for an arbitrary direction is not generally parallel to this direction, although this parallelism has occurred in some of the examples studied in this chapter. The aim of the following study is to establish the general conditions of occurrence of this singularity and examine the behavior of the *value* of the moment of inertia as a function of the direction of the axis.

When the inertia vector of a body C with respect to a point O for a given direction $\mathbf{a}$ is parallel to $\mathbf{a}$, it is said to be a *principal direction of inertia* of the body with respect to the point. So, if $\mathbf{a}$ is a unit vector parallel to a principal direction of inertia, then, according to Eq. (3.4),

$$\mathbf{I}_a^{C/O} = \mathbf{II}^{C/O} \cdot \mathbf{a} = \lambda \mathbf{a}, \tag{6.1}$$

where, according to Eqs. (3.10) and (3.14), the module of the inertia vector will be the corresponding moment of inertia itself, that is,

$$\lambda = I_{aa}^{O}. \tag{6.2}$$

In other words, the unit vector describing a principal direction of inertia, if any, is a *normalized autovector* of the inertia tensor, and the moment of inertia for this same direction is the *associated eigenvalue* (see Appendix A). Rewriting Eq. (6.1) as

$$(\mathbb{II}^{C/O} - I_{aa}^{O}\mathbb{1}) \cdot \mathbf{a} = 0, \tag{6.3}$$

and recalling that $\mathbf{a}$ is not null, the result is that the determinant of the tensor should vanish, that is,

$$\left| \mathbb{II}^{C/O} - I_{aa}^{O}\mathbb{1} \right| = 0. \tag{6.4}$$

Equation (6.4), called the *characteristic equation*, will have as solutions the eigenvalues I_{aa}^{O}. Adopting a system of Cartesian axes $\{x_1, x_2, x_3\}$, with origin in point O and arbitrary orientation, Eq. (6.4) becomes

$$\begin{vmatrix} (I_{11}^{O} - I_{aa}^{O}) & I_{12}^{O} & I_{13}^{O} \\ I_{21}^{O} & (I_{22}^{O} - I_{aa}^{O}) & I_{23}^{O} \\ I_{31}^{O} & I_{32}^{O} & (I_{33}^{O} - I_{aa}^{O}) \end{vmatrix} = 0. \tag{6.5}$$

The inertia tensor is real and symmetric with dimension 3. Therefore, it admits three real eigenvalues (see Appendix A). This means, then, that for every body C and every point O, there are three real eigenvalues, called *principal moments of inertia* of the body with respect to the point, to which three normalized eigenvectors are associated, that is, unit vectors that characterize the principal directions of inertia. Equation (6.5), when expanded, results in a cubic equation for I_{aa}^{O}, whose roots, λ_1, λ_2, and λ_3, all real, are the desired eigenvalues. Let us then suppose that the unit vectors $\mathbf{a}_1$, $\mathbf{a}_2$, and $\mathbf{a}_3$ are the respective associated normalized eigenvectors. The following vectorial relations must be fulfilled:

$$\mathbb{II}^{C/O} \cdot \mathbf{a}_1 = \lambda_1 \mathbf{a}_1; \quad \mathbb{II}^{C/O} \cdot \mathbf{a}_2 = \lambda_2 \mathbf{a}_2; \quad \mathbb{II}^{C/O} \cdot \mathbf{a}_3 = \lambda_3 \mathbf{a}_3. \tag{6.6}$$

By dot-multiplying the first equation by $\mathbf{a}_2$ and the second by $\mathbf{a}_1$, subtracting the results, and recalling that, given the symmetry of the inertia tensor, $\mathbf{a}_j \cdot \mathbf{II}^{C/O} \cdot \mathbf{a}_k = \mathbf{a}_k \cdot \mathbf{II}^{C/O} \cdot \mathbf{a}_j$, we have

$$0 = (\lambda_1 - \lambda_2)\,\mathbf{a}_1 \cdot \mathbf{a}_2, \tag{6.7}$$

and, doing the same with the other two pairs of equations, we have

$$0 = (\lambda_2 - \lambda_3)\,\mathbf{a}_2 \cdot \mathbf{a}_3, \tag{6.8}$$

$$0 = (\lambda_3 - \lambda_1)\,\mathbf{a}_3 \cdot \mathbf{a}_1. \tag{6.9}$$

If the three roots of Eq. (6.5) are distinct, that is, $\lambda_1 \neq \lambda_2 \neq \lambda_3$, then Eqs. (6.7-6.9) guarantee that $\mathbf{a}_1 \cdot \mathbf{a}_2 = \mathbf{a}_2 \cdot \mathbf{a}_3 = \mathbf{a}_3 \cdot \mathbf{a}_1 = 0$ and *the three principal directions of inertia are orthogonal to each other* (see Fig. 6.1a).

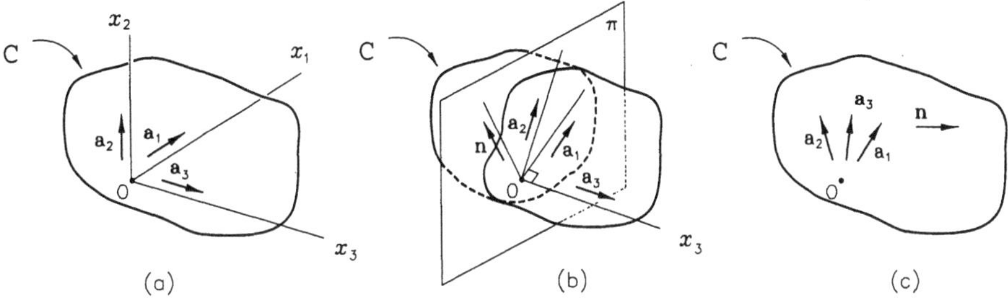

Figure 6.1

When there are two identical roots and one distinct, that is, $\lambda_1 = \lambda_2 = \lambda \neq \lambda_3$, then $\mathbf{a}_1 \cdot \mathbf{a}_3 = \mathbf{a}_2 \cdot \mathbf{a}_3 = 0$ and the principal direction of inertia associated to the distinct root is orthogonal to the other two. When this occurs, all directions orthogonal to the unit vectors associated to the distinct root are principal directions of inertia. In fact, if π is a plane orthogonal to $\mathbf{n}_3$ and $\mathbf{n}$ is an arbitrary unit parallel to π (see Fig. 6.1b), $\mathbf{n} = n_1\mathbf{a}_1 + n_2\mathbf{a}_2$, then

$$\mathbf{I}_n^{C/O} = \mathbf{II}^{C/O} \cdot \mathbf{n} = n_1\lambda\mathbf{a}_1 + n_2\lambda\mathbf{a}_2 = \lambda\mathbf{n}. \tag{6.10}$$

Lastly, if the three roots are equal, that is, $\lambda_1 = \lambda_2 = \lambda_3 = \lambda$, given a unit vector of an arbitrary direction in space (see Fig. 6.1c), $\mathbf{n} = n_1\mathbf{a}_1 + n_2\mathbf{a}_2 + n_3\mathbf{a}_3$, we have

$$\mathbf{I}_n^{C/O} = \mathbf{II}^{C/O} \cdot \mathbf{n} = n_1\lambda\mathbf{a}_1 + n_2\lambda\mathbf{a}_2 + n_3\lambda\mathbf{a}_3 = \lambda\mathbf{n}, \tag{6.11}$$

meaning that any direction in space is a principal direction of inertia of the body with respect to the point.

The notation adopted for moments of inertia usually indicates the body C, point O, and a double index for the direction of the axis in the form of $I_{nn}^{C/O}$. For the sake of simplicity, if there is a single body in question, the indication of this has been abandoned in the elements of inertia matrix in the form of I_{nn}^{O}. Maintaining the repeated indices plays a formal rather than an actual operational role within the matrix. To emphasize a principal moment of inertia, however, we will adopt a simple index in the notation. Therefore, the form $I_{a}^{C/O}$, or when there is no ambiguity, I_{a}^{O}, will indicate a principal moment of inertia of the body with respect to the point, and it is understood that **a** is a principal direction of inertia for that point.

Example 6.1   Let us consider a homogeneous parallelpiped C, with mass m and dimensions a, b, c, and a system of axes $\{x_1, x_2, x_3\}$, parallel to its edges, with origin in its center O (see Fig. 6.2). The moments of inertia with respect to the coordinate axes are (see Appendix C)

$$I_{11}^{O} = \frac{1}{12}m(b^2 + c^2), \quad I_{22}^{O} = \frac{1}{12}m(c^2 + a^2), \quad I_{33}^{O} = \frac{1}{12}m(a^2 + b^2).$$

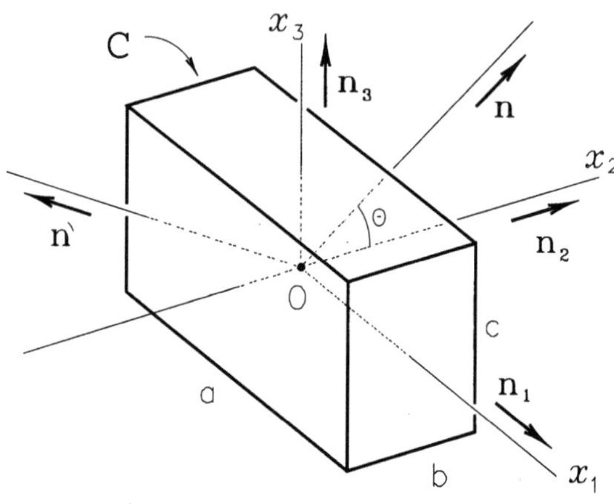

Figure 6.2

Given the symmetry, all products of inertia for these axes will be null. The inertia tensor may be expressed, then, by the matrix

$$\mathbf{II}^{C/O} = \frac{1}{12}m \begin{pmatrix} b^2 + c^2 & 0 & 0 \\ 0 & c^2 + a^2 & 0 \\ 0 & 0 & a^2 + b^2 \end{pmatrix}.$$

The inertia vectors for the coordinate directions are, therefore,

$$\mathbf{I}_1^O = \frac{1}{12}m(b^2 + c^2)\,\mathbf{n}_1,$$

$$\mathbf{I}_2^O = \frac{1}{12}m(c^2 + a^2)\,\mathbf{n}_2,$$

$$\mathbf{I}_3^O = \frac{1}{12}m(a^2 + b^2)\,\mathbf{n}_3.$$

As expected, the inertia vectors obtained are parallel to the respective directions, consisting, therefore, of three principal directions of inertia orthogonal to each other. Let us now assume that $c = b$. In this case, the inertia tensor is reduced to

$$\mathbf{II}^{C/O} = \frac{1}{12}m \begin{pmatrix} 2b^2 & 0 & 0 \\ 0 & a^2 + b^2 & 0 \\ 0 & 0 & a^2 + b^2 \end{pmatrix}.$$

Now, let $\mathbf{n}$ be an arbitrary unit vector parallel to plane $x_2 x_3$, that is, $\mathbf{n} = \cos\theta\,\mathbf{n}_2 + \sin\theta\,\mathbf{n}_3$, θ being an arbitrary angle, the inertia vector of the body with respect to point O for the direction $\mathbf{n}$ is

$$\mathbf{I}_n^O = \mathbf{II}^{C/O} \cdot \mathbf{n} = \frac{1}{12}m(a^2 + b^2)(\cos\theta\,\mathbf{n}_2 + \sin\theta\,\mathbf{n}_3) = \frac{1}{12}m(a^2 + b^2)\mathbf{n},$$

indicating that $\mathbf{n}$ is a normalized autovector associated, therefore, to a principal direction of inertia. As θ is arbitrary, the presence of two identical roots $I_{22}^O = I_{33}^O = \frac{1}{12}m(a^2 + b^2)$ guarantees that every direction parallel to the plane $x_2 x_3$ is a principal direction of inertia. Lastly, if the solid is reduced to a cube, with $c = b = a$, the inertia tensor is

$$\mathbf{II}^{C/O} = \frac{1}{6}ma^2\,\mathbb{1}$$

and any direction $\mathbf{n}'$ will be a principal direction of inertia, since

$$\mathbf{I}_{n'}^O = \frac{1}{6}ma^2\,\mathbb{1} \cdot \mathbf{n}' = \frac{1}{6}ma^2\mathbf{n}'.$$

The above example also illustrates the existing relationship between planes of symmetry and the principal directions of inertia. As shown in Section 6.3, if a plane π is of symmetry for a body C, the inertia vector with respect to any point O of the plane for the direction $\mathbf{n}$ orthogonal to the plane is parallel to $\mathbf{n}$. In other words, this means that, for any point of a plane of symmetry, the direction orthogonal to it is a principal direction of inertia for the point. If there are two orthogonal planes of symmetry, then two principal directions of inertia are established also orthogonal to each other, for any point belonging to the line of intersection of the planes. Naturally there will be, in this case, a third principal direction of inertia orthogonal to the previous ones, completing the trihedron of the principal directions.

It is very worthwhile to identify planes of symmetry when establishing principal directions of inertia. Nevertheless, it has been demonstrated that for each and every body and point in space, the symmetry of the inertia tensor guarantees the existence of three principal directions of inertia orthogonal to each other, although there is no evident geometric symmetry in their mass distribution. In the latter case, establishing the principal moments of inertia requires solving Eq. (6.5). Having established the eigenvalues, the normalized eigenvectors are obtained from Eq. (6.1).

Example 6.2 The moments and products of inertia of the homogeneous rectangular plate of mass m with respect to its center O for axes $\{z_1, z_2, z_3\}$ (see Fig. 6.3) are equal to (see Appendix C)

$$I_{11}^O = \frac{1}{12}ma^2, \quad I_{22}^O = \frac{1}{3}ma^2, \quad I_{33}^O = \frac{5}{12}ma^2, \quad I_{12}^O = I_{23}^O = I_{31}^O = 0.$$

The axes $\{z_1, z_2, z_3\}$ define three mutually orthogonal planes of symmetry and are, therefore, principal directions of inertia of the plate for point O. Transferring the inertia properties for the axes $\{y_1, y_2, y_3\}$, parallel to the previous ones and with origin in vertex A, then we have, using Eqs. (5.7) and (5.8),

$$I_{11}^A = \frac{1}{3}ma^2, \quad I_{22}^A = \frac{4}{3}ma^2, \quad I_{33}^A = \frac{5}{3}ma^2,$$

$$I_{12}^A = -\frac{1}{2}ma^2, \quad I_{23}^A = 0, \quad I_{31}^A = 0.$$

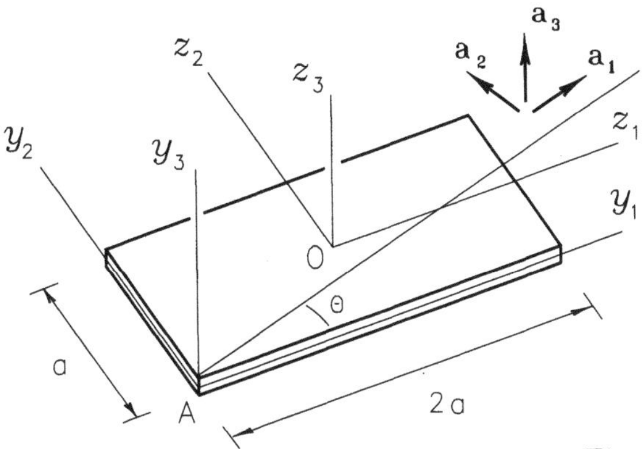

$$Figure\ 6.3$$

The inertia tensor of the plate with respect to point A may be expressed then by

$$\mathbf{II}^{P/A} = \frac{1}{3}ma^2 \begin{pmatrix} 1 & -3/2 & 0 \\ -3/2 & 4 & 0 \\ 0 & 0 & 5 \end{pmatrix}.$$

The characteristic equation to determine the eigenvalues, Eq. (6.5), is of the form (note that, to simplify, we will leave out the constant $\frac{1}{3}ma^2$; once the roots are found, the principal moments of inertia will be obtained by multiplying it by the constant put in evidence)

$$\begin{vmatrix} 1-\lambda & -3/2 & 0 \\ -3/2 & 4-\lambda & 0 \\ 0 & 0 & 5-\lambda \end{vmatrix} = 0,$$

which may be factored as

$$\big((1-\lambda)(4-\lambda) - 9/4\big)\big(5-\lambda\big) = 0$$

and whose roots are

$$\lambda_1 = \frac{1}{2}(5 - 3\sqrt{2}), \quad \lambda_2 = \frac{1}{2}(5 + 3\sqrt{2}), \quad \lambda_3 = 5.$$

Since three different roots were obtained, the principal directions of inertia for vertex A are orthogonal to each other, with two normalized eigenvectors, $\mathbf{a}_1$ and $\mathbf{a}_2$, parallel to the plane $y_1\,y_2$ and $\mathbf{a}_3$ orthogonal to it, since the body is flat and, in fact, $\lambda_3 = \lambda_1 + \lambda_2$. The principal moments of inertia will then be $I_j^A = \frac{1}{3}ma^2\lambda_j$, that is,

$$I_1^A = \frac{5 - 3\sqrt{2}}{6}ma^2, \quad I_2^A = \frac{5 + 3\sqrt{2}}{6}ma^2, \quad I_3^A = \frac{5}{3}ma^2.$$

The eigenvectors must fulfill Eq. (6.1). Expressing then $\mathbf{a}_1$ in coordinates $\{y_1, y_2, y_3\}$, $\mathbf{a}_1 = (\cos\theta, \sin\theta, 0)$, and replacing in Eq. (6.1), we have

$$\begin{pmatrix} 1 & -3/2 & 0 \\ -3/2 & 4 & 0 \\ 0 & 0 & 5 \end{pmatrix} \begin{pmatrix} \cos\theta \\ \sin\theta \\ 0 \end{pmatrix} = \lambda_1 \begin{pmatrix} \cos\theta \\ \sin\theta \\ 0 \end{pmatrix},$$

whose solution is (check it out)

$$\theta = 22^0\, 30'.$$

The other principal direction of inertia is naturally defined by (see Fig. 6.3)

$$\mathbf{a}_2 = \mathbf{a}_3 \times \mathbf{a}_1 = (-\sin\theta, \cos\theta, 0).$$

The principal directions of inertia of a body C with respect to a point O define three directions orthogonal to each other. If the roots of the characteristic equation are all different from each other, these directions are, as already shown, unique; when there are two repeated roots, we can always choose two directions orthogonal to each other and orthogonal to the autovector associated with the distinct root to form a Cartesian trihedron; in the case of repeating the three roots, every mutually orthogonal trihedron consists of a set of principal directions of inertia. So, by adopting a system of Cartesian axes with origin in O and parallel to the principal directions of inertia of the body with respect to O, the inertia matrix becomes diagonal. In fact, if $\mathbf{a}_1, \mathbf{a}_2, \mathbf{a}_3$ is an orthonormal basis parallel to the principal directions of inertia, all products of inertia for these coordinate directions, called *principal coordinates*, cancel each other out, since

$$I_{jk}^{C/O} = \mathbf{I}_j^{C/O} \cdot \mathbf{a}_k = I_{jj}^{C/O} \mathbf{a}_j \cdot \mathbf{a}_k = 0, \qquad j \neq k. \tag{6.12}$$

So, when adopting the principal directions of inertia as coordinate directions with origin on the point, the inertia tensor is expressed by

$$\mathbf{II}^{C/O} = \begin{pmatrix} I_1^O & 0 & 0 \\ 0 & I_2^O & 0 \\ 0 & 0 & I_3^O \end{pmatrix}, \tag{6.13}$$

where I_j^O, $j = 1, 2, 3$, are the principal moments of inertia of the body with respect to the point. One of the major advantages, therefore, of adopting principal coordinates lies in reducing the operations involving the inertia tensor. In fact, if $\mathbf{u}$ is a unit vector characterizing a given arbitrary direction with components (u_1, u_2, u_3) in the principal basis $\mathbf{a}_1, \mathbf{a}_2, \mathbf{a}_3$, the inertia vector of C with respect to O for direction $\mathbf{u}$ is, according to Eq. (3.4),

$$\mathbf{I}_u^{C/O} = \sum_{j=1}^{3} I_j^O u_j \mathbf{a}_j. \tag{6.14}$$

The moment of inertia of the body with respect to an axis E passing through O and parallel to the unit vector $\mathbf{u}$ is, according to Eq. (3.10),

$$I_E^C = \sum_{j=1}^{3} I_j^O u_j^2. \tag{6.15}$$

Lastly, if $\mathbf{v} = (v_1, v_2, v_3)$ is another arbitrary unit vector, the product of inertia of the body with respect to O for directions $\mathbf{u}$ and $\mathbf{v}$, is, according to Eq. (3.16),

$$I_{uv}^{C/O} = \sum_{j=1}^{3} I_j^O u_j v_j. \tag{6.16}$$

Example 6.3   Reverting to the previous example, choosing axes $\{x_1, x_2, x_3\}$, with origin in vertex A and parallel to the principal directions of inertia with respect to this point (see Fig. 6.4), the corresponding moments and products of inertia are

$$I_1^A = \frac{5 - 3\sqrt{2}}{6} ma^2, \quad I_2^A = \frac{5 + 3\sqrt{2}}{6} ma^2, \quad I_3^A = \frac{5}{3} ma^2,$$
$$I_{12}^A = I_{23}^A = I_{31}^A = 0.$$

The inertia tensor with respect to point A may then be expressed by the diagonal matrix

$$\mathbf{II}^{P/A} = \frac{1}{6} ma^2 \begin{pmatrix} 5 - 3\sqrt{2} & 0 & 0 \\ 0 & 5 + 3\sqrt{2} & 0 \\ 0 & 0 & 10 \end{pmatrix}.$$

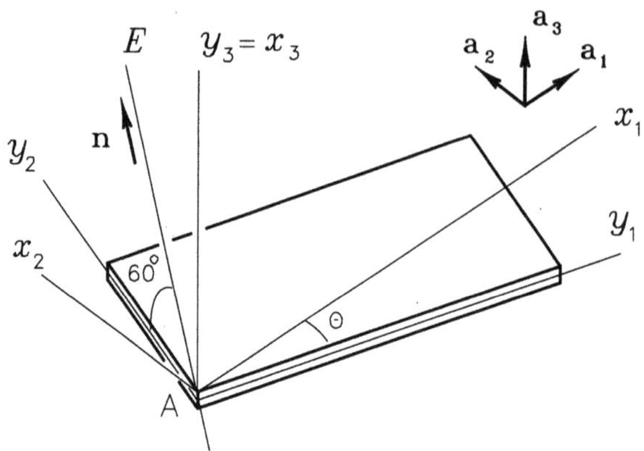

Figure 6.4

The unit vector $\mathbf{n}$, expressed on the principal basis, is

$$\mathbf{n} = \frac{1}{2}(\sin\theta\mathbf{a}_1 + \cos\theta\mathbf{a}_2 + \sqrt{3}\,\mathbf{a}_3).$$

The inertia vector of the plate with respect to point A for direction $\mathbf{n}$, according to Eq. (6.14), is

$$\mathbf{I}_n^{P/A} = \frac{1}{12}ma^2\left[(5-3\sqrt{2})\sin\theta\mathbf{a}_1 + (5+3\sqrt{2})\cos\theta\mathbf{a}_2 + 10\sqrt{3}\,\mathbf{a}_3\right].$$

The moment of inertia of the plate with respect to axis E is, according to Eq. (6.15),

$$\begin{aligned}
I_E^P &= \frac{1}{24}\left[(5-3\sqrt{2})\sin^2\theta + (5+3\sqrt{2})\cos^2\theta + 30\right]ma^2 \\
&= \frac{1}{24}(35 + 3\sqrt{2}\cos 2\theta)ma^2 \\
&= \frac{19}{12}ma^2.
\end{aligned}$$

We will now consider a body C and an arbitrary point O, and let $\{x_1, x_2, x_3\}$ be Cartesian axes with origin O and parallel to the principal directions of inertia of C with respect to O and E an axis passing through O parallel to the unit vector $\mathbf{a} = a_1\mathbf{n}_1 + a_2\mathbf{n}_2 + a_3\mathbf{n}_3$ (see Fig. 6.5). Also let P: (x_1, x_2, x_3) be a point of E whose distance d to O is defined by

$$d = \frac{k}{\sqrt{I_{aa}}}, \tag{6.17}$$

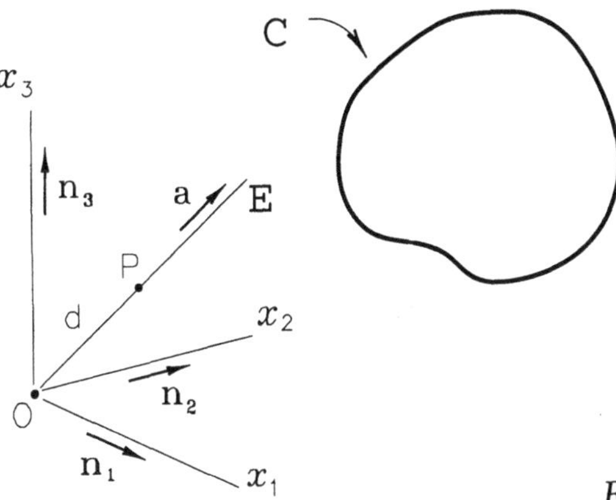

Figure 6.5

where k is an arbitrary constant with dimension $[M^{\frac{1}{2}}L^2]$. For each selected orientation, therefore, the quadratic distance from P to O will be inversely proportional to the moment of inertia for the direction; particularly, when $\mathbf{a} = \mathbf{n}_j$, P$=$ P$_j$, $j = 1, 2, 3$. Therefore, varying the orientation of $\mathbf{a}$, P describes a surface around point O, whose nature we will now investigate.

The moment of inertia of the body with respect to point O for direction $\mathbf{a}$ may be expressed as a function of the principal moments of inertia, according to Eq. (6.15), as

$$I_{aa} = I_1 a_1^2 + I_2 a_2^2 + I_3 a_3^2. \tag{6.18}$$

The coordinates x_j of point P and the components a_j of unit vector $\mathbf{a}$ are related by

$$a_j = \frac{x_j}{d}; \quad \text{therefore,} \quad a_j^2 = x_j^2 \frac{I_{aa}}{k^2}, \qquad j = 1, 2, 3. \tag{6.19}$$

Replacing Eq. (6.19) in Eq. (6.18) and dividing both members by I_{aa}, we get

$$1 = \frac{I_1}{k^2} x_1^2 + \frac{I_2}{k^2} x_2^2 + \frac{I_3}{k^2} x_3^2, \tag{6.20}$$

and, as $d_j = k/\sqrt{I_j}$ is the distance to O from point P$_j$, $j = 1, 2, 3$, then

$$1 = \frac{x_1^2}{d_1^2} + \frac{x_2^2}{d_2^2} + \frac{x_3^2}{d_3^2}, \tag{6.21}$$

an equation of an ellipsoid of semi-axes d_1, d_2, and d_3, called the *inertia ellipsoid* of the body with respect to the point (see Fig. 6.6). It is worth mentioning that the actual dimensions of the inertia ellipsoid will depend on a scale, depending on the arbitrated value for the constant k.

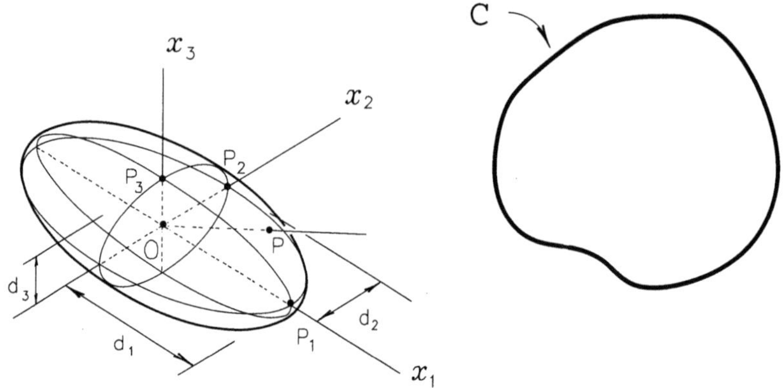

Figure 6.6

The inertia ellipsoid of a body C with respect to a point O has center in O and is oriented according to the principal directions of inertia of C with respect to O. The distance d from any point P of the surface of an ellipsoid to its center lies between its maximum and minimum semi-axes, that is, if the order of the axes is chosen so that $d_1 \geq d_2 \geq d_3$, then

$$d_1 \geq d \geq d_3. \tag{6.22}$$

In other words, since the ellipsoid has been built from quadratic distances to its center in inverse proportion to the moment of inertia associated with the direction, the minimum and maximum moments of inertia of the body with respect to the point will occur in principal directions of inertia and the moment of inertia for an arbitrary direction **a** will necessarily be between these values, that is,

$$I_1 \leq I_{aa} \leq I_3. \tag{6.23}$$

We can, therefore, say that every body is, from the viewpoint of its inertia of rotation relative to any point, equivalent to an ellipsoid, with mass equal to that of the body and semi-axes directed according to the principal directions of inertia of the body for that point.

Example 6.4 The inertia ellipsoid for rectangular plate P, analyzed in the two preceding examples, with respect to point O is oriented according to the main axes $\{z_1, z_2, z_3\}$ for the point (see Fig. 6.7a). Arbitrating, for convenience, $k = \sqrt{m/12}\,a^2$, the semi-axes of the ellipsoid will be, according to Eq. (6.17),

$$d_1 = \frac{k}{\sqrt{I_1^O}} = a, \quad d_2 = \frac{k}{\sqrt{I_2^O}} = \frac{1}{2}a, \quad d_3 = \frac{k}{\sqrt{I_3^O}} = \frac{1}{\sqrt{5}}a.$$

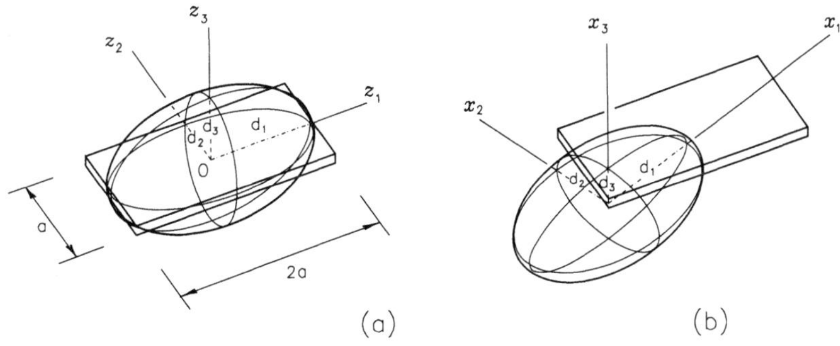

Figure 6.7

Note that d_1 is the largest semi-axis, associated with the minimum moment of inertia of the plate with respect to point O, $I_1^O = ma^2/12$, and that d_3 is the smallest semi-axis, associated with the maximum moment of inertia of the plate with respect to O, $I_3^O = 5ma^2/12$. The inertia ellipsoid of the plate with respect to its vertex A is oriented according to axes $\{x_1, x_2, x_3\}$ (see Fig. 6.7b), with semi-axes (adopting the same k)

$$d_1 = \frac{a}{\sqrt{2(5 - 3\sqrt{2})}}, \quad d_2 = \frac{a}{\sqrt{2(5 + 3\sqrt{2})}}, \quad d_3 = \frac{a}{2\sqrt{5}},$$

with $d_1 > d_2 > d_3$. As expected, when dealing with a flat body, the maximum moment of inertia occurs for the direction orthogonal to the plane, whatever the chosen point of the plane may be.

The moment of inertia of a body C with respect to a given axis E is always a nonnegative scalar that measures a distribution involving the mass and the average quadratic distance to the axis. The farther the average distance from the body to the axis, the greater the corresponding moment of inertia, and there will, therefore, be no upper limit for the moment of inertia of a body with respect to an axis. Approximating, inversely, the axis from the body, we find a minimum value for the moment of inertia. In fact, Eq. (5.3) establishes that, given an arbitrary direction **a**, the moment of inertia of the body with respect to any point O, for that direction, is the algebraic sum of the moment of inertia of the body with respect to its mass center, for the same direction, with the moment of inertia of mass center with respect to O. This means that, whatever the chosen direction, the point with respect to which the moment of inertia is minimum is the mass center of the body. On the other hand, one of the principal directions of inertia of the body with respect to any point is, as seen above, the direction of the smallest value for the moment of inertia of the body with respect to the point. So the conclusion is that the smallest absolute value for the moments of inertia of any body is that of the smallest moment of inertia for one of the three principal directions of inertia with respect to the mass center of the body.

Example 6.5   Returning once again to Example 6.2 (also see Examples 6.3 and 6.4), the smallest moment of inertia of the plate with respect to point A is the principal moment of inertia associated with axis x_1,

$$I_1^A = \frac{5 - 3\sqrt{2}}{6}\, ma^2 = 0.1262\, ma^2,$$

and the smallest absolute moment of inertia of the plate is the principal moment of inertia with respect to point O, associated with axis z_1,

$$I_1^O = \frac{1}{12}\, ma^2 = 0.0833\, ma^2.$$

As expected, $I_1^O < I_1^A$.

Exercise Series #8   (Sections 6.1 to 6.4)

P8.1   Establish the position of the centroid of the homogeneous body shown.

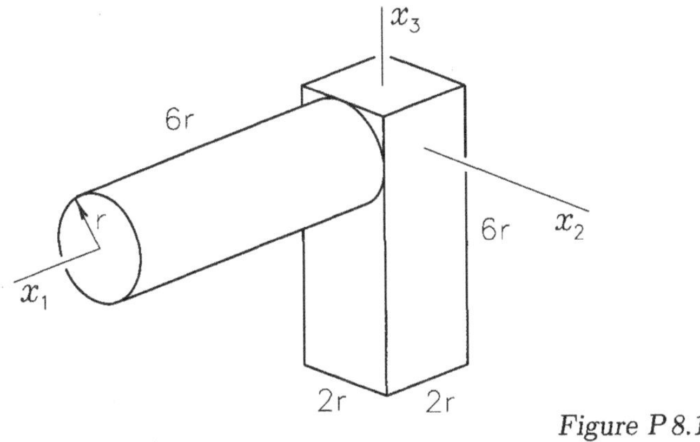

Figure P 8.1

P8.2   Establish the position of the centroid of the homogeneous plate.

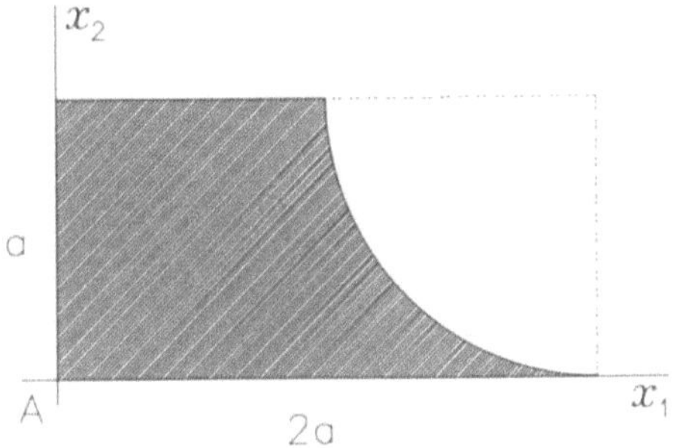

Figure P 8.2

P8.3 Establish the distance of the centroid of the homogeneous angle plate to axis x_1.

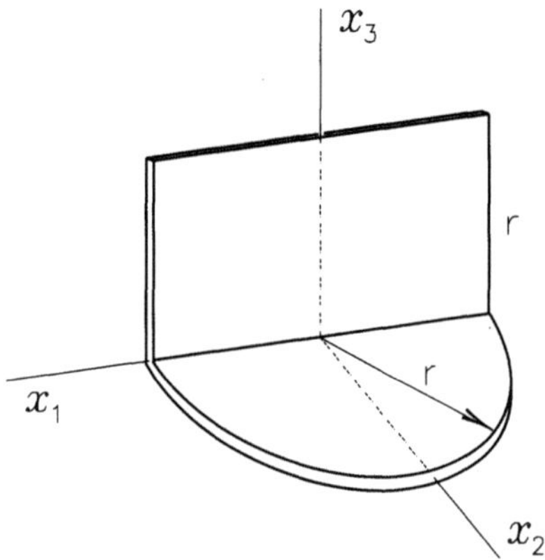

Figure P 8.3

P8.4 Establish the coordinates of the centroid of the disk with an off-center circular hole.

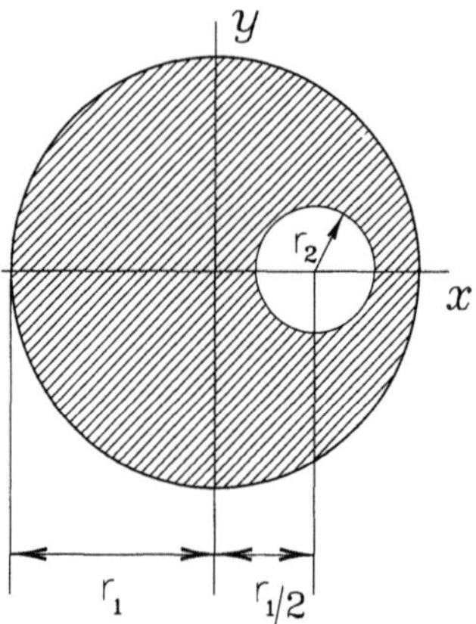

Figure P 8.4

P8.5 Compute the distance from point E to point A so that, when cutting
the rectangle according to segment CE and hanging the ABCE trapezoid by
vertex E, the side BC stays horizontal.

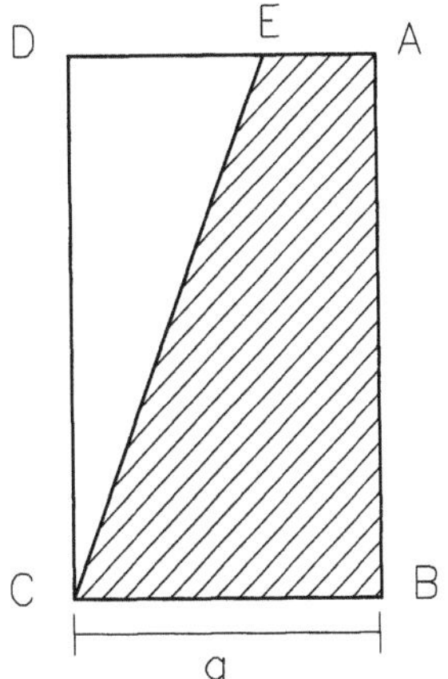

Figure P 8.5

P8.6 Find the coordinates of the mass center of a system of particles located
on the vertices of a rectangular parallelpiped, masses C, D, and F being equal
to 3 kg, masses E and G equal to 4 kg, mass A equal to 1 kg, mass B equal
to 2 kg, and mass H equal to 5 kg.

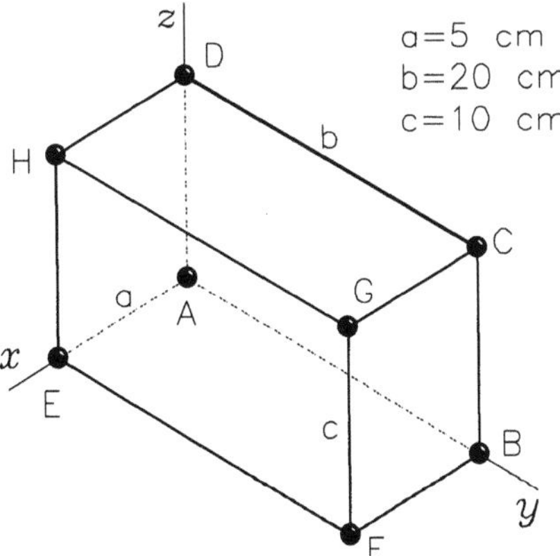

Figure P 8.6

P8.7 Find the coordinates of the centroid of the truss consisting of homogeneous rods of the same density. Quotas are in meters.

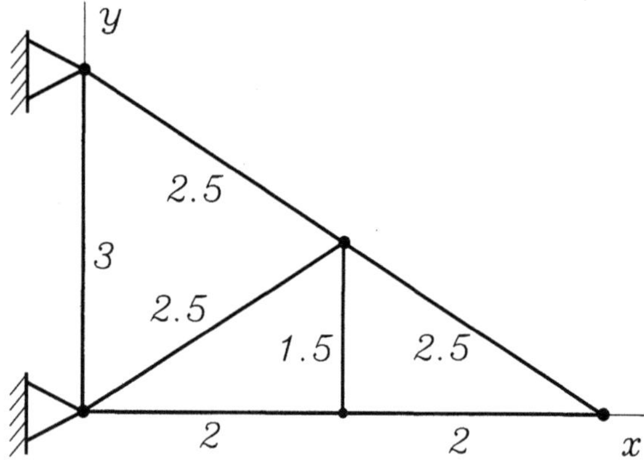

Figure P 8.7

P8.8 Two halves of a homogeneous cylinder with mass m are kept together using a rope with two equal masses at the end. Calculate the minimum value of each of these masses to keep the halves in contact, if there is no friction.

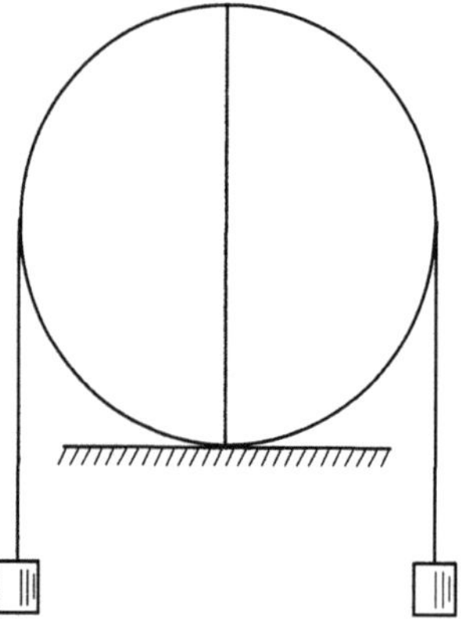

Figure P 8.8

P8.9 The body consists of a hemisphere and a cylinder of equal density, coupled together. Find the relation between the maximum height a of the cylinder and the radius r of the hemisphere so that the body stays in stable equilibrium in the position indicated.

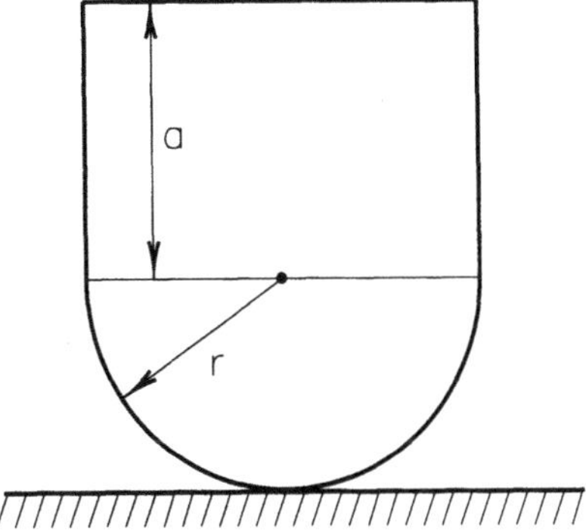

Figure *P 8.9*

P8.10 A small ball E, with mass equal to 5 kg, is set in the position shown. Given the vectors $\mathbf{a} = (1,1,0)$ and $\mathbf{b} = (0,3,4)$, find the moments of inertia of E with regard to point O for the directions defined by vectors $\mathbf{a}$ and $\mathbf{b}$ and the product of inertia of E with regard to O for the same directions.

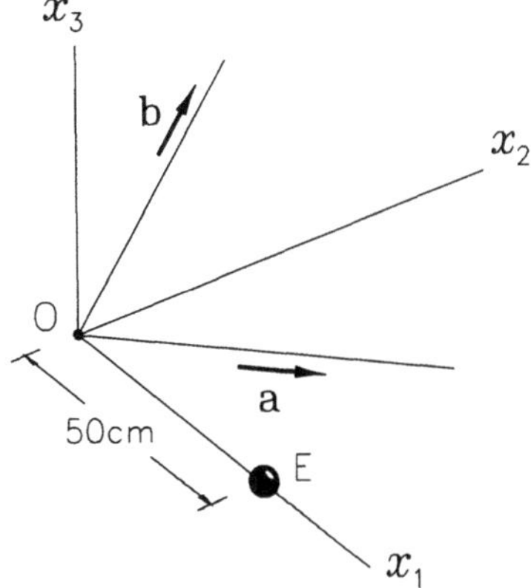

Figure *P 8.10*

P8.11 Consider the thread of beads hanging between two points with the same elevation, forming a catenary $\left(y = \frac{a}{2}(e^{x/a} + e^{-x/a} - 4)\right)$ with the arrow a, as shown. There are 62 identical beads of the same mass m distributed along the thread. Estimate the moment of inertia of the set with regard to the axis x.

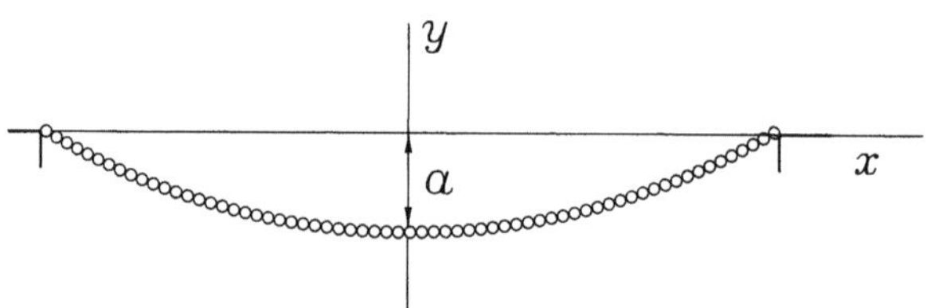

Figure P 8.11

P8.12 Given a system of particles $\mathcal{S}$ and an arbitrary point O, show that the sum of the moments of inertia of $\mathcal{S}$ with regard to three mutually orthogonal axes passing through O is an invariant.

P8.13 Show that the moment of inertia of a homogeneous sphere with regard to an axis that contains its center is smaller than the arithmetic average of the moments of inertia of any other body of the same mass with regard to three mutually orthogonal axes, passing through its mass center.

P8.14 Find the moments and products of inertia of the homogeneous bar, of mass equal to 3 kg and length equal to $\sqrt{2}/2$ m, with regard to point A, for the coordinate directions.

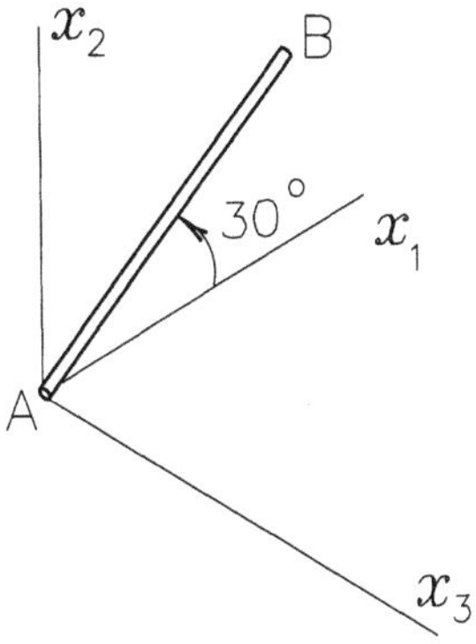

Figure P 8.14

P8.15 A homogeneous wire, with mass m, is shaped around a cylinder, forming a helix with constant pace that completes a whole turn around the cylinder, as shown. Find the inertia vector of the wire with regard to point O, the center of the cylinder base, for direction $\mathbf{n}_1$.

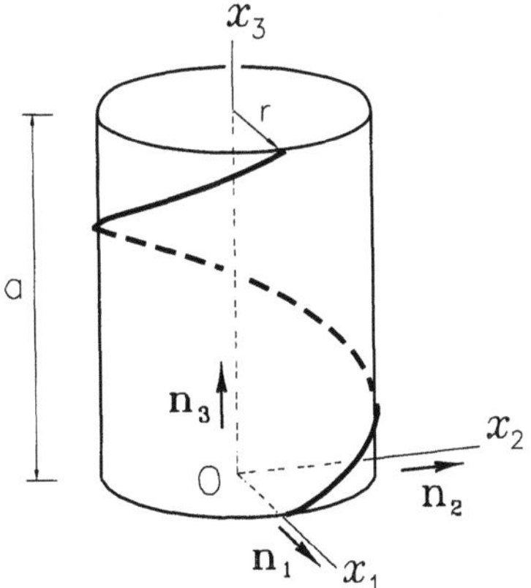

Figure P 8.15

P8.16 Consider the homogeneous rod B, with mass m, situated in relation to the system of coordinates $\{x_1, x_2, x_3\}$, as shown. Find the inertia vector of the rod with respect to O for direction $\mathbf{n}$. Also, find its product of inertia with respect to O for directions $\mathbf{n}$ and $\mathbf{n}_1$.

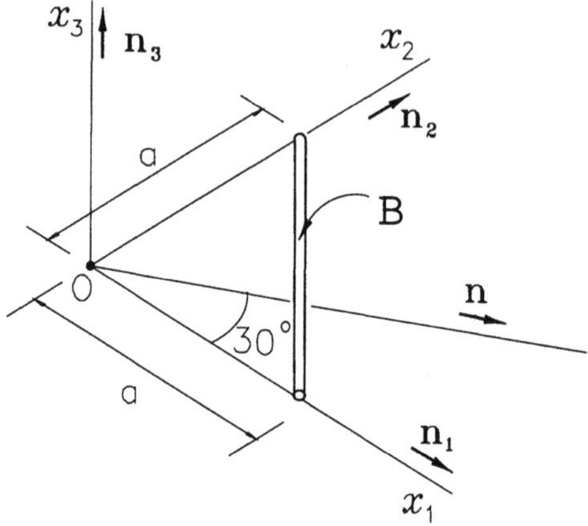

Figure P 8.16

P8.17 Find the moments of inertia of the homogeneous angle plate with density ρ with respect to the coordinate axes.

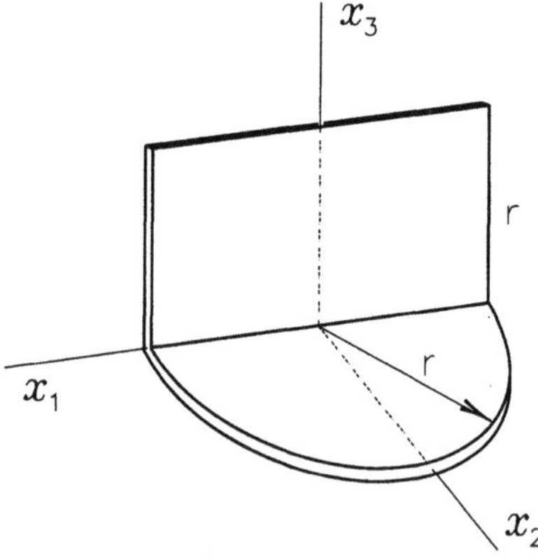

Figure P 8.17

P8.18 When using an eraser to rub out a mistake in his notebook, the student discovered that the heavy table shook much less when the movement was in the direction $b - b$ than in $a - a$. Why do you think this happens?

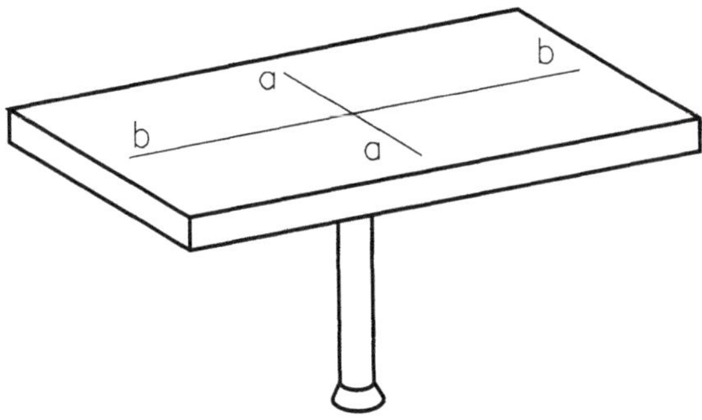

Figure P 8.18

P8.19 Show that for any body and any of its points, the moment of inertia with respect to a given direction cannot be greater than the sum of the moments of inertia with respect to two directions orthogonal to the former.

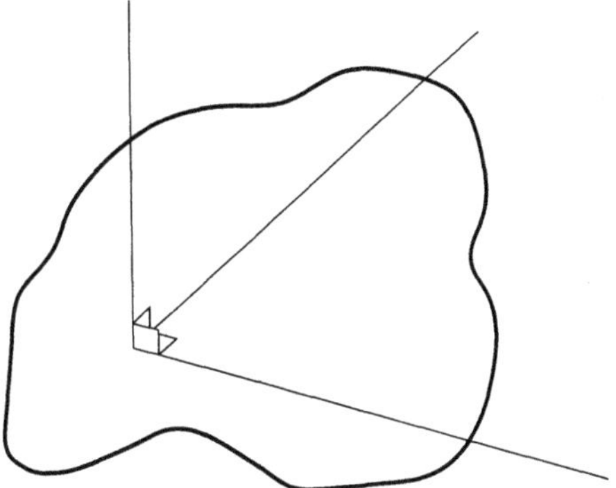

Figure P 8.19

Exercise Series #9   (Sections 6.5 to 6.6)

P9.1 Consider the system comprising four particles of the same mass m, in the configuration shown. If (u_1, u_2, u_3) are the director cosines of axis E, passing through P_0, find the moment of inertia of the system with regard to E. Also, calculate the product of inertia of the system with regard to its mass center for the directions parallel to x_1 and x_2.

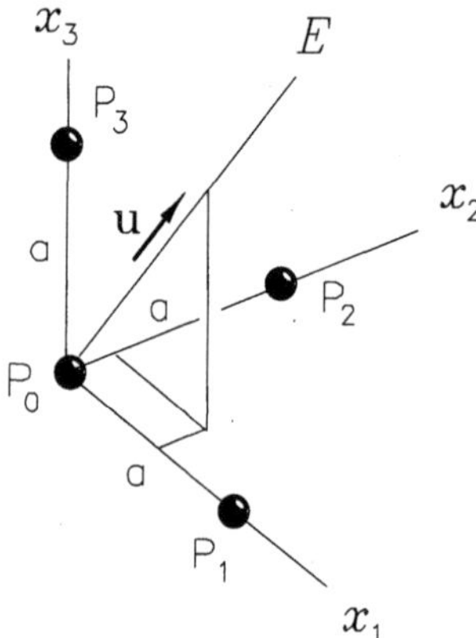

Figure P 9.1

P9.2 Find the moment of inertia of the homogeneous ring of mass m with regard to axis E.

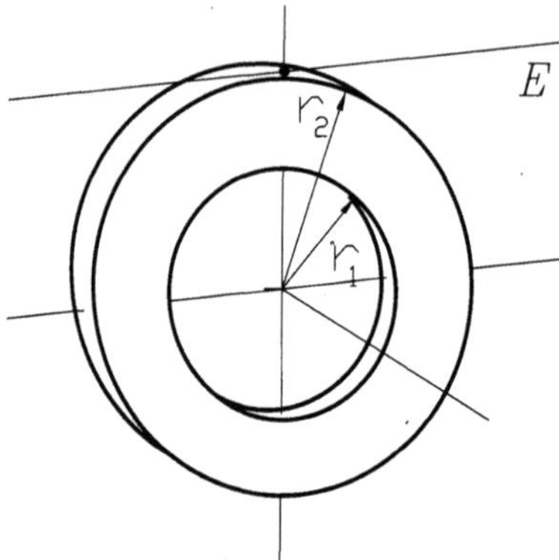

Figure P 9.2

P9.3 The solid consists of a cone of mass equal to 2.5 kg, welded to a hemisphere of mass equal to 5 kg. Find the moment of inertia of the body with regard to axis x_1 and the product of inertia with regard to point A for directions $\mathbf{n}$ and $\mathbf{n}_1$.

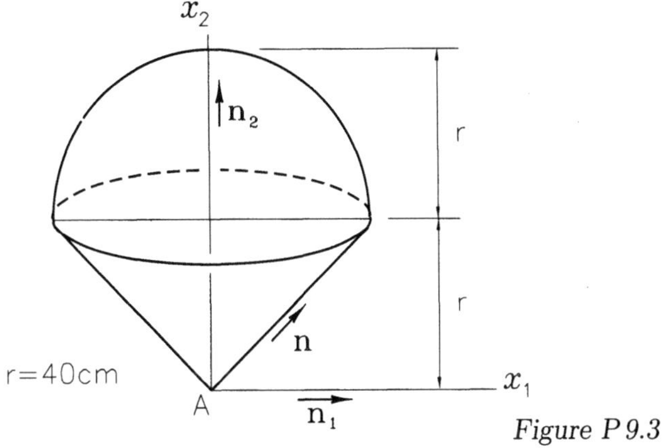

Figure P 9.3

P9.4 Two squares and a circle were cut from a thin steel plate, with density equal to 8 kg/m^2, and placed one on top of the other, as shown. Find the moment of inertia of the body with regard to axis E.

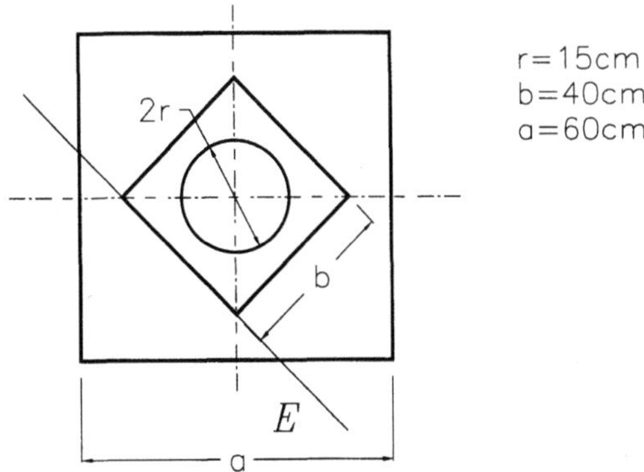

Figure P 9.4

P9.5 A homogeneous rod of mass m and length $3a$ is folded as shown. Establish its inertia matrix with regard to point O for the Cartesian axes indicated.

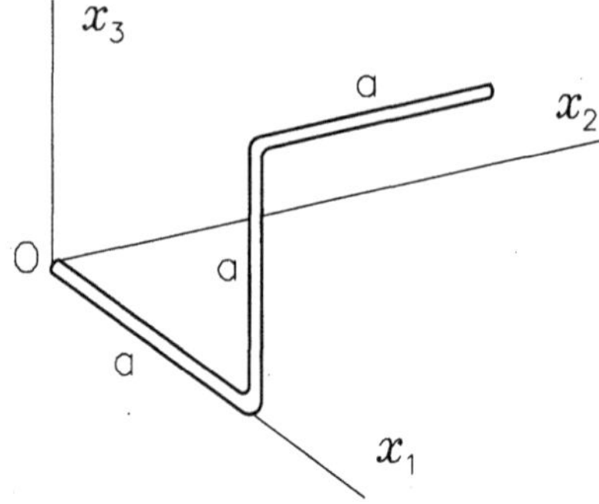

Figure P 9.5

P9.6 Consider the homogeneous annular cylinder with an external radius R. Find the value of the inner radius r that maximizes the moment of inertia of the cylinder with regard to axis z', parallel to the axis of symmetry z and tangent to the inner edge.

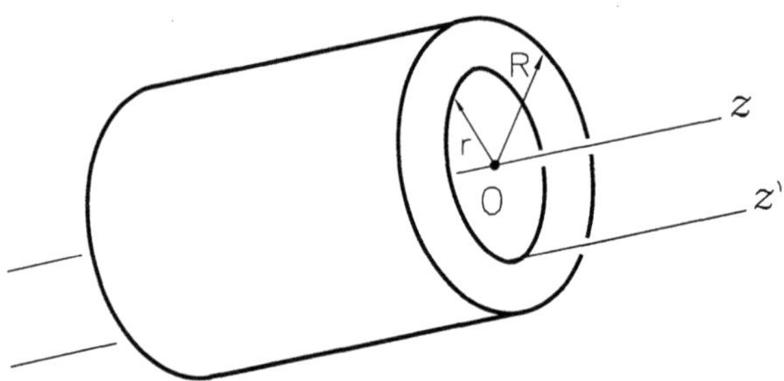

Figure P 9.6

P9.7 Consider a thin prismatic body, with length c and square cross section with edge a, whose longitudinal axis of symmetry is parallel to axis x_1, d being the distance between them, as shown. Analyze the error made in computing the moments of inertia with regard to the Cartesian axes indicated when dimension a is neglected, considering the body to be a *rod* rather than a *parallelpiped*. Estimate the maximum values of the ratios a/c and a/d so that the percentual error made when calculating any of the moments of inertia with respect to the Cartesian axes shown in the figure is below 0.1%.

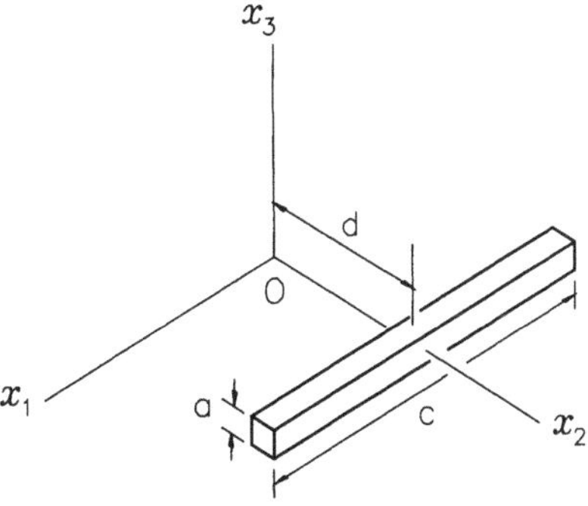

Figure P 9.7

P9.8 Three identical square homogeneous plates will be welded as outlined in Fig. P9.8. Find the value to be chosen for the angle θ so that the moment of inertia of the set with regard to axis E is minimum.

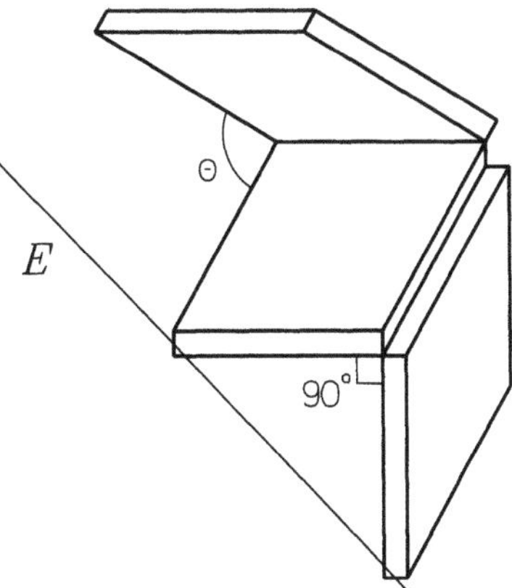

Figure P 9.8

P9.9 Consider a spherical shell with mass m and radius r. Find the inertia vector with respect to point Q for the direction of the unit vector **n**, and compute the moment of inertia with respect to axis E.

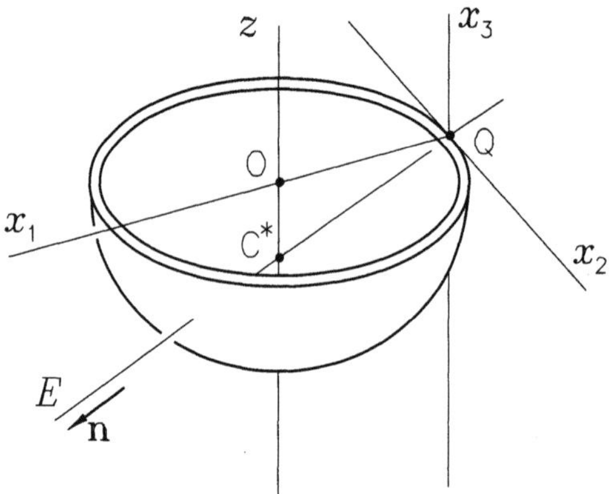

Figure *P 9.9*

P9.10 Calculate the moments of inertia of the homogeneous body with density ρ with respect to the coordinate axis.

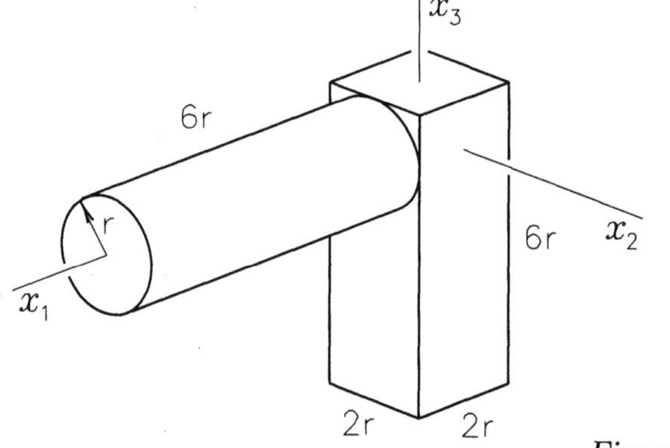

Figure *P 9.10*

P9.11 Find the moment of inertia of the homogeneous body of density ρ with respect to axis E.

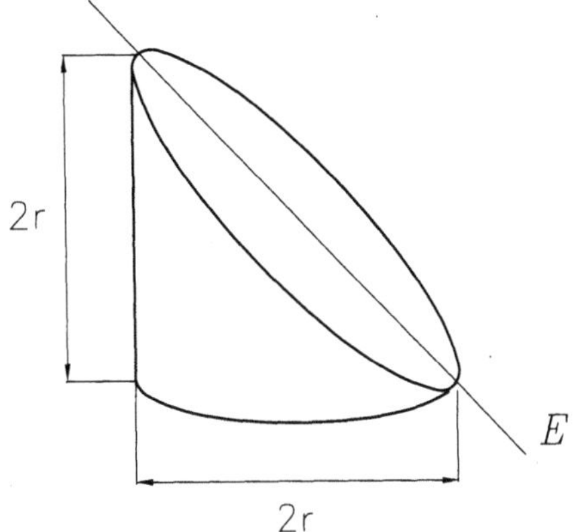

Figure P 9.11

P9.12 The particles P_0, P_1, P_2, and P_3, each with mass m, occupy the vertices of two rectangular triangles, as shown. Find the minimum and maximum moment of inertia of this system with regard to P_0.

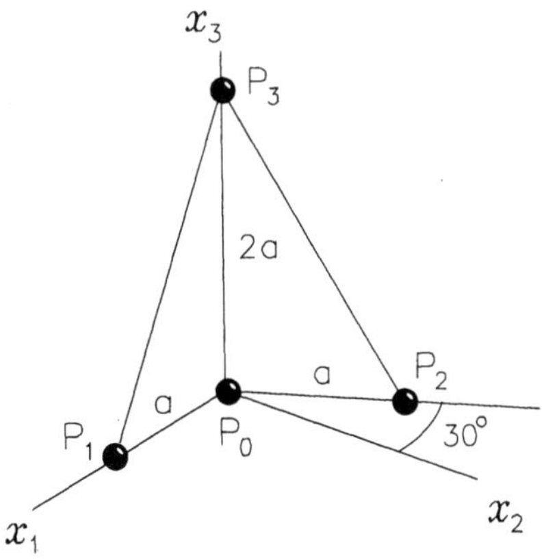

Figure P 9.12

P9.13   The homogeneous rod of mass m and length $2a$ is folded as shown. Find its principal moments of inertia with regard to point O.

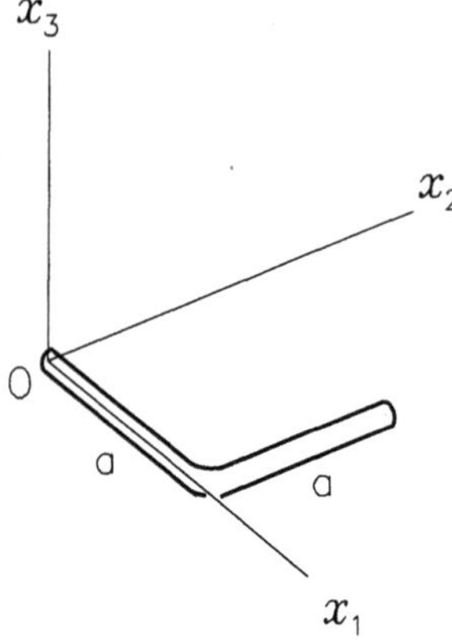

Figure P 9.13

P9.14   Find the distance from vertex A to a point O on the hypotenuse AB of the triangular plate, so that the line passing through A and B is a principal axis of inertia of the plate with regard to O.

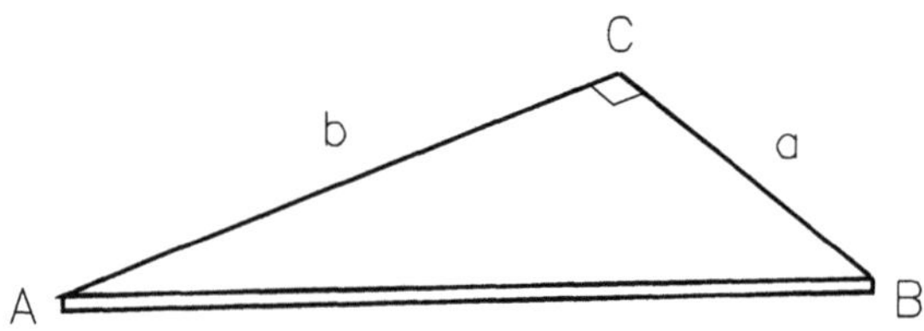

Figure P 9.14

P9.15 Find the smallest angle between axis x_1 and the axis of minimum moment of inertia of the homogeneous perforated plate, with regard to point A.

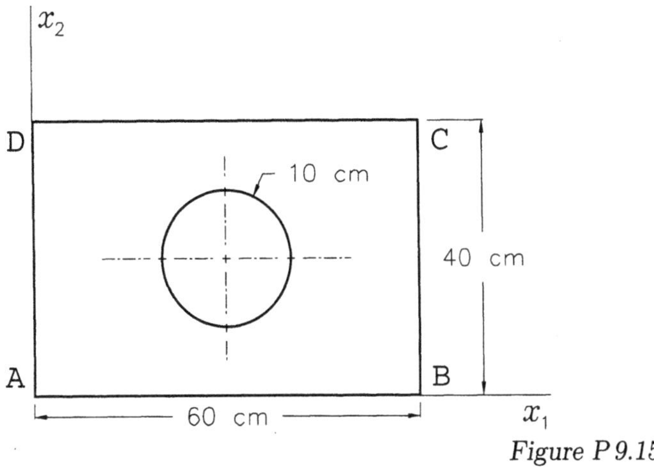

Figure P 9.15

P9.16 Find the ratio between the height and radius of a homogeneous cylinder so that the inertia ellipsoid with regard to the center of one of the bases is reduced to a sphere.

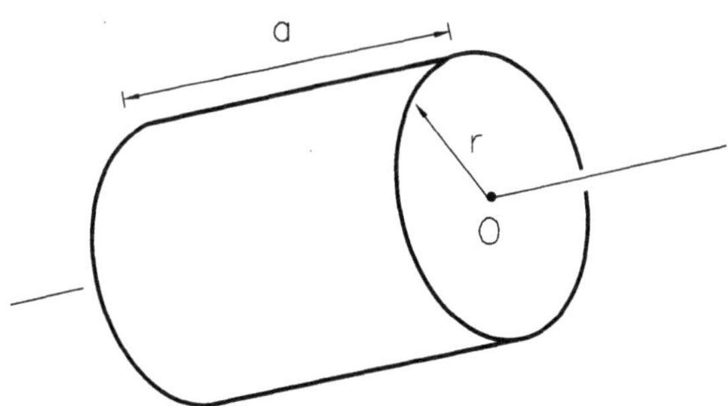

Figure P 9.16

P9.17 Show that the inertia ellipsoid of a homogeneous cube with regard to a vertex is of revolution. Then find the principal moments of inertia of a cube with mass m and edge a with regard to a vertex.

P9.18 The homogeneous disk, with mass $2m$, is welded to the half-wheel of mass m, as shown. Find the moment of inertia of the set with regard to axis E. What is the proportion between the semi-axes of the inertia ellipsoid with regard to point O?

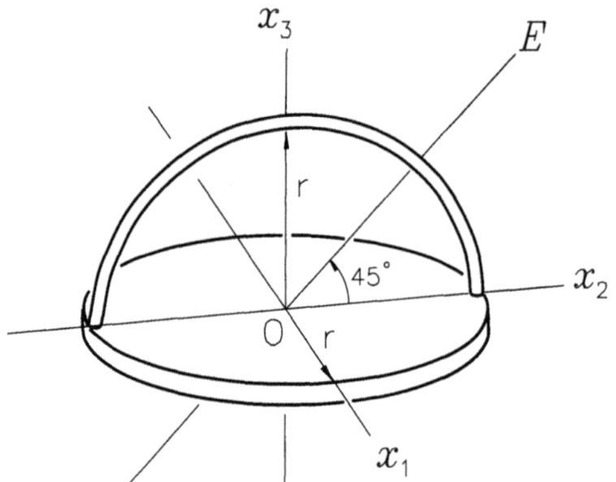

Figure P 9.18

P9.19 The body C, with mass m, is a solid of revolution about x_3. Due to a manufacturing flaw, the distribution of mass is imperfect, so that the principal axis of inertia with a maximum moment of inertia with regard to the mass center O undergoes a slight deviation θ in plane $x_1 x_3$, as shown.

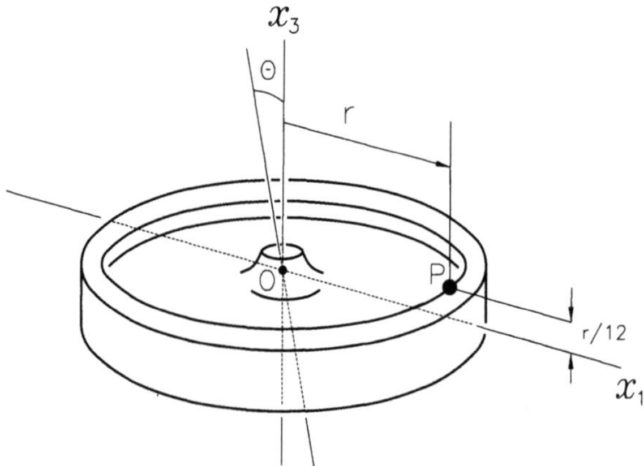

Figure P 9.19

Knowing that $I_{13}^{C/O} = mr^2/720$, estimate the amount of mass to be added in point P, as shown, in order to correctly balance the body, bringing x_3 into a principal axis of inertia.

P9.20 Demonstrate the following inertia properties, valid for every rigid body C with mass center P*:

- A principal axis of inertia for P* is also a principal axis of inertia for all points of the axis.
- If a line is a principal axis of inertia of the body for two of its points, then it contains the mass center of the body and is a principal axis of inertia for P*.
- If P is a point belonging to one of the principal axes of inertia with regard to P*, then the principal directions of inertia for P and P* are parallel.

P9.21 Demonstrate that the inertia ellipsoids of the five regular polyhedrons (tetrahedron, hexahedron, octahedron, dodecahedron, and icosahedron) with respect to their geometric centers are all spherical.

P9.22 Let **a** be a unit vector of any direction and P be the intersection point of the line passing through an arbitrary point O and is parallel to **a**, with the surface of the ellipsoid of inertia of a rigid body with regard to O, as illustrated. Show that the inertia vector of the body with regard to O for direction **a** is orthogonal to the ellipsoid surface at point P.

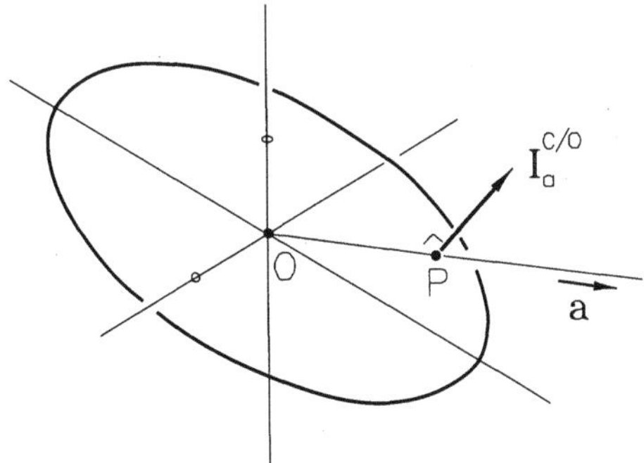

Figure P 9.22

P9.23   The china vase has mass m and mass center O, located on its axis of symmetry x_3. Axes $\{x_1, x_2, x_3\}$ are, therefore, one of the possible orientations for the principal directions of inertia of the vase with respect to O. Q is the intersection point of axis x_2 with the outer surface of the vase. What would you expect with regard to the principal directions of inertia of the vase with respect to Q? If I and J are the transverse and longitudinal moments of inertia, respectively, with respect to the mass center, compute the inertia tensor with respect to point Q.

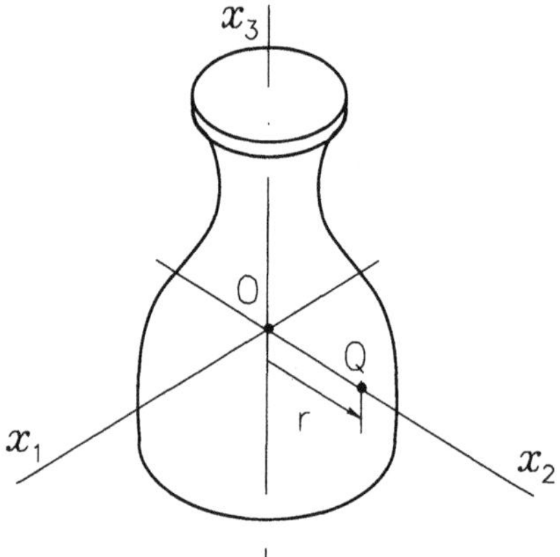

Figure P 9.23

P9.24   Find the principal moments of inertia of the homogeneous equilateral triangular plate with mass m, with respect to vertex A.

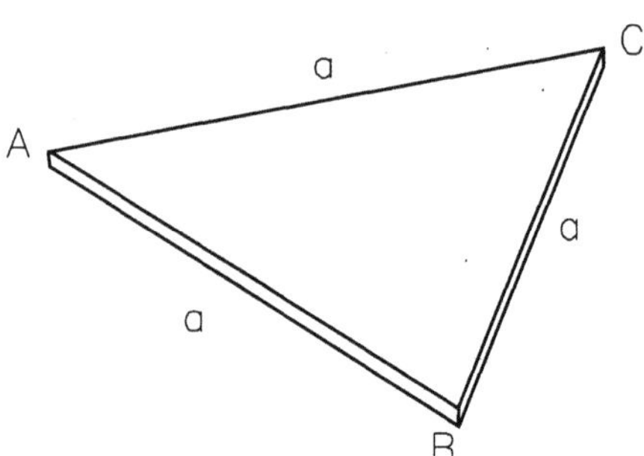

Figure P 9.24

P9.25 Find the principal moments of inertia of the homogeneous wire with mass m with respect to point O.

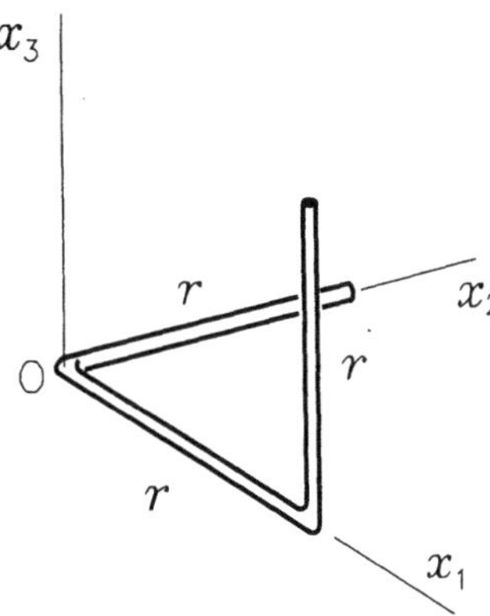

Figure P 9.25

P9.26 Find the smallest moment of inertia of the semicircular wire of mass m and radius r with respect to point A.

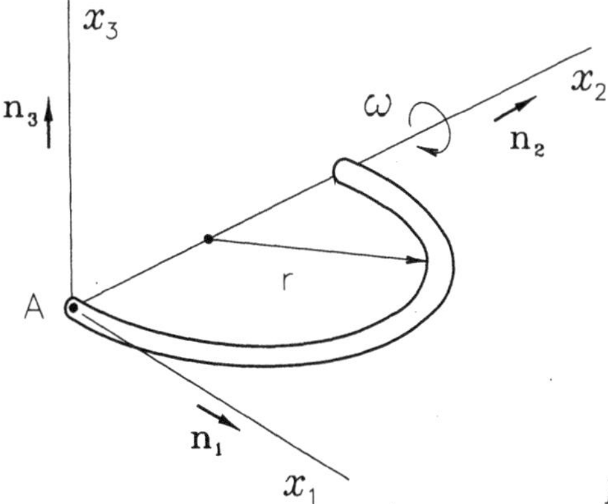

Figure P 9.26

P9.27　Consider the spherical shell with mass m and radius r. Calculate the principal moments of inertia for point Q.

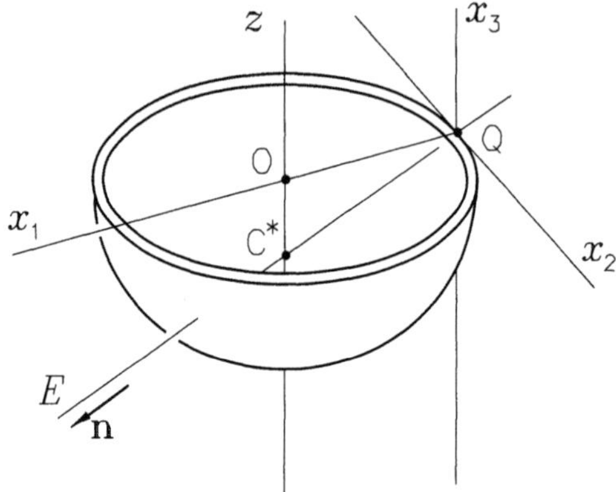

Figure P 9.27

P9.28　A letter **G** is shaped from a homogeneous wire with the indicated geometry. Find the smallest angle between the axis with minimum moment of inertia of the body with respect to point O and the coordinate axis x_1.

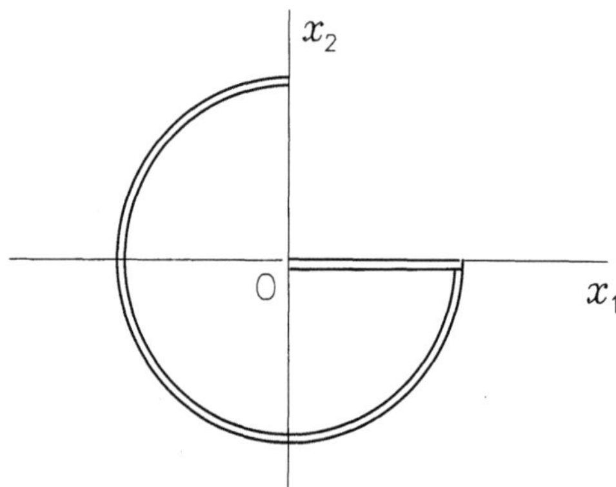

Figure 9.28

P9.29 Calculate the smallest moment of inertia of the homogeneous plate with a surface density ρ with respect to point A.

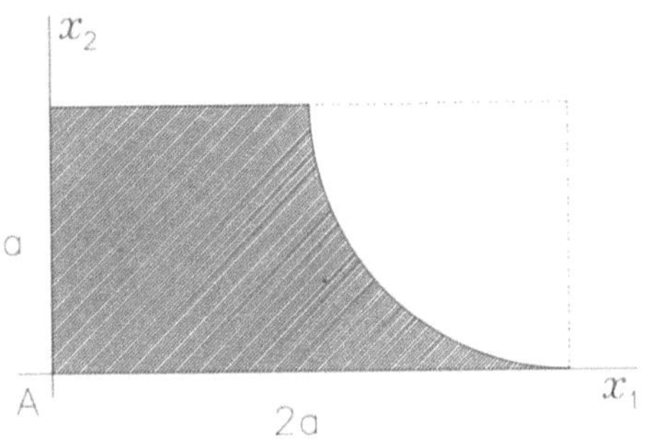

Figure P 9.29

P9.30 Find the moment of inertia of the hollow homogeneous cylinder, of mass m, with respect to axis E. Once radii a and b are fixed, what should the height c be so that the ellipsoid of inertia of the body with respect to its center O degenerates into a sphere?

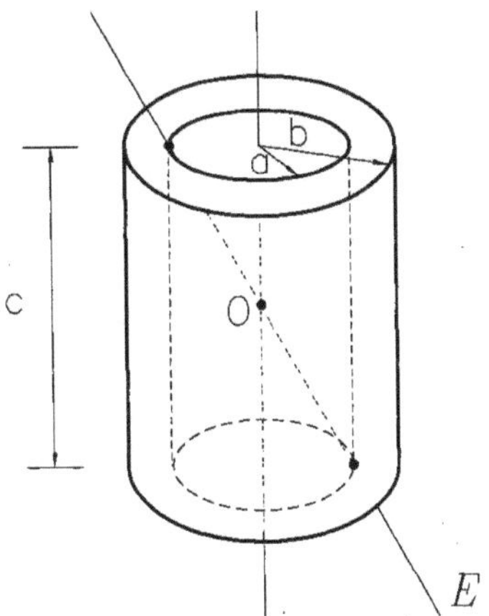

Figure P 9.30

Dynamics of the Rigid Body

Chapter 7

This chapter completes a study that has been taking shape in the earlier chapters. In fact, the analysis of the motion of a rigid body assumes the mastering of forces, kinematics, principles ruling the variation of its dynamic properties, and the inertia properties of the body. To determine the motion of a rigid body it is first necessary to carefully analyze the system of forces acting on the body. After identifying the forces and torques applied — the corresponding free body diagram having therefore been made — it is necessary to reduce the system to the mass center of the body or to another specific point, where applicable; to do so, the methods studied in Chapter 2 will be used. Next, we must identify the degrees of freedom of the body, characterize its holonomicity, and choose a set of coordinates that describe its configuration in a given reference frame. The determination of its angular velocity and acceleration, as well as the velocity and acceleration of its mass center — each of these vectors expressed in terms of the chosen coordinates and their time derivatives — is the following task. The use of kinematic theorems, studied in Chapter 3, especially facilitates this task. The basic principles of dynamics are introduced, from the conceptual viewpoint, in Chapter 1 and applied to the motion of the particle in Chapter 4. The principal methods of dynamic analysis are discussed therein. Chapter 5 generalizes dynamic concepts and principles to be employed in discrete systems of particles and continuous bodies. The same general equations for the motion of the mass center of a body, the motion of the body around the mass center, and the relations between work and energy will be adopted in the study of the motion of the rigid body, with notable simplifications resulting from the hypothesis of rigidity. Chapter 6, lastly, studies the inertia of the rigid body. Translation inertia — the mass — brings nothing new in relation to a general system; inertia of rotation, characterized by the inertia tensor, will, on the contrary, bring the equations to a new way, simplifying the study of the motion of the rigid body, as will be seen in this chapter.

Section 7.1 introduces the dynamic properties for a rigid body. Here it is shown that the angular momentum and kinetic energy of the body can be expressed as a function of its inertia tensor with respect to the mass center and the angular velocity vector of the body. Section 7.2 discusses the central theme of the chapter: the equations of motion for a rigid body. It shows that there are two kind of equations. The equations of the first kind are equations governing the movement of the mass center of the body, similar to the equations of motion for a particle, relating the time rate of the momentum vector to the resultant force applied. The equations of the second kind establish the rotating motion of the body around its mass center, involving the inertia tensor, and relate the time rate of the angular momentum vector to the resultant torque. Section 7.3 generalizes the work concept of a general system of forces, showing how to evaluate the work done by a torque applied to a rigid body. In Section 7.4, the resultant work on a rigid body is related to the variation of its kinetic energy in a given interval, establishing the energy balances. Section 7.5 is devoted to the study of plane motion; it shows that three scalar equations are generally available, and the conditions for maintaining plane motion are analyzed. These five sections provide the basic tools to describe, in principle, each and every situation involving the motion of a rigid body. More specific and interesting applications, such as gyroscopes, some aspects of stability of motion, and situations involving collision between bodies, will be discussed in Chapter 8.

7.1 Dynamic Properties

A body is said to be rigid when all its elements maintain invariant mutual distances in time. The hypothesis of rigidity, therefore, guarantees that if P and Q are two points of a rigid body C moving arbitrarily in a reference frame $\mathcal{R}$ (see Fig. 1.1), the velocity component of P relative to Q in $\mathcal{R}$ in the direction of line (r) containing P and Q is necessarily null; in other words, the following kinematic constraint must be fulfilled for any two points P and Q of C:

$$^{\mathcal{R}}\mathbf{v}^{P/Q} \cdot \mathbf{p}^{P/Q} = 0, \tag{1.1}$$

where $\mathbf{p}^{P/Q}$ is the position vector of P with respect to Q, parallel to line r. In fact, the relative velocity between two points of a rigid body may, according to Eq. (3.8.4), be expressed by $^{\mathcal{R}}\mathbf{v}^{P/Q} = {}^{\mathcal{R}}\boldsymbol{\omega}^{C} \times \mathbf{p}^{P/Q}$, a vector orthogonal to $\mathbf{p}^{P/Q}$. The result, then, is that the dot product of relative velocity with the position vector will be null for any instant. Equation (1.1) may be interpreted, therefore, as the kinematic translation of the hypothesis of rigidity.

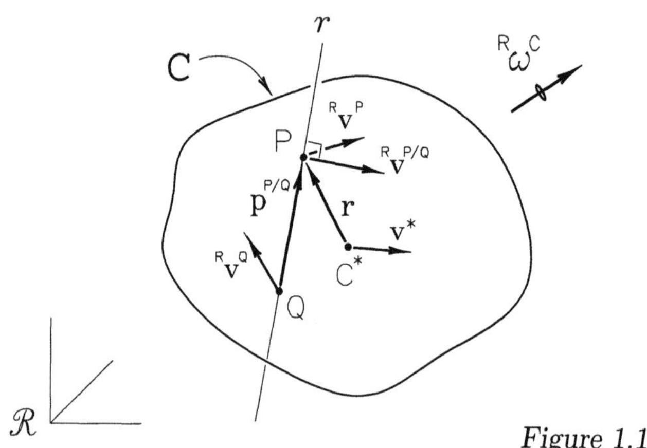

Figure 1.1

The kinematic theorem for velocities applied to a rigid body establishes, as shown in Section 3.8, that if P and Q are any two points of a rigid body C moving in a given reference frame $\mathcal{R}$, with angular velocity $^{\mathcal{R}}\boldsymbol{\omega}^{C}$, then the velocities of P and Q in $\mathcal{R}$ satisfy the relationship

$$^{\mathcal{R}}\mathbf{v}^{P} = {}^{\mathcal{R}}\mathbf{v}^{Q} + {}^{\mathcal{R}}\boldsymbol{\omega}^{C} \times \mathbf{p}^{P/Q}. \tag{1.2}$$

In particular, if C^* is the mass center of C and $\mathbf{v}^* = {}^{\mathcal{R}}\mathbf{v}^{C^*}$ is its velocity in $\mathcal{R}$ (see Fig. 1.1), then the velocities of P and C^* in $\mathcal{R}$ will be related according to

$$^{\mathcal{R}}\mathbf{v}^{P} = \mathbf{v}^* + {}^{\mathcal{R}}\boldsymbol{\omega}^{C} \times \mathbf{r}, \tag{1.3}$$

where $\mathbf{r} = \mathbf{p}^{P/C^*}$ is the position vector of point P with respect to the mass center of the body.

When a body C, with mass m, moves in a reference frame $\mathcal{R}$, each point P of the body, representative of an infinitesimal mass element dm, will have velocity $\mathbf{v} = {}^{\mathcal{R}}\mathbf{v}^P$ and momentum $d\mathbf{G} = \mathbf{v}\,dm$ (see Fig. 1.2). The set of vectors $d\mathbf{G}$ is, as shown in Section 5.1, a simple distributed system $\mathcal{G}$ of momentum vectors, whose resultant is the *momentum vector of the rigid body*, in the reference frame $\mathcal{R}$, given by

$$ {}^{\mathcal{R}}\mathbf{G}^C \rightleftharpoons \int_C d\mathbf{G} = \int_C \mathbf{v}\,dm. \tag{1.4} $$

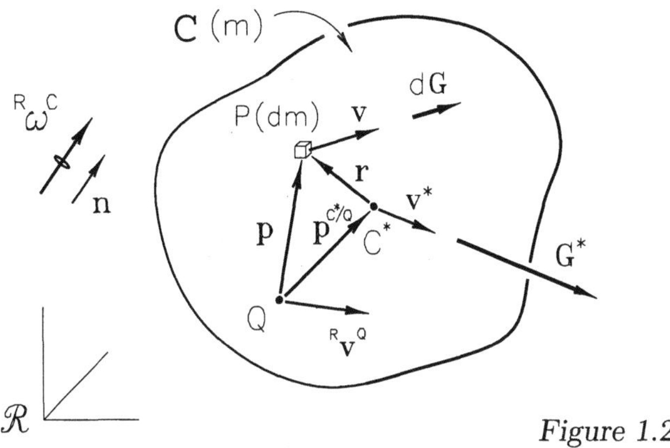

Figure 1.2

We call the vector

$$ \mathbf{G}^* = {}^{\mathcal{R}}\mathbf{G}^{C^*} \rightleftharpoons m\mathbf{v}^* \tag{1.5} $$

the *momentum of the mass center of the body*, that is, the momentum vector, in $\mathcal{R}$, of a (fictitious) particle with mass equal to that of the body and moving in $\mathcal{R}$ as its mass center. Of course, given the definition of the mass center of the body, Eq. (6.1.4), then

$$ {}^{\mathcal{R}}\mathbf{G}^C = \mathbf{G}^* = m\mathbf{v}^*. \tag{1.6} $$

Example 1.1   Body C consists of a homogeneous cylinder, with mass m, welded to a light rod, with the dimensions indicated (see Fig. 1.3).

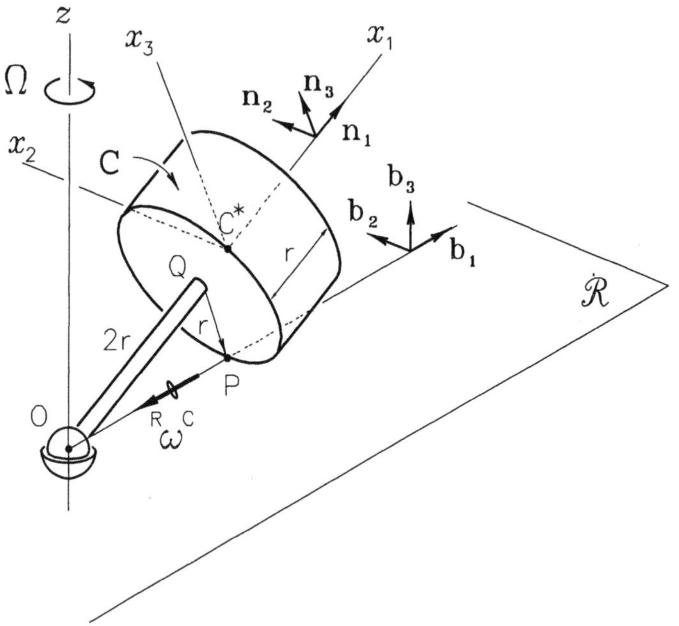

Figure 1.3

The free end of the rod is connected to point O, fixed to the horizontal plane by means of a ball joint and the cylinder rolls over the plane, the reference frame $\mathcal{R}$, a point P of its bottom edge always being in contact with the latter. The motion of C in $\mathcal{R}$ is prescribed and may be characterized as follows. Consider a reference frame B, defined by the vertical plane containing axis z and mass center C^* of C. Orthonormal bases $\mathbf{b}_1, \mathbf{b}_2, \mathbf{b}_3$ and $\mathbf{n}_1, \mathbf{n}_2, \mathbf{n}_3$ are fixed in B, with the directions indicated in the figure. B moves in relation to $\mathcal{R}$ with vertical simple angular velocity of constant module Ω, as shown. All points on axis x_1, therefore, also have a constant speed in reference frame $\mathcal{R}$. As points O and P are fixed in C, O is at the same time fixed in $\mathcal{R}$ and, at each instant, the rolling conditions ensure that the velocity of P in $\mathcal{R}$ is null, that is, $^{\mathcal{R}}\mathbf{v}^P = {}^{\mathcal{R}}\mathbf{v}^O = 0$. Then from Eq. (1.2), we have

$$^{\mathcal{R}}\boldsymbol{\omega}^C \times \mathbf{p}^{P/O} = 0,$$

so

$$^{\mathcal{R}}\boldsymbol{\omega}^C = \omega\mathbf{b}_1.$$

But, from the additiveness relation of angular velocities, Eq. (3.3.10),

$$^{\mathcal{R}}\boldsymbol{\omega}^C = {}^{B}\boldsymbol{\omega}^C + {}^{\mathcal{R}}\boldsymbol{\omega}^B = \omega_1\mathbf{n}_1 + \Omega\mathbf{b}_3.$$

By comparing the last two expressions, it is then easy to see that (check)

$$\omega = -2\Omega \quad \text{and} \quad \omega_1 = -\sqrt{5}\,\Omega$$

and, therefore, that

$$^{\mathcal{R}}\boldsymbol{\omega}^C = -2\Omega\mathbf{b}_1 = \frac{2}{\sqrt{5}}\Omega(-2\mathbf{n}_1 + \mathbf{n}_3).$$

The velocity of the mass center may then be obtained from the kinematic theorem of velocities, resulting in

$$\begin{aligned}
\mathbf{v}^* &= {}^{\mathcal{R}}\mathbf{v}^O + {}^{\mathcal{R}}\boldsymbol{\omega}^C \times \mathbf{p}^{C^*/O} \\
&= 0 + \frac{2}{\sqrt{5}}\Omega(-2\mathbf{n}_1 + \mathbf{n}_3) \times \frac{5}{2}r\mathbf{n}_1 \\
&= \sqrt{5}\,\Omega r\mathbf{n}_2.
\end{aligned}$$

The momentum vector of body C may then be obtained from Eq. (1.6), that is,

$$^{\mathcal{R}}\mathbf{G}^C = m\mathbf{v}^* = \sqrt{5}\,m\Omega r\mathbf{n}_2.$$

See the corresponding animation.

The resultant moment with respect to a point Q, fixed in a body C, of the distributed system $\mathcal{G}$ of momentum vectors of the elements of C, in a reference frame $\mathcal{R}$, is the *angular momentum vector of the rigid body with respect to the point*, in $\mathcal{R}$, defined, therefore, as

$$^{\mathcal{R}}\mathbf{H}^{C/Q} \rightleftharpoons \int_C \mathbf{p} \times d\mathbf{G} = \int_C \mathbf{p} \times \mathbf{v}\, dm, \tag{1.7}$$

where $\mathbf{p} = \mathbf{p}^{P/Q}$ is the position vector of a general point P of C with respect to the chosen point Q (see Fig. 1.2). As C is a rigid body, the velocity field of the points of the body is defined by Eq. (1.2). Therefore,

$$\mathbf{p} \times \mathbf{v} = \mathbf{p} \times {}^{\mathcal{R}}\mathbf{v}^Q + \mathbf{p} \times ({}^{\mathcal{R}}\boldsymbol{\omega}^C \times \mathbf{p}). \tag{1.8}$$

Integrating the first part of the right side of Eq. (1.8) throughout the body, then, according to the definition of mass center, Eq. (6.1.4),

$$\int_C \mathbf{p} \times {}^{\mathcal{R}}\mathbf{v}^Q\, dm = \int_C \mathbf{p}\, dm \times {}^{\mathcal{R}}\mathbf{v}^Q = \mathbf{p}^* \times m\,{}^{\mathcal{R}}\mathbf{v}^Q, \tag{1.9}$$

where $\mathbf{p}^* = \mathbf{p}^{C^*/Q}$. Now expressing the angular velocity vector of the body in the reference frame as $^{\mathcal{R}}\boldsymbol{\omega}^C = \omega\mathbf{n}$, where $\mathbf{n}$ is a unit vector with the same direction as $^{\mathcal{R}}\boldsymbol{\omega}^C$ (see Fig. 1.2) and ω is a positive scalar, the integration in the body of the second part of the right side of Eq. (1.8) will, using Eqs. (6.3.8) and (6.3.4), give (see Section 6.3)

$$\begin{aligned}
\int_C \mathbf{p} \times (^{\mathcal{R}}\boldsymbol{\omega}^C \times \mathbf{p})\, dm &= \omega \int_C \mathbf{p} \times (\mathbf{n} \times \mathbf{p})\, dm \\
&= \omega\, \mathbf{I}_n^{C/Q} \\
&= \omega\, \mathbb{II}^{C/Q} \cdot \mathbf{n} \\
&= \mathbb{II}^{C/Q} \cdot {}^{\mathcal{R}}\boldsymbol{\omega}^C.
\end{aligned} \qquad (1.10)$$

By combining the terms, then, we have the following expression for the angular momentum vector:

$$^{\mathcal{R}}\mathbf{H}^{C/Q} = \mathbb{II}^{C/Q} \cdot {}^{\mathcal{R}}\boldsymbol{\omega}^C + \mathbf{p}^* \times m\,{}^{\mathcal{R}}\mathbf{v}^Q. \qquad (1.11)$$

Equation (1.11) expresses the general form for establishing the angular momentum vector of a rigid body, with respect to an arbitrary point Q of the body, in a given reference frame. This expression will take a simpler form, canceling out the second term on the right side, if the chosen point Q is the mass center of the body — in this case, $\mathbf{p}^* = 0$ — or if point Q is fixed in $\mathcal{R}$ — canceling $^{\mathcal{R}}\mathbf{v}^Q$. The first hypothesis is valid whatever the motion of the body in the reference frame. Thus, by replacing the general point Q for C^* in Eq. (1.11), we have

$$^{\mathcal{R}}\mathbf{H}^{C/C^*} = \mathbb{II}^{C/C^*} \cdot {}^{\mathcal{R}}\boldsymbol{\omega}^C. \qquad (1.12)$$

Equation (1.12) expresses quite a simple and extremely useful relation to find the angular momentum vector of a rigid body with respect to its mass center in a given reference frame, being equal to the product of the inertia tensor of the body with respect to the mass center by the angular velocity vector of the body in the reference frame.

If the rigid body C moves in a reference frame $\mathcal{R}$ so that a point O of C stays fixed in $\mathcal{R}$, then, as defined in Section 3.8, there is a *motion with a fixed point* and Eq. (1.11) takes the simplified form

$$^{\mathcal{R}}\mathbf{H}^{C/O} = \mathbb{II}^{C/O} \cdot {}^{\mathcal{R}}\boldsymbol{\omega}^C, \qquad \text{O fixed in } \mathcal{R}. \qquad (1.13)$$

In short, if Q is an arbitrary point of a rigid body C moving in a reference frame $\mathcal{R}$ with angular velocity $^{\mathcal{R}}\boldsymbol{\omega}^{C}$, the angular momentum of C with respect to Q in $\mathcal{R}$ depends on the inertia tensor of the body with respect to the point, on the angular velocity of the body, and on the velocity of the point in the reference frame, as generally expressed by Eq. (1.11). If C* is the mass center of C, the angular momentum of the body with respect to C* in $\mathcal{R}$ depends on the inertia tensor of the body with respect to the mass center and on the angular velocity of the body in the reference frame, as expressed in Eq. (1.12). Lastly, if body C moves in $\mathcal{R}$ so that a point O of C remains fixed in the reference frame, then the angular momentum of C with respect to O, in $\mathcal{R}$, may be expressed as a function of the inertia tensor of the body with respect to O and the angular velocity vector of C in $\mathcal{R}$, as given in Eq. (1.13). The first two equations mentioned are absolutely general while the last is restricted to a particular kind of motion of the body. It is worth noting that there is also another possibility for the second term of Eq. (1.11) to vanish, that is, when Q moves parallel to vector $\mathbf{p}^{*}$, vanishing the corresponding vectorial product; this is, however, too unusual a situation to dwell upon it further.

Example 1.2 Returning to the previous example (see Fig. 1.3), the inertia tensor of body C with respect to its mass center C* can be expressed in the coordinate system $\{x_1, x_2, x_3\}$ by the matrix (see Appendix C)

$$\mathbf{II}^{C/C^*} = mr^2 \begin{pmatrix} \frac{1}{2} & 0 & 0 \\ 0 & \frac{1}{3} & 0 \\ 0 & 0 & \frac{1}{3} \end{pmatrix}.$$

The inertia tensor with respect to point Q can then be obtained from Eq. (6.5.6), resulting in

$$\mathbf{II}^{C/Q} = \frac{1}{2}mr^2 \begin{pmatrix} 1 & 0 & 0 \\ 0 & \frac{7}{6} & 0 \\ 0 & 0 & \frac{7}{6} \end{pmatrix}.$$

Likewise, the inertia tensor with respect to point O may be obtained, using the same relation, being given by (check it)

$$\mathbf{II}^{C/O} = \frac{1}{2}mr^2 \begin{pmatrix} 1 & 0 & 0 \\ 0 & \frac{79}{6} & 0 \\ 0 & 0 & \frac{79}{6} \end{pmatrix}.$$

The angular velocity vector of the cylinder, expressed on the basis $\mathbf{n}_1, \mathbf{n}_2, \mathbf{n}_3$, is (see preceding example)

$$^{\mathcal{R}}\boldsymbol{\omega}^C = \frac{2}{\sqrt{5}}\Omega(-2\mathbf{n}_1 + \mathbf{n}_3).$$

The velocity of point Q in the reference frame $\mathcal{R}$ is

$$^{\mathcal{R}}\mathbf{v}^Q = {}^{\mathcal{R}}\mathbf{v}^O + {}^{\mathcal{R}}\boldsymbol{\omega}^C \times \mathbf{p}^{Q/O} = \frac{4}{\sqrt{5}}\Omega r\mathbf{n}_2.$$

The angular momentum vector of the cylinder with respect to point Q is then, according to Eq. (1.11),

$$^{\mathcal{R}}\mathbf{H}^{C/Q} = \mathbf{II}^{C/Q} \cdot {}^{\mathcal{R}}\boldsymbol{\omega}^C + \mathbf{p}^{C^*/Q} \times m\,{}^{\mathcal{R}}\mathbf{v}^Q$$

$$= \frac{1}{2}mr^2 \begin{pmatrix} 1 & 0 & 0 \\ 0 & \frac{7}{6} & 0 \\ 0 & 0 & \frac{7}{6} \end{pmatrix} \cdot \frac{2}{\sqrt{5}}\Omega(-2\mathbf{n}_1 + \mathbf{n}_3) + \frac{1}{2}r\mathbf{n}_1 \times m\frac{4}{\sqrt{5}}\Omega r\mathbf{n}_2$$

$$= \frac{1}{\sqrt{5}}mr^2\Omega \left(-2\mathbf{n}_1 + \frac{19}{6}\mathbf{n}_3\right).$$

The angular momentum vector with respect to the mass center is, according to Eq. (1.12),

$$^{\mathcal{R}}\mathbf{H}^{C/C^*} = \mathbf{II}^{C/C^*} \cdot {}^{\mathcal{R}}\boldsymbol{\omega}^C = \frac{2}{\sqrt{5}}mr^2\Omega \left(-\mathbf{n}_1 + \frac{1}{3}\mathbf{n}_3\right).$$

Lastly, the angular momentum of C with respect to point O, fixed in the reference frame, is, according to Eq. (1.13),

$$^{\mathcal{R}}\mathbf{H}^{C/O} = \mathbf{II}^{C/O} \cdot {}^{\mathcal{R}}\boldsymbol{\omega}^C = \frac{1}{\sqrt{5}}mr^2\Omega \left(-2\mathbf{n}_1 + \frac{79}{6}\mathbf{n}_3\right).$$

Note that the three computed angular momentum vectors are not parallel to each other, nor are any of them parallel to the angular velocity vector of the body in the reference frame. See the corresponding animation.

The angular momentum vectors of a rigid body C with respect to two points P and Q, in the same reference frame $\mathcal{R}$, are related by a simple expression, as follows:

$$^{\mathcal{R}}\mathbf{H}^{C/P} = {}^{\mathcal{R}}\mathbf{H}^{C/Q} + \mathbf{p}^{Q/P} \times {}^{\mathcal{R}}\mathbf{G}^C. \tag{1.14}$$

In fact, as the angular momentum vector of a body with respect to a point is the resultant moment, with respect to the point, of the distributed system $\mathcal{G}$, angular momenta with respect to two points must satisfy the moments transport theorem, Eq. (2.3.4), where the resultant of the system, when applicable, is exactly the momentum vector of the body in the respective reference frame. By substituting the resultant moments, therefore, in the moments transport theorem, for the respective angular momentum vectors and the resultant for the momentum vector of the body, Eq. (1.14) is obtained.

Example 1.3   Returning to Example 1.1 (see Fig. 1.3), the angular momentum vector of cylinder C with respect to point Q could be calculated from its angular momentum with respect to, say, C* and its momentum in the reference frame $\mathcal{R}$. So, from Eq. (1.14), we have

$$
\begin{aligned}
{}^{\mathcal{R}}\mathbf{H}^{C/Q} &= {}^{\mathcal{R}}\mathbf{H}^{C/C^*} + \mathbf{p}^{C^*/Q} \times {}^{\mathcal{R}}\mathbf{G}^C \\
&= \frac{2}{\sqrt{5}} m r^2 \Omega \left(-\mathbf{n}_1 + \frac{1}{3}\mathbf{n}_3 \right) + \frac{1}{2} r \mathbf{n}_1 \times \sqrt{5}\, m \Omega r \mathbf{n}_2 \\
&= \frac{1}{\sqrt{5}} m r^2 \Omega \left(-2\mathbf{n}_1 + \frac{19}{6}\mathbf{n}_3 \right),
\end{aligned}
$$

as obtained in Example 1.2, using Eq. (1.11). The reader must pay attention to the fact that Eq. (1.14) is valid for any two points of the body, and it is, therefore, unnecessary for one of these points to be the mass center, as happens by chance in this example. See the corresponding animation.

The *kinetic energy of a rigid body* C in a given reference frame $\mathcal{R}$ is defined, as seen in Section 5.1 for bodies in general, as the integral in the body of the kinetic energy of its elements, that is,

$$
{}^{\mathcal{R}}K^C = \int_C \frac{1}{2} \mathbf{v} \cdot \mathbf{v}\, dm, \tag{1.15}
$$

where $\mathbf{v} = {}^{\mathcal{R}}\mathbf{v}^P$ is the velocity in $\mathcal{R}$ of a general point P of the body (see Fig. 1.2). Here, too, the nature of the velocity field of the points of a rigid body will bring to its kinetic energy a peculiar expression. By then using Eq. (1.3), the quadratic velocity is

$$
\mathbf{v} \cdot \mathbf{v} = \mathbf{v}^* \cdot \mathbf{v}^* + \left({}^{\mathcal{R}}\boldsymbol{\omega}^C \times \mathbf{r} \right) \cdot \left({}^{\mathcal{R}}\boldsymbol{\omega}^C \times \mathbf{r} \right) + 2\mathbf{v}^* \cdot {}^{\mathcal{R}}\boldsymbol{\omega}^C \times \mathbf{r}, \tag{1.16}
$$

where $\mathbf{r}$ is the position vector of the general point P with respect to the mass center of the body. The kinetic energy of the body will then be

$$
\begin{aligned}
{}^{\mathcal{R}}K^C &= \int_C \frac{1}{2}\mathbf{v}^* \cdot \mathbf{v}^* \, dm + \int_C \frac{1}{2}({}^{\mathcal{R}}\boldsymbol{\omega}^C \times \mathbf{r}) \cdot ({}^{\mathcal{R}}\boldsymbol{\omega}^C \times \mathbf{r}) \, dm \\
&\quad + \int_C \mathbf{v}^* \cdot {}^{\mathcal{R}}\boldsymbol{\omega}^C \times \mathbf{r} \, dm \\
&= \frac{1}{2}m\mathbf{v}^* \cdot \mathbf{v}^* + \frac{1}{2}{}^{\mathcal{R}}\boldsymbol{\omega}^C \cdot \int_C \mathbf{r} \times ({}^{\mathcal{R}}\boldsymbol{\omega}^C \times \mathbf{r}) \, dm \\
&\quad + \mathbf{v}^* \cdot {}^{\mathcal{R}}\boldsymbol{\omega}^C \times \int_C \mathbf{r} \, dm.
\end{aligned}
\tag{1.17}
$$

The *kinetic energy of the mass center of a body*, as defined in Section 5.1, is

$$
{}^{\mathcal{R}}K^{C^*} = \frac{1}{2}m\mathbf{v}^* \cdot \mathbf{v}^*,
\tag{1.18}
$$

equal, therefore, to the first term of Eq. (1.17). The second term may be developed, as in Eq. (1.10), as

$$
\begin{aligned}
\frac{1}{2}{}^{\mathcal{R}}\boldsymbol{\omega}^C \cdot \int_C \mathbf{r} \times ({}^{\mathcal{R}}\boldsymbol{\omega}^C \times \mathbf{r}) \, dm &= \frac{1}{2}{}^{\mathcal{R}}\boldsymbol{\omega}^C \cdot \omega \int_C \mathbf{r} \times (\mathbf{n} \times \mathbf{r}) \, dm \\
&= \frac{1}{2}{}^{\mathcal{R}}\boldsymbol{\omega}^C \cdot \omega \, I_n^{C/C^*} \\
&= \frac{1}{2}{}^{\mathcal{R}}\boldsymbol{\omega}^C \cdot \omega \, \mathbf{II}^{C/C^*} \cdot \mathbf{n} \\
&= \frac{1}{2}{}^{\mathcal{R}}\boldsymbol{\omega}^C \cdot \mathbf{II}^{C/C^*} \cdot {}^{\mathcal{R}}\boldsymbol{\omega}^C.
\end{aligned}
\tag{1.19}
$$

Lastly, the third and last term of Eq. (1.17) will be null, since, according to Eq. (5.1.9), $\int_C \mathbf{r} \, dm = 0$.

In short, the kinetic energy of a rigid body may be expressed as

$$
{}^{\mathcal{R}}K^C = \frac{1}{2}m\mathbf{v}^* \cdot \mathbf{v}^* + \frac{1}{2}{}^{\mathcal{R}}\boldsymbol{\omega}^C \cdot \mathbf{II}^{C/C^*} \cdot {}^{\mathcal{R}}\boldsymbol{\omega}^C,
\tag{1.20}
$$

where the first term, called the *kinetic energy of translation* or the *kinetic energy of the mass center*, ${}^{\mathcal{R}}K^{C^*}$, is defined by Eq. (1.18) and the second term, called the *kinetic energy of rotation* or the *kinetic energy around the mass center*, is given by

$$
{}^{\mathcal{R}}K^{C/C^*} = \frac{1}{2}{}^{\mathcal{R}}\boldsymbol{\omega}^C \cdot \mathbf{II}^{C/C^*} \cdot {}^{\mathcal{R}}\boldsymbol{\omega}^C.
\tag{1.21}
$$

Note that the lowering of the kinetic energy of a rigid body results in the general breaking down of the kinetic energy of a body, expressed in Eq. (5.1.26). In the case of the rigid body, in particular, the kinetic energy around the mass center is reduced to the kinetic energy of rotation, as defined by Eq. (1.21) above. In fact, for a rigid body $\dot{\mathbf{r}} = \boldsymbol{\omega} \times \mathbf{r}$ and, from Eq. (5.1.25),

$$
\begin{aligned}
{}^{\mathcal{R}}K^{C/C^*} &= \int_C \frac{1}{2}\dot{\mathbf{r}} \cdot \dot{\mathbf{r}}\, dm \\
&= \int_C \frac{1}{2}({}^{\mathcal{R}}\boldsymbol{\omega}^C \times \mathbf{r}) \cdot ({}^{\mathcal{R}}\boldsymbol{\omega}^C \times \mathbf{r})\, dm \\
&= \frac{1}{2}{}^{\mathcal{R}}\boldsymbol{\omega}^C \cdot \int_C \mathbf{r} \times ({}^{\mathcal{R}}\boldsymbol{\omega}^C \times \mathbf{r})\, dm \\
&= \frac{1}{2}{}^{\mathcal{R}}\boldsymbol{\omega}^C \cdot \mathbf{II}^{C/C^*} \cdot {}^{\mathcal{R}}\boldsymbol{\omega}^C.
\end{aligned}
\tag{1.22}
$$

Example 1.4   Let us now consider the motion of a coin M, in the shape of a homogeneous and thin disk, with mass m and radius r, rolling over a flat surface (see Fig. 1.4). Let us find its kinetic energy at a certain instant when the coin's plane is vertical, that is, when $\theta = 0$. The angular velocity of the coin in the reference frame $\mathcal{R}$, where it is fixed to the surface, expressed in terms of coordinates $\phi(t)$, $\theta(t)$, and $\psi(t)$, is (see Example 3.3.4)

$$
{}^{\mathcal{R}}\boldsymbol{\omega}^M = \dot{\psi}\cos\theta\,\mathbf{n}_1 + \dot{\theta}\mathbf{n}_2 + (\dot{\phi} + \dot{\psi}\sin\theta)\mathbf{n}_3.
$$

In the condition $\theta = 0$, we have

$$
{}^{\mathcal{R}}\boldsymbol{\omega}^M = \dot{\psi}\mathbf{n}_1 + \dot{\theta}\mathbf{n}_2 + \dot{\phi}\mathbf{n}_3.
$$

The velocity of center O of the coin, at the instant in question, may be obtained from the rolling condition, ${}^{\mathcal{R}}\mathbf{v}^C = 0$, resulting in

$$
\begin{aligned}
{}^{\mathcal{R}}\mathbf{v}^O &= {}^{\mathcal{R}}\mathbf{v}^C + {}^{\mathcal{R}}\boldsymbol{\omega}^M \times \mathbf{p}^{O/C} \\
&= 0 + (\dot{\psi}\mathbf{n}_1 + \dot{\theta}\mathbf{n}_2 + \dot{\phi}\mathbf{n}_3) \times r\mathbf{n}_1 \\
&= r(\dot{\phi}\mathbf{n}_2 - \dot{\theta}\mathbf{n}_3).
\end{aligned}
$$

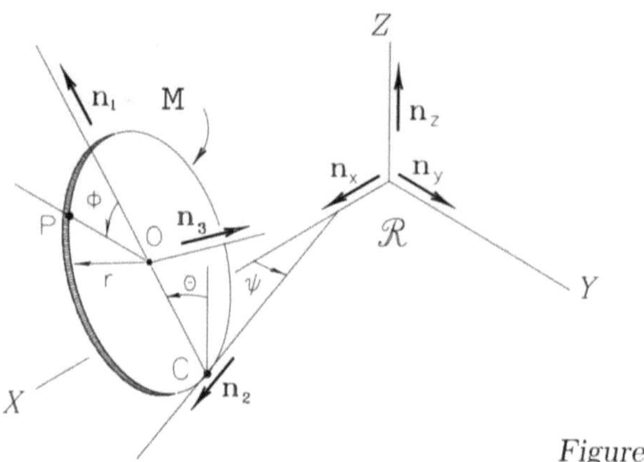

Figure 1.4

The kinetic energy of translation of the coin is then, according to Eq. (1.18),

$$^{\mathcal{R}}K^O = \frac{1}{2}m\,{}^{\mathcal{R}}\mathbf{v}^O \cdot {}^{\mathcal{R}}\mathbf{v}^O = \frac{1}{2}mr^2(\dot{\phi}^2 + \dot{\theta}^2).$$

The inertia tensor of the coin with respect to its mass center is (see Appendix C)

$$\mathbf{II}^{M/O} = \frac{1}{2}mr^2 \begin{pmatrix} \frac{1}{2} & 0 & 0 \\ 0 & \frac{1}{2} & 0 \\ 0 & 0 & 1 \end{pmatrix}.$$

The kinetic energy of rotation of the coin is then, according to Eq. (1.21),

$$
\begin{aligned}
^{\mathcal{R}}K^{M/O} &= \frac{1}{2}{}^{\mathcal{R}}\boldsymbol{\omega}^M \cdot \mathbf{II}^{M/O} \cdot {}^{\mathcal{R}}\boldsymbol{\omega}^M \\
&= \frac{1}{2}(\dot{\psi},\dot{\theta},\dot{\phi}) \cdot \frac{1}{2}mr^2 \begin{pmatrix} \frac{1}{2} & 0 & 0 \\ 0 & \frac{1}{2} & 0 \\ 0 & 0 & 1 \end{pmatrix} \cdot (\dot{\psi},\dot{\theta},\dot{\phi}) \\
&= \frac{1}{8}mr^2\left(\dot{\psi}^2 + \dot{\theta}^2 + 2\dot{\phi}^2\right).
\end{aligned}
$$

The overall kinetic energy of the coin in this instant is therefore, according to Eq. (1.20),

$$^{\mathcal{R}}K^M = {}^{\mathcal{R}}K^O + {}^{\mathcal{R}}K^{M/O} = \frac{1}{8}mr^2\left(\dot{\psi}^2 + 5\dot{\theta}^2 + 6\dot{\phi}^2\right).$$

When a rigid body C moves in a reference frame $\mathcal{R}$ with a fixed point O, it is convenient to define its *kinetic energy of rotation around the point*, so that

$$^{\mathcal{R}}K^{C/O} \rightleftharpoons \frac{1}{2}{}^{\mathcal{R}}\boldsymbol{\omega}^C \cdot \mathbf{II}^{C/O} \cdot {}^{\mathcal{R}}\boldsymbol{\omega}^C. \tag{1.23}$$

If P is a general point of C, whose position with respect to the fixed point O is given by vector $\mathbf{p} = \mathbf{p}^{P/O}$, its velocity in $\mathcal{R}$ is $\mathbf{v} = {}^{\mathcal{R}}\mathbf{v}^P = {}^{\mathcal{R}}\boldsymbol{\omega}^C \times \mathbf{p}$ and the kinetic energy of the body in $\mathcal{R}$ is, according to Eq. (1.15),

$$\begin{aligned}
^{\mathcal{R}}K^C &= \int_C \frac{1}{2}\mathbf{v} \cdot \mathbf{v}\, dm \\
&= \int_C \frac{1}{2}({}^{\mathcal{R}}\boldsymbol{\omega}^C \times \mathbf{p}) \cdot ({}^{\mathcal{R}}\boldsymbol{\omega}^C \times \mathbf{p})\, dm \\
&= \frac{1}{2}{}^{\mathcal{R}}\boldsymbol{\omega}^C \cdot \int_C \mathbf{p} \times ({}^{\mathcal{R}}\boldsymbol{\omega}^C \times \mathbf{p})\, dm \\
&= \frac{1}{2}{}^{\mathcal{R}}\boldsymbol{\omega}^C \cdot \mathbf{II}^{C/O} \cdot {}^{\mathcal{R}}\boldsymbol{\omega}^C \\
&= {}^{\mathcal{R}}K^{C/O}.
\end{aligned} \tag{1.24}$$

In short, when there is a motion of a rigid body with a fixed point, a kinetic energy of rotation around the point may be defined — similar to the kinetic energy of rotation in the more general case, but with the inertia tensor with respect to the point instead of the inertia tensor with respect to the mass center — the latter being equal to the total kinetic energy of the body in the reference frame, that is,

$$^{\mathcal{R}}K^C = \frac{1}{2}{}^{\mathcal{R}}\boldsymbol{\omega}^C \cdot \mathbf{II}^{C/O} \cdot {}^{\mathcal{R}}\boldsymbol{\omega}^C, \qquad \text{O fixed in } \mathcal{R}. \tag{1.25}$$

Example 1.5 Going back to Example 1.1 (see Fig. 1.3), the kinetic energy of the body in the reference frame may be calculated, according to Eq. (1.25), as

$$\begin{aligned}
^{\mathcal{R}}K^C &= \frac{1}{2}{}^{\mathcal{R}}\boldsymbol{\omega}^C \cdot \mathbf{II}^{C/O} \cdot {}^{\mathcal{R}}\boldsymbol{\omega}^C \\
&= \frac{1}{2}\frac{2}{\sqrt{5}}\Omega(-2\mathbf{n}_1 + \mathbf{n}_3) \cdot \frac{1}{2}mr^2 \begin{pmatrix} 1 & 0 & 0 \\ 0 & \frac{79}{6} & 0 \\ 0 & 0 & \frac{79}{6} \end{pmatrix} \cdot \frac{2}{\sqrt{5}}\Omega(-2\mathbf{n}_1 + \mathbf{n}_3) \\
&= \frac{103}{30}mr^2\Omega^2.
\end{aligned}$$

7.2 Equations of Motion

The equations governing the motion of a rigid body C in an inertial reference frame $\mathcal{R}$ are, at least in principle, the same as those governing a continuous system, or body, as shown in Section 5.4. So, if $\mathbf{F}$ is the resultant of the external forces acting on C and ${}^{\mathcal{R}}\mathbf{G}^{C}$ is its momentum in $\mathcal{R}$, Eq. (5.4.1) will, for C, be as follows:

$$ {}^{\mathcal{R}}\dot{\mathbf{G}}^{C} = \mathbf{F}, \tag{2.1} $$

that is, the time rate in an inertial reference frame $\mathcal{R}$ of the momentum vector of rigid body C, in this same reference frame, is, at each instant, equal to the resultant $\mathbf{F}$ of the system of external forces acting on the body. Likewise, the first equation of motion for a rigid body can be expressed in terms of the motion of its mass center, as stated in Eq. (5.4.2), that is,

$$ \dot{\mathbf{G}}^{*} = \mathbf{F}, \tag{2.2} $$

where $\mathbf{G}^{*}$ is the momentum of the mass center of the body, and as Eq. (1.6) shows, equal to the momentum of the body in the reference frame $\mathcal{R}$. Equation (2.2) is, of course, equivalent to

$$ m\mathbf{a}^{*} = \mathbf{F}, \tag{2.3} $$

where m is the mass of the body and $\mathbf{a}^{*}$ is the acceleration of its mass center in the inertial reference frame.

Choosing a system of Cartesian axes $\{x_1, x_2, x_3\}$, in the directions in which the vectors involved are broken down, Eq. (2.3) develops into a set of three scalar equations governing the motion of the mass center of the body, that is,

$$ \begin{aligned} ma_1^{*} &= F_1, \\ ma_2^{*} &= F_2, \\ ma_3^{*} &= F_3. \end{aligned} \tag{2.4} $$

Equation (2.1) [or (2.2), or (2.3)] is also called the *equation of translation of the rigid body*, to the extent that it governs the translation of the mass center of the body in the inertial reference frame. We will especially refer to Eqs. (2.4) also as *equations of motion of the first type*. It is worth mentioning that the Cartesian axes adopted to break Eq. (2.3) down into Eqs. (2.4) may be fixed in any reference frame.

Example 2.1 Consider a homogeneous rod B, with mass m and length a, resting on a smooth horizontal table $\mathcal{R}$, when a horizontal force with module H and direction orthogonal to the rod's axis is applied to end P (see Fig. 2.1). Adopting the orthonormal basis $\mathbf{b}_1, \mathbf{b}_2, \mathbf{b}_3$, fixed in B and in the indicated direction, the external forces applied to the rod are $\mathbf{H} = H\mathbf{b}_2$, $\mathbf{N} = N\mathbf{b}_3$, and $\mathbf{P} = -mg\mathbf{b}_3$ (assuming that the normal force exerted by the table and weight are both parallel and uniformly distributed forces, reduceable, therefore, to the mass center).

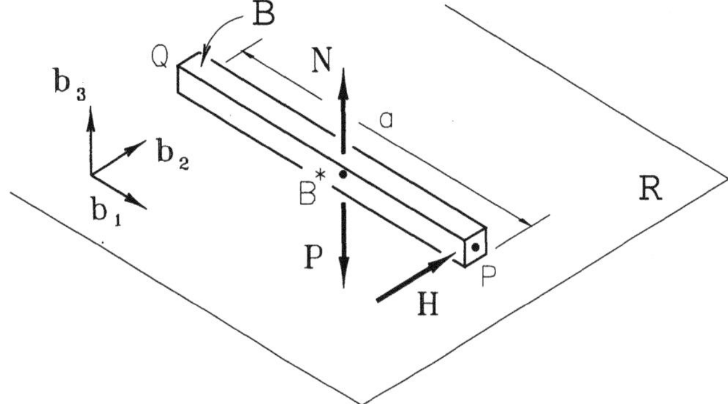

Figure 2.1

The external resultant force is then

$$\mathbf{F} = H\mathbf{b}_2 + (N - mg)\mathbf{b}_3.$$

The rod will move and continue in contact with the table; the acceleration of the mass center in relation to the latter is then

$$\mathbf{a}^* = a_1\mathbf{b}_1 + a_2\mathbf{b}_2.$$

Assuming that the table is an inertial reference frame, the equations of motion for the mass center, Eqs. (2.4), are

$$ma_1 = 0,$$
$$ma_2 = H,$$
$$0 = N - mg,$$

resulting in

$$a_1 = 0, \quad a_2 = \frac{H}{m}, \quad N = mg.$$

Note that the first two equations of motion are second-order ordinary differential equations for the coordinates of the mass center of the rod (whose time derivatives, up to second-order, are present in the expressions of a_1 and a_2), while the third is an algebraic equation that leads to the unknown force component, N. Also, pay attention to the fact that the equations only govern the motion of point B^*, informing nothing about the rotation of the rod. For instance, it is not possible only with the result obtained so far to know something about the motion of point P. Lastly, it is worth noting that the chosen basis to write the equations of motion is not fixed in the inertial reference frame, but is convenient for addressing the problem. See the corresponding animation.

As already mentioned, the equations of motion for a rigid body are, in principle, the same as those governing the motion of a continuous body, discussed in Section 5.4. The most important of them, among the equations of motion of the second kind, is that which governs the motion of the body around its mass center, Eq. (5.4.7). If C is, therefore, a rigid body moving in an inertial reference frame $\mathcal{R}$ under the action of a system of external forces $\mathcal{F}_e$, whose resultant moment with respect to the mass center C^* of the body is $\mathbf{M}^{\mathcal{F}_e/C^*}$, the time derivative in $\mathcal{R}$ of the angular momentum vector of the body with respect to its mass center, $^\mathcal{R}\mathbf{H}^{C/C^*}$, satisfies the relation

$$^\mathcal{R}\dot{\mathbf{H}}^{C/C^*} = \mathbf{M}^{\mathcal{F}_e/C^*}. \tag{2.5}$$

For a rigid body, however, the angular momentum vector with respect to the mass center is as described in Eq. (1.12), and its time rate in $\mathcal{R}$ is, then,

$$\begin{aligned}
^\mathcal{R}\dot{\mathbf{H}}^{C/C^*} &= \frac{^\mathcal{R}d}{dt}\,^\mathcal{R}\mathbf{H}^{C/C^*} \\
&= \frac{^Cd}{dt}\,^\mathcal{R}\mathbf{H}^{C/C^*} + {}^\mathcal{R}\boldsymbol{\omega}^C \times {}^\mathcal{R}\mathbf{H}^{C/C^*} \\
&= \frac{^Cd}{dt}\left(\mathbf{II}^{C/C^*} \cdot {}^\mathcal{R}\boldsymbol{\omega}^C\right) + {}^\mathcal{R}\boldsymbol{\omega}^C \times \mathbf{II}^{C/C^*} \cdot {}^\mathcal{R}\boldsymbol{\omega}^C \\
&= \mathbf{II}^{C/C^*} \cdot {}^\mathcal{R}\boldsymbol{\alpha}^C + {}^\mathcal{R}\boldsymbol{\omega}^C \times \mathbf{II}^{C/C^*} \cdot {}^\mathcal{R}\boldsymbol{\omega}^C.
\end{aligned} \tag{2.6}$$

Note that, when obtaining the above result, it was necessary to transpose the time rate in $\mathcal{R}$ to time rate in C, where the inertia tensor is an invariant. It is worth recalling that the position vector $\mathbf{p}$ appears in the expression of the inertia tensor of the body, Eq. (6.3.2) (in this case, $\mathbf{r}$, the position of a general point of the body with respect to the mass center), whose time rate in $\mathcal{R}$ is different from zero but whose variation in C is null; hence the inertia tensor of the body has a null derivative in reference frame C, a fact exploited in the deduction of Eq. (2.6) above.

So substituting Eq. (2.6) in Eq. (2.5), then the *equation of motion of the second kind for a rigid body*, also called the *equation of rotation of the rigid body*, is obtained.

$$\mathbf{II}^{C/C^*} \cdot {}^{\mathcal{R}}\boldsymbol{\alpha}^C + {}^{\mathcal{R}}\boldsymbol{\omega}^C \times \mathbf{II}^{C/C^*} \cdot {}^{\mathcal{R}}\boldsymbol{\omega}^C = \mathbf{M}^{\mathcal{F}_e/C^*}. \qquad (2.7)$$

Example 2.2 Returning to Example 2.1, we are now going to analyze the rotation of the rod around point B^* (see Fig. 2.1). The inertia tensor of the rod with respect to its mass center, expressed in the directions of the chosen basis $\mathbf{b}_1, \mathbf{b}_2, \mathbf{b}_3$, is (see Appendix C)

$$\mathbf{II}^{B/B^*} = \frac{1}{12} ma^2 \begin{pmatrix} 0 & 0 & 0 \\ 0 & 1 & 0 \\ 0 & 0 & 1 \end{pmatrix}.$$

As the rod remains in contact with the plane, vector $\mathbf{b}_3$ will be fixed simultaneously in B and $\mathcal{R}$, characterizing a simple angular velocity, that is,

$${}^{\mathcal{R}}\boldsymbol{\omega}^B = \omega \mathbf{b}_3, \qquad {}^{\mathcal{R}}\boldsymbol{\alpha}^B = \dot{\omega} \mathbf{b}_3.$$

The resultant moment with respect to point B^* of the forces acting on the rod is

$$\mathbf{M}^{\mathcal{F}_e/B^*} = \frac{1}{2} Ha\mathbf{b}_3.$$

So by substituting it in Eq. (2.7), then we get

$$\frac{1}{12} ma^2 \begin{pmatrix} 0 & 0 & 0 \\ 0 & 1 & 0 \\ 0 & 0 & 1 \end{pmatrix} \cdot \dot{\omega} \mathbf{b}_3 + \omega \mathbf{b}_3 \times \frac{1}{12} ma^2 \begin{pmatrix} 0 & 0 & 0 \\ 0 & 1 & 0 \\ 0 & 0 & 1 \end{pmatrix} \cdot \omega \mathbf{b}_3 = \frac{1}{2} Ha\mathbf{b}_3,$$

which results in

$$\dot{\omega} = \frac{6H}{am},$$

that is, a constant angular acceleration (assuming that **H** remains orthogonal to the rod and with constant module H). Starting from rest, the angular velocity of the rod at an instant t will, therefore, be $\omega = (6H/am)\,t$. Now we can, with the help of the kinematic theorem for accelerations, find the acceleration of other points of the rod. The initial acceleration of point P, on which the force is applied, is (see Fig. 2.2)

$$
\begin{aligned}
{}^{\mathcal{R}}\mathbf{a}^P &= {}^{\mathcal{R}}\mathbf{a}^{B^*} + {}^{\mathcal{R}}\boldsymbol{\omega}^B \times ({}^{\mathcal{R}}\boldsymbol{\omega}^B \times \mathbf{p}^{P/B^*}) + {}^{\mathcal{R}}\boldsymbol{\alpha}^B \times \mathbf{p}^{P/B^*} \\
&= \frac{H}{m}\mathbf{b}_2 + 0 + \frac{6H}{am}\mathbf{b}_3 \times \frac{a}{2}\mathbf{b}_1 \\
&= 4\frac{H}{m}\mathbf{b}_2.
\end{aligned}
$$

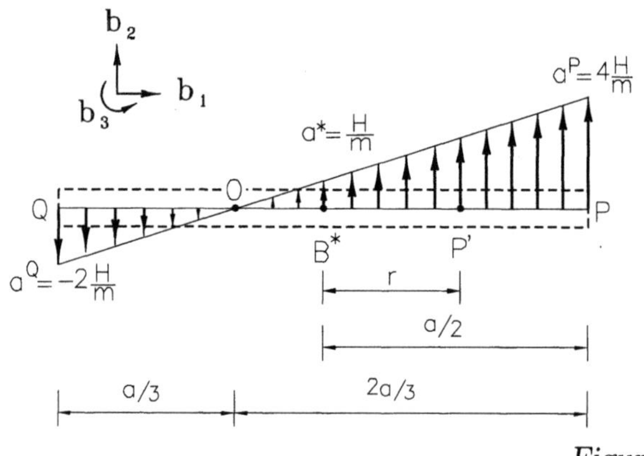

Figure 2.2

The acceleration of the other end of the rod, at the same instant, is

$$
\begin{aligned}
{}^{\mathcal{R}}\mathbf{a}^Q &= {}^{\mathcal{R}}\mathbf{a}^{B^*} + {}^{\mathcal{R}}\boldsymbol{\omega}^B \times ({}^{\mathcal{R}}\boldsymbol{\omega}^B \times \mathbf{p}^{Q/B^*}) + {}^{\mathcal{R}}\boldsymbol{\alpha}^B \times \mathbf{p}^{Q/B^*} \\
&= -2\frac{H}{m}\mathbf{b}_2.
\end{aligned}
$$

Exploring the example a little more, note that there is a linear variation in the value of the initial acceleration of the points of the rod, from $-2H/m$ in point Q to $4H/m$ in point P. In fact, if P' is a general point of the rod, its position vector with respect to B* may be expressed as $\mathbf{p}^{P'/B^*} = r\mathbf{b}_1$, with $|r| \leq a/2$, (see Fig. 2.2), and the initial acceleration of P' is

$$
{}^{\mathcal{R}}\mathbf{a}^{P'} = \frac{H}{m}\mathbf{b}_2 + \frac{6H}{am}\mathbf{b}_3 \times r\mathbf{b}_1 = \frac{H}{m}\left(1 + 6\frac{r}{a}\right)\mathbf{b}_2,
$$

a linear function of r. There must, therefore, be a point O whose initial acceleration is null, which happens for $r = -a/6$. Point O, at a distance of $a/3$ from the free end Q, is called the *center of percussion* relative to P, to the extent that, when the rod hits on P, we have in O a point unsensitive to the action, with null initial acceleration in $\mathcal{R}$. See the corresponding animation.

If a system of Cartesian axes $\{x_1, x_2, x_3\}$ and an orthonormal basis $\mathbf{n}_1, \mathbf{n}_2, \mathbf{n}_3$, both *fixed in body* C, are chosen for breaking down the vectors $\boldsymbol{\omega}$, $\boldsymbol{\alpha}$, and $\mathbf{M}$ and the inertia tensor II^{C/C^*}, expressed as a matrix of inertia according to the same axes, we have

$$\mathcal{R}\boldsymbol{\omega}^C = \omega_1 \mathbf{n}_1 + \omega_2 \mathbf{n}_2 + \omega_3 \mathbf{n}_3,$$

$$\mathcal{R}\boldsymbol{\alpha}^C = \alpha_1 \mathbf{n}_1 + \alpha_2 \mathbf{n}_2 + \alpha_3 \mathbf{n}_3,$$

$$\mathbf{M}^{\mathcal{F}_e/C^*} = M_1 \mathbf{n}_1 + M_2 \mathbf{n}_2 + M_3 \mathbf{n}_3, \tag{2.8}$$

$$\mathrm{II}^{C/C^*} = \begin{pmatrix} I_{11}^* & I_{12}^* & I_{13}^* \\ I_{21}^* & I_{22}^* & I_{23}^* \\ I_{31}^* & I_{32}^* & I_{33}^* \end{pmatrix}.$$

Equation (2.7) may now be divided into a system of three scalar equations. In their most general form, these equations assume a relatively extensive expression, given below (check it):

$$I_{11}^*\alpha_1 + I_{12}^*\alpha_2 + I_{13}^*\alpha_3 + I_{23}^*(\omega_2^2 - \omega_3^2)$$
$$+ (I_{33}^* - I_{22}^*)\omega_3\omega_2 + (I_{13}^*\omega_2 - I_{12}^*\omega_3)\omega_1 = M_1;$$

$$I_{22}^*\alpha_2 + I_{23}^*\alpha_3 + I_{21}^*\alpha_1 + I_{31}^*(\omega_3^2 - \omega_1^2) \tag{2.9}$$
$$+ (I_{11}^* - I_{33}^*)\omega_1\omega_3 + (I_{21}^*\omega_3 - I_{23}^*\omega_1)\omega_2 = M_2;$$

$$I_{33}^*\alpha_3 + I_{31}^*\alpha_1 + I_{32}^*\alpha_2 + I_{12}^*(\omega_1^2 - \omega_2^2)$$
$$+ (I_{22}^* - I_{11}^*)\omega_2\omega_1 + (I_{32}^*\omega_1 - I_{31}^*\omega_2)\omega_3 = M_3.$$

Equations (2.9) are somewhat inconvenient, given their complexity. They may, however, be substantially simplified if the Cartesian axes chosen are the principal axes of inertia of the body with respect to its mass center. In fact, as the products of inertia associated with the principal directions of inertia are all null, the equations of motion of the second kind will be reduced to (confirm)

$$I_1^* \alpha_1 + (I_3^* - I_2^*)\omega_3\omega_2 = M_1,$$
$$I_2^* \alpha_2 + (I_1^* - I_3^*)\omega_1\omega_3 = M_2, \qquad (2.10)$$
$$I_3^* \alpha_3 + (I_2^* - I_1^*)\omega_2\omega_1 = M_3.$$

Equations (2.10) are known as *Euler's equations*. We must not forget that, in order to use these much simpler equations than Eqs. (2.9), we must know the principal moments of inertia of the body with respect to the mass center and break the resultant moment and angular velocity and acceleration vectors down into the principal directions of inertia.

Equations. (2.7), (2.9), and (2.10) come from Eqs. (2.5) and (2.6). When deducting Eq. (2.6), in turn (have another look at it), it was implicitly assumed that the inertia tensor with respect to the mass center is a constant in body C, which is always true. When breaking down into a system of Cartesian axes, however, the inertia matrix arises, whose components — which are the moments and products of inertia — *depend* on the direction of the axes, that is, on the motion of the reference frame to which these axes are associated. This is the reason why, in the decomposition given by Eqs. (2.8), it is expressly stated that the chosen axes shall be fixed *in the body*. To use Euler's equations, in particular, this is also an imperative constraint since the principal directions of inertia are also fixed in the body. A special case where this constraint may be relaxed occurs when the body is of revolution, that is, it has an axis of symmetry. Every axisymmetric body has an inertia ellipsoid, relative to its mass center, also axisymmetric. One of its principal axes is the axis of symmetry and the others are any axis orthogonal to it passing through the center of the body (see Section 6.6). So choosing, say, the unit vector $\mathbf{n}_3$ parallel to the axis of symmetry and arbitrary directions (maintaining, of course, the orthogonality) for $\mathbf{n}_1$ and $\mathbf{n}_2$, there might be the basis fixed in a reference frame in relation to which the body has a simple angular velocity, parallel to the axis of symmetry, the principal moments of inertia staying constant.

Equations. (2.4), which are the equations of motion of the first type, and Eqs. (2.10), which are the simplest form for the equations of motion of the second kind, are mutually independent and form a set of six scalar equations that govern the motion of a rigid body. When the forces and torques acting on the body are known, these equations, when integrated, are generally sufficient to establish the six generalized coordinates that define the configuration of the body in the reference frame. It is more usual, however, due to the presence of links, not to know some part of the forces and torques applied to the body; in this case, the number of degrees of freedom of the body is reduced and part of those six equations will be used to find the unknown efforts.

Example 2.3 Let us consider the disk D, with mass m and radius r, welded to a light rod with length $3r$, whose end is connected to point O, fixed on plane $\mathcal{R}$, inertial, by means of a ball and socket joint, as shown (see Fig. 2.3).

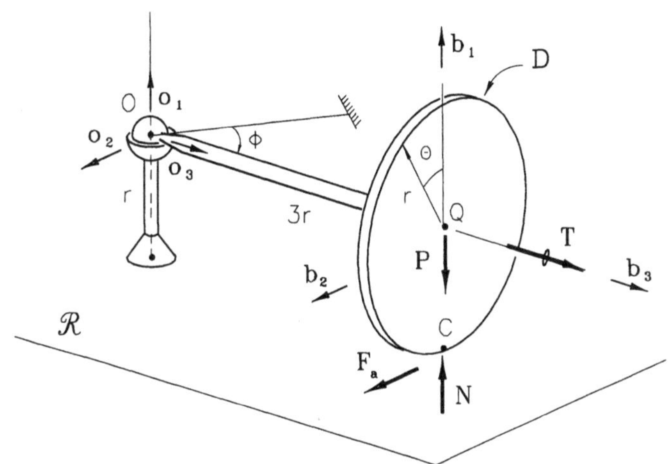

Figure 2.3

The disk is at rest, supported by the plane, when torque **T**, parallel to its axis of symmetry, is applied. We wish to find the initial acceleration of point Q, mass center of the body, and the initial values of the forces applied to it. It is assumed that the coefficient of friction present is enough to prevent sliding on C. This same system is studied in Chapter 2, from the viewpoint of the analysis of forces and torques, and in Chapter 3, from the viewpoint of its kinematics. For a more complete approach, however, we will now reexamine every aspect here. The figure indicates the forces and torques to be considered: a force applied to O with, in principle, three components; the weight, applied to Q; two components of force on the point of contact C (there is no tendency to slide in the direction parallel to $\mathbf{b}_3$); and the known torque $\mathbf{T}$. Reducing this system to point Q gives

$$\mathbf{F} = (O_1 + N - mg)\mathbf{b}_1 + (O_2 + F_a)\mathbf{b}_2 + O_3\mathbf{b}_3,$$
$$\mathbf{M}^{\mathcal{F}e/Q} = 3rO_2\mathbf{b}_1 - 3rO_1\mathbf{b}_2 + (T - rF_a)\mathbf{b}_3.$$

The angular velocity of the body may be expressed in terms of the angular coordinates ϕ and θ, such as

$$^{\mathcal{R}}\boldsymbol{\omega}^D = -\dot{\phi}\mathbf{b}_1 + \dot{\theta}\mathbf{b}_3.$$

The angular acceleration may then be obtained through differentiation, resulting in

$$^{\mathcal{R}}\boldsymbol{\alpha}^D = -\ddot{\phi}\mathbf{b}_1 + \dot{\theta}\dot{\phi}\mathbf{b}_2 + \ddot{\theta}\mathbf{b}_3.$$

The velocity of point O in $\mathcal{R}$ is null, the same happening with the velocity of point C. From the kinematic theorem $^{\mathcal{R}}\mathbf{v}^C = {}^{\mathcal{R}}\mathbf{v}^O + {}^{\mathcal{R}}\boldsymbol{\omega}^D \times \mathbf{p}^{C/O}$ results the relation

$$\dot{\theta} = 3\dot{\phi}$$

and, consequently,

$$\ddot{\theta} = 3\ddot{\phi}.$$

The components of the angular velocity and acceleration on the chosen basis will, therefore, be

$$\omega_1 = -\dot{\phi}, \quad \omega_2 = 0, \quad \omega_3 = 3\dot{\phi},$$
$$\alpha_1 = -\ddot{\phi}, \quad \alpha_2 = 3\dot{\phi}^2, \quad \alpha_3 = 3\ddot{\phi}.$$

The velocity of point Q may be obtained from

$$^{\mathcal{R}}\mathbf{v}^Q = {}^{\mathcal{R}}\mathbf{v}^O + {}^{\mathcal{R}}\boldsymbol{\omega}^D \times \mathbf{p}^{Q/O} = 3r\dot{\phi}\mathbf{b}_2,$$

and its acceleration is, by differentiation in $\mathcal{R}$,

$$^{\mathcal{R}}\mathbf{a}^Q = 3r(\ddot{\phi}\mathbf{b}_2 - \dot{\phi}^2\mathbf{b}_3).$$

The chosen directions are principal directions of inertia for the body (of revolution) with regard to Q, and the principal moments of inertia are (see Appendix C)

$$I_1^Q = I_2^Q = \frac{1}{4}mr^2, \quad I_3^Q = \frac{1}{2}mr^2.$$

We then have all three elements to comprise the six equations of motion for the body in question. In fact, substituting the above results in Eqs. (2.4) and (2.10), we have

$$0 = O_1 + N - mg, \tag{a}$$

$$3mr\ddot{\phi} = O_2 + F_a, \tag{b}$$

$$-3mr\dot{\phi}^2 = O_3, \tag{c}$$

$$-\frac{1}{4}mr^2\ddot{\phi} = 3rO_2, \tag{d}$$

$$\frac{3}{4}mr^2\dot{\phi}^2 + \frac{3}{4}mr^2\dot{\phi}^2 = -3rO_1, \tag{e}$$

$$\frac{3}{2}mr^2\ddot{\phi} = T - rF_a. \tag{f}$$

Here we have a holonomic system with a single degree of freedom, described by coordinate $\phi(t)$, which appears in Eqs. (b–f) with their first and second time rates. For the unknown components of force, that is, O_1, O_2, O_3, N, and F_a, there is a system of algebraic equations. Substituting (d) in (b) and the result in (f) gives

$$\ddot{\phi} = \frac{12}{55}\frac{T}{mr^2},$$

a relation that may be integrated directly for T constant, resulting in

$$\dot{\phi}(t) = \frac{12}{55}\frac{T}{mr^2}t$$

and

$$\phi(t) = \frac{6}{55}\frac{T}{mr^2}t^2 + \phi(0).$$

In the initial condition, immediately after applying the torque, $\dot{\phi} = 0$ and the initial acceleration of point Q is

$$^{\mathcal{R}}\mathbf{a}^Q = \frac{36}{55}\frac{T}{mr}\mathbf{b}_2.$$

The components of force, at the first instant, are, therefore,

$$O_1 = O_3 = 0, \qquad O_2 = -\frac{1}{55}\frac{T}{r}, \qquad N = mg, \qquad F_a = \frac{37}{55}\frac{T}{r}.$$

Note that the equations of motion were expressed in the directions of the orthonormal basis $\mathbf{b}_1, \mathbf{b}_2, \mathbf{b}_3$, that is not fixed in the body nor in the inertial reference frame. This was only possible due to the symmetry of the body, whose inertia ellipsoid is of revolution. Therefore, any direction parallel to the plane of $\mathbf{b}_1$ and $\mathbf{b}_2$ are principal directions of inertia for the disk with respect to its center. When this particular condition fails to occur, the reader must remember to choose a basis fixed in the body and — wishing to use Euler's equations — parallel to the principal directions of inertia for the mass center. See the corresponding animation.

If $\mathbf{n}_1, \mathbf{n}_2, \mathbf{n}_3$ is an orthonormal basis, fixed in a rigid body C, and parallel to its principal directions of inertia for the mass center, then the components of the angular acceleration vector of the body in an inertial reference frame are the time rates of the respective components of the angular velocity vector of C in the same reference frame, that is,

$$\alpha_j = \dot{\omega}_j, \qquad j = 1, 2, 3. \tag{2.11}$$

In fact, as the time rate of the angular velocity vector in the reference frame is equal to its time rate in the body and the basis is fixed in C, the result is Eq. (2.11). The substitution of Eq. (2.11) in Eq. (2.10) gives an alternative form for Euler's equations:

$$\begin{aligned}
I_1^*\dot{\omega}_1 + (I_3^* - I_2^*)\omega_3\omega_2 &= M_1; \\
I_2^*\dot{\omega}_2 + (I_1^* - I_3^*)\omega_1\omega_3 &= M_2; \\
I_3^*\dot{\omega}_3 + (I_2^* - I_1^*)\omega_2\omega_1 &= M_3.
\end{aligned} \tag{2.12}$$

Example 2.4 A homogeneous rod, with mass m and length $2r$, is pivoted in its mass center O to a fork, and may turn freely around the horizontal axis x_2, while the fork may, in turn, rotate freely around the vertical axis z, as shown (see Fig. 2.4a). The rod is at rest in the horizontal position, when force $\mathbf{F}$, of module $10\,mg$ and direction indicated in the figure, is applied to end Q. The initial acceleration of the opposite end P is to be found. Figure 2.4b indicates the forces and torques applied to the rod, broken down in the principal coordinate directions, of inertia for the rod with respect to its mass center. Torque $\mathbf{T}_1$ results, at least in principle, from the rotation constraint of the link in the corresponding direction. The resultant moment of the system of external forces with respect to O is

$$\mathbf{M}^{\mathcal{F}_{e/O}} = T_1\mathbf{b}_1 - rF_3\mathbf{b}_2 + rF_2\mathbf{b}_3.$$

The angular velocity of the rod in the inertial reference frame $\mathcal{R}$, supporting the fork, may be expressed by

$$^{\mathcal{R}}\boldsymbol{\omega}^{B} = {}^{S}\boldsymbol{\omega}^{B} + {}^{\mathcal{R}}\boldsymbol{\omega}^{S} = -\dot{\theta}\mathbf{b}_2 + \dot{\phi}(\sin\theta\,\mathbf{b}_1 + \cos\theta\,\mathbf{b}_3),$$

where θ is the angle between axis x_1 and the horizontal plane containing poin O and ϕ measures the rotation of the fork.

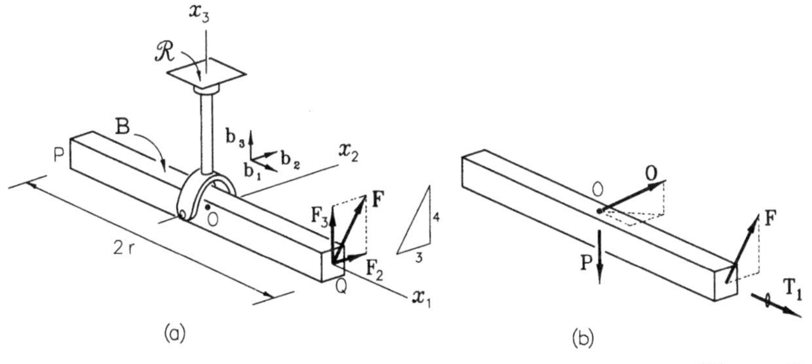

$$\textit{Figure 2.4}$$

As basis $\mathbf{b}_1, \mathbf{b}_2, \mathbf{b}_3$ is fixed in B, the expression for the angular acceleration is

$$^{\mathcal{R}}\boldsymbol{\alpha}^{B} = (\ddot{\phi}\sin\theta + \dot{\phi}\dot{\theta}\cos\theta)\mathbf{b}_1 - \ddot{\theta}\mathbf{b}_2 + (\ddot{\phi}\cos\theta - \dot{\phi}\dot{\theta}\sin\theta)\mathbf{b}_3.$$

The principal moments of inertia of the rod for its mass center are

$$I_1^{O} = 0, \quad I_2^{O} = I_3^{O} = \frac{1}{3}mr^2.$$

So substituting the above terms in Eq. (2.12), we have

$$0 = T_1,$$

$$-\frac{1}{3}mr^2\ddot{\theta} - \frac{1}{3}mr^2\dot{\phi}^2\sin\theta\cos\theta = -rF_3,$$

$$\frac{1}{3}mr^2(\ddot{\phi}\cos\theta - \dot{\phi}\dot{\theta}\sin\theta) - \frac{1}{3}mr^2\dot{\phi}\dot{\theta}\sin\theta = rF_2,$$

which can be reduced to

$$T_1 = 0,$$

$$\ddot{\theta} + \dot{\phi}^2\sin\theta\cos\theta = \frac{3F_3}{mr},$$

$$\ddot{\phi}\cos\theta - 2\dot{\phi}\dot{\theta}\cos\theta = \frac{3F_2}{mr},$$

and whose solution is (note that, at $t = 0$, $\dot{\phi} = \dot{\theta} = \phi = \theta = 0$)

$$\ddot{\theta} = 24\frac{g}{r} \quad \text{and} \quad \ddot{\phi} = 18\frac{g}{r}.$$

As point O is fixed in $\mathcal{R}$ and, at the initial instant $^{\mathcal{R}}\boldsymbol{\omega}^B = 0$, then

$$^{\mathcal{R}}\mathbf{a}^P = {}^{\mathcal{R}}\boldsymbol{\alpha}^B \times \mathbf{p}^{P/O} = 6\frac{g}{r}(-4\mathbf{b}_2 + 3\mathbf{b}_3) \times (-r)\mathbf{b}_1 = -6g(3\mathbf{b}_2 + 4\mathbf{b}_3).$$

See the corresponding animation.

7.3 Work on a Rigid Body

Consider that C is a rigid body moving in a reference frame $\mathcal{R}$, and let $^{\mathcal{R}}\boldsymbol{\omega}^C$ be its angular velocity vector in $\mathcal{R}$. If $\mathbf{T}$ is a torque applied on C during a certain interval of time (t_1, t_2), the *work of the torque on the body* in the interval, in the reference frame $\mathcal{R}$, is defined by (see Fig. 3.1)

$$^{\mathcal{R}}\mathcal{T}_{12}^T \rightleftharpoons \int_{t_1}^{t_2} \mathbf{T} \cdot {}^{\mathcal{R}}\boldsymbol{\omega}^C \, dt. \tag{3.1}$$

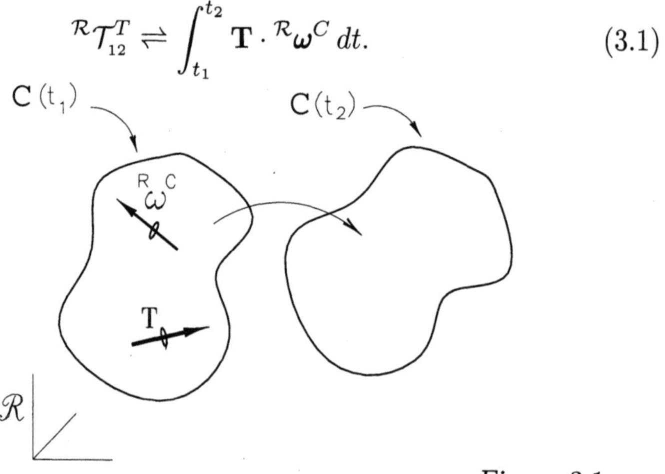

Figure 3.1

The work of a torque applied on a rigid body is a scalar with the same physical dimension of work of a force applied to a point, that is, $[ML^2T^{-2}]$.

If M torques $\mathbf{T}_j$, $j = 1, 2, \ldots, M$, act on a body C, the *work of the torques* will be the algebraic sum of the work, being naturally equal to the work done by the vector sum of the torques involved, in the same interval. In fact, if $\mathbf{T} = \mathbf{T}_1 + \mathbf{T}_2 + \ldots + \mathbf{T}_M$, then,

$$
\begin{aligned}
{}^{\mathcal{R}}\mathcal{T}_{12}^{T} &= \int_{t_1}^{t_2} \mathbf{T} \cdot {}^{\mathcal{R}}\boldsymbol{\omega}^C \, dt \\
&= \int_{t_1}^{t_2} \sum_{j=1}^{M} \mathbf{T}_j \cdot {}^{\mathcal{R}}\boldsymbol{\omega}^C \, dt \\
&= \sum_{j=1}^{M} \int_{t_1}^{t_2} \mathbf{T}_j \cdot {}^{\mathcal{R}}\boldsymbol{\omega}^C \, dt \\
&= \sum_{j=1}^{M} {}^{\mathcal{R}}\mathcal{T}_{12}^{T_j}.
\end{aligned}
\tag{3.2}
$$

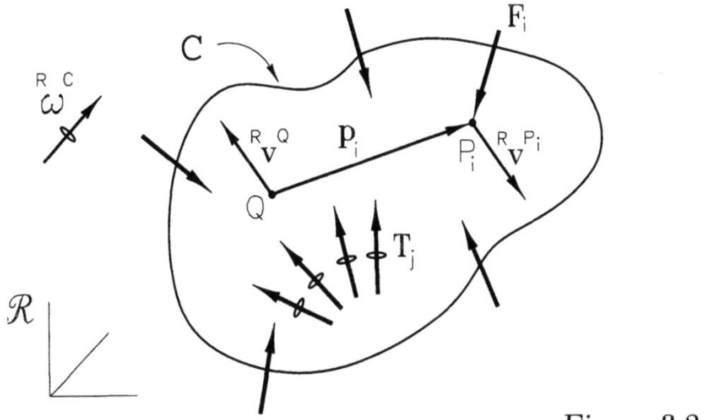

Figure 3.2

Let us now assume that a system $\mathcal{F}$ of forces act on the rigid body C, comprising N forces F_i, $i = 1, 2, \ldots, N$, each applied to a certain point of the body, and M torques T_j, $j = 1, 2, \ldots, M$ (see Fig. 3.2). We will call as the *resultant work of the system* $\mathcal{F}$ on the body C in the interval (t_1, t_2), in a given reference frame $\mathcal{R}$, to the algebraic sum of the work done by the components of $\mathcal{F}$, that is,

$$
{}^{\mathcal{R}}\mathcal{T}_{12}^{\mathcal{F}} \rightleftharpoons \sum_{i=1}^{N} {}^{\mathcal{R}}\mathcal{T}_{12}^{F_i} + \sum_{j=1}^{M} {}^{\mathcal{R}}\mathcal{T}_{12}^{T_j}.
\tag{3.3}
$$

The reader should note that, when calculating the resultant work of a system $\mathcal{F}$ on a rigid body, the work of *all* forces applied, in their respective points, is added and the work of *all* torques applied. This sum is only possible because, as previously commented, both works are scalars with the same physical dimension.

Arbitrating a point Q, *fixed in the body*, to reduce the system $\mathcal{F}$, we get a resultant vector, $\mathbf{R}$, and a resultant moment vector, $\mathbf{M}^{\mathcal{F}/Q}$. If P_i is the point of C where force $\mathbf{F}_i$ is applied, its velocity in $\mathcal{R}$ may, according to Eq. (3.8.3), be expressed by ${}^{\mathcal{R}}\mathbf{v}^{P_i} = {}^{\mathcal{R}}\mathbf{v}^Q + {}^{\mathcal{R}}\boldsymbol{\omega}^C \times \mathbf{p}_i$, where $\mathbf{p}_i$ is the position vector of the point P_i with respect to Q. The work done, therefore, by force $\mathbf{F}_i$ in the considered interval is, according to Eq. (4.5.5),

$$
\begin{aligned}
{}^{\mathcal{R}}\mathcal{T}_{12}^{F_i} &= \int_{t_1}^{t_2} \mathbf{F}_i \cdot {}^{\mathcal{R}}\mathbf{v}^{P_i}\, dt \\
&= \int_{t_1}^{t_2} \mathbf{F}_i \cdot ({}^{\mathcal{R}}\mathbf{v}^Q + {}^{\mathcal{R}}\boldsymbol{\omega}^C \times \mathbf{p}_i)\, dt \\
&= \int_{t_1}^{t_2} \mathbf{F}_i \cdot {}^{\mathcal{R}}\mathbf{v}^Q\, dt + \int_{t_1}^{t_2} \mathbf{p}_i \times \mathbf{F}_i \cdot {}^{\mathcal{R}}\boldsymbol{\omega}^C\, dt \\
&= \int_{t_1}^{t_2} \mathbf{F}_i \cdot {}^{\mathcal{R}}\mathbf{v}^Q\, dt + \int_{t_1}^{t_2} \mathbf{M}^{F_i/Q} \cdot {}^{\mathcal{R}}\boldsymbol{\omega}^C\, dt,
\end{aligned}
\tag{3.4}
$$

where $\mathbf{M}^{F_i/Q}$ is the moment of force $\mathbf{F}_i$ with respect to point Q. The resultant work of system $\mathcal{F}$ on body C can be calculated, then, by substituting Eqs. (3.1) and (3.4) in Eq. (3.3), resulting in

$$
\begin{aligned}
{}^{\mathcal{R}}\mathcal{T}_{12}^{\mathcal{F}} &= \sum_{i=1}^{N}\left(\int_{t_1}^{t_2} \mathbf{F}_i \cdot {}^{\mathcal{R}}\mathbf{v}^Q\, dt + \int_{t_1}^{t_2} \mathbf{M}^{F_i/Q} \cdot {}^{\mathcal{R}}\boldsymbol{\omega}^C\, dt \right) \\
&\quad + \sum_{j=1}^{M} \int_{t_1}^{t_2} \mathbf{T}_j \cdot {}^{\mathcal{R}}\boldsymbol{\omega}^C\, dt \\
&= \int_{t_1}^{t_2} \sum_{i=1}^{N} \mathbf{F}_i \cdot {}^{\mathcal{R}}\mathbf{v}^Q\, dt + \int_{t_1}^{t_2} \left(\sum_{i=1}^{N} \mathbf{M}^{F_i/Q} + \sum_{j=1}^{M} \mathbf{T}_j \right) \cdot {}^{\mathcal{R}}\boldsymbol{\omega}^C\, dt \\
&= \int_{t_1}^{t_2} \mathbf{R} \cdot {}^{\mathcal{R}}\mathbf{v}^Q\, dt + \int_{t_1}^{t_2} \mathbf{M}^{\mathcal{F}/Q} \cdot {}^{\mathcal{R}}\boldsymbol{\omega}^C\, dt.
\end{aligned}
\tag{3.5}
$$

We conclude then that, given a general system of forces $\mathcal{F}$ acting on rigid body C moving in a reference frame $\mathcal{R}$, to establish the resultant work of $\mathcal{F}$ on C, in $\mathcal{R}$, in a given interval, it is sufficient to reduce the system to an *arbitrary point of the body*, calculate the work of a force equal to the resultant applied to the point, and add the work of the resultant moment of the system with respect to the point, as if a torque were applied to the body. In other words, the work of a system of forces $\mathcal{F}$ on a rigid body C is equal to the work of its reduction to an arbitrary point of the body.

Example 3.1 A homogeneous disk D, with mass m and radius r, is at rest, supported by a smooth flat surface, fixed in a reference frame $\mathcal{R}$, when a system of forces is applied, consisting of forces $\mathbf{A} = 4mg\mathbf{n}_1$, applied to a string rolled around it, and $\mathbf{B} = 3mg\mathbf{n}_2$, applied to its center O, as well as by torque $\mathbf{T} = mgr\mathbf{n}_3$ (see Fig. 3.3). The resultant work of this system during a time interval τ is to be found.

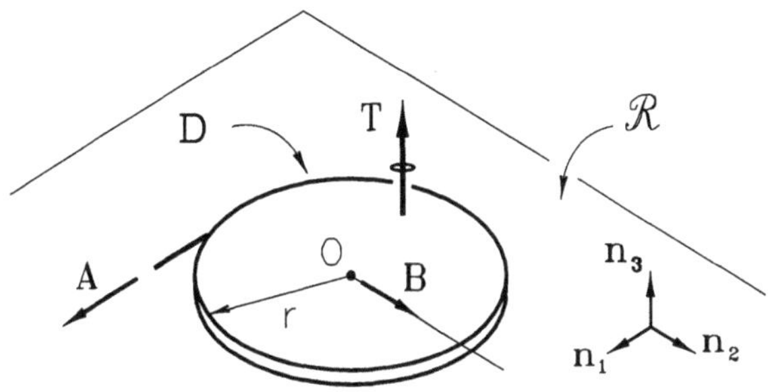

Figure 3.3

By reducing the system, say, to the center of the disk, then

$$\mathbf{R} = A\mathbf{n}_1 + B\mathbf{n}_2,$$
$$\mathbf{M}^{\mathcal{F}/O} = (rA + T)\mathbf{n}_3.$$

Since the disk is in plane motion, the acceleration of the mass center has the form $^{\mathcal{R}}\mathbf{a}^O = a_1\mathbf{n}_1 + a_2\mathbf{n}_2$, where the basis $\mathbf{n}_1, \mathbf{n}_2, \mathbf{n}_3$ is fixed in $\mathcal{R}$, with the indicated orientation. The angular velocity of the disk may be expressed by $^{\mathcal{R}}\boldsymbol{\omega}^D = \omega\mathbf{n}_3$ and its angular acceleration is $^{\mathcal{R}}\boldsymbol{\alpha}^D = \dot{\omega}\mathbf{n}_3$. The equations of motion for this body are then

$$ma_1 = A,$$
$$ma_2 = B,$$
$$I_3^O \dot{\omega} = rA + T,$$

which, integrated, result in

$$^{\mathcal{R}}\mathbf{v}^O = \frac{1}{m}(A\mathbf{n}_1 + B\mathbf{n}_2)\, t, \qquad \omega = \frac{rA + T}{I_3^O}\, t.$$

The resultant work in the considered interval is then, according to Eq. (3.5),

$$
\begin{aligned}
^{\mathcal{R}}\mathcal{T}_{0\tau}^{\mathcal{F}} &= \int_0^\tau (A\mathbf{n}_1 + B\mathbf{n}_2) \cdot \frac{1}{m}(A\mathbf{n}_1 + B\mathbf{n}_2)\, t\, dt \\
&\quad + \int_0^\tau (rA + T)\mathbf{n}_3 \cdot \frac{rA + T}{I_3^O}\, t\, \mathbf{n}_3\, dt \\
&= \left(\frac{A^2 + B^2}{m} + \frac{(rA + T)^2}{\frac{1}{2}mr^2} \right) \frac{\tau^2}{2} \\
&= \frac{75}{2} mg^2 \tau^2.
\end{aligned}
$$

See the corresponding animation.

A force $\mathbf{F}_i$ applied on a certain point P_i of a rigid body will do no work, in a given time interval (t_1, t_2), if, as we already know, the velocity of the point in the reference frame is, in this interval, orthogonal to the force (or null). Likewise, a torque $\mathbf{T}_j$ applied to a rigid body will do no work in a given interval if, in that interval, the angular velocity of the body in the reference frame is orthogonal to the torque (or null), as indicated in Eq. (3.1). Therefore, forces and torques that do not work do not contribute to the resultant work of the system $\mathcal{F}$. When there are, therefore, elements of $\mathcal{F}$ that do not contribute, the resultant work expressed in Eq. (3.5) may be calculated by ignoring these terms. In other words, the resultant work may be calculated by previously suppressing from the resultant $\mathbf{R}$ and from the resultant moment $\mathbf{M}^{\mathcal{F}/Q}$ the forces and torques that are known not to work. In fact, we will consider a force $\mathbf{F}_i$, component of system $\mathcal{F}$, so that, during the interval (t_1, t_2), velocity $\mathbf{v}_i$ of point P_i of the body where it is applied is orthogonal to the force (or null), that is, the dot product $\mathbf{F}_i \cdot \mathbf{v}_i = 0$ at any instant of the considered interval. Let us also assume a torque $\mathbf{T}_j$, also a component of $\mathcal{F}$, orthogonal to $^{\mathcal{R}}\boldsymbol{\omega}^C$ during the same interval, that is, $\mathbf{T}_j \cdot {}^{\mathcal{R}}\boldsymbol{\omega}^C = 0$, in the interval. So decomposing the reduction of system $\mathcal{F}$ to point Q as

$$\mathbf{R} = \overline{\mathbf{R}} + \mathbf{F}_i, \qquad \mathbf{M}^{\mathcal{F}/Q} = \overline{\mathbf{M}} + \mathbf{M}^{\mathbf{F}_i/Q} + \mathbf{T}_j, \qquad (3.6)$$

and substituting in Eq. (3.5), we have

$$
^{\mathcal{R}}\mathcal{T}_{12}^{\mathcal{F}} = \int_{t_1}^{t_2} \overline{\mathbf{R}} \cdot {}^{\mathcal{R}}\mathbf{v}^Q \, dt + \int_{t_1}^{t_2} \overline{\mathbf{M}} \cdot {}^{\mathcal{R}}\boldsymbol{\omega}^C \, dt
$$
$$
+ \int_{t_1}^{t_2} \mathbf{F}_i \cdot {}^{\mathcal{R}}\mathbf{v}^Q \, dt + \int_{t_1}^{t_2} \mathbf{M}^{\mathbf{F}_i/Q} \cdot {}^{\mathcal{R}}\boldsymbol{\omega}^C \, dt \qquad (3.7)
$$
$$
+ \int_{t_1}^{t_2} \mathbf{T}_j \cdot {}^{\mathcal{R}}\boldsymbol{\omega}^C \, dt.
$$

But, from the kinematic theorem for velocities, $^{\mathcal{R}}\mathbf{v}^Q = \mathbf{v}_i - {}^{\mathcal{R}}\boldsymbol{\omega}^C \times \mathbf{p}_i$, and the third integral in Eq. (3.7) is reduced to

$$
\int_{t_1}^{t_2} \mathbf{F}_i \cdot {}^{\mathcal{R}}\mathbf{v}^Q \, dt = \int_{t_1}^{t_2} \mathbf{F}_i \cdot \mathbf{v}_i \, dt - \int_{t_1}^{t_2} \mathbf{F}_i \cdot {}^{\mathcal{R}}\boldsymbol{\omega}^C \times \mathbf{p}_i \, dt
$$
$$
= - \int_{t_1}^{t_2} \mathbf{p}_i \times \mathbf{F}_i \cdot {}^{\mathcal{R}}\boldsymbol{\omega}^C \, dt \qquad (3.8)
$$
$$
= - \int_{t_1}^{t_2} \mathbf{M}^{\mathbf{F}_i/Q} \cdot {}^{\mathcal{R}}\boldsymbol{\omega}^C \, dt.
$$

So substituting Eq. (3.8) in Eq. (3.7) and noting that, by hypothesis, $\mathbf{T}_j \cdot {}^{\mathcal{R}}\boldsymbol{\omega}^C = 0$, then

$$
^{\mathcal{R}}\mathcal{T}_{12}^{\mathcal{F}} = \int_{t_1}^{t_2} \overline{\mathbf{R}} \cdot {}^{\mathcal{R}}\mathbf{v}^Q \, dt + \int_{t_1}^{t_2} \overline{\mathbf{M}} \cdot {}^{\mathcal{R}}\boldsymbol{\omega}^C \, dt, \qquad (3.9)
$$

where $\overline{\mathbf{R}}$ is the resultant of the *working forces* and $\overline{\mathbf{M}}$ is the resultant moment with respect to point Q, including only *the working forces and torques*, in the interval. Hence the result is that, in fact, the reduction $\{\overline{\mathbf{R}}, \overline{\mathbf{M}}\}$ may always be used for calculating the resultant work of system $\mathcal{F}$. It may happen that some force or torque, included in the reduction

$\{\overline{\mathbf{R}}, \overline{\mathbf{T}}\}$, does not work in the interval, this fact not being evident a priori. This happens because the integration of the dot product between the force and velocity (or the integration of the dot product between the torque and angular velocity) may vanish in the interval, even though the integrand is not null throughout. Of course, even when this eventuality occurs, there will be no *error* involved in calculating the resultant work, but only effectively unnecessary calculations.

Example 3.2   Figure 3.4 reproduces the system studied in Example 2.3. The applied system $\mathcal{F}$ consists of the components of force O_1, O_2, O_3, P, N, and F_a, applied to the indicated points, and torque $\mathbf{T}$.

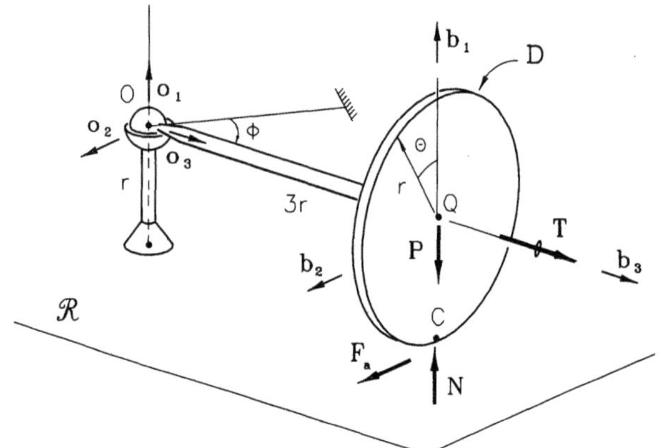

Figure 3.4

Force $\mathbf{O}$ does not work, since its application point is fixed in $\mathcal{R}$. Force $\mathbf{P}$ does not work, since it is always orthogonal to the velocity of point Q. Nor do forces $\mathbf{N}$ and $\mathbf{F}_a$ work, due to the fact that they always act on the point of contact, whose velocity in $\mathcal{R}$ is null. The only component of the system $\mathcal{F}$ that will contribute to the resultant work will, therefore, be torque $\mathbf{T}$. We then have $\overline{\mathbf{R}} = 0$ and $\overline{\mathbf{M}} = \mathbf{T}$. The resultant work in an interval, say, of a complete revolution of the arm, is, therefore, according to Eq. (3.9),

$$
^{\mathcal{R}}\mathcal{T}_{12}^{\mathcal{F}} = \int_{t_1}^{t_2} \mathbf{T} \cdot {}^{\mathcal{R}}\boldsymbol{\omega}^C \, dt = \int_{t_1}^{t_2} T\mathbf{b}_3 \cdot (-\dot{\phi}\mathbf{b}_1 + 3\dot{\phi}\mathbf{b}_3) \, dt
$$

$$
= 3T \int_0^{2\pi} d\phi = 6\pi T.
$$

It has been clearly demonstrated in Section 5.5 that, as Eq. (5.5.8) indicates, the resultant work of the system consisting of the forces of mutual interaction between elements of a mechanical system that keep constant distances in time is null. Now, a rigid body is exactly a system where *all* its elements keep invariant distances between each other with time. Consequently, the resultant work of the internal forces, for a rigid body, will always be null, that is, whatever the initial and final configurations of the body may be,

$$^{\mathcal{R}}\mathcal{T}_{12}^{\mathcal{F}_i} = 0. \tag{3.10}$$

This result means that only the resultant work of the *external forces* applied on a rigid body requires consideration, that is,

$$^{\mathcal{R}}\mathcal{T}_{12}^{\mathcal{F}} = {}^{\mathcal{R}}\mathcal{T}_{12}^{\mathcal{F}_e}. \tag{3.11}$$

Concerning, therefore, the calculation of the resultant work done by a system of forces acting on a rigid body, we will hereinafter always be referring to a *system of external forces*.

Returning once again to Section 5.5, it is shown that, as Eq. (5.5.13) indicates, the resultant work of the external forces acting on any mechanical system may be broken down into the sum of a resultant work on the mass center of the body with a resultant work around the mass center, that is,

$$^{\mathcal{R}}\mathcal{T}_{12}^{\mathcal{F}_e} = {}^{\mathcal{R}}\mathcal{T}_{12}^{F} + {}^{\mathcal{R}}\mathcal{T}_{12}^{\mathcal{F}_e/C^*}, \tag{3.12}$$

where the first term on the right represents the work done by a force $\mathbf{F}$ equal to the resultant of the system of external forces, applied to the mass center of the body, while the second term represents what we usually call resultant work around the mass center.

Equation (3.5), on the other hand, establishes that the resultant work of a system of forces applied to a rigid body C moving in a reference frame $\mathcal{R}$ can be calculated by reducing it to an arbitrary point Q of the body. If, in particular, the mass center, C^*, itself is chosen to be that point, the resultant work of the external forces may be expressed as

$$^{\mathcal{R}}\mathcal{T}_{12}^{\mathcal{F}_e} = \int_{t_1}^{t_2} \mathbf{F} \cdot \mathbf{v}^* \, dt + \int_{t_1}^{t_2} \mathbf{M}^{\mathcal{F}_e/C^*} \cdot {}^{\mathcal{R}}\boldsymbol{\omega}^C \, dt, \tag{3.13}$$

where $\mathbf{v}^*$ is the velocity, in $\mathcal{R}$, of the mass center of the body.

Now comparing the two decompositions, it is evident that, for a given rigid body, the resultant work of the (external) forces on the mass center, also more simply called the *resultant work of translation of the body*, is

$$^{\mathcal{R}}\mathcal{T}_{12}^{F} = \int_{t_1}^{t_2} \mathbf{F} \cdot \mathbf{v}^* \, dt \tag{3.14}$$

and the resultant work of the (external) forces around the mass center, also called, for the sake of simplicity, the *resultant work of rotation of the body*, will be

$$^{\mathcal{R}}\mathcal{T}_{12}^{\mathcal{F}_e/C^*} = \int_{t_1}^{t_2} \mathbf{M}^{\mathcal{F}_e/C^*} \cdot {}^{\mathcal{R}}\boldsymbol{\omega}^C \, dt. \tag{3.15}$$

Example 3.3 Let us consider a slim prismatic body, with mass m and length r, supported by a smooth, horizontal flat surface, and being hinged in end B (see Fig. 3.5). The body is at rest when a force, of constant module A, which always remains orthogonal to the axis of the body, as indicated, is applied to the free end A.

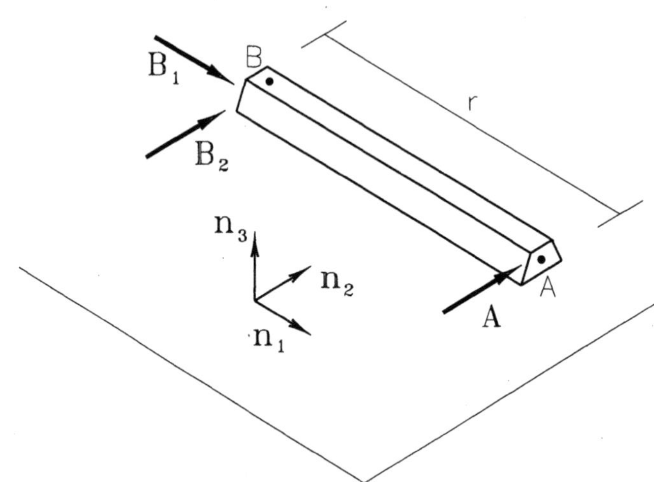

Figure 3.5

This is, therefore, a plane motion with a fixed point. The acting forces will be (those parallel to the plane): A, B_1, and B_2. Reducing this system to point B, we have

$$\mathbf{F} = B_1 \mathbf{n}_1 + (A + B_2)\mathbf{n}_2,$$
$$\mathbf{M}^{\mathcal{F}e/B} = rA\mathbf{n}_3.$$

The equations of motion for the problem are

$$ma_1 = B_1, \tag{a}$$
$$ma_2 = A + B_2, \tag{b}$$
$$\frac{1}{3}mr^2\ddot{\theta} = rA, \tag{c}$$

where a_1 and a_2 are the scalar components of the acceleration of the mass center and $\theta(t)$ describes the rotation of the bar. Assuming that the body leaves the resting position with $\theta = 0$, then, from Eq. (c),

$$\ddot{\theta} = \frac{3A}{mr}; \qquad \text{then} \qquad \dot{\theta} = \frac{3A}{mr}\,t \qquad \text{and} \qquad \theta = \frac{3A}{2mr}\,t^2.$$

The components of the acceleration of the mass center may be expressed as

$$a_1 = -\frac{r}{2}\dot{\theta}^2, \qquad a_2 = \frac{r}{2}\ddot{\theta}.$$

The unknown components of force may then be obtained from Eqs. (a) and (b), resulting in

$$B_1 = -\frac{9}{2}\frac{A^2}{mr}\,t^2, \qquad B_2 = \frac{1}{2}A.$$

The velocity of the mass center is given by

$$\mathbf{v}^* = \frac{r}{2}\dot{\theta}\mathbf{n}_2,$$

and the resultant work of translation during, say, a quarter of a turn is, according to Eq. (3.14),

$$
\begin{aligned}
{}^{\mathcal{R}}\mathcal{T}^F_{0\frac{\pi}{2}} &= \int_{t_1}^{t_2} \left(B_1\mathbf{n}_1 + (A + B_2)\mathbf{n}_2\right) \cdot \frac{r}{2}\dot{\theta}\mathbf{n}_2 \, dt \\
&= \int_0^{\pi/2} (A + B_2)\frac{r}{2}\, d\theta \\
&= \frac{3\pi}{8}Ar.
\end{aligned}
$$

The resultant moment of the system of forces with respect to the mass center is

$$\mathbf{M}^{\mathcal{F}_e/C^*} = \frac{r}{2}(A - B_2)\mathbf{n}_3 = \frac{Ar}{4}\,\mathbf{n}_3,$$

and the resultant work of rotation, in the same interval, is, according to Eq. (3.15),

$$\begin{aligned}
^{\mathcal{R}}\mathcal{T}_{0\frac{\pi}{2}}^{\mathcal{F}_e/C^*} &= \int_{t_1}^{t_2} \frac{Ar}{4}\,\mathbf{n}_3 \cdot \dot{\theta}\mathbf{n}_3\,dt \\
&= \frac{Ar}{4}\int_0^{\pi/2} d\theta \\
&= \frac{\pi}{8}Ar.
\end{aligned}$$

The overall resultant work, in the interval, is, therefore, according to Eq. (3.12),

$$^{\mathcal{R}}\mathcal{T}_{0\frac{\pi}{2}}^{\mathcal{F}_e} = \frac{3\pi}{8}Ar + \frac{\pi}{8}Ar = \frac{\pi}{2}Ar.$$

Considering now only force $\mathbf{A}$, the reduction to, say, point A consists of

$$\overline{\mathbf{R}} = A\mathbf{n}_2, \qquad \overline{\mathbf{M}} = 0$$

and the resultant work is, according to Eq. (3.9),

$$^{\mathcal{R}}\mathcal{T}_{0\frac{\pi}{2}}^{\mathcal{F}_e} = \int_{t_1}^{t_2} \overline{\mathbf{R}} \cdot {}^{\mathcal{R}}\mathbf{v}^A\,dt = \int_{t_1}^{t_2} A \cdot r\dot{\theta}\,dt = Ar\int_0^{\pi/2} d\theta = \frac{\pi}{2}Ar,$$

as obtained above.

7.4 Work and Energy

Some general equations are established in Section 5.6, valid for any mechanical system, relating a change in some kind of energy of the system in an inertial reference frame, between two configurations, to the resultant work applied. These equations may easily be particularized now for the rigid body, taking into account Eq. (3.11). So, Eq. (5.6.1), applied to a rigid body C moving in an inertial reference frame $\mathcal{R}$, is reduced to

$$^{\mathcal{R}}K^C(2) - {}^{\mathcal{R}}K^C(1) = {}^{\mathcal{R}}\mathcal{T}_{12}^{\mathcal{F}_e}, \tag{4.1}$$

that is, the change in the kinetic energy of a rigid body in an inertial reference frame, between two configurations of its motion, is equal to the resultant work of the applied external forces, in the same interval.

Example 4.1 Fiure. 4.1 illustrates the system analyzed in Example 2.3 (take another look at it). Assuming $\phi(0) = 0$, the rod will have given a complete turn in instant t which satisfies

$$2\pi = \frac{6}{55}\frac{T}{mr^2}t^2, \qquad \text{that means,} \qquad t^2 = \frac{55\pi}{3}\frac{mr^2}{T}.$$

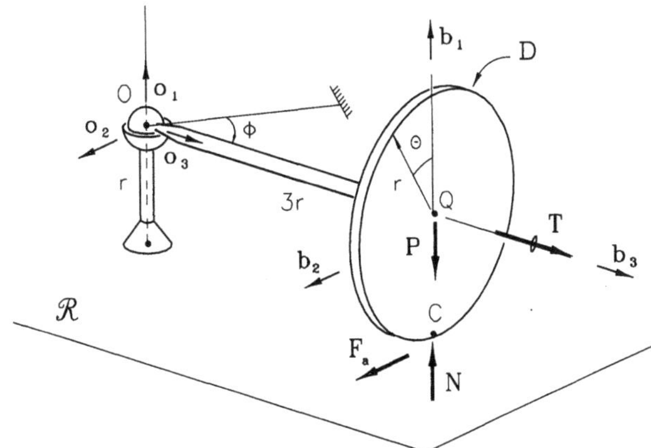

Figure 4.1

In this instant, then, we have (check)

$$\dot{\phi}^2 = \frac{48\pi}{55}\frac{T}{mr^2}.$$

The kinetic energy of translation of the disk is, according to Eq. (1.18),

$$\begin{aligned}
{}^{\mathcal{R}}K^Q &= \frac{1}{2}m\,{}^{\mathcal{R}}\mathbf{v}^Q \cdot {}^{\mathcal{R}}\mathbf{v}^Q \\
&= \frac{1}{2}m(3r\dot{\phi})^2 \\
&= \frac{9}{2}mr^2\dot{\phi}^2.
\end{aligned}$$

The kinetic energy of rotation of the disk, in the same instant, is, according to Eq. (1.21),

$$\begin{aligned}
{}^{\mathcal{R}}K^{D/Q} &= \frac{1}{2}{}^{\mathcal{R}}\boldsymbol{\omega}^D \cdot \mathbf{\Pi}^{D/Q} \cdot {}^{\mathcal{R}}\boldsymbol{\omega}^D \\
&= \frac{1}{2}(-\dot{\phi}\mathbf{b}_1 + 3\dot{\phi}\mathbf{b}_3) \cdot mr^2 \begin{pmatrix} \frac{1}{4} & 0 & 0 \\ 0 & \frac{1}{4} & 0 \\ 0 & 0 & \frac{1}{2} \end{pmatrix} \cdot (-\dot{\phi}\mathbf{b}_1 + 3\dot{\phi}\mathbf{b}_3) \\
&= \frac{19}{8}mr^2\dot{\phi}^2.
\end{aligned}$$

The kinetic energy of the disk, after a complete turn of the arm, is, therefore,

$$^{\mathcal{R}}K^D = {}^{\mathcal{R}}K^Q + {}^{\mathcal{R}}K^{D/Q}$$

$$= \frac{55}{8}mr^2\dot{\phi}^2$$

$$= 6\pi T.$$

If the reader now goes back to Example 3.2, it will be found that this is exactly the value obtained for the resultant work done by the system of applied external forces, during a complete rotation, confirming what is established in Eq. (4.1). Try calculating the kinetic energy of the disk via Eq. (1.25), comparing the result.

When, from among the external forces acting on a body, conservative forces contribute, their respective work can be calculated, as shown earlier, by the change of a potential function. The sum of potential functions of all conservative forces present is called the *potential energy of the body*, $^{\mathcal{R}}\Phi^C$, depending on its configuration. Equation (5.6.3), expressed for a rigid body, is then as follows:

$$^{\mathcal{R}}K^C(2) + {}^{\mathcal{R}}\Phi^C(2) - {}^{\mathcal{R}}K^C(1) - {}^{\mathcal{R}}\Phi^C(1) = {}^{\mathcal{R}}T_{12}^{\mathcal{F}_N}, \qquad (4.2)$$

where, now, $^{\mathcal{R}}T_{12}^{\mathcal{F}_N}$ is the resultant work of the nonconservative *external* forces applied to the body in the interval.

Defining the *mechanical energy* of a rigid body in a reference frame $\mathcal{R}$ as the sum, at each instant, of its kinetic energy with its potential energy, in the same reference frame, that is,

$$^{\mathcal{R}}E^C \rightleftharpoons {}^{\mathcal{R}}K^C + {}^{\mathcal{R}}\Phi^C, \qquad (4.3)$$

Equation (4.2) can be expressed alternatively as

$$^{\mathcal{R}}E^C(2) - {}^{\mathcal{R}}E^C(1) = {}^{\mathcal{R}}T_{12}^{\mathcal{F}_N}. \qquad (4.4)$$

Equation (4.4) establishes, therefore, the equality between the change of the mechanical energy of a rigid body, in an inertial reference frame, and the resultant work of the nonconservative external forces, in the considered interval.

Example 4.2 The homogeneous rod B, with mass m and length r, has its end O articulated by a pivot on a fixed support, and can move freely around the axis x_2 (see Fig. 4.2). A wire is fixed to end P, passing through a guide without friction in Q, to which a force with constant module F is applied, as indicated. The ends of two linear springs, with elastic constants k and $3k$, are fixed to the mass center of the rod, both with a natural length r, and whose opposite ends are fixed to points M and N, respectively.

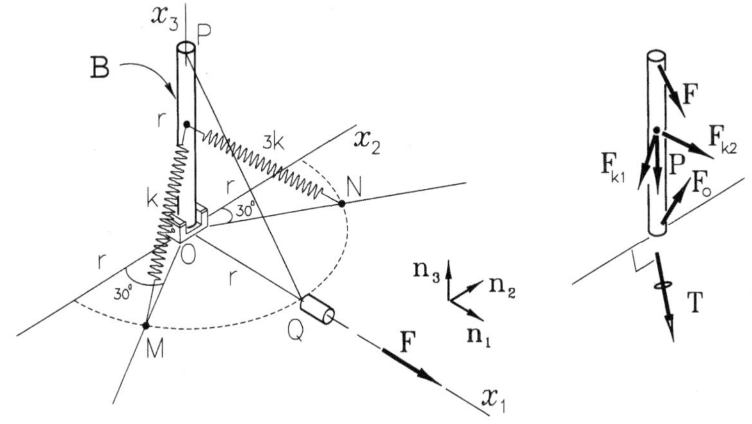

Figure 4.2

The rod is initially at rest in the vertical position, as shown. The speed of end P when passing close to the guide is to be found. The rod's weight, the force applied by the wire, the forces applied by the springs, an unknown force applied on the pivot (three components), and a torque, also due to the pivot, orthogonal to the axis x_2 (two components) all act on the rod. The force on the pivot does not work, since point O does not move. The angular velocity of the rod is necessarily parallel to axis x_2, which means that the applied torque does not contribute either to the resultant work. The work of the force applied by the wire, between the initial (vertical) and final (horizontal) positions of the rod, may be simply calculated as the product of the module of force, constant, by the total displacement of the wire, that is,

$$\mathcal{T}_{12}^F = \sqrt{2}Fr.$$

The weight is a conservative force and its potential function, in each of the positions considered, will be (arbitrating the reference in $x_3 = 0$)

$$\Phi_g(1) = \frac{1}{2}mgr, \qquad \Phi_g(2) = 0.$$

The force applied by the spring is also conservative. The potential function associated to the pair of springs, for each position, will be

$$\Phi_k(1) = \frac{1}{2}(k+3k)\left(\frac{\sqrt{5}}{2}r - r\right)^2 = \frac{\left(\sqrt{5}-2\right)^2}{2}kr^2,$$

$$\Phi_k(2) = \frac{1}{2}(k+3k)\left(\frac{\sqrt{3}}{2}r - r\right)^2 = \frac{\left(\sqrt{3}-2\right)^2}{2}kr^2.$$

As point O is fixed, the kinetic energy of the rod may be calculated from Eq. (1.25), that is,

$$K^B(1) = 0,$$

$$K^B(2) = \frac{1}{2}\omega(0,1,0)\cdot\frac{1}{3}mr^2\begin{pmatrix}1 & 0 & 0\\ 0 & 1 & 0\\ 0 & 0 & 0\end{pmatrix}\omega\begin{pmatrix}0\\ 1\\ 0\end{pmatrix} = \frac{1}{6}mr^2\omega^2.$$

So by substituting the above results in Eq. (4.4), we have

$$\frac{1}{6}mr^2\omega^2 + \frac{1}{2}\left(\left(\sqrt{3}-2\right)^2 - \left(\sqrt{5}-2\right)^2\right)kr^2 - \frac{1}{2}mgr = \sqrt{2}Fr.$$

Solving for ω and substituting in the kinematic relation $v = \omega r$, the velocity is found to be

$$v = r\left[6\sqrt{2}\,\frac{F}{mr} + 3\frac{g}{r} + 3\left(\left(\sqrt{5}-2\right)^2 - \left(\sqrt{3}-2\right)^2\right)\frac{k}{m}\right]^{\frac{1}{2}}.$$

See the corresponding animation.

As the reader has already had the opportunity to see, the equations that relate resultant work with a change in energy are extremely useful in finding velocities and angular velocities as a function of a given configuration of the body. There is also the advantage of bypassing the calculation of unknown forces whenever they do not contribute to the resultant work.

Example 4.3　　A drum T, with outer radius $2r$, is coupled to two disks with radius r. The set, with a total mass m and moment of inertia $I = 1.3\,mr^2$ with respect to the axis of symmetry, is supported, at rest, on a sloping rail, under the action of a force applied to a wound wire on the edge of the drum, as shown in Fig. 4.3. Suddenly, the force applied to the wire increased to the value $A = 4mg$, so that the drum accelerates up the rail. Assuming that the coefficient of friction with the rail is sufficient to prevent the relative sliding, we want to find the velocity of the center of the drum and its angular velocity when a complete turn has been made around its axis.

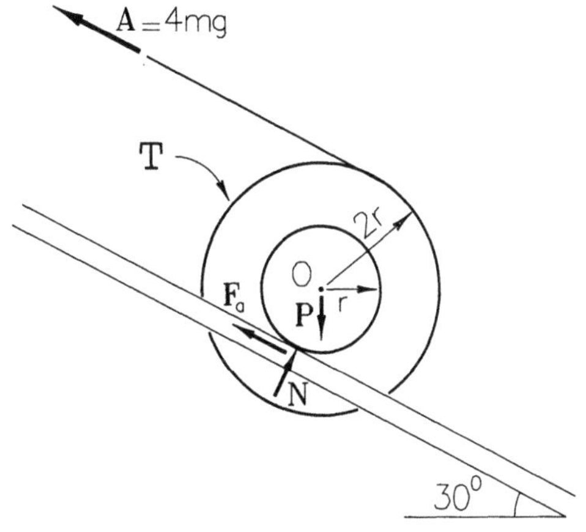

Figure 4.3

The figure shows the acting forces, among them only the weight $\mathbf{P}$ and the force $\mathbf{A}$ do work in the considered interval. As the weight is a conservative force, its contribution to the resultant work is given by the change (negative) in its potential function, that is, assuming that the rail is fixed in an inertial reference frame $\mathcal{R}$,

$$^{\mathcal{R}}\mathcal{T}^P_{0\,2\pi} = \Phi_g(0) - \Phi_g(2\pi) = -\pi mgr.$$

The work of force $\mathbf{A}$, constant, can be calculated simply by the product of the module of force by the overall displacement of the wire, in the direction of the force and, therefore, is

$$^{\mathcal{R}}\mathcal{T}^A_{0\,2\pi} = 4mg \cdot 6\pi r = 24\pi mgr.$$

The resultant work will, therefore, be

$$^{\mathcal{R}}\mathcal{T}^{\mathcal{F}_e}_{0\,2\pi} = 24\pi mgr - \pi mgr = 23\pi mgr.$$

The initial kinetic energy of translation is null, and the final is

$$^{\mathcal{R}}K^O(2\pi) = \frac{1}{2}mv^2 = \frac{1}{2}mr^2\omega^2.$$

The intial kinetic energy of rotation is also null, and the final is

$$^{\mathcal{R}}K^{T/O}(2\pi) = \frac{1}{2}I\omega^2.$$

The final kinetic energy of the drum is, therefore,

$$^{\mathcal{R}}K^{T}(2\pi) = {}^{\mathcal{R}}K^{O}(2\pi) + {}^{\mathcal{R}}K^{T/O}(2\pi) = \frac{1}{2}\left(mr^2 + I\right)\omega^2.$$

So, from Eq. (4.1), we have

$$\frac{1}{2}\left(mr^2 + I\right)\omega^2 = 23\pi mgr$$

and, therefore,

$$\omega = 2\sqrt{5\pi g/r},$$

and the velocity of the mass center is

$$v^* = 2\sqrt{5\pi gr}.$$

Note that the velocity was obtained without needing to calculate the unknown components of the contact force. See the corresponding animation.

Returning once again to Section 5.6, it is shown, according to Eq. (5.6.7), that the change of the kinetic energy of the mass center of any mechanical system in an inertial reference frame $\mathcal{R}$, in a given interval, is equal to the resultant work on its mass center, in this interval. In terms of a rigid body C, the result then is that the change of the kinetic energy of translation of the body in $\mathcal{R}$, in a given interval, will be equal to the resultant work of translation, that is,

$$^{\mathcal{R}}K^{C^*}(2) - {}^{\mathcal{R}}K^{C^*}(1) = {}^{\mathcal{R}}\mathcal{T}_{12}^{F}. \tag{4.5}$$

Likewise, Eq. (5.6.8), which establishes the relation between the change of the kinetic energy around the mass center, in an inertial reference frame, and the resultant work around the mass center of the system, will, for a rigid body, be reduced to the form

$$^{\mathcal{R}}K^{C/C^*}(2) - {}^{\mathcal{R}}K^{C/C^*}(1) = {}^{\mathcal{R}}\mathcal{T}_{12}^{\mathcal{F}_e/C^*}, \tag{4.6}$$

that is, the change, in an inertial reference frame, of the kinetic energy of rotation of the body, in a given interval, is equal to the resultant work of rotation applied to the body, in the interval.

Example 4.4 Returning to the preceding example (see Fig. 4.3), the resultant work of translation after a turn of the drum is, according to Eq. (4.5),

$$^{\mathcal{R}}\mathcal{T}_{0\,2\pi}^{F} = {}^{\mathcal{R}}K^{O}(2\pi) = \frac{1}{2}mv^2 = 10\pi mgr.$$

The resultant work of rotation is, according to Eq. (4.6),

$$^{\mathcal{R}}\mathcal{T}_{0\,2\pi}^{\mathcal{F}_e/C^*} = {}^{\mathcal{R}}K^{T/O}(2\pi) = \frac{1}{2}I\omega^2 = 13\pi mgr.$$

The sum of the works is, naturally, equal to the overall resultant work. In fact,

$$^{\mathcal{R}}\mathcal{T}_{0\,2\pi}^{F} + {}^{\mathcal{R}}\mathcal{T}_{0\,2\pi}^{\mathcal{F}_e/C^*} = 10\pi mgr + 13\pi mgr = 23\pi mgr.$$

Note that both works include the contribution of the unknown forces of contact. The major utility of Eqs. (4.5) and (4.6) to find the velocity of the mass center and angular velocity is, therefore, restricted to those situations where the forces applied to the body are all known.

Equations (4.1) to (4.6), relating the change of the (kinetic, rotation kinetic, mechanical, etc.) energy of a rigid body in an inertial reference frame with the resultant work (of the outer forces, around the mass center, nonconservative, etc.) are useful to find the angular velocity of the body and velocities of its points, in the reference frame. As it is already known, when wishing to know the *configuration* of the body (its orientation in the reference frame, the position of a point of the body with respect to a fixed point in reference frame), as a time function, there is no way to avoid establishing and solving the equations of motion. In some cases, a hybrid approach is advisable, that is, the solution is facilitated by using energy methods, Newton's second law, and Euler's equations simultaneously. In the following section, these methods will be employed in solving some examples involving the plane motion of a rigid body.

7.5 Plane Motion

When a rigid body C has a *plane motion* in an inertial reference frame $\mathcal{R}$ (see Section 3.8), the equations ruling its evolution take a much simpler form, if compared with those ruling the general motion of a body, as will be seen below.

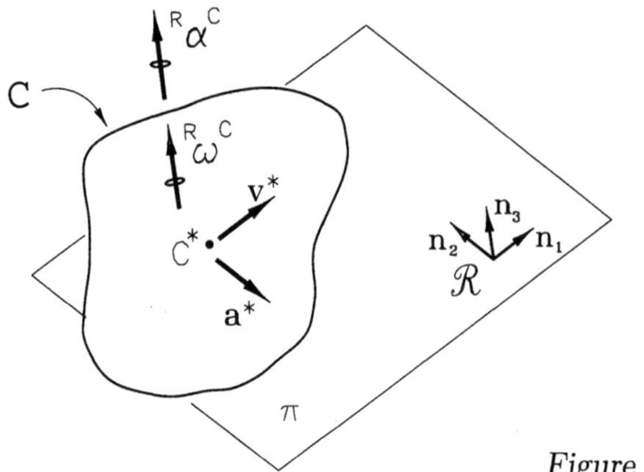

Figure 5.1

Consider, then, π as a plane fixed in an inertial reference frame $\mathcal{R}$, $\mathbf{n}_1, \mathbf{n}_2, \mathbf{n}_3$ an orthonormal basis, with $\mathbf{n}_3$ orthogonal to the plane, fixed therefore simultaneously in the body and in the inertial reference frame (see Fig. 5.1). If C^* is the mass center of C, moving parallel to π, with velocity $\mathbf{v}^*$ and acceleration $\mathbf{a}^*$ in $\mathcal{R}$, then

$$\mathbf{v}^* \cdot \mathbf{n}_3 = 0, \qquad \mathbf{a}^* \cdot \mathbf{n}_3 = 0. \tag{5.1}$$

As there is simple angular velocity, then

$$^R\boldsymbol{\omega}^C = \omega\mathbf{n}_3, \qquad ^R\boldsymbol{\alpha}^C = \dot{\omega}\mathbf{n}_3. \tag{5.2}$$

The following scalar components are, therefore, null:

$$v_3^* = 0; \quad a_3^* = 0; \quad \omega_1 = \omega_2 = 0; \quad \alpha_1 = \alpha_2 = 0. \tag{5.3}$$

The equations of motion of the first kind, Eqs. (2.4), for the condition of plane motion, are reduced, then, to

$$\begin{aligned} ma_1^* &= F_1, \\ ma_2^* &= F_2, \\ 0 &= F_3. \end{aligned} \tag{5.4}$$

Note that, generally, a_1^* and a_2^* are expressed in terms of the coordinates describing the motion of the body and, therefore, only two differential equations will be available to find those coordinates. The third equation establishes that a necessary condition for plane motion is that the scalar component of the resultant force, in the direction orthogonal to the plane, must be null.

Now Eqs. (5.1) to (5.3) being substituted in the equations of motion of the second kind, Eqs. (2.9), we have (check)

$$I_{13}^*\dot{\omega} - I_{23}^*\omega^2 = M_1,$$
$$I_{23}^*\dot{\omega} + I_{31}^*\omega^2 = M_2, \tag{5.5}$$
$$I_{33}^*\dot{\omega} = M_3.$$

Note the significant simplification introduced in the equations of motion of the second kind by the kinematic constraints due to plane motion. We have a set of three first-order nonlinear differential equations for $\omega(t)$ (or, depending on the available data, three algebraic equations for the scalar components of the resultant moment applied). Note also that only three components of the inertia tensor are present in the equations: $I_{13}^* = I_{31}^*$, I_{23}^*, and I_{33}^*.

Let us now assume that the direction orthogonal to the plane is a principal direction of inertia for the body in question with respect to its mass center (see Section 6.6). Now, if $I_{33}^* = I_3^*$ is a principal moment of inertia, then the associated products of inertia will be null, that is,

$$I_{13}^* = 0, \qquad I_{23}^* = 0. \tag{5.6}$$

In this case, then, Eqs. (5.5) are reduced to

$$0 = M_1,$$
$$0 = M_2, \tag{5.7}$$
$$I_3^*\dot{\omega} = M_3.$$

Equations (5.7) are nothing but *Euler's equations* for the condition of plane motion, when one of the principal directions of inertia coincides with the direction orthogonal to it. We now have a first-order linear differential equation to find $\omega(t)$ (assuming that M_3 does not depend on ω) and two other algebraic equations that establish null values for the scalar components of the applied resultant moment, in the directions parallel to the plane.

In practice, knowing the forces and torques applied to a rigid body that has a plane motion in an inertial reference frame, the direction orthogonal to the plane being a principal direction of inertia of the body with respect to the mass center, then we have two equations of motion of the first type (the first two of Eqs. (5.4), in directions parallel to the plane) and an equation of motion of the second kind (the third of Eqs. (5.7), in the direction orthogonal to the plane) available to find the coordinates. It is worth recalling, however, that a rigid body with a plane motion has, at the most, three degrees of freedom.

Example 5.1   A cylinder C, with mass m and radius r, is at rest, sustained by a flexible tape that has one end fixed and another passing through a pulley, as shown in Fig. 5.2. At a given instant, a constant vertical force $V = 4mg$ is applied to the free end, putting the cylinder in motion. We want to find the speed of its center O, after a time interval τ.

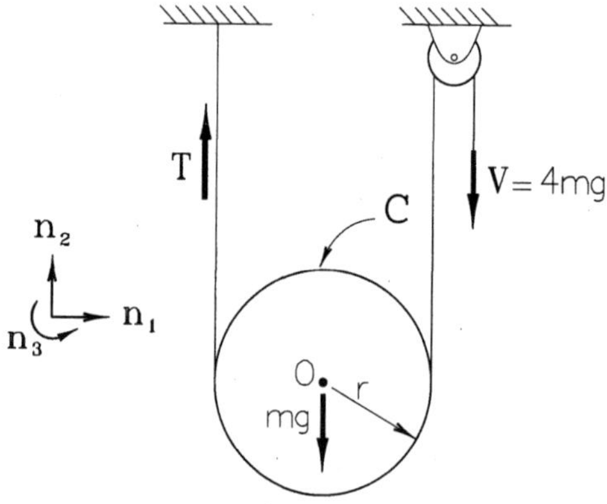

Figure 5.2

Adopting the basis indicated, the resultant force applied to the cylinder is

$$\mathbf{F} = (T + 3mg)\mathbf{n}_2.$$

The resultant moment with respect to the mass center is

$$\mathbf{M}^{\mathcal{F}_e/O} = (4mg - T)r\,\mathbf{n}_3.$$

The cylinder describes a vertical plane motion with a principal direction of inertia orthogonal to the plane. Substituting, then, in the first two of

Eqs. (5.4) and in the last of Eqs. (5.7), we have the system of equations:

$$ma_1^* = 0; \tag{a}$$

$$ma_2^* = T + 3mg; \tag{b}$$

$$\frac{1}{2}mr^2\dot{\omega} = (4mg - T)r. \tag{c}$$

From Eq. (a) we have $a_1^* = 0$; therefore, $v_1^* = 0$, that is, the center of the cylinder moves on the vertical. Everything happens, then, as if the cylinder were to roll over the fixed part of the tape, which gives us the kinematic relation

$$v^* = v_2^* = r\omega; \qquad \text{then,} \qquad a^* = a_2^* = r\dot{\omega}.$$

Substituting the above relation in Eq. (b) and resolving for T, then

$$T = m(r\dot{\omega} - 3g)$$

and, substituting in Eq. (c) and resolving for $\dot{\omega}$, gives

$$\dot{\omega} = \frac{14}{3}\frac{g}{r}, \qquad \text{so,} \qquad \omega(\tau) = \frac{14}{3}\frac{g}{r}\tau.$$

The desired velocity is, therefore,

$$v^*(\tau) = \frac{14}{3}g\tau.$$

Note that the principal kinematic unknown, in this case, is $\omega(t)$, since the finding of a_1^* has proven to be trivial and a_2^* could be expressed as a function of $\dot{\omega}$. The other unknown present is the traction T, which will be constant in time, that is,

$$T = m(r\dot{\omega} - 3g) = \frac{5}{3}mg.$$

Example 5.2 A rod B, with mass m and length $2r$, has a little wheel at its end Q that slides, without friction therefore, over the horizontal plane, while end P slides over the vertical plane, as shown in Fig. 5.3.

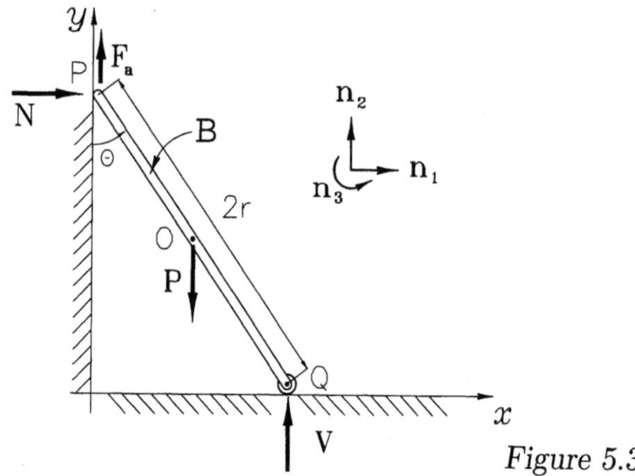

Figure 5.3

Assuming that the rod is left at rest in an essentially vertical position ($\theta = 0$) and that the friction coefficient μ, at the point of contact P, is insufficient to prevent the sliding, then we wish to study the motion of the rod and, particularly, find the velocity of end P when it collides with the horizontal plane. The rod describes a plane motion, parallel to xy, inertial. Assuming that the forces indicated are the only acting forces, points P and Q are restricted to moving along the axes y and x, respectively, the latter then consisting of two holonomic kinematic constraints. The body has, therefore, a single degree of freedom, which may be conveniently described by the coordinate θ. The forces applied are the weight, **P**, in O; force **V**, in Q; and the normal, **N**, and friction $\mathbf{F}_a$ in P. The reduction to the rod mass center of the system of external forces is

$$\mathbf{F} = N\mathbf{n}_1 + (V + F_a - mg)\mathbf{n}_2,$$
$$\mathbf{M}^{\mathcal{F}_e/O} = r\left(V \sin\theta - F_a \sin\theta - N\cos\theta\right)\mathbf{n}_3.$$

The angular velocity of the rod may be expressed by $\boldsymbol{\omega}^B = \dot\theta\mathbf{n}_3$ and, from the kinematic constraints, we have

$$\mathbf{v}^P = \dot{y}\mathbf{n}_2, \qquad \mathbf{v}^Q = \dot{x}\mathbf{n}_1.$$

Using, then, the kinematic theorem of velocities for points of a rigid body, it is easy to see that

$$\dot{x} = 2r\dot\theta\cos\theta, \qquad \dot{y} = -2r\dot\theta\sin\theta,$$

and that

$$\mathbf{v}^O = r\dot\theta\left(\cos\theta\,\mathbf{n}_1 - \sin\theta\,\mathbf{n}_2\right).$$

The angular acceleration of the rod is $\boldsymbol{\alpha}^B = \ddot{\theta}\mathbf{n}_3$, and the acceleration of the mass center may be obtained by differentiating its velocity, that is,

$$\mathbf{a}^O = r\left((\ddot{\theta}\cos\theta - \dot{\theta}^2\sin\theta)\mathbf{n}_1 - (\ddot{\theta}\sin\theta + \dot{\theta}^2\cos\theta)\mathbf{n}_2\right).$$

Substituting in the equations of motion (as always, two of the first type, in the directions parallel to the plane, and one of the second kind, in the direction orthogonal to the plane), we have

$$mr(\ddot{\theta}\cos\theta - \dot{\theta}^2\sin\theta) = N, \tag{a}$$

$$-mr(\ddot{\theta}\sin\theta + \dot{\theta}^2\cos\theta) = V + F_a - mg, \tag{b}$$

$$\frac{1}{3}mr^2\ddot{\theta} = r\left(V\sin\theta - F_a\sin\theta - N\cos\theta\right). \tag{c}$$

Note that in this system of equations, the following are unknown: θ; V; and N ($|F_a| = \mu|N|$, since there is sliding). Solving Eqs. (a) and (b) for V and N and substituting in Eq. (c), we get a differential equation for $\theta(t)$, as follows:

$$\left(\frac{2}{3} \pm \mu\cos\theta\sin\theta\right)\ddot{\theta} \mp \mu\,\mathrm{sen}^2\theta\,\dot{\theta}^2 - \frac{g}{2r}\sin\theta = 0,$$

where the signals depend on the direction of the friction force. This equation is, evidently, too complex to admit an analytical solution, requiring numerical integration. Let us, therefore, see how the system behaves if the friction at the contact point P is negligible (imagine that there is another small wheel at end P). In this case, the above equation is reduced to

$$\ddot{\theta} - \frac{3g}{4r}\sin\theta = 0,$$

still a nonlinear equation but involving less complex handling. (Note that there is a similarity between the latter equation and that governing the motion of the simple pendulum (see Section 4.2); the minus sign, however, will give the solution a nonperiodic character, as was to be expected.) Integrating in the variable θ, with the initial conditions prescribed, we have

$$\dot{\theta}^2 = \frac{3g}{2r}(1 - \cos\theta).$$

In particular, for $\theta = \pi/2$, we have, immediately before the impact of the rod with the horizontal plane,

$$\omega = \sqrt{\frac{3g}{2r}},$$

and the searched velocity is

$$\mathbf{v}^P = -2r\omega\mathbf{n}_2 = -\sqrt{6gr}\,\mathbf{n}_2.$$

See the corresponding animation.

When a rigid body C has a plane motion, as described at the beginning of this section, its kinetic energy also assumes a simpler expression. In fact, as $\mathbf{v}^* = v_1^* \mathbf{n}_1 + v_2^* \mathbf{n}_2$, the kinetic energy of translation of the body in reference frame $\mathcal{R}$ is, according to Eq. (1.18),

$$^{\mathcal{R}}K^{C^*} = \frac{1}{2} m \left(v_1^{*2} + v_2^{*2} \right). \tag{5.8}$$

On the other hand, since $^{\mathcal{R}}\boldsymbol{\omega}^C = \omega \mathbf{n}_3$, the kinetic energy of rotation of the body in $\mathcal{R}$ is, according to Eq. (1.21) (check),

$$^{\mathcal{R}}K^{C/C^*} = \frac{1}{2} I_{33}^* \omega^2. \tag{5.9}$$

Note that, particularly for the rotation term, the condition of plane motion resulted in a significant simplification of the expression, solely with the intervention of the component I_{33}^* of the inertia tensor. When the direction orthogonal to the plane is a principal of the inertia for the body with respect to its mass center, this component is exactly the principal moment of inertia I_3^*, associated to this direction.

The overall kinetic energy of the body is, of course, the sum of the above components, that is,

$$^{\mathcal{R}}K^C = \frac{1}{2} m \left(v_1^{*2} + v_2^{*2} \right) + \frac{1}{2} I_{33}^* \omega^2. \tag{5.10}$$

Example 5.3 Returning to Example 5.1 (see Fig. 5.2), we will now use the energy balance to find the velocity of the center of the cylinder at the instant when it has completed a turn around its axis of symmetry. Now, the kinetic energy of the cylinder, at a general instant, is, according to Eq. (5.10),

$$\begin{aligned}
^{\mathcal{R}}K^C &= \frac{1}{2} m v^{*2} + \frac{1}{2} I_3^* \omega^2 \\
&= \frac{1}{2} \left(m r^2 \omega^2 + \frac{1}{2} m r^2 \omega^2 \right) \\
&= \frac{3}{4} m r^2 \omega^2.
\end{aligned}$$

Force $\mathbf{T}$ does not work, since, at each instant, its application point has a null velocity in the reference frame; forces $\mathbf{V}$ and $\mathbf{P}$ both have a constant module and are parallel to the velocities of the respective application points. The resultant work of the external forces after a complete turn is, then,

$$\mathcal{R}\mathcal{T}^{\mathcal{F}_e}_{0\,2\pi} = 4mg \cdot 4\pi r - mg \cdot 2\pi r = 14\pi mgr.$$

From the balance of energy, Eq. (4.1), we then have

$$\frac{3}{4}mr^2\left(\omega^2(2\pi) - \omega^2(0)\right) = 14\pi mgr;$$

therefore,

$$\omega^2(2\pi) = \frac{56\pi}{3}\frac{g}{r},$$

and the searched velocity is

$$v^* = 2\sqrt{\frac{14\pi}{3}gr}.$$

Note that, from the analysis of Example 5.1, we had, from the last equation of motion,

$$\ddot{\theta} = \dot{\omega} = \frac{14}{3}\frac{g}{r}.$$

Now integrating not in time but in θ, in the interval $(0, 2\pi)$ and with the initial condition $\dot{\theta}(0) = 0$, we have

$$\frac{\dot{\theta}^2}{2} = \frac{14}{3}\frac{g}{r}\theta;$$

therefore,

$$\omega^2(2\pi) = \frac{56\pi}{3}\frac{g}{r} \qquad \text{and} \qquad v^* = 2\sqrt{\frac{14\pi}{3}gr},$$

as obtained above. See the corresponding animation.

Example 5.4 Going back to Example 5.2 (see Fig. 5.3), forces $\mathbf{V}$ and $\mathbf{N}$ do not work and, when there is no friction, the mechanical energy of the rod is conserved. The initial kinetic energy $(\theta = 0)$ is null and the gravitational potential energy, assuming the reference in $y = 0$, is mgr. In the final position $(\theta = \pi/2)$, the kinetic energy of translation of the rod is

$$K^O = \frac{1}{2}mv^2 = \frac{1}{2}mr^2\omega^2,$$

the kinetic energy of rotation is

$$K^{B/O} = \frac{1}{2}\frac{1}{3}mr^2\omega^2,$$

and the gravitational potential energy is null. Substituting Eq. (4.4), then, in the energy balance, we have

$$\frac{1}{2}mr^2\omega^2 + \frac{1}{6}mr^2\omega^2 - mgr = 0,$$

so,

$$\omega^2 = \frac{3g}{2r},$$

and the velocity of end P, immediately before the impact, is

$$\mathbf{v}^P = -2r\omega\mathbf{n}_2 = -\sqrt{6gr}\,\mathbf{n}_2,$$

as obtained in Example 5.2. See the corresponding animation.

Equations (5.7) are, as already mentioned, applicable to the situations of plane motion where the direction orthogonal to the plane is a principal direction of inertia for the body with respect to its mass center. In this case, then, the scalar components of the resultant moment with respect to the mass center in the directions parallel to the plane are null and the body stays naturally in plane motion. When, however, the direction orthogonal to the plane of the motion is *not* a principal direction of inertia of the body with respect to its mass center, Eqs. (5.7) are no longer applicable; in this case the pertinent equations of motion of the second kind will be Eqs. (5.5). It is then evident that, for a rigid body to continue in a plane motion with a given angular velocity not parallel to a principal direction of inertia of the body with respect to its mass center, it is necessary to apply a torque parallel to the plane of motion.

Example 5.5   Consider a homogeneous rectangular plate A, with mass m, welded with a slope angle θ to a light rod B, which revolves with a vertical angular velocity with constant module ω in relation to an inertial reference frame $\mathcal{R}$ (see Fig. 5.4). The Cartesian axes $\{x_1, x_2, x_3\}$ and $\{y_1, y_2, y_3\}$ are fixed simultaneously in A and B, with the orientations indicated. Note that $\{y_1, y_2, y_3\}$ are principal axes of inertia for the plate, with respect to point O, and that the latter has a horizontal plane motion, with angular velocity parallel to x_3.

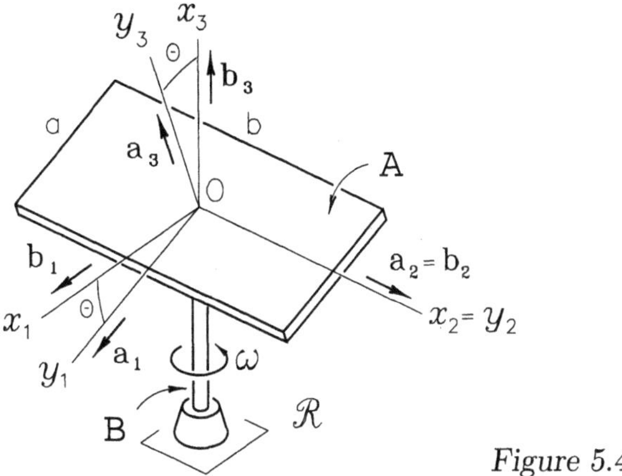

Figure 5.4

The forces system acting on the plate consists of its weight and the reactions on the link at O that, since it is a welding, comprises, at least
in principle, three force components, F_1, F_2, F_3, and three torque components, M_1, M_2, M_3. The inertia tensor of the plate with respect to point
O, expressed on the basis of the principal directions, is the matrix (see
Appendix C)

$$\mathbb{I}^{A/O} = \frac{1}{12} m \begin{pmatrix} b^2 & 0 & 0 \\ 0 & a^2 & 0 \\ 0 & 0 & b^2 + a^2 \end{pmatrix}.$$

The inertia vector with respect to point O, associated with direction $\mathbf{b}_3$,
is, according to Eq. (6.2.2),

$$\mathbf{I}_{\mathbf{b}_3}^{A/O} = \mathbb{I}^{A/O} \cdot \mathbf{b}_3 = \frac{1}{12} m \left(-b^2 \sin\theta \mathbf{a}_1 + (b^2 + a^2) \cos\theta \mathbf{a}_3 \right).$$

The components of inertia present in Eq. (5.5) may, then, be calculated
(note that the indices now refer to axes $\{x_1, x_2, x_3\}$):

$$I_{13}^O = \mathbf{I}_{\mathbf{b}_3}^{A/O} \cdot \mathbf{b}_1 = \frac{1}{12} ma^2 \sin\theta \cos\theta;$$

$$I_{23}^O = \mathbf{I}_{\mathbf{b}_3}^{A/O} \cdot \mathbf{b}_2 = 0;$$

$$I_{33}^O = \mathbf{I}_{\mathbf{b}_3}^{A/O} \cdot \mathbf{b}_3 = \frac{1}{12} m(b^2 + a^2 \cos^2\theta).$$

The mass center of the plate does not move in relation to $\mathcal{R}$, and the
angular acceleration is null. The equations of motion of the first type,
Eqs. (5.4), and of the second kind, Eqs. (5.5), will then be, in the case, as

follows (check):

$$0 = F_1; \tag{a}$$

$$0 = F_2; \tag{b}$$

$$0 = F_3 - mg; \tag{c}$$

$$0 = M_1; \tag{d}$$

$$\frac{1}{12}ma^2 \sin\theta \cos\theta \omega^2 = M_2; \tag{e}$$

$$0 = M_3. \tag{f}$$

We conclude that the efforts applied to the link are a vertical force $\mathbf{F} = mg\mathbf{b}_3$ and a horizontal torque $\mathbf{T} = \frac{1}{12}ma^2 \sin\theta \cos\theta \omega^2 \mathbf{b}_2$. The latter is the torque required to keep the plate turning with an angular velocity of constant module ω, in the vertical direction. Note that the torque is a vector fixed in the plate; it therefore turns uniformly in relation to the fixed reference frame. The torque will vanish, naturally, if $\theta = k\pi/2$, k an integer, when the angular velocity vector is parallel to a principal direction of inertia of the plate with respect to its mass center.

Example 5.6   Let us now consider a rigid body C consisting of two rings, with radius r and mass m each, welded to a homogeneous rod, with length $2\sqrt{2}\,r$ and mass also equal to m, with the configuration shown in Fig. 5.5. The body is rolling over an inertial horizontal flat surface $\mathcal{R}$, in such a way that point O, the mass center of C, moves in $\mathcal{R}$ with a constant speed v (direction of x_2). We want to study how the forces that the horizontal plane applies on C behave. The Cartesian axes $\{y_1, y_2; y_3\}$ and the orthonormal basis $\mathbf{b}_1, \mathbf{b}_2, \mathbf{b}_3$ are fixed in C; axes $\{x_1, x_2, x_3\}$ and basis $\mathbf{n}_1, \mathbf{n}_2, \mathbf{n}_3$ are fixed in a reference frame that translates in $\mathcal{R}$ with a constant speed (basis $\mathbf{n}_1, \mathbf{n}_2, \mathbf{n}_3$ is also, of course, fixed in $\mathcal{R}$). The body describes a vertical plane motion, that is, parallel to plane x_1, x_2. The acting forces consist of the weight, $\mathbf{P} = -3mg\mathbf{n}_1$, applied to O, and the vertical, $\mathbf{V}_1$ and $\mathbf{V}_2$, and horizontal (friction), $\mathbf{H}_1$ and $\mathbf{H}_2$ components applied on the points of contact of each ring with the horizontal plane, as shown.

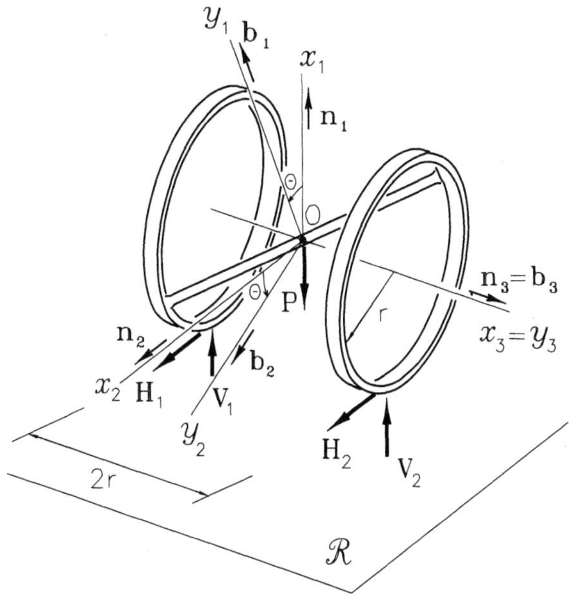

Figure 5.5

The reduction to O of this system of forces is

$$\mathbf{F} = (V_1 + V_2 - 3mg)\,\mathbf{n}_1 + (H_1 + H_2)\,\mathbf{n}_2,$$

$$\mathbf{M}^{\mathcal{F}e/O} = (H_1 - H_2)r\,\mathbf{n}_1 + (V_2 - V_1)r\,\mathbf{n}_2 - (H_1 + H_2)r\,\mathbf{n}_3$$

$$= \big((H_1 - H_2)\cos\theta + (V_2 - V_1)\sin\theta\big)r\,\mathbf{b}_1$$

$$+ \big(-(H_1 - H_2)\sin\theta + (V_2 - V_1)\cos\theta\big)r\,\mathbf{b}_2$$

$$- (H_1 + H_2)r\,\mathbf{b}_3.$$

The velocity of the mass center is $^{\mathcal{R}}\mathbf{v}^O = v\mathbf{n}_2$, constant, and the acceleration, therefore, will be null. The first two equations of motion of the first kind for the problem are, then,

$$0 = V_1 + V_2 - 3mg, \tag{a}$$

$$0 = H_1 + H_2. \tag{b}$$

(Note that the third of Eqs. (5.4) is, in this case, identically null.) We will now adopt axes $\{y_1, y_2, y_3\}$ to find the inertia matrix of the body with respect to point O. In order to calculate the moments of inertia we will use the expressions available in Appendix C and the relationships for transposition of axis and of inertia properties of composite bodies (see Section 6.5):

$$I^{C/O}_{y_1 y_1} = \frac{1}{12} m (2\sqrt{2}r)^2 + 2 \left(\frac{1}{2} mr^2 + mr^2 \right) = \frac{11}{3} mr^2;$$

$$I^{C/O}_{y_2 y_2} = \frac{1}{12} m (2r)^2 + 2 \left(\frac{1}{2} mr^2 + mr^2 \right) = \frac{10}{3} mr^2;$$

$$I^{C/O}_{y_3 y_3} = \frac{1}{12} m (2r)^2 + 2 mr^2 = \frac{7}{3} mr^2.$$

Plane $y_2 y_3$ is of symmetry for the body; therefore,

$$I^{C/O}_{y_1 y_2} = I^{C/O}_{y_1 y_3} = 0.$$

The remaining product of inertia may be found by integrating along the rod, since, for the couple of rings, plane $y_1 y_2$ is also of symmetry. Now, taking the variable of length s along the rod, we have

$$I^{C/O}_{y_2 y_3} = - \int_{-\sqrt{2}r}^{\sqrt{2}r} y_2 y_3 \, dm$$

$$= \frac{m}{4\sqrt{2}r} \int_{-\sqrt{2}r}^{\sqrt{2}r} s^2 \, ds$$

$$= \frac{1}{3} mr^2.$$

The inertia tensor of C with respect to O, expressed according to axes $\{y_1, y_2, y_3\}$, consists, then, of the matrix

$$\mathbf{II}^{C/O} = \frac{1}{3} mr^2 \begin{pmatrix} 11 & 0 & 0 \\ 0 & 10 & 1 \\ 0 & 1 & 7 \end{pmatrix}.$$

The angular velocity of C in $\mathcal{R}$ is, from the condition of plane motion and rolling condition,

$$^{\mathcal{R}}\boldsymbol{\omega}^C = \omega \mathbf{n}_3 = \frac{v}{r} \mathbf{n}_3,$$

and the angular acceleration is, therefore, null. So, by substituting in Eqs. (5.5), we have

$$-\frac{1}{3} mr^2 \omega^2 = (H_1 - H_2) r \cos\theta + (V_1 - V_2) r \sin\theta, \tag{c}$$

$$0 = -(H_1 - H_2) r \sin\theta + (V_2 - V_1) r \cos\theta, \tag{d}$$

$$0 = -(H_1 + H_2) r. \tag{e}$$

Thus solving the system of algebraic equations (a–d) (note that Eqs. (e) and (b) are dependent), the unknown forces are

$$H_1 = -\frac{1}{6}\frac{mv^2}{r}\cos\theta,$$

$$H_2 = \frac{1}{6}\frac{mv^2}{r}\cos\theta,$$

$$V_1 = \frac{3}{2}mg + \frac{1}{6}\frac{mv^2}{r}\sin\theta,$$

$$V_2 = \frac{3}{2}mg - \frac{1}{6}\frac{mv^2}{r}\sin\theta.$$

We also see here, as in the preceding example, the periodical character of the solution. More precisely, as, in both cases, the body moves with angular velocity of constant module, the solution is harmonic. Note that the forces of friction are always in opposition (said to be in *phase opposition*), both consisting of functions of the type $A\cos\theta$, where, in the case, $|A| = mv^2/6r$. Note also that the friction components vanish when the rod passes through the vertical plane. On the other hand, the normal components consist of harmonic functions with an *offset*, that is, both have a constant term, equal to half the weight of the body, around which they oscillate, also harmonically, and with the same range of the friction force. Also note that the normal components are equal when the rod is parallel to the horizontal plane. See the corresponding animation.

Exercise Series #10   (Sections 7.1 to 7.2)

P10.1 The homogeneous cone, with mass m, is linked in vertix A to an inertial reference frame by means of a frictionless ball joint. If, throughout its motion, the angle β, the spin $\dot{\phi} = p$, and precession $\dot{\psi} = n$ remain constant, estimate the kinetic energy of the cone in the reference frame.

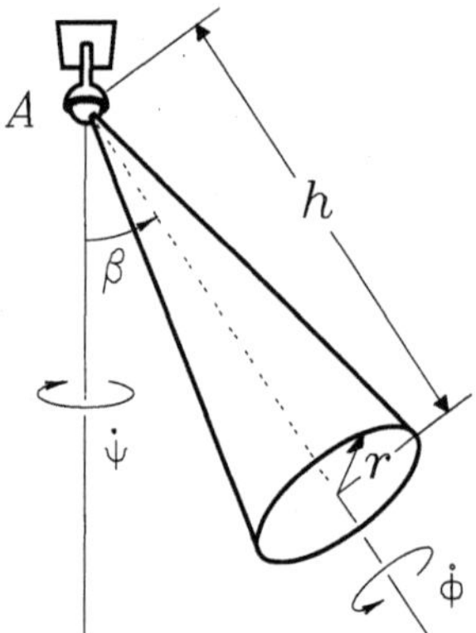

Figure P 10.1

P10.2 The homogeneous cylinder C, with mass m, turns around the support S with angular velocity of constant module ω, while the latter turns around base B with angular acceleration of constant module α, starting from rest. Find, at the instant when the module of the angular velocity vector of S in B is 4ω, the angular momentum vector of C with respect to its mass center, in referential B. Also calculate, at the same instant, the time rate in B of this vector.

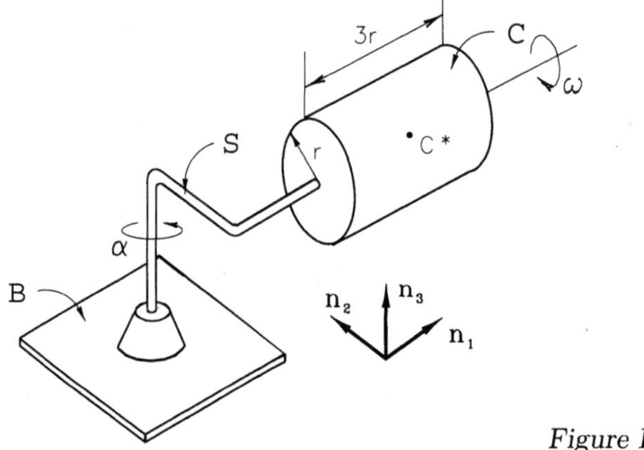

Figure P 10.2

P10.3 Consider a body C moving in relation to two different reference frames $\mathcal{R}$ and $\mathcal{R}'$. If m is the total mass of C, ${}^{\mathcal{R}}\boldsymbol{\omega}^{\mathcal{R}'}$ is the angular velocity vector of $\mathcal{R}'$ in $\mathcal{R}$ and Q is an arbitrary point fixed in $\mathcal{R}'$, show that the momentum vectors of C in $\mathcal{R}$ and $\mathcal{R}'$ are related according to ${}^{\mathcal{R}}\mathbf{G}^C = {}^{\mathcal{R}'}\mathbf{G}^C + m({}^{\mathcal{R}}\mathbf{v}^Q + {}^{\mathcal{R}}\boldsymbol{\omega}^{\mathcal{R}'} \times \mathbf{q}^*)$, where $\mathbf{q}^*$ is the position vector of the mass center of C with respect to point Q. Use this result to calculate ${}^A\mathbf{G}^S$ in Example 5.1.3.

P10.4 A planetary mechanism moves on the horizontal plane, consisting of a crank D, with mass $2m$, supporting the axes of three identical disks, with mass m and radius r. Disk A is fixed, disk B rolls over A, and disk C rolls over B, while the crank turns at a constant angular velocity ω, as shown. Calculate the kinetic energy of the set.

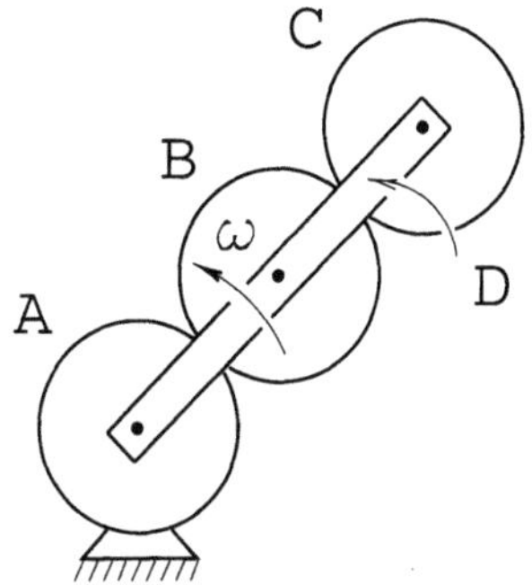

Figure P 10.4

P10.5 Demonstrate that if ${}^{\mathcal{R}}\mathbf{G}^C$ and ${}^{\mathcal{R}}\mathbf{H}^{C/C^*}$ are the momentum vector of a rigid body C in a reference frame $\mathcal{R}$ and angular momentum vector of the same body with respect to its mass center, in the same reference frame, the kinetic energy of C in $\mathcal{R}$ may be expressed by

$$
{}^{\mathcal{R}}K^C = \frac{1}{2}\left(\mathbf{v}^* \cdot {}^{\mathcal{R}}\mathbf{G}^C + {}^{\mathcal{R}}\boldsymbol{\omega}^C \cdot {}^{\mathcal{R}}\mathbf{H}^{C/C^*}\right),
$$

where $\mathbf{v}^*$ is the velocity, in $\mathcal{R}$, of the mass center of C and ${}^{\mathcal{R}}\boldsymbol{\omega}^C$ is its angular velocity in $\mathcal{R}$.

P10.6   The corner plate illustrated consists of two homogeneous square flaps,
very thin if compared to its dimensions. The corner plate is made to turn with
a constant angular velocity around an axis orthogonal to one of the flaps,
passing through its center, as shown. Find the value of the angle θ between
the flaps that minimizes the kinetic energy of the corner plate.

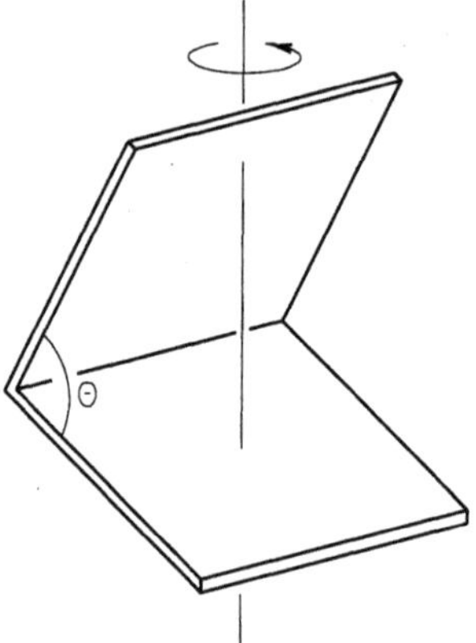

Figure P 10.6

P10.7   The homogenous wire with mass m and radius r turns, in relation to
a given reference frame $\mathcal{R}$, around axis x_2, with an angular velocity of module
ω, in the direction shown. Find its angular momentum vector with respect to
point A, in this reference frame. Also calculate its angular momentum with
respect to axis x_1, in the same reference frame. Basis $\mathbf{n}_1, \mathbf{n}_2, \mathbf{n}_3$ and Cartesian
axes $\{x_1, x_2, x_3\}$ are fixed in the wire.

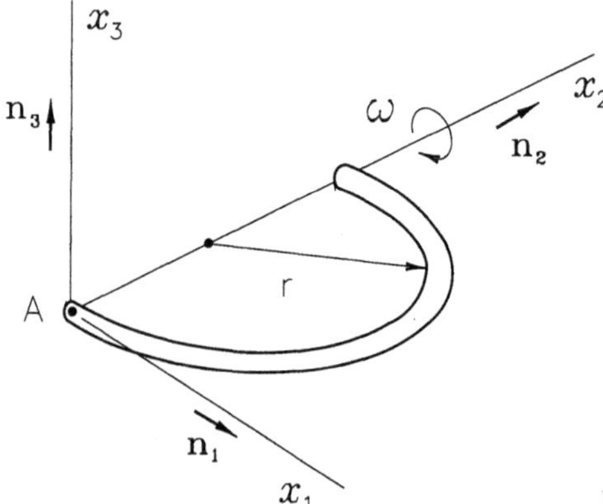

Figure P 10.7

P10.8 A homogeneous disk D, with radius $r = 0.1$ m and mass $m = 3$ kg, turns in relation to the fork with a constant angular velocity of module $\omega = 120$ rpm, in the direction shown, while the light fork turns around axis x_1 in relation to a given reference frame $\mathcal{R}$ with an angular velocity also constant of module $\Omega = 60$ rpm in the direction shown. Find the angular momentum of the disk with respect to point O, the origin of system of coordinates, in $\mathcal{R}$. Also calculate its kinetic energy in $\mathcal{R}$.

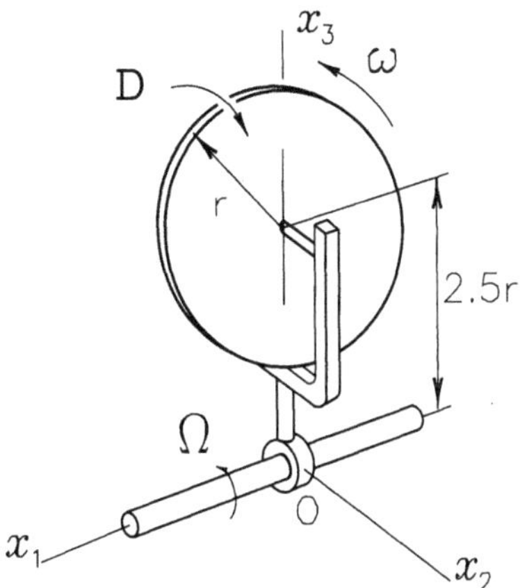

Figure P 10.8

P10.9 A homogeneous rectangular plate A, with mass m, is welded at a slope angle θ to a light shaft B turning with a vertical angular velocity of constant module ω in relation to the reference frame $\mathcal{R}$. Find the angular momentum vector of the plate with respect to O, in reference frame $\mathcal{R}$. Also calculate its kinetic energy in $\mathcal{R}$.

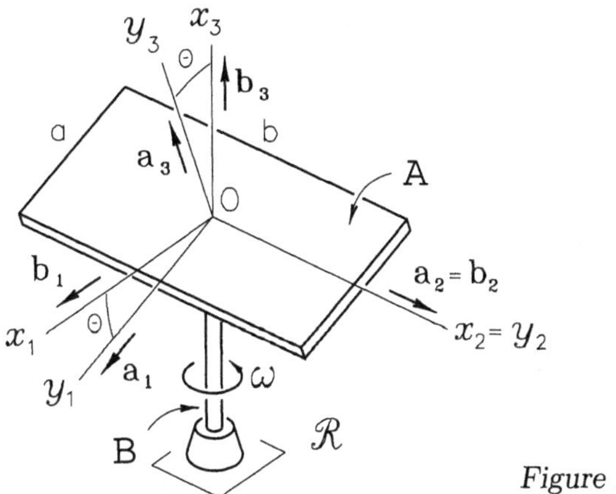

Figure P 10.9

P10.10   An element of a transmission chain consists of two pulleys, the larger with moment of inertia with respect to the axis of rotation equal to I_1 and the smaller with moment of inertia with respect to the axis of rotation equal to I_2, and a belt with mass m, which interconnect the pulleys. If the larger pulley is turning at an angular velocity ω, calculate the kinetic energy of the set.

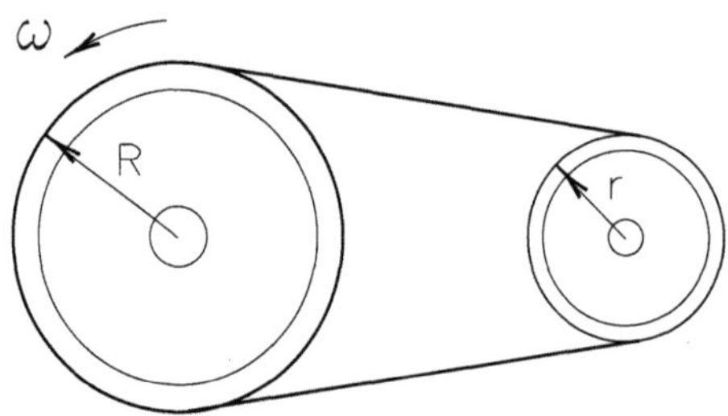

Figure P 10.10

P10.11   The crankshaft mechanism consists of arm B, with mass m and length r, joined at O, turning around the latter at a constant angular velocity ω, and of a cursor C, with mass $3m$, sliding along the guide according to a periodic motion, guaranteed by a connecting element, arm A, with length a, light and joined to the other two, as shown. Find the kinetic energy of the set when $\theta = 0$ and when $\theta = \pi/2$.

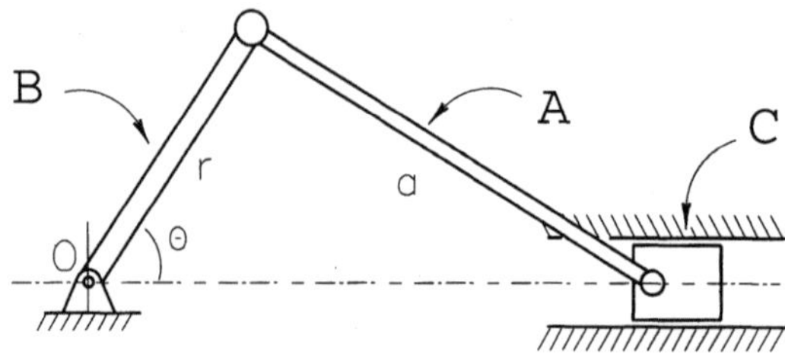

Figure P 10.11

P10.12 Consider a body C moving in relation to two different reference frames $\mathcal{R}$ and $\mathcal{R}'$. If m is the total mass of C and $^{\mathcal{R}}\boldsymbol{\omega}^{\mathcal{R}'}$ is the angular velocity vector of $\mathcal{R}'$ in $\mathcal{R}$, show that the angular momentum vectors of C with respect to a given point O in reference frames $\mathcal{R}$ and $\mathcal{R}'$ are related according to

$$^{\mathcal{R}}\mathbf{H}^{C/O} = {}^{\mathcal{R}'}\mathbf{H}^{C/O} + \mathbf{II}^{C/O}\,{}^{\mathcal{R}}\boldsymbol{\omega}^{\mathcal{R}'} + m\mathbf{p}^* \times \left({}^{\mathcal{R}}\mathbf{v}^O - {}^{\mathcal{R}'}\mathbf{v}^O\right),$$

where $\mathbf{p}^*$ is the position vector of the mass center of C with respect to O. Use this result to calculate $^A\mathbf{H}^{S/Q}$ and $^A\mathbf{H}^{S/O}$ in Example 5.1.3.

P10.13 The rigid body C consists of a thin disk with mass $2m$ and radius r, welded to a slim rod with mass m and length $6r$, with a common axis of symmetry. The other end of the rod is connected by a ball joint to a fixed vertical shaft, in point O, while the disk is supported on a flat surface, as shown. The body is at rest when a horizontal torque $T = 12mgr$ is applied, always in the direction of axis x_2, then starting the motion of C, with the disk rolling over the plane and point O fixed. Adopting the directions of the mobile axes $\{x_1, x_2, x_3\}$ and the orthonormal basis $\mathbf{n}_1, \mathbf{n}_2, \mathbf{n}_3$ associated with them, calculate the force applied by the plane on the periphery of the disk after the first complete turn of the rod around x_3.

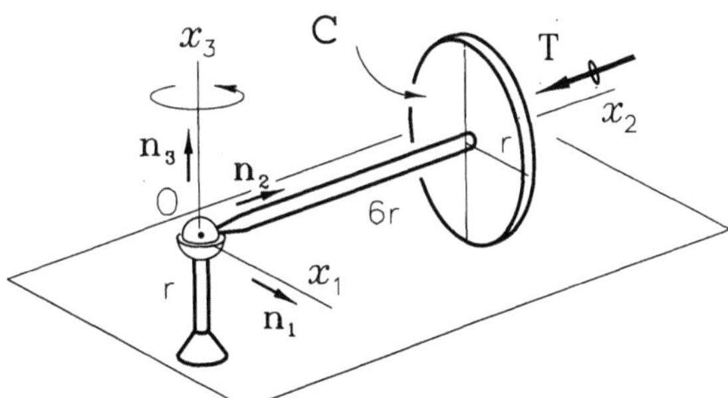

Figure P 10.13

P10.14   A homogeneous cube, with mass m and edge a, is initially at rest on a horizontal plane, merely supported on the vertices B, C, and D and joined by a ball joint in vertix A. Suddenly a torque of module $T = 2\sqrt{2}mga$ is applied in the direction shown. Find the initial acceleration or vertix C. Also calculate the module of force in the ball joint, at this same instant.

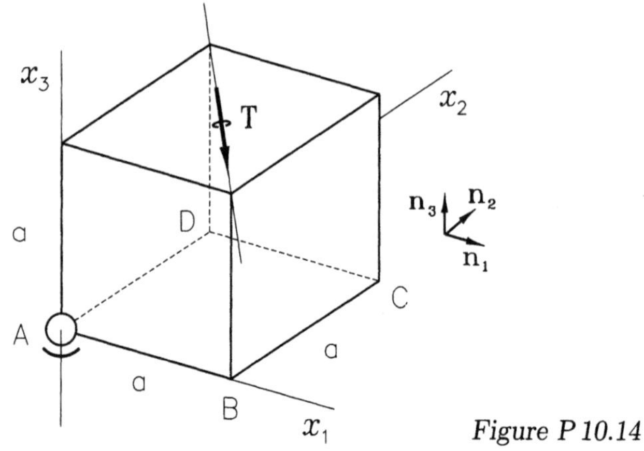

Figure P 10.14

P10.15   A homogeneous cone, with mass m, has its vertix connected by a ball joint at point O and is initially at rest, supported on a horizontal plane, when the vertical torque **T**, with constant module $T = 5mgr$, is applied. Thanks to the friction present in the contact between the periphery of the base of the cone and the plane, rolling occurs. Calculate the force exerted by the plane on the cone in the beginning and after a complete turn around the axis x_1.

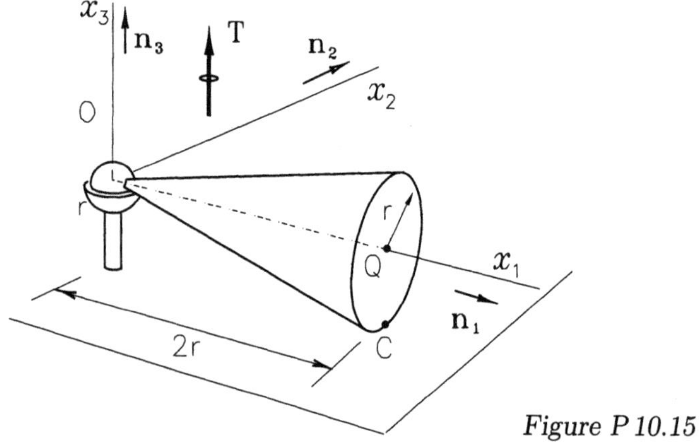

Figure P 10.15

P10.16 The homogeneous disk D, with mass m and radius r, rolls over the conical surface and can turn freely around the light arm B that, in turn, is pivoted on the vertical shaft, fixed, and may also turn around it, in the vertical direction. The system is at rest when the vertical torque of module $T = 4mgr$ is applied to it. Calculate the initial value of the force that the surface applies on the disk.

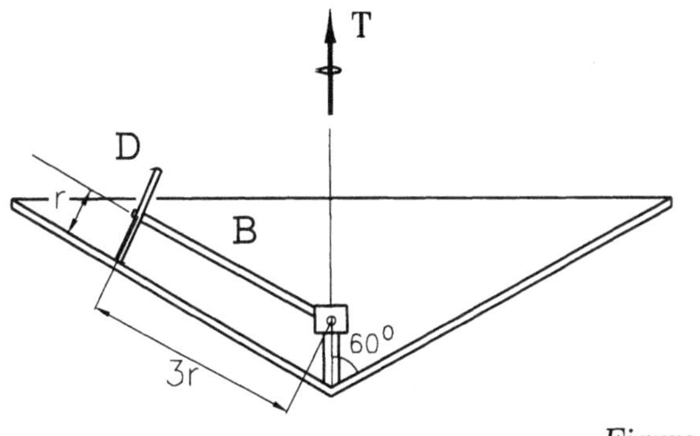

Figure P 10.16

P10.17 A cylinder, with mass m and radius $2r$, rolls over the cylindrical surface with radius $6r$, being initially at rest in the position shown. Under the action of a horizontal force of constant module H, applied to a cable surrounding a cylindrical flange with radius r and negligible mass, as shown, the cylinder moves. Calculate the initial angular acceleration of the cylinder.

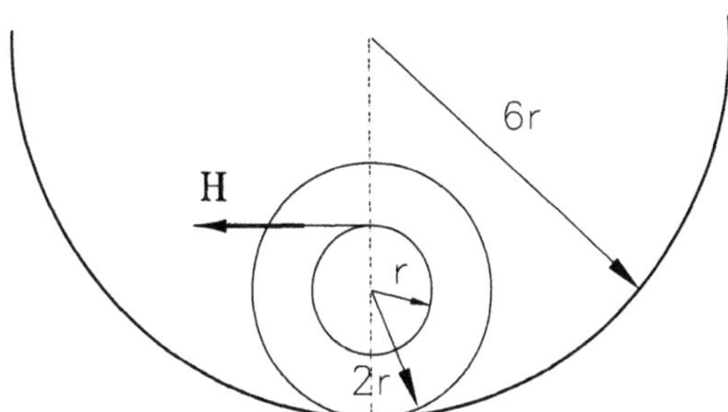

Figure P 10.17

P10.18 A body with circular geometry is at rest, supported on a horizontal plane, as shown. A horizontal force H is then applied on its periphery. Calculate the value of the friction force, knowing that the body rolls over the plane, in the case of a cylinder. Now redo the calculations for the case of a ring.

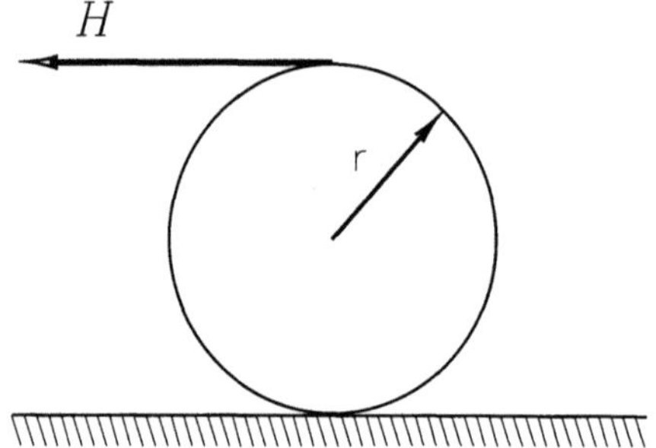

Figure P 10.18

P10.19 The thin semidisk D, with mass m and radius r, is mounted on a shaft, making an angle of 20^0, as shown, turning at an angular velocity of constant module $\omega = 30$ rad/s. Find the angular momentum vector of D with respect to point O. Also calculate its kinetic energy.

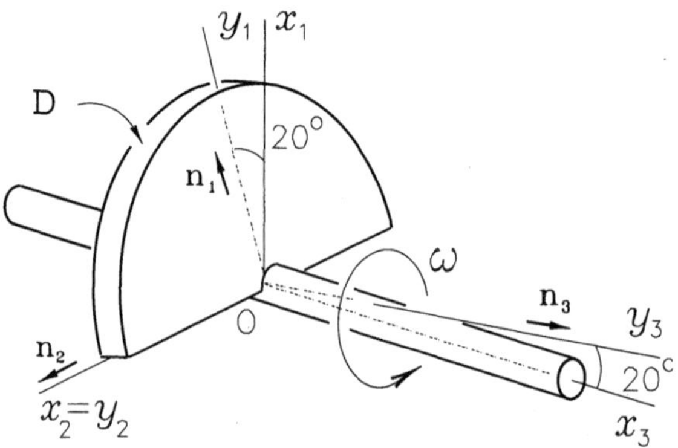

Figure P 10.19

P10.20 A homogeneous bar, with mass m and length r, turns with a constant angular velocity ω around a vertical shaft, as shown. Calculate the module of the resulting force acting on point P.

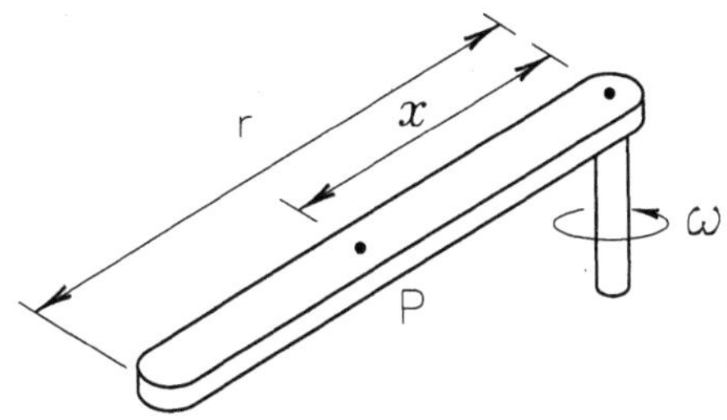

Figure P 10.20

P10.21 A radar sensor consists of a semicylindrical shell, with mass m, radius r, and height a, turning in relation to a reference frame $\mathcal{R}$ around a vertical axis, as shown. Axes $\{x_1, x_2, x_3\}$ and the orthonormal basis $\mathbf{b}_1, \mathbf{b}_2, \mathbf{b}_3$ are fixed in the sensor. At a certain instant, the angular velocity of the shell has module Ω in the positive direction of x_1. Calculate, in this same instant, the kinetic energy of the sensor in $\mathcal{R}$.

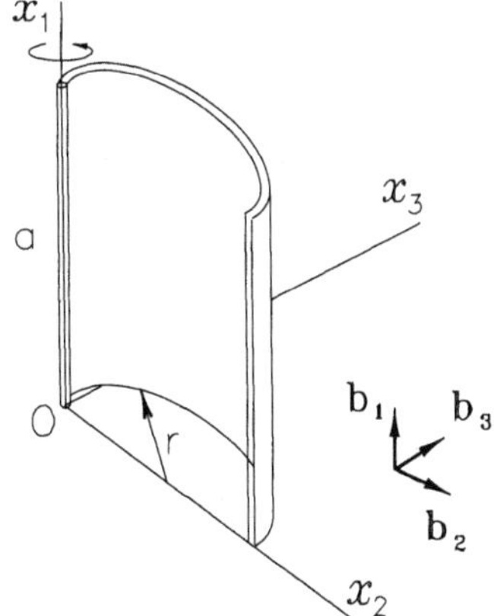

Figure P 10.21

Exercise Series #11   (Sections 7.3 to 7.5)

P11.1 Four identical homogeneous bars, with mass m and length a each, are joined to each other and move in the vertical plane, mantaining the same angle with the vertical, as shown. The set leaves its state of rest with $\theta = \pi/2$, moving under the action of the constant torque T applied to the right-hand bar, joined at the fixed point O. Assuming the friction in the joints and in the small wheels to be negligible, calculate the velocity of point A immediately before the impact ($\theta = 0$).

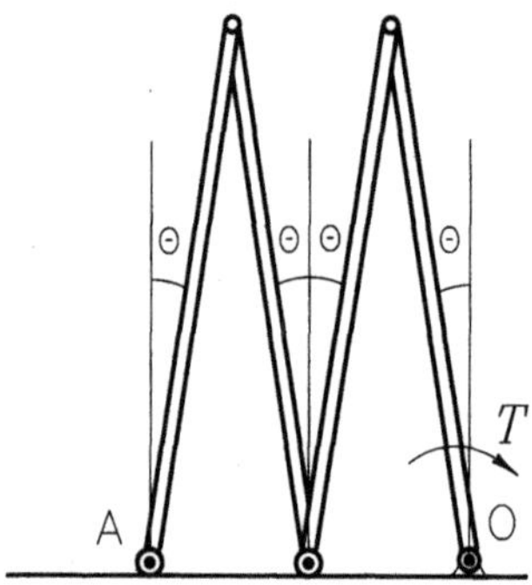

Figure P 11.1

P11.2 A homogeneous cone, with mass m, has its vertex fixed by a ball joint at point O and is initially at rest, supported on a horizontal plane, when vertical torque **T**, with constant module $T = 5mgr$, is applied. Rolling occurs because of the friction present in the contact between the periphery of the cone base and the plane. What is the work done by torque **T** from the beginning up to point Q to complete a circumference with center O?

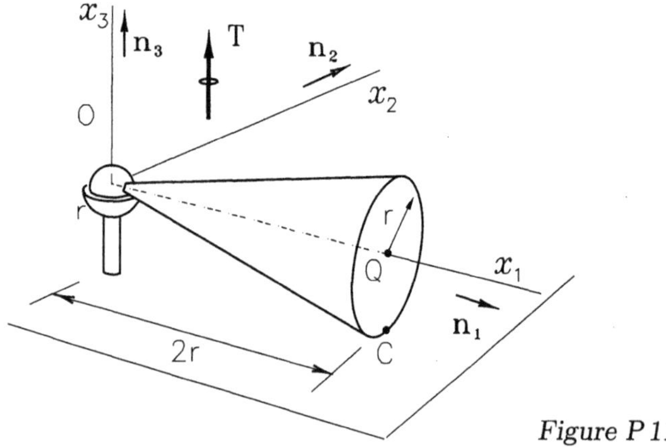

Figure P 11.2

P11.3 Rod B has an end (P) fixed in the horizontal plane by a ball joint and another end (Q), merely supported on the smooth vertical plane, as shown. The rod is abandoned from rest essentially on the vertical plane, sliding until it collides with the horizontal plane. Calculate the velocity of point Q immediately before the impact.

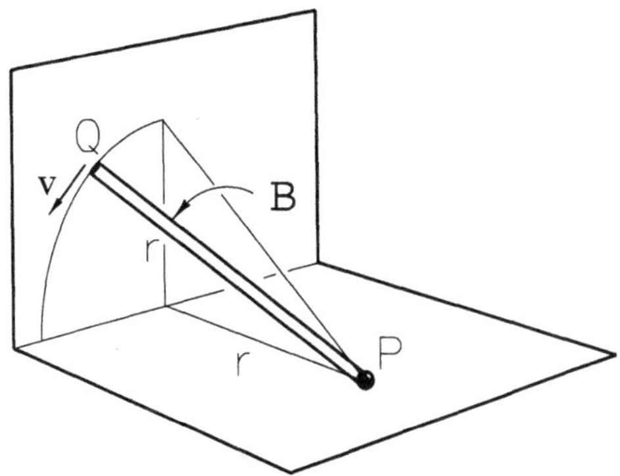

Figure P 11.3

P11.4 Disk D, with mass m and radius r, rolls over the conical surface and can turn freely around the light arm B that, in turn, can rotate around the fixed vertical shaft. The system is at rest when torque $T = 4mgr$ is applied to the disk, in the direction shown. Calculate the velocity of the center of the disk after a complete turn of the arm around the vertical shaft.

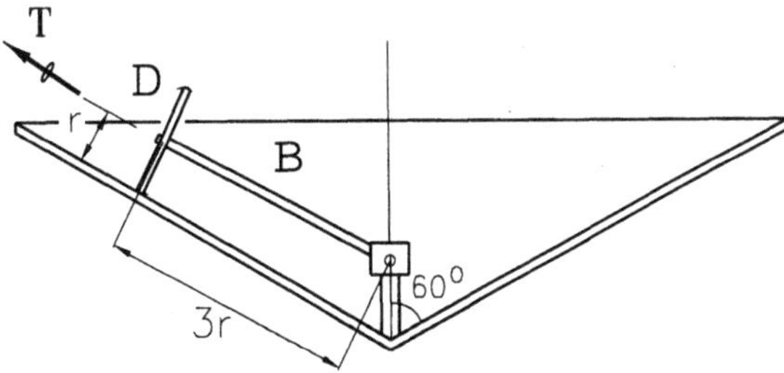

Figure P 11.4

P11.5   A cylinder, with mass m and radius $2r$, rolls inside the cylindrical surface of radius $6r$, leaving its initial state of rest in the position shown, under the action of a horizontal force with constant module $H = 4mg$, applied to a wire surrounding a cylindrical flange with radius r, as shown. Calculate the velocity of the center of the cylinder after it completes a half-turn around its axis of symmetry.

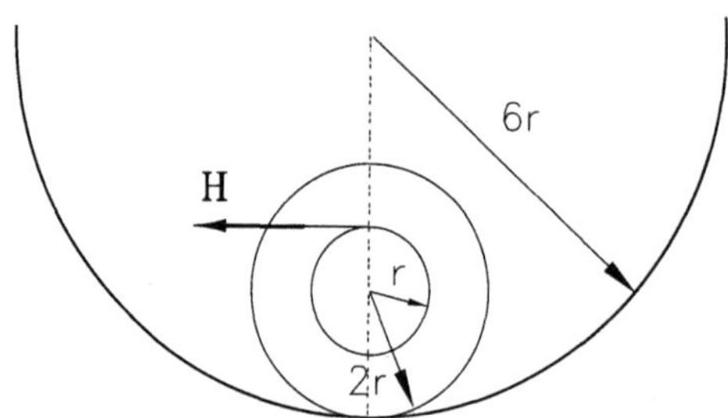

Figure P 11.5

P11.6   A homogeneous rod, with mass m and length b, is freely pivoted at end Q and hung on the horizontal position by a thread fixed at the other end P, as shown. The thread is cut, and the rod now moves on the vertical plane with the end Q fixed. Find the ratio between the cutting effort on the pin of the pivot immediately after cutting the thread, and this same force in the condition of original equilibrium. Next calculate this ratio when the rod is at an angle of 45^0 with the vertical.

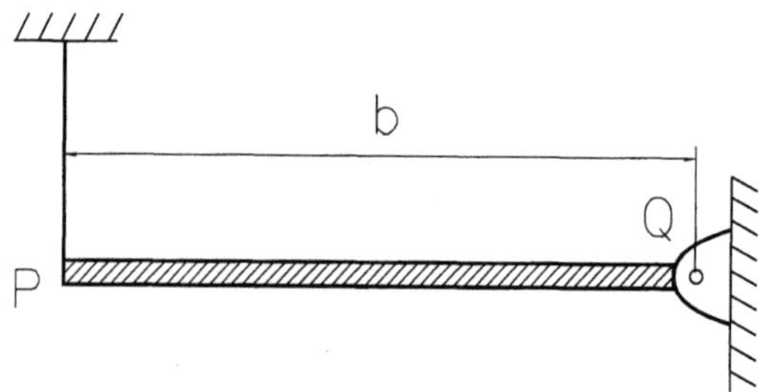

Figure P 11.6

P11.7   Before the start of a soccer game the referee should toss a coin to choose the attacking field. The coin is tossed with an angular velocity ω_0, orthogonal to the plane of the figure and with its center having a velocity v_0, making an angle θ_0 with the vertical, as shown. Find the values of v, w, and θ with which the coin is picked up, as a result of the difference between elevations d, shown. Ignore the air's resistance.

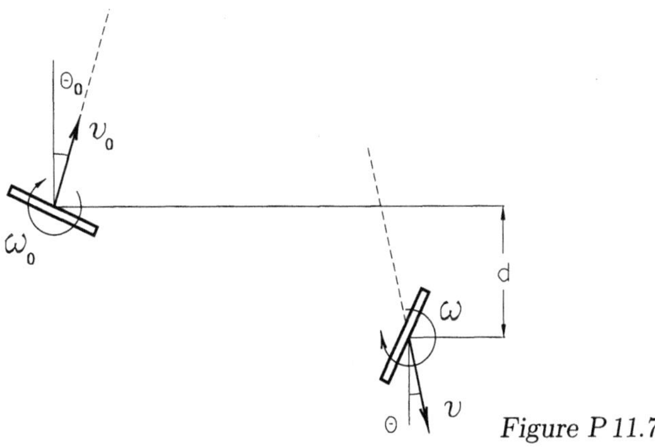

Figure P 11.7

P11.8   A cylinder C, with mass m, is at rest, supported by the corners A and B, when the latter is suddenly removed. Find the module of the force that the cylinder exerts on corner A at this instant, knowing that no sliding occurs in the mutual contact.

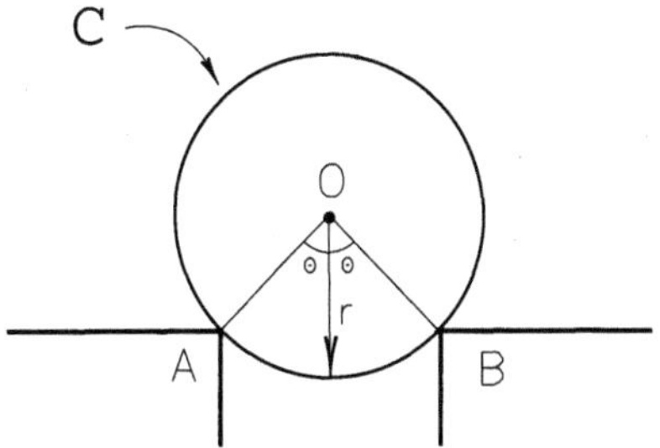

Figure P 11.8

P11.9 The homogeneous rod, with mass m and length a, has a small light wheel fixed at its end A, which is supported by the sloping plane. The set is released from rest in the position shown. Find the initial value of the force of contact N on the wheel's axis.

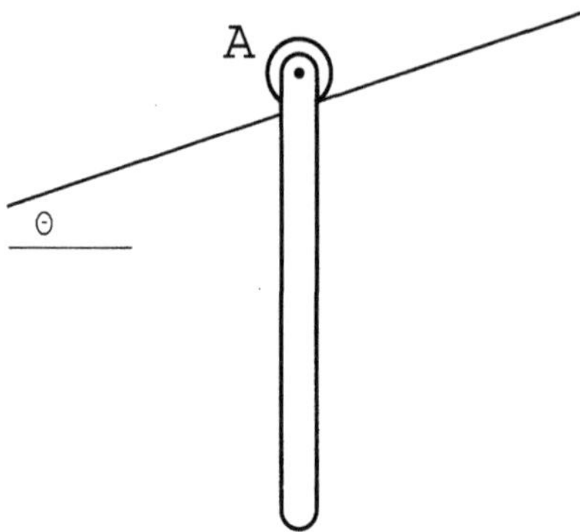

Figure P 11.9

P11.10 A homogeneous cylinder, with mass m and radius r, is left at rest on a cylindrical surface with radius $4r$, in the position shown. Knowing that the cylinder rolls over the surface, find the magnitude of the friction force at the initial instant and the magnitude of the velocity of point Q in the position of minimum elevation.

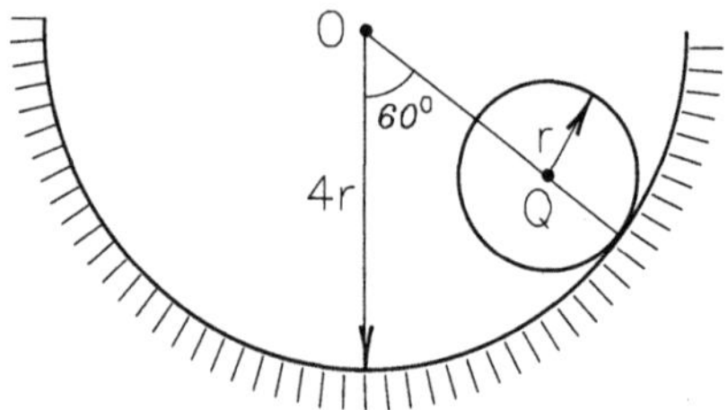

Figure P 11.10

P11.11 A slender homogeneous rod, with mass m and length a, consists of a dense but fragile agglomerate of wood shavings (a lot of pressure and little glue), resisting a maximum bending moment of $mga/54$. The rod can only move on the vertical plane, pivoted at one end on point O, as shown. Given an initial impulse from the position of vertical equilibrium, find the slope θ where the rod breaks and the coordinate x of the breaking point P.

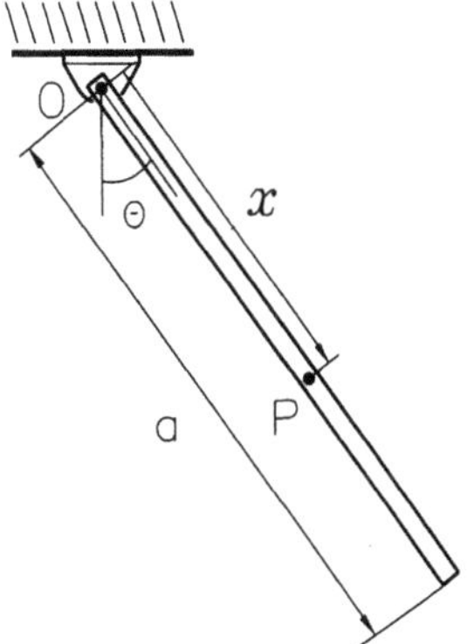

Figure P 11.11

P11.12 Gear A is left from its state of rest on the surface of fixed gear B, as shown. A small imbalance starts the motion of A to the left. Analyzing the gear as, essentially, a homogeneous cylinder with radius r, find the angular position θ where A loses contact with B.

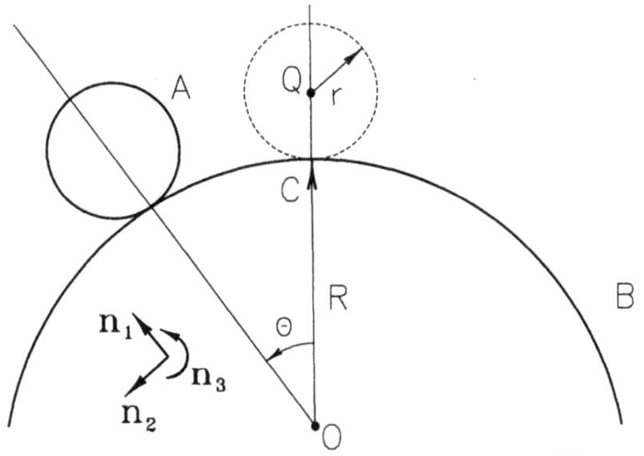

Figure P 11.12

P11.13 A slender homogeneous rod, with mass m and length a, is at rest, on a horizontal flat surface, as shown. Suddenly, a force orthogonal to its axis and with magnitude equal to its weight is applied to one end of the rod. Knowing that the coefficient of dynamic friction between the surfaces in contact is $\mu = 0.5$, find the distance to the opposite end of the point of the rod that will have a null acceleration at the beginning. *Hint:* Find the first- and second-kind equations of motion for the rod, arbitrating a point P, s far from the opposite end, as a fixed point and solve the equations for s.

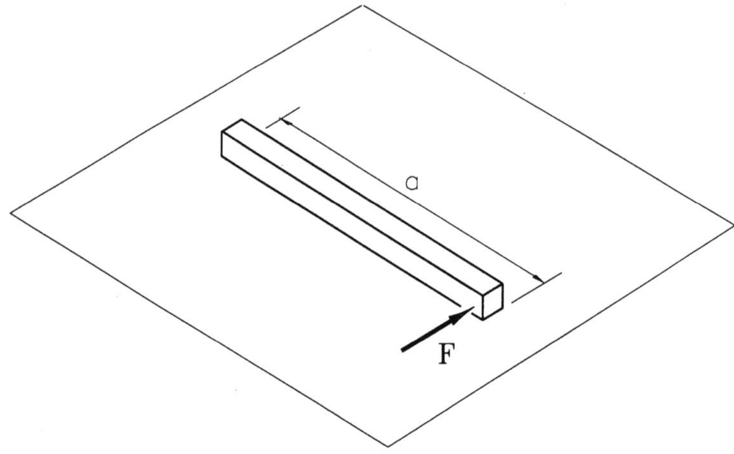

Figure P 11.13

P11.14 Referring to the previous problem, find the initial angular acceleration of the rod and acceleration, also at the same instant, of the point of application of force F.

P11.15 The system illustrated in Fig. P 11.15 consists of a fixed crown C and a pair of identical gears, A and B, whose centers are fixed to a rod, jointly with a vertical axis. The rod has a mass of 3 kg, while the gears have a mass of 6 kg each. The system is at rest when a torque $T = 6$ Nm is applied to the axis, in the direction shown. Considering the gears as homogeneous cylinders with the same radius, calculate how long the angular velocity of the rod will take to reach 25 rad/s.

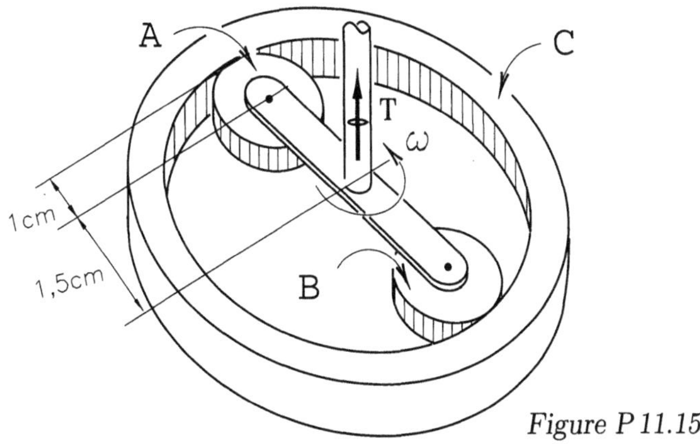

Figure P 11.15

P11.16 Two pulleys, with masses m and $3m$, are connected by a light belt, as shown. The set is at rest when a torque T is applied to the smaller pulley. Calculate the angular acceleration of this pulley, assuming that its mass is essentially concentrated on the periphery (both) and that the friction on the bearings can be disregarded.

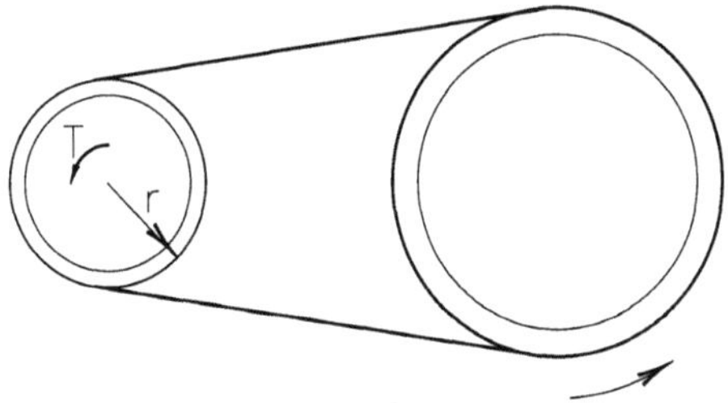

Figure P 11.16

P11.17 A rod, with mass m, is at rest over three identical cylinders, with mass $4m$ each. When applying a force F on the end of the rod, the system moves without sliding. Find the acceleration of the rod.

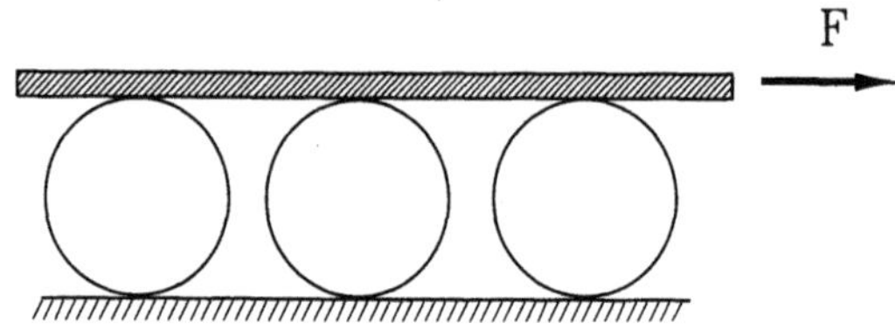

Figure P 11.17

P11.18 Ring A, homogeneous and with mass m, is supported by the sloping plane, initially at rest under the action of a force with magnitude T_0 and parallel to the plane, applied to a thread wound around the ring, as shown. The force magnitude suddenly increases to the value T and remains as such, causing an upward motion of the ring on the plane. Find the angular velocity of the ring at the instant when it is completing a turn around its axis of symmetry.

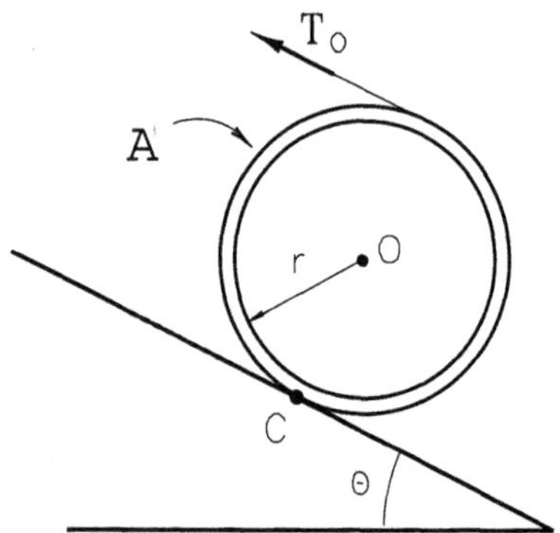

Figure P 11.18

P11.19 A rigid body is revolving around a vertical axis under the action of a torque T. A resistant torque then occurs on the bearings, being proportional to the square of the instantaneous angular velocity, so that $T_R = k\omega^2$. Find the angular velocity as a function of time knowing the moment of inertia of the body with respect to the rotation axis, I.

P11.20 Consider a rigid body, consisting of two rings, with radius r and mass m each, welded to a homogeneous rod, with length $2\sqrt{2}\,r$ and mass m, rolling over a horizontal plane, with its mass center, O, moving with a horizontal speed v, as described in Example 5.6. If the coefficient of friction between the plane and rings is $\mu = 0.5$, what is the greatest possible value for speed v before sliding occurs? What do you imagine will happen if the velocity is faster than this limit value?

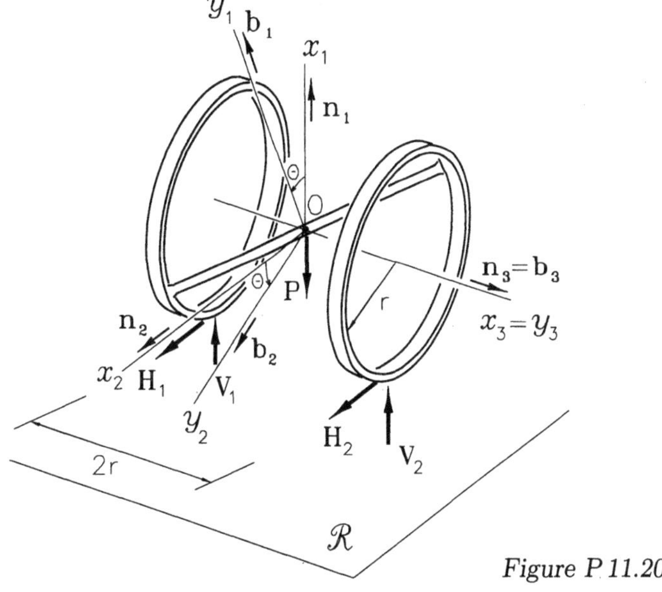

Figure P 11.20

P11.21 A uniform rod is supported by two identical pulleys, whose axes
are $2r$ apart from each other, turning at the same constant angular velocity
of module ω, but in opposite directions. With the rod placed with its mass
center exactly in $x = 0$, the system stays in equilibrium, under the action
of the perfectly balanced normal and friction forces. However, the rod being
inserted with its mass center displaced by x_0, as illustrated in Fig. P 11.21,
the friction force on the right-hand pulley is greater than the friction force
on the left-hand pulley, accelerating the rod in this direction until the mass
center returns to the centered position, to then start acceleration to the left. A
periodic motion is thus established (strictly speaking, harmonic) for the rod.
If the coefficient of friction between the pulleys and the rod is μ, calculate the
frequency of this motion.

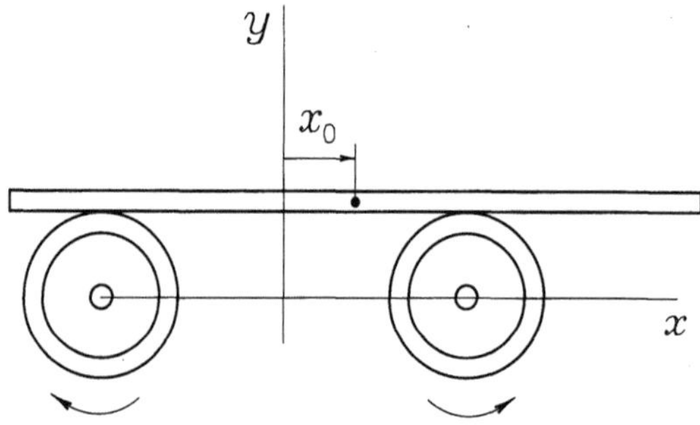

Figure P 11.21

Advanced Topics

Chapter 7 provides the concepts and equations required to describe the motion of a rigid body. The principles of conservation of momentum, angular momentum, and energy are also presented and the concept of work done by a torque applied to a rigid body is introduced. These expressions are applied to several examples of plane motion. In this chapter, slightly more complex situations involving three-dimensional rigid body dynamics are analyzed in detail. Some categories of motion, deserve special attention. For instance, the *gyroscopic motion*, with relevant applications in different branches of engineering, will be studied in detail in this chapter.

Section 8.1 is devoted to the study of motion of a rigid body with a fixed point. Special attention is given to the motion of bodies of revolution and to the application of Euler's equations in this class of problem. Section 8.2 discusses gyroscopic motion, closely related to what is discussed in the preceding section. Emphasis is given to the so-called gyroscopic effect and its applications. It also studies the free gyroscope and the motion of he top. In Section 8.3 the general motion of a rigid body is studied, providing some classic examples of dynamics. Some simple aspects involving approximate solutions are discussed. Comments on the rigid body dynamic stability are also made based on specific examples. Finally, Section 8.4 addresses situations of impact, discussing the usual methodology in the study of the collision between bodies.

8.1 Motion with a Fixed Point

When a rigid body C moves in relation to an inertial reference frame $\mathcal{R}$ in such a way that a point O, fixed in the body, throughout the motion remains also fixed in $\mathcal{R}$, it is called *motion with a fixed point*. This is quite a common occurrence in the various possible kinds of motion of a rigid body; this is what happens with a body bound by a pivot or a ball and socket joint, with a top supported by a horizontal plane with friction, with the rotor of a gyroscope, for example.

As discussed in Section 7.3.8 (have another look at it), if C is a rigid body moving with a point fixed in $\mathcal{R}$ (see Fig. 1.1), then there will be, in principle, three degrees of freedom in $\mathcal{R}$. In fact, when the position of a point of the body is determined, only three angular coordinates are required to define its orientation in relation to the reference frame. The result, then, is that three dynamic equations of the *second kind* are, in general, sufficient to describe this kind of motion.

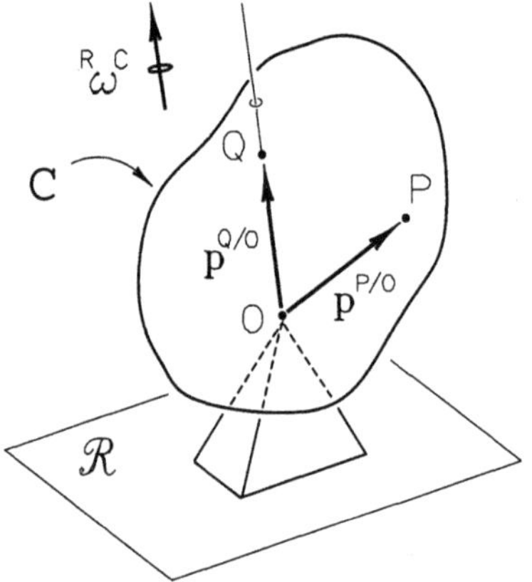

Figure 1.1

The general equations of motion for a rigid body, Eqs. (7.2.3) and (7.2.7), are naturally applicable to the situation of motion with a fixed point. When, however, only three angular coordinates are to be determined, it is more convenient to analyze the motion of the body from equations of motion of the second kind referring to the point. As seen in Section 7.5.4, if O is a fixed point in an inertial reference frame $\mathcal{R}$, $\mathbf{M}^{\mathcal{F}_e/O}$ is the resultant moment of the external forces applied to a body with respect to O and $^{\mathcal{R}}\mathbf{H}^{C/O}$ is the angular momentum vector of the body with respect to the point, then,

$$^{\mathcal{R}}\dot{\mathbf{H}}^{C/O} = \mathbf{M}^{\mathcal{F}e/O}. \tag{1.1}$$

Now, if C is a rigid body and O is a point of the body fixed in the inertial reference frame $\mathcal{R}$, then, according to Eq. (7.1.13),

$$^{\mathcal{R}}\mathbf{H}^{C/O} = \mathbb{I}^{C/O} \cdot {}^{\mathcal{R}}\boldsymbol{\omega}^{C}, \tag{1.2}$$

and the time rate in $\mathcal{R}$ of the angular momentum vector of the body with respect to O is, in the same way as the development that led to Eq. (7.2.6),

$$
\begin{aligned}
^{\mathcal{R}}\dot{\mathbf{H}}^{C/O} &= \frac{^{\mathcal{R}}d}{dt} {}^{\mathcal{R}}\mathbf{H}^{C/O} \\
&= \frac{^{C}d}{dt} {}^{\mathcal{R}}\mathbf{H}^{C/O} + {}^{\mathcal{R}}\boldsymbol{\omega}^{C} \times {}^{\mathcal{R}}\mathbf{H}^{C/O} \\
&= \frac{^{C}d}{dt} \left(\mathbb{I}^{C/O} \cdot {}^{\mathcal{R}}\boldsymbol{\omega}^{C} \right) + {}^{\mathcal{R}}\boldsymbol{\omega}^{C} \times \mathbb{I}^{C/O} \cdot {}^{\mathcal{R}}\boldsymbol{\omega}^{C} \\
&= \mathbb{I}^{C/O} \cdot {}^{\mathcal{R}}\boldsymbol{\alpha}^{C} + {}^{\mathcal{R}}\boldsymbol{\omega}^{C} \times \mathbb{I}^{C/O} \cdot {}^{\mathcal{R}}\boldsymbol{\omega}^{C}.
\end{aligned} \tag{1.3}
$$

So by substituting Eq. (1.3) in Eq. (1.1), the vector equation that governs the motion with a fixed point is as follows:

$$\mathbb{I}^{C/O} \cdot {}^{\mathcal{R}}\boldsymbol{\alpha}^{C} + {}^{\mathcal{R}}\boldsymbol{\omega}^{C} \times \mathbb{I}^{C/O} \cdot {}^{\mathcal{R}}\boldsymbol{\omega}^{C} = \mathbf{M}^{\mathcal{F}e/O}. \tag{1.4}$$

Note that Eq. (1.4) is completely similar to Eq. (7.2.7), differing by the inertia tensor of the body, now with respect to point O, and by the resultant moment, with respect also to the fixed point O.

Example 1.1 Figure 1.2 reproduces the system already studied previously in Examples 7.1.1 to 7.1.3 (take another look at them). The forces applied to C comprise its weight, $\mathbf{P} = -mg\mathbf{b}_3$, applied in C^*; a normal force, $\mathbf{N} = N\mathbf{b}_3$, and a friction component, $\mathbf{F}_a = F_a\mathbf{b}_2$, both applied in the point of contact P; and three components of force, $\mathbf{F}_1$, $\mathbf{F}_2$, and $\mathbf{F}_3$, applied by the link in O. Here there is a motion with a fixed point; therefore, the main equation for analyzing the problem is the equation of motion of the second kind with respect to point O. The resultant moment of the system of forces applied with respect to point O is (check)

$$
\begin{aligned}
\mathbf{M}^{\mathcal{F}e/O} &= \sqrt{5}r\,(mg - N)\mathbf{b}_2 + \sqrt{5}rF_a\mathbf{b}_3 \\
&= rF_a\mathbf{n}_1 + \sqrt{5}r\,(mg - N)\mathbf{n}_2 + 2rF_a\mathbf{n}_3.
\end{aligned}
$$

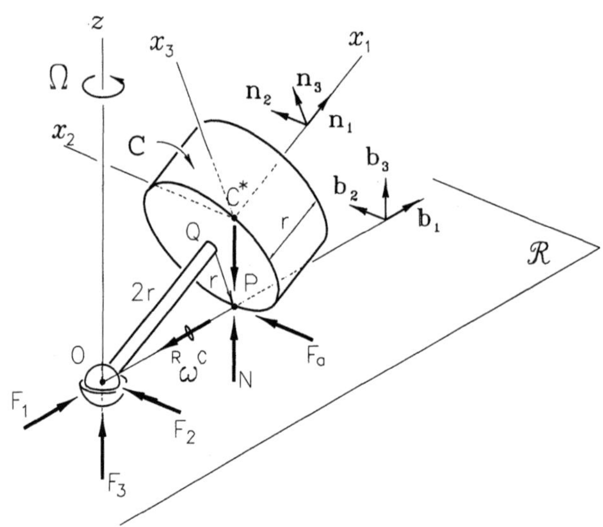

Figure 1.2

(Note that the basis $\mathbf{n}_1, \mathbf{n}_2, \mathbf{n}_3$ is parallel to the principal directions of inertia of the body with respect to point O and is, therefore, more convenient for breaking down the resultant moment vector.) The angular velocity vector of the body, obtained in Example 7.1.1, is

$$^{\mathcal{R}}\boldsymbol{\omega}^C = -2\Omega\mathbf{b}_1 = \frac{2}{\sqrt{5}}\Omega(-2\mathbf{n}_1 + \mathbf{n}_3),$$

and the angular acceleration vector may be obtained then by differentiation, that is,

$$^{\mathcal{R}}\boldsymbol{\alpha}^C = {}^{\mathcal{R}}\dot{\boldsymbol{\omega}}^C = -2\Omega\dot{\mathbf{b}}_1 = -2\Omega^2\mathbf{b}_2 = -2\Omega^2\mathbf{n}_2.$$

The inertia tensor of the body with respect to point O is expressed in the basis $\mathbf{n}_1, \mathbf{n}_2, \mathbf{n}_3$ by the matrix (see Example 7.1.2)

$$\mathbf{II}^{C/O} = \frac{1}{2}mr^2 \begin{pmatrix} 1 & 0 & 0 \\ 0 & \frac{79}{6} & 0 \\ 0 & 0 & \frac{79}{6} \end{pmatrix}.$$

We then have (check it out, it is good to practice)

$$\mathbf{II}^{C/O} \cdot {}^{\mathcal{R}}\boldsymbol{\alpha}^C = -\frac{79}{6}\Omega^2 mr^2\mathbf{n}_2,$$

$$^{\mathcal{R}}\boldsymbol{\omega}^C \times \mathbf{II}^{C/O} \cdot {}^{\mathcal{R}}\boldsymbol{\omega}^C = \frac{146}{15}\Omega^2 mr^2\mathbf{n}_2.$$

By substituting in Eq. (1.4), the equations obtained are

$$0 = rF_a,$$

$$-\frac{103}{30}\Omega^2 mr^2 = \sqrt{5}r(mg - N),$$

$$0 = 2rF_a.$$

The result, therefore, is that the friction force, for the condition under study, is null and the normal force is

$$N = m\left(g + \frac{103}{30\sqrt{5}}\,\Omega^2 r\right).$$

A glance at the result shows that it is not surprising that the friction force is null; this happens because the motion of the body is *uniform*, that is, C moves in $\mathcal{R}$ with a constant module angular velocity. Its mass center also has a constant speed. Note that the normal force is also constant and greater than the body weight. If we now wish to determine the force applied by the linkage in O (usually there is no interest in it), we need only to resort to the equations of motion of the first kind. In fact, the resultant force applied to the body is

$$\mathbf{F} = F_1\mathbf{b}_1 + F_2\mathbf{b}_2 + (F_3 + N - mg)\mathbf{b}_3,$$

and the acceleration of the mass center, which describes a uniform circular trajectory in $\mathcal{R}$, is

$$\mathbf{a}^* = -\sqrt{5}r\Omega^2\mathbf{b}_1.$$

Substituting then in Eqs. (7.2.4), we have

$$-\sqrt{5}mr\Omega^2 = F_1,$$

$$0 = F_2,$$

$$0 = (F_3 + N - mg);$$

therefore:

$$F_1 = -\sqrt{5}mr\Omega^2; \qquad F_2 = 0; \qquad F_3 = -\frac{103}{30\sqrt{5}}\,m\Omega^2 r.$$

It is worth noting that forces $\mathbf{N}$ and $\mathbf{F}_a$ were determined without needing to refer to the equations of motion of the first kind, used solely to determine the force in the link in O. Finally, note that, for the sake of convenience, different bases were used for the equations of rotation and translation of the body.

Adopting a basis of orthonormal vectors $\mathbf{n}_1, \mathbf{n}_2, \mathbf{n}_3$ to break down the angular velocity, angular acceleration, and resultant moment vectors and, also, expressing the inertia tensor of the body with respect to the point fixed in the same basis, Eq. (1.4) will be expanded, in the same way as when breaking down Eq. (7.2.7), which resulted in Eqs. (7.2.9). The difference, of course, lies in the moments and products of inertia, now referring to the fixed point O, and in the scalar components of the resultant moment, also now with respect to point O.

These equations, such as Eqs. (7.2.9), are too complex and, again, it is desirable to choose a basis parallel to the principal directions of inertia of the body with respect to the fixed point, thereby going back to *Euler's equations*, now for motion with a fixed point:

$$I_1^O \alpha_1 + (I_3^O - I_2^O)\omega_3\omega_2 = M_1^O;$$

$$I_2^O \alpha_2 + (I_1^O - I_3^O)\omega_1\omega_3 = M_2^O; \qquad (1.5)$$

$$I_3^O \alpha_3 + (I_2^O - I_1^O)\omega_2\omega_1 = M_3^O.$$

It is worth mentioning that, usually, to ensure that the basis stays parallel to the principal directions of inertia, it must be fixed in the body; an exception to this constraint happens when the body is of revolution and the fixed point belongs to the axis of symmetry, as shown in some of the following examples.

Example 1.2   A homogeneous bar B, with mass m and length r, is pivoting freely on its end O at the support S, which, in turn, rotates with a vertical simple angular velocity, with constant module ω, in relation to the inertial reference frame $\mathcal{R}$. We want to study its general motion in $\mathcal{R}$ and, in particular, to determine the value of ω so that the inclination θ remains constant with time; see Fig. 1.3. The axes $\{y_1, y_2, y_3\}$ are principal axes of inertia of the bar with respect to point O. The corresponding principal moments of inertia are (see Appendix C)

$$I_1^O = \frac{1}{3}mr^2, \qquad I_2^O = \frac{1}{3}mr^2, \qquad I_3^O = 0.$$

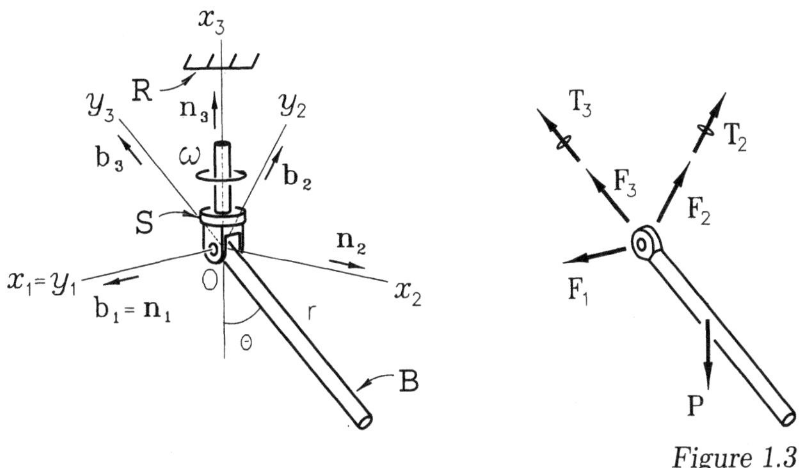

Figure 1.3

The system of external forces acting on the bar consists of its weight, $\mathbf{P}$; three components of force on the pivot, $\mathbf{F}_1$, $\mathbf{F}_2$, and $\mathbf{F}_3$, and two components of torque, also due to the link in O, $\mathbf{T}_2$ and $\mathbf{T}_3$. The resultant moment with respect to point O is, then,

$$\mathbf{M}^{\mathcal{F}_e/O} = -\frac{1}{2}mgr\sin\theta\,\mathbf{b}_1 + T_2\mathbf{b}_2 + T_3\mathbf{b}_3.$$

The angular velocity of the bar in $\mathcal{R}$ may be expressed as

$$^{\mathcal{R}}\boldsymbol{\omega}^B = {}^{\mathcal{R}}\boldsymbol{\omega}^S + {}^{S}\boldsymbol{\omega}^B = \omega\mathbf{n}_3 + \dot{\theta}\mathbf{n}_1 = \dot{\theta}\mathbf{b}_1 + \omega\sin\theta\,\mathbf{b}_2 + \omega\cos\theta\,\mathbf{b}_3.$$

The angular acceleration of the bar in $\mathcal{R}$ is, therefore,

$$^{\mathcal{R}}\boldsymbol{\alpha}^B = \ddot{\theta}\mathbf{n}_1 + \dot{\theta}\omega\mathbf{n}_2 = \ddot{\theta}\mathbf{b}_1 + \dot{\theta}\omega\cos\theta\,\mathbf{b}_2 - \dot{\theta}\omega\sin\theta\,\mathbf{b}_3.$$

Substituting, then, the kinematic, momentum, and inertia terms in Eqs. (1.5), we have (check)

$$\ddot{\theta} - \omega^2\sin\theta\cos\theta = -\frac{3}{2}\frac{g}{r}\sin\theta, \tag{a}$$

$$\frac{2}{3}mr^2\omega\dot{\theta}\cos\theta = T_2, \tag{b}$$

$$0 = T_3. \tag{c}$$

We then have a differential equation that governs the motion of the bar, so that

$$\ddot{\theta} - \left(\omega^2\cos\theta - \frac{3g}{2r}\right)\sin\theta = 0, \tag{d}$$

whose integration (non-trivial, in time) leads to the solution of the problem. The components of the torque are determined, naturally, from Eqs. (b) and (c). Assuming an initial condition of the kind $\dot{\theta}\,(\theta_0) = 0$, we can easily integrate Eq. (d) in the variable θ, resulting in

$$\dot{\theta}^2(\theta) = \omega^2\left(\sin^2\theta - \sin^2\theta_0\right) + 3\frac{g}{r}\left(\cos\theta - \cos\theta_0\right).$$

We then have from Eqs. (b) and (c)

$$T_2 = \frac{2}{3}mr^2\omega\cos\theta\left(\omega^2(\sin^2\theta - \sin^2\theta_0) + 3\frac{g}{r}(\cos\theta - \cos\theta_0)\right)^{\frac{1}{2}},$$

$$T_3 = 0.$$

Now assuming θ constant, the bar and support move together; therefore,

$$^{\mathcal{R}}\boldsymbol{\omega}^B = \omega\mathbf{n}_3 = \omega(\sin\theta\mathbf{b}_2 + \cos\theta\mathbf{b}_3) \qquad \text{and} \qquad ^{\mathcal{R}}\boldsymbol{\alpha}^B = 0.$$

The equations of motion are then reduced to

$$\omega^2\cos\theta = \frac{3}{2}\frac{g}{r}, \tag{e}$$

$$0 = T_2, \tag{f}$$

$$0 = T_3. \tag{g}$$

The last two equations establish that, for this uniform motion with θ constant, the torque on the pivot is null. The first one immediately gives the desired relation:

$$\omega = \sqrt{\frac{3g}{2r\cos\theta}}.$$

Note that the same result could be obtained by zeroing the second term of Eq. (d). Would you know how to explain? See the corresponding animation.

The balance of energy is also easier when principal directions of inertia are adopted. In particular, the kinetic energy of a body C that moves with a point fixed in an inertial reference frame $\mathcal{R}$, whose general expression is given by Eq. (7.1.25), is reduced, when the products of inertia are null, to

$$^{\mathcal{R}}K^C = \frac{1}{2}\,^{\mathcal{R}}\boldsymbol{\omega}^C \cdot \mathbf{II}^{C/O} \cdot {}^{\mathcal{R}}\boldsymbol{\omega}^C$$

$$= \frac{1}{2}\left(I_1^O\omega_1^2 + I_2^O\omega_2^2 + I_3^O\omega_3^2\right). \tag{1.6}$$

Example 1.3 Returning to the previous example (see Fig. 1.3), let us now consider that the bar B is abandoned from rest in relation to support S with $\theta = \pi/2$, while the latter moves in relation to the inertial reference frame $\mathcal{R}$ with a prescribed angular velocity $^{\mathcal{R}}\boldsymbol{\omega}^S = \omega\mathbf{n}_3$, as before. Let us now analyze how its energy varies. Taking the quota of point O as a reference, the initial gravitational potential energy ($\theta = \pi/2$) is null and the kinetic energy is, according to Eq. (1.6),

$$^{\mathcal{R}}K^B(1) = \frac{1}{2}I_2^O\,\omega_2^2 = \frac{1}{6}mr^2\omega^2.$$

The initial mechanical energy of the bar is, therefore,

$$^{\mathcal{R}}E^B(1) = \frac{1}{6}mr^2\omega^2.$$

In a generic position θ, the gravitational potential energy is $^{\mathcal{R}}\Phi^B(2) = -\frac{1}{2}mgr\cos\theta$ and the kinetic energy is, also according to Eq. (1.6),

$$^{\mathcal{R}}K^B(2) = \frac{1}{2}\left(\frac{1}{3}mr^2\dot{\theta}^2 + \frac{1}{3}mr^2\omega^2\mathrm{sen}^2\theta\right) = \frac{1}{6}mr^2(\dot{\theta}^2 + \omega^2\sin^2\theta).$$

For the prescribed initial conditions, we have (see the previous example)

$$\dot{\theta}^2(\theta) = \omega^2(\sin^2\theta - 1) + \frac{3g}{r}\cos\theta.$$

The mechanical energy of the bar in a general position is, then,

$$\begin{aligned}
^{\mathcal{R}}E^B(2) &= -\frac{1}{2}mgr\cos\theta + \frac{1}{6}mr^2\left(\omega^2(2\sin^2\theta - 1) + \frac{3g}{r}\cos\theta\right) \\
&= \frac{1}{6}mr^2\omega^2(2\sin^2\theta - 1).
\end{aligned}$$

The change in mechanical energy of the body in the interval is

$$^{\mathcal{R}}E^B(2) - {}^{\mathcal{R}}E^B(1) = -\frac{1}{3}mr^2\omega^2\cos^2\theta.$$

There is, therefore, a reduction in the mechanical energy of the bar, as θ decreases. What is responsible for consuming this energy? Well, most of the work is already computed in the variation of mechanical energy and the forces in the linkage in O do not contribute, obviously, to the resultant work. The explanation rests on the effect of torque $\mathbf{T}_2$, whose work, in the considered interval, is, according to Eq. (7.3.1),

$$
\begin{aligned}
{}^{\mathcal{R}}\mathcal{T}_{12}^{T_2} &= \int_0^t \mathbf{T}_2 \cdot {}^{\mathcal{R}}\boldsymbol{\omega}^B \, dt \\
&= \int_0^t \frac{2}{3} m r^2 \omega \cos\theta \, \dot{\theta} \, \omega \sin\theta \, dt \\
&= \frac{2}{3} m r^2 \omega^2 \int_{\pi/2}^{\theta} \cos\theta \sin\theta \, d\theta \\
&= -\frac{1}{3} m r^2 \omega^2 \cos^2\theta,
\end{aligned}
$$

as expected. See the corresponding animation.

Whenever a point O stays fixed simultaneously in a rigid body C and in an inertial reference frame $\mathcal{R}$, the situation created is called motion with a fixed point, satisfying Eq. (1.4), even when there is no point effectively belonging to the body that stays fixed in $\mathcal{R}$ during the motion. In other terms, as every rigid body forms a reference frame, any point in space that does not move in relation to the body can be understood as an extension of it, the corresponding kinetic relations, thereby being valid; if the point in consideration is also fixed in an inertial reference frame, then the body is describing a motion with a fixed point.

Example 1.4   Consider a rigid body C consisting of two rings A and B, with masses m and $2m$, respectively, rigidly connected by means of a light structure, showing the configuration in Fig. 1.4. The body is at rest, supported by a horizontal flat surface, when the torque $\mathbf{T}$ of module $T = 13mgr$ is applied, as shown. We want to determine the initial acceleration of point Q, the center of the larger ring, assuming that the coefficient of friction between the body and the surface is sufficient to prevent relative sliding. Well, the rolling condition of the rings on the horizontal plane ensures that axis x_3, of symmetry for the body, will describe a conical surface with the vertex in O, that is, point O, fixed in the inertial reference frame, also remains fixed in relation to C. We have, therefore, a situation of motion with a fixed point, which may be conveniently analyzed from Eqs. (1.5). The applied system of forces consists of the weights of the rings, the components in the points of contact with the surface (three, in principle, for each), and the applied torque. The resultant moment with respect to O is then (check)

$$\mathbf{M}^{\mathcal{F}_e/O} = \left(\frac{15}{2}mg - 2(A_2 + 2B_2)\right) r\,\mathbf{n}_1 + \sqrt{3}\left(A_1 + 2B_1\right) r\,\mathbf{n}_2$$
$$+ \left(T + (A_1 + 2B_1)\,r\right)\mathbf{n}_3,$$

where A_j and B_j, $j = 1, 2, 3$, are the components, in the basis $\mathbf{n}_1, \mathbf{n}_2, \mathbf{n}_3$, of the forces applied to the support points on the plane.

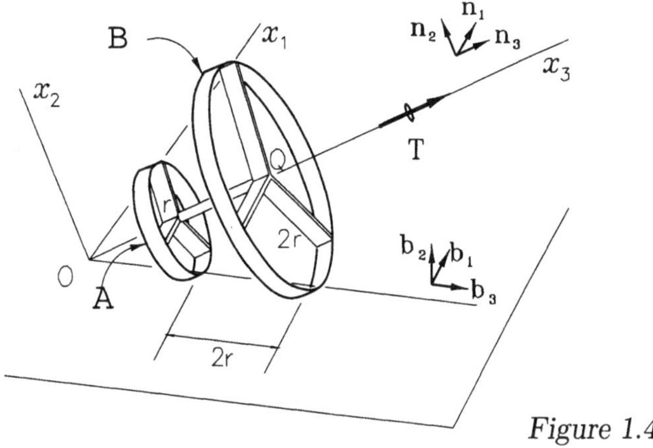

Figure 1.4

The principal moments of inertia of the body with respect to coordinate axes $\{x_1, x_2, x_3\}$ are

$$I_1^O = I_2^O = \frac{1}{2}mr^2 + 3mr^2 + 4mr^2 + 24mr^2 = \frac{63}{2}mr^2,$$
$$I_3^O = mr^2 + 8mr^2 = 9mr^2.$$

The angular velocity vector of the body in the inertial reference frame may be expressed by

$$^{\mathcal{R}}\boldsymbol{\omega}^C = \omega\mathbf{n}_3 + \Omega\mathbf{b}_2 = \left(\frac{1}{2}\omega + \Omega\right)\mathbf{b}_2 + \frac{\sqrt{3}}{2}\omega\mathbf{b}_3.$$

However, from the rolling condition in the two points of contact, it guarantees that the angular velocity vector is necessarily parallel to $\mathbf{b}_3$; therefore, $\Omega = -\frac{1}{2}\omega$ and

$$^{\mathcal{R}}\boldsymbol{\omega}^C = \frac{\sqrt{3}}{2}\omega\mathbf{b}_3.$$

The angular acceleration of the body can then be obtained by differentiating the angular velocity, to get

$$^{\mathcal{R}}\boldsymbol{\alpha}^C = \frac{\sqrt{3}}{2}(\dot{\omega}\mathbf{b}_3 + \omega\dot{\mathbf{b}}_3)$$

$$= \frac{\sqrt{3}}{2}(\dot{\omega}\mathbf{b}_3 - \frac{1}{2}\omega^2\mathbf{b}_1).$$

Initially,

$$^{\mathcal{R}}\boldsymbol{\omega}^C = 0 \qquad \text{and} \qquad ^{\mathcal{R}}\boldsymbol{\alpha}^C = \frac{\sqrt{3}}{4}\dot{\omega}(-\mathbf{n}_2 + \sqrt{3}\mathbf{n}_3).$$

So substituting in Eqs. (1.5) leads to

$$0 = \frac{15}{2}mg - 2(A_2 + 2B_2), \tag{a}$$

$$-\frac{63\sqrt{3}}{8}mr^2\dot{\omega} = \sqrt{3}\,r(A_1 + 2B_1), \tag{b}$$

$$\frac{27}{4}mr^2\dot{\omega} = T + r(A_1 + 2B_1). \tag{c}$$

Substituting Eq. (b) in Eq. (c) and solving for $\dot{\omega}$, we get

$$\dot{\omega} = \frac{8}{117}\frac{T}{mr^2}.$$

Therefore, initially,

$$^{\mathcal{R}}\mathbf{a}^Q = {}^{\mathcal{R}}\boldsymbol{\alpha}^C \times \mathbf{p}^{Q/O} = -\frac{4}{39}\frac{T}{mr}\mathbf{n}_1 = -\frac{4}{3}g\,\mathbf{n}_1.$$

8.2 Gyroscopic Motion

The *inertia* is a basic property of mechanics that can be used to detect motion. Inside a closed elevator, we know if we are going up or down because, due to the inertia, we can perceive the direction of the acceleration to which we are submitted. With the known mass m of a body with small dimensions and measuring, albeit indirectly, the resultant force applied to it, we can determine from Eq. (7.2.3) how the body moves in relation to an inertial reference frame. This is, in general, the principle of the operation of a *vibrometer* or *accelerometer*.

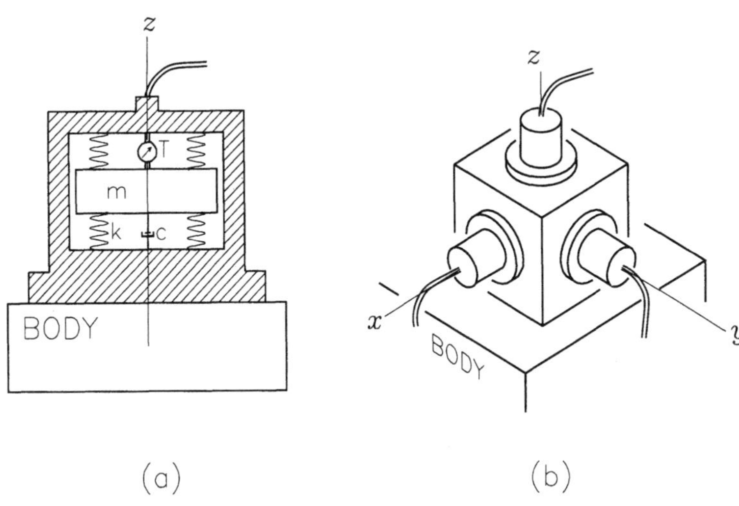

Figure 2.1

Figure 2.1a simply illustrates in a diagram a one-dimensional accelerometer; mass m is suspended elastically to a casing, which, in turn, is fixed in the body whose motion we want to determine. The model comprises the inertia, characterized by the mass m; rigidity, given by the elastic constant k; and absorption, defined by parameter c. An electromechanical transducer T — converting displacement in electric signal — measures, therefore, the relative displacement between the suspended mass and the casing, which, in turn, is related to the resultant force applied on the mass, which will indirectly provide a measure of acceleration of the casing in the direction of the axis of the accelerometer. (Many technological details are involved to, among other things, ensure the linearity of the system, taking into account the damping, and guarantee good electromechanical transduction, the study of which is of no interest to us for the time being; what matters here is to perceive the general idea that an acceleration may be assessed by indirectly measuring the associated resultant force, as established by Newton's second law.) Measuring, then, the acceleration at each instant, the absolute velocity and displacement in the inertial reference frame may be easily obtained by integrating in time.

Figure 2.1b illustrates a layout of three accelerometers arranged according to mutually orthogonal axes. So, with this layout, the acceleration *vector* can be obtained, which, when integrated, will give us the position vector of the body. Since its initial position is known, the position, in an inertial reference frame, may be known at any time. This is the basic principle of *inertial navigation*. There is, however, one very important detail: In order to integrate the acceleration vector, it is necessary for the axes of the accelerometers to keep their orientation fixed in the inertial reference frame. Maintaining the spatial orientation is one of the applications of the *gyroscope*.

Just as the mass is the inertia of translation of a rigid body, its inertia tensor with respect to the mass center is its inertia of rotation. In the same way as when measuring the applied resultant force is used to assess the acceleration of the mass center of the body, according to Eq. (7.2.3), monitoring the applied resultant torque may provide information on the rotational behavior of the body in an inertial reference frame, according to Eq. (7.2.5). Another alternative in using the inertia of rotation as a guide for spatial orientation is precisely the use of the free gyroscope, that is, the one whose rotor can remain turning around a fixed direction in space, as is discussed below. In this case, relative rotations instead of the applied torque are measured.

An inertial platform consists, then, of a layout of three accelerometers, such as shown in the layout in Fig. 2.1b, whose orientation is kept fixed in space, in relation to an inertial reference frame, with the help of two gyroscopes. The reasons for this are that each gyroscope is put to turn at an angular velocity prescribed around the axis of symmetry of its rotor, two degrees of freedom therefore remaining for its angular orientation. Since three angular coordinates are required to fully characterize the orientation relative to the inertial reference frame, two gyroscopes are used, with noncoinciding rotation axes, which makes one of the coordinates redundant.

Figure 2.2 shows a drawing of a gyroscope with two degrees of freedom, consisting of a rotor C, which is just a well-balanced axisymmetric body (that is, built with particular accuracy to guarantee that its principal axes of inertia coincide with the geometric symmetry axes), which turns at a constant angular velocity in relation to an inertial reference frame. To guarantee this condition, the rotor is suspended by rings in its support. See the corresponding animation.

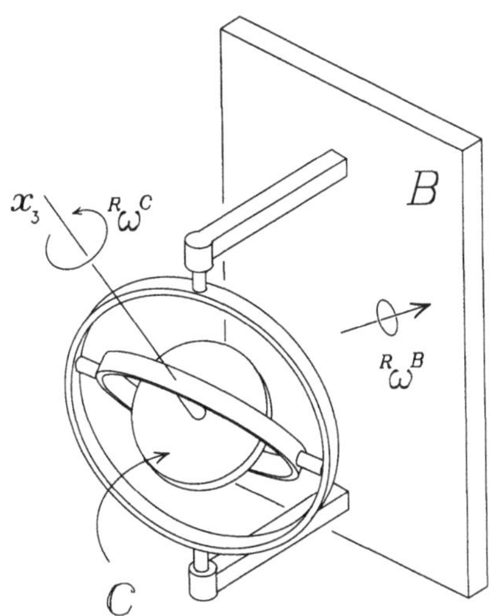

Figure 2.2

Assuming ideal bearings, that is, without friction, the resultant moment with respect to the mass center applied to the rotor is null and, from Eq. (7.2.5),

$$^R\dot{\mathbf{H}}^{C/C^*} = 0, \tag{2.1}$$

that is, the angular momentum vector of the gyroscope with respect to its mass center in an inertial reference frame is conserved. Since the axis x_3 around which the rotor is turning is a principal axis of inertia, then, if $J = I_{x_3}$ is the corresponding principal moment of inertia, the angular momentum vector will be expressed, according to Eq. (7.1.12), by

$$^R\mathbf{H}^{C/C^*} = J\omega\mathbf{n}_3 = J\,^R\boldsymbol{\omega}^C. \tag{2.2}$$

Now, since, from Eq. (2.1), the angular momentum vector of the rotor in the inertial reference frame is constant, the result is that the angular velocity vector of the rotor also remains constant in the reference frame. A system with these characteristics is called a *free gyroscope*. When the support of a free gyroscope is fixed in a body moving arbitrarily in an inertial reference frame, the change in orientation of the body in the reference frame may be measured by taking as reference the orientation of the rotor shaft.

Gyroscopes are extremely useful for navigation, as simple guides for spatial orientation, as in the case of the gyroscopic compasses, as a component of an inertial navigation system, or as an auxiliary element for rotational stabilization. They are found in seagoing vessels, aircraft, launcher rockets, space satellites, oil-drilling equipment, and war applications, among others. For example, gyroscopic systems can keep the solar panels of a spacecraft aligned with the sun, thereby optimizing the capture of the energy required to operate them. Of course, the ideal gyroscope is merely fictitious and many technological details omitted herein are required to compensate the friction in the bearings, discount the earth's rotation, and reduce the aerodynamic drag, among other minor effects. Let us now analyze the main dynamic phenomena associated with this system, commonly known generally as *gyroscopic effect*. The reader must bear in mind, however, that the gyroscopic effect is not a new dynamic principle, resulting solely from the behavior described by Eq. (7.2.5) or Eq. (1.1), as shown ahead.

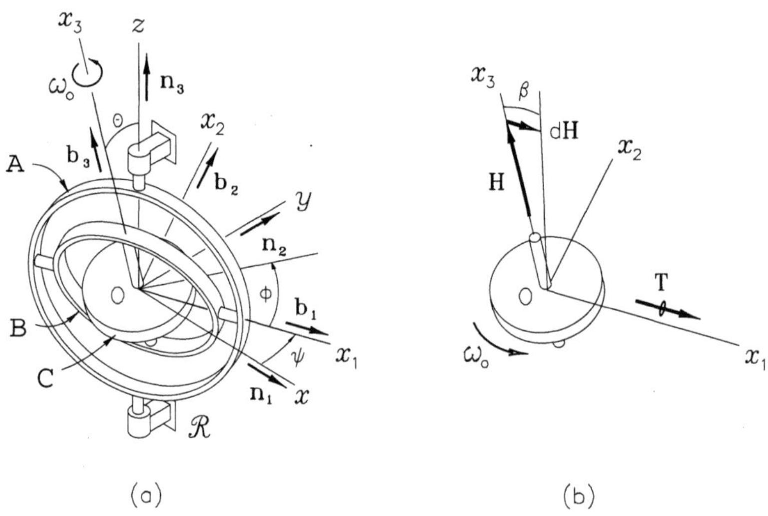

Figure 2.3

Let us, then, consider rotor C, with mass m and axial and transversal moments of inertia, with respect to the mass center, J and I, respectively, suspended by rings in the inertial reference frame $\mathcal{R}$ (see Fig. 2.3a). The Cartesian axes $\{x, y, z\}$ are fixed in the reference frame $\mathcal{R}$. The rotation of ring A, which has a simple angular velocity in $\mathcal{R}$ in the direction of z, is measured by the angle ψ. The Cartesian axes $\{x_1, x_2, x_3\}$ and the orthonormal basis $\mathbf{b}_1, \mathbf{b}_2, \mathbf{b}_3$ are fixed in ring B, which has a simple angular velocity in relation to A in the direction of x_1. The rotation between B and A is measured by the angle θ. Last, the rotation between C and B is given by the angle ϕ. The angular coordinates that, therefore, describe the attitude (orientation) of the rotor in relation to $\mathcal{R}$ are the *Euler angles* (see Section 3.8): $\phi(t)$; $\theta(t)$; and $\psi(t)$, as shown in the figure. The spin of the body is, therefore, $\dot{\phi}$, corresponding to the angular velocity of C in B; the *nutation* is $\dot{\theta}$, which describes the angular velocity of ring B in A; and the *precession* is $\dot{\psi}$, which measures the angular velocity of ring A in the reference frame $\mathcal{R}$.

The angular velocity in $\mathcal{R}$ of ring B is

$$\begin{aligned}
{}^{\mathcal{R}}\boldsymbol{\omega}^{B} &= {}^{\mathcal{R}}\boldsymbol{\omega}^{A} + {}^{A}\boldsymbol{\omega}^{B} \\
&= \dot{\psi}\mathbf{n}_3 + \dot{\theta}\mathbf{b}_1 \\
&= \dot{\theta}\mathbf{b}_1 + \dot{\psi}\sin\theta\,\mathbf{b}_2 + \dot{\psi}\cos\theta\,\mathbf{b}_3.
\end{aligned} \tag{2.3}$$

The angular velocity of C in relation to B is, in turn,

$$ {}^{B}\boldsymbol{\omega}^{C} = \dot{\phi}\mathbf{b}_3. \tag{2.4}$$

The angular velocity vector of the rotor in the reference frame $\mathcal{R}$ is, therefore,

$$ {}^{\mathcal{R}}\boldsymbol{\omega}^{C} = \omega_1\mathbf{b}_1 + \omega_2\mathbf{b}_2 + \omega_3\mathbf{b}_3, \tag{2.5}$$

where

$$ \omega_1 = \dot{\theta}, \qquad \omega_2 = \dot{\psi}\sin\theta, \qquad \omega_3 = \dot{\psi}\cos\theta + \dot{\phi}. \tag{2.6}$$

The angular acceleration of C in $\mathcal{R}$ may then be obtained by differentiation, resulting in

$$ {}^{\mathcal{R}}\boldsymbol{\alpha}^{C} = \alpha_1\mathbf{b}_1 + \alpha_2\mathbf{b}_2 + \alpha_3\mathbf{b}_3, \tag{2.7}$$

where (check: the basis $\mathbf{b}_1, \mathbf{b}_2, \mathbf{b}_3$ varies in $\mathcal{R}$, with $\dot{\mathbf{b}}_j = {}^{\mathcal{R}}\boldsymbol{\omega}^{B} \times \mathbf{b}_j$, $j = 1, 2, 3$)

$$\begin{aligned}
\alpha_1 &= \dot{\omega}_1 + \dot{\phi}\omega_2 = \ddot{\theta} + \dot{\psi}\dot{\phi}\sin\theta, \\
\alpha_2 &= \dot{\omega}_2 - \dot{\phi}\omega_1 = \ddot{\psi}\sin\theta + \dot{\psi}\dot{\theta}\cos\theta - \dot{\phi}\dot{\theta}, \\
\alpha_3 &= \dot{\omega}_3 = \ddot{\psi}\cos\theta + \ddot{\phi} - \dot{\psi}\dot{\theta}\sin\theta.
\end{aligned} \tag{2.8}$$

If we apply a torque $\mathbf{T}$, say, in the direction of x_1, there will be a variation of the angular momentum vector, as shown in Fig. 2.3b. If no torques are applied, the resultant moment with respect to the mass center of the rotor is null, that is,

$$M_1^* = M_2^* = M_3^* = 0. \tag{2.9}$$

Substituting, then, Eqs. (2.5), (2.7), and (2.9) in Euler's equations, the system of differential equations governing the motion of the free gyroscope, in terms of the components of the angular velocity vector, is

$$I(\dot{\omega}_1 + \dot{\phi}\omega_2) + (J - I)\omega_3\omega_2 = 0,$$
$$I(\dot{\omega}_2 - \dot{\phi}\omega_1) + (I - J)\omega_1\omega_3 = 0, \tag{2.10}$$
$$J\dot{\omega}_3 = 0.$$

Alternatively, expressing the equations of motion in terms of Euler's angles, using Eqs. (2.6) and (2.8), leads to

$$I(\ddot{\theta} - \dot{\psi}^2 \sin\theta \cos\theta) + J(\dot{\psi}\cos\theta + \dot{\phi})\dot{\psi}\sin\theta = 0,$$
$$I(\ddot{\psi}\sin\theta + 2\dot{\psi}\dot{\theta}\cos\theta) - J(\dot{\psi}\cos\theta + \dot{\phi})\dot{\theta} = 0, \tag{2.11}$$
$$\ddot{\psi}\cos\theta - \dot{\psi}\dot{\theta}\sin\theta + \ddot{\phi} = 0.$$

As an initial value, therefore, is prescribed for the angular velocity, of module ω_0, around the axis of symmetry, the solution (verify; it is easy), is

$$\omega_1 = \omega_2 = 0, \qquad \omega_3 = \omega_0, \tag{2.12}$$

that is,

$$\dot{\phi} = \omega_0, \qquad \dot{\psi} = \dot{\theta} = 0, \tag{2.13}$$

thereby keeping reference frames A and B fixed in $\mathcal{R}$ with Eq. (2.1) being satisfied. In short, given an initial spin condition with null nutation and precession, the rotor thus remains throughout, therefore preserving the spatial orientation of its axis of symmetry. In other words, the angular velocity vector and angular momentum vector of the rotor in the inertial reference frame stay constant in time.

Equation (2.12) expresses a possible solution for Eqs (2.10), but other solutions may be investigated. There is, for example, an interesting solution with constant precession and spin and null nutation, with $\theta \neq 0$, that is,

$$\dot{\phi} = \omega_0, \qquad \dot{\psi} = \Omega, \qquad \dot{\theta} = 0, \tag{2.14}$$

where Ω and ω_0 are constants. The components of the angular velocity vector, in this case, would be

$$\omega_1 = 0, \qquad \omega_2 = \Omega \sin \theta, \qquad \omega_3 = \Omega \cos \theta + \omega_0. \tag{2.15}$$

It is easy to check that the above conditions satisfy Eqs. (2.10) provided that (check)

$$\Omega \sin \theta \big(J\omega_0 + (J - I)\Omega \cos \theta \big) = 0. \tag{2.16}$$

This relation is satisfied for $\Omega = 0$, returning to the previous case of null precession and nutation, or to

$$\omega_0 = \frac{I - J}{J} \cos \theta \, \Omega. \tag{2.17}$$

The angular velocity vector of the rotor in the inertial reference frame, therefore, is

$$^{\mathcal{R}}\boldsymbol{\omega}^C = \Omega \left(\sin \theta \mathbf{b}_2 + \frac{I}{J} \cos \theta \mathbf{b}_3 \right). \tag{2.18}$$

The angular momentum vector of the rotor with respect to its mass center is then

$$\begin{aligned}
^{\mathcal{R}}\mathbf{H}^{C/C^*} &= I\omega_1 \mathbf{b}_1 + I\omega_2 \mathbf{b}_2 + J\omega_3 \mathbf{b}_3 \\
&= I\Omega \sin \theta \mathbf{b}_2 + I\Omega \cos \theta \mathbf{b}_3 \\
&= I\Omega \mathbf{n}_3.
\end{aligned} \tag{2.19}$$

It is therefore found that, for this solution of constant precession and spin with null nutation, the angular velocity vector of the rotor in the inertial reference frame varies with time, although its angular momentum vector remains invariant.

Example 2.1 Consider a homogeneous hemispherical rotor with mass m and radius r rotating around its axis of symmetry at the rate $p = 120$ rpm, with a constant inclination of 60° with respect to the vertical (see Fig. 2.4).

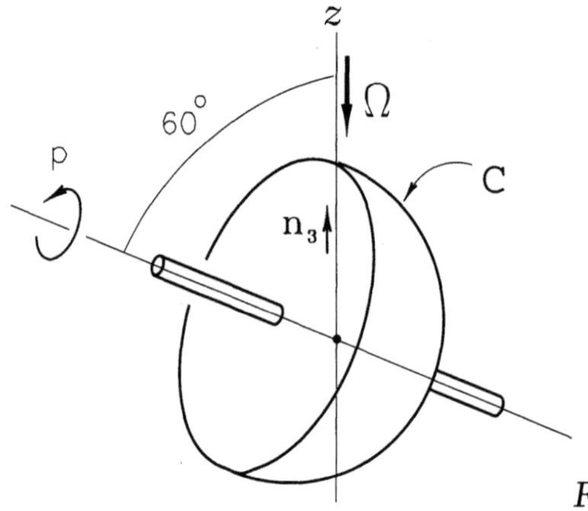

Figure 2.4

Let us look for the corresponding nonnull precession. The tables in Appendix C give us the following values for the principal moments of inertia with respect to the mass center of the rotor:

$$I = \frac{83}{320} mr^2; \qquad J = \frac{2}{5} mr^2.$$

The given spin is

$$\omega_0 = \frac{120 \times 2\pi}{60} = 4\pi \text{ rad/s}.$$

Resolving Eq. (2.17) for precession Ω, then we have

$$\Omega = \frac{J}{(I - J)\cos\theta}\, \omega_0 = -\frac{1024}{45}\pi \text{ rad/s} = -682.667 \text{ rpm}.$$

See the corresponding animation.

The above example resulted in a negative precession; let us see what that means. The motion made with null nutation leads to (see Fig. 2.3)

$$^{\mathcal{R}}\boldsymbol{\omega}^A = \dot\psi \mathbf{n}_3 \qquad \text{and} \qquad ^{B}\boldsymbol{\omega}^C = \dot\phi \mathbf{b}_3. \tag{2.20}$$

So, when the scalar product of these two vectors is positive, we say that there is *direct precession*; when, on the contrary, the scalar product is negative, we say that there is *retrograde precession*. Now,

$$^{\mathcal{R}}\boldsymbol{\omega}^A \cdot {}^{B}\boldsymbol{\omega}^C = \dot\psi \mathbf{n}_3 \cdot \dot\phi \mathbf{b}_3 = \Omega\omega_0 \cos\theta = \frac{I - J}{J} \cos^2\theta\, \Omega^2, \tag{2.21}$$

so the precession will be direct if $I > J$ and the body is said, then, to be *slender* (see Fig. 2.5a); otherwise, the precession will be retrograde if $I < J$, in which case the body is said to be *nonslender* or *blunt* (see Fig. 2.5b). The rotor in Example 7.1 is, therefore, nonslender. See also the corresponding animation

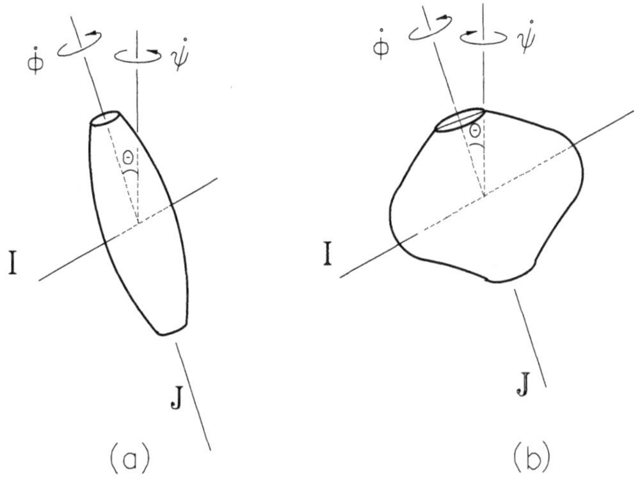

Figure 2.5

The solutions for Eqs. (2.11) expressed by Eqs. (2.13) and (2.14) are, of course, quite simple particular cases. The determination of a general analytical solution, given initial arbitrary conditions, is not an easy task. We can, however, take valuable information about the angular behavior of the rotor, using the principles of conservation applicable to this specific situation of motion with the null resultant torque with respect to the mass center, as discussed ahead.

We will then consider a free gyroscope for which initial arbitrary conditions are given, that is, with the three components of the nonnull angular velocity vector. Let us first note that the third of Eqs. (2.10) can be directly integrated in time, offering the solution

$$\omega_3 = \omega_0, \tag{2.22}$$

where ω_0 is a scalar constant. The angular momentum vector with respect to the mass center is, at each instant,

$$^{\mathcal{R}}\mathbf{H}^{C/C^*} = I\omega_1\mathbf{b}_1 + I\omega_2\mathbf{b}_2 + J\omega_3\mathbf{b}_3, \tag{2.23}$$

and its scalar component in the direction of axis z is

$$H_z = {}^{\mathcal{R}}\mathbf{H}^{C/C^*} \cdot \mathbf{n}_3 = I\omega_2 \sin\theta + J\omega_3 \cos\theta. \tag{2.24}$$

Now, the angular momentum vector of the rotor is conserved and, in particular, its scalar component in the direction of axis z is also conserved. So if H_{z_0} is the initial value of the angular momentum of the rotor with respect to axis z, then at any instant we have

$$I\omega_2 \sin\theta + J\omega_3 \cos\theta = H_{z_0}. \tag{2.25}$$

Then, for $\theta \neq 0$,

$$\omega_2 = \frac{1}{I\sin\theta}\left(H_{z_0} - J\cos\theta\,\omega_3\right). \tag{2.26}$$

Substituting then in the second of Eqs. (2.6) leads to a general expression for the precession as a function of inclination θ and, naturally, of the initial conditions,

$$\dot{\psi}(\theta) = \frac{1}{I\sin^2\theta}\left(H_{z_0} - J\cos\theta\,\omega_3\right). \tag{2.27}$$

Now substituting the above expression in the third of Eqs. (2.6) leads to an expression for the spin, also as a function of inclination θ and the initial conditions,

$$\dot{\phi}(\theta) = \left(1 + \frac{J\cos^2\theta}{I\sin^2\theta}\right)\omega_3 - \frac{H_{z_0}\cos\theta}{I\sin^2\theta}. \tag{2.28}$$

The system of forces acting on the rotor is, as we saw, null, so the resultant work applied during any considered time interval is also null. Equation (7.4.1) establishes, therefore, that the kinetic energy of the rotor is conserved, that is,

$$\frac{1}{2}\left(I\omega_1^2 + I\omega_2^2 + J\omega_3^2\right) = K_0, \tag{2.29}$$

where the scalar constant K_0 is the kinetic energy of the rotor in the given initial condition. So solving Eq. (2.29) for ω_1 leads to

$$\omega_1 = \sqrt{\frac{1}{I}\left(2K_0 - I\omega_2^2 - J\omega_3^2\right)}. \tag{2.30}$$

Equations (2.22), (2.26), and (2.30) therefore provide the solution for the components of the angular velocity vector as a function of the coordinate θ and initial conditions of the motion, represented by ω_3, H_{z_0}, and K_0. Substituting then Eq. (2.30) in the first of Eqs. (2.6) leads to an expression for the nutation, also as a function of the inclination θ, that is,

$$\dot{\theta}(\theta) = \sqrt{\frac{1}{I}\left(2K_0 - I\omega_2^2 - J\omega_3^2\right)}, \tag{2.31}$$

where ω_2 is expressed by Eq. (2.26). Equations (2.27), (2.28), and (2.31) then provide the general expressions for the precession, the spin, and the nutation as a function of the inclination θ of the rotor axis and, of course, of the prescribed initial conditions. Note that the initial conditions are implicitly expressed by the initial kinetic energy, K_0, and by the value, also initial, of the scalar component of the angular momentum vector of the body with respect to its mass center, in the direction of the vertical axis, H_{z_0}.

Example 2.2 Consider a free gyroscope whose rotor C has the shape of a disk with mass m and radius r (see Fig. 2.6). Given an initial condition, with $\theta = 30^o$, of spin $\dot{\phi}_0 = \sqrt{3g/r}$, nutation $\dot{\theta}_0 = 5\sqrt{g/r}$ and null precession, how will the gyroscope be moving when the axis of symmetry passes through the horizontal position? The prescribed initial conditions, with $\theta = \pi/6$, are

$$\omega_1 = 5\sqrt{\frac{g}{r}}, \qquad \omega_2 = 0, \qquad \omega_3 = \sqrt{\frac{3g}{r}}.$$

The vertical component of the angular momentum vector is

$$H_{z_0} = J\omega_3 \cos\theta = \frac{3}{4}mr^2\sqrt{\frac{g}{r}}.$$

The initial kinetic energy of the rotor is, in turn,

$$K_0 = \frac{1}{2}\left(\frac{1}{4}mr^2\frac{25g}{r} + \frac{1}{2}mr^2\frac{3g}{r}\right) = \frac{31}{8}mgr.$$

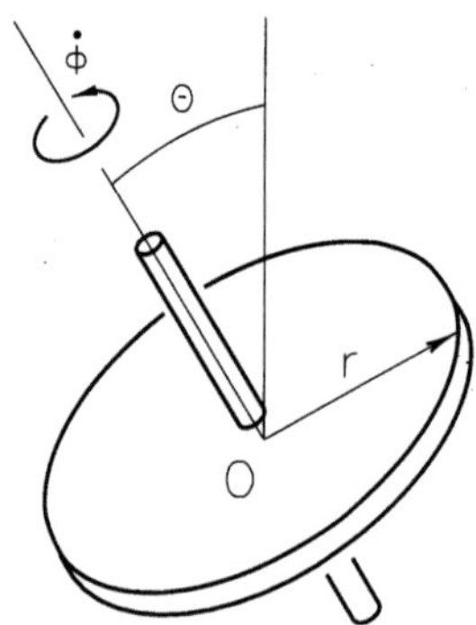

Figure 2.6

If $\theta = \pi/2$, according to Eq. (2.26), it leads, then, to

$$\omega_2 = 3\sqrt{\frac{g}{r}};$$

therefore,

$$\dot{\psi} = 3\sqrt{\frac{g}{r}}$$

and, from Eq. (2.28),

$$\dot{\phi} = \sqrt{\frac{3g}{r}}.$$

Last, from Eq. (2.31),

$$\dot{\theta} = 4\sqrt{\frac{g}{r}}.$$

The components of the angular velocity vector in this position are, then,

$$\omega_1 = 4\sqrt{\frac{g}{r}}, \qquad \omega_2 = 3\sqrt{\frac{g}{r}}, \qquad \omega_3 = \sqrt{\frac{3g}{r}}.$$

Note that, in the initial condition, we have

$$^{\mathcal{R}}\mathbf{H}^{C/C^*} = \frac{1}{4}mr\sqrt{gr}(5\mathbf{b}_1 + 2\sqrt{3}\mathbf{b}_3),$$

while in the final condition we get

$$^{\mathcal{R}}\mathbf{H}^{C/C^*} = \frac{1}{4}mr\sqrt{gr}(4\mathbf{b}_1 + 3\mathbf{b}_2 + 2\sqrt{3}\mathbf{b}_3),$$

both, therefore, with the same module $H = (\sqrt{37}/4)\,mr\sqrt{gr}$, as to be expected. In fact, resolving basis $\mathbf{b}_1, \mathbf{b}_2, \mathbf{b}_3$ in basis $\mathbf{n}_1, \mathbf{n}_2, \mathbf{n}_3$ (see Fig. 2.3), we have

$$\mathbf{b}_1 = \cos\psi\,\mathbf{n}_1 + \sin\psi\,\mathbf{n}_2,$$
$$\mathbf{b}_2 = -\cos\theta\sin\psi\,\mathbf{n}_1 + \cos\theta\cos\psi\,\mathbf{n}_2 + \sin\theta\,\mathbf{n}_3,$$
$$\mathbf{b}_3 = \sin\theta\sin\psi\,\mathbf{n}_1 - \sin\theta\cos\psi\,\mathbf{n}_2 + \cos\theta\,\mathbf{n}_3.$$

We may choose the orientation of axes $\{x, y, z\}$ so that, for the initial condition of $\theta = \pi/6$, we have $\psi = 0$ and

$$\mathbf{b}_1 = \mathbf{n}_1, \qquad \mathbf{b}_2 = \frac{\sqrt{3}}{2}\mathbf{n}_2 + \frac{1}{2}\mathbf{n}_3, \qquad \mathbf{b}_3 = -\frac{1}{2}\mathbf{n}_2 + \frac{\sqrt{3}}{2}\mathbf{n}_3.$$

Then, the angular momentum vector in the initial condition, expressed in the base fixed in the inertial reference frame, would be

$$^{\mathcal{R}}\mathbf{H}^{C/C^*} = \frac{1}{4}mr\sqrt{gr}\,(5\mathbf{n}_1 - \sqrt{3}\mathbf{n}_2 + 3\mathbf{n}_3).$$

If $\theta = (\pi/2)$, we have

$$\mathbf{b}_1 = \cos\psi\,\mathbf{n}_1 + \sin\psi\,\mathbf{n}_2, \qquad \mathbf{b}_2 = \mathbf{n}_3, \qquad \mathbf{b}_3 = \sin\psi\,\mathbf{n}_1 - \cos\psi\,\mathbf{n}_2.$$

The angular momentum vector, then, is

$$^{\mathcal{R}}\mathbf{H}^{C/C^*} = \frac{1}{4}mr\sqrt{gr}\left((4\cos\psi + 2\sqrt{3}\sin\psi)\mathbf{n}_1\right.$$
$$\left. + (4\sin\psi - 2\sqrt{3}\cos\psi)\mathbf{n}_2 + 3\mathbf{n}_3\right).$$

Now, not only must the module of the angular momentum vector remain constant, but the vector itself is also invariant; therefore:

$$4\cos\psi + 2\sqrt{3}\sin\psi = 5;$$
$$4\sin\psi - 2\sqrt{3}\cos\psi = -\sqrt{3}.$$

The result then is that $\psi = 0.38025$ rad (check). See the corresponding animation.

Now let us see what happens if a torque with constant module T and parallel to axis x_1 is applied, during a time interval τ, to a rotor that is turning with spin ω_0, null precession, and also null nutation. When there is a nonnull torque component in this direction, Euler's equations become as follows:

$$I(\dot{\omega}_1 + \dot{\phi}\omega_2) + (J - I)\omega_3\omega_2 = T;$$
$$I(\dot{\omega}_2 - \dot{\phi}\omega_1) + (I - J)\omega_1\omega_3 = 0; \qquad (2.32)$$
$$J\dot{\omega}_3 = 0.$$

There is, in this case, no conservation of the angular momentum vector of the rotor with respect to its mass center, since the resultant moment is no longer null. There will, then, be a variation in the angular momentum vector, in the direction of the applied torque, as established by the equation of motion of the second kind. Figure 2.7 illustrates what happens in a short time interval $\Delta\tau$. The original angular momentum vector is parallel to the rotor axis; the application of the torque in the x_1-direction generates a small component $\Delta\mathbf{H}$, additional to the original angular momentum vector, in the $\mathbf{b}_1$-direction. See the corresponding animation.

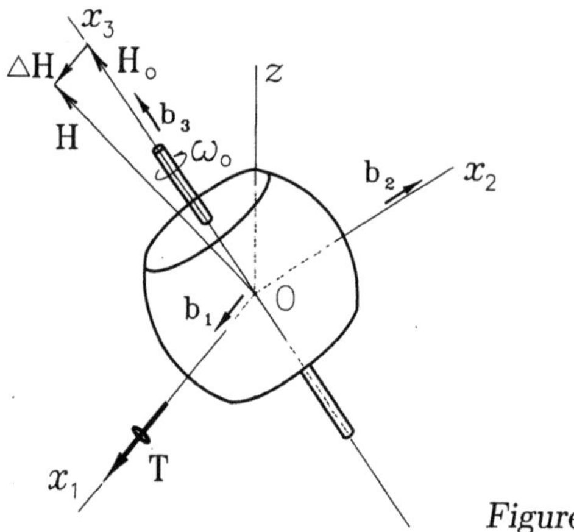

Figure 2.7

Taking a look, then, at the expression of the angular momentum vector, Eq. (2.23), it is evident that a component ω_1 will appear in the angular velocity vector. This is an apparently natural result: The presence of a torque component orthogonal to the original angular velocity vector produces, after a time interval, a new angular velocity component in the direction of the applied torque.

So let us return to Eqs. (2.32). The third equation is identical to that studied previously and, integrated directly, ensures that the component ω_3 remains constant in time. The second is also identical to the former in the case of a null resultant moment. Algebraically resolving for $\dot{\omega}_2$ leads to

$$\dot{\omega}_2 = \left(\dot{\phi} + \frac{J - I}{I}\omega_3 \right) \omega_1, \qquad (2.33)$$

showing that an angular velocity component will also appear in the direction of axis x_2, since ω_1 is now different from zero. This phenomenon, surprising at first sight, is nothing but the fact that, given a body that turns with constant angular velocity in a given fixed direction when a torque is applied in a direction orthogonal to the former, generates an angular velocity component in the direction orthogonal to the other two; it is the so-called gyroscopic effect.

The complete analysis of the gyroscopic effect, of course, requires the integration of Eqs. (2.32). Yet, as in the case of the free gyroscope, we can extract useful information from conservation equations. If torque $\mathbf{T}$ continues to be applied in the direction of axis x_1, there will be a conservation of the rotor angular momentum with respect to axis z. This means that Eqs (2.26), (2.27), and (2.28) remain valid, and the precession and spin may be easily determined as a function of the inclination θ. On the other hand, if the torque has a constant module T, its work done on the rotor in a given interval is, according to Eq. (7.3.1),

$$^{\mathcal{R}}\mathcal{T}_{12}^{T} = \int_{t_1}^{t_2} \mathbf{T} \cdot {}^{\mathcal{R}}\boldsymbol{\omega}^{C}\, dt = T \int_{t_1}^{t_2} \omega_1\, dt = T \int_{\theta_0}^{\theta} d\theta = T(\theta - \theta_0). \quad (2.34)$$

If K_0 is the initial kinetic energy of the rotor, the energy balance is, according to Eq. (7.4.1),

$$\frac{1}{2}\left(I\omega_1^2 + I\omega_2^2 + J\omega_3^2 \right) - K_0 = T(\theta - \theta_0). \qquad (2.35)$$

Resolving, then, for ω_1 and substituting in the first of Eqs. (2.6), we obtain an expression for the nutation as a function of the inclination θ, as follows:

$$\dot{\theta}(\theta) = \sqrt{\frac{1}{I}\left(2K_0 + 2T(\theta - \theta_0) - I\omega_2^2 - J\omega_3^2\right)}, \qquad (2.36)$$

where ω_2 is given by Eq. (2.26).

Example 2.3 Returning to the previous example (see Fig. 2.6), consider the same given initial conditions; now a torque $\mathbf{T} = (3/\pi)\, mgr\mathbf{b}_1$, however, is applied to the rotor. In this case, what is the rotor angular velocity when the axis of symmetry passes through the horizontal position? Well, since Eqs. (2.22), (2.26), (2.27), and (2.28) continue to be valid, we have, as in the previous example,

$$\dot{\psi}(\pi/2) = 3\sqrt{\frac{g}{r}} \qquad \text{and} \qquad \dot{\phi}(\pi/2) = \sqrt{\frac{3g}{r}},$$

therefore,

$$\omega_2(\pi/2) = 3\sqrt{\frac{g}{r}} \qquad \text{and} \qquad \omega_3(\pi/2) = \sqrt{\frac{3g}{r}}.$$

In order to determine the nutation, we now employ Eq. (2.36), resulting in (verify)

$$\dot{\theta}(\pi/2) = \omega_1(\pi/2) = 2\sqrt{\frac{6g}{r}}.$$

The angular velocity vector in the required situation is, then,

$$^{\mathcal{R}}\boldsymbol{\omega}^C = \sqrt{\frac{3g}{r}}\left(\sqrt{8}\mathbf{b}_1 + \sqrt{3}\mathbf{b}_2 + \mathbf{b}_3\right).$$

See the corresponding animation.

Quite an interesting situation of gyroscopic motion occurs when a body of revolution C has any point O of its axis of symmetry, non-coinciding with its mass center, fixed in an inertial reference frame $\mathcal{R}$, moving under the action of its own weight. This is, for example, the case of a top moving while keeping its tip fixed on a horizontal plane with friction. This is, then, a case of motion with a fixed point with three degrees of freedom and nonnull resultant moment with respect to the fixed point. The best description for the motion of a top or a similar object is also that given by Euler's coordinates.

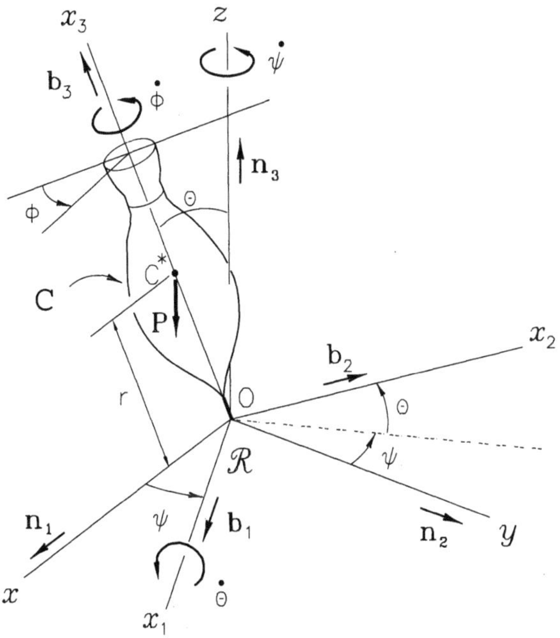

Figure 2.8

Figure 2.8 illustrates a body of revolution C moving with a point O of its axis of symmetry fixed in an inertial reference frame $\mathcal{R}$. The system of Cartesian axes $\{x, y, z\}$, with origin in O, and the orthonormal basis $\mathbf{n}_1, \mathbf{n}_2, \mathbf{n}_3$ are fixed in $\mathcal{R}$. The attitude (orientation) of C with respect to the fixed axes in $\mathcal{R}$ is entirely determined by the angular coordinates $\phi(t)$, $\psi(t)$, and $\theta(t)$; ϕ measures the rotation of C around x_3, its axis of symmetry; ψ measures the rotation of the vertical plane containing axes x_1 and z, around the latter; and θ measures the inclination of axis x_3 with respect to the vertical. As mentioned previously, the rate $\dot{\phi}$ is called *spin*, the rate $\dot{\psi}$ is called *precession*, and the rate $\dot{\theta}$ is called *nutation*.

The kinematic description of the top is, thus, identical to that of the free gyroscope. Its angular velocity vector may be expressed, therefore, according to Eqs. (2.5) and (2.6). Likewise, its angular acceleration vector may be described according to Eqs. (2.7) and (2.8). The resultant moment with respect to point O is the moment applied by the weight force, that is,

$$\mathbf{M}^{\mathcal{F}_e/O} = r\mathbf{b}_3 \times (-mg)\mathbf{n}_3 = mgr\sin\theta\,\mathbf{b}_1. \qquad (2.37)$$

So, adopting I and J as the transversal and longitudinal principal moments of inertia, respectively, with respect to point O, we have Euler's equations for a body of revolution in motion with a fixed point under the action of the weight itself:

$$I(\dot{\omega}_1 + \dot{\phi}\omega_2) + (J - I)\omega_3\omega_2 = mgr\sin\theta;$$
$$I(\dot{\omega}_2 - \dot{\phi}\omega_1) + (I - J)\omega_1\omega_3 = 0; \qquad (2.38)$$
$$J\dot{\omega}_3 = 0.$$

Note that Eqs. (2.38) are similar to Eqs. (2.32), with the difference from the latter that the torque in the direction of x_1 is no longer constant, varying with the inclination θ, and by the moments of inertia, now with respect to point O.

The same equations can be expressed in terms of Euler's coordinates. Indeed, substituting Eqs. (2.6) and (2.8) in Eqs. (2.38) leads to

$$I(\ddot{\theta} - \dot{\psi}^2\sin\theta\cos\theta) + J(\dot{\psi}\cos\theta + \dot{\phi})\dot{\psi}\sin\theta = mgr\sin\theta,$$
$$I(\ddot{\psi}\sin\theta + 2\dot{\psi}\dot{\theta}\cos\theta) - J(\dot{\psi}\cos\theta + \dot{\phi})\dot{\theta} = 0, \qquad (2.39)$$
$$\ddot{\psi}\cos\theta - \dot{\psi}\dot{\theta}\sin\theta + \ddot{\phi} = 0.$$

The determination of the time evolution of Euler's coordinates for initial arbitrary conditions is extremely difficult, given the complexity of the coupled system of equations, requiring numerical integration. However, we can extract valuable information on the general behavior of the body, such as the evolution of the spin, precession and nutation components as a function of the inclination θ, as was similarly done for the free gyroscope.

Returning to Eqs. (2.38), the third equation is obviously uncoupled and can be integrated in time, resulting in

$$\omega_3 = \omega_0, \qquad (2.40)$$

that is, the scalar component of the angular velocity vector in the direction of the axis of symmetry for the body remains constant during motion.

The resultant moment with respect to axis z is null; then, the angular momentum of the body with respect to the vertical axis is conserved. The balance expressed by Eq. (2.25) is, therefore, applicable, resulting in ω_2, according to Eq. (2.26), $\dot{\psi}$, according to Eq. (2.27), and $\dot{\phi}$, according to Eq. (2.28).

To obtain an expression also for the nutation, we can use the energy balance. In the absence of nonconservative forces, the mechanical energy of the body remains constant in time. Admitting that E_0 is the initial mechanical energy, then in a generic instant, we have

$$\frac{1}{2}\left(I\omega_1^2 + I\omega_2^2 + J\omega_3^2\right) + mgr\cos\theta = E_0. \tag{2.41}$$

So resolving for $\dot{\theta} = \omega_1$, it leads to

$$\dot{\theta}(\theta) = \sqrt{\frac{1}{I}\left(2E_0 - 2mgr\cos\theta - J\omega_3^2\right) - \omega_2^2}. \tag{2.42}$$

Example 2.4 A homogeneous cone C, with mass m and dimensions indicated, has its vertex bound by a spherical joint to point O, fixed in an inertial reference frame $\mathcal{R}$ (see Fig. 2.9).

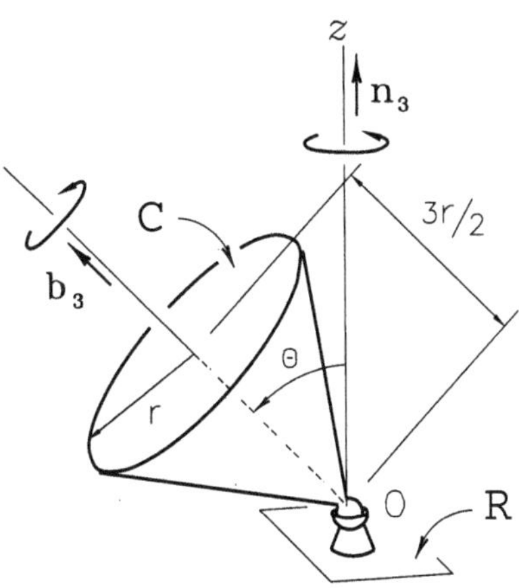

Figure 2.9

The cone is left turning with an angular velocity ω_0 around its axis of symmetry, with the latter in the vertical position. A small disturbance induces the cone to incline slightly, leading it to acquire motion of precession and nutation. It was also found that the nutation motion is periodic, with amplitude $\pi/2$, that means the cone starts the motion with its axis of symmetry in the vertical, sloping until the horizontal position, returning to the initial position, and so on. It is to be expected that the amplitude of this oscillation depends on the initial conditions. We are then going to investigate the value of the initial angular velocity ω_0, which results in the observed behavior. The principal moments of inertia with respect to O for the cone are (see Appendix C)

$$I = \frac{3}{20}mr^2 + \frac{3}{80}m\left(\frac{3}{2}r\right)^2 + m\left(\frac{3}{4}\frac{3}{2}r\right)^2 = \frac{3}{2}mr^2, \qquad J = \frac{3}{10}mr^2.$$

In the initial condition, $\theta = 0$, the time rates of Euler's angles are

$$\dot{\theta} = 0, \qquad \dot{\psi} = 0, \qquad \dot{\phi} = \omega_0.$$

The scalar component of the angular momentum of the body with respect to the vertical axis is

$$H_{z_0} = \mathbf{n}_3 \cdot \mathbf{\Pi}^{C/O} \cdot {}^{\mathcal{R}}\boldsymbol{\omega}^C = \frac{3}{10}mr^2\omega_0.$$

Also in the initial condition, the kinetic energy is

$$K_0 = \frac{1}{2}{}^{\mathcal{R}}\mathbf{H}^{C/O} \cdot {}^{\mathcal{R}}\boldsymbol{\omega}^C = \frac{1}{2}J\omega_3^2 = \frac{3}{20}mr^2\omega_0^2,$$

the potential energy is

$$\Phi_0 = mgz^* = \frac{9}{8}mgr,$$

and the mechanical energy is

$$E_0 = \frac{3}{20}mr^2\omega_0^2 + \frac{9}{8}mgr.$$

When $\theta = \pi/2$, then we have, from Eq. (2.26) (verify),

$$\omega_2 = \frac{1}{5}\omega_0.$$

Substituting the above results in Eq. (2.42) leads to (check)

$$\dot{\theta} = \sqrt{\frac{3g}{2r} - \frac{\omega_0^2}{25}}.$$

Now, if $\theta = \pi/2$ corresponds to the amplitude of the nutation motion, then, necessarily, $\dot{\theta} = 0$ in this position. Hence the result is that

$$\omega_0 = 5\sqrt{\frac{3g}{2r}}$$

is the initial value that must be given to the angular velocity of the cone to provide the nutation motion with the described amplitude. The precession will also be of an oscillating nature; in fact, starting from a null initial value, Eq. (2.27) shows that, for the position with the horizontal axis of symmetry,

$$\dot{\psi} = \frac{1}{5}\omega_0 = \sqrt{\frac{3g}{2r}}.$$

Last, in the horizontal position, Eq. (2.28) provides

$$\dot{\phi} = \omega_0.$$

Note that the value obtained for the spin when $\theta = \pi/2$ is equal to its initial value. Do you think that the spin remains constant during motion? Clear up the doubt by calculating it for an arbitrary position. See the corresponding animation.

The general solution of Eqs. (2.38) or (2.39) depends heavily on the initial conditions prescribed. It may be demonstrated that, as in the case of the free gyroscope, there is a solution in which the body moves with *null nutation*. We are then going to study, below, how this motion occurs and verify which initial conditions are necessary. So, admitting that $\theta = \theta_0$ is a possible solution for Eqs. (2.38), we have $\dot{\theta} = \ddot{\theta} = 0$; therefore, $\omega_1 = \dot{\omega}_1 = 0$. So, substituting in those equations yields

$$I\dot{\phi}\omega_2 + (J - I)\omega_3\omega_2 = mgr\sin\theta,$$
$$I\dot{\omega}_2 = 0, \tag{2.43}$$
$$J\dot{\omega}_3 = 0.$$

Integrating the two last equations leads to

$$\omega_3 = \omega_0 \quad \text{and} \quad \omega_2 = \Omega, \tag{2.44}$$

where ω_0 and Ω are constants depending on the initial conditions. It then results that, for constant θ, $\dot{\psi}$ is constant and, consequently, $\dot{\phi}$ is also constant. In other words, there is a possible solution for the motion of the top with precession and spin constants and null nutation. The relation between the rates that provide this solution may be extracted from the first of the Eqs. (2.43). So rearranging the equation, we have, for $\theta \neq 0$,

$$(J - I)\cos\theta\,\dot{\psi}^2 + J\dot{\phi}\dot{\psi} - mgr = 0, \tag{2.45}$$

a second-order algebraic equation (don't forget that θ, $\dot{\phi}$, and $\dot{\psi}$ are now constants in time), whose solutions we are now going to analyze.

Let us first assume that $J = I$; in this special condition Eq. (2.45) is reduced to

$$\dot{\psi} = \frac{mgr}{J\dot{\phi}}, \tag{2.46}$$

that is, precession is inversely proportional to the spin, the relation between both being indpendent of θ. This occurs, and in this special case alone, because the body is inertially equivalent to a sphere, the direction around which it turns being indifferent.

If, however, $J \neq I$, there is a pair of roots, as follows:

$$\dot{\psi} = \frac{-J\dot{\phi} \pm \sqrt{\Delta}}{2(J - I)\cos\theta}, \tag{2.47}$$

where

$$\Delta = J^2\dot{\phi}^2 + 4(J - I)mgr\cos\theta. \tag{2.48}$$

Then there are two real roots if $\Delta \geq 0$, that is,

$$|\dot{\phi}| \geq \frac{2}{J}\sqrt{(I - J)mgr\cos\theta}. \tag{2.49}$$

Note that, according to Eq. (2.48), Δ is always nonnegative if $(J - I)\cos\theta \geq 0$, in which case there is no restriction for the spin value. When, on the contrary, $(J - I)\cos\theta < 0$, the spin shall satisfy the minimum condition established in Eq. (2.49).

Example 2.5 A washer C consists of a disk with mass m and radius r welded to a light slender shaft with length $2a$ (see Fig. 2.10). The end O of the shaft is fixed by a rotula to an inertial support S. Admitting that the body moves with constant precession and spin, we would like to determine the possible values for precession $\dot{\psi}$ if the spin is $\dot{\phi} = 2\sqrt{g/r}$ and the inclination of the axis of symmetry is $\theta = \pi/3$.

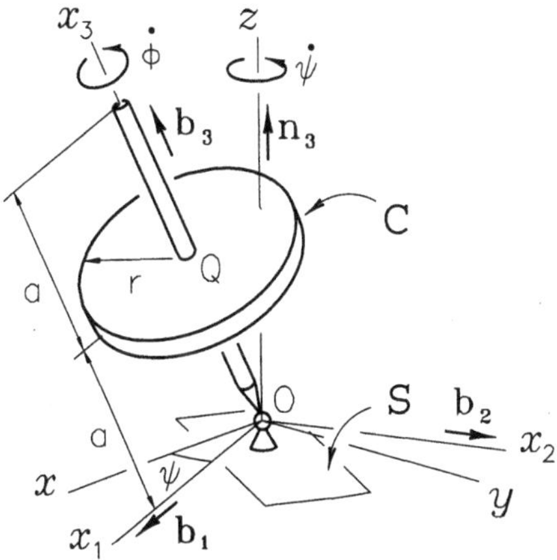

Figure 2.10

The only force that generates moment with respect to point O is the weight of the washer; therefore,

$$\mathbf{M}^{\mathcal{F}_e/O} = \mathbf{p}^{Q/O} \times \mathbf{P} = a\mathbf{b}_3 \times (-mg)\mathbf{n}_3 = mga\sin\theta\,\mathbf{b}_1.$$

The principal moments of inertia of the washer with respect to point O are

$$I = I_1^O = I_2^O = \frac{1}{4}mr^2 + ma^2 \qquad \text{and} \qquad J = I_3^O = \frac{1}{2}mr^2.$$

Let us initially look at what happens if $a = r/2$. In this case, $J = I$ and the inertia tensor is reduced to $\mathbf{II}^{C/O} = \frac{1}{2}mr^2\,\mathbb{1}$. Therefore, from Eq. (2.46), we have

$$\dot{\psi} = \frac{mg\frac{r}{2}}{\frac{1}{2}mr^2 2\sqrt{g/r}} = \frac{1}{2}\sqrt{\frac{g}{r}}.$$

The washer then presents a constant spin of module $2\sqrt{g/r}$ and constant precession of module $\frac{1}{2}\sqrt{g/r}$, regardless of the angle θ. Note that, in fact,

$$^S\mathbf{H}^{C/O} = \mathbf{II}^{C/O} \cdot {}^S\boldsymbol{\omega}^C = \frac{1}{2}mr^2(\dot{\phi}\mathbf{b}_3 + \dot{\psi}\mathbf{n}_3),$$

so

$$^{S}\dot{\mathbf{H}}^{C/O} = \frac{1}{2}mr^2\dot{\phi}\dot{\psi}\mathbf{n}_3 \times \mathbf{b}_3 = \frac{1}{2}mgr\sin\theta\mathbf{b}_1 = mga\sin\theta\mathbf{b}_1 = \mathbf{M}^{\mathcal{F}_e/O}.$$

Assuming now that $a = r/4$, then $I = \frac{5}{16}mr^2$, and so $J - I = \frac{3}{16}mr^2$. Since $(J - I)\cos\theta > 0$, from Eq. (2.48) we then have

$$\Delta = \frac{1}{4}m^2r^4 \cdot 4\frac{g}{r} + 4 \cdot \frac{3}{16}mr^2 \cdot mg\frac{r}{4} \cdot \frac{1}{2} = \frac{35}{32}(mr^2)^2\frac{g}{r}$$

and, from Eq. (2.47), we have

$$\dot{\psi} = \frac{16}{3}\left(-1 \pm \sqrt{\frac{35}{32}}\right)\sqrt{\frac{g}{r}}.$$

The roots, for this geometric relationship, are, therefore,

$$\dot{\psi} = 0.2444\sqrt{\frac{g}{r}} \qquad \text{and} \qquad \dot{\psi} = -10.911\sqrt{\frac{g}{r}}.$$

Note that there is a solution with a precession module lower than that of the spin and another of a greater module in the opposite direction. Let us now see what happens with, say, $a = 3r/4$. In this case, $I = \frac{13}{16}mr^2$ and $J - I = -\frac{5}{16}mr^2$. Now, $(J - I)\cos\theta < 0$, and Eq. (2.49) must be verified. In fact,

$$\frac{2}{J}\sqrt{(I - J)mgr\cos\theta} = \frac{2}{\frac{13}{16}mr^2}\sqrt{\frac{5}{16}mr^2 \cdot mg\frac{3r}{4}\frac{1}{2}} = 0.84265\sqrt{\frac{g}{r}} < \dot{\phi}.$$

So, we have

$$\Delta = \frac{3}{8}(mr^2)^2\frac{g}{r},$$

therefore,

$$\dot{\psi} = \frac{16}{5}\left(1 \pm \sqrt{\frac{3}{8}}\right)\sqrt{\frac{g}{r}},$$

resulting in the following values for precession:

$$\dot{\psi} = 1.2404\sqrt{\frac{g}{r}} \qquad \text{and} \qquad \dot{\psi} = 5.1596\sqrt{\frac{g}{r}}.$$

In this case, therefore, we also have a "low" and another "high" precesssion, but both with the same sign. See the corresponding animation.

As mentioned before, Eqs. (2.39) require numerical integration, their solution depending on the initial conditions of the problem. We saw in Example 2.5 possible solutions, depending on the geometry, for the permanent motion of the top, that is, the motion with null nutation and constant precession and spin. The reader who has already noticed, however, the motion of a top knows that it does not behave as described in that example. This difference in behavior is due to the existence of a friction torque at the tip of the top, resulting in no conservation of the angular momentum of the body with respect to the vertical axis and, since this resistive torque does work, the kinetic energy of the top is also not conserved. In this case, therefore, Eqs. (2.25) and (2.41) do not apply and it is no longer possible to analytically determine $\dot{\phi}$, $\dot{\theta}$, and $\dot{\psi}$ as a function of θ.

Thus, assuming the presence of a resistive torque $\mathbf{T} = -T\mathbf{b}_3$, and of constant module (see Fig. 2.8), Eqs. (2.39) are converted into

$$I(\ddot{\theta} - \dot{\psi}^2 \sin\theta \cos\theta) + J(\dot{\psi}\cos\theta + \dot{\phi})\dot{\psi}\sin\theta = mgr\sin\theta,$$

$$I(\ddot{\psi}\sin\theta + 2\dot{\psi}\dot{\theta}\cos\theta) - J(\dot{\psi}\cos\theta + \dot{\phi})\dot{\theta} = 0, \qquad (2.50)$$

$$J(\ddot{\psi}\cos\theta - \dot{\psi}\dot{\theta}\sin\theta + \ddot{\phi}) = -T.$$

The numerical solution of these equations will lead to the observed behavior, as the following example illustrates.

Example 2.6 We will now consider the washer in the previous example (see Fig. 2.10), with $a = 3r/4$, as analyzed at the end of that example, and let us now suppose a small friction torque $T = 0.1$ Nm, applied toward $\mathbf{b}_3$, in the opposite direction to that of the spin. Admitting then radius $r = 0.1$ m, mass $m = 1$ kg, the transversal and longitudinal moments of inertia, $I = \frac{13}{16}mr^2$ and $J = \frac{1}{2}mr^2$, respectively, we can numerically integrate the equations. Establishing initial conditions for the inclination angle of the washer axis and for the spin, identical to those in the previous example ($\theta = \pi/3$, $\dot{\phi} = 2\sqrt{g/r}$), as well as fixing the low precession value as an initial condition for the permanent motion of the washer, $\dot{\psi} = 1.2404\sqrt{g/r}$, Eqs. (2.50) can be numerically integrated, obtaining, for the time evolution of θ, $\dot{\phi}$ and $\dot{\psi}$, the result illustrated in the figures ahead.

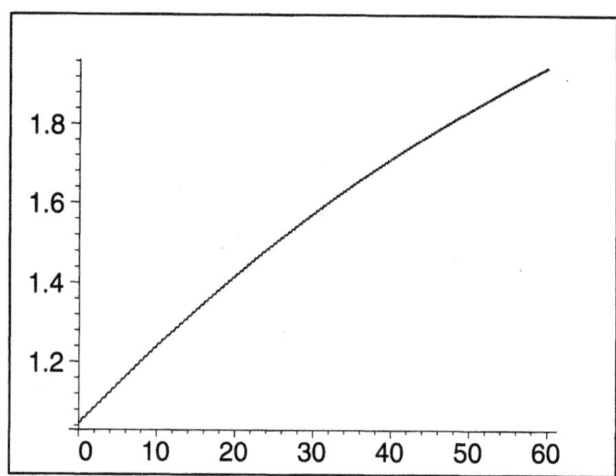

Figure 2.11

Figure 2.11 shows the evolution of the angle θ during the first minute of motion. Note that the inclination increases continuously with time, starting from the value $\theta = \pi/3 = 1.047$ rad in $t = 0$ until approximately $\theta = 1.31$ rad in $t = 60$ s. Also note that this increase does not occur according to a linear function. There is, therefore, an almost constant nutation that derived from the function $\theta(t)$, as shown, that is, due to the friction torque present in point O, in the direction of the body axis of symmetry, the top "drops."

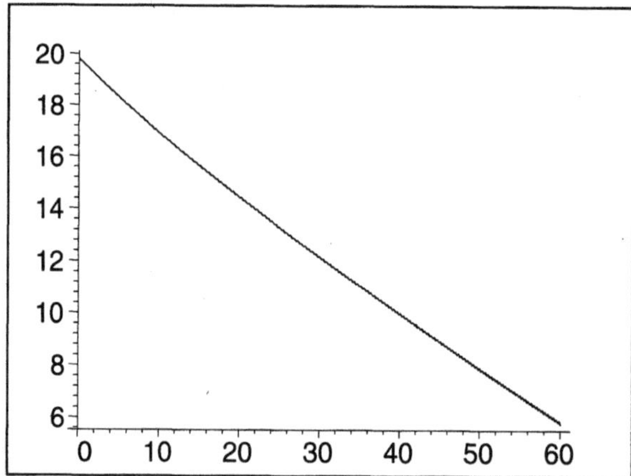

Figure 2.12

Figure 2.12 illustrates the evolution of the spin, showing that this continuously decreases, although not in a linear way, from its initial value,

$\dot{\phi} = 2\sqrt{g/r} \approx 20$ rad/s, until a value of approximately $\dot{\phi} = 16.6$ rad/s, in $t = 60$ s. Last, Fig. 2.13 shows the behavior of the precession during one minute, showing that it increases continuously with time, from its initial value, $\dot{\psi} = 1.2404\sqrt{g/r} \approx 12.404$ rad/s in $t = 0$, until an approximate value of $\dot{\psi} = 13.66$ rad/s in $t = 60$ s. The simulated behavior is according to that observed in the motion of a real top, where there is indeed a small nutation and where the precession is increasing as the spin drops, all these effects provided by the small friction torque present in point O.

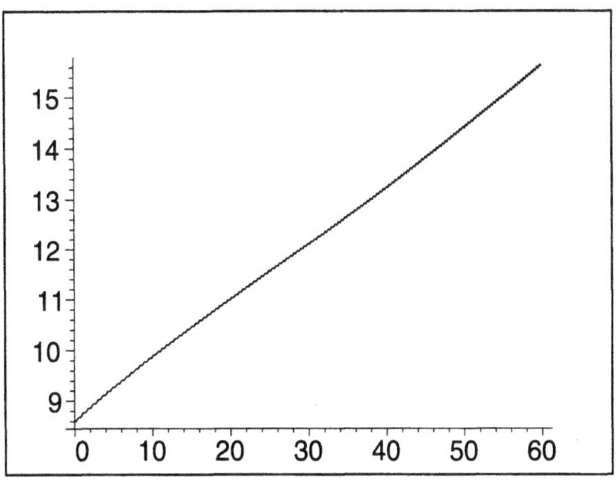

Figure 2.13

See the corresponding animation.

8.3 General Motion

When the motion of a rigid body in an inertial reference frame does not fit any of the categories discussed herein above, that is, if there is no plane motion or motion with a fixed point nor any particular case of such, it is said that the body describes a *general motion* in the reference frame. The equations governing the general motion are the equation of motion of the first kind, Eq. (7.2.2) or Eq. (7.2.3), and the equation of motion of the second kind, Eq. (7.2.7).

A system of Cartesian coordinates with an arbitrary orientation having been chosen, the equation of motion of the first kind can be resolved, as we saw, in three scalar equations, such as Eqs. (7.2.4). Having arbitrated a system of Cartesian coordinates *fixed* in the body

— which may, eventually, be the same as that chosen for the decomposition of the equation of motion of the first kind —, Eq. (7.2.7) may also be resolved in three scalar equations, such as Eqs. (7.2.9), or, in the case of axes parallel to the principal directions of inertia with respect to the mass center, Eqs. (7.2.10). In short, we have a set of six nonlinear dynamic equations, generally independent of each other.

A rigid body has six degrees of freedom, there being no constraints to its motion. So, choosing a set of six generalized coordinates for the description of this motion, we have a parity between the number of equations and the unknown quantities.

Example 3.1   Consider a coin M, modeled as a homogenous slim disk with mass m and radius r, thrown in space turning with angular velocity $^{\mathcal{R}}\boldsymbol{\omega}^M = \omega_0 \mathbf{n}_1$, around its cross axis y_1, coinciding, at this instant, with axis x_1, and its center O having a speed $\mathbf{v}_0 = u_1\mathbf{n}_1 + u_2\mathbf{n}_2 + u_3\mathbf{n}_3$, where basis $\mathbf{n}_1, \mathbf{n}_2, \mathbf{n}_3$ is fixed in an inertial reference frame (see Fig. 3.1). This is a holonomic system with six degrees of freedom, since there are no applicable kinematic constraints.

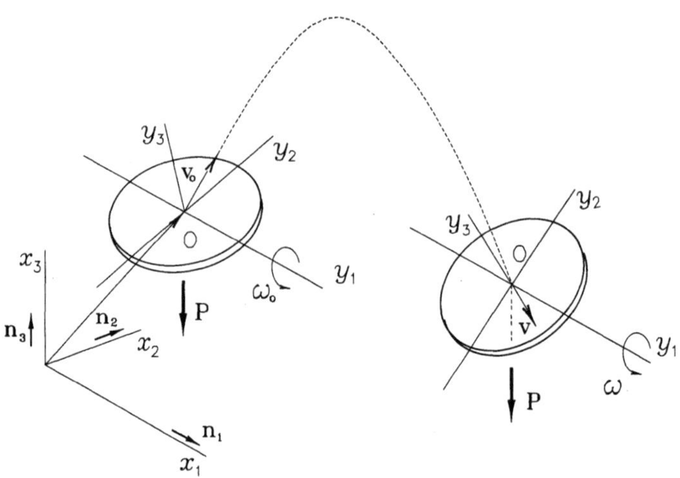

Figure 3.1

The coordinates $x_1(t)$, $x_2(t)$, and $x_3(t)$, describing the position of the coin center with respect to the origin of the Cartesian system, and $\theta_1(t)$, $\theta_2(t)$, and $\theta_3(t)$, describing successive rotations of the coin around the axes $\{x_1, x_2, x_3\}$, respectively, are sufficient to characterize the configuration of the body at any instant. The only force applied is the weight, $\mathbf{P} = -mg\mathbf{n}_3$. The resultant moment with respect to point O is, therefore, null. The equations of motion of the first kind are

$$m\ddot{x}_1 = 0,$$
$$m\ddot{x}_2 = 0,$$
$$m\ddot{x}_3 = -mg,$$

whose solution, given the initial conditions, is

$$x_1(t) = x_1(0) + u_1 t,$$
$$x_2(t) = x_2(0) + u_2 t,$$
$$x_3(t) = x_3(0) + u_3 t - \frac{1}{2}gt^2,$$

that is, the coin mass center moves with constant speed, in the directions parallel to x_1 and x_2, and is uniformly decelerated, in the direction parallel to x_3, describing, thereby, a parabolic trajectory in a vertical plane. As the resultant moment is null, the angular momentum vector is conserved and, the initial angular velocity being parallel to a principal direction of inertia, then, at any instant,

$$^{\mathcal{R}}\mathbf{H}^{M/O}(t) = {}^{\mathcal{R}}\mathbf{H}^{M/O}(0) = I\omega_0\mathbf{n}_1,$$

where I is the moment of inertia of the disk with respect to axis y_1. The coin continues, therefore, to turn at the speed ω_0 around axis y_1, which remains parallel to its original direction. Thus,

$$\theta_1(t) = \theta_1(0) + \omega_0 t, \qquad \theta_2(t) = \theta_3(t) = 0.$$

Each constraint imposed on the general motion of a rigid body reduces its number of degrees of freedom in the exact proportion in which it introduces, as unknown quantities, components of force or torque present in the linkage corresponding to the constraint. When the constraint is holonomic, it is always possible to express a coordinate as a function of the others, thus reducing the number of unknown quantities. On the other hand, every kinematic constraint, being the result of a linkage, introduces, in turn, an unknown force or torque. It is therefore necessary to have six dynamic equations to determine the mutually independent coordinates and the unknown linkage components. When, on the contrary, the kinematic constraint is not holonomic, it is not possible to directly obtain a coordinate as a function of the others and, moreover, new unknown quantities are introduced by the nonholonomic linkages. We then have a coupled system of equations, in which the six second-order dynamic equations appear, plus a set of first-order equations, in equal number to that of the nonholonomic constraints present. In many cases, nevertheless, it is possible to choose coordinates that partially uncouple the system of equations, in order only to solve a subsystem of greater interest. In short, when only one portion of the coordinates is of interest, a subset of equations may normally be solved, provided that they do not involve the other unknown equations.

Example 3.2   A homogeneous sphere E, with mass m and radius r, moves inside a vertical pipe with radius R, rolling on its inside surface (see Fig. 3.2).

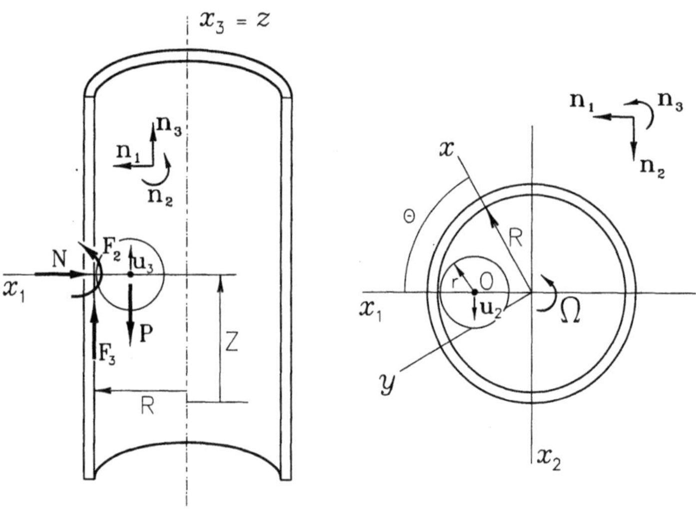

Figure 3.2

Given an initial condition of velocity for the sphere center, we want to study how its position progresses in time. This is a nonholonomic system, in principle, since it involves a rolling condition, with three degrees of freedom. In fact, we have a holonomic constraint requiring contact between the surfaces and two non-holomonic constraints to guarantee the nonsliding between them. Axes $\{x_1, x_2, x_3\}$ are fixed in the plane that contains the pipe axis of symmetry and, at each instant, center O of the sphere; basis $\mathbf{n}_1, \mathbf{n}_2, \mathbf{n}_3$ is parallel to these axes. We may choose, as coordinates for describing the motion, variables $z(t)$ and $\theta(t)$, which determine the position of O, and three angles to characterize their attitude. Since the body is spherical, its spatial orientation is indifferent or, at least, irrelevant. It is interesting for us, therefore, to essentially determine the position of the sphere mass center. The applied external forces are the weight, $\mathbf{P} = -mg\mathbf{n}_3$, and the components in the contact, N, F_2, and F_3, as shown. The reduction to the mass center of this force system is

$$\mathbf{F} = -N\mathbf{n}_1 + F_2\mathbf{n}_2 + (F_3 - mg)\mathbf{n}_3,$$
$$\mathbf{M}^{\mathcal{F}_{e}/O} = -rF_3\mathbf{n}_2 + rF_2\mathbf{n}_3.$$

The inertia tensor of the sphere with respect to O is (see Appendix C)

$$\mathbb{II}^{E/O} = I\,\mathbb{1}, \qquad \text{where} \quad I = \frac{2}{5}mr^2.$$

The angular velocity vector of the sphere in relation to the pipe, expressed in the same basis, is

$$\boldsymbol{\omega}^E = \omega_1\mathbf{n}_1 + \omega_2\mathbf{n}_2 + \omega_3\mathbf{n}_3.$$

The time rates of the basis are

$$\dot{\mathbf{n}}_1 = \dot{\theta}\mathbf{n}_2, \qquad \dot{\mathbf{n}}_2 = -\dot{\theta}\mathbf{n}_1, \qquad \dot{\mathbf{n}}_3 = 0.$$

The angular acceleration vector is then

$$\boldsymbol{\alpha}^E = (\dot{\omega}_1 - \dot{\theta}\omega_2)\mathbf{n}_1 + (\dot{\omega}_2 + \dot{\theta}\omega_1)\mathbf{n}_2 + \dot{\omega}_3\mathbf{n}_3.$$

The velocity of point O, expressed in terms of the point coordinates, is

$$\mathbf{v}^O = (R - r)\dot{\theta}\mathbf{n}_2 + \dot{z}\mathbf{n}_3.$$

The acceleration of the mass center then is

$$\mathbf{a}^O = (R - r)(-\dot{\theta}^2\mathbf{n}_1 + \ddot{\theta}\mathbf{n}_2) + \ddot{z}\mathbf{n}_3.$$

The velocity of the point of contact C is

$$\mathbf{v}^C = \mathbf{v}^O + \boldsymbol{\omega}^E \times \mathbf{p}^{C/O}$$
$$= \big((R - r)\dot{\theta} + r\omega_3\big)\mathbf{n}_2 + (\dot{z} - r\omega_2)\mathbf{n}_3.$$

Now, the rolling condition requires that this velocity is null, which results in the following relationships:

$$\omega_2 = \frac{\dot{z}}{r}; \qquad \omega_3 = -\frac{R - r}{r}\dot{\theta}.$$

We can then write the equations of motion in terms of the selected coordinates. Substituting the above results in Eqs. (7.2.4), we have

$$-m(R - r)\dot{\theta}^2 = -N, \tag{a}$$
$$m(R - r)\ddot{\theta} = F_2, \tag{b}$$
$$m\ddot{z} = F_3 - mg. \tag{c}$$

Since the principal moments of inertia are all equal, the terms of Eqs. (7.2.10) involving differences in moments of inertia are canceled, reducing the equations of motion of the second kind to

$$\dot{\omega}_1 - \dot{\theta}\frac{\dot{z}}{r} = 0, \tag{d}$$
$$\frac{2}{5}m\left(\ddot{z} + r\dot{\theta}\omega_1\right) = -F_3, \tag{e}$$
$$-\frac{2}{5}m\left(R - r\right)\ddot{\theta} = F_2. \tag{f}$$

Note that components ω_2 and ω_3 of the angular velocity vector are expressed in terms of coordinates z and θ, from the rolling condition, and this is why they do not appear in the equations of motion. Component ω_1, however, does not depend on those coordinates, being present in Eqs. (d) and (e). It is interesting to solve the equation system for $z(t)$ and $\theta(t)$, the unknowns ω_1, F_2, F_3, and N being secondary. Let us then consider the following initial conditions as given:

$$\omega_1(0) = 0; \qquad \mathbf{v}^O(0) = u_2\mathbf{n}_2 + u_3\mathbf{n}_3,$$

that is, the sphere rolls initially without pivoting, having a component of velocity in the horizontal direction and another velocity component in the vertical direction. The second condition may also be expressed equivalently by

$$\dot\theta(0) = \Omega, \qquad \text{where} \quad \Omega = \frac{u_2}{R-r} \qquad \text{and} \qquad \omega_2(0) = \frac{\dot z}{r} = \frac{u_3}{r}.$$

From Eqs. (b) and (f), we have

$$\ddot\theta = 0; \qquad \text{then,} \quad \dot\theta = \Omega.$$

Substituting then in Eq. (d), we get

$$\dot\omega_1 = \frac{\Omega}{r}\dot z; \qquad \text{then,} \quad \omega_1 = \frac{\Omega}{r}z.$$

Now solving Eq. (c) for F_3 and substituting in Eq. (e), the following linear differential equation is obtained for coordinate $z(t)$ (check):

$$\ddot z + \frac{2}{7}\Omega^2\, z + \frac{5}{7}g = 0,$$

the general solution of which is (verify: just substitute)

$$z(t) = \frac{r}{\omega_0}\Big(A\sin\omega_0 t + B(1 - \cos\omega_0 t)\Big),$$

where

$$\omega_0 = \sqrt{\frac{2}{7}}\,\Omega, \qquad A = \frac{u_3}{r}, \qquad B = -\frac{5g}{7r\omega_0}.$$

We have, therefore, a periodic solution with frequency

$$f_0 = \frac{1}{2\pi}\sqrt{\frac{2}{7}}\,\Omega$$

for the vertical component of the position of the sphere mass center and a linear solution,

$$\theta(t) = \theta(0) + \Omega t,$$

for its angular displacement. The velocity of the sphere center at a generic instant is, then,

$$\begin{aligned}
\mathbf{v}^O(t) &= (R-r)\dot\theta(t)\mathbf{n}_2 + \dot z(t)\mathbf{n}_3 \\
&= (R-r)\Omega\mathbf{n}_2 + r(A\cos\omega_0 t + B\sin\omega_0 t)\mathbf{n}_3 \\
&= u_2\mathbf{n}_2 + \left(u_3\cos\omega_0 t - \frac{5g}{7\omega_0}\sin\omega_0 t\right)\mathbf{n}_3,
\end{aligned}$$

where ω_0 is defined above. It is interesting to observe that the dynamic solution for the motion of the sphere, in the absence of dissipative forces, is in fact periodic. The work of the weight force is sometimes positive, sometimes negative. If, however, null initial conditions were given, that is, if the sphere is abandoned from rest, we will have $u_2(0) = u_3(0) = 0$ and $\Omega = 0$. The differential equation governing the motion will, in this case, be

$$\ddot{z} + \frac{5}{7}g = 0,$$

whose solution is

$$z(t) = -\frac{5}{14}gt^2,$$

that is, a free-fall motion delayed due to the rotation inertia of the sphere and the work of the weight force will always be positive. See the corresponding animation.

When a rigid body has a general motion, involving nonholonomic kinematic constraints, the equation system may be too complex due, essentially, to its nonlinearity. Moreover, the number of equations — besides the six equations of motion, we will have the nonholonomic constraint equations — makes it even more difficult to obtain a general solution. We then reinforce the importance of selecting a subset of coordinates, when possible, to facilitate the solution. Even so, the equation system might remain inaccessible to an analytical solution, and the only possible way is the numerical integration. Some approximate solutions may be obtained analytically, by admitting, for instance, small values for one or more coordinates. In this case, as done in the study of a simple pendulum, seen in Example 4.2.2, the equations may be *linearized*, to help obtain an analytical solution.

Example 3.3   Let us now consider a coin M, modeled as a thin homogeneous disk with mass m and radius r, which rolls on a flat horizontal surface (see Fig. 3.3). The constraints to the motion of the coin establish that it is always in contact with the surface (a holonomic constraint) and that there is rolling in the contact point (two nonholonomic constraints). Choosing as generalized coordinates the Cartesian coordinates of the coin center, $x(t)$, $y(t)$, and $z(t)$, and the angles describing its spatial orientation, $\phi(t)$, $\theta(t)$, and $\psi(t)$, the holonomic constraint may be expressed by

$$z = r\cos\theta.$$

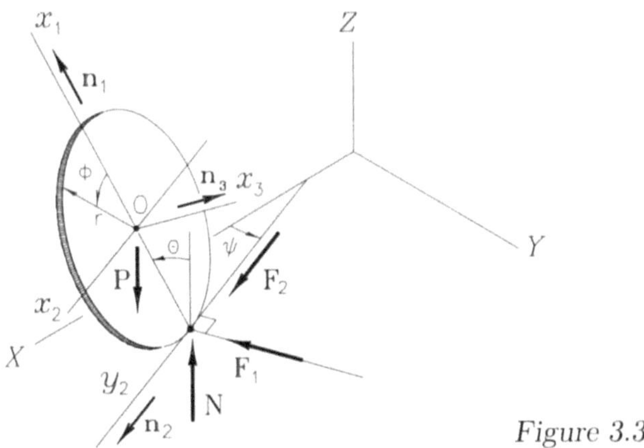

Figure 3.3

The angular velocity of the coin in $\mathcal{R}$ may be expressed by (see Example 3.3.4)

$$^{\mathcal{R}}\boldsymbol{\omega}^M = \dot{\psi}\cos\theta\,\mathbf{n}_1 + \dot{\theta}\,\mathbf{n}_2 + (\dot{\psi}\sin\theta + \dot{\phi})\mathbf{n}_3.$$

The velocity of point O is

$$^{\mathcal{R}}\mathbf{v}^O = \dot{x}\mathbf{n}_x + \dot{y}\mathbf{n}_y - r\dot{\theta}\sin\theta\,\mathbf{n}_z.$$

Writing the velocity of the contact point C as

$$^{\mathcal{R}}\mathbf{v}^C = {}^{\mathcal{R}}\mathbf{v}^O + {}^{\mathcal{R}}\boldsymbol{\omega}^M \times \mathbf{p}^{C/O}$$

and zeroing it, given the rolling condition, the relations obtained are as follows:

$$\dot{x}\sin\psi - \dot{y}\cos\psi - r\dot{\theta}\cos\theta = 0; \tag{a}$$

$$\dot{x}\cos\psi + \dot{y}\sin\psi - r(\dot{\psi}\sin\theta + \dot{\phi}) = 0. \tag{b}$$

(If the reader decides to check the above calculation, it will be found that the scalar component of the rolling condition in the direction of $\mathbf{n}_3$ is identical to the scalar component in the direction of $\mathbf{n}_1$, expressed by Eq. (a), only two equations for nonholonomic constraint remaining, as foreseen.) The external forces acting on the coin are its weight, $\mathbf{P}$, and the force of contact applied by the flat surface. The latter may be opened in a normal component, $\mathbf{N}$, and two friction components, $\mathbf{F}_1$ and $\mathbf{F}_2$, as shown. The reduction of this system of forces to the mass center of the coin is

$$\mathbf{F} = \big((N - mg)\cos\theta + F_1\sin\theta\big)\mathbf{n_1} + F_2\mathbf{n_2}$$
$$+ \big((N - mg)\sin\theta - F_1\cos\theta\big)\mathbf{n_3},$$
$$\mathbf{M}^{\mathcal{F}_e/O} = r(N\sin\theta - F_1\cos\theta)\mathbf{n}_2 - rF_2\mathbf{n}_3.$$

The reader should note that the coin rolling on the surface is a nonholonomic system with $l = 3$ degrees of freedom. The number of generalized coordinates required to describe its motion is $r = 5$ (the z-coordinate was eliminated by a holonomic constraint equation). Moreover, the rolling condition introduced three unknown force components, increasing to eight the number of scalar unknowns. The full analysis of the coin motion for arbitrary initial conditions would involve, therefore, solving a coupled system of first-order differential equations consisting of Eqs. (a) and (b), plus six second-order dynamic equations (first and second kind). The general solution of such a system is too complex; if we wish to know the angular behavior of the body, without bothering to know the location at each instant of the mass center, we may choose to express the acceleration of point O solely in terms of the angular coordinates, reducing the system to the six equations of motion. In fact, the angular acceleration of the coin in reference frame $\mathcal{R}$ can be expressed by

$$\mathcal{R}\boldsymbol{\alpha}^M = \alpha_1 \mathbf{n}_1 + \alpha_2 \mathbf{n}_2 + \alpha_3 \mathbf{n}_3,$$

where (see Example 3.9.3)

$$\alpha_1 = \dot{\omega}_1 + \dot{\phi}\omega_2, \qquad \alpha_2 = \dot{\omega}_2 - \dot{\phi}\omega_1, \qquad \alpha_3 = \dot{\omega}_3.$$

The acceleration of point O can then be expressed by (see, once again, the aforementioned example)

$$\mathcal{R}\mathbf{a}^O = r\left[-\left(\omega_2^2 + \omega_3(\omega_3 - \dot{\phi})\right)\mathbf{n}_1 + (\dot{\omega}_3 + \omega_1\omega_2)\mathbf{n}_2 + (-\dot{\omega}_2 + \omega_3\omega_1)\mathbf{n}_3\right].$$

The inertia tensor of the coin with respect to point O is

$$\mathbf{II}^{M/O} = \begin{pmatrix} I & 0 & 0 \\ 0 & I & 0 \\ 0 & 0 & J \end{pmatrix},$$

where (see Appendix C)

$$I = \frac{1}{4}mr^2 \qquad \text{and} \qquad J = \frac{1}{2}mr^2.$$

Substituting, then, the above results in Eqs. (7.2.4) and Eqs. (7.2.10), the following equation system that governs the coin motion is obtained:

$$-mr\left(\omega_2^2 + \omega_3(\omega_3 - \dot{\phi})\right) = (N - mg)\cos\theta + F_1 \sin\theta; \qquad (c)$$

$$mr\left(\dot{\omega}_3 + \omega_1\omega_2\right) = F_2; \tag{d}$$

$$mr\left(-\dot{\omega}_2 + \omega_3\omega_1\right) = (N - mg)\sin\theta - F_1\cos\theta; \tag{e}$$

$$\dot{\omega}_1 + \omega_2(\omega_3 + \dot{\phi}) = 0; \tag{f}$$

$$\frac{1}{4}mr\left(\dot{\omega}_2 - \omega_1(\omega_3 + \dot{\phi})\right) = N\sin\theta - F_1\cos\theta; \tag{g}$$

$$\frac{1}{2}mr\,\dot{\omega}_3 = -F_2. \tag{h}$$

This is a compact form to express the equations of motion; to obtain the general form of the equations, just substitute the angular velocity vector components and its derivatives in terms of $\theta(t)$, $\phi(t)$, and $\psi(t)$. Of course, a general analytical solution for this equation system from arbitrary initial conditions is quite difficult. We can, however, obtain very interesting particular solutions relatively easily. Let us then assume that the coin begins pure rolling on the plane, that is, with initial angular velocity $^{\mathcal{R}}\boldsymbol{\omega}^M(0) = \Omega\mathbf{n}_3$, with $\mathbf{n}_3$ essentially horizontal (see Fig. 3.4).

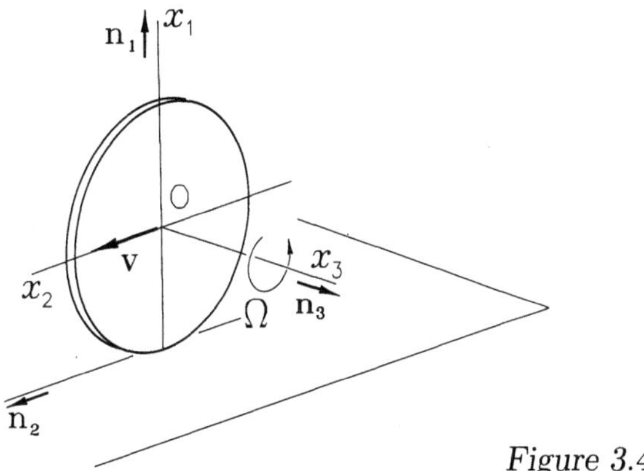

Figure 3.4

The initial conditions given are

$$\theta(0) = \theta_0, \ \dot{\theta}(0) = 0, \ \psi(0) = \psi_0, \ \dot{\psi}(0) = 0, \ \phi(0) = \phi_0, \ \dot{\phi}(0) = \Omega.$$

Let us see if, for this given initial condition, there is a solution in which the coin remains approximately in a vertical plane with invariant orientation, that is, a solution with small θ, $\dot{\theta}$, and $\dot{\psi}$. Neglecting, then, the second-order terms in these variables, we have

$$\omega_1 \approx \dot{\psi}, \qquad \omega_2 = \dot{\theta}, \qquad \omega_3 \approx \dot{\phi},$$

and the system of equations of motion may be linearized, resulting in (verify; it is important to master the linearization technique)

$$0 = N - mg + F_1\theta, \tag{i}$$

$$mr\,\ddot{\phi} = F_2, \tag{j}$$

$$mr\,(\dot{\phi}\dot{\psi} - \ddot{\theta}) = (N - mg)\theta - F_1, \tag{k}$$

$$\ddot{\psi} + 2\dot{\theta}\dot{\phi} = 0, \tag{l}$$

$$\frac{1}{4}mr\,(\ddot{\theta} - 2\dot{\psi}\dot{\phi}) = N\theta - F_1, \tag{m}$$

$$mr\,\ddot{\phi} = -2F_2. \tag{n}$$

From Eqs. (j) and (n) we get

$$\ddot{\phi} = 0; \qquad \text{so,} \qquad \dot{\phi} = \Omega.$$

Equation (l) can then be integrated, resulting in

$$\dot{\psi} = 2\Omega(\theta_0 - \theta).$$

So, removing the term $N\theta - F_1$ from Eqs. (k) and (m) leads to

$$\ddot{\theta} + \frac{4}{5}\left(3\Omega^2 - \frac{g}{r}\right)\theta = \frac{12}{5}\Omega^2\theta_0. \tag{o}$$

Now, Eq. (o) is of the kind $\ddot{u} + \kappa u = \lambda$, the solution of which is harmonic for $\kappa > 0$, that means, for the angular velocity satisfying the condition

$$\Omega > \sqrt{\frac{g}{3r}}, \tag{p}$$

and hyperbolic for $\kappa < 0$. Admitting then that, say, $\Omega = \sqrt{2g/r}$, then

$$\ddot{\theta} + 4\frac{g}{r}\theta = \frac{12}{5}\Omega^2\theta_0,$$

the solution of which is

$$\theta(t) = -\frac{1}{5}\theta_0\cos\left(2\sqrt{\frac{g}{r}}\,t\right) + \frac{6}{5}\theta_0.$$

Substituting in the expression for $\dot{\psi}$ and integrating, then

$$\psi(t) = \psi_0 + \frac{\sqrt{2}}{5}\theta_0\left[\sin\left(2\sqrt{\frac{g}{r}}\,t\right) - 2\sqrt{\frac{g}{r}}\,t\right].$$

The coin behavior is, therefore, oscillatory with natural frequency $f = (1/\pi)\sqrt{g/r}$, around axis y_2 and also oscillatory (with the same frequency) plus a linear term around axis x_1. This motion is said to be *stable*. When the initial angular velocity is insufficient, not satisfying Eq. (p), the motion is *unstable*, the coin plane deviating from the almost-vertical initial orientation.

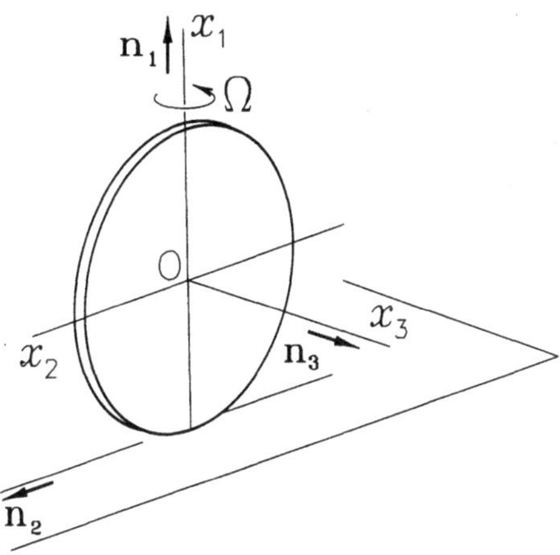

Figure 3.5

Let us now examine another hypothesis of motion, as that illustrated in Fig. 3.5, in which the coin is abandoned pivoting around axis x_1, essentially vertical, that is, the initial angular velocity being given as $^{\mathcal{R}}\boldsymbol{\omega}^M = \Omega\mathbf{n}_1$. It is reasonable to presume, in this case, a motion with small θ, $\dot{\theta}$, and $\dot{\phi}$. The initial conditions, for this case, are

$$\theta(0) = \theta_0,\ \dot{\theta}(0) = 0,\ \psi(0) = \psi_0,\ \dot{\psi}(0) = \Omega,\ \phi(0) = \phi_0,\ \dot{\phi}(0) = 0.$$

The linearized scalar components of the angular velocity vector are

$$\omega_1 \approx \dot{\psi}, \qquad \omega_2 = \dot{\theta}, \qquad \omega_3 \approx \dot{\psi}\theta + \dot{\phi}.$$

The linearized equations of motion are then (once more, it is worth checking)

$$0 = N - mg + F_1\theta; \tag{q}$$

$$mr\left(\dot{\omega}_3 + \dot{\psi}\dot{\theta}\right) = F_2; \tag{r}$$

$$mr\left(-\ddot{\theta} + \dot{\psi}\omega_3\right) = (N - mg)\theta - F_1; \tag{s}$$

$$\ddot{\psi} = 0; \tag{t}$$

$$\frac{1}{4}mr\left(\ddot{\theta} - \dot{\psi}(\omega_3 + \dot{\phi})\right) = N\theta - F_1; \tag{u}$$

$$mr\,\dot{\omega}_3 = -2F_2. \tag{v}$$

Equation (t) can be integrated, resulting in

$$\dot{\psi} = \Omega.$$

So, eliminating F_2 from Eqs. (r) and (v) leads to

$$\dot{\omega}_3 = -\frac{2}{3}\Omega\dot{\theta},$$

which may also be directly integrated, providing

$$\omega_3 = \frac{\Omega}{3}(5\theta_0 - \theta).$$

Then, eliminating the term $N\theta - F_1$ from Eqs. (s) and (u), we obtain a differential equation for $\theta(t)$, as follows:

$$\ddot{\theta} + \left(\Omega^2 - \frac{4g}{5r}\right)\theta = 2\Omega^2\theta_0. \tag{w}$$

We then again have an equation of the kind $\ddot{u} + \kappa u = \lambda$, the solution of which is harmonic for $\kappa > 0$, that is, for the initial angular velocity satisfying the condition

$$\Omega > 2\sqrt{\frac{g}{5r}}, \tag{x}$$

the solution being hyperbolic for the case of $\kappa < 0$. The angular velocity expressed in (x) is then the minimum to prevent the coin from dropping. Then let $\Omega = \sqrt{g/r}$. For this angular velocity, Eq. (w) becomes

$$\ddot{\theta} + \frac{g}{5r}\theta = 2\frac{g}{r}\theta_0,$$

with the solution

$$\theta(t) = -9\,\theta_0 \cos\left(\sqrt{\frac{g}{5r}}\,t\right) + 10\,\theta_0, \tag{y}$$

a periodic solution with natural frequency $f = (1/\pi)\sqrt{g/5r}$. Replacing now Eq. (y) in the expression for the linearized ω_3 and solving for $\dot{\phi}$, we get

$$\dot{\phi} = \frac{5\Omega}{2}(\theta_0 - \theta),$$

which, integrated, results in

$$\phi(t) = 15\,\theta_0 \left[\sqrt{5}\sin\left(\sqrt{\frac{g}{5r}}\,t\right) - \sqrt{\frac{g}{5r}}\,t\right] + \phi_0, \qquad (z)$$

a solution with a periodic behavior of the same frequency $f = (1/\pi)\sqrt{g/5r}$ plus a linear term. The reader has certainly already watched a coin in essentially pure rolling, as in the first hypothesis studied, or in essentially pure pivoting, as in the second, and knows that, in fact, the permanent solutions found here are not obtained. This happens in both cases due to the presence of a slight friction torque in the rolling, which, in practice, does not happen in a single point of contact, as in the adopted model. The result is thus that $\dot{\phi}$, in the first case, and $\dot{\psi}$, in the second, are slightly decreasing. Given, however, an initial value Ω for one or the other, the oscillatory solution for dropping θ is basically the same and the values obtained for the stability condition will still be valid. In other words, the ideal model studied correctly shows us the angular velocities from which the coin begins to irremediably drop. See the corresponding animation.

Example 3.3 shows *stable* solutions — harmonic in this case — and frankly *unstable* solutions. The dynamic stability of the motion of a rigid body requires a separate chapter, so is not discussed herein. We may, nevertheless, introduce, albeit superficially, this concept so that the reader at least acquires some idea on the subject.

The static concept of stability is certainly known to the reader (see Fig. 3.6).

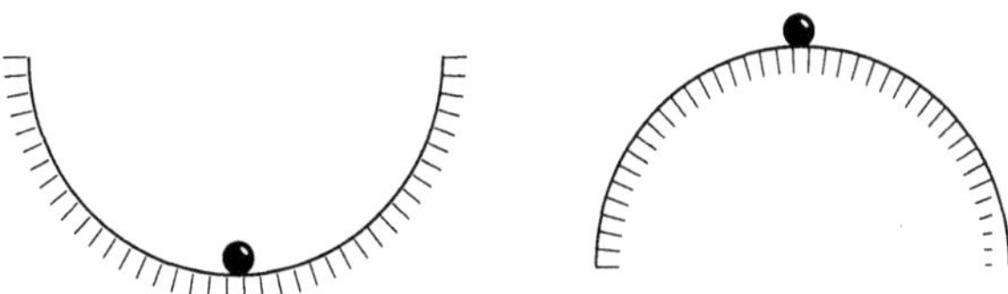

Figure 3.6

The little ball on the left is at rest, at the bottom of the hemispherical cavity. A small disturbance, that is, a slight displacement to which it is submitted, removing it from the equilibrium position, takes it back to the former position; the equilibrium is then said to be *stable*. Now the little ball on the right, when, in turn, disturbed slightly, moves increasingly farther from the original position; the equilibrium is said, in this case, to be *unstable*.

Dynamic stability is, to a certain extent, similar. A rigid body C, subject to a certain system of external forces, has some motion condition, say, periodic motion, with respect to an inertial reference frame $\mathcal{R}$. If a small disturbing force is introduced, the body motion may return, after a certain interval of time, to its original condition — and, in this case, it is said that its motion is *dynamically stable* — or may be definitively removed from that motion condition, in which case it is said to be *dynamically unstable*. For example, the motion of the coin rolling on a horizontal plane, studied in Example 3.3, shows that the stability condition depends on the angular velocity of the body.

Usually, the stability of a motion can be analyzed by studying the behavior of the solutions for the equations of motion. In the coin example, it is seen that one of the equations (for the variable θ) is in the $\ddot{u} + ku = 0$ form, the solution of which is harmonic for $k > 0$, meaning that θ remains oscillating around the position of *dynamic equilibrium*. On the other hand, for $k < 0$, the solution is of the exponential kind, moving away, therefore, from the original position of dynamic equilibrium. The reader has certainly already found it difficiult to stay balancing on a standing bicycle.

When a rigid body moves with null moment with respect to its mass center and angular velocity parallel to a principal direction of inertia (say, the direction x_3), Euler's equations are reduced to

$$\dot{\omega}_1 = 0, \qquad \dot{\omega}_2 = 0, \qquad \dot{\omega}_3 = 0. \tag{3.1}$$

Therefore, the components of the angular velocity vector remain invariant in time and the body keeps the original angular velocity. Now try and perform a very simple experiment: Take a rectangular cardboard box or some other light material, with three notably different dimensions, and experiment throwing it up, impressing, at the same time, an angular velocity around one of the principal axes of inertia with respect to the mass center. Do this twice or three times and then repeat the operation, turning it around each of the other two principal axes of inertia. What did you find? (If you continued reading without performing the experiment, you are missing a great opportunity to appreciate a truly interesting effect; why don't you stop reading now and test it before continuing?)

Example 3.4 Consider a homogeneous rectangular block B, with mass m and three different dimensions, so that the principal directions of inertia for its mass center are parallel to the Cartesian axes $\{x_1, x_2, x_3\}$, with the three principal moments of inertia different to each other (see Fig. 3.7).

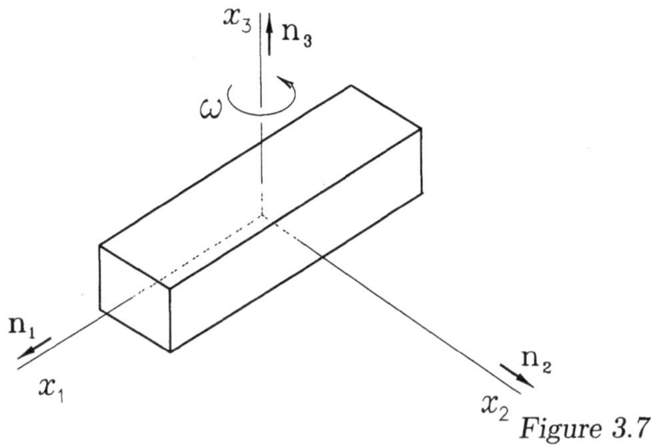

Figure 3.7

Now assume that the block is moving in space, solely under the action of its weight, so that the resultant moment with respect to the mass center is null. Also, let the angular velocity vector of the block in an inertial reference frame $\mathcal{R}$ be

$$^{\mathcal{R}}\boldsymbol{\omega}^{B} = \omega_1 \mathbf{n}_1 + \omega_2 \mathbf{n}_2 + \omega_3 \mathbf{n}_3,$$

where ω_1 and ω_2 are much smaller than ω_3. In other words, although the intention was to turn the block around one of its principal directions of inertia, either due to the slight nonhomogeneity of the block or due to our incapacity to turn it closely around axis x_3, there are small components of angular velocity in the two other directions. Euler's equations, for null moment, are, then,

$$I_1^* \dot{\omega}_1 + (I_3^* - I_2^*)\omega_3 \omega_2 = 0, \tag{a}$$

$$I_2^* \dot{\omega}_2 + (I_1^* - I_3^*)\omega_1\omega_3 = 0, \qquad\qquad\text{(b)}$$

$$I_3^* \dot{\omega}_3 + (I_2^* - I_1^*)\omega_2\omega_1 = 0. \qquad\qquad\text{(c)}$$

Admitting ω_1 and ω_2 as very small, their product may be neglected and Eq. (c) may be integrated, giving

$$\omega_3 = \omega_0,$$

where ω_0 is a constant. Now differentiating Eq. (b) with respect to time, we have

$$I_2^* \ddot{\omega}_2 + (I_1^* - I_3^*)\omega_0\dot{\omega}_1 = 0.$$

Next, solving Eq. (a) for $\dot{\omega}_1$, substituting in the previous equation, and rearranging lead to (check)

$$\ddot{\omega}_2 + \frac{(I_1^* - I_3^*)(I_2^* - I_3^*)}{I_1^* I_2^*}\omega_0^2\omega_2 = 0.$$

Operating inversely with Eqs. (a) and (b) leads likewise to,

$$\ddot{\omega}_1 + \frac{(I_3^* - I_2^*)(I_3^* - I_1^*)}{I_2^* I_1^*}\omega_0^2\omega_1 = 0.$$

Both equations are (again) of the kind

$$\ddot{u} + ku = 0,$$

leading to a periodic solution (harmonic) for $k > 0$, that is, for $I_1^* < I_3^*$ and $I_2^* < I_3^*$ or also for $I_1^* > I_3^*$ and $I_2^* > I_3^*$. In short, the motion is stable only if I_3^* is the maximum or minimum principal moment of inertia. Placing the block, however, to turn around the principal direction of inertia with the principal moment of inertia of intermediary value, we have $k < 0$ and the initial values of ω_1 and ω_2 increase exponentially, making the motion unstable. (The author now suggests that the reader (re)do the experiment.)

8.4 Impulse and Impact

Section 4.7 gives the concept of impulse of a force in a given interval, as the integration of the force during the interval, that is, given by the vector

$$\mathbf{I}_{12}^F = \int_{t_1}^{t_2} \mathbf{F}\, dt. \tag{4.1}$$

If a rigid body is subject to a system of external forces $\mathcal{F}_e$, with resultant $\mathbf{F}$, it is called the *resultant impulse*, in a given interval, to the vector sum of the impulses of each component of $\mathcal{F}_e$ in the same interval. It is, therefore, equal to the impulse of the resultant force, that is,

$$\mathbf{I}_{12}^{\mathcal{F}_e} = \sum_{i=1}^{n} \int_{t_1}^{t_2} \mathbf{F}_i\, dt = \int_{t_1}^{t_2} \sum_{i=1}^{n} \mathbf{F}_i\, dt = \int_{t_1}^{t_2} \mathbf{F}\, dt = \mathbf{I}_{12}^F. \tag{4.2}$$

If a rigid body C moves in an inertial reference frame $\mathcal{R}$ under the action of a system of forces (and, possibly, torques) $\mathcal{F}_e$, the change of the momentum vector of C in $\mathcal{R}$ in a given interval (t_1, t_2) is equal to the resultant impulse in this interval, that is,

$$^{\mathcal{R}}\mathbf{G}^C(t_2) - {}^{\mathcal{R}}\mathbf{G}^C(t_1) = \mathbf{I}_{12}^{\mathcal{F}_e}. \tag{4.3}$$

In fact, integrating in time the equation of motion of the first kind, Eq. (7.2.1), yields

$$\int_{t_1}^{t_2} {}^{\mathcal{R}}\dot{\mathbf{G}}^C\, dt = \int_{t_1}^{t_2} \mathbf{F}\, dt, \tag{4.4}$$

which immediately leads to Eq. (4.3). Section 4.7 also introduces the idea of angular impulse of the moment of a force with respect to a point, in a given interval, defined as the integration of this moment during the interval, that is,

$$\mathbf{I}_{12}^{F/O} = \int_{t_1}^{t_2} \mathbf{M}^{F/O}\, dt = \int_{t_1}^{t_2} \mathbf{p} \times \mathbf{F}\, dt, \tag{4.5}$$

where $\mathbf{p}$ is the position vector of any point of the line of action of $\mathbf{F}$ with respect to O.

If a torque $\mathbf{T}$ acts on a rigid body, its angular impulse in a given interval is, likewise, defined by

$$\mathbf{I}_{12}^{T} \rightleftharpoons \int_{t_1}^{t_2} \mathbf{T}\, dt. \tag{4.6}$$

Now let rigid body C, which moves under the action of a system of external forces $\mathcal{F}_e$, consist of n forces $\mathbf{F}_i$, $i = 1, 2, \ldots, n$, and m torques $\mathbf{T}_j$, $j = 1, 2, \ldots, m$. Choosing an arbitrary point Q of C to reduce the system, the resultant angular impulse with respect to the point in a given interval of time (t_1, t_2) is equal to the angular impulse with respect to Q of the resultant moment with respect to the point, that is,

$$\begin{aligned}
\mathbf{I}_{12}^{\mathcal{F}_e/Q} &= \sum_{i=1}^{n} \int_{t_1}^{t_2} \mathbf{p}_i \times \mathbf{F}_i\, dt + \sum_{j=1}^{m} \int_{t_1}^{t_2} \mathbf{T}_j\, dt \\
&= \int_{t_1}^{t_2} \left(\sum_{i=1}^{n} \mathbf{p}_i \times \mathbf{F}_i + \sum_{j=1}^{m} \mathbf{T}_j \right) dt \\
&= \int_{t_1}^{t_2} \mathbf{M}^{\mathcal{F}_e/Q}\, dt.
\end{aligned} \tag{4.7}$$

In particular, if C* is the mass center of C, reducing the forces system $\mathcal{F}_e$ to C* will lead to a resultant angular impulse with respect to the mass center

$$\mathbf{I}_{12}^{\mathcal{F}_e/C^*} = \int_{t_1}^{t_2} \mathbf{M}^{\mathcal{F}_e/C^*}\, dt. \tag{4.8}$$

If a rigid body C moves in an inertial reference frame $\mathcal{R}$ under the action of a system of external forces $\mathcal{F}_e$, the change in $\mathcal{R}$ of the angular momentum vector of C with respect to its mass center C*, in a given interval of time (t_1, t_2), is equal to the resultant angular impulse with respect to C*, in this interval, that is,

$$^{\mathcal{R}}\mathbf{H}^{C/C^*}(t_2) - {}^{\mathcal{R}}\mathbf{H}^{C/C^*}(t_1) = \mathbf{I}_{12}^{\mathcal{F}_e/C^*}. \tag{4.9}$$

In fact, being integrated in time between the instants t_1 and t_2, Eq. (7.2.5) yields

$$\int_{t_1}^{t_2} {}^{\mathcal{R}}\dot{\mathbf{H}}^{C/C^*}\, dt = \int_{t_1}^{t_2} \mathbf{M}^{\mathcal{F}_e/C^*}\, dt, \tag{4.10}$$

which, with Eq. (4.8), results in Eq. (4.9).

Example 4.1 A homogeneous cylinder C, with mass m and radius r, rolls freely on a horizontal plane, its center describing a straight trajectory at a constant speed v. When it reaches a slope with inclination θ (see Fig. 4.1), the cylinder begins to roll on the new surface, but now with the velocity of point O no longer constant, due to the work done by the weight. We wish to determine the maximum elevation reached by its center. Now, the cylinder undergoes an impact upon colliding with the slope. Then analyzing the motion before the collision, there is a uniform motion with conservation of the kinetic energy, since the forces involved do not work. As the velocity of the mass center is v, the rolling condition leads to an angular velocity $\omega = v/r$, and the kinetic energy of the cylinder is

$$^{\mathcal{R}}K^{C} = \frac{1}{2}mv^2 + \frac{1}{2}\frac{1}{2}mr^2\left(\frac{v}{r}\right)^2 = \frac{3}{4}mv^2.$$

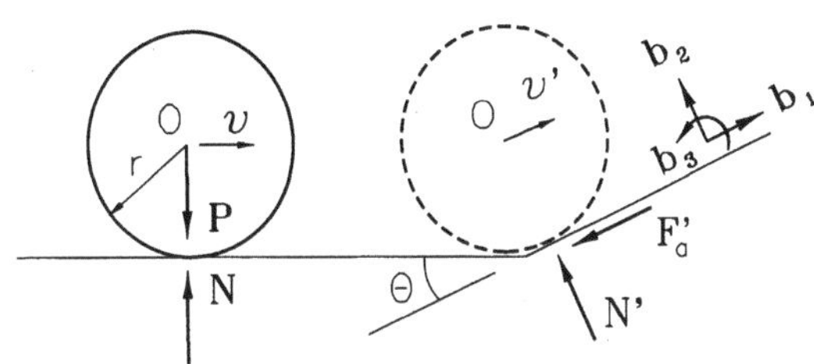

Figure 4.1

During the collision, only the impact forces need to be taken into account to calculate the impulse (see Section 4.7): in this case, the normal in the slope, $\mathbf{N}'$, and the friction force in the slope, $\mathbf{F}'_a$. The momentum of the cylinder immediately before and after the impact with the slope must satisfy Eq. (4.3), that is,

$$\mathbf{G}^{C}(t_2) - \mathbf{G}^{C}(t_1) = mv'\mathbf{b}_1 - mv(\cos\theta\mathbf{b}_1 - \sin\theta\mathbf{b}_2)$$

$$= \int_{t_1}^{t_2}(-F'_a\mathbf{b}_1 + N'\mathbf{b}_2)\,dt.$$

Then,

$$m(v' - v\cos\theta) = -\int_{t_1}^{t_2}F'_a\,dt, \tag{a}$$

$$mv \sin \theta = \int_{t_1}^{t_2} N' \, dt. \tag{b}$$

The resultant moment with respect to the mass center, during the impact, is

$$\mathbf{M}^{\mathcal{F}_e/C^*} = -F_a' r \, \mathbf{b}_3.$$

The angular momentum of the cylinder with respect to its mass center shall satisfy Eq. (4.9), that is,

$$\mathbf{H}^{C/C^*}(t_2) - \mathbf{H}^{C/C^*}(t_1) = -\frac{1}{2} mrv' \mathbf{b}_3 + \frac{1}{2} mrv \mathbf{b}_3$$

$$= -\int_{t_1}^{t_2} F_a' r \, \mathbf{b}_3 \, dt. \tag{c}$$

Substituting, then, the impulse of the friction force of Eq. (c) in Eq. (a) leads to

$$v' = \frac{1}{3}(1 + 2\cos\theta)\, v.$$

Admitting that there is no sliding after the collision, there will be no work of nonconservative forces, the mechanical energy being conserved. Taking, then, the reference for the potential energy in the elevation r leads to

$$\begin{aligned}
{}^{\mathcal{R}}E^C(t_2) &= {}^{\mathcal{R}}K^C(t_2) + 0 \\
&= {}^{\mathcal{R}}K^{C^*}(t_2) + {}^{\mathcal{R}}K^{C/C^*}(t_2) \\
&= \frac{1}{2} mv'^2 + \frac{1}{2}\frac{1}{2} mr^2 \left(\frac{v'}{r}\right)^2 \\
&= \frac{3}{4} mv'^2 \\
&= \frac{(1 + 2\cos\theta)^2}{12} mv^2.
\end{aligned}$$

At the maximum elevation, the kinetic energy is null, having been entirely converted into potential energy

$$ {}^{\mathcal{R}}E^C(t_3) = {}^{\mathcal{R}}\Phi^C(t_3) = mga.$$

The height a reached by the cylinder center (in relation to the initial elevation, r) then it is obtained by equaling the energies, to result in

$$a = \frac{(1 + 2\cos\theta)^2}{12} \frac{v^2}{g}.$$

See the corresponding animation.

Let a rigid body be C, which moves in an inertial reference frame $\mathcal{R}$ with a fixed point O (see Section 8.1), on which a system of external forces $\mathcal{F}_e$ acts, consisting of forces and torques. The reduction to the fixed point of this system leads to a resultant, $\mathbf{F}$, and to a resultant moment, $\mathbf{M}^{\mathcal{F}_e/O}$. The change in the angular momentum vector of C with respect to O in $\mathcal{R}$, in a given interval (t_1, t_2), is equal to the resultant impulse with respect to O, in this interval, that is,

$$^{\mathcal{R}}\mathbf{H}^{C/O}(t_2) - {}^{\mathcal{R}}\mathbf{H}^{C/O}(t_1) = \mathbf{I}_{12}^{\mathcal{F}_e/O}. \tag{4.11}$$

In fact, Eq. (4.7) for a fixed point O is

$$\mathbf{I}_{12}^{\mathcal{F}_e/O} = \int_{t_1}^{t_2} \mathbf{M}^{\mathcal{F}_e/O}\, dt, \tag{4.12}$$

and, Eq. (1.1) being integrated in time, we get

$$\int_{t_1}^{t_2} {}^{\mathcal{R}}\dot{\mathbf{H}}^{C/O}\, dt = \int_{t_1}^{t_2} \mathbf{M}^{\mathcal{F}_e/O}\, dt, \tag{4.13}$$

resulting immediately in Eq. (4.11).

Example 4.2   Two slender rods, with the same mass m and length r, are at rest in the configuration shown in Fig. 4.2.

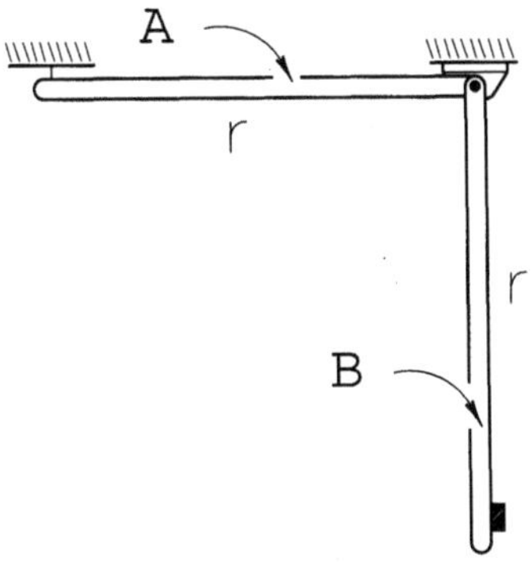

Figure 4.2

With the thread holding rod A cut, it moves until it collides with rod B (there is a small catch so that rod A pushes the other). After the impact, the two rods stay together in motion. We want to determine how much energy is lost in the impact and what is the maximum angle θ that the rods achieve. We will consider t_0 as the instant at which the thread is cut. At this instant, the kinetic energy of rod A is null and the potential energy, taking into account point O as a reference, is also null. Until the instant immediately before the impact (t_1), the only forces acting on A are the weight, conservative, and the force in the pivot, which does not work. The mechanical energy of rod A between these instants is, therefore, conserved. At the instant t_1 we then have

$$^{\mathcal{R}}K^A(t_1) = \frac{1}{2}\frac{1}{3}mr^2\omega^2, \qquad ^{\mathcal{R}}\Phi^A(t_1) = -\frac{1}{2}mgr.$$

Since the energy is conserved, then

$$\frac{1}{6}mr^2\omega^2 - \frac{1}{2}mgr = 0; \qquad \text{so,} \qquad \omega(t_1) = \sqrt{\frac{3g}{r}}.$$

During impact, forces arise in the mutual contact between the rods and other forces in the respective pivots. The latter do not produce any angular impulse with respect to O, while the former generate two angular impulses that are mutually canceled (the forces are equal and opposite). We can then conclude from Eq. (4.11) that the angular momentum of the pair of rods is conserved during the collision. The angular momentum of the rods with respect to O before the impact is

$$^{\mathcal{R}}\mathbf{H}^{A/O}(t_1) + {}^{\mathcal{R}}\mathbf{H}^{B/O}(t_1) = \frac{1}{3}mr^2\omega(t_1) + 0.$$

The angular momentum of the rods after the impact (instant t_2) is

$$^{\mathcal{R}}\mathbf{H}^{A/O}(t_2) + {}^{\mathcal{R}}\mathbf{H}^{B/O}(t_2) = \frac{1}{3}mr^2\omega(t_2) + \frac{1}{3}mr^2\omega(t_2).$$

The angular momenta being equal, therefore, the angular velocity of both rods immediately after the collision is

$$\omega(t_2) = \frac{1}{2}\sqrt{\frac{3g}{r}}.$$

The kinetic energy of the set before the collision is

$$^{\mathcal{R}}K^{AB}(t_1) = \frac{1}{6}mr^2\omega^2(t_1) = \frac{1}{2}mgr.$$

The kinetic energy of the set after the collision is

$$^{\mathcal{R}}K^{AB}(t_2) = \frac{1}{2}\omega(t_2)\frac{2}{3}mr^2\omega(t_2) = \frac{1}{4}mgr.$$

Since, during the collision, there is no change in the potential energy of the set, the change in the mechanical energy is

$$^{\mathcal{R}}E^{AB}(t_2) - {}^{\mathcal{R}}E^{AB}(t_1) = -\frac{1}{4}mgr,$$

a loss of energy, therefore, equal to $\frac{1}{4}mgr$. Between instants t_2 and t_3, when the rods reach the maximum angle θ, the mechanical energy of the set of rods is conserved again. At this instant, all energy is in the form of gravitational potential energy; therefore,

$$^{\mathcal{R}}\Phi^{AB}(t_3) = 2\frac{1}{2}mgr\cos\theta = mgr\cos\theta,$$

so,

$$\cos\theta = \frac{1}{4}, \qquad \text{that is,} \qquad \theta = 75^\circ\,31'.$$

Exercise Series #12　　(Sections 8.1 to 8.4)

P12.1　A spherical shell, with mass m and radius r, is welded to a light rod with length $2r$, whose end is linked by a ball and socket joint to the fixed point O. The body moves so that the inclination of the rod with respect to the vertical remains constant, with $\theta = 45^0$, while point Q maintains a horizontal velocity of constant module v. Determine the rate of rotation ω of the shell around its axis of symmetry. Calculate the value for v in such a way that the body moves without rotating around the axis of symmetry.

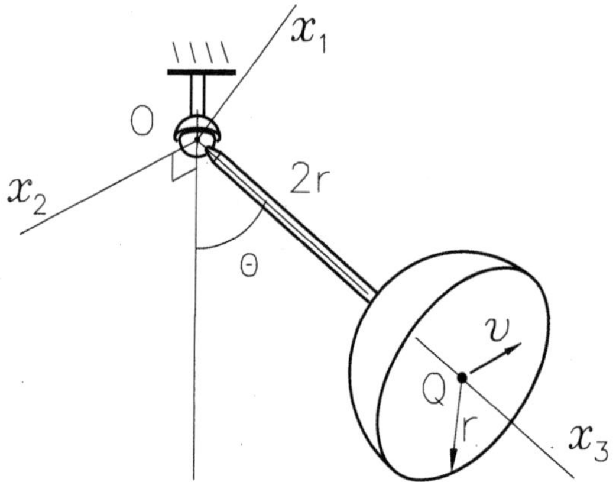

Figure P 12.1

P12.2　Rod B, with mass m, is pivoting in the fork A, which rotates around its vertical axis with angular velocity of constant module ω. The set moves, keeping constant the rod inclination with the vertical axis. Determine the value of angle θ. What is the new angular velocity, in rpm, to keep $\theta=60^0$?

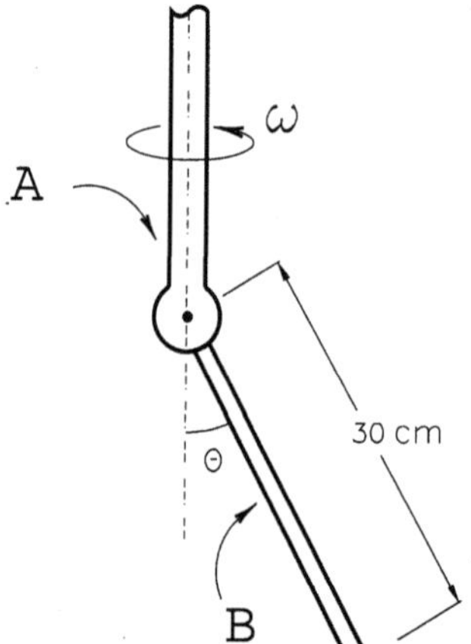

Figure P 12.2

P12.3 The vertical rod B can rotate freely around the vertical axis x_3 and have negligible inertia of rotation around this axis. Body C consists of two identical rods, each with mass m and length r, welded orthogonally, freely pivoting in B, at end O, as shown, and therefore being free to rotate around axis x_1. What is the value of the angular velocity of B as a function of angle θ if this remains constant?

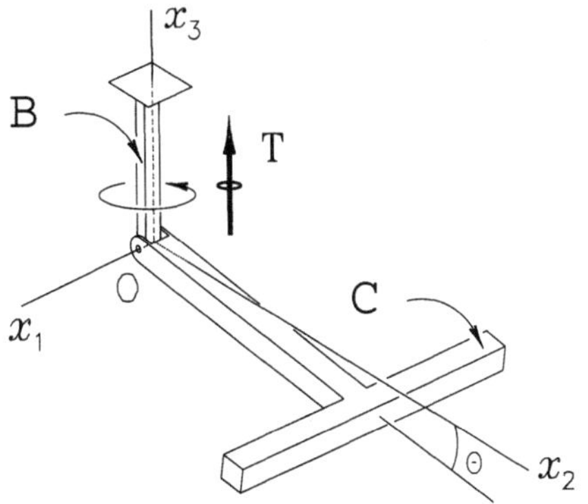

Figure P 12.3

P12.4 The homogeneous slender rod, with length $4r$, freely pivots on the support A, fixed in the disk B, with radius r, while the disk rotates around the vertical axis with a constant angular velocity of module ω, as shown. Determine the value of ω that maintains the rod fixed in relation to the disk, with $\theta = 60^0$.

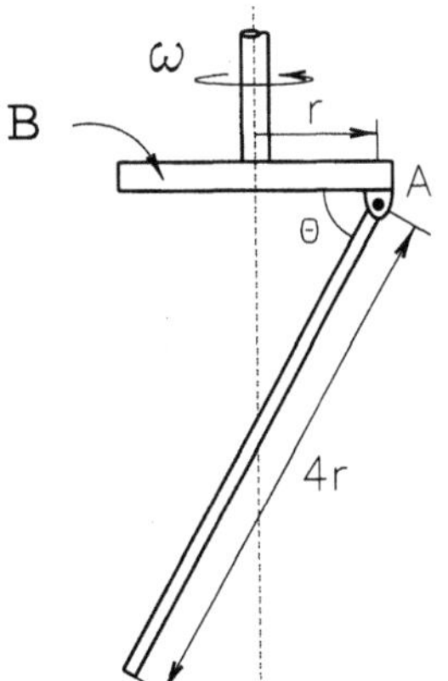

Figure P 12.4

P12.5   The alien spacecraft moves under the action of the earth's gravitational field so that, while its mass center G describes an orbital trajectory, the spacecraft, a body of revolution, turns with constant precession of four revolutions per hour around axis z', fixed in an inertial reference frame. The angle between z' and its axis of symmetry, z, remains constant. Determine the rate of rotation (constant) of the spacecraft around z (the spin), knowing that its transversal moment of inertia with respect to the mass center is double the longitudinal moment of inertia. *Suggestion:* First show that $\mathbf{H}^{C/G}$ is fixed in the inertial reference frame and is parallel to z'.

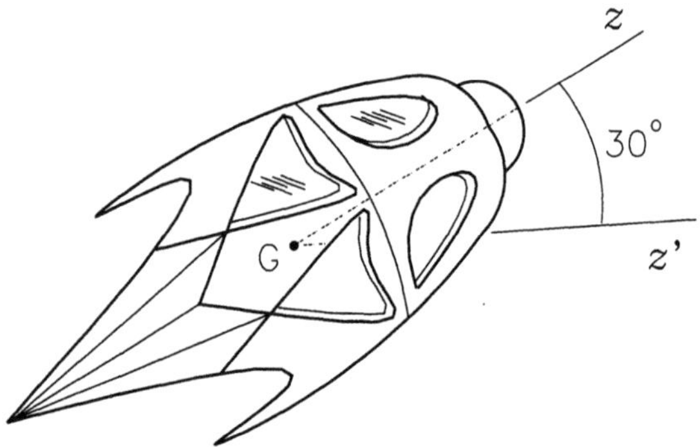

Figure P 12.5

P12.6   Consider a satellite as an axisymmetrical body moving in space with a null resultant external force. Show that the angular velocity vector ω of the satellite in relation to an inertial reference frame moves in relation to the satellite, rotating around the axis of symmetry at the constant rate

$$\Omega = \frac{J - I}{J}\omega_z,$$

where J is the moment of inertia with respect to the axis of symmetry, I is the transversal moment of inertia relative to the mass center, and ω_z is the component of the angular velocity of the satellite in the direction of its axis of symmetry.

P12.7 Consider a cone frustum C with mass m and transversal and longitudinal moments of inertia with respect to its mass center of I and J, respectively, connected to the support $\mathcal{R}$ by means of a flexible hose, as shown. The hose offers very little resistance to flexion (consider null) and very high resistance to torsion (consider that it does not twist). This body can move so that a point P of its axis of symmetry stays fixed in the vertical axis, while its mass center moves in a circular trajectory on the horizontal plane with constant speed v. Determine v as a function of distance r between the mass center and point P, angle θ between the axis of symmetry and vertical axis, and the properties of inertia of the cone frustum.

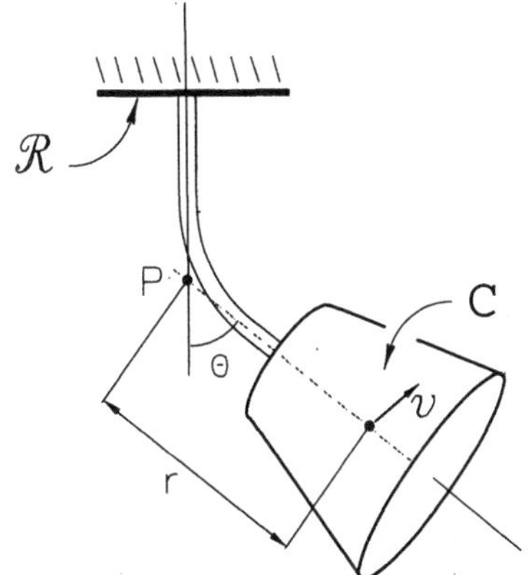

Figure P 12.7

P12.8 A homogeneous slender rod B is pivoting in a vertical axis that rotates with constant angular velocity, as shown. Determine the inclination angle θ of the rod.

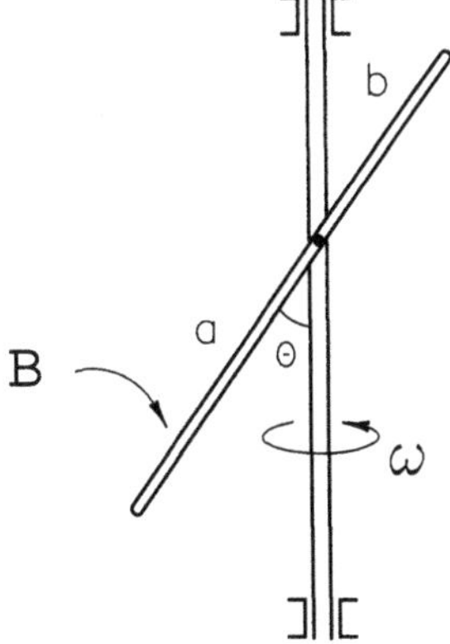

Figure P 12.8

P12.9 The homogeneous disk D, with mass m, is welded to the light horizontal shaft, which can turn freely around the axes x_2 and x_3. Play in the coupling in O also permits small rotations around x_1. The set is moving so that the axis system turns around x_3 with constant angular velocity of module Ω, while the disk rolls on the fixed horizontal plane. Determine the force that the plane applies on the disk.

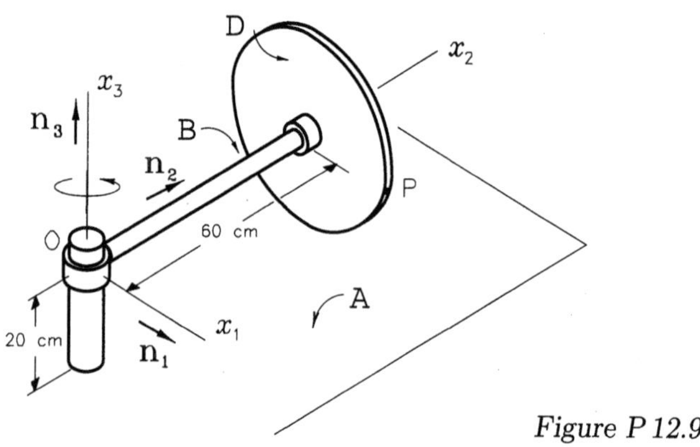

Figure *P 12.9*

P12.10 The homogeneous disk D, with radius r, can turn freely around axis z, while the light rod B is articulated by means of a ball and socket joint to point O, fixed in an inertial reference frame, and turns in relation to the latter around axis Z. Determine the relationship between the spin, ω, and precession, Ω, for the system to stay in motion with $\theta = \pi/2$.

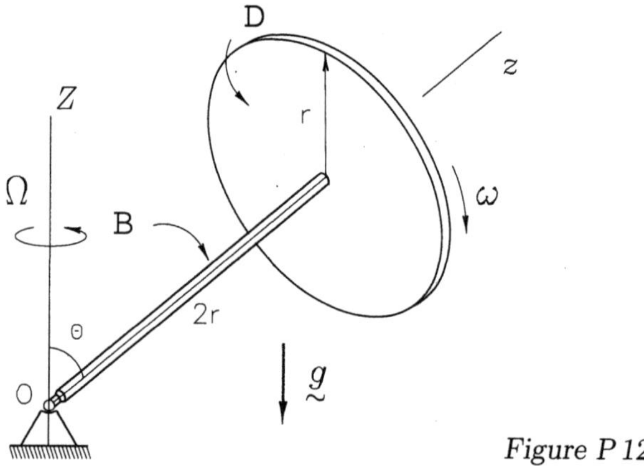

Figure *P 12.10*

P12.11 The body consists of two identical homogeneous disks, with radius r, connected together by a telescopic shaft with negligible mass and variable length. The body is made to turn around its axis of symmetry and is disturbed slightly enough to present a precession. Determine the shortest length of the telescopic shaft to guarantee direct precession.

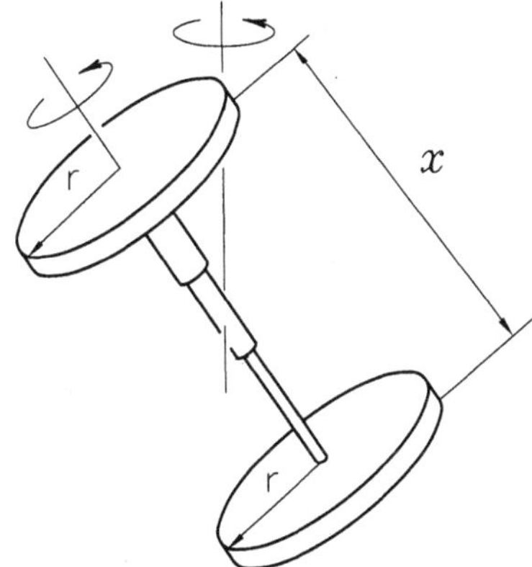

Figure P 12.11

P12.12 A slender rod, with mass m and length r, has one fixed end, articulated by a ball and socket joint in point O, fixed in the inertial reference frame, and turns with constant angular velocity of module ω around the vertical axis. Determine the angle θ with the vertical and calculate the force in joint O.

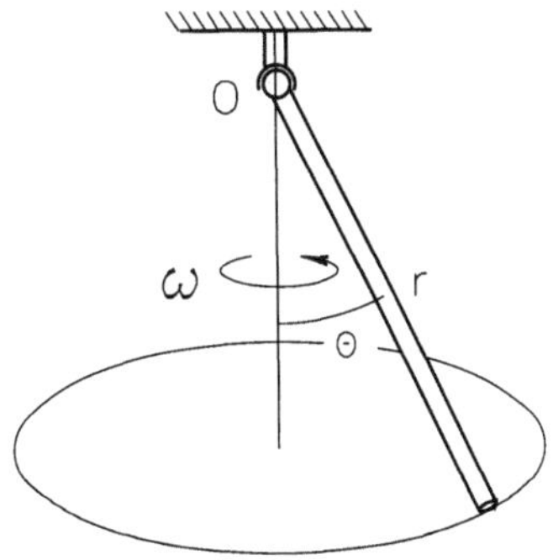

Figure P 12.12

P12.13 The homogeneous slender rod, with length $4r$, is pivoting freely in the support in A, fixed in disk B, with radius r, while the disk turns around the vertical axis with constant angular velocity of module ω, as shown. Determine the highest value of angle θ for which it is possible to keep the rod fixed in relation to the disk if $\omega = \sqrt{3g/r}$.

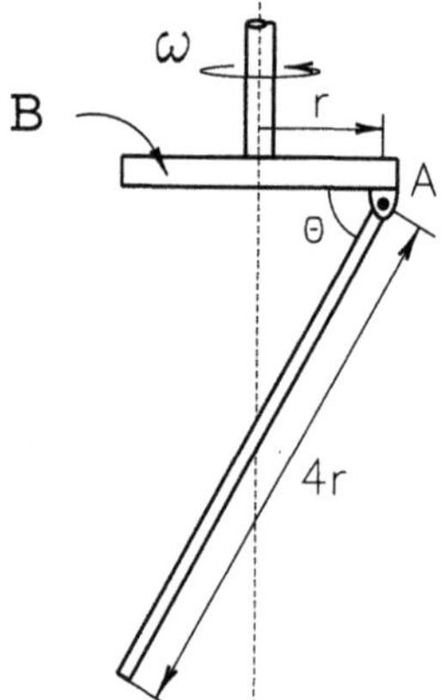

Figure P 12.13

P12.14 A homogeneous rod, with mass m and length r, is pivoted in one end, moving from the horizontal position of rest and colliding with the object with mass m' when it reaches the vertical position, as shown. Knowing the friction coefficient μ between the object and horizontal plane and knowing that the collision is nonelastic, determine to what extent the object is displaced until it stops again.

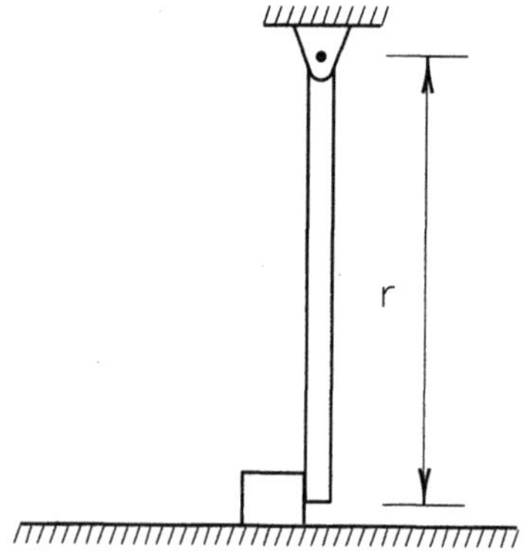

Figure P 12.14

P12.15   A homogeneous cube is moving on the horizontal plane with velocity v when it collides with the bottom step. What is the value of v from which the cube topples over?

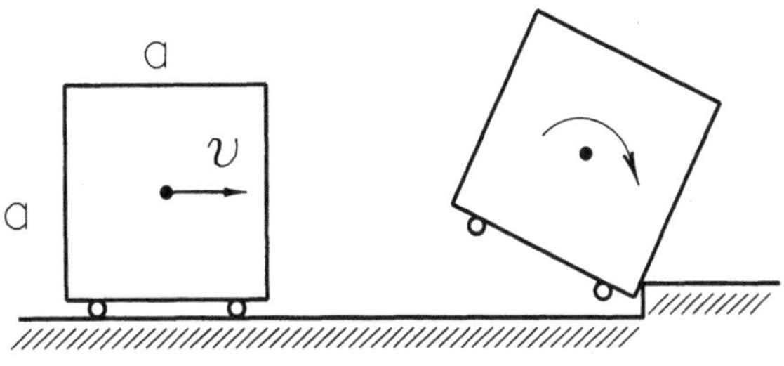

Figure P 12.15

P12.16   A radar sensor consists of a semicylindrical shell, with mass m, radius r, and height a, which turns in relation to an inertial reference frame $\mathcal{R}$ around a vertical axis, as shown. Axes $\{x_1, x_2, x_3\}$ and the orthonormal basis $\mathbf{b}_1, \mathbf{b}_2, \mathbf{b}_3$ are fixed in the sensor. At a certain instant, the shell angular velocity has module Ω and the angular acceleration module $2\Omega^2$, both in the positive direction of axis x_1. Calculate, at this same instant, the resultant force and torque in point O.

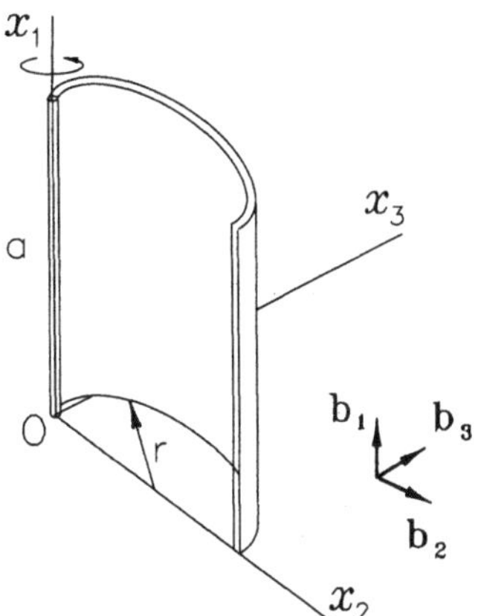

Figure P 12.16

P12.17 Figure P12.17 illustrates in diagram form the motion of a *dancing bottle*. The bottle is placed with its axis of symmetry slightly inclined in relation to the vertical and made to roll on a horizontal plane with an angular velocity in order to ensure a periodic motion. The motion may be modeled as follows. The bottle G, with mass center Q at a distance z from the base, is instantaneously supported on point C, at a distance r from its axis of symmetry x_3. The geometric place of the successive points of contact in the plane describes a circle with radius R and center O$'$. Everything happens as if the bottle moves with point O, intersection of axes x_3 and Z, fixed in the plane. Bases $\mathbf{n}_1, \mathbf{n}_2, \mathbf{n}_3$ and $\mathbf{b}_1, \mathbf{b}_2, \mathbf{b}_3$ are fixed in a vertical plane that, at each instant, contains axes x_3 and Z. The transversal and longitudinal moments of inertia of the bottle with respect to its mass center can be expressed by $I = p\,mr^2$ and $J = q\,mr^2$, respectively, where p and q are real positives. Calculate the precession Ω of the bottle. Next, verify this value for $R = 2r$, $z = r$, $p = 7/8$, $q = 3/4$, and $\theta = \tan^{-1}(1/5)$.

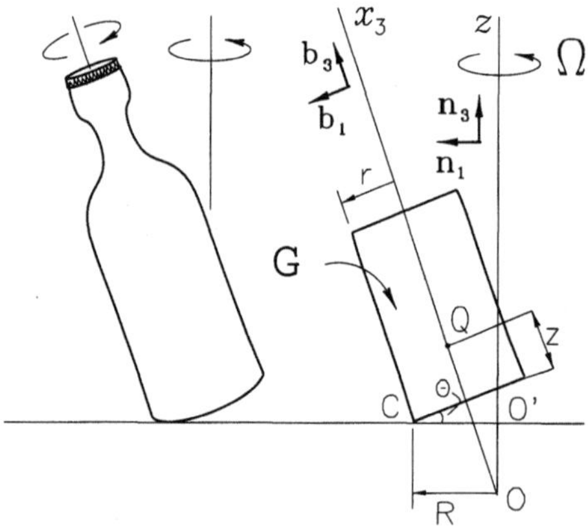

Figure P 12.17

P12.18 A homogeneous ring, with radius r, is made to roll on a horizontal plane, its center keeping a constant velocity v, with the ring moving in a vertical plane, with pure rolling. What is the lowest value for v in order to keep this motion stable?

*L*inear Algebra

Appendix A

The purpose of this appendix is to make the text more self-contained. It actually only discusses a tiny part of the subject known as *linear algebra* and, even then, very briefly. The aim here is for the reader who already has some knowledge on the subject to recall the main operations of algebra of real numbers, vectors, and tensors, giving a reference that can be immediately used to facilitate some aspects of the main text. Attention is centered on the workability of this algebra, without caring about the related mathematical formalism. Therefore, several results are given without demonstration, giving the reader the choice to look for further detail in the vast bibliography available. The algebra of matrices is not explicitly discussed here, but in order for the operations with tensors to be more accessible to the reader, the association is duly made between tensors and their matrix representation in a system of Cartesian axes.

Section A.1 addresses scalars, their sum and product, and the properties satisfied by these operations. Section A.2 discusses vectors. Operations, such as adding, multiplying by a scalar, dot product, vector product, mixed product, double vector product, and their geometric meanings are presented. Section A.3 addresses tensors, giving their definition and presenting the dyadic product or tensor product of two vectors, the product between tensors, the inverse tensor, transposed tensor, symmetric tensor, orthogonal tensor, and trace of a tensor. Last, Section A.4 discusses eigenvalues and eigenvectors, being particularly useful for a better understanding of the notion of principal directions of inertia, discussed in Section 6.6.

A.1 Scalars

Scalar properties are those wholly determined by a real number. For example, the mass of a body, the distance between two points, and the moment of inertia of a body with respect to an axis are scalar properties. Real numbers follow an *algebra* where two operations are defined — addition and product — that satisfy the properties established below.

A.1.1 Addition of Two Real Numbers

1. Commutativity:
$$a + b = b + a. \tag{1.1}$$

2. Associativity:
$$(a + b) + c = a + (b + c). \tag{1.2}$$

3. Existence of the null element:

There is a real number 0 so that, for every real number a,

$$a + 0 = a. \tag{1.3}$$

4. Existence of the symmetric:

For every real number a there is another real number, called the symmetric of a and represented by $-a$, so that

$$a + (-a) = 0. \tag{1.4}$$

A.1.2 Product of Two Real Numbers

1. Commutativity:
$$ab = ba. \tag{1.5}$$

2. Associativity:
$$(ab)c = a(bc). \tag{1.6}$$

3. Distributivity with respect to the addition:

$$(a + b)\,c = a\,c + b\,c. \tag{1.7}$$

4. Existence of the identity element:

There is a real number 1 so that, for every real number a,

$$a \cdot 1 = a. \tag{1.8}$$

5. Existence of the inverse:

For every real number $a \neq 0$, there is another real number, called the inverse of a and represented by $1/a$, so that

$$a \cdot \frac{1}{a} = 1. \tag{1.9}$$

A.2 Vectors

Properties of a vector nature are those that, to be defined, require the specification of two parameters: module and direction. Examples of vector properties are the position of a material point in relation to the origin of a coordinate system, the angular velocity of a rigid body, and the force applied by one particle on another. Vector properties are represented geometrically by arrows. The length of the arrow is proportional to the module of the property. The direction of the arrow is the same as that of the property. It is usual to write vectors in boldface font and their modules in the corresponding letters in italic or boldface between vertical lines, that is, if $\mathbf{v}$ represents a vector, its module will be represented by

$$|\mathbf{v}| = v. \tag{2.1}$$

In general, in three dimensions, any vector $\mathbf{u}$ may be expressed as a linear combination of any three noncoplanar vectors, $\mathbf{v}_1, \mathbf{v}_2, \mathbf{v}_3$. Therefore, $\mathbf{u} = u_1\mathbf{v}_1 + u_2\mathbf{v}_2 + u_3\mathbf{v}_3$, with u_1, u_2, and u_3 being scalars. The scalars $u_1|\mathbf{v}_1|$, $u_2|\mathbf{v}_2|$, and $u_3|\mathbf{v}_3|$ are called the *scalar components* of $\mathbf{u}$ in the directions of $\mathbf{v}_1$, $\mathbf{v}_2$, and $\mathbf{v}_3$, respectively. Vectors $\mathbf{v}_1, \mathbf{v}_2, \mathbf{v}_3$ are a *basis* for the three-dimensional space. Although it is *unnecessary*, it is convenient to choose the mutually orthogonal vectors of the basis. The basis, in this case, is called an *orthogonal basis*. A particularly useful set of basis vectors is what is called a *Cartesian basis*, made up of three mutually orthogonal unit vectors $\mathbf{n}_1, \mathbf{n}_2, \mathbf{n}_3$, that is, all with a module equal to 1. This basis is also called an *orthonormal basis*. A geometric representation of $\mathbf{v}$ in such a basis is shown in Fig. 2.1.

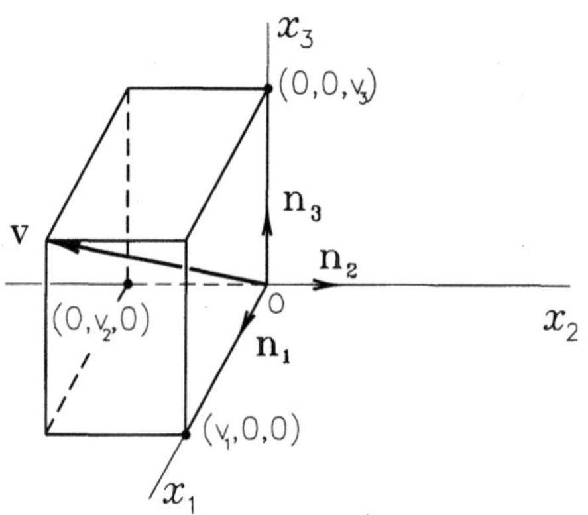

Figure 2.1

Vector $\mathbf{v}$, in the coordinate system $\{x_1, x_2, x_3\}$, is represented as

$$\mathbf{v} = v_1\mathbf{n}_1 + v_2\mathbf{n}_2 + v_3\mathbf{n}_3, \tag{2.2}$$

or, compactly, by the triad of its scalar components v_1, v_2, and v_3 (see Fig. 2.1)

$$\mathbf{v} = (v_1, v_2, v_3). \tag{2.3}$$

The module of $\mathbf{v}$ is given by

$$|\mathbf{v}| = v = \sqrt{v_1^2 + v_2^2 + v_3^2}. \tag{2.4}$$

The null vector,

$$\mathbf{0} = (0, 0, 0), \tag{2.5}$$

has a module equal to zero and indeterminate direction.

   Two vectors $\mathbf{u}$ and $\mathbf{v}$ are *equal* when they have the same module and direction.

If $\mathbf{u} = \mathbf{v}$, then necessarily $u_1 = v_1$, $u_2 = v_2$ and $u_3 = v_3$, that is, $(u_1, u_2, u_3) = (v_1, v_2, v_3)$. The vector $\mathbf{v} = (v_1, v_2, v_3)$ is called *symmetric* of $\mathbf{u} = (u_1, u_2, u_3)$ if $v_1 = -u_1$, $v_2 = -u_2$, and $v_3 = -u_3$. Vector $\mathbf{v}$, symmetric of $\mathbf{u}$, may be represented by $-\mathbf{u}$, having the same module but in the opposite direction to that of $\mathbf{u}$.

A.2.1 Multiplication by a Scalar

Let α be a scalar and $\mathbf{v}$ a vector. The vector $\alpha\mathbf{v}$, called the *scalar multiple* of $\mathbf{v}$, is defined by

$$\alpha\mathbf{v} = (\alpha v_1, \alpha v_2, \alpha v_3) \tag{2.6}$$

and has a module equal to $\alpha|\mathbf{v}|$ and the same direction as $\mathbf{v}$. Figure 2.2 shows the geometric representation of the vector $\alpha\mathbf{v}$, with $\alpha < 0$ and $|\alpha| > 1$. It is easy to see that if $\alpha > 0$, $\alpha\mathbf{v}$ will have the same direction as $\mathbf{v}$ and, if $\alpha < 0$, $\alpha\mathbf{v}$ will have the opposite direction to $\mathbf{v}$.

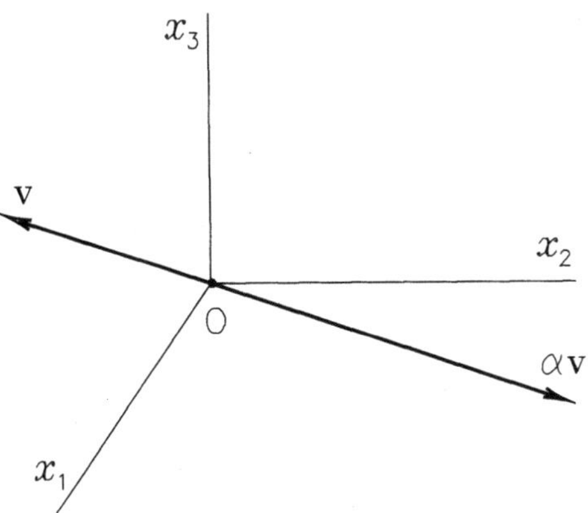

Figure 2.2

The multiplication of a vector by a scalar has the following properties:

$$\alpha(\beta\mathbf{v}) = (\alpha\beta)\mathbf{v} \qquad \text{(associative)}; \tag{2.7}$$

$$(\alpha + \beta)\mathbf{v} = \alpha\mathbf{v} + \beta\mathbf{v} \qquad \text{(distributive)}. \tag{2.8}$$

A.2.2   Sum of Two Vectors

The sum of vectors $\mathbf{u}$ and $\mathbf{v}$ is the vector

$$\mathbf{u} + \mathbf{v} = (u_1 + v_1, u_2 + v_2, u_3 + v_3). \tag{2.9}$$

The subtraction of two vectors may be obtained by adding $-\mathbf{v}$ to vector $\mathbf{u}$:

$$\mathbf{u} - \mathbf{v} = (u_1 - v_1, u_2 - v_2, u_3 - v_3). \tag{2.10}$$

The sum of vectors is commutative, associative, and distributive with respect to the multiplication by a scalar:

$$\mathbf{u} + \mathbf{v} = \mathbf{v} + \mathbf{u} \qquad \text{(commutative);} \tag{2.11}$$

$$(\mathbf{u} + \mathbf{v}) + \mathbf{w} = \mathbf{u} + (\mathbf{v} + \mathbf{w}) \quad \text{(associative);} \tag{2.12}$$

$$a(\mathbf{u} + \mathbf{v}) = a\mathbf{u} + a\mathbf{v} \qquad \text{(distributive).} \tag{2.13}$$

Figure 2.3 shows the geometric representation of the sum of two vectors $\mathbf{u}$ and $\mathbf{v}$, also known as the *parallelogram rule*, for $\mathbf{u}$ and $\mathbf{v}$ on the plane $x_1 x_2$.

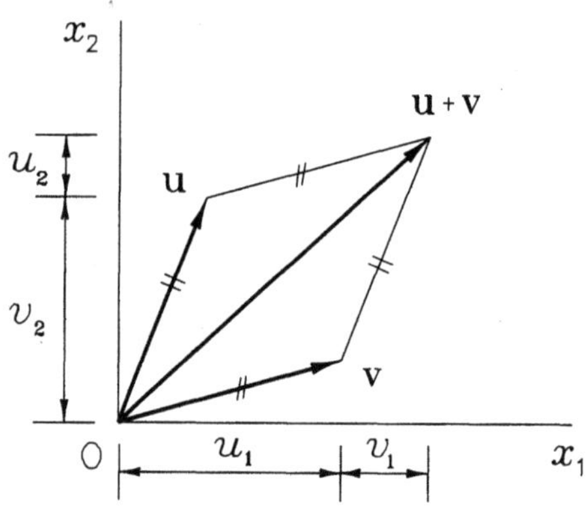

Figure 2.3

Example 2.1   Given the two-dimensional vectors $\mathbf{u} = (5, 2)$ and $\mathbf{v} = (1, -2)$, calculate the result of $3\mathbf{u} - \mathbf{v}$. Now, according to Eq. (2.6), $3\mathbf{u} = (15, 6)$ and $-\mathbf{v} = (-1, 2)$. Therefore, according to Eq. (2.10),

$$3\mathbf{u} - \mathbf{v} = 3\mathbf{u} + (-\mathbf{v}) = (14, 8).$$

A.2.3 Dot Product

The *dot product* of two vectors $\mathbf{u}$ and $\mathbf{v}$ is the scalar defined by

$$\mathbf{u} \cdot \mathbf{v} = uv \cos \theta, \tag{2.14}$$

where θ indicates the angle between the directions of $\mathbf{u}$ and $\mathbf{v}$. Rewriting Eq. (2.14) as

$$\mathbf{u} \cdot \mathbf{v} = u(v \cos \theta), \tag{2.15}$$

it may be found that $\mathbf{u} \cdot \mathbf{v}$ is the product of the module of $\mathbf{u}$ by the orthogonal projection of $\mathbf{v}$ in the direction of $\mathbf{u}$, as shown in Fig. 2.4.

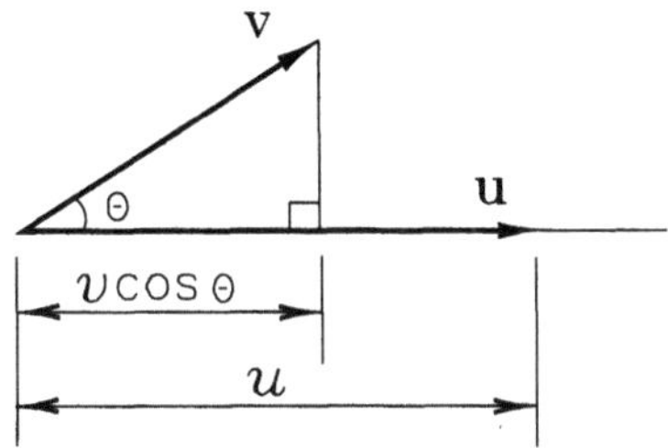

Figure 2.4

The dot product of vectors is commutative,

$$\mathbf{u} \cdot \mathbf{v} = \mathbf{v} \cdot \mathbf{u}, \tag{2.16}$$

distributive with respect to addition,

$$\mathbf{u} \cdot (\mathbf{v} + \mathbf{w}) = \mathbf{u} \cdot \mathbf{v} + \mathbf{u} \cdot \mathbf{w}, \tag{2.17}$$

and distributive with respect to multiplication by a scalar,

$$(a\mathbf{u}) \cdot \mathbf{v} = \mathbf{u} \cdot (a\mathbf{v}) = a(\mathbf{u} \cdot \mathbf{v}). \tag{2.18}$$

Two nonnull vectors $\mathbf{u}$ and $\mathbf{v}$ whose directions are orthogonal are called orthogonal vectors and will have a null dot product, since the projection of one in the direction of the other is null. Therefore,

$$\mathbf{u} \cdot \mathbf{v} = 0 \qquad \text{if} \qquad \mathbf{u} \perp \mathbf{v}. \tag{2.19}$$

If $\mathbf{u}$ and $\mathbf{v}$ have the same direction,

$$\mathbf{u} \cdot \mathbf{v} = \pm uv, \tag{2.20}$$

with the positive sign if $\mathbf{u}$ and $\mathbf{v}$ have the same direction and negative in the opposite case. From Eqs. (2.19) and (2.20) we see that

$$\mathbf{n}_1 \cdot \mathbf{n}_1 = \mathbf{n}_2 \cdot \mathbf{n}_2 = \mathbf{n}_3 \cdot \mathbf{n}_3 = 1 \tag{2.21}$$

and that

$$\mathbf{n}_1 \cdot \mathbf{n}_2 = \mathbf{n}_2 \cdot \mathbf{n}_3 = \mathbf{n}_3 \cdot \mathbf{n}_1 = 0. \tag{2.22}$$

Therefore, using these results and Eq. (2.2), we obtain (check)

$$\mathbf{u} \cdot \mathbf{v} = \sum_{j=1}^{3} u_j v_j = u_1 v_1 + u_2 v_2 + u_3 v_3. \tag{2.23}$$

This result allows us to write

$$|\mathbf{v}|^2 = v^2 = v_1^2 + v_2^2 + v_3^2 = \mathbf{v} \cdot \mathbf{v}. \tag{2.24}$$

The result of dividing $\mathbf{v}$ by $(\mathbf{v} \cdot \mathbf{v})^{1/2}$ is a vector with module equal to 1 and with the same direction as $\mathbf{v}$. This vector is called the *unit vector of direction* $\mathbf{v}$, also called the *versor* of $\mathbf{v}$,

$$\mathbf{n}_v = \frac{\mathbf{v}}{\sqrt{\mathbf{v} \cdot \mathbf{v}}}. \tag{2.25}$$

Once $\mathbf{n}_v$ is obtained, any vector $\mathbf{w}$, of module w and with the same direction as $\mathbf{v}$, may be represented as

$$\mathbf{w} = w\mathbf{n}_v. \tag{2.26}$$

Considering three mutually orthogonal vectors $\mathbf{u}$, $\mathbf{v}$, and $\mathbf{w}$ and their respective versors $\mathbf{n}_u$, $\mathbf{n}_v$, and $\mathbf{n}_w$, any vector $\mathbf{r}$ may be represented by

$$\mathbf{r} = \mathbf{r} \cdot \mathbf{n}_u \, \mathbf{n}_u + \mathbf{r} \cdot \mathbf{n}_v \, \mathbf{n}_v + \mathbf{r} \cdot \mathbf{n}_w \, \mathbf{n}_w. \tag{2.27}$$

A.2.4 Vector Product

The *vector product* of two vectors $\mathbf{u}$ and $\mathbf{v}$, represented by $\mathbf{u} \times \mathbf{v}$, is the vector orthogonal to the plane formed by $\mathbf{u}$ and $\mathbf{v}$, with a module numerically equal to the area of the parallelogram with sides u and v and with the direction given by the motion of translation that a screw with a right-hand thread would have when turning from $\mathbf{u}$ to $\mathbf{v}$, by the smallest angle (see Fig. 2.5).

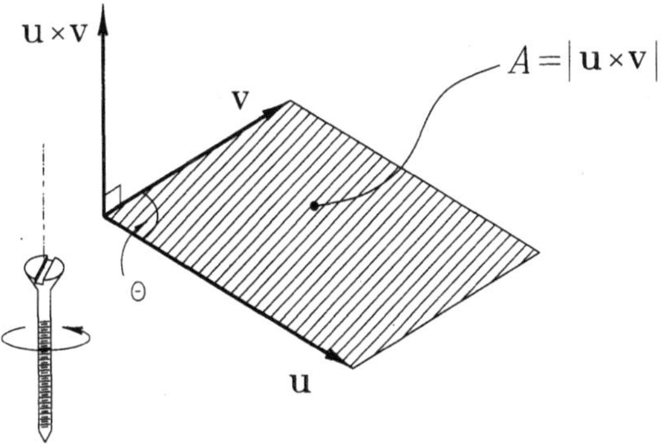

Figure 2.5

The area of the parallelogram with sides u and v is given by $uv \sin\theta$ (see Fig. 2.6) and, therefore,

$$|\mathbf{u} \times \mathbf{v}| = uv \sin\theta. \tag{2.28}$$

Since the direction of $\mathbf{u} \times \mathbf{v}$ is given by the rotation oriented from $\mathbf{u}$ to $\mathbf{v}$, it follows that

$$\mathbf{u} \times \mathbf{v} \neq \mathbf{v} \times \mathbf{u}, \tag{2.29}$$

that is, the vector product is not commutative. More specifically,

$$\mathbf{u} \times \mathbf{v} = -(\mathbf{v} \times \mathbf{u}). \tag{2.30}$$

Nor is the vector product associative, that is,

$$\mathbf{u} \times (\mathbf{v} \times \mathbf{w}) \neq (\mathbf{u} \times \mathbf{v}) \times \mathbf{w}, \tag{2.31}$$

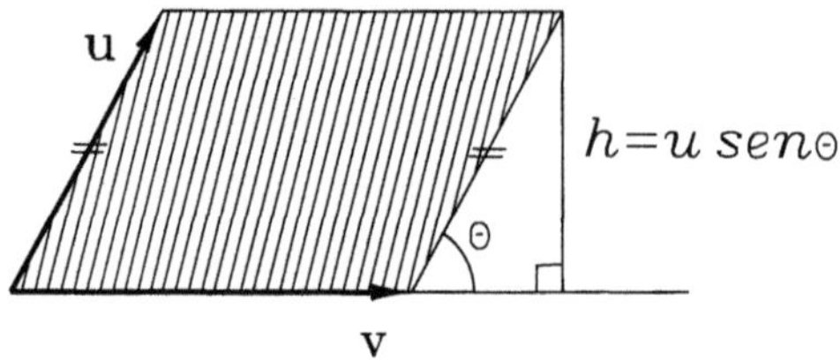

Figure 2.6

But it is distributive in relation to the vector sum

$$\mathbf{u} \times (\mathbf{v} + \mathbf{w}) = \mathbf{u} \times \mathbf{v} + \mathbf{u} \times \mathbf{w}. \tag{2.32}$$

Note that if $\mathbf{u}$ and $\mathbf{v}$ are parallel vectors, by the definition of vector product,

$$\mathbf{u} \times \mathbf{v} = \mathbf{0}. \tag{2.33}$$

For an orthonormal basis, the following properties are valid:

$$\mathbf{n}_1 \times \mathbf{n}_1 = \mathbf{n}_2 \times \mathbf{n}_2 = \mathbf{n}_3 \times \mathbf{n}_3 = \mathbf{0}; \tag{2.34}$$

$$\mathbf{n}_1 \times \mathbf{n}_2 = \mathbf{n}_3; \quad \mathbf{n}_2 \times \mathbf{n}_3 = \mathbf{n}_1; \quad \mathbf{n}_3 \times \mathbf{n}_1 = \mathbf{n}_2. \tag{2.35}$$

Writing vectors $\mathbf{u}$ and $\mathbf{v}$ in terms of their components and performing the vector product, we obtain (check)

$$\mathbf{u} \times \mathbf{v} = (u_2 v_3 - u_3 v_2)\mathbf{n}_1 + (u_3 v_1 - u_1 v_3)\mathbf{n}_2 + (u_1 v_2 - u_2 v_1)\mathbf{n}_3. \tag{2.36}$$

The above result may be represented compactly as

$$\mathbf{u} \times \mathbf{v} = \begin{vmatrix} \mathbf{n}_1 & \mathbf{n}_2 & \mathbf{n}_3 \\ u_1 & u_2 & u_3 \\ v_1 & v_2 & v_3 \end{vmatrix}, \tag{2.37}$$

where $|\ \ |$ indicates a determinant. (Although this is not actually a determinant, it acts as a memorizing formula, according to the usual rules for calculating determinants.)

Example 2.2 Consider a known force $\mathbf{F} = 3(1, 1, -1)$ Newtons (N). Its module is, according to Eq. (2.4), $3\sqrt{3}$ N. Let P be a point of coordinates, in meters (m), P:$(2, 3, 5)$ and let O be another point of coordinates O:$(1, 2, 5)$. The position vector of P with respect to O is, then, vector $\mathbf{p} = (1, 1, 0)$ m. If $\mathbf{F}$ is applied on P, its moment with respect to point O, $\mathbf{M}^{F/O}$, is then

$$\mathbf{M}^{F/O} = \mathbf{p} \times \mathbf{F} = (-1, 1, 0) \text{ Nm.}$$

A.2.5 Mixed Product

By making the dot product of $\mathbf{u}$ with the vector $\mathbf{v} \times \mathbf{w}$, we obtain the scalar $\mathbf{u} \cdot (\mathbf{v} \times \mathbf{w})$, called the *mixed product* of $\mathbf{u}$, $\mathbf{v}$, and $\mathbf{w}$. It may be shown that this product (a scalar) is numerically equal to the volume of the parallelpiped with sides u, v, and w, since, as Fig. (2.7) shows, $|\mathbf{u} \cdot (\mathbf{v} \times \mathbf{w})| = u \sin \theta A$, where A is the area of the parallelogram with sides v and w and $u \sin \theta$ is the height of the parallelpiped formed by vectors $\mathbf{u}$, $\mathbf{v}$, and $\mathbf{w}$.

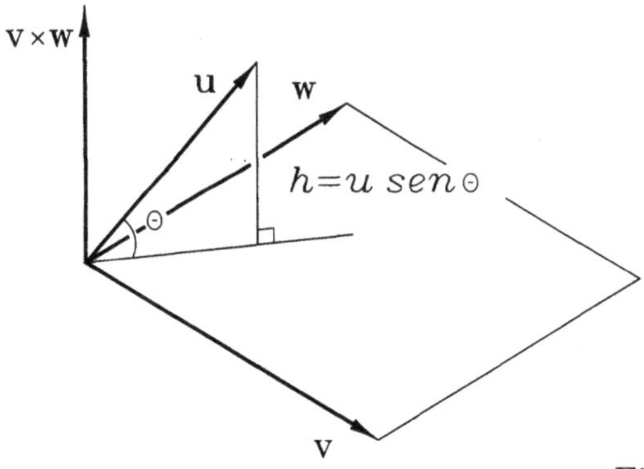

Figure 2.7

The mixed product may be written as

$$\mathbf{u} \cdot (\mathbf{v} \times \mathbf{w}) = u_1(v_2 w_3 - v_3 w_2) + u_2(v_3 w_1 - v_1 w_3) + u_3(v_1 w_2 - v_2 w_1), \quad (2.38)$$

which is Laplace's development according to the first line of the determinant, which has scalar components of $\mathbf{u}$, $\mathbf{v}$, and $\mathbf{w}$ as its lines, that is,

$$\mathbf{u} \cdot (\mathbf{v} \times \mathbf{w}) = \begin{vmatrix} u_1 & u_2 & u_3 \\ v_1 & v_2 & v_3 \\ w_1 & w_2 & w_3 \end{vmatrix}.$$ (2.39)

Note that the mixed product $\mathbf{u} \cdot (\mathbf{v} \times \mathbf{w})$ will be positive if $\mathbf{u}$ and $(\mathbf{v} \times \mathbf{w})$ are on the same side of the plane given by $\mathbf{v}$ and $\mathbf{w}$ and negative when the contrary. The mixed product has the cyclical property

$$\mathbf{u} \cdot (\mathbf{v} \times \mathbf{w}) = \mathbf{v} \cdot (\mathbf{w} \times \mathbf{u}) = \mathbf{w} \cdot (\mathbf{u} \times \mathbf{v}).$$ (2.40)

The result of this relation is that if two of the three vectors $\mathbf{u}$, $\mathbf{v}$, or $\mathbf{w}$ are parallel, its mixed product will be null.

A.2.6 Double Vector Product

The *double vector product* $\mathbf{u} \times (\mathbf{v} \times \mathbf{w})$ defines, of course, a vector that is orthogonal to $\mathbf{u}$ and to vector $\mathbf{v} \times \mathbf{w}$. Since $\mathbf{v} \times \mathbf{w}$ is perpendicular to the plane defined by $\mathbf{v}$ and $\mathbf{w}$, then the result of the double vector product is necessarily parallel to this plane (see Fig. 2.8).

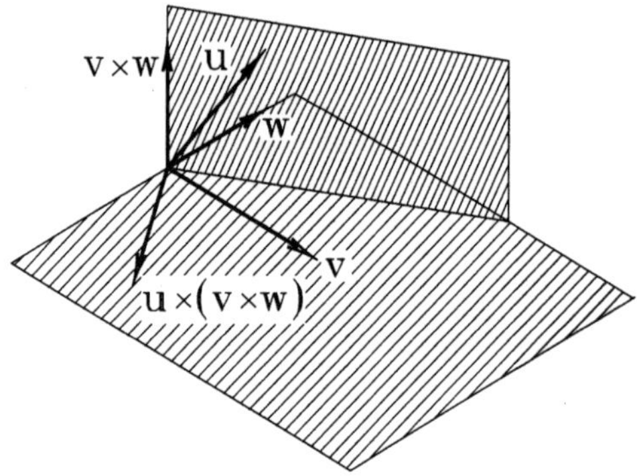

Figure 2.8

It may be shown that the double vector product has the algebraic representation

$$\mathbf{u} \times (\mathbf{v} \times \mathbf{w}) = (\mathbf{u} \cdot \mathbf{w})\mathbf{v} - (\mathbf{u} \cdot \mathbf{v})\mathbf{w}.$$ (2.41)

From the above relation, when $\mathbf{u} = \mathbf{w} = \mathbf{n}_r$, the unit vector in direction r, we have

$$\mathbf{n}_r \times (\mathbf{v} \times \mathbf{n}_r) = (\mathbf{n}_r \cdot \mathbf{n}_r)\mathbf{v} - (\mathbf{n}_r \cdot \mathbf{v})\mathbf{n}_r = \mathbf{v} - (\mathbf{n}_r \cdot \mathbf{v})\mathbf{n}_r.$$ (2.42)

Therefore, $\mathbf{n}_r \times (\mathbf{v} \times \mathbf{n}_r)$ is the component of $\mathbf{v}$ in the direction orthogonal to r (see Fig. 2.9).

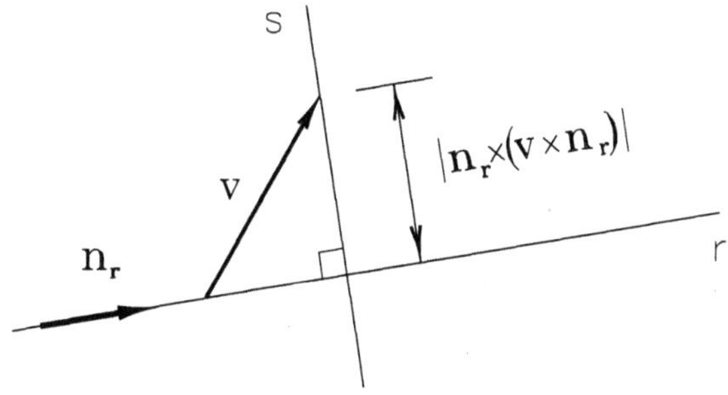

Figure 2.9

Example 2.3 Let O:$(0,0,0)$, P:$(1,2,3)$, and Q:$(2,1,2)$ be three given points (coordinates in meters) of a plane. Now let $\mathbf{p}$, $\mathbf{q}$, and $\mathbf{r}$ be the position vectors from O to P, from O to Q, and from Q to P, respectively. The sides of the triangle OPQ will, of course, be the modules of those vectors p, q, and r, respectively. We want to determine the height a of the triangle, relative to side OQ. Now, the height relative to one side is the perpendicular lowered from the opposite vertex. Therefore, this means determining the module of component $\mathbf{p}_\perp$ of vector $\mathbf{p}$ orthogonal to vector $\mathbf{q}$. Since $\mathbf{p} = (1,2,3)$, $\mathbf{q} = (2,1,2)$, and $\mathbf{q}$ is not a unit vector, then

$$\mathbf{p}_\perp = \frac{1}{q^2} \mathbf{q} \times (\mathbf{p} \times \mathbf{q})$$
$$= \frac{1}{9}(-11, 8, 7).$$

The height to be found, then, will be equal to the module of the vector obtained:

$$a = |\mathbf{p}_\perp| = \frac{1}{9}\sqrt{(-11)^2 + 8^2 + 7^2} = \sqrt{26}\ \text{m}.$$

A.3 Tensors

A tensor is an operator that associates one given vector $\mathbf{v}$ with another vector $T\mathbf{v}$, according to the following properties:

$$T(\mathbf{v} + \mathbf{w}) = T\mathbf{v} + T\mathbf{w}; \tag{3.1}$$

$$T(a\mathbf{v}) = aT\mathbf{v}. \tag{3.2}$$

As defined, the tensor is equivalent to a *linear transformation* or *linear operator* that leads one vector to another. The *null tensor* 0 associates with any vector $\mathbf{v}$ in space the null vector

$$0\mathbf{v} = \mathbf{0}. \tag{3.3}$$

The *unit tensor* $\mathbb{1}$ transforms any vector $\mathbf{v}$ in itself, that is,

$$\mathbb{1}\mathbf{v} = \mathbf{v}. \tag{3.4}$$

The sum of two tensors T and S is the tensor $T + S$ given by

$$(T + S)\mathbf{v} = T\mathbf{v} + S\mathbf{v}, \tag{3.5}$$

and the product of the scalar a by the tensor T is defined by

$$(aT)\mathbf{v} = a(T\mathbf{v}). \tag{3.6}$$

The *diadic product* or *tensor product* of vectors $\mathbf{u}$ and $\mathbf{v}$ is defined as the tensor $T = \mathbf{u} \otimes \mathbf{v}$ so that, for any vector $\mathbf{w}$,

$$T\mathbf{w} = (\mathbf{u} \otimes \mathbf{v})\mathbf{w} = (\mathbf{v} \cdot \mathbf{w})\mathbf{u}. \tag{3.7}$$

It can be shown that that operators $T + S$, aT, and $\mathbf{u} \otimes \mathbf{v}$ meet properties (3.1) and (3.2). In the same way as vectors can be expressed through their scalar components with respect to a system of orthogonal coordinates $\{x_1, x_2, x_3\}$, a tensor can be represented in this system, considering T as the linear transformation that leads $\mathbf{u} = (u_1, u_2, u_3)$ in $T\mathbf{u} = \mathbf{v} = (v_1, v_2, v_3)$, so that

$$T_{11}u_1 + T_{12}u_2 + T_{13}u_3 = v_1,$$

$$T_{21}u_1 + T_{22}u_2 + T_{23}u_3 = v_2, \tag{3.8}$$

$$T_{31}u_1 + T_{32}u_2 + T_{33}u_3 = v_3.$$

This transformation of the components of $\mathbf{u}$ in the respective components of $\mathbf{v}$ can be compactly represented by the system of linear equations

$$T\mathbf{u} = \mathbf{v}, \tag{3.9}$$

where $\mathbf{u} = \operatorname{col}[u_1, u_2, u_3]$ and $\mathbf{v} = \operatorname{col}[v_1, v_2, v_3]$ indicate column matrices and

$$T = \begin{pmatrix} T_{11} & T_{12} & T_{13} \\ T_{21} & T_{22} & T_{23} \\ T_{31} & T_{32} & T_{33} \end{pmatrix} \tag{3.10}$$

is the square matrix of third order whose elements T_{ij} are called *tensor components* T in the given system of coordinates. The matrix $[T] = [T_{ij}]$ is called the matrix of tensor T with respect to the system of coordinates $\{x_1, x_2, x_3\}$.

A.3.1 Product of Tensors

The product of two tensors T and S may be defined as the tensor TS so that

$$(TS)\mathbf{v} = T(S\mathbf{v}). \tag{3.11}$$

Equation (3.11) establishes that the composition of T and S is obtained by the successive application of S and T to vector $\mathbf{v}$, in that order. It may be shown that if T and S fulfill properties (3.1) and (3.2), TS, defined by Eq. (3.11), also obeys those rules of linearity, that is, TS is also a tensor. The product of two tensors obeys these properties:

$$(TS)R = T(SR);$$

$$T(R + S) = TR + TS;$$

$$(R + S)T = RT + ST; \tag{3.12}$$

$$a(TS) = T(aS);$$

$$\mathbb{1}T = T\mathbb{1} = T.$$

In terms of the matrix representation associated with a system of orthogonal coordinates, the product of two tensors, TS, and the rules given by Eqs. (3.12) are equivalent to the properties of matrix algebra. In this context, it is found that the product of two tensors is not commutative, that is,

$$TS \neq ST. \tag{3.13}$$

When equality is valid for two particular tensors T and S, it is said that they *commute*.

A.3.2   Inverse Tensor

A tensor is *invertible* when there is another tensor T^{-1}, called the *inverse* of T, so that

$$T^{-1}T = TT^{-1} = \mathbb{1}. \tag{3.14}$$

When applying the tensor T^{-1} to vector $\mathbf{v} = T\mathbf{u}$, of Eq. (3.9), we see that

$$T^{-1}\mathbf{v} = T^{-1}T\mathbf{u} = \mathbb{1}\mathbf{u} = \mathbf{u}. \tag{3.15}$$

Therefore, the linear system $T\mathbf{u} = \mathbf{v}$ can be solved through the inverse tensor to obtain the solution vector $\mathbf{u} = T^{-1}\mathbf{v}$. It can be shown that the necessary and sufficient condition for the existence of T^{-1} is that $|T| \neq 0$. The inverse tensor, if any, is unique. It may be shown that the product TS is invertible if T and S are also invertible, the following relation being valid:

$$(TS)^{-1} = S^{-1}T^{-1}. \tag{3.16}$$

Tensors that do not have inverses are called *singular tensors*. The null tensor does not have an inverse, since $|0| = 0$. Similarly, the dyadic product $\mathbf{u} \otimes \mathbf{v}$ is not invertible. This may be seen when developing Eq. (3.7) in terms of components, in order to obtain the matrix representation

$$\mathbf{u} \otimes \mathbf{v} = \begin{pmatrix} u_1 v_1 & u_1 v_2 & u_1 v_3 \\ u_2 v_1 & u_2 v_2 & u_2 v_3 \\ u_3 v_1 & u_3 v_2 & u_3 v_3 \end{pmatrix}. \tag{3.17}$$

The above equation shows that the matrix of $\mathbf{u} \otimes \mathbf{v}$ has proportional lines and, therefore,

$$|\mathbf{u} \otimes \mathbf{v}| = 0. \tag{3.18}$$

A.3.3 Transposed Tensor

The tensor T^T, called the *transposed* of tensor T, is defined by

$$(T^T \mathbf{u}) \cdot \mathbf{v} = \mathbf{u} \cdot (T\mathbf{v}) \tag{3.19}$$

and obeys the following properties:

$$(T + S)^T = T^T + S^T; \tag{3.20}$$

$$(aT)^T = aT^T; \tag{3.21}$$

$$(TS)^T = S^T T^T. \tag{3.22}$$

In terms of the matrix representation, let $[T] = [T_{ij}]$. By the representation of the dot product of vectors, Eq. (2.23) and by Eq. (3.8), it may be shown that $T^T = [T_{ji}]$, that is, the operation of transposition is equivalent to transforming the columns of T in the lines of T^T.

A.3.4 Symmetric and Antisymmetric Tensor

A tensor is called *symmetric* when

$$T^T = T, \tag{3.23}$$

that is, $T_{ij} = T_{ji}$ for every i, j. From Eq. (3.19), if T is symmetric, then

$$\mathbf{u} \cdot T\mathbf{v} = \mathbf{v} \cdot T\mathbf{u}. \tag{3.24}$$

The tensor so that

$$T = -T^T \tag{3.25}$$

is called *antisymmetric*, resulting hence that, in this case,

$$T_{ij} = -T_{ji} \tag{3.27}$$

and

$$T_{ii} = 0. \tag{3.28}$$

It may be shown that tensor $T+T^T$ is symmetric (just transpose this sum to check the property) and tensor $T - T^T$ is antisymmetric. Therefore, any tensor T can be broken down into the sum of a symmetric tensor $S = \frac{1}{2}(T + T^T)$ and antisymmetric tensor $A = \frac{1}{2}(T - T^T)$.

A.3.5 Orthogonal Tensor

A tensor is called *orthogonal* if it is invertible, that is, if

$$|T| \neq 0 \tag{3.29}$$

and

$$T^{-1} = T^T. \tag{3.30}$$

Therefore, if T is an orthogonal tensor, then the following relation is satisfied:

$$T^T T = \mathbb{1}. \tag{3.31}$$

Now consider an orthogonal tensor T and vectors $\mathbf{u}_0$, $\mathbf{v}_0$, $\mathbf{u}$, and $\mathbf{v}$, so that

$$T\mathbf{u} = \mathbf{u}_0 \tag{3.32}$$

and

$$T\mathbf{v} = \mathbf{v}_0. \tag{3.33}$$

The dot product $\mathbf{u}_0 \cdot \mathbf{v}_0$, in view of the above equations, is written as

$$\mathbf{u}_0 \cdot \mathbf{v}_0 = T\mathbf{u} \cdot T\mathbf{v}, \tag{3.34}$$

which, by Eq. (3.19), becomes

$$\mathbf{u}_0 \cdot \mathbf{v}_0 = \mathbf{u} \cdot T^T(T\mathbf{v}). \tag{3.35}$$

Since T is orthogonal, it follows that

$$\mathbf{u}_0 \cdot \mathbf{v}_0 = \mathbf{u} \cdot \mathbb{1}\mathbf{v} = \mathbf{u} \cdot \mathbf{v}. \qquad (3.36)$$

The result, therefore, is that the dot product of vectors $\mathbf{u}_0$ and $\mathbf{v}_0$, transformed linearly by the orthogonal tensor T, is identical to the dot product of vectors $\mathbf{u}$ and $\mathbf{v}$. Since the angle between the directions of two vectors or the module of a given vector can be determined through the dot product, the result obtained shows that angles and magnitudes are preserved by an orthogonal linear transformation. In particular, mutually orthogonal unit vectors remain orthonormal when transformed by an orthogonal tensor.

Example 3.1   Consider the vectors $\mathbf{u} = (5,0)$ and $\mathbf{v} = (4,3)$ and the tensor T, expressed in the same basis by

$$T = \begin{pmatrix} 1 & 0 \\ 0 & -1 \end{pmatrix}.$$

Then $|\mathbf{u}| = 5$, $|\mathbf{v}| = 5$, and $\mathbf{u} \cdot \mathbf{v} = 20$. By now applying the linear transformation T on $\mathbf{u}$ and $\mathbf{v}$, we have

$$\mathbf{u}' = T\mathbf{u} = \begin{pmatrix} 1 & 0 \\ 0 & -1 \end{pmatrix} \begin{Bmatrix} 5 \\ 0 \end{Bmatrix} = (5,0)$$

and

$$\mathbf{v}' = T\mathbf{v} = \begin{pmatrix} 1 & 0 \\ 0 & -1 \end{pmatrix} \begin{Bmatrix} 4 \\ 3 \end{Bmatrix} = (4,-3),$$

with

$$|\mathbf{u}'| = 5, \quad |\mathbf{v}'| = 5, \quad \text{and} \quad \mathbf{u}' \cdot \mathbf{v}' = 20.$$

Therefore, transformation T keeps angles and distances unchanged. In fact,

$$T^T = \begin{pmatrix} 1 & 0 \\ 0 & -1 \end{pmatrix} = T \quad \text{and} \quad T^T T = \mathbb{1},$$

therefore being an orthogonal tensor.

A.3.6　Trace of a Tensor

The *trace* of a tensor T is the real number defined as

$$\operatorname{tr}(T) = \sum_{i=1}^{3} T_{ii}.$$
(3.37)

The following properties are valid:

$$\operatorname{tr}(T + S) = \operatorname{tr}(T) + \operatorname{tr}(S);$$
(3.38)

$$\operatorname{tr}(aT) = a\operatorname{tr}(T);$$
(3.39)

$$\operatorname{tr}(\mathbf{u} \otimes \mathbf{v}) = \mathbf{u} \cdot \mathbf{v};$$
(3.40)

$$\operatorname{tr}(T^{T}) = \operatorname{tr}(T);$$
(3.41)

$$\operatorname{tr}(TS) = \operatorname{tr}(ST).$$
(3.42)

All those properties are easily shown from definition (3.37) and the respective tensor operations.

A.3.7　Linear Systems

Consider a system of linear equations of the type

$$
\begin{aligned}
a_{11}x_1 + a_{12}x_2 + \cdots + a_{1n}x_n &= b_1 \\
a_{21}x_1 + a_{22}x_2 + \cdots + b_{2n}x_n &= b_1 \\
&\vdots \\
a_{n1}x_1 + a_{n2}x_2 + \cdots + a_{nn}x_n &= b_n,
\end{aligned}
$$
(3.43)

where $x_1, x_2, \ldots, x_n$ are unknown, $b_1, b_2, \ldots, b_n$ are known values, and a_{ij}, $i, j = 1, 2, \ldots, n$, are the *coefficients*, also known. This system can be represented more compactly as

$$A\mathbf{x} = \mathbf{b},$$
(3.44)

where A is the *coefficient matrix*, defined by

$$
A = \begin{pmatrix}
a_{11} & a_{12} & \cdots & a_{1n} \\
a_{21} & a_{22} & \cdots & a_{2n} \\
& & \ddots & \\
a_{n1} & a_{n2} & \cdots & a_{nn}
\end{pmatrix},
\tag{3.45}
$$

$\mathbf{x} = (x_1, x_2, \ldots, x_n)$ is the unknown vector, and $\mathbf{b} = (b_1, b_2, \ldots, b_n)$ is a given vector.

Solving the system means to determine $\mathbf{x}$, given tensor A and vector $\mathbf{b}$. Of course, the key to the general solution of this problem lies in determining the inverse tensor A^{-1}, which, applied to vector $\mathbf{b}$, will give vector $\mathbf{x}$, as shown in Section A.3.2. An easy way to obtain the desired solution is *Cramer's rule*, which consists of calculating the determinant of the coefficient matrix, Δ, and the determinants characteristic of the unknown quantities, Δ_j, which are the determinants of the matrices obtained by substituting, in the coefficien matrix, the jth column by the vector $\mathbf{b}$, that is,

$$
\Delta_j = \begin{vmatrix}
a_{11} & a_{12} & \cdots & a_{1j-1} & b_1 & a_{1j+1} & \cdots & a_{1n} \\
a_{21} & a_{22} & \cdots & a_{2j-1} & b_2 & a_{2j+1} & \cdots & a_{2n} \\
& & \ddots & & & & & \\
a_{n1} & a_{n2} & \cdots & a_{nj-1} & b_n & a_{nj+1} & \cdots & a_{nn}
\end{vmatrix}, \quad j = 1, 2, \ldots, n. \tag{3.46}
$$

Each unknown quantity is then obtained from the quotient between its characteristic determinant and the determinant of the coefficient matrix, that is,

$$
x_j = \frac{\Delta_j}{\Delta}, \qquad j = 1, 2, \ldots, n. \tag{3.47}
$$

Example 3.2 Let a linear system consist of the three equations

$$
2x + y - 2z = 3,
$$
$$
-x + 2y + z = 11,
$$
$$
5x - 3y + 5z = 0,
$$

so that it may be expressed alternatively by

$$
\begin{pmatrix}
2 & 1 & -2 \\
-1 & 2 & 1 \\
5 & -3 & 5
\end{pmatrix}
\begin{Bmatrix} x \\ y \\ z \end{Bmatrix}
= \begin{Bmatrix} 3 \\ 11 \\ 0 \end{Bmatrix}.
$$

The determinant of the coefficient matrix is

$$\Delta = \begin{vmatrix} 2 & 1 & -2 \\ -1 & 2 & 1 \\ 5 & -3 & 5 \end{vmatrix} = 50.$$

The characteristic determinant for the coordinate x is

$$\Delta_x = \begin{vmatrix} 3 & 1 & -2 \\ 11 & 2 & 1 \\ 0 & -3 & 5 \end{vmatrix} = 50; \qquad \text{therefore,} \qquad x = \frac{\Delta_x}{\Delta} = 1.$$

Likewise,

$$\Delta_y = \begin{vmatrix} 2 & 3 & -2 \\ -1 & 11 & 1 \\ 5 & 03 & 5 \end{vmatrix} = 250; \qquad \text{therefore,} \qquad y = \frac{\Delta_y}{\Delta} = 5.$$

And, last,

$$\Delta_z = \begin{vmatrix} 2 & 1 & 32 \\ -1 & 2 & 11 \\ 5 & -3 & 0 \end{vmatrix} = 100; \qquad \text{therefore,} \qquad z = \frac{\Delta_z}{\Delta} = 2.$$

The roots are, therefore, $x = 1$, $y = 5$, and $z = 2$.

A.4 Eigenvalues and Eigenvectors

A tensor T applied to a nonnull vector $\mathbf{u}$ can result in another vector that is a scalar multiple of the original vector $\mathbf{u}$. If

$$T\mathbf{u} = \lambda\mathbf{u}, \qquad \mathbf{u} \neq \mathbf{0}, \tag{4.1}$$

then λ is called an *eigenvalue* of T and $\mathbf{u}$ is called an *eigenvector* of T associated with this eigenvalue. Rewriting Eq. (4.1) as

$$(T - \lambda\mathbb{1})\mathbf{u} = \mathbf{0}, \tag{4.2}$$

considering the matrix $[T_{ij}]$ associated with T and breaking down the vector $\mathbf{u}$ in the same system of coordinates, we can see that the eigenvalues T can be obtained from the homogeneous system of linear equations, defined by the equation

$$\begin{pmatrix} (T_{11} - \lambda) & T_{12} & T_{13} \\ T_{21} & (T_{22} - \lambda) & T_{23} \\ T_{31} & T_{32} & (T_{33} - \lambda) \end{pmatrix} \begin{pmatrix} u_1 \\ u_2 \\ u_3 \end{pmatrix} = \mathbf{0}, \qquad (4.3)$$

which has a nontrivial solution $\mathbf{u} \neq \mathbf{0}$ provided that

$$|T - \lambda \mathbb{1}| = 0. \qquad (4.4)$$

The algebraic expansion of this determinant leads to a third-degree equation, called the *characteristic equation*. Its roots are the eigenvalues of T. Once the eigenvalues are obtained, a eigenvector $\mathbf{u}_i$ associated with the eigenvalue λ_i, $i = 1, 2, 3$, is obtained, thereby solving the linear system

$$T\mathbf{u}_i = \lambda_i \mathbf{u}_i \qquad (4.5)$$

to obtain the components of $\mathbf{u}_i$. It should be noted that the eigenvalue equation is a homogeneous equation and, therefore, the scalar components of the eigenvector $\mathbf{u}_i$ can only be determined with an arbitrary multiplicative constant, that is, if $\mathbf{u}_i$ is an eigenvector of T associated with the eigenvalue λ_i, then $k\mathbf{u}_i$ $(k \neq 0)$ is also an eigenvector of T associated with the eigenvalue λ_i. This is the same as saying that only the *directions* of the eigenvectors can be determined and not their *magnitudes*. As we are free to choose any multiplicative constant k, it is very often convenient to make the eigenvectors of T also have a module equal to 1. To do so, just divide each eigenvector $\mathbf{u}_i$ by its module, obtaining the eigenvectors

$$\mathbf{e}_i = \frac{\mathbf{u}_i}{\sqrt{\mathbf{u}_i \cdot \mathbf{u}_i}}, \qquad i = 1, 2, 3, \qquad (4.6)$$

which are called *normalized eigenvectors* of T. As we are especially interested in symmetric tensors, we discuss below some properties of the eigenvalues and eigenvectors of real and symmetric matrices.

The eigenvalues of a real and symmetric matrix are real.

In fact, let

$$T\mathbf{u}_i = \lambda_i \mathbf{u}_i, \qquad (4.7)$$

with λ_i being a complex number. Conjugating this equation, that is, taking the conjugated complex of both members, we obtain

$$T\mathbf{u}_i^* - \lambda_i^*\mathbf{u}_i^*, \tag{4.8}$$

since T is a real matrix. By dot-multiplying Eq. (4.7) by $\mathbf{u}_i^*$ and Eq. (4.8) by $\mathbf{u}_i$, we obtain

$$\mathbf{u}_i^* \cdot T\mathbf{u}_i = \lambda_i\mathbf{u}_i^* \cdot \mathbf{u}_i \tag{4.9}$$

and

$$\mathbf{u}_i \cdot T\mathbf{u}_i^* = \lambda_i^*\mathbf{u}_i \cdot \mathbf{u}_i^*. \tag{4.10}$$

Subtracting member by member Eqs. (4.9) and (4.10), then

$$\mathbf{u}_i^* \cdot T\mathbf{u}_i - \mathbf{u}_i \cdot T\mathbf{u}_i^* = (\lambda_i - \lambda_i^*)\,\mathbf{u}_i \cdot \mathbf{u}_i^*, \tag{4.11}$$

since $\mathbf{u}_i^* \cdot \mathbf{u}_i = \mathbf{u}_i \cdot \mathbf{u}_i^*$. As T is symmetric, it is natural that $\mathbf{u}_i^* \cdot T\mathbf{u}_i = \mathbf{u}_i \cdot T^T\mathbf{u}_i^* = \mathbf{u}_i \cdot T\mathbf{u}_i^*$, and Eq. (4.11) becomes

$$(\lambda_i - \lambda_i^*)\mathbf{u}_i \cdot \mathbf{u}_i^* = 0. \tag{4.12}$$

Since $\mathbf{u}_i \cdot \mathbf{u}_i^* > 0$ for $\mathbf{u}_i \neq \mathbf{0}$, it so happens that

$$\lambda_i = \lambda_i^*; \tag{4.13}$$

therefore, λ_i is *real*.

Eigenvectors associated with the different eigenvalues of a real and symmetric matrix are orthogonal.

Let λ_1 and λ_2 be different eigenvalues of real and symmetric T, with the associated respective eigenvectors $\mathbf{u}_1$ and $\mathbf{u}_2$. Then, since

$$T\mathbf{u}_1 = \lambda_1\mathbf{u}_1 \tag{4.14}$$

and

$$T\mathbf{u}_2 = \lambda_2\mathbf{u}_2, \tag{4.15}$$

we can dot-multiply the first equation by $\mathbf{u}_2$ and the second by $\mathbf{u}_1$, obtaining

$$\mathbf{u}_2 \cdot T\mathbf{u}_1 = \lambda_1\mathbf{u}_2 \cdot \mathbf{u}_1 \tag{4.16}$$

and

$$\mathbf{u}_1 \cdot T\mathbf{u}_2 = \lambda_2 \mathbf{u}_1 \cdot \mathbf{u}_2. \tag{4.17}$$

Subtracting member by member Eqs. (4.16) and (4.17), then

$$\mathbf{u}_2 \cdot T\mathbf{u}_1 - \mathbf{u}_1 \cdot T\mathbf{u}_2 = (\lambda_1 - \lambda_2)(\mathbf{u}_2 \cdot \mathbf{u}_1) \tag{4.18}$$

and, since T is symmetric,

$$\mathbf{u}_2 \cdot T\mathbf{u}_1 = \mathbf{u}_1 \cdot T^T\mathbf{u}_2 = \mathbf{u}_1 \cdot T\mathbf{u}_2; \tag{4.19}$$

therefore, Eq. (4.18) becomes

$$(\lambda_1 - \lambda_2)(\mathbf{u}_1 \cdot \mathbf{u}_2) = 0. \tag{4.20}$$

Since, by hypothesis, $\lambda_1 \neq \lambda_2$, it results that

$$\mathbf{u}_1 \cdot \mathbf{u}_2 = 0 \qquad \text{and, therefore,} \quad \mathbf{u}_1 \perp \mathbf{u}_2. \tag{4.21}$$

Hence the result that if the eigenvalues of real and symmetric T are different, for each eigenvalue of T there will be an associated eigenvector that will be orthogonal to the other eigenvectors. Therefore, the three eigenvectors of T form an orthogonal set that is a basis for the Cartesian space. Let $\mathbf{n}_1, \mathbf{n}_2, \mathbf{n}_3$ be the orthogonal normalized eigenvectors of T. Any vector of space can be expressed in this basis as

$$\mathbf{u} = \sum_{i=1}^{3} a_i \mathbf{n}_i \tag{4.22}$$

and the scalar components of $\mathbf{u}$ in the basis $\mathbf{n}_1, \mathbf{n}_2, \mathbf{n}_3$ are easily obtained by

$$a_i = \mathbf{u} \cdot \mathbf{n}_i. \tag{4.23}$$

The vector $\mathbf{v} = T\mathbf{u}$ can also be easily represented in this basis:

$$\mathbf{v} = T\left(\sum_{i=1}^{3} a_i \mathbf{n}_i\right) = \sum_{i=1}^{3} a_i T\mathbf{n}_i = \sum_{i=1}^{3} a_i \lambda_i \mathbf{n}_i. \tag{4.24}$$

If an eigenvalue λ_j of a real and symmetric T have multiplicity $k \geq 2$, there are k orthonormal eigenvectors associated with λ_j.

These k eigenvectors form a basis for the k-dimensional subspace and there are countless ways of selecting them. This result, which we will not demonstrate here, allows us to say that there is at least one basis of eigenvectors of a real and symmetric T for the three-dimensional space.

Example 4.1   Supposing that the tensor T is given, whose matrix, in a given system of coordinates is

$$T = \begin{pmatrix} 1 & 1 & 0 \\ 2 & 1 & 0 \\ 0 & 0 & 3 \end{pmatrix},$$

we want to determine, in the same system of coordinates, its lowest eigenvalue and corresponding normalized eigenvector. The characteristic equation for the given tensor is

$$(3 - \lambda)[(1 - \lambda)^2 - 2] = 0,$$

whose solutions are $\lambda_1 = 1 - \sqrt{2}$, $\lambda_2 = 1 + \sqrt{2}$, and $\lambda_3 = 3$. The lowest eigenvalue is, therefore, $\lambda_1 = 1 - \sqrt{2}$. The corresponding eigenvector, $\mathbf{u}$, will be the one that satisfies the equation

$$\begin{pmatrix} 1 & 1 & 0 \\ 2 & 1 & 0 \\ 0 & 0 & 3 \end{pmatrix} \begin{Bmatrix} u_1 \\ u_2 \\ u_3 \end{Bmatrix} = (1 - \sqrt{2}) \begin{Bmatrix} u_1 \\ u_2 \\ u_3 \end{Bmatrix},$$

equivalent to

$$u_1 + u_2 = (1 - \sqrt{2})\, u_1,$$
$$2u_1 + u_2 = (1 - \sqrt{2})\, u_2,$$
$$3u_3 = (1 - \sqrt{2})\, u_3.$$

Of course, vector $\mathbf{u}$ is determined by these equations less a constant, since every vector parallel to a given direction will be a eigenvector of the tensor T associated with eigenvalue λ_1. To obtain the normalized eigenvector, the relation $u_1^2 + u_2^2 + u_3^2 = 1$ must be fulfilled, thereby providing the solution

$$\mathbf{u} = \frac{1}{\sqrt{3}}(1, -\sqrt{2}, 0).$$

Exercise Series #13   (Sections A.1 to A.4)

P13.1 Given the vectors $\mathbf{a} = (1, 1, 1)$, $\mathbf{b} = (1, -1, \alpha)$, and $\mathbf{c} = (1, \beta, \gamma)$, determine α, β, and γ so that $\mathbf{a}$, $\mathbf{b}$, and $\mathbf{c}$ form an orthogonal basis.

P13.2 Calculate the area of the parallelogram whose adjacent sides are the vectors $\mathbf{a} = (-1, 2, 0)$ and $\mathbf{b} = (0, -2, -3)$ (components in cm).

P13.3 Resolve the vector $\mathbf{v}$, with module equal to five units, in three mutually orthogonal components so that their modules are in the proportion $1 : 2 : 2$. Determine the angles between $\mathbf{v}$ and each of its three components.

P13.4 To which values of α are the vectors $\mathbf{a} = (5, \alpha, 2)$, $\mathbf{b} = (1, 0, \alpha)$, and $\mathbf{c} = (3, 4, -1)$ coplanar?

P13.5 Determine a unit vector $\mathbf{n}$ parallel to the intersection of the planes defined by equations $x + y - 2z = 5$ and $4x - 3y + 2z = 0$.

P13.6 Determine p and q so that the vector $\mathbf{v} = (p, q, -1)$ is parallel to the planes defined by $4x + y + 6z = -2$ and $x + y + z = 0$.

P13.7 Calculate the angle between planes $x + 2y + z = 0$ and $2x - y + 3z = 0$.

P13.8 Given the four points $A(2, 1, 3)$, $B(1, 2, 1)$, $C(-1, -2, -2)$, and $D(1, -4, 0)$, determine the shortest distance between the line passing through A and B and the line passing through C and D.

P13.9 Let $\mathbf{a}$, $\mathbf{b}$, and $\mathbf{c} = -\mathbf{b} - \mathbf{a}$ be the sides of a triangle. Demonstrate that $\mathbf{c} \times \mathbf{a} = \mathbf{a} \times \mathbf{b} = \mathbf{b} \times \mathbf{c}$, thereby proving the *sine law* of trigonometry.

P13.10 Determine a unit vector $\mathbf{n}$ orthogonal to the plane defined by $3x + 4y + 5\sqrt{3}z = 17$.

P13.11 Given the vectors $\mathbf{a} = (1, 1, -1)$, $\mathbf{b} = (1, 0, \alpha)$, and $\mathbf{c} = (1, 2, 1)$, determine α so that $\mathbf{a} \times \mathbf{b} \cdot \mathbf{c}$ is null.

P13.12 Given $\mathbf{u} = (1, 1, 1)$, determine the component of $\mathbf{v} = (-1, 2, 0)$ in the direction orthogonal to $\mathbf{u}$.

P13.13 Let

$$T = \begin{pmatrix} \frac{1}{2} & \frac{\sqrt{3}}{2} & 0 \\ -\frac{\sqrt{3}}{2} & \frac{1}{2} & 0 \\ 0 & 0 & 1 \end{pmatrix}$$

be an orthogonal tensor and $\mathbf{u}_0 = \frac{1}{2}(\sqrt{2}, \sqrt{2}, 0)$ and $\mathbf{v}_0 = \frac{1}{2}(-\sqrt{2}, \sqrt{2}, 0)$ two orthonormal unit vectors. Show that vectors $\mathbf{u} = T\mathbf{u}_0$ and $\mathbf{v} = T\mathbf{v}_0$ are also orthonormal.

P13.14 Given the vectors $\mathbf{a} = (1,1,0)$, $\mathbf{b} = (0,1,1)$, and $\mathbf{c} = (1,1,-2)$, determine a unit vector $\mathbf{n}$ so that $(\mathbf{a} \otimes \mathbf{b})\mathbf{c} \cdot \mathbf{n} = 0$ and, simultaneously, $(\mathbf{b} \otimes \mathbf{c})(\mathbf{a} \times \mathbf{n}) = \mathbf{b}$.

P13.15 Are there vectors $\mathbf{a}$ and $\mathbf{b}$ so that $\mathbf{a} \otimes \mathbf{b} = \mathbb{1}$?

P13.16 Show that, given three vectors $\mathbf{a}$, $\mathbf{b}$, and $\mathbf{c}$, linearly independent, then

$$(\mathbf{a} \otimes \mathbf{b} - \mathbf{b} \otimes \mathbf{a})\mathbf{c} = \mathbf{c} \times (\mathbf{a} \times \mathbf{b}).$$

P13.17 Solve the system

$$x + y + z = 2,$$
$$3x + y + \frac{1}{2}z = 3,$$
$$2x - z = 0.$$

P13.18 Let the symmetric tensor be

$$T = \begin{pmatrix} 0 & 1 \\ 1 & 0 \end{pmatrix}.$$

Determine the normalized eigenvectors of T.

Linkages

This appendix presents a small set of models for *linkages*. It illustrates the most common models for mechanical linkages rather than covering all possible cases of interaction between two bodies through contact. They are presented in a logical sequence, starting with a *clamp* or *welding*, which, as it does not admit any displacement or rotation, responds with three force and three torque components. Each model is then reached by loosening each constraint in succession, by allowing either a displacement in or a rotation around a certain direction. The usual models for the *linear elastic spring*, *viscous damper*, and *fluctuation force* are also indicated at the end.

The reader is recommended to carefully study in detail the models for and the compositions of such linkages. It is fundamental to master this modeling to be able to correctly build what are called *free body diagrams*, without which, in turn, any analysis in dynamics will be irretrievably prejudiced.

The left column indicates the most common name for the type of linkage. The second column illustrates what the physical model of the linkage is like and the associated coordinated axes. The third column indicates the force and torque components present in the linkage, for the chosen system of Cartesian coordinates. Last, the fourth column points out the *kinematic constraints* associated to the linkage and other pertinent information.

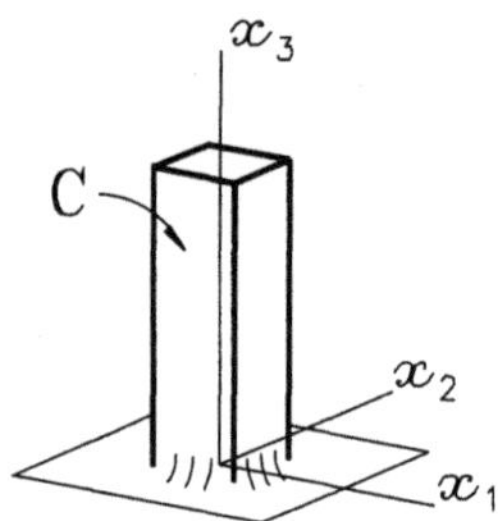

B.01
Clamp
or
welding

$F_1,\ F_2,\ F_3$

$T_1,\ T_2,\ T_3$

Does not admit
displacements or
rotatations.

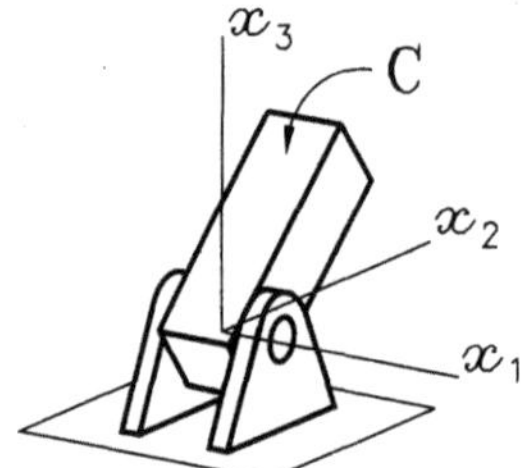

B.02
Pin
or
pivot

$F_1,\ F_2,\ F_3$

$T_2,\ T_3$

Only admits
rotation in the
x_1-direction.

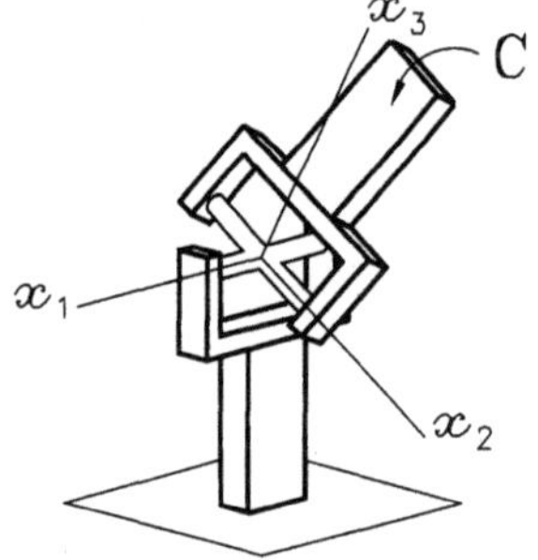

B.03
Universal
joint

$F_1,\ F_2,\ F_3$

T_3

Admits rotation
in the x_1- and
x_2-directions.

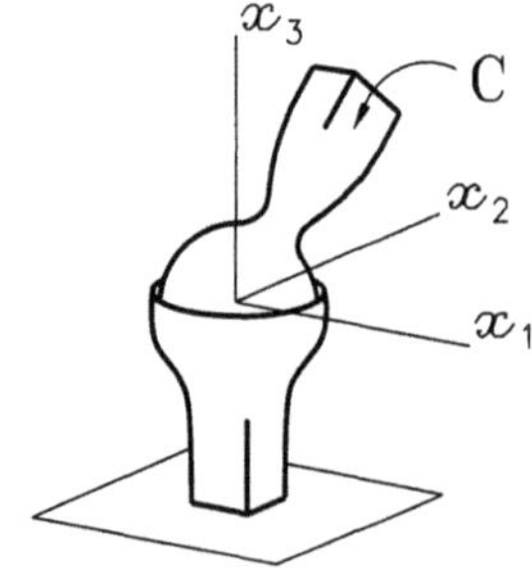

B.04
Ball and
socket
joint

$F_1,\ F_2,\ F_3$

Admits rotation in the
three directions. Does not
admit displacements.

B.05
Sliding
cursor

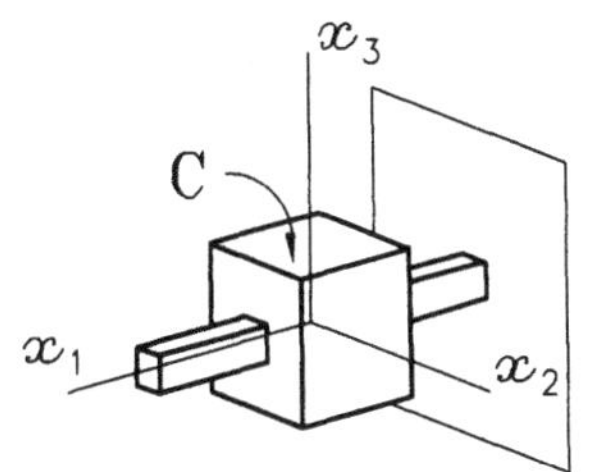

$F_2,\ F_3$

$T_1,\ T_2,\ T_3$

Only admits
displacement in the
x_1-direction.

B.06
Double
cursor

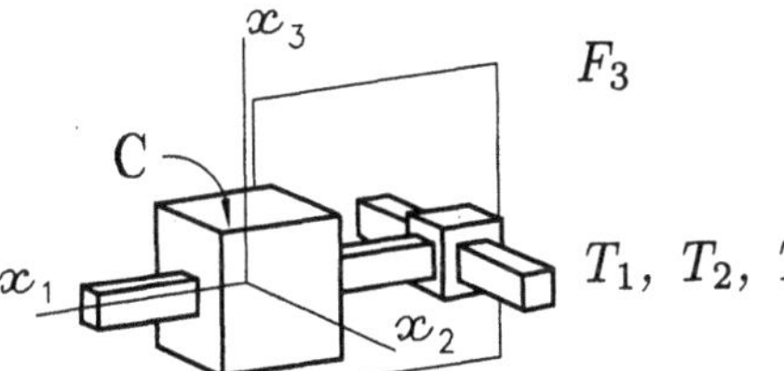

F_3

$T_1,\ T_2,\ T_3$

Admits displacement
in the x_1- and
x_2-directions.

B.07
Triple
cursor

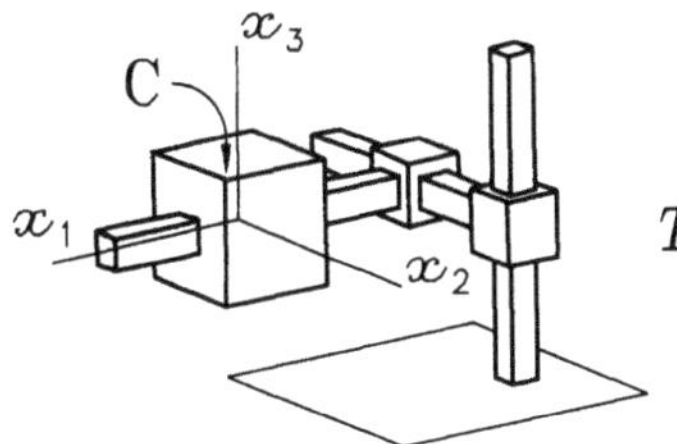

$T_1,\ T_2,\ T_3$

Admits displacement
in all directions.
Does not admit
rotations.

B.08
Sliding cur-
sor in cylin-
drical guide

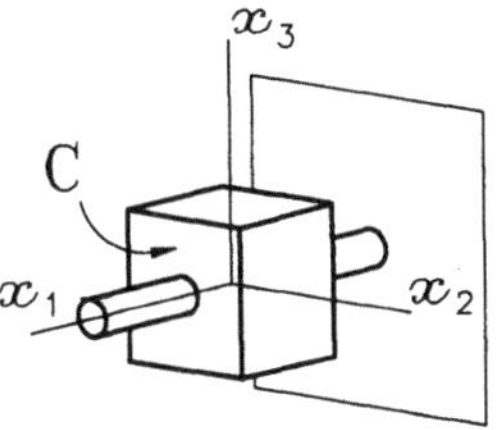

$F_2,\ F_3$

$T_2,\ T_3$

Admits displacement
and rotation, both
in the x_1-direction.

B.09
Flexible
hose

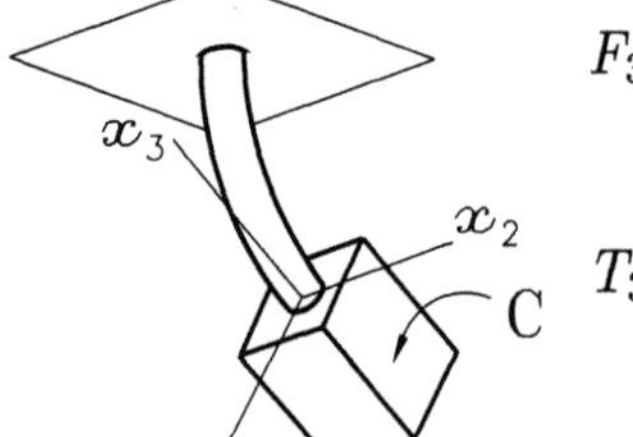

F_3

T_3

Admits bending,
but not
twisting.

B.10
Support
with
friction

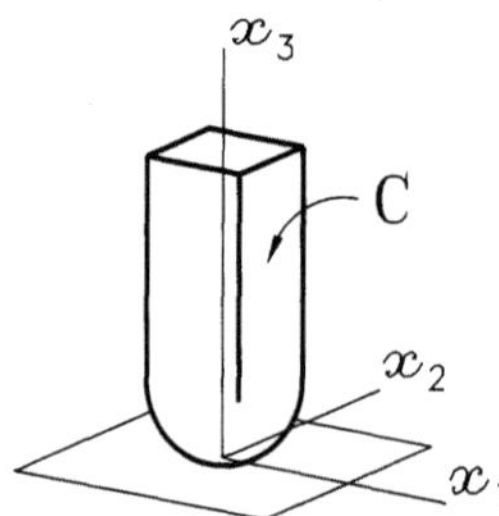

F_1, F_2, F_3

A single point
of contact.
Only admits force.

B.11
Cylindrical
swivel

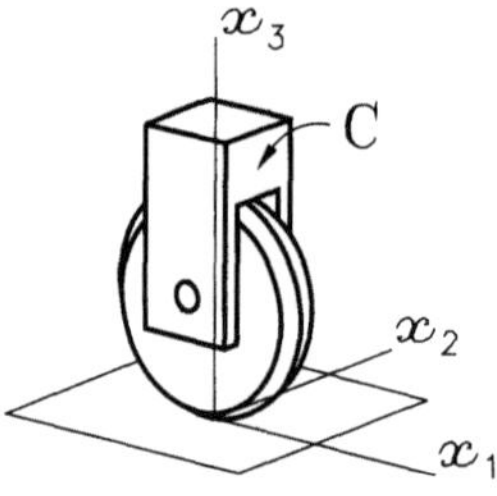

F_2, F_3

Force of contact
with friction
exclusively in the
x_2-direction.

B.12
Spherical
swivel

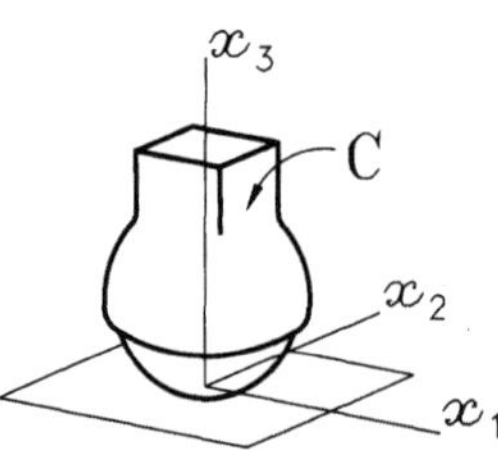

F_3

Support without
friction.

B.13

Cable

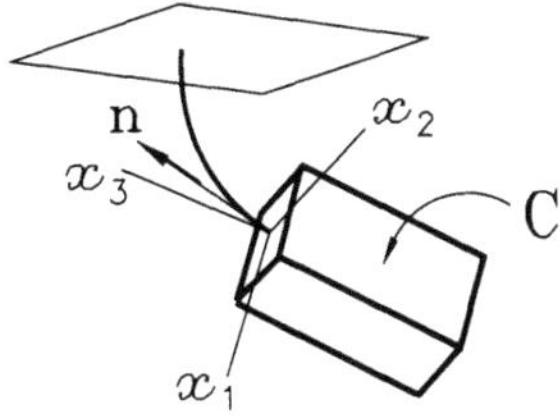

$\mathbf{F} = F\mathbf{n}$

Force in the direction of the tangent to the cable at the end.

B.14
Linear
elastic
spring

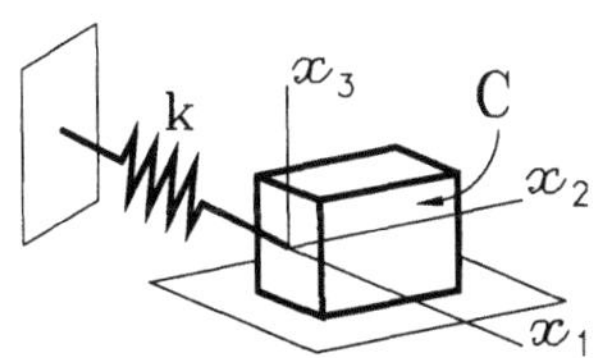

$F = k\delta$

Force proportional (k) to displacement (δ), in the direction of spring.

B.15
Viscous
damper

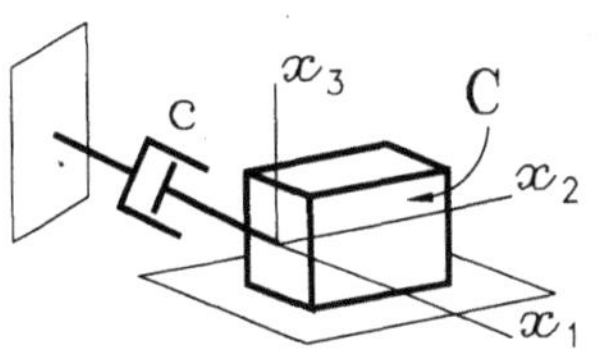

$F = c\dot{\delta}$

Force proportional (c) to the rate of displacement $(\dot{\delta})$.

B.16
Fluid
at rest

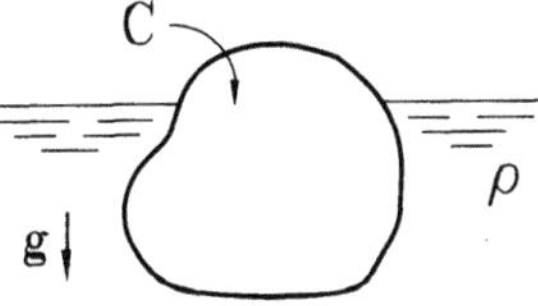

$\mathbf{F} = -\rho \mathbf{g} V$

Thrust: a vertical force equal to the weight of the displaced fluid.

Properties of Inertia

Appendix C

This is a supplementary appendix to Chapter 6, grouping reasonably complete tables with the properties of inertia of a large number of geometric figures. It also summarizes the main formulas for calculating moments and products of inertia and the most important relations — such as the theorem of parallel axes — that allow us to determine the inertial properties of a body with respect to a given point, knowing the same properties for the mass center of the body.

The appendix consists of four sections, as follows: C.1, *lines or bars*; C.2, *plates or crosssections*; C.3, *surfaces or shells*; and C.4, *volumes or solids*.

The tables include perimeters, areas, or volumes, whether the figure is a line, a cross section, (or a shell), or even a solid, respectively; the coordinates of its centroid in a system of Cartesian axes shown in the respective figure; and the moments and products of inertia with respect to the coordinate axes and axes parallel to them, passing through the centroid of the figure. When, therefore, the moments and products of inertia with respect to the axes passing through the centroid and their coordinates are known, it becomes easy to determine the moments and products of inertia with respect to any other point, for parallel directions, using the relations presented here.

Moments of inertia with respect to a point O, the origin of a system of Cartesian axes, are calculated according to the integrals (see

Chapter 6):

$$I_{11}^{O} = \int_{F} (x_2^2 + x_3^2)\, dm; \tag{1}$$

$$I_{22}^{O} = \int_{F} (x_3^2 + x_1^2)\, dm; \tag{2}$$

$$I_{33}^{O} = \int_{F} (x_1^2 + x_2^2)\, dm. \tag{3}$$

Products of inertia with respect to a point O, the origin of the same system of axes, can be calculated by the integrals (see Chapter 6):

$$I_{12}^{O} = -\int_{F} x_1 x_2\, dm; \tag{4}$$

$$I_{23}^{O} = -\int_{F} x_2 x_3\, dm; \tag{5}$$

$$I_{31}^{O} = -\int_{F} x_3 x_1\, dm, \tag{6}$$

where F is the figure (region) in which the integration is performed and x_j, $j = 1, 2, 3$, are the coordinates of the one-, two-, or three-dimensional mass element, dm, whichever the case. The moments and products of inertia with respect to axes, parallel to the previous ones, but with the origin in the centroid of the body, are obtained from the relationships of axis transposition (see Chapter 6), so that

$$I_{11}^{*} = I_{11}^{O} - m(x_2^{*2} + x_3^{*2}), \tag{7}$$

$$I_{22}^{*} = I_{22}^{O} - m(x_3^{*2} + x_1^{*2}), \tag{8}$$

$$I_{33}^{*} = I_{33}^{O} - m(x_1^{*2} + x_2^{*2}), \tag{9}$$

$$I_{12}^{*} = I_{12}^{O} + m x_1^{*} x_2^{*}, \tag{10}$$

$$I_{23}^{*} = I_{23}^{O} + m x_2^{*} x_3^{*}, \tag{11}$$

$$I_{31}^{*} = I_{31}^{O} + m x_3^{*} x_1^{*}. \tag{12}$$

The table gives for every plane line in Section C.1 its *perimeter*, the coordinates of the *centroid* on the plane of the figure, and its *moments* *of inertia* with respect to two Cartesian axes $\{x_1, x_2\}$, belonging to the plane, with the origin in a geometrically convenient point O. The table also includes the *moments of inertia* with respect to Cartesian axes parallel to the previous ones, $\{y_1, y_2\}$, with the origin in the centroid C^*, and the corresponding *product of inertia*. When discussing plane figures, the moments of inertia with respect to axes x_3 and y_3 may be obtained from the relations (see Chapter 6)

$$I^O_{33} = I^O_{11} + I^O_{22}, \qquad I^*_{33} = I^*_{11} + I^*_{22}, \tag{13}$$

and every product of inertia involving the direction orthogonal to the plane will be null (see Chapter 6), that is,

$$I^O_{23} = I^O_{31} = I^*_{23} = I^*_{31} = 0. \tag{14}$$

For each plane figure in Section C.2, the table provides its *area*, the coordinates of the *centroid* on the plane of the figure, its *moments of inertia of area* with respect to two Cartesian axes belonging to the plane, $\{x_1, x_2\}$, with the origin in a geometrically convenient point O, its *moments of inertia of area* with respect to Cartesian axes parallel to the previous ones, $\{y_1, y_2\}$, with the origin in the centroid C^*, and the corresponding *products of inertia of area*.

As in the case of the plane lines, the plane surfaces will have their moments of inertia of area with respect to axes x_3 and y_3 obtained from the relations (see Chapter 6)

$$J^O_{33} = J^O_{11} + J^O_{22}, \qquad J^*_{33} = J^*_{11} + J^*_{22}, \tag{15}$$

and all their products of inertia of area involving the direction orthogonal to the plane of the figure will be null (see Chapter 6), that is,

$$J^O_{23} = J^O_{31} = J^*_{23} = J^*_{31} = 0. \tag{16}$$

When the figure represents a homogeneous flat plate with mass m, its moments and products of inertia can be easily obtained from the expressions for the moments and products of inertia of area, respectively, according to the relations (see Chapter 6)

$$I^O_{jj} = \frac{m}{A} J^O_{jj}, \qquad I^*_{jj} = \frac{m}{A} J^*_{jj}, \qquad j = 1, 2, 3, \tag{17}$$

and

$$I^O_{jk} = \frac{m}{A}\, J^O_{jk}, \qquad I^*_{jk} = \frac{m}{A}\, J^*_{jk}, \qquad j, k = 1, 2, 3. \tag{18}$$

For each shell or surface in Section C.3, the table provides its area, the coordinates of the *centroid* of the figure, its *moments of inertia* with respect to three Cartesian axes, $\{x_1, x_2, x_3\}$, with the origin in a geometrically convenient point O, its *moments of inertia* with respect to Cartesian axes parallel to the previous ones, with the origin in the centroid C^*, and the corresponding *products of inertia*.

For each solid in Section C.4, the table provides its *volume*, the coordinates of its *centroid*, its *moments of inertia* with respect to three Cartesian axes, $\{x_1, x_2, x_3\}$, with the origin in a geometrically convenient point O, its *moments of inertia* with respect to Cartesian axes parallel to the previous ones, with the origin in the centroid C^*, and the corresponding *products of inertia*.

For those figures in which the algebraic expression of moments or products of inertia proved to be too long, it was decided to present the results relative to the system of coordinates $\{x_1, x_2, x_3\}$, the moments and products of inertia with respect to the centroid being expressed as in Eqs. (7–12), taking into account the values given for the respective coordinates of the centroid. Thus, for example, for the frustum of a rectangular pyramid, Fig. C.4.04, we have the entry $I^*_{11} = I^O_{11} - mx^{*2}_3$, since the expression for I^O_{11} is now too long and, in this case, $x^*_2 = 0$.

Line	*Geometry*	*Perimeter, Centroid Position*	*Moments of Inertia*	*Products of Inertia*
C.1.01 Slender rod	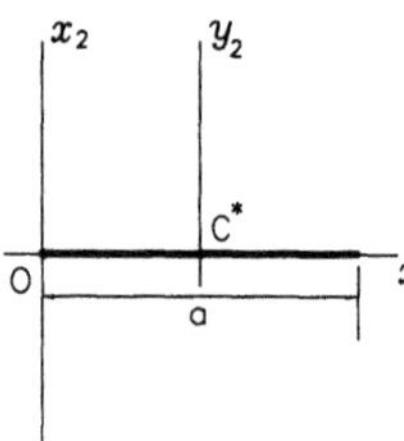	$C = a$ $x_1^* = \frac{a}{2}$ $x_2^* = 0$	$I_{11}^O = 0$ $I_{22}^O = \frac{ma^2}{3}$ $I_{11}^* = 0$ $I_{22}^* = \frac{ma^2}{12}$	$I_{12}^O = 0$ $I_{12}^* = 0$
C.1.02 Circular ring	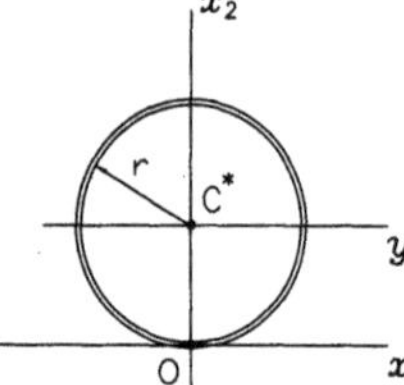	$C = 2\pi r$ $x_1^* = 0$ $x_2^* = r$	$I_{11}^O = \frac{3mr^2}{2}$ $I_{22}^O = \frac{mr^2}{2}$ $I_{11}^* = \frac{mr^2}{2}$ $I_{22}^* = \frac{mr^2}{2}$	$I_{12}^O = 0$ $I_{12}^* = 0$
C.1.03 Semi-circular ring	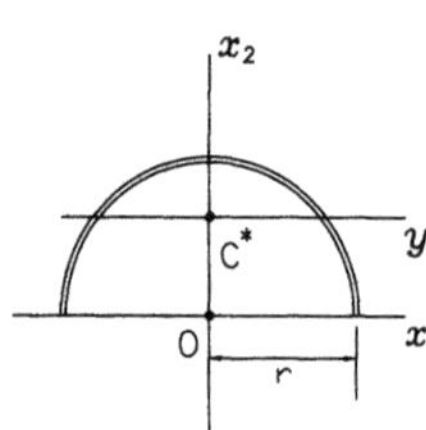	$C = \pi r$ $x_1^* = 0$ $x_2^* = \frac{2r}{\pi}$	$I_{11}^O = \frac{mr^2}{2}$ $I_{22}^O = \frac{mr^2}{2}$ $I_{11}^* = \frac{(\pi^2-8)mr^2}{2\pi^2}$ $I_{22}^* = \frac{mr^2}{2}$	$I_{12}^O = 0$ $I_{12}^* = 0$
C.1.04 Arc of a circle	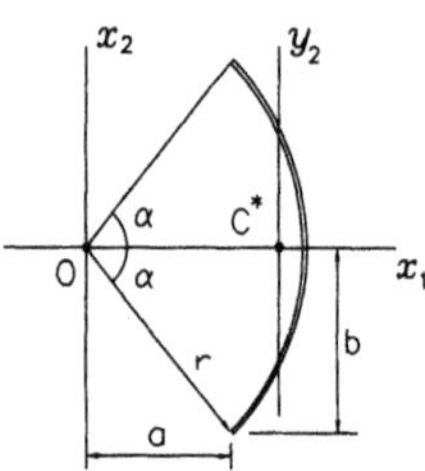	$C = 2\alpha r$ $x_1^* = \frac{b}{\alpha}$ $x_2^* = 0$	$I_{11}^O = \frac{m(\alpha r^2 - ab)}{2\alpha}$ $I_{22}^O = \frac{m(\alpha r^2 + ab)}{2\alpha}$ $I_{11}^* = \frac{m(\alpha r^2 - ab)}{2\alpha}$ $I_{22}^* = \frac{m(\alpha^2 r^2 + ab\alpha - 2b^2)}{2\alpha^2}$	$I_{12}^O = 0$ $I_{12}^* = 0$

Section	Geometry	Area, Centroid Position	Moments of Inertia	Products of Inertia
C.2.01 Rectangle	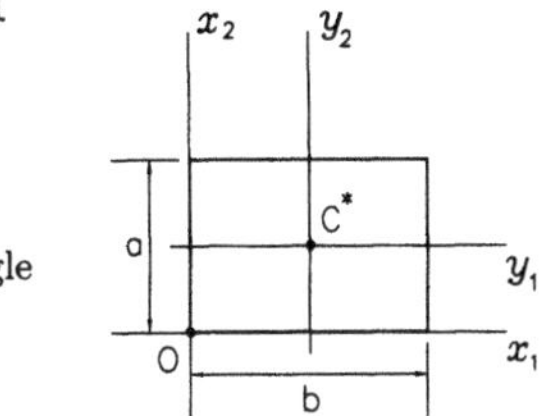	$A = ab$ $x_1^* = \dfrac{b}{2}$ $x_2^* = \dfrac{a}{2}$	$J_{11}^O = \dfrac{a^3 b}{3}$ $J_{22}^O = \dfrac{ab^3}{3}$ $J_{11}^* = \dfrac{a^3 b}{12}$ $J_{22}^* = \dfrac{ab^3}{12}$	$J_{12}^O = -\dfrac{a^2 b^2}{4}$ $J_{12}^* = 0$
C.2.02 Square	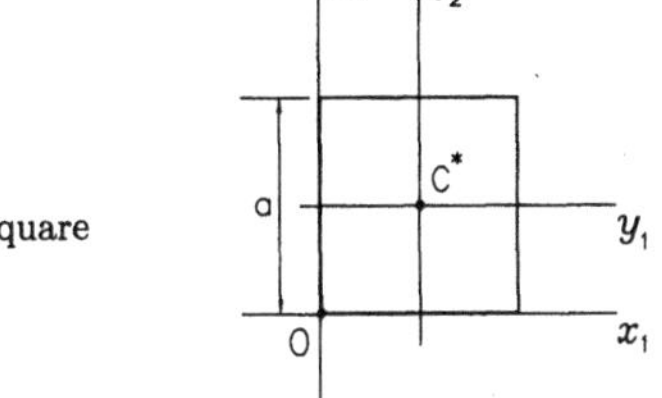	$A = a^2$ $x_1^* = \dfrac{a}{2}$ $x_2^* = \dfrac{a}{2}$	$J_{11}^O = \dfrac{a^4}{3}$ $J_{22}^O = \dfrac{a^4}{3}$ $J_{11}^* = \dfrac{a^4}{12}$ $J_{22}^* = \dfrac{a^4}{12}$	$J_{12}^O = -\dfrac{a^4}{4}$ $J_{12}^* = 0$
C.2.03 Triangle	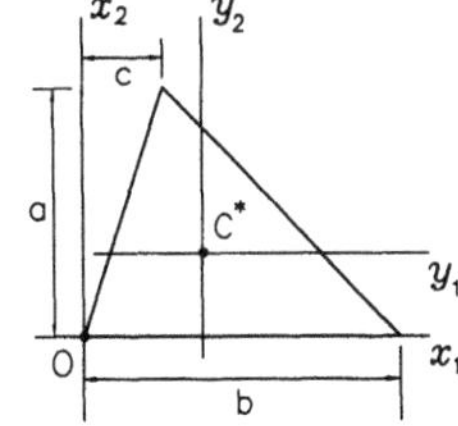	$A = \dfrac{ab}{2}$ $x_1^* = \dfrac{(b+c)}{3}$ $x_2^* = \dfrac{a}{3}$	$J_{11}^O = \dfrac{a^3 b}{12}$ $J_{22}^O = \dfrac{ab(b^2 + bc + c^2)}{12}$ $J_{11}^* = \dfrac{a^3 b}{36}$ $J_{22}^* = \dfrac{ab(b^2 - bc + c^2)}{36}$	$J_{12}^O = -\dfrac{a^2 b(b+2c)}{24}$ $J_{12}^* = \dfrac{a^2 b(b-2c)}{72}$
C.2.04 Right-angle triangle	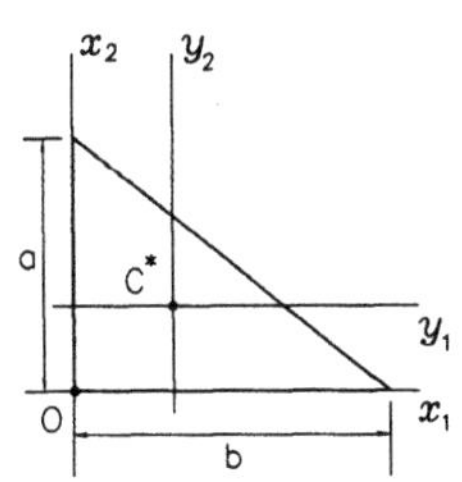	$A = \dfrac{ab}{2}$ $x_1^* = \dfrac{b}{3}$ $x_2^* = \dfrac{a}{3}$	$J_{11}^O = \dfrac{a^3 b}{12}$ $J_{22}^O = \dfrac{ab^3}{12}$ $J_{11}^* = \dfrac{a^3 b}{36}$ $J_{22}^* = \dfrac{ab^3}{36}$	$J_{12}^O = -\dfrac{a^2 b^2}{24}$ $J_{12}^* = \dfrac{a^2 b^2}{72}$

Section	Geometry	Area, Centroid Position	Moments of Inertia	Products of Inertia
C.2.05 Equilateral triangle	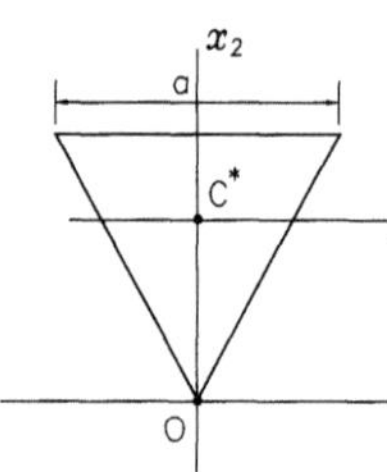	$A = \frac{\sqrt{3}a^2}{4}$ $x_1^* = 0$ $x_2^* = \frac{\sqrt{3}a}{3}$	$J_{11}^O = \frac{3\sqrt{3}a^4}{32}$ $J_{22}^O = \frac{\sqrt{3}a^4}{96}$ $J_{11}^* = \frac{\sqrt{3}a^4}{96}$ $J_{22}^* = \frac{\sqrt{3}a^4}{96}$	$J_{12}^O = 0$ $J_{12}^* = 0$
C.2.06 Isosceles trapezoid	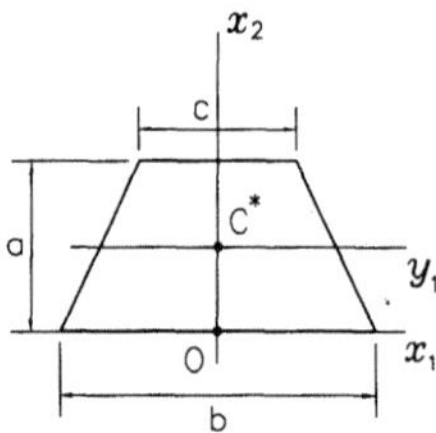	$A = \frac{a(b+c)}{2}$ $x_1^* = 0$ $x_2^* = \frac{a(b+2c)}{3(b+c)}$	$J_{11}^O = \frac{a^3(b+3c)}{12}$ $J_{22}^O = \frac{a(b^3+b^2c+bc^2+c^3)}{48}$ $J_{11}^* = \frac{a^3(b^2+4bc+c^2)}{36(b+c)}$ $J_{22}^* = \frac{a(b^3+b^2c+bc^2+c^3)}{48}$	$J_{12}^O = 0$ $J_{12}^* = 0$
C.2.07 Right-angle trapezoid	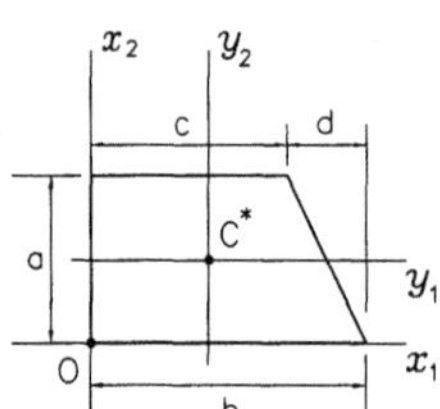	$A = \frac{a(b+c)}{2}$ $x_1^* = \frac{b^2+bc+c^2}{3(b+c)}$ $x_2^* = \frac{a(b+2c)}{3(b+c)}$	$J_{11}^O = \frac{a^3(b+3c)}{12}$ $J_{22}^O = \frac{a(b^3+b^2c+bc^2+c^3)}{12}$ $J_{11}^* = \frac{a^3(b^2+4bc+c^2)}{36(b+c)}$ $J_{22}^* = \frac{a(b^4+2b^3c+2bc^3+c^4)}{36(b+c)}$	$J_{12}^O = \frac{a^2(b^2+2bc+3c^2)}{-24}$ $J_{12}^* = J_{12}^O + Ax_1^*x_2^*$
C.2.08 Parallel-ogram	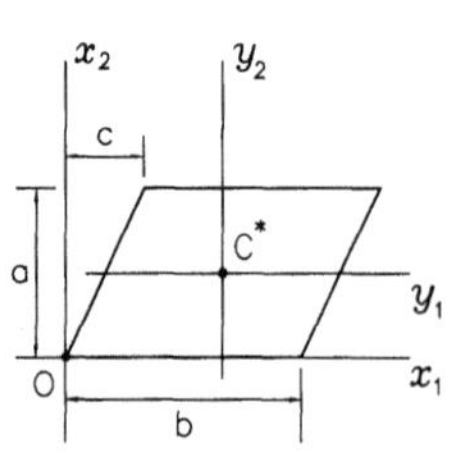	$A = ab$ $x_1^* = \frac{(b+c)}{2}$ $x_2^* = \frac{a}{2}$	$J_{11}^O = \frac{a^3b}{3}$ $J_{22}^O = \frac{ab(2b^2+3bc+2c^2)}{6}$ $J_{11}^* = \frac{a^3b}{12}$ $J_{22}^* = \frac{ab(b^2+c^2)}{12}$	$J_{12}^O = -\frac{a^2b(4c+3b)}{12}$ $J_{12}^* = -\frac{a^2bc}{12}$

Section	Geometry	Area, Centroid Position	Moments of Inertia	Products of Inertia
C.2.09 Lozenge	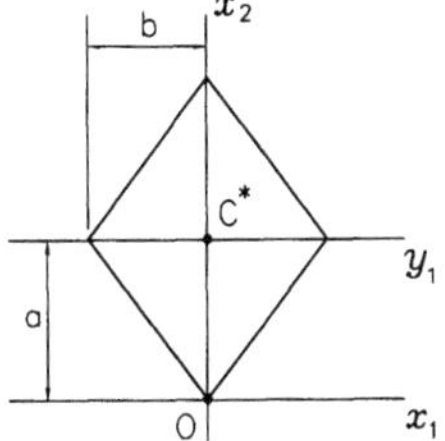	$A = 2ab$ $x_1^* = 0$ $x_2^* = a$	$J_{11}^O = \frac{7a^3 b}{3}$ $J_{22}^O = \frac{ab^3}{3}$ $J_{11}^* = \frac{a^3 b}{3}$ $J_{22}^* = \frac{ab^3}{3}$	$J_{12}^O = 0$ $J_{12}^* = 0$
C.2.10 Circle	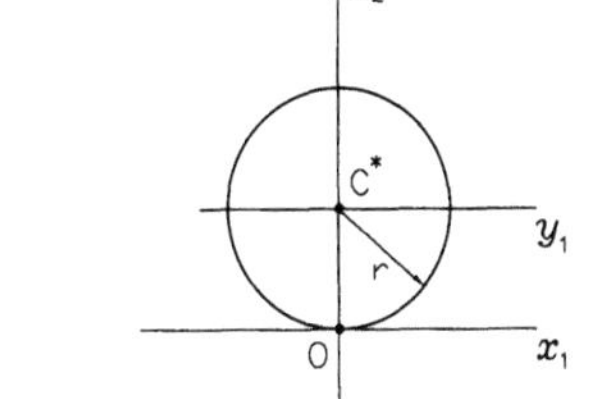	$A = \pi r^2$ $x_1^* = 0$ $x_2^* = r$	$J_{11}^O = \frac{5\pi r^4}{4}$ $J_{22}^O = \frac{\pi r^4}{4}$ $J_{11}^* = \frac{\pi r^4}{4}$ $J_{22}^* = \frac{\pi r^4}{4}$	$J_{12}^O = 0$ $J_{12}^* = 0$
C.2.11 Semi-circle	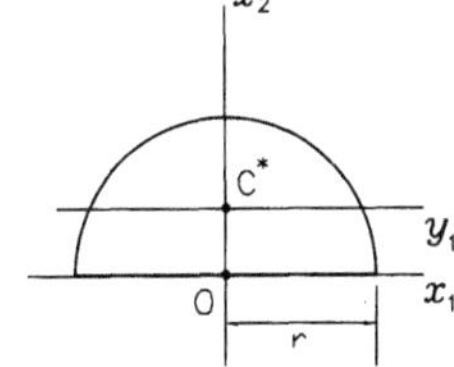	$A = \frac{\pi r^2}{2}$ $x_1^* = 0$ $x_2^* = \frac{4r}{3\pi}$	$J_{11}^O = \frac{\pi r^4}{8}$ $J_{22}^O = \frac{\pi r^4}{8}$ $J_{11}^* = \frac{(9\pi^2 - 64)r^4}{72\pi}$ $J_{22}^* = \frac{\pi r^4}{8}$	$J_{12}^O = 0$ $J_{12}^* = 0$
C.2.12 Quarter circle	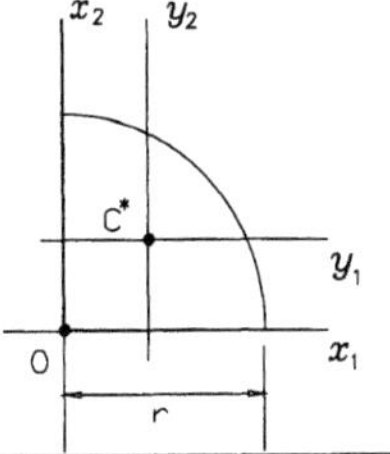	$A = \frac{\pi r^2}{4}$ $x_1^* = \frac{4r}{3\pi}$ $x_2^* = \frac{4r}{3\pi}$	$J_{11}^O = \frac{\pi r^4}{16}$ $J_{22}^O = \frac{\pi r^4}{16}$ $J_{11}^* = \left(\frac{\pi}{16} - \frac{4}{9\pi}\right) r^4$ $J_{22}^* = \left(\frac{\pi}{16} - \frac{4}{9\pi}\right) r^4$	$J_{12}^O = -\frac{r^4}{8}$ $J_{12}^* = \left(\frac{4}{9\pi} - \frac{1}{8}\right) r^4$

Section	Geometry	Area, Centroid Position	Moments of Inertia	Products of Inertia
C.2.13 Circular sector	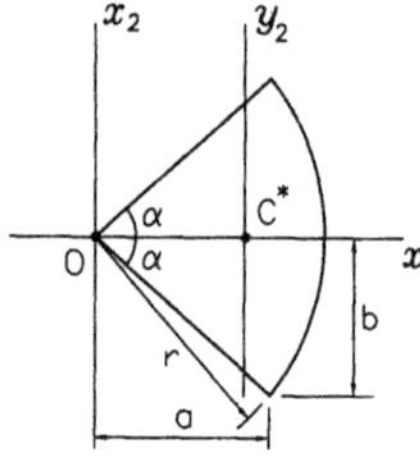	$A = \alpha r^2$ $x_1^* = \frac{2b}{3\alpha}$ $x_2^* = 0$	$J_{11}^O = \frac{(\alpha r^2 - ab)r^2}{4}$ $J_{22}^O = \frac{(\alpha r^2 + ab)r^2}{4}$ $J_{11}^* = \frac{(\alpha r^2 - ab)r^2}{4}$ $J_{22}^* = \frac{((3\alpha r)^2 + 9\alpha ab - 16b^2)r^2}{36\alpha}$	$J_{12}^O = 0$ $J_{12}^* = 0$
C.2.14 Circular segment	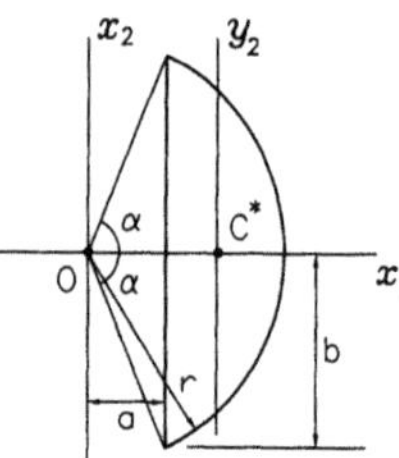	$A = \alpha r^2 - ab$ $x_1^* = \frac{2b^3}{3(\alpha r^2 - ab)}$ $x_2^* = 0$	$J_{11}^O = \frac{3\alpha r^4 - 3abr^2 - 2ab^3}{12}$ $J_{22}^O = \frac{\alpha r^4 - abr^2 + 2ab^3}{4}$ $J_{11}^* = \frac{3\alpha r^4 - 3abr^2 - 2ab^3}{12}$ $J_{22}^* = J_{22}^O - A x_1^{*2}$	$J_{12}^O = 0$ $J_{12}^* = 0$
C.2.15 Ellipse	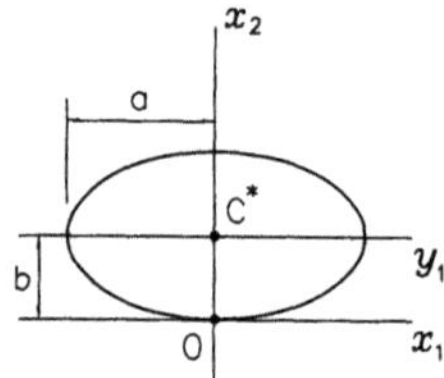	$A = \pi ab$ $x_1^* = 0$ $x_2^* = b$	$J_{11}^O = \frac{5\pi ab^3}{4}$ $J_{22}^O = \frac{\pi a^3 b}{4}$ $J_{11}^* = \frac{\pi ab^3}{4}$ $J_{22}^* = \frac{\pi a^3 b}{4}$	$J_{12}^O = 0$ $J_{12}^* = 0$
C.2.16 Semi-ellipse	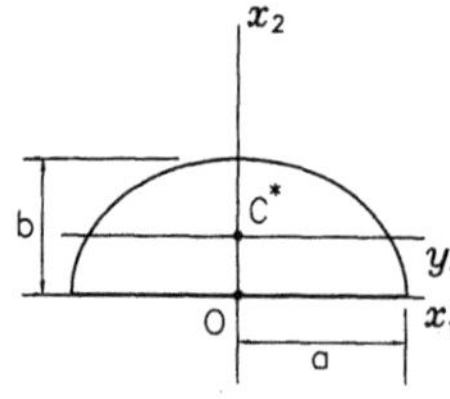	$A = \frac{\pi ab}{2}$ $x_1^* = 0$ $x_2^* = \frac{4b}{3\pi}$	$J_{11}^O = \frac{\pi ab^3}{8}$ $J_{22}^O = \frac{\pi a^3 b}{8}$ $J_{11}^* = \frac{(9\pi^2 - 64)ab^3}{72\pi}$ $J_{22}^* = \frac{\pi a^3 b}{8}$	$J_{12}^O = 0$ $J_{12}^* = 0$

Section	Geometry	Area, Centroid Position	Moments of Inertia	Products of Inertia
C.2.17 Parabolic segment	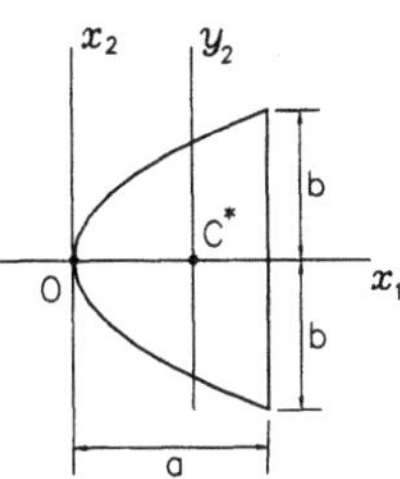	$A = \frac{4ab}{3}$ $x_1^* = \frac{3a}{5}$ $x_2^* = 0$	$J_{11}^O = \frac{4ab^3}{15}$ $J_{22}^O = \frac{4a^3b}{7}$ $J_{11}^* = \frac{4ab^3}{15}$ $J_{22}^* = \frac{16a^3b}{175}$	$J_{12}^O = 0$ $J_{12}^* = 0$
C.2.18 Semi-parabolic segment	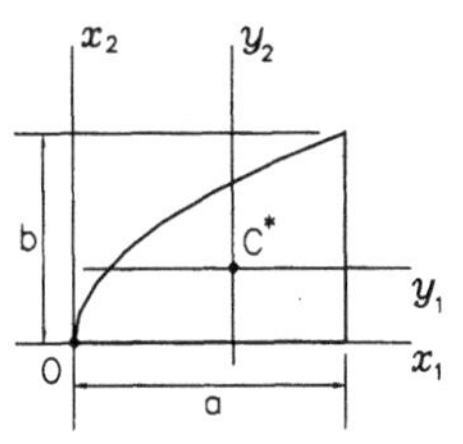	$A = \frac{2ab}{3}$ $x_1^* = \frac{3a}{5}$ $x_2^* = \frac{3b}{8}$	$J_{11}^O = \frac{2ab^3}{15}$ $J_{22}^O = \frac{2a^3b}{7}$ $J_{11}^* = \frac{19ab^3}{480}$ $J_{22}^* = \frac{8a^3b}{175}$	$J_{12}^O = -\frac{a^2b^2}{6}$ $J_{12}^* = -\frac{a^2b^2}{60}$
C.2.19 Regular pentagon	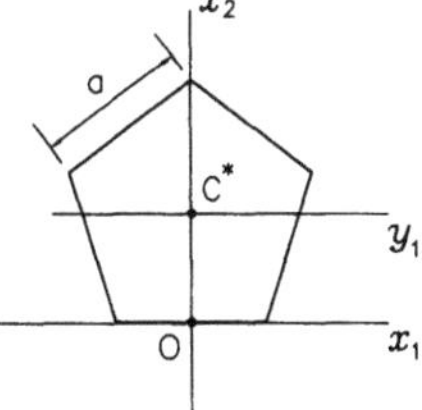	$A = \frac{5a^2 \tan \frac{3\pi}{10}}{4}$ $x_1^* = 0$ $x_2^* = \frac{a \tan \frac{3\pi}{10}}{2}$	$J_{11}^O = 1{,}05438a^4$ $J_{22}^O = 0{,}23955a^4$ $J_{11}^* = 0{,}23955a^4$ $J_{22}^* = 0{,}23955a^4$	$J_{12}^O = 0$ $J_{12}^* = 0$
C.2.20 Regular hexagon	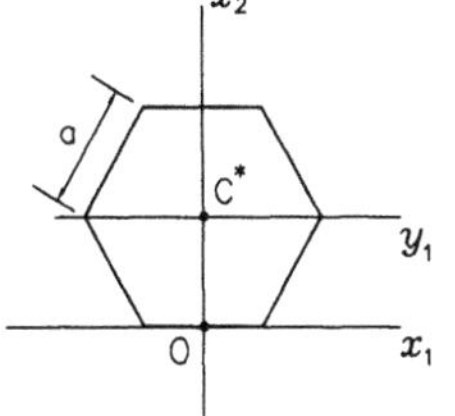	$A = \frac{3\sqrt{3}a^2}{2}$ $x_1^* = 0$ $x_2^* = \frac{\sqrt{3}a}{2}$	$J_{11}^O = \frac{23\sqrt{3}a^4}{16}$ $J_{22}^O = \frac{7\sqrt{3}a^4}{16}$ $J_{11}^* = \frac{5\sqrt{3}a^4}{16}$ $J_{22}^* = \frac{7\sqrt{3}a^4}{16}$	$J_{12}^O = 0$ $J_{12}^* = 0$

Section	Geometry	Area, Centroid Position	Moments of Inertia	Products of Inertia
C.2.21 Square frame	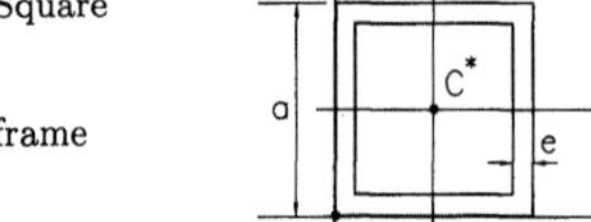	$A = 4e(a-e)$ $x_1^* = \dfrac{a}{2}$ $x_2^* = \dfrac{a}{2}$	$J_{11}^O = \dfrac{e(5a^3 - 9a^2 e + 8ae^2 - 4e^3)}{3}$ $J_{22}^O = \dfrac{e(5a^3 - 9a^2 e + 8ae^2 - 4e^3)}{3}$ $J_{11}^* = \dfrac{a^4 - (a-2e)^4}{12}$ $J_{22}^* = \dfrac{a^4 - (a-2e)^4}{12}$ $J_{12}^O = a^2 e(e-a)$ $J_{12}^* = 0$
C.2.22 Rectangular frame	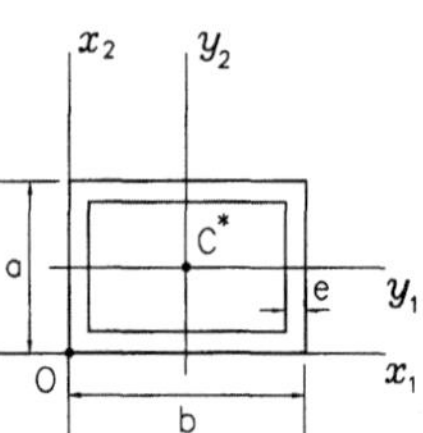	$A = 2e(b+a-2e)$ $x_1^* = \dfrac{b}{2}$ $x_2^* = \dfrac{a}{2}$	$J_{11}^O = \dfrac{e(3a(b-2e)(a-e) + 2be^2 + 2a^3 - 4e^3)}{3}$ $J_{22}^O = \dfrac{e(3b(a-2e)(b-e) + 2ae^2 + 2b^3 - 4e^3)}{3}$ $J_{11}^* = \dfrac{a^3 b - (b-2e)(a-2e)^3}{12}$ $J_{22}^* = \dfrac{ab^3 - (a-2e)(b-2e)^3}{12}$ $J_{12}^O = \dfrac{abe(2e-a-b)}{2}$ $J_{12}^* = 0$
C.2.23 Circular crown	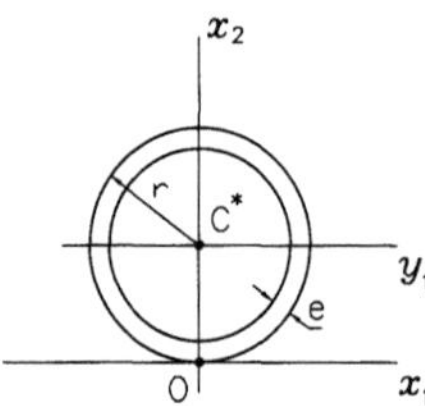	$A = \pi e(2r-e)$ $x_1^* = 0$ $x_2^* = r$	$J_{11}^O = \dfrac{\pi[r^4 - (r-e)^4 + 8r^3 e - 4r^2 e^2]}{4}$ $J_{22}^O = \dfrac{\pi[r^4 - (r-e)^4]}{4}$ $J_{11}^* = \dfrac{\pi[r^4 - (r-e)^4]}{4}$ $J_{22}^* = \dfrac{\pi[r^4 - (r-e)^4]}{4}$ $J_{12}^O = 0$ $J_{12}^* = 0$
C.2.24 Semi-circular crown	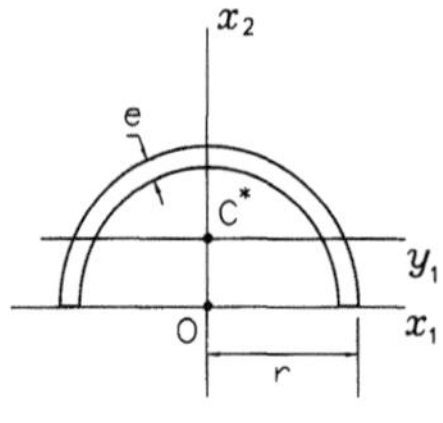	$A = \dfrac{\pi e(2r-e)}{2}$ $x_1^* = 0$ $x_2^* = \dfrac{4r^3 - 4(r-e)^3}{3\pi(2re - e^2)}$	$J_{11}^O = \dfrac{\pi[r^4 - (r-e)^4]}{8}$ $J_{22}^O = \dfrac{\pi[r^4 - (r-e)^4]}{8}$ $J_{11}^* = J_{11}^O - A x_2^{*2}$ $J_{22}^* = \dfrac{\pi[r^4 - (r-e)^4]}{8}$ $J_{12}^O = 0$ $J_{12}^* = 0$

Section	Geometry	Area, Centroid Position	Moments of Inertia	Products of Inertia
C.2.25 Elliptic crown	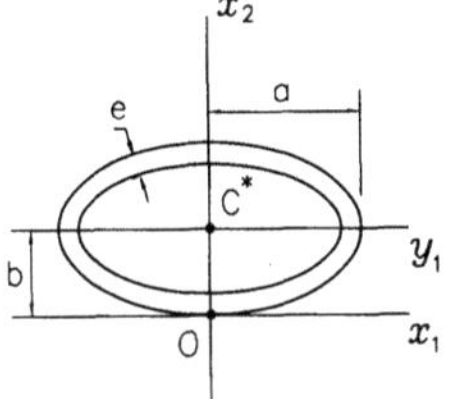	$A = \pi e(a+b-e)$ $x_1^* = 0$ $x_2^* = b$	$J_{11}^O = J_{11}^* + A x_2^{*\,2}$ $J_{22}^O = \dfrac{\pi[a^3 b - (a-e)^3(b-e)]}{4}$ $J_{11}^* = \dfrac{\pi[ab^3 - (b-e)^3(a-e)]}{4}$ $J_{22}^* = \dfrac{\pi[a^3 b - (a-e)^3(b-e)]}{4}$ $J_{12}^O = 0$ $J_{12}^* = 0$	
C.2.26 Semi- elliptic crown	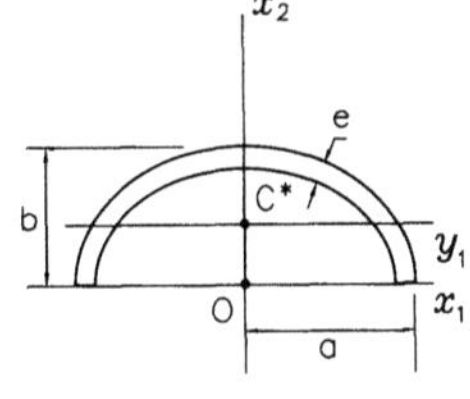	$A = \dfrac{\pi e(a+b-e)}{2}$ $x_1^* = 0$ $x_2^* = \dfrac{4[ab^2-(a-e)(b-e)^2]}{3\pi e(a+b-e)}$	$J_{11}^O = \dfrac{\pi[ab^3 - (b-e)^3(a-e)]}{8}$ $J_{22}^O = \dfrac{\pi[a^3 b - (a-e)^3(b-e)]}{8}$ $J_{11}^* = J_{11}^O - A x_2^{*\,2}$ $J_{22}^* = \dfrac{\pi[a^3 b - (a-e)^3(b-e)]}{8}$ $J_{12}^O = 0$ $J_{12}^* = 0$	
C.2.27 L profile (equal legs)	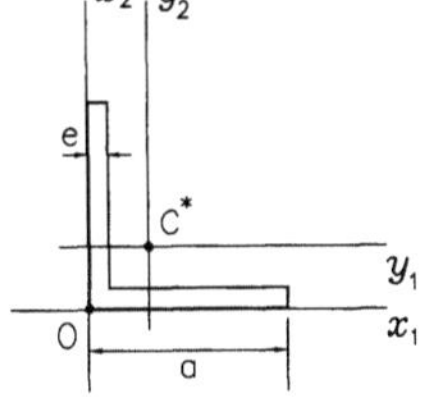	$A = e(2a-e)$ $x_1^* = \dfrac{a^2+(a-e)e}{2(2a-e)}$ $x_2^* = \dfrac{a^2+(a-e)e}{2(2a-e)}$	$J_{11}^O = \dfrac{e(a^3+ae^2-e^3)}{3}$ $J_{22}^O = \dfrac{e(a^3+ae^2-e^3)}{3}$ $J_{11}^* = \dfrac{e(a^2-ae+e^2)(5a^2-5ae+e^2)}{12(2a-e)}$ $J_{22}^* = \dfrac{e(a^2-ae+e^2)(5a^2-5ae+e^2)}{12(2a-e)}$ $J_{12}^O = -\dfrac{e^2(2a^2-e^2)}{4}$ $J_{12}^* = \dfrac{a^2 e(a-e)^2}{4(2a-e)}$	
C.2.28 L profile (unequal legs)	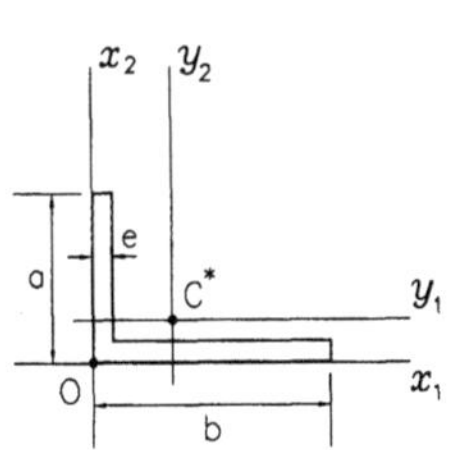	$A = e(a+b-e)$ $x_1^* = \dfrac{e[2(b-e)+a]+(b-e)^2}{2(a+b-e)}$ $x_2^* = \dfrac{e[b+2(a-e)]+(a-e)^2}{2(a+b-e)}$	$J_{11}^O = \dfrac{e(a^3+be^2-e^3)}{3}$ $J_{22}^O = \dfrac{e(ae^2+b^3-e^3)}{3}$ $J_{11}^* = \dfrac{e(a^3+be^2-e^3)}{3} - \dfrac{e[a^2+e(b-e)]^2}{4(a+b-e)}$ $J_{22}^* = \dfrac{e(b^3+ae^2-e^3)}{3} - \dfrac{e[b^2+e(a-e)]^2}{4(a+b-e)}$ $J_{12}^O = -\dfrac{e^2(a^2+b^2-e^2)}{4}$ $J_{12}^* = \dfrac{abe(b-e)(a-e)}{4(a+b-e)}$	

Section	Geometry	Area, Centroid Position	Moments of Inertia	Products of Inertia
C.2.29 I profile	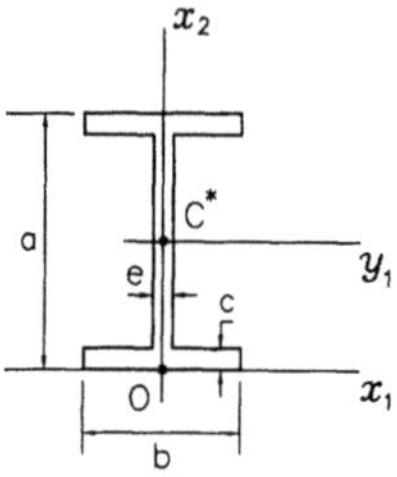	$A = 2bc + (a-2c)e$ $x_1^* = 0$ $x_2^* = \dfrac{a}{2}$	$J_{11}^O = \dfrac{a^3 e + (3a^2 c - 3ac^2 + 2c^3)(b-e)}{3}$ $J_{22}^O = \dfrac{2b^3 c + (a-2c)e^3}{12}$ $J_{11}^* = \dfrac{a^3 b - (a-2c)^3(b-e)}{12}$ $J_{22}^* = \dfrac{2b^3 c + (a-2c)e^3}{12}$ $J_{12}^O = 0$ $J_{12}^* = 0$	
C.2.30 Regular I profile	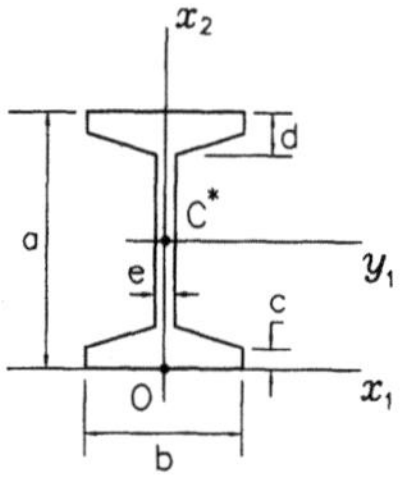	$A = ae + (d+c)(b-e)$ $x_1^* = 0$ $x_2^* = \dfrac{a}{2}$	$J_{11}^O = J_{11}^* + A x_2^{*2}$ $J_{22}^O = \dfrac{2b^3 c + (a-2d)e^3}{12} + \dfrac{(d-c)(b^4 - e^4)}{24(b-e)}$ $J_{11}^* = \dfrac{a^3 b}{12} - \dfrac{(b-e)[(a-2c)^4 - (a-2d)^4]}{96(d-c)}$ $J_{22}^* = \dfrac{2b^3 c + (a-2d)e^3}{12} + \dfrac{(d-c)(b^4 - e^4)}{24(b-e)}$ $J_{12}^O = 0$ $J_{12}^* = 0$	
C.2.31 U profile	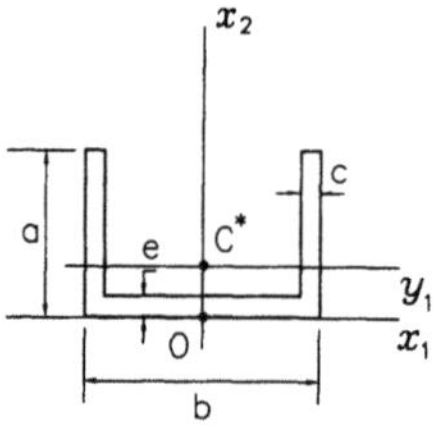	$A = 2ac + be - 2ce$ $x_1^* = 0$ $x_2^* = \dfrac{2a^2 c + (b-2c)e^2}{2[2ac + (b-2c)e]}$	$J_{11}^O = \dfrac{2a^3 c + be^3 - 2ce^3}{3}$ $J_{22}^O = \dfrac{ab^3 - (a-e)(b-2c)^3}{12}$ $J_{11}^* = J_{11}^O - A x_2^{*2}$ $J_{22}^* = \dfrac{ab^3 - (a-e)(b-2c)^3}{12}$ $J_{12}^O = 0$ $J_{12}^* = 0$	
C.2.32 Regular U profile	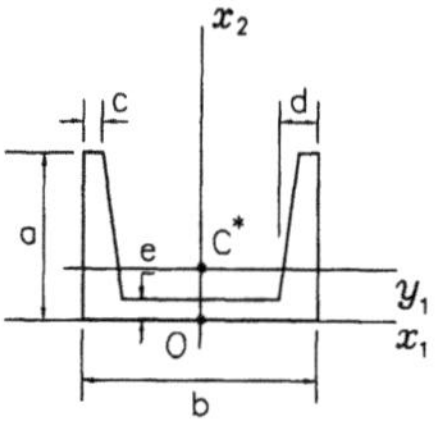	$A = be + (d+c)(a-e)$ $x_1^* = 0$ $x_2^* = -\dfrac{2a^2(c-4d) + 2ae(c-d) - e^2[3b + 2(2c-5d)]}{6[a(c+d) + e(b-c-d)]}$	$J_{11}^O = \dfrac{2ca^3 + (b-2d)e^3}{3} + \dfrac{(d-c)(a^4 - e^4)}{6(a-e)}$ $J_{22}^O = \dfrac{ab^3}{12} - \dfrac{(a-e)[(b-2c)^4 - (b-2d)^4]}{96(d-c)}$ $J_{11}^* = J_{11}^O - A x_2^{*2}$ $J_{22}^* = \dfrac{ab^3}{12} - \dfrac{(a-e)[(b-2c)^4 - (b-2d)^4]}{96(d-c)}$ $J_{12}^O = J_{12}^* = 0$	

Section	Geometria	Area, Centroid Position	Moments of Inertia	Products of Inertia
C.2.33 Z profile		$A = e(2b+a-2e)$ $x_1^* = 0$ $x_2^* = \dfrac{a}{2}$	$J_{11}^O = \dfrac{e[a^3+(3a^2-3ae+2e^2)(b-e)]}{3}$ $J_{22}^O = \dfrac{e[ae^2+2(b-e)(4b^2-2be+e^2)]}{12}$ $J_{11}^* = \dfrac{a^3b-(b-e)(a-2e)^3}{12}$ $J_{22}^* = \dfrac{e[ae^2+2(b-e)(4b^2-2be+e^2)]}{12}$ $J_{12}^O = \dfrac{be(a-e)(b-e)}{2}$ $J_{12}^* = \dfrac{be(a-e)(b-e)}{2}$	
C.2.34 T profile		$A = bc+ae-ce$ $x_1^* = 0$ $x_2^* = \dfrac{a^2e+bc^2-c^2e}{2(ae+bc-ce)}$	$J_{11}^O = \dfrac{a^3e+bc^3-c^3e}{3}$ $J_{22}^O = \dfrac{b^3c+(a-c)e^3}{12}$ $J_{11}^* = J_{11}^O - Ax_2^{*2}$ $J_{22}^* = \dfrac{b^3c+(a-c)e^3}{12}$ $J_{12}^O = 0$ $J_{12}^* = 0$	

Surface	Geometry	Area, Centroid Position	Moments of Inertia	Products of Inertia
C.3.01 Spherical shell		$A = 4\pi r^2$ $x_1^* = 0$ $x_2^* = 0$ $x_3^* = 0$	$I_{11}^O = \frac{2mr^2}{3}$ $I_{22}^O = \frac{2mr^2}{3}$ $I_{33}^O = \frac{2mr^2}{3}$ $I_{11}^* = \frac{2mr^2}{3}$ $I_{22}^* = \frac{2mr^2}{3}$ $I_{33}^* = \frac{2mr^2}{3}$	$I_{12}^O = I_{23}^O = I_{31}^O = 0$ $I_{12}^* = I_{23}^* = I_{31}^* = 0$
C.3.02 Hemi- spherical shell		$A = 2\pi r^2$ $x_1^* = 0$ $x_2^* = 0$ $x_3^* = -\frac{r}{2}$	$I_{11}^O = \frac{2mr^2}{3}$ $I_{22}^O = \frac{2mr^2}{3}$ $I_{33}^O = \frac{2mr^2}{3}$ $I_{11}^* = \frac{5mr^2}{12}$ $I_{22}^* = \frac{5mr^2}{12}$ $I_{33}^* = \frac{2mr^2}{3}$	$I_{12}^O = I_{23}^O = I_{31}^O = 0$ $I_{12}^* = I_{23}^* = I_{31}^* = 0$
C.3.03 Circular cylindrical shell		$A = 2\pi ra$ $x_1^* = 0$ $x_2^* = 0$ $x_3^* = \frac{a}{2}$	$I_{11}^O = \frac{mr^2}{2} + \frac{ma^2}{3}$ $I_{22}^O = \frac{mr^2}{2} + \frac{ma^2}{3}$ $I_{33}^O = mr^2$ $I_{11}^* = \frac{mr^2}{2} + \frac{ma^2}{12}$ $I_{22}^* = \frac{mr^2}{2} + \frac{ma^2}{12}$ $I_{33}^* = mr^2$	$I_{12}^O = I_{23}^O = I_{31}^O = 0$ $I_{12}^* = I_{23}^* = I_{31}^* = 0$
C.3.04 Semicirc. cylindrical shell		$A = \pi ra$ $x_1^* = 0$ $x_2^* = 0$ $x_3^* = -\frac{2r}{\pi}$	$I_{11}^O = mr^2$ $I_{22}^O = \frac{mr^2}{2} + \frac{ma^2}{12}$ $I_{33}^O = \frac{mr^2}{2} + \frac{ma^2}{12}$ $I_{11}^* = \frac{(\pi^2-4)mr^2}{\pi^2}$ $I_{22}^* = \frac{(\pi^2-8)mr^2}{2\pi^2} + \frac{ma^2}{12}$ $I_{33}^* = \frac{mr^2}{2} + \frac{ma^2}{12}$	$I_{12}^O = I_{23}^O = I_{31}^O = 0$ $I_{12}^* = I_{23}^* = I_{31}^* = 0$

Surface	Geometry	Area, Centroid Position	Moments of Inertia	Products of Inertia
C.3.05 Conical shell		$A = \pi r \sqrt{a^2 + r^2}$ $x_1^* = 0$ $x_2^* = 0$ $x_3^* = \dfrac{a}{3}$	$I_{11}^O = \dfrac{mr^2}{4} + \dfrac{ma^2}{6}$ $I_{22}^O = \dfrac{mr^2}{4} + \dfrac{ma^2}{6}$ $I_{33}^O = \dfrac{mr^2}{2}$ $I_{11}^* = \dfrac{mr^2}{4} + \dfrac{ma^2}{18}$ $I_{22}^* = \dfrac{mr^2}{4} + \dfrac{ma^2}{18}$ $I_{33}^* = \dfrac{mr^2}{2}$	$I_{12}^O = I_{23}^O = I_{31}^O = 0$ $I_{12}^* = I_{23}^* = I_{31}^* = 0$
C.3.06 Semi-conical shell		$A = \dfrac{\pi r \sqrt{a^2 + r^2}}{2}$ $x_1^* = 0$ $x_2^* = \dfrac{a}{3}$ $x_3^* = -\dfrac{4r}{3\pi}$	$I_{11}^O = \dfrac{mr^2}{4} + \dfrac{ma^2}{6}$ $I_{22}^O = \dfrac{mr^2}{2}$ $I_{33}^O = \dfrac{mr^2}{4} + \dfrac{ma^2}{6}$ $I_{11}^* = \dfrac{(9\pi^2 - 64)mr^2}{36\pi^2} + \dfrac{ma^2}{18}$ $I_{22}^* = \dfrac{(9\pi^2 - 32)mr^2}{18\pi^2}$ $I_{33}^* = \dfrac{mr^2}{4} + \dfrac{ma^2}{18}$	$I_{12}^O = I_{31}^O = 0$ $I_{23}^O = \dfrac{mra^2}{3\pi\sqrt{a^2 + r^2}}$ $I_{12}^* = I_{31}^* = 0$ $I_{23}^* = \dfrac{mra(3a - 4\sqrt{a^2 + r^2})}{9\pi\sqrt{a^2 + r^2}}$

Solid	Geometry	Volume, Centroid Position	Moments of Inertia	Products of Inertia
C.4.01 Rectang. parallel-piped		$V = abc$ $x_1^* = \frac{c}{2}$ $x_2^* = \frac{b}{2}$ $x_3^* = \frac{a}{2}$	$I_{11}^O = \frac{m(a^2+b^2)}{3}$ $I_{22}^O = \frac{m(a^2+c^2)}{3}$ $I_{33}^O = \frac{m(b^2+c^2)}{3}$ $I_{11}^* = \frac{m(a^2+b^2)}{12}$ $I_{22}^* = \frac{m(a^2+c^2)}{12}$ $I_{33}^* = \frac{m(b^2+c^2)}{12}$	$I_{12}^O = -\frac{mbc}{4}$ $I_{23}^O = -\frac{mab}{4}$ $I_{31}^O = -\frac{mac}{4}$ $I_{12}^* = I_{23}^* = I_{31}^* = 0$
C.4.02 Cube		$V = a^3$ $x_1^* = \frac{a}{2}$ $x_2^* = \frac{a}{2}$ $x_3^* = \frac{a}{2}$	$I_{11}^O = \frac{2ma^2}{3}$ $I_{22}^O = \frac{2ma^2}{3}$ $I_{33}^O = \frac{2ma^2}{3}$ $I_{11}^* = \frac{ma^2}{6}$ $I_{22}^* = \frac{ma^2}{6}$ $I_{33}^* = \frac{ma^2}{6}$	$I_{12}^O = -\frac{ma^2}{4}$ $I_{23}^O = -\frac{ma^2}{4}$ $I_{31}^O = -\frac{ma^2}{4}$ $I_{12}^* = I_{23}^* = I_{31}^* = 0$
C.4.03 Rec-tangular pyramid		$V = \frac{abc}{3}$ $x_1^* = 0$ $x_2^* = 0$ $x_3^* = \frac{a}{4}$	$I_{11}^O = \frac{m(b^2+2a^2)}{20}$ $I_{22}^O = \frac{m(c^2+2a^2)}{20}$ $I_{33}^O = \frac{m(b^2+c^2)}{20}$ $I_{11}^* = \frac{m(4b^2+3a^2)}{80}$ $I_{22}^* = \frac{m(4c^2+3a^2)}{80}$ $I_{33}^* = \frac{m(b^2+c^2)}{20}$	$I_{12}^O = I_{23}^O = I_{31}^O = 0$ $I_{12}^* = I_{23}^* = I_{31}^* = 0$
C.4.04 Frustum of rectang. pyramid	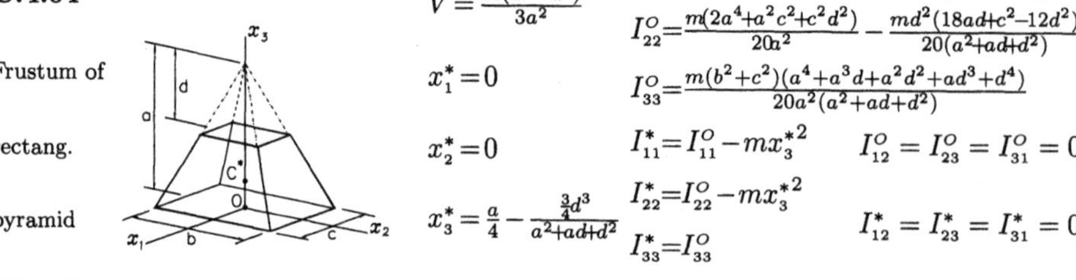	$V = \frac{bc(a^3-d^3)}{3a^2}$ $x_1^* = 0$ $x_2^* = 0$ $x_3^* = \frac{a}{4} - \frac{\frac{3}{4}d^3}{a^2+ad+d^2}$	$I_{11}^O = \frac{m(2a^4+a^2b^2+b^2d^2)}{20a^2} - \frac{md^2(18ad+b^2-12d^2)}{20(a^2+ad+d^2)}$ $I_{22}^O = \frac{m(2a^4+a^2c^2+c^2d^2)}{20a^2} - \frac{md^2(18ad+c^2-12d^2)}{20(a^2+ad+d^2)}$ $I_{33}^O = \frac{m(b^2+c^2)(a^4+a^3d+a^2d^2+ad^3+d^4)}{20a^2(a^2+ad+d^2)}$ $I_{11}^* = I_{11}^O - mx_3^{*2}$ $I_{22}^* = I_{22}^O - mx_3^{*2}$ $I_{33}^* = I_{33}^O$	$I_{12}^O = I_{23}^O = I_{31}^O = 0$ $I_{12}^* = I_{23}^* = I_{31}^* = 0$

Solid	Geometry	Volume, Centroid Position	Moments of Inertia	Products of Inertia
C.4.05 Regular tetra- hedron		$V=\dfrac{\sqrt{2}a^3}{12}$ $x_1^*=0$ $x_2^*=0$ $x_3^*=\dfrac{\sqrt{6}a}{12}$	$I_{11}^O=\dfrac{11ma^2}{120}$ $I_{22}^O=\dfrac{11ma^2}{120}$ $I_{33}^O=\dfrac{ma^2}{20}$ $I_{11}^*=\dfrac{ma^2}{20}$ $I_{22}^*=\dfrac{ma^2}{20}$ $I_{33}^*=\dfrac{ma^2}{20}$	$I_{12}^O=I_{23}^O=I_{31}^O=0$ $I_{12}^*=I_{23}^*=I_{31}^*=0$
C.4.06 Rectang. tetra- hedron		$V=\dfrac{abc}{6}$ $x_1^*=\dfrac{c}{4}$ $x_2^*=\dfrac{b}{4}$ $x_3^*=\dfrac{a}{4}$	$I_{11}^O=\dfrac{m(a^2+b^2)}{10}$ $I_{22}^O=\dfrac{m(a^2+c^2)}{10}$ $I_{33}^O=\dfrac{m(b^2+c^2)}{10}$ $I_{11}^*=\dfrac{3m(a^2+b^2)}{80}$ $I_{22}^*=\dfrac{3m(a^2+c^2)}{80}$ $I_{33}^*=\dfrac{3m(b^2+c^2)}{80}$	$I_{12}^O=-\dfrac{mbc}{20}$ $I_{23}^O=-\dfrac{mab}{20}$ $I_{31}^O=-\dfrac{mac}{20}$ $I_{12}^*=\dfrac{mbc}{80}$ $I_{23}^*=\dfrac{mab}{80}$ $I_{31}^*=\dfrac{mac}{80}$
C.4.07 Right angle prism		$V=\dfrac{abc}{2}$ $x_1^*=\dfrac{a}{3}$ $x_2^*=\dfrac{b}{3}$ $x_3^*=\dfrac{c}{2}$	$I_{11}^O=\dfrac{mb^2}{6}+\dfrac{mc^2}{3}$ $I_{22}^O=\dfrac{ma^2}{6}+\dfrac{mc^2}{3}$ $I_{33}^O=\dfrac{ma^2}{6}+\dfrac{mb^2}{6}$ $I_{11}^*=\dfrac{mb^2}{18}+\dfrac{mc^2}{12}$ $I_{22}^*=\dfrac{ma^2}{18}+\dfrac{mc^2}{12}$ $I_{33}^*=\dfrac{ma^2}{18}+\dfrac{mb^2}{18}$	$I_{12}^O=-\dfrac{mab}{12}$ $I_{23}^O=-\dfrac{mbc}{6}$ $I_{31}^O=-\dfrac{mac}{6}$ $I_{12}^*=\dfrac{mab}{36}$ $I_{23}^*=0$ $I_{31}^*=0$
C.4.08 Equilateral triangular prism		$V=\dfrac{\sqrt{3}ab^2}{4}$ $x_1^*=0$ $x_2^*=\dfrac{\sqrt{3}b}{6}$ $x_3^*=\dfrac{a}{2}$	$I_{11}^O=\dfrac{ma^2}{3}+\dfrac{mb^2}{8}$ $I_{22}^O=\dfrac{ma^2}{3}+\dfrac{mb^2}{24}$ $I_{33}^O=\dfrac{mb^2}{6}$ $I_{11}^*=\dfrac{ma^2}{12}+\dfrac{mb^2}{24}$ $I_{22}^*=\dfrac{ma^2}{12}+\dfrac{mb^2}{24}$ $I_{33}^*=\dfrac{mb^2}{12}$	$I_{12}^O=0$ $I_{23}^O=-\dfrac{\sqrt{3}mab}{12}$ $I_{31}^O=0$ $I_{12}^*=I_{23}^*=I_{31}^*=0$

Solid	Geometry	Volume, Centroid Position	Moments of Inertia	Products of Inertia
C.4.09 Regular octa- hedron		$V = \frac{\sqrt{2}a^3}{3}$ $x_1^* = 0$ $x_2^* = 0$ $x_3^* = 0$	$I_{11}^O = \frac{ma^2}{10}$ $I_{22}^O = \frac{ma^2}{10}$ $I_{33}^O = \frac{ma^2}{10}$ $I_{11}^* = \frac{ma^2}{10}$ $I_{22}^* = \frac{ma^2}{10}$ $I_{33}^* = \frac{ma^2}{10}$	$I_{12}^O = I_{23}^O = I_{31}^O = 0$ $I_{12}^* = I_{23}^* = I_{31}^* = 0$
C.4.10 Sphere		$V = \frac{4\pi r^3}{3}$ $x_1^* = 0$ $x_2^* = 0$ $x_3^* = 0$	$I_{11}^O = \frac{2mr^2}{5}$ $I_{22}^O = \frac{2mr^2}{5}$ $I_{33}^O = \frac{2mr^2}{5}$ $I_{11}^* = \frac{2mr^2}{5}$ $I_{22}^* = \frac{2mr^2}{5}$ $I_{33}^* = \frac{2mr^2}{5}$	$I_{12}^O = I_{23}^O = I_{31}^O = 0$ $I_{12}^* = I_{23}^* = I_{31}^* = 0$
C.4.11 Hemi- sphere		$V = \frac{2\pi r^3}{3}$ $x_1^* = 0$ $x_2^* = 0$ $x_3^* = -\frac{3r}{8}$	$I_{11}^O = \frac{2mr^2}{5}$ $I_{22}^O = \frac{2mr^2}{5}$ $I_{33}^O = \frac{2mr^2}{5}$ $I_{11}^* = \frac{83mr^2}{320}$ $I_{22}^* = \frac{83mr^2}{320}$ $I_{33}^* = \frac{2mr^2}{5}$	$I_{12}^O = I_{23}^O = I_{31}^O = 0$ $I_{12}^* = I_{23}^* = I_{31}^* = 0$
C.4.12 Spherical sector		$V = \frac{2\pi a r^2}{3}$ $x_1^* = 0$ $x_2^* = 0$ $x_3^* = \frac{6r-3a}{8}$	$I_{11}^O = \frac{m(a^2-3ra+6r^2)}{10}$ $I_{22}^O = \frac{m(a^2-3ra+6r^2)}{10}$ $I_{33}^O = \frac{3mra}{5} - \frac{ma^2}{5}$ $I_{11}^* = \frac{m(12r^2+84ra-13a^2)}{320}$ $I_{22}^* = \frac{m(12r^2+84ra-13a^2)}{320}$ $I_{33}^* = \frac{3mra}{5} - \frac{ma^2}{5}$	$I_{12}^O = I_{23}^O = I_{31}^O = 0$ $I_{12}^* = I_{23}^* = I_{31}^* = 0$

Solid	Geometry	Volume, Centroid Position	Moments of Inertia	Products of Inertia
C.4.13 Spherical segment	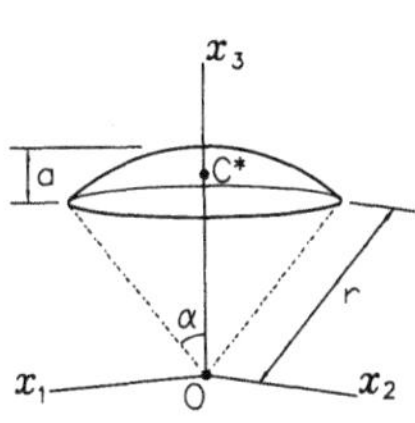	$V = \dfrac{\pi a^2(3r-a)}{3}$ $x_1^* = 0$ $x_2^* = 0$ $x_3^* = \dfrac{3(2r-a)^2}{4(3r-a)}$	$I_{11}^O = \dfrac{m(60r^3 - 80r^2a + 45ra^2 - 9a^3)}{20(3r-a)}$ $I_{22}^O = \dfrac{m(60r^3 - 80r^2a + 45ra^2 - 9a^3)}{20(3r-a)}$ $I_{33}^O = \dfrac{m(20ar^2 - 15ra^2 + 3a^3)}{10(3r-a)}$ $I_{11}^* = I_{11}^O - mx_3^{*2}$ $I_{22}^* = I_{22}^O - mx_3^{*2}$ $I_{33}^* = I_{33}^O$	$I_{12}^O = I_{23}^O = I_{31}^O = 0$ $I_{12}^* = I_{23}^* = I_{31}^* = 0$
C.4.14 Circular cylinder	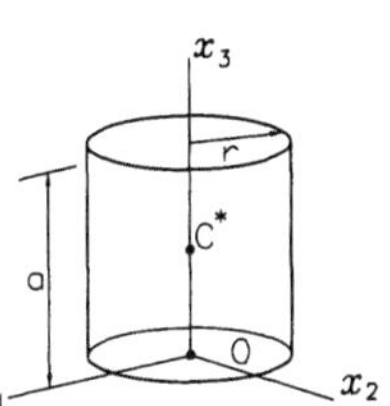	$V = \pi a r^2$ $x_1^* = 0$ $x_2^* = 0$ $x_3^* = \dfrac{a}{2}$	$I_{11}^O = \dfrac{mr^2}{4} + \dfrac{ma^2}{3}$ $I_{22}^O = \dfrac{mr^2}{4} + \dfrac{ma^2}{3}$ $I_{33}^O = \dfrac{mr^2}{2}$ $I_{11}^* = \dfrac{mr^2}{4} + \dfrac{ma^2}{12}$ $I_{22}^* = \dfrac{mr^2}{4} + \dfrac{ma^2}{12}$ $I_{33}^* = \dfrac{mr^2}{2}$	$I_{12}^O = I_{23}^O = I_{31}^O = 0$ $I_{12}^* = I_{23}^* = I_{31}^* = 0$
C.4.15 Semi- circular cylinder	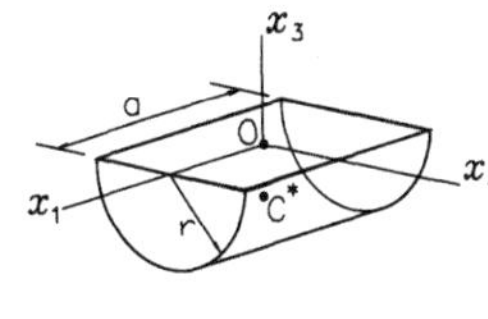	$V = \dfrac{\pi a r^2}{2}$ $x_1^* = 0$ $x_2^* = 0$ $x_3^* = -\dfrac{4r}{3\pi}$	$I_{11}^O = \dfrac{mr^2}{2}$ $I_{22}^O = \dfrac{mr^2}{4} + \dfrac{ma^2}{12}$ $I_{33}^O = \dfrac{mr^2}{4} + \dfrac{ma^2}{12}$ $I_{11}^* = \dfrac{(9\pi^2 - 32)mr^2}{18\pi^2}$ $I_{22}^* = \dfrac{(9\pi^2 - 64)mr^2}{36\pi^2} + \dfrac{ma^2}{12}$ $I_{33}^* = \dfrac{mr^2}{4} + \dfrac{ma^2}{12}$	$I_{12}^O = I_{23}^O = I_{31}^O = 0$ $I_{12}^* = I_{23}^* = I_{31}^* = 0$
C.4.16 Cylindrical sector	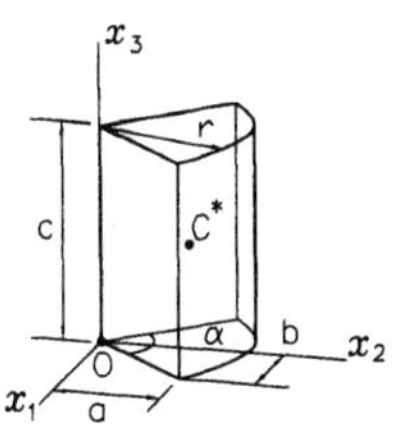	$V = c a r^2$ $x_1^* = 0$ $x_2^* = \dfrac{2b}{3\alpha}$ $x_3^* = \dfrac{c}{2}$	$I_{11}^O = \dfrac{mr^2}{4} + \dfrac{mab}{4\alpha} + \dfrac{mc^2}{3}$ $I_{22}^O = \dfrac{mr^2}{4} - \dfrac{mab}{4\alpha} + \dfrac{mc^2}{3}$ $I_{33}^O = \dfrac{mr^2}{2}$ $I_{22}^* = \dfrac{mr^2}{4} - \dfrac{mab}{4\alpha} + \dfrac{mc^2}{12}$ $I_{33}^* = \dfrac{mr^2}{2} - \dfrac{4mb^2}{9\alpha^2}$ $I_{11}^* = \dfrac{mr^2}{4} + \dfrac{mab}{4\alpha} + \dfrac{mc^2}{12} - \dfrac{4mb^2}{9\alpha^2}$	$I_{12}^O = I_{31}^O = 0$ $I_{23}^O = -\dfrac{mbc}{3\alpha}$ $I_{12}^* = I_{23}^* = I_{31}^* = 0$

Solid	Geometry	Volume, Centroid Position	Moments of Inertia	Products of Inertia
C.4.17 Cylindrical segment	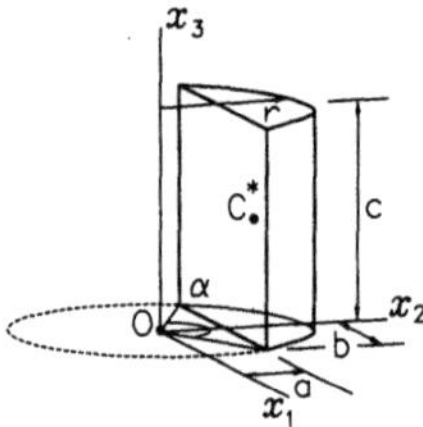	$V = \alpha c r^2 - abc$ $x_1^* = 0$ $x_2^* = \dfrac{2b^3}{3\alpha r^2 - 3ab}$ $x_3^* = \dfrac{c}{2}$	$I_{11}^O = \dfrac{m(\alpha r^4 - abr^2 + 2ab^3)}{4(\alpha r^2 - ab)} + \dfrac{mc^2}{3}$ $I_{22}^O = \dfrac{m(3\alpha r^4 - 3abr^2 - 2ab^3)}{12(\alpha r^2 - ab)} + \dfrac{mc^2}{3}$ $I_{33}^O = \dfrac{m(3\alpha r^4 - 3abr^2 + 2ab^3)}{6(\alpha r^2 - ab)}$ $I_{11}^* = I_{11}^O - m(x_2^{*2} + x_3^{*2})$ $I_{22}^* = I_{22}^O - m x_3^{*2}$ $I_{33}^* = I_{33}^O - m x_2^{*2}$	$I_{12}^O = I_{31}^O = 0$ $I_{23}^O = -\dfrac{mb^3 c}{3\alpha r^2 - 3ab}$ $I_{12}^* = I_{23}^* = I_{31}^* = 0$
C.4.18 Elliptic cylinder	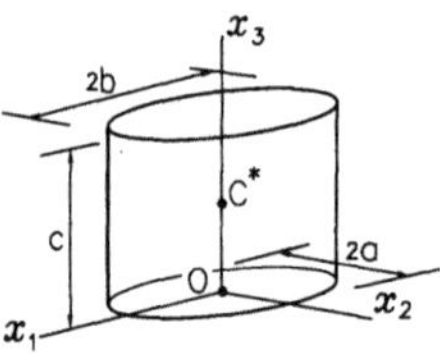	$V = \pi abc$ $x_1^* = 0$ $x_2^* = 0$ $x_3^* = \dfrac{c}{2}$	$I_{11}^O = \dfrac{ma^2}{4} + \dfrac{mc^2}{3}$ $I_{22}^O = \dfrac{mb^2}{4} + \dfrac{mc^2}{3}$ $I_{33}^O = \dfrac{ma^2}{4} + \dfrac{mb^2}{4}$ $I_{11}^* = \dfrac{ma^2}{4} + \dfrac{mc^2}{12}$ $I_{22}^* = \dfrac{mb^2}{4} + \dfrac{mc^2}{12}$ $I_{33}^* = \dfrac{ma^2}{4} + \dfrac{mb^2}{4}$	$I_{12}^O = I_{23}^O = I_{31}^O = 0$ $I_{12}^* = I_{23}^* = I_{31}^* = 0$
C.4.19 Semi-elliptic cylinder	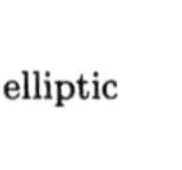	$V = \dfrac{\pi abc}{2}$ $x_1^* = 0$ $x_2^* = 0$ $x_3^* = -\dfrac{4a}{3\pi}$	$I_{11}^O = \dfrac{ma^2}{4} + \dfrac{mc^2}{12}$ $I_{22}^O = \dfrac{ma^2}{4} + \dfrac{mb^2}{4}$ $I_{33}^O = \dfrac{mb^2}{4} + \dfrac{mc^2}{12}$ $I_{11}^* = \dfrac{ma^2}{4} + \dfrac{mc^2}{12} - \dfrac{16ma^2}{9\pi^2}$ $I_{22}^* = \dfrac{ma^2}{4} + \dfrac{mb^2}{4} - \dfrac{16ma^2}{9\pi^2}$ $I_{33}^* = \dfrac{mb^2}{4} + \dfrac{mc^2}{12}$	$I_{12}^O = I_{23}^O = I_{31}^O = 0$ $I_{12}^* = I_{23}^* = I_{31}^* = 0$
C.4.20 Circular cone	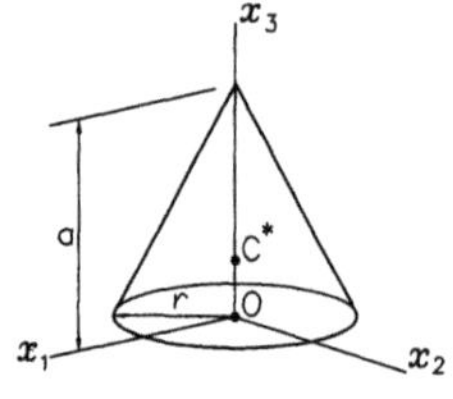	$V = \dfrac{\pi a r^2}{3}$ $x_1^* = 0$ $x_2^* = 0$ $x_3^* = \dfrac{a}{4}$	$I_{11}^O = \dfrac{3mr^2}{20} + \dfrac{ma^2}{10}$ $I_{22}^O = \dfrac{3mr^2}{20} + \dfrac{ma^2}{10}$ $I_{33}^O = \dfrac{3mr^2}{10}$ $I_{11}^* = \dfrac{3mr^2}{20} + \dfrac{3ma^2}{80}$ $I_{22}^* = \dfrac{3mr^2}{20} + \dfrac{3ma^2}{80}$ $I_{33}^* = \dfrac{3mr^2}{10}$	$I_{12}^O = I_{23}^O = I_{31}^O = 0$ $I_{12}^* = I_{23}^* = I_{31}^* = 0$

Solid	Geometry	Volume, Centroid Position	Moments of Inertia	Products of Inertia
C.4.21 Frustum of circular cone	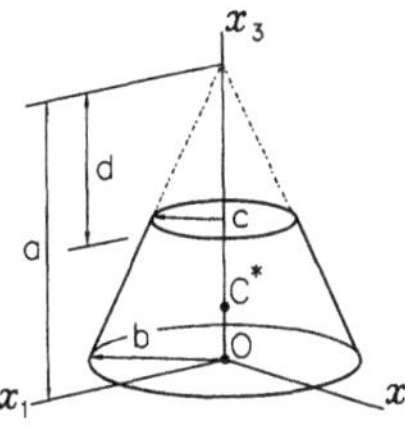	$V = \dfrac{\pi(ab^2 - dc^2)}{3}$ $x_1^* = 0$ $x_2^* = 0$ $x_3^* = \dfrac{\frac{3}{4}c^2 d(d-a)}{ab^2 - dc^2} + \dfrac{a}{4}$	$I_{11}^O = \dfrac{m(2a^4 + 3a^2 b^2 + 3b^2 d^2)}{20a^2} - \dfrac{3md^2(6ad + b^2 - 4d^2)}{20(a^2 + ad + d^2)}$ $I_{22}^O = \dfrac{m(2a^4 + 3a^2 b^2 + 3b^2 d^2)}{20a^2} - \dfrac{3md^2(6ad + b^2 - 4d^2)}{20(a^2 + ad + d^2)}$ $I_{33}^O = \dfrac{3m(ab^4 - dc^4)}{10(ab^2 - dc^2)}$ $I_{11}^* = I_{11}^O - mx_3^{*2}$ $I_{22}^* = I_{22}^O - mx_3^{*2}$ $I_{33}^* = I_{33}^O$	$I_{12}^O = I_{23}^O = I_{31}^O = 0$ $I_{12}^* = I_{23}^* = I_{31}^* = 0$
C.4.22 Semi-circular cone	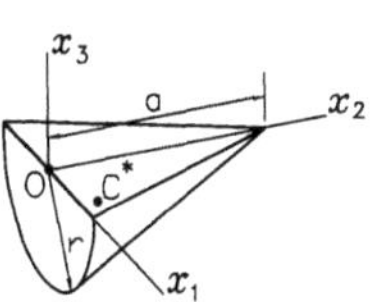	$V = \dfrac{\pi a r^2}{6}$ $x_1^* = 0$ $x_2^* = \dfrac{a}{4}$ $x_3^* = -\dfrac{r}{\pi}$	$I_{11}^O = \dfrac{3mr^2}{20} + \dfrac{ma^2}{10}$ $I_{22}^O = \dfrac{3mr^2}{10}$ $I_{33}^O = \dfrac{3mr^2}{20} + \dfrac{ma^2}{10}$ $I_{11}^* = \dfrac{3mr^2}{20} + \dfrac{3ma^2}{80} - \dfrac{mr^2}{\pi^2}$ $I_{22}^* = \dfrac{3mr^2}{10} - \dfrac{mr^2}{\pi^2}$ $I_{33}^* = \dfrac{3mr^2}{20} + \dfrac{3ma^2}{80}$	$I_{12}^O = 0$ $I_{23}^O = \dfrac{mra}{5\pi}$ $I_{31}^O = 0$ $I_{12}^* = 0$ $I_{23}^* = -\dfrac{mra}{20\pi}$ $I_{31}^* = 0$
C.4.23 Ellipsoid	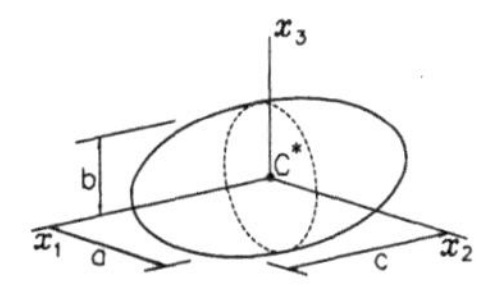	$V = \dfrac{4\pi abc}{3}$ $x_1^* = 0$ $x_2^* = 0$ $x_3^* = 0$	$I_{11}^O = \dfrac{m(a^2 + b^2)}{5}$ $I_{22}^O = \dfrac{m(b^2 + c^2)}{5}$ $I_{33}^O = \dfrac{m(a^2 + c^2)}{5}$ $I_{11}^* = \dfrac{m(a^2 + b^2)}{5}$ $I_{22}^* = \dfrac{m(b^2 + c^2)}{5}$ $I_{33}^* = \dfrac{m(a^2 + c^2)}{5}$	$I_{12}^O = I_{23}^O = I_{31}^O = 0$ $I_{12}^* = I_{23}^* = I_{31}^* = 0$
C.4.24 Semi-ellipsoid	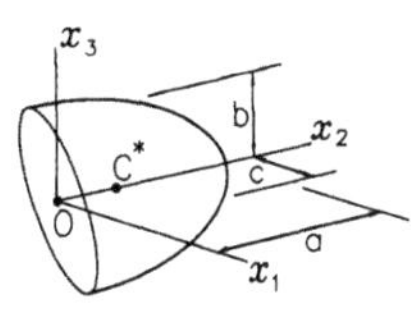	$V = \dfrac{2\pi abc}{3}$ $x_1^* = 0$ $x_2^* = \dfrac{3a}{8}$ $x_3^* = 0$	$I_{11}^O = \dfrac{ma^2}{5} + \dfrac{mb^2}{5}$ $I_{22}^O = \dfrac{mb^2}{5} + \dfrac{mc^2}{5}$ $I_{33}^O = \dfrac{ma^2}{5} + \dfrac{mc^2}{5}$ $I_{11}^* = \dfrac{mb^2}{5} + \dfrac{19ma^2}{320}$ $I_{22}^* = \dfrac{mb^2}{5} + \dfrac{mc^2}{5}$ $I_{33}^* = \dfrac{mc^2}{5} + \dfrac{19ma^2}{320}$	$I_{12}^O = I_{23}^O = I_{31}^O = 0$ $I_{12}^* = I_{23}^* = I_{31}^* = 0$

Solid	Geometry	Volume, Centroid Position	Moments of Inertia	Products of Inertia
C.4.25 Paraboloid of revolution		$V = \frac{\pi a r^2}{2}$ $x_1^* = 0$ $x_2^* = \frac{a}{3}$ $x_3^* = 0$	$I_{11}^O = \frac{mr^2}{6} + \frac{ma^2}{6}$ $I_{22}^O = \frac{mr^2}{3}$ $I_{33}^O = \frac{mr^2}{6} + \frac{ma^2}{6}$ $I_{11}^* = \frac{mr^2}{6} + \frac{ma^2}{18}$ $I_{22}^* = \frac{mr^2}{3}$ $I_{33}^* = \frac{mr^2}{6} + \frac{ma^2}{18}$	$I_{12}^O = I_{23}^O = I_{31}^O = 0$ $I_{12}^* = I_{23}^* = I_{31}^* = 0$
C.4.26 Elliptic paraboloid		$V = \frac{\pi a b c}{2}$ $x_1^* = 0$ $x_2^* = \frac{a}{3}$ $x_3^* = 0$	$I_{11}^O = \frac{ma^2}{6} + \frac{mb^2}{6}$ $I_{22}^O = \frac{mb^2}{6} + \frac{mc^2}{6}$ $I_{33}^O = \frac{ma^2}{6} + \frac{mc^2}{6}$ $I_{11}^* = \frac{ma^2}{18} + \frac{mb^2}{6}$ $I_{22}^* = \frac{mb^2}{6} + \frac{mc^2}{6}$ $I_{33}^* = \frac{ma^2}{18} + \frac{mc^2}{6}$	$I_{12}^O = I_{23}^O = I_{31}^O = 0$ $I_{12}^* = I_{23}^* = I_{31}^* = 0$
C.4.27 Circular torus		$V = 2\pi^2 a^2 r$ $x_1^* = 0$ $x_2^* = 0$ $x_3^* = 0$	$I_{11}^O = \frac{5ma^2}{8} + \frac{mr^2}{2}$ $I_{22}^O = \frac{5ma^2}{8} + \frac{mr^2}{2}$ $I_{33}^O = \frac{3ma^2}{4} + mr^2$ $I_{11}^* = \frac{5ma^2}{8} + \frac{mr^2}{2}$ $I_{22}^* = \frac{5ma^2}{8} + \frac{mr^2}{2}$ $I_{33}^* = \frac{3ma^2}{4} + mr^2$	$I_{12}^O = I_{23}^O = I_{31}^O = 0$ $I_{12}^* = I_{23}^* = I_{31}^* = 0$
C.4.28 Semi-circular torus		$V = \pi^2 a^2 r$ $x_1^* = 0$ $x_2^* = \frac{a^2 + 4r^2}{2\pi r}$ $x_3^* = 0$	$I_{11}^O = \frac{5ma^2}{8} + \frac{mr^2}{2}$ $I_{22}^O = \frac{5ma^2}{8} + \frac{mr^2}{2}$ $I_{33}^O = \frac{3ma^2}{4} + mr^2$ $I_{22}^* = \frac{5ma^2}{8} + \frac{mr^2}{2}$ $I_{11}^* = \frac{5ma^2}{8} - \frac{m(a^2+4r^2)^2}{4\pi^2 r^2} + \frac{mr^2}{2}$ $I_{33}^* = \frac{3ma^2}{4} - \frac{m(a^2+4r^2)^2}{4\pi^2 r^2} + mr^2$	$I_{12}^O = I_{23}^O = I_{31}^O = 0$ $I_{12}^* = I_{23}^* = I_{31}^* = 0$

Answers to the Exercises

The answers provided here are for almost all of the problems proposed in the series of exercises at the end of each chapter. Of course, answers are not given to those exercises that require a demonstration or to those requiring to model a set of forces and torques acting on a mechanical system.

The purpose of the exercises is to help the reader assimilate the concepts presented in the text, sometimes to complement some ideas not explicitly addressed in the body of the text. It is suggested that the reader work *each* exercise proposed as the corresponding sections of the text are being studied. The answers are available to check if the reader has the right solution to the exericse. Do not be discouraged if this does not happen right away; try again. Some exercises are quite simple, while others require a little more work, and a few propose a slight challenge to the reader. They are distributed in a more or less random manner, within each series, not necessarily in increasing order of difficulty. Many of the exercises were chosen to stress a different aspect of the theory and its applications. This is why the study of the text, accompanied by an *individual effort* in solving each proposed problem, will certainly result in the desired mastery of dynamics.

1.1   Galileo was right. If your conclusion was not this one, then try to think about it a little more.

1.2   Aristotle did not clearly realize the presence of the friction forces.

1.3   For example, the statement: "I am lying."

1.4   The first one is a discrete system and the second a continuous one; the field ρ is defined over a continuous domain.

1.5 $$M = \sum_{i=1}^{n} m_i + \sum_{j=1}^{m} \int_{C_j} \rho_j \, dv.$$

1.7   With this more general formulation, mass variation is allowed.

1.9   0.001%.

1.10   781.237 N.

1.11 r or $\dfrac{5}{3}r$.

1.12 $\dfrac{4r}{3\pi}$, r the radius.

1.13   If the applied forces and torques are not correctly identified, the entire formulation of the problem will be jeopardized.

1.14 $^{\mathcal{R}}K^{C}$; $\mathbf{M}^{\mathcal{V}/x_1}$.

2.1   12.65 Nm; 12 Nm.

2.2   50 N; $20(3\mathbf{n}_1 - 2\mathbf{n}_3)$ N, $13.5\mathbf{n}_1$ Nm.

2.3   4 N and 236 N; 31 N.

2.4   7.874; 12.247; 9.144.

2.5 $\dfrac{1}{31}(-16, 22, 17)$.

2.6 $u\big(\alpha, \beta, (1 + 5\beta)\big)$, α and β real.

2.7 $20\sqrt{2}$ N; $10((3 + \sqrt{2}), 2, 0)$ N; 50 N e 60 Nm.

2.8   1.257 m.

2.9 $\dfrac{1}{2}\rho g l z^2$

2.10 $2G\rho^2 r\dfrac{p-r}{p}$, $p=\sqrt{a^2+r^2}$; yes.

2.11 1.28 rad.

2.14 $\dfrac{GmM}{r^2}$; $\dfrac{GmMa^2}{2r^3}\sin 2\psi$.

2.15 26 cm.

2.16 No; 25% of binding.

2.17 $\dfrac{1}{109}(70\mathbf{n}_1-55\mathbf{n}_2-72\mathbf{n}_3)$.

2.20 14/27.

2.22 $V_1=\dfrac{1}{6}Qc$, $V_2=\dfrac{1}{3}Qc$; yes; $\dfrac{Qc^2}{3(V_1'-\frac{1}{2}Qc)}\mathbf{n}_1+\lambda\mathbf{n}_2$, λ real.

2.23 Two force and two torque; two degrees of freedom.

2.24 $a,\ a/2,\ a/3,\ \ldots,\ a/n$.

2.25 $\dfrac{mgr}{2\sqrt{R^2+2rR}}$.

2.26 $\theta=\tan^{-1}\dfrac{r^2\mu}{r^2-\frac{c^2}{4}(1+\mu^2)}$.

2.27 13.34 N.

2.28 $\dfrac{17}{48}\,a$.

3.1 $c_1+c_2+c_3$; c_1-c_2; no.

3.2 $-\dot\phi\sin\theta\,\mathbf{n}_1+(\Omega+\dot\phi\cos\theta)\mathbf{n}_2+\dot\theta\mathbf{n}_3$.

3.3 $\dfrac{c}{\cos\theta}$, c constant.

3.5 0; $-\omega_2\omega_3\mathbf{a}_1+\omega_3\omega_1\mathbf{a}_2+\omega_1\omega_2\mathbf{a}_3$.

3.7 $(\dot\phi\dot\theta+\ddot\psi\cos\theta-\dot\psi\dot\theta\sin\theta)\mathbf{n}_1+(\ddot\theta-\dot\phi\dot\psi\cos\theta)\mathbf{n}_2$
$+(\ddot\phi+\ddot\psi\sin\theta+\dot\psi\dot\theta\cos\theta)\mathbf{n}_3$.

3.10 300.03 rad/s; 1341.64 rad/s^2.

3.11 $\dfrac{\sqrt{30}}{2}\Omega;\ \dfrac{9+\sqrt{3}}{4}\Omega^2$.

3.12 $(3\mathbf{n}_1 - \pi\sqrt{2}\mathbf{n}_3)$ rad/s; $3\pi(\sqrt{2}\mathbf{n}_2 - \frac{1}{2}\mathbf{n}_3)$ rad/s^2.

3.13 $-u\dot{\phi}\mathbf{n}_1;\ (r\dot{\phi}^2 - u\ddot{\phi})\mathbf{n}_1 - u\dot{\phi}^2\mathbf{n}_2,\ u = c - r\phi$.

3.14 $(\ddot{r} - r\dot{\theta}^2 - r\dot{\phi}^2\cos^2\theta)\mathbf{b}_1 + (r\ddot{\theta} + 2\dot{r}\dot{\theta} + r\dot{\phi}^2\cos\theta\sin\theta)\mathbf{b}_2$
$+(r\ddot{\phi}\cos\theta + 2\dot{r}\dot{\phi}\cos\theta - 2r\dot{\theta}\dot{\phi}\sin\theta)\mathbf{b}_3$.

3.15 $-\omega_0^2 l\mathbf{n}_2;\ 2\omega_0^2 l(-3\mathbf{n}_1 - 2\mathbf{n}_2 + 2\mathbf{n}_3)$.

3.16 $(r\theta_0\omega_0 + a\Omega)\mathbf{n}_2 - v\mathbf{n}_3;\ \theta_0\omega_0(2v\mathbf{n}_2 + r\theta_0\omega_0\mathbf{n}_3)$.

3.17 0; 48 rad/s^2; 2.824 rad/s; -4.484 rad/s^2.

3.18 $\dfrac{1}{6z}\big(-3v\mathbf{n}_1 + (v - 2u)\mathbf{n}_2 - (v - 2u)\mathbf{n}_3\big)$.

3.19 $-\dfrac{1}{2\sqrt{2}z}\big(\mathbf{n}_2 + \mathbf{n}_3\big);\ -\dfrac{a}{6z}(3\mathbf{n}_1 + \mathbf{n}_2 - \mathbf{n}_3)$.

3.20 2.792 m/s; 18.6 m/s^2.

3.21 $\dfrac{v}{2r};\ 0;\ v\sin\theta;\ \dfrac{v^2}{2r}(1 + 3\sin^2\theta)^{\frac{1}{2}}$.

3.22 $-\pi\omega^2\mathbf{n}_3;\ 2c\omega\mathbf{n}_1$.

3.23 $\dfrac{v}{2a}(2\mathbf{n}_2 + \mathbf{n}_3);\ 0$.

3.24 $2.027\,\omega^2 r$.

4.1 $r\dot{\theta}\mathbf{a}_1 + \dot{r}\mathbf{a}_2 - r\omega\sin\theta\mathbf{a}_3;\ \dfrac{\sqrt{5}}{5}l\omega^2$.

4.2 $8\sqrt{2}$ m/s^2.

4.3 $\dfrac{a^2 + b^2}{a}$ and $\dfrac{a^2}{\sqrt{a^2 + b^2}}$.

4.4 $2\pi a\sqrt{1 + \sin^2(4\pi t)};\ 4\pi^2 a\sqrt{1 + 4\cos^2(4\pi t)}$.

4.5 $\dfrac{l\omega_0^2}{\sqrt{2}}.$

4.6 $0.$

4.7 The ellipse $\dfrac{x^2}{A^2} + \dfrac{y^2}{B^2} - \dfrac{2xy}{AB}\cos(\phi - \psi) = \sin^2(\phi - \psi).$

4.8 2 m.

4.9 $4r\sin\dfrac{\phi}{2}.$

4.10 400 m/s^2, directed to the disk center.

4.11 $\dfrac{g}{v}(v^2 - a^2).$

4.12 $\dfrac{g}{v}a.$

5.1 $\dfrac{\sqrt{4a^2 + 3b^2}}{\sqrt{3}\,ab}\,v;\ \dfrac{\sqrt{16a^2 + 3b^2}}{2\sqrt{3}\,a}\,v.$

5.2 $\dfrac{b}{2a}v.$

5.3 10 m/s^2.

5.4 $\omega\sin\theta\cos\psi\,\mathbf{b}_1 + \omega\sin\theta\sin\psi\,\mathbf{b}_2 + (\Omega + \dot{\psi})\mathbf{b}_3;$
$\omega(\dot{\psi} + \Omega)\sin\theta\sin\psi(-\mathbf{b}_1 + \mathbf{b}_2);$
$-r(\dot{\psi} + \Omega)\sin\theta\cos\psi\,\mathbf{b}_1 + (R\Omega + r(\dot{\psi} + \Omega)\sin\theta\sin\psi)\mathbf{b}_2.$

5.5 $\Omega\left(\dfrac{\pi}{2}\sin(\Omega t)\mathbf{n}_1 + \mathbf{n}_3\right);$
$\dfrac{\pi}{2}\Omega^2\left[\cos(\Omega t)\mathbf{n}_1 + \sin(\Omega t)\mathbf{n}_2\right].$

5.6 4.1356 m/s

5.7 $\dfrac{v}{\sqrt{2}\,r};\ \dfrac{v}{\sqrt{2}\,r}.$

5.8 $\dfrac{v^2}{2r^2};\ \dfrac{v^2}{2r^2}.$

5.9 $4\pi(\sqrt{3}\mathbf{n}_2 - \mathbf{n}_3);\ 237.51\,\mathbf{n}_1.$

5.10 $\dfrac{1 + \tan^2 \theta}{\cos \beta + \tan^2 \theta / \cos \beta}\, \Omega \mathbf{b}_3;\ \Omega / \cos \beta;\ \Omega \cos \beta.$

5.11 $\Omega / \cos \beta;\ \Omega.$

5.12 $\dfrac{\pi}{5}(-3\mathbf{n}_1 + \mathbf{n}_2 + 3\mathbf{n}_3)\ \text{m/s};\ \dfrac{\pi^2}{5}(\mathbf{n}_1 + 3\mathbf{n}_2)\ \text{m/s}^2.$

5.13 $\text{Yes};\ -\dfrac{v}{r \cos \beta}\mathbf{n}_1;\ -\dfrac{v^2}{r^2}\tan \beta \mathbf{n}_2.$

5.14 $\Omega \mathbf{n}_2 + \dfrac{1}{6}(\Omega - \omega)\mathbf{n}_3;\ -\dfrac{1}{6}\Omega(\Omega - \omega)\mathbf{n}_1;\ r\Omega \mathbf{n}_1;\ \dfrac{1}{6}r\Omega(\Omega - \omega)\mathbf{n}_2.$

5.15 $\dfrac{2u}{\sqrt{3}a}.$

5.16 $\dfrac{a}{r} = \dfrac{1 + \sin \beta}{\cos \beta - \sin \beta}.$

5.17 $\dfrac{\omega}{\cos \theta}(\sin^2 \theta \sin^2 \phi + \cos^2 \theta)\mathbf{n}_3;$
$\omega \sin \theta(\cos \phi \mathbf{n}_1 + \sin \phi \mathbf{n}_2 + \sin^2 \phi \tan \theta \mathbf{n}_3).$

5.18 $\Omega\left(-\dfrac{11}{12}\sin \theta \mathbf{n}_1 + \cos \theta \mathbf{n}_3\right);\ -\dfrac{11}{144}\Omega^2 \cos \theta \sin \theta \mathbf{n}_3.$
$\dfrac{\Omega R}{24}\big(\cos \theta(\sin \theta - 2\cos \theta) - 22\big)\mathbf{n}_2.$

5.19 61.538 rpm.

5.20 $-\dfrac{v^2 \sin^2 \theta}{r^2 \sin \beta}\,\mathbf{n}_2;\ -\dfrac{v^2 \sin \theta}{r \sin \beta \cos \beta}(\cos \theta\, \mathbf{n}_1 - \sin \theta\, \mathbf{n}_3).$

5.21 $(2\cos \theta - \sin \theta)\}\cos \theta\, \Omega^2 r\big((\cos \theta + 2\sin \theta)\mathbf{n}_1 + (2\cos \theta - \sin \theta)\mathbf{b}_3\big).$

5.22 $2(b - r)\,\Omega.$

5.23 $\omega \mathbf{n}_2 + \dfrac{\omega - \omega'}{2}\,\mathbf{n}_3;\ -\dfrac{\omega(\omega - \omega')}{2}\,\mathbf{n}_1.$

5.24 $-\omega_0\left(\dfrac{1}{\sqrt{3}}\,\mathbf{n}_1 + \mathbf{n}_2\right);\ \dfrac{\omega_0^2}{3\sqrt{3}}\,n_3.$

5.25 $(\dot{\phi} + \dot{\theta})(\sin \psi \mathbf{c}_1 - \cos \psi \mathbf{c}_2) + \dot{\psi}\mathbf{c}_3;\ -\phi_0 \lambda^2 \mathbf{c}_3;$
$-3\omega_0(\sin \psi_0 \mathbf{c}_1 - \cos \psi_0 \mathbf{c}_2).$

5.26 4π rad/s.

5.27 Three.

5.28 $\dfrac{\omega + \omega'}{2}r$; $\dfrac{\omega^2 - \omega'^2}{4}r$.

5.29 $\dfrac{v\sin\theta(s-r)}{2rs + \cos\theta(a + 2b\cos\theta)(s+r)}$.

5.30 $-2\Omega\mathbf{b}_1$; $-2\Omega^2\mathbf{b}_2$; $-\sqrt{5}\Omega^2 r\mathbf{b}_1$; $2\sqrt{5}\,\Omega^2 r\mathbf{b}_3$.

5.31 $\dfrac{N_D}{2N_E}(\Omega + \Omega')$, N_D and N_E being the number of teeth in D and E, respectively.

6.1 $-m\dfrac{v^2 b^4}{a^2 y^3}$, in the y-direction.

6.2 19.55 m; 2.61 s.

6.3 $\dfrac{ma^2}{r^3}$; a logarithmic spiral.

6.4 17 times.

6.5 The runway is on the line of the equator, in the direction $\mathcal{N} - \mathcal{S}$.

6.6 4.02 m/s.

6.7 3.

6.8 12,768 km.

6.9 $m\sqrt{4\omega^4 r^2 + g^2}$; no.

6.10 $\theta = \cos^{-1}\dfrac{1}{3}\left(2 + \dfrac{v^2}{gr}\right)$; $\sqrt{gr}$; 0.841 rad.

6.11 0.11 a; 0.90 mg.

6.12 1.642 s.

6.13 $\dfrac{3(3 - \theta)\sin\theta + 2(1 - \cos\theta)}{3 - \theta}mg$.

6.14 $m\big[(2\dot{r}\omega + r\omega^2 \sin\theta\cos\theta - g\cos\theta)\,\mathbf{b}_2 + 2\,(\dot{r}\omega\cos\theta - r\omega^2\sin\theta)\,\mathbf{b}_3\big]$,
where $\dot{r} = \big(\omega^2(1 + \cos^2\theta)r^2 + 2gr\sin\theta\big)^{\frac{1}{2}}$.

6.15 $\theta = \dfrac{a_0 p^2}{r(q^2 - p^2)}\left(\sin pt - \dfrac{p}{q}\sin qt\right), \quad q = \sqrt{g/r}.$

6.16 5.5365.

6.17 $\dfrac{39\pi}{4}\rho g a^4$.

6.18 $\sqrt{(8 - \pi)gr}$.

6.19   The parabola $y = \dfrac{1}{2}\dfrac{\omega^2}{g}x^2 + a$.

6.20 $\dfrac{a}{\sqrt{a^2 + R^2}}\sqrt{2gb - \left(\left(\dfrac{a}{a-b}\right)^2 - 1\right)v_0^2}$.

6.21 $\sqrt{g(a + 2r)};\ 2mg\dfrac{a + 2r}{a - \pi r}$.

6.22 $\sqrt{u^2 - \dfrac{k}{m}(r - r_0)^2}$.

6.23 1.1864 rad.

6.24 $\frac{\pi}{6}$ rad; $\frac{\pi}{4}$ rad.

6.25 $\sqrt{2}\sin\dfrac{\theta}{2};\ x = 2r/3$.

6.26 $m\sin\theta\alpha r$.

6.27 $\dfrac{1}{2}\left(r - \dfrac{g\cos\theta}{\Omega^2\sin^2\theta}\right)\left(e^{\Omega\sin\theta t} + e^{-\Omega\sin\theta t}\right) + \dfrac{g\cos\theta}{\Omega^2\sin^2\theta};\ 0.$

6.28 $\dfrac{2}{\sin\theta}\sqrt{\dfrac{g\cos\theta}{3r}}$.

6.29   No; $-m\Omega^2\sin^2\theta r^2$;   The horizontal force does negative work, draining energy from P.

7.1 $\dfrac{m}{a}(gx + v^2)$.

7.2 $\dfrac{F}{m(3 - 2\sin^2\theta)}$.

7.3 $\dfrac{F}{2m}$.

7.4 $g\tan\dfrac{\alpha-\beta}{2}$.

7.5 $\dfrac{3}{5}r$.

7.6 $1.210\,u$; $2.545\,u$.

7.7 ω.

7.8 $\dfrac{1}{2}mv^2$.

7.9 36.4^0.

7.10 $a\sqrt{\dfrac{km_1}{(m_1+m_2)m_2}}$; $2\sqrt{3}\,a\dfrac{m_2}{m_1+m_2}$.

7.11 $2\pi\sqrt{\dfrac{3m_1m_2}{k(m_1+m_2)}}$.

7.12 $v_1=v_2=0.79\,v$; $\theta_1=\theta_2=0.32$ rad.

7.13 $mg\left(3+\dfrac{4(a+r)}{c-\pi r}\right)$.

7.14 $\dfrac{1}{2}\sqrt{gr-9v^2}$.

7.15 $\dfrac{6}{5}(1+2\sqrt{3})$ m/s.

7.16 $0.2195\sqrt{\dfrac{k}{m}}\,a$.

7.17 $v_1=v_2=v\cos\theta$.

7.18 $v_A=2.37$ m/s; $v_B=2.73$ m/s; $v_C=4.90$ m/s.

7.19 $\sqrt{m_1m_3}$.

7.20 $\dfrac{13}{7}\left(\text{or }\dfrac{7}{13}\right)$.

7.21 $\dfrac{8}{21}\dfrac{F}{m}$.

7.22   138.56 N.

7.23 $\rho Q(v_1 + v_2 \cos \alpha)$; $\rho Q(v_1 + v_2 \cos \alpha - 2v_0)$.

7.24 $\dfrac{1 + \sin \theta}{2}\, ve$; $\dfrac{1 - \sin \theta}{2}\, ve$.

7.25 $F = \rho v^2 e \cos \theta$; $T = \dfrac{1}{2}\rho v^2 e^2 \sin \theta$.

7.26 $\dfrac{1}{4r}\sqrt{v_0^2 - 4(\sqrt{3} - 1)gr}$; 87.5%.

7.27 $\alpha = \tan^{-1} \dfrac{2F/g}{m_1 + 2m_2 + 2m_3}$; $\beta = \tan^{-1} \dfrac{2F/g}{m_2 + 2m_3}$;

        $\gamma = \tan^{-1} \dfrac{2F/g}{m_3}$.

7.28 $3.561\sqrt{ga}$.

7.29 $2\dfrac{r}{c}\sin^2\left(\dfrac{c}{2r}\right) g$.

8.1 $(1.7596, 0, -1.1202)\, r$.

8.2 $(0.6278, 0.4511)\, a$.

8.3 $0.3366\, r$.

8.4 $-\dfrac{r_1 r_2^2}{2(r_1^2 - r_2^2)}, 0$.

8.5 $0.366\, a$.

8.6 $(3.2, 9.6, 6)$ cm.

8.7 $(1.47, 0.94)$ m.

8.8 $\dfrac{2m}{3\pi}$.

8.9 $\dfrac{a}{r} = \dfrac{1}{\sqrt{2}}$.

8.10 0.625 kg m^2; 1.25 kg m^2; 0.530 kg m^2.

8.11 $29.7\, ma^2$; $43.0\, ma^2$.

8.14 0.125; 0.375; 0.5; -0.2165; 0; 0 $(\mathrm{kg\,m^2})$.

8.15 $\dfrac{1}{2}m\left(r^2 + \dfrac{2a^2}{3}\right)\mathbf{n}_1$.

8.16 $\dfrac{1}{12}ma^2\big((2\sqrt{3}-1)\mathbf{n}_1 + (2-\sqrt{3})\mathbf{n}_2\big)$; $\dfrac{2\sqrt{3}-1}{12}\,ma^2$.

8.17 $1.0594\rho r^4$; $1.7260\rho r^4$; $1.4521\rho r^4$.

8.18 To the highest moment of inertia with respect to direction $a - a$.

9.1 $2\,\mathrm{ma^2}$; $\dfrac{1}{4}\,\mathrm{ma^2}$.

9.2 $\dfrac{1}{4}m(5r_2^2 + r_1^2)$.

9.3 $2.02\ \mathrm{kgm^2}$; $1.428\ \mathrm{kgm^2}$.

9.4 $0.2957\ \mathrm{kgm^2}$.

9.5
$$\frac{1}{3}ma^2 \begin{pmatrix} 5/3 & -1/2 & -3/2 \\ -1/2 & 11/3 & -1/2 \\ -3/2 & -1/2 & 8/3 \end{pmatrix}.$$

9.6 $R/\sqrt{3}$.

9.7 $\dfrac{a}{c} < 0.0316$; $\dfrac{a}{d} < 0.0775$.

9.8 340^0.

9.9 $\dfrac{1}{6\sqrt{5}}\,mr^2(5\mathbf{n}_1 - 4\mathbf{n}_3)$; $\dfrac{7}{15}\,mr^2$.

9.10 $185.42\,\rho r^5$, $538.85\,\rho r^5$, $378.85\,\rho r^5$.

9.11 $\dfrac{13}{24}\pi\rho r^5$.

9.12 $5.5\,\mathrm{ma^2}$; $2\,\mathrm{ma^2}$.

9.13 $\dfrac{5-3\sqrt{2}}{12}ma^2$, $\dfrac{5+3\sqrt{2}}{12}ma^2$, $\dfrac{5}{6}ma^2$.

9.14 $\dfrac{a^2 + 3b^2}{4(a^2 + b^2)^{\frac{1}{2}}}.$

9.15 $30^0.$

9.16 $\sqrt{3}/2.$

9.17 $\dfrac{2}{3}ma^2.$

9.18 $\dfrac{5}{4}mr^2;\ \dfrac{\sqrt{2}}{\sqrt{3}}.$

9.19 $\dfrac{m}{60}.$

9.23   They are parallel to $\{x_1, x_2, x_3\}$; $I + mr^2$, I, $J + mr^2$, $I_{jk} = 0.$

9.24 $\dfrac{3}{8}ma^2;\ \dfrac{1}{24}ma^2;\ \dfrac{5}{12}ma^2.$

9.25 $\dfrac{5}{9}mr^2;\ \dfrac{7 - 3\sqrt{2}}{18}mr^2;\ \dfrac{7 + 3\sqrt{2}}{18}mr^2.$

9.26 $0.1905\, mr^2.$

9.27 $\dfrac{5}{3}mr^2;\ \dfrac{7 - 3\sqrt{2}}{6}mr^2;\ \dfrac{7 + 3\sqrt{2}}{6}mr^2.$

9.28 $35.78^0.$

9.29 $1.458\, \rho a^4.$

9.30 $m\dfrac{a^2(2a^2 + 2b^2 + 5c^2/3) + b^2c^2}{2(4a^2 + c^2)};\ \sqrt{3(a^2 + b^2)}.$

10.1 $\dfrac{3}{40}mr^2\left(2p^2 + n^2 + 2pn\cos\beta + 4\dfrac{h^2}{r^2}n^2\sin^2\beta\right).$

10.2 $mr^2\omega\left(\dfrac{1}{2}\mathbf{n}_1 + 4\mathbf{n}_3\right);\ mr^2(2\omega^2\mathbf{n}_2 + \alpha\mathbf{n}_3).$

10.4 $\dfrac{49}{3}mr^2\omega^2.$

10.6 0.723 rad.

10.7 $mr^2\omega\left(-\dfrac{2}{\pi}\mathbf{n}_1 + \dfrac{1}{2}\mathbf{n}_2\right);\ -\dfrac{2}{\pi}mr^2\omega\mathbf{n}_1.$

10.8 $\dfrac{3\pi}{100}(13, 2, 0)\ \mathrm{kg\,m^2/s};\ 5.0335\ \mathrm{Joules}.$

10.9 $\dfrac{m\omega}{12}\left(-b^2\sin\theta\,\mathbf{a}_1 + (b^2 + a^2)\cos\theta\,\mathbf{a}_3\right);\ \dfrac{m\omega^2}{24}(b^2 + a^2\cos^2\theta).$

10.10 $\dfrac{\omega^2}{2}\left(I_1 + \dfrac{R^2}{r^2}I_2 + mR^2\right).$

10.11 $\dfrac{1}{6}mr^2\omega^2;\ \dfrac{5}{3}mr^2\omega^2.$

10.13 $13.077\,mg.$

10.14 $\dfrac{36}{11}\,g\mathbf{n}_3;\ 2.88\,mg.$

10.15 $mg\left(-\dfrac{4}{5}\mathbf{n}_2 + \dfrac{3}{4}\mathbf{n}_3\right);\ mg\left(-\dfrac{4}{5}\mathbf{n}_2 + \dfrac{15 + 32\pi}{20}\mathbf{n}_3\right).$

10.16 $1.88\,mg.$

10.17 $\dfrac{H}{2mr}.$

10.18 $\dfrac{H}{3};\ 0.$

10.19 $\dfrac{mr^2}{4}(-10.26\mathbf{n}_1 + 56.38\mathbf{n}_3);\ 211.84\,mr^2.$

10.20 $m\dfrac{r - x}{r}\left[\left(\dfrac{r + x}{2}\right)^2\omega^4 + g^2\right]^{\frac{1}{2}}.$

10.21 $mr^2\Omega^2.$

11.1 $\sqrt{\dfrac{3(\pi T - 4mga)}{4m}}.$

11.2 $10\pi mgr.$

11.3 $\sqrt{3gr}.$

11.4 $2.941\sqrt{gr}.$

11.5 $6.354\sqrt{gr}$.

11.6 $1/2;\ \dfrac{1}{2}\sqrt{\dfrac{101}{2}}$.

11.7 $\sqrt{v_0^2 + 2gd};\ \omega_0;\ \sin^{-1}\left(\dfrac{v_0\sin\theta_0}{\sqrt{v_0^2 + 2gd}}\right)$.

11.8 $\dfrac{1}{3}mg\sqrt{1 + 8\cos^2\theta}$.

11.9 $\dfrac{\cos\theta}{1 + 3\sin^2\theta}mg$.

11.10 $\dfrac{\sqrt{3}}{6}mg;\ \sqrt{2gr}$.

11.11 $30^0;\ a/3$.

11.12 $\cos^{-1}\dfrac{4}{7}$.

11.13 $k - \dfrac{11}{36}\dfrac{a^2}{k} + \dfrac{a}{2}$, where $k = \dfrac{1}{6}\left(\sqrt{1655} - 18\right)^{\frac{1}{3}}a$.

11.14 $4.6856\,\dfrac{g}{a};\ 3.16692\,g$.

11.15 0.01781 s.

11.16 $\dfrac{T}{4mr^2}$.

11.17 $\dfrac{2F}{11m}$.

11.18 $2\left(\dfrac{\pi}{r}\left(\dfrac{T}{m} - \dfrac{1}{2}g\sin\theta\right)\right)^{\frac{1}{2}}$.

11.19 $\sqrt{\dfrac{T}{k}}\,\dfrac{e^{\beta t} - 1}{e^{\beta t} + 1},\ \beta = \dfrac{2}{I}\sqrt{kT}$.

11.20 $3\sqrt{gr}$.

11.21 $\sqrt{\dfrac{\mu g}{r}}$.

12.1 $\dfrac{3}{2}\left(\dfrac{3g}{\sqrt{2}v} - \dfrac{v}{r}\right);\ \sqrt{\dfrac{3gr}{\sqrt{2}}}$.

12.2 $\cos^{-1}\dfrac{5g}{\omega^2}$; 94.58 rpm.

12.3 $3\sqrt{\dfrac{g}{8r\sin\theta}}.$

12.4 $\sqrt{\dfrac{\sqrt{3}g}{r}}.$

12.5 $\dfrac{\sqrt{3}\pi}{900}$ rad/s.

12.7 $r\sin\theta\sqrt{\dfrac{mgr}{mr^2\cos\theta+I(2-\cos\theta)-J(1-\cos\theta)}}.$

12.8 $\cos^{-1}\dfrac{3g}{2\omega^2}\dfrac{a-b}{a^2-ab+b^2}.$

12.9 $m(g+\dfrac{r\Omega^2}{2}).$

12.10 $\omega\Omega=\dfrac{4g}{r}.$

12.11 $r.$

12.12 $\cos^{-1}\dfrac{3g}{2r\omega^2};\ \dfrac{1}{2}mr\omega^2\sqrt{1+\dfrac{7g^2}{4r^2\omega^4}}.$

12.13 3.0506 rad.

12.14 $\dfrac{3r}{2\mu}\dfrac{m^2}{(m+3m')^2}.$

12.15 $1.051\sqrt{ga}.$

12.16 $\mathbf{F}=m\left(g\mathbf{b}_1-\dfrac{\pi+4}{\pi}r\Omega^2\mathbf{b}_2+2\dfrac{\pi-1}{\pi}r\Omega^2\mathbf{b}_3\right);$

$\mathbf{T}=mr\left(4r\Omega^2\mathbf{b}_1+\dfrac{1}{\pi}\left(2g-(\pi-1)a\Omega^2\right)\mathbf{b}_2-\left(g+\dfrac{4-\pi}{2\pi}a\Omega^2\right)\mathbf{b}_3\right).$

12.17 $\left(\dfrac{(r\cos\theta-z\sin\theta)}{(R-r\cos\theta+z\sin\theta)\beta}\dfrac{g}{r}\right)^{\frac{1}{2}},$

where $\beta=\left(\sin\theta+\dfrac{z}{r}\cos\theta-(qz\sin\theta-pr\cos\theta)\sin\theta\right)$; $0.707\sqrt{\dfrac{g}{r}}.$

12.18 $\dfrac{1}{2}\sqrt{gr}.$

13.1 $\alpha = 0$; $\beta = 1$; $\gamma = -2$.

13.2 7 cm^2.

13.3 $\cos^{-1}\frac{1}{3}$; $\cos^{-1}\frac{2}{3}$; $\cos^{-1}\frac{2}{3}$.

13.4 $\frac{1}{6}(-19 \pm \sqrt{265})$.

13.5 $\pm\dfrac{1}{\sqrt{165}}(4, 10, 7)$.

13.6 $p = 5/3$; $q = -2/3$.

13.7 1.237 rad.

13.8 $3\sqrt{2}$.

13.10 $\dfrac{1}{10}(3, 4, 5\sqrt{3})$.

13.11 $\alpha = -3$.

13.12 $\dfrac{1}{3}(-4, 5, -1)$.

13.14 $\dfrac{1}{4}(1, -1, \sqrt{14})$.

13.15   No.

13.17 $x = 1$, $y = -1$, $z = 2$.

13.18 $\mathbf{n}_1 = \pm\dfrac{\sqrt{2}}{2}(1, 1)$, $\mathbf{n}_2 = \mp\dfrac{\sqrt{2}}{2}(1, -1)$.

Index

vector systems, 27, 35
velocity of approach, 270
velocity of separation, 270
velocity vector, 12
versor of a direction, 642
vibrometer, 574
viscosity of a fluid, 70
viscous damper, 663
viscous friction, 70

work and energy, 361
work of a force, 214, 248
work of a torque on a rigid body, 510
work of torques, 510
wrench, 53